Lecture Notes in Mathematics

A collection of informal reports and seminars
Edited by A. Dold, Heidelberg and B. Eckmann, Zürich

T0236967

337

Cambridge Summer School in Mathematical Logic

Held in Cambridge/England, August 1–21, 1971

Edited by A. R. D. Mathias, Cambridge/England
H. Rogers, Massachusetts Institute of Technology, Cambridge, MA/USA

Springer-Verlag
Berlin · Heidelberg · New York 1973

AMS Subject Classifications (1970): 02-02, 02C15, 02C20, 02E05, 02F10, 02F25, 02F99, 02H10, 02H13, 02H99, 02K20, 02K35, 02K99, 04-02, 04A20, 04A25, 05C15

ISBN 3-540-05569-X Springer-Verlag Berlin · Heidelberg · New York
ISBN 0-387-05569-X Springer-Verlag New York · Heidelberg · Berlin

Offsetdruck: Julius Beltz, Hemsbach/Bergstr.

PREFACE

This volume is the tangible residue of the Summer School in Mathematical Logic that was held in Cambridge in 1971 and which lasted from the first to the twenty-first of August. It was recognised as an European meeting of the Association for Symbolic Logic, and was financed by grants from the Scientific Affairs Division of the North Atlantic Treaty Organization, who listed it as an Advanced Study Institute, from the Science Research Council of Great Britain, the Division of Logic Methodology and Philosophy of Science of the International Union of History and Philosophy of Science, the British Council, and the IBM Corporation. To all those the Organizing Committee express their gratitude. They record here their thanks to the Master and Fellows of Pembroke College and the Master and Fellows of Trinity College for board and lodging; the secretariat of the Department of Pure Mathematics; the University of Cambridge for the use of its lecture rooms; and the administrative secretary of the Summer School, Miss Catherine Braithwaite.

Addresses were invited on five topics: model theory, intuitionism, the priority method, infinitary combinatorics, and finite automata. In addition H.M. Friedmann gave some further lectures on realizability and other topics, there was a session devoted to problems in model theory, there were contributed papers, and a panal of experts under the chairmanship of S.C. Kleene debated intuitionism. A list of participants is appended, and a fuller report will be given to the Association for Symbolic Logic.

The papers presented here are in the main versions of addresses actually delivered during the Summer School except for those of N. Goodman and C.C. Chang, who had accepted invitations to speak but were in the event unable to come, and of K. Kunen who spoke on another topic. The content of the lectures by J.H. Conway and D. Pilling on finite automata may be found in the book by J.H. Conway entitled: "Regular Algebra and Finite Machines", published by Chapman and Hall.

Hartley Rogers
A.R.D. Mathias

TABLE OF CONTENTS

SET THEORY AND COMBINATORICS

MODEL THEORY

LIST OF PARTICIPANTS

John A.H. Anderson
John G. Anderson
Daniel Andler
Paul Bacsich
Henk Barendregt
Jon Barwise
Sanat Basu
Görg Belger
Robert Bonnet
Marie-Claire Bonnet
David Bryars
Christopher Brickhill
Leslie Burkholder
Stan Burris
Doug Bush
Denis Capatos
Rafael Casas
Claude Christen
Jean-Claude Collet
John Conway
Barry Cooper
John Cornwell
Bruno Courcelle
Nigel Cutland
Dirk van Dalen
Maryvonne Daquenet
Keith Devlin
Walter Deuber
Frank Drake
Diana Dubrovsky
Michael Dummett
Raymond Durand
Susan Eisenbach
Paul Eklof
Paul Erdös
Michael Falkoff
Ulrich Felgner
Ed Fisher
Jeanne Forrante
Harvey Friedman
Sy Friedman
Robin Gandy
Joe Gielen
Wim Gielen
Jean-Yves Girard
Klaus Gloede
Derek Goldrei
Richard Gostanian
W.E. Gould
David Gudjonsson
Andras Hajnal
Bob Hale
Janusz Halicki
Alan Hamilton
Chris Harding
Martha F. Harrell

Leo Harrington
Alex Häussler
Allan Hayes
Johannes Heidema
J. van Heljenoort
Klaus Heidler
Arend Heyting
Denis Higgs
Peter Hiller
Roger Hindley
Louis Hodes
Susan Hodes
Wilfried Hodges
Michel Holz
Albert Hoogewijs
Paul Howard
Martin Hyland
J.J. de Iongh
Stephen Jackson
Martin Janta-Polczynski
Michael Jean
Ronald Jensen
James Jones
Istvan Juhasz
Aki Kanamori
H. Jerome Keisler
Steve Kleene
E.M. Kleinberg
Piotr Kossowski
Georg Kreisel
Frank Kriwacsek

Ken Kunen
Alistair Lachlan
Peter Ladkin
Jay John Tuthill Lagemann
Jean Larson
Reginald Lawson
Robert Lebeuf
D.H. Lehmer
Manuel Lerman
Francis Lowenthal
Nancy Lynch
Angus MacIntyre
John MacIntyre
David MacQueen
Menachem Magidor
M. Makkai
Johann Makowsky
Viktor Marek
Attila Maté
A.R.D. Mathias
Mathieu Meyer
Eric Milner
Gadi Moran
Yiannis Moschovakis
Michael Moss
Cathérine Muhlrad-Greif
Gert Müller
John Myhill
S.A. Negrepontis
H. Joachim Neuhaus
Eva Nosal

Kempachiro Ohashi
Leszek Pacholski
Donald Pelletier
Jerzy Perzanowski
Robert Phillips
Don Pilling
Donald Potts
Birgit Poulsen
Marian Pour-El
Alex Prestel
Karel Prikry
Richard Rado
Peter Rado
William Raines
Ken Rasmussen
Cecylia Rauszer
Joachim Reineke
Dave C. Rine
Hartley Rogers
Par Rogers
Harvey Rose
George Sacerdote
Gerald Sacks
Henrik Sahlqvist
Hidehisa Sakai
Ken Salomon
Arturo Sangalli
Leonard Sasso
James Schmerl
Philip Scott
Krister Segerberg
Saharon Shelah
J.R. Shoenfield
Richard Shore
Jack Silver
Lewis Simonoff
Harry Simmons
Stephen G. Simpson
John Skvoretz
James Smith
Leonard Smith
R. Solomon
Ippolito Spadafora
Marian Srebrny
John Staples
Staunton Steen
Jacques Stern
Gabriele Stolzenberg
Michael Stone
Adrian Tang
Steven Thomason
Anne Troelstra
Bob Vaught
Wim Veldman
Guy Vidal-Naquet
François-Yves Villemin
Roel de Vrijer
Stan Wainer
Philip Walters
Bogden Weglorz
Alec Wilkie
Agnieska Wojciechowska
Pawel Zbierski
Martin Ziegler

I. LECTURES ON INTUITIONISM

D. van Dalen

CONTENTS

0. INTRODUCTION

In the following lectures we have tried to present an introduction to intuitionistic mathematics in the more traditional sense and also a modest survey of some topics in the foundations of intuitionistic mathematics of more recent date. The list of subjects is far from complete. In particular no attempt has been made to incorporate a formal treatment of intuitionistic logic.

Intuitionism was created by L.E.J.Brouwer in the beginning of this century. Without slighting the efforts of earlier constructivists, one can say that he virtually singlehanded undertook to reconstruct mathematics according to his views. For a mathematical-philosophical exposition we refer to [Kreisel-Newman, 70] and [Van Stigt, 71].

The subject matter of intuitionism consists of constructive activities of the mind, in particular mental constructions. The question as to which objects, processes, etc. are constructive cannot be answered as it would require a knowledge of all (possible) constructions. We consider the domain of the constructive as essentially open and approach the notion accordingly, i.e. any time we may add new (classes of) constructive objects that we recognize as such.

The actual mathematical practice of intuitionism employs these constructive objects and arguments in the cultivation of such well-known disciplines as analysis, algebra, topology, etc. In the foundations of intuitionistic mathematics it is exactly the subject matter of intuitionism that must be analyzed. And therefore a study of the extent of the universe of the constructive is a central issue in foundational work. For the purpose of these lectures it is assumed that the reader has sufficient understanding of constructions and constructivity to follow the exposition.

The following examples will illustrate the kind of constructions studied in finitistic mathematics and in intuitionistic mathematics.

(i) A simple example of a (mental) construction is provided by the natural numbers. The basic construction here is the junction of units. To be precise, we start out with unit and when a natural number has been constructed, the next number n' is obtained as the junction of n and unit into a new object. The natural numbers are obtained by iterating the above construction: $0, 0', 0'', 0''', \ldots$.

Note that in the construction no properties of the natural numbers beyond equality and inequality are used, the construction does not use the 'internal structure' so to speak.

The natural numbers are simple in still another way: suppose a natural number n is given, i.e. the construction of n is given, the natural number is a construction which is at the same time a proof that the object constructed is a natural number. Natural numbers carry their own proof.

Natural numbers certainly belong to the domain of finitism, which applies constructions to concrete, spatio-temporal, objects (cf. [Kreisel,65], p.119). The natural numbers are insofar typical for the combinatorial operations of finitism, that the latter can be coded into the natural numbers. The primitive recursive functions provide a familiar example. Relations between primitive recursive functions translate into simple relations between their encodings, e.g. the relation 'f is the composition of g and h ' translates into a primitive recursive relation between the encodings of f, g and h.

(ii) Under (i) we discussed some constructions and objects which are completed, in the sense that they are heriditarily constructed and e.g. encodable in natural numbers, or finite configurations in a production system over some alphabet.

Intuitionism, however, allows a wider range of constructions and objects. One may think of such abstract objects as functions, functionals, species, constructions on constructions, etc.

We will mention one concrete example of a non-finitist construction: the operation $\lambda ab(\lambda x.a(b(x)))$, where a and b are variables for constructive mappings of N into N.

In the case of the primitive recursive functions one could still recognise the finitistic nature by the coding of their recursion equations in the natural numbers. In the case of the constructive mappings we have left the domain of finitism, as there is no apparant way to represent them as finite configurations of some sort (neither is there a coding in the natural numbers).

Once we have introduced our objects we must answer the question how to prove statements about these objects. In intuitionistic mathematics these proofs are constructions, in particular mental constructions of the intuitionist.

In simple cases these constructions turn out to be finitistic objects, e.g. the proof of $2+3 = 5$ consists of constructions of 2, 3, 2+3, 5 and a comparison.

Characteristic for statements concerning natural numbers is the principle of complete induction:

$$A0 \wedge \forall x\ (Ax \rightarrow Ax') \rightarrow \forall x\ Ax.$$

In order to prove An for a natural number n one has to prove A0, A1, ..., An parallel to the construction of n.

Even for proving simple identities, such as $m+n = n+m$, one employs the induction principle, i.e. by paralleling the construction of m+n and n+m (in the specific order indicated by the double induction), one obtains a proof for the identity $m+n = n+m$.

Without introducing more sophisticated objects, we can consider more complex 'properties'. One way to do so is to define propositions by means of the operations of first order logic. As, however, the logical particals were used in mathematics in their non-constructive sense, the intuitionist needs a constructive reinterpretation of the logical connectives. We will return to that problem in section 2. From such a reinterpretation it appears plausible that the formal properties of assertions valid for the classical and for the constructive interpretation diverge.

We will give an illustration, involving the disjunction, which is of interest also without subtle analysis.

Consider the following problem: are there irrational numbers a and b such that a^b is rational?

Non-constructive solution: $\sqrt{2}^{\sqrt{2}}$ is either rational or irrational. In the first case $a = b = \sqrt{2}$, in the second case $a = \sqrt{2}^{\sqrt{2}}$ and $b = \sqrt{2}$.

Constructive solution: Show that $\sqrt{2}^{\sqrt{2}}$ is irrational. (E.g. apply Gelfand's theorem, that states that a^b is transcedent if a and b are algebraic irrational, [Gelfand], p.106, Thm 2.)

The first solution would also disappoint a classical mathematician, he might not have realised that, strictly speaking, this kind of solution to the problem would be possible. He may have wanted a and b, maybe in some specific notation, without introducing undecided case distinctions. This is fully in accordance with the intuitionistic meaning of the logical connectives. As the second solution shows, there is such an intuitionistic solution.

For more examples see [Heyting, 56].

We cannot hope to get such a clear view of the total possibilities of the mathematical imagination, that we could give an explicit description of the abstract objects to be considered in intuitionism. We may, however, indicate some legitimate procedures for generating (creating, or 'constructing') abstract objects.

In particular we consider those objects created by 'the second act of intuitionism', according to Brouwer: "infinitely proceeding sequences whose terms are chosen more or less freely from mathematical entities previously acquired" and "species, i.e. properties supposable for mathematical entities previously acquired." We shall return to these notions later. For the moment let us restrict our attention to the so-called <u>lawlike</u> sequences of natural numbers (i.e. lawlike mappings from N in N).

Lawlike sequences are given by a law (i.e. algorithm or effective procedure) which associates to each natural number a natural number together with a proof of this fact.

For lawlike sequences (and for abstract objects in general) we distinguish two kinds of equality: (i) an intensional (or definitional) equality - two sequences are equal, if they are given by the same law,

(ii) an extensional equality - $a=b$ iff

$$\forall x\ (ax = bx).$$

Examples of lawlike sequences are well-known from recursion theory.

<u>Primitive recursive functions</u>. A primitive recursive function is given by its recursion equations and a proof of its being defined. The recursion equations usually are coded into the natural numbers

and the coding of f is called its index.

The law determining f is the set of recursion equations and it is completely determined by the index of f. Hence we say that f and g are intensionally equal if the indices of f and g are equal.

We know from recursion theory that considerable use is made of indices of primitive recursive (and general recursive, of course) functions and not only of their graphs. Therefore the introduction of intensional equality makes perfect sense.

<u>Heriditarily recursive operations.</u> (HRO, see [Troelstra, 71]). Heriditarily recursive operations (hro's for short) may serve as a model for intuitionistic arithmetic of finite types, in the intensional version.

Let types be inductively defined by (i) $0 \in T$, (ii) $\sigma,\tau \in T \Rightarrow (\sigma)\tau \in T$. As usual 0 is the type of N and if σ,τ are types of K,L, then $(\sigma)\tau$ is the type of the mappings of K into L.

We define the following sets V_σ for all $\sigma \in T$:

$$V_0 = N$$

$$V_{(\sigma)\tau} = \{x \mid (\forall y \in V_\sigma) \exists n (T(x,y,n) \wedge Un \in V_\tau)\}$$

(T is Kleene's predicate and U is the resultfunction)

In a natural way the elements of $V_{(\sigma)\tau}$ represent (recursive) operations of type $(\sigma)\tau$.

For these operations Φ_x the law is given by the index x (which codes the recursion equations).

From their definition it is necessary to consider an intensional equality relation, since Φ_x may map extensionally identical operations on distinct operations.

In general it is meaningful to consider intensional equality, as laws are taken into consideration and not only graphs of sequences.

Another kind of abstract object, which is not at first sight suggested by the second act of intuitionism was introduced by Brouwer when reflecting on the concept of the idealised mathematician or creative subject. This concept arises naturally from the basic aim of intuitionism, of considering mental constructions, when one ignores (or 'abstracts from') the interactions of different subjects. The position, taken by intuitionists in conformity with the above views may be characterised as 'mathematical solipsism'. Formal consequences of the solipsist principles, cf. Brouwer's papers on essentially negative properties (see e.g. [Heyting, 56] Ch.VIII; [Troelstra, 69] , §16), extend earlier material. However, one has to keep in mind that the impossibility of certain proofs, which is shown with the help of these strong solipsist principles, is not sufficient to establish the negative properties considered when one refers only to those objects which are defined without reference to the creative subject.

A systematic approach to the study of the creative subject was made by Kripke, Kreisel, Myhill and others (cf [Troelstra, 69], p.95 ff.) In the present lectures we will not touch on the subject of the creative subject.

General reading material on intuitionism can be found in a number of expositions, in particular [Heyting, 56], [Kreisel, 65], [Fraenkel - Bar-Hillel, 72], [Troelstra, 69].

Part of the material of these lectures overlaps with Troelstra's 'Principles of Intuitionism', the reader who is familiar with the Principles may skip the introductory material.

In particular the sections 2 and 4 are not covered in Troelstra's notes.

In preparing these lectures we have drawn upon a number of sources, mentioned in the text. Moreover we are indebted to a number of logicians for conversations, in particular G.Kreisel and A.S.Troelstra have put their stamp on the material presented.

1. SOME CONCEPTS OF INTUITIONISTIC MATHEMATICS

In this section we will discuss a number of concepts that play a role in the literature. As most of them have already been extensively treated we will not elaborate the subjects.

1.1. Species

Suppose that we have at our disposal a well-defined collection of mathematical objects. For example the natural numbers are potentially given by their way of generation, so that we can view their totality as a conceptional unit. The well-defined properties of these collections are species.

Objects with the required property are called elements of the species and we will denote "a is element of the species S" as usual by $a \in S$.
Thus species are not constructed in the sense that their elements are constructively, or even freely, generated, rather they are results of a comprehension principle. Although we need not actually know for an object a and species S whether $a \in S$ holds, we know what it means for $a \in S$ to hold; namely that we have a proof that a has the property S.

The answer to the question whether impredicative species are allowed depends on ones willingness to recognize the powerspecies (i.e. the collection of all subspecies) of a basic collection (such as the natural numbers) as a basic collection.

In Brouwer's original definition ([Brouwer, 24], p. 245) there is a hierarchy of species:

Species of first order is a property of mathematical entities only (which are introduced in [Brouwer, 24] as spreads and elements of spreads, cf. section 3). Species of order n+1 are properties of mathematical entities or species of order i $(i = 1,\ldots,n)$.

In Brouwer's writings not too much attention is paid to species and their nature. it is not clear for example how far impredicative species were allowed. Closely connected are the problems of the existence of the powerspecies and of quantification over species. Brouwer did not make himself explicit on these points. It may be noted that for Brouwer's mathematical practice a weak kind of impredicativity suffices, namely the kind represented by (single) generalized inductive definitions.

Lately the metamathematical aspects of the theory of species were considered in a number of papers, specifically [Martin-Löf 71], [Prawitz 71], [Troelstra 71,71A, this volume], [Kreisel-Troelstra 70], [Friedman, this volume].

In practical intuitionistic mathematics the use of the notion of species has mainly been unproblematic. As a matter of fact most applications of the comprehension principle have been predicative. Impredicative applications have been used in the case of the intuitionistic ordinals and of K (see section 3.).

Although the definition of species is intensional it is useful to consider an extensional equality relation for species:
$X = Y := \forall x(x \in X \leftrightarrow x \in Y)$.

1.1.1.

Definition: (i) X is <u>inhabited</u> if $\exists x(x \in X)$ ($\mathrm{Ч}(X)$)

(ii) Y is <u>detachable</u> in X if $(\forall x \in X)(x \in Y \vee x \notin Y)$

(iii) X is <u>empty</u> if $\neg \exists x\ (x \in X)$, $(X = \phi)$

(iv) X is <u>discrete</u> if $\forall x \in X \forall y \in X\ (x = y \vee x \neq y)$

Note that $X \neq \phi \rightarrow \mathrm{Ч}(X)$ is not valid, as it is equivalent to $\neg\neg\exists x A x \rightarrow \exists x A x$.

The usual set-theoretic operations and relations will be used in the sequel.

In addition to the equality relation on species there is the so-called "apartness relation" which is the positive analogue of the inequality.

Notation $x \# y$.

The apartness relation is governed by the following laws:

1.1.2.

(a) $\neg x \# y \leftrightarrow x = y$

(b) $x \# y \leftrightarrow y \# x$

(c) $x \# y \rightarrow x \# z \vee y \# z$

Note that the equality relation on a species with apartness relation is stable, i.e. $\neg\neg x = y \rightarrow x = y$. It has been shown however that the presence of an apartness relation is stronger than stability of the equality relation in the sense that a Kripke model has been constructed on which the equality relation is stable, but no apartness relation can be introduced [van Dalen-Gordon, 71]. De Iongh and Troelstra have examined the relation between several notions of finiteness of species, see [Troelstra, 67].

1.2. Sequences

A central notion is that of mapping or function, which, contrary to classical mathematical practice, is introduced intensionally.

1.2.1.

Definition: A mapping ϕ from a species X into a species Y is a process which assigns to each $x \in X$ an element $y \in Y$, such that $x = x' \rightarrow \phi x = \phi x'$.

Note that this condition is necessary in case the equality relation is not the intensional (definitional) one. Because, then it must be shown that the (coarser) extensional equality relation is preserved, e.g. if X and Y themselves consist of species.

Note that the equality relation on X need not be the original intensional one, e.g. if X consists of species then the equality relation considered can be the extensional one and we want ϕ to carry extensionally equal elements into the same object.

Notation: we will write $\phi : X \to Y$ or $\phi \in (X)Y$.

1.2.2.

ϕ is <u>bi-unique</u> (or one-one, or an injection) if

$$\forall x \in X \forall x' \in X\ (\phi x = \phi x' \to x = x').$$

ϕ is weakly <u>bi-unique</u> if $\forall x \in X \forall x' \in X\ (x \neq x' \to \phi x \neq \phi x')$.

If the equality relation on X is stable, then a weakly bi-unique ϕ is bi-unique. A bi-unique ϕ has an inverse ϕ^{-1}.

If $\phi \in (X)Y$ and X and Y possess apartness relations $\#$ and $\#'$, then ϕ is called <u>strongly bi-unique</u> if

$$\forall x \in X \forall x' \in X\ (x \# x' \to \phi x \#' \phi x')\ldots$$

Note that in case ϕ is bi-jection and X has an apartness relation, ϕ induces in a natural way an apartness relation on Y.

Notions like homomorphism, isomorphism can be introduced as usual.

A special place is taken by mappings of type (N)X, where N is the species of natural numbers. One is familiar with this fact e.g. through the introduction of real numbers.
Mappings of type (N)X are called <u>sequences</u>, we will use $\xi, \eta, \zeta, \xi_1, \eta_1, \zeta_1, \ldots$ to denote sequences in general.

A host of notions of sequences is known and we will return to these in section 4. Here we will consider the so-called lawlike sequences, i.e. those that are given by a law or algorithm. Along with the law we must provide a proof of its applicability to all natural numbers.

We will denote lawlike sequences by a,b,c,... With respect to this notion we can express Ch ch's Thesis: All lawlike sequences are recursive, or in symbols $\forall a \, \exists z \, \forall x \, (ax = \{z\}x)$.

One should keep in mind that Church's Thesis pertains to "mechanically computable" functions. There is no reason to expect Church's Thesis to hold for a wider class, e.g. one containing "subjectively computable" functions (see [Troelstra, 69A] § 16).

Kreisel has initiated the study of formal systems for analysis with variables for lawlike sequences (besides the coice sequences). For a comprehensive treatment see [Kreisel-Troelstra, 70]. In a number of these formal systems the assumption of Church's Thesis can be shown consistent.

Lawlike functions are, in contrast to the various notions of choice sequences, complete objects, i.e. objects which allow a complete description (the law + the proof of applicability). As explained in section 0 lawlike sequences are provided with an intensional equality relation. In general we will denote intensional equality by $\equiv$ and extensional equality by $=$.

On the basis of the intuistionistic notion of validity the existence of lawlike sequences is often guaranteed by choice principles.

Suppose we have a proof of $\forall x \, \exists y \, A(x,y)$ where x and y range over natural numbers and A is a lawlike statement (i.e. does not contain parameters for incomplete objects), then the proof contains a complete description how to associate a y to a given x. Therefore we may conclude the existence of a lawlike sequence such that

$$\forall x \, A(x,ax).$$

So $\forall x \exists y A(x,y) \rightarrow \exists a \, \forall x \, A(x,ax)$ holds.

Completely analogously we arrive at selection principles for more general cases.

Suppose that X is a species with intensional equality relation such that for each construction it is decidable whether or not it produces an element of X then

$$\forall x \in X \exists y \in Y\, A(x,y) \rightarrow \exists \Phi \in (X)Y \forall x \in X\, A(x,\Phi x).$$

Φ need not be lawlike, in particular A may contain "incomplete parameters", in that case Φ is lawlike <u>in</u> the parameters.
With each species X with extensional equality relation we can associate a "lifted" species X' of elements of X together with their definitions. On this species the selection principle can be applied. The result does however in general not carry over to original species.

We list some common choice principles

(1) $\forall x\, \exists \xi\, B(x,\xi,\xi_0,\ldots) \rightarrow \exists \eta\, \forall x\, B(x,\lambda y\cdot\eta\{x,y\},\xi_0,\ldots)$

(2) $\forall x\, \exists y\, B(x,y,\xi_0,\ldots) \rightarrow \exists \xi\, \forall x\, B(x,\xi x,\xi_0,\ldots)$

(3) $\forall x\, \exists a\, A(x,a) \rightarrow \exists b\, \forall x\, A(x,\lambda y\cdot b\{x,y\})$

(4) $\forall x\, \exists y\, A(x,y) \rightarrow \exists a\, \forall x\, A(x,ax)$

where x,y range over N, A does not contain non-lawlike parameters. In (1) and (3) B and A are extensional w.r.t. ξ and a, i.e.

$$\xi = \xi' \wedge B(x,\xi,\ldots) \rightarrow B(x,\xi',\ldots)$$

$$a = a' \wedge A(x,a) \rightarrow A(x,a').$$

The selection principle needs justification for the various notions of sequence as it is not a priori clear whether the resulting sequence is of the required kind.

1.3. Some notions from analysis

Once we have at our disposal the natural numbers, the construction of rational numbers is unproblematic. The classical procedure is acceptable. The introduction of real numbers is however a different matter, here one essentially needs sequences (of rationals or of rational intervals) and the question arises which notion of sequence has to be used. We will return to this matter later and for the moment we introduce the reals without specifying the notion of sequence.

In this text we will introduce real numbers by the device of Cauchy-sequence, any other approach would do just as well (cf. [Brouwer, 18], [Heyting, 53]).

We will denote a sequence $\lambda n \cdot x_n$ by $\langle x_n \rangle_n$, or simply $\langle x_n \rangle$.

1.3.1.

<u>Definition</u>: (i) a Cauchy-sequence is a sequence $\langle r_n \rangle$ of rationals such that $\forall k\, \exists n\, \forall m\, (|r_n - r_{n+m}| < 2^{-k})$

(ii) $\langle r_n \rangle \sim \langle s_n \rangle$ if

$$\forall k\, \exists n\, \forall m\, (|r_{n+m} - s_{n+m}| < 2^{-k}).$$

It is easily shown that $\sim$ is an equivalence relation. The equivalence classes under $\sim$ are called <u>real numbers</u>. To simplify the notation we will denote the real numbers also by $\langle r_n \rangle$.

1.3.2.

<u>Definition</u>: (i) $\langle r_n \rangle + \langle s_n \rangle = \langle r_n + s_n \rangle$

(ii) $\langle r_n \rangle \cdot \langle s_n \rangle = \langle r_n \cdot s_n \rangle$.

As usual one shows addition and multiplication are compatible with the equivalence relation. Therefore we can define addition and multiplication for reals. The definition of the inverse requires more care. In classical mathematics one proceeds as follows: let r_n determine the real x_0, we have either $x_0 = 0$ or $x_0 \neq 0$ (where $0 = \langle 0 \rangle_n$), accordingly we define

$$x_0^{-1} = 0 \text{ or } x_0^{-1} = \langle s_n \rangle, \text{ where } s_n = \begin{cases} 0 \text{ if } r_n = 0 \\ r_n^{-1} \text{ else} \end{cases}$$

It is a matter of routine to check that x_0^{-1} classically is a Cauchy-

sequence. The above dichotomy is by definition equivalent to
$\forall k\, \exists n\, \forall m\, (|r_{n+m}| < 2^{-k})$ or $\neg \forall k\, \exists n\, \forall m\, (|r_{n+m}| < 2^{-k})$.
Unfortunately there is no way to check in general which of the two disjuncts holds.

Following Brouwer, intuitionists have motivated their refusal to accept these dubious applications of the principle of the excluded third by reducing these to some unsolved problem in the realm of the natural numbers.

We will outline the method so that the reader can apply it if desired. We take it for granted that the real number π possesses a decimal expansion (for a proof see e.g. [Brouwer, 20]). Consider the statement Πn := n is the number of the last decimal of a sequence of 10 consecutive numerals 7 in the decimal expansion of π.

At present we have no means to decide the problem $\forall n \neg \Pi n \vee \neg \forall n \neg \Pi n$. Using the property Πn, we define a Cauchy-sequence:

$$r_n = \begin{cases} 2^{-n} & \text{if } \neg \exists m \leq n\, \Pi m \\ 2^{-m} & \text{if } m \leq n,\ \Pi m \text{ and } k < m \rightarrow \neg \Pi k \end{cases}$$

Evidently $\langle r_n \rangle$ is a Cauchy-sequence, moreover $\langle r_n \rangle$ is lawlike (even primitive recursive).

Now $\langle r_n \rangle = 0$ is equivalent to $\forall k\, \exists n\, \forall m\, (|r_{n+m}| < 2^{-k})$ or $\neg \exists n\, \Pi n$, or $\forall n \neg \Pi n$ (1)

and $\langle r \rangle \neq 0$ is equivalent to $\neg \forall k\, \exists n\, \forall m\, (|r_{n+m}| < 2^{-k})$ or $\neg \forall n \neg \Pi n$. (2)

But we have no grounds to assert (1) or (2), hence there is no justification for $\langle r_n \rangle = 0 \vee \langle r_n \rangle \neq 0$.

Counterexamples of the above kind do not reduce the statement under consideration to a contradiction (an absurdity), they show that at present there is no evidence to assert them.

In this particular case it is shown that we cannot hope to define the inverse for all reals.

Another example, due to Heyting [Heyting, 56], p. 17, shows that not for all reals x $\neg\neg Rat(x) \to Rat(x)$, where Rat(x) stands for "x is rational".

$$\text{Define } r_n = \begin{cases} \sum_{k=1}^{n} 3\cdot 10^{-k} & \text{if } \neg\, \exists m \leqslant n\ \Pi m \\ \sum_{k=1}^{m} 3\cdot 10^{-k} & \text{if } m \leqslant n,\ \Pi m \text{ and } p < m \to \neg\, \Pi p \end{cases}$$

Let $x = \langle r_n \rangle$.

Rat(x) implies that there exist natural numbers p and q such that p = qx. Given p and q we can decide whether $\exists m\ \Pi m$ or $\neg\exists m\ \Pi m$. So Rat(x) cannot be asserted.

Now $\neg Rat(x)$ implies in particular $\forall n \neg\exists m \leqslant n\ \Pi m$ and hence $x = 3^{-1}$, which contradicts $\neg Rat(x)$.

Conclusion $\neg\neg Rat(x)$.

From (1) and (2) we conclude that $\neg\neg Rat(x) \to Rat(x)$ cannot be asserted.

Using the real number x constructed above we can exhibit a mapping which is weakly bi-unique but not bi-unique.

Let $\mathbb{R}$ be the species of reals, $\mathbb{Q}$ the species of rationals and $\tilde{\mathbb{Q}}$ the species of the non-non-rationals

$$\mathbb{Q} \subset \tilde{\mathbb{Q}} \subset \mathbb{R}.$$

Consider the natural mapping p: $\mathbb{R}/\mathbb{Q} \to \mathbb{R}/\tilde{\mathbb{Q}}$. p is weakly bi-unique. Now consider the real number x with the property $\neg\neg Rat(x)$ and Rat(x) unknown.

Clearly $p(x+\mathbb{Q}) = p(\mathbb{Q})$, but it is unknown whether $x+\mathbb{Q} = \mathbb{Q}$, so we are not allowed to say that p is bi-unique.

Note that the identity relation on $\mathbb{R}/\mathbb{Q}$ is not stable, otherwise p would be bi-unique.

Let us return to the problem of the inverse. Consider a Cauchy-sequence $\langle r_n \rangle$. It is no restriction to suppose $r_n \neq 0$ for all n. A simple calculation tells us that if $\langle r_n^{-1} \rangle$ is a Cauchy-sequence the following holds: $\exists k\, \exists n\, \forall m\, (|r_{n+m}| > 2^{-k})$. (*)

Conversely if (*) holds the inverse exists. This leads us to the following definition for Cauchy-sequences.

1.3.3.

<u>Definition</u>: (i) $\langle r_n \rangle \# \langle s_n \rangle := \exists k\, \exists n\, \forall m\, (|r_{n+m} - s_{n+m}| > 2^{-k}$

(ii) $\langle r_n \rangle < \langle s_n \rangle := \exists k\, \exists n\, \forall m\, (s_{n+m} - r_{n+m} > 2^{-k})$

In a natural way this induces relations on the reals.

1.3.4.

<u>Theorem</u>: $\neg\, x \# y \leftrightarrow x = y$

$x \# y \leftrightarrow y \# x$

$x \# y \rightarrow x \# z \vee y \# z$

$x \# y \rightarrow x + z \# y + z$

$x \# y \wedge z \# 0 \rightarrow xz \# yz$

The proofs are straightforward, see [Heyting, 56], [Troelstra, 69A].
In particular $\#$ is an apartness relation. From this we conclude that the equality relation on the reals is stable (cf. 1.1.2.).
We can now abbreviate (*): x^{-1} is defined if $x \# 0$.
The ordering and the apartness relation are connected as is shown by the

1.3.5.

<u>Theorem</u>: $x \# y \leftrightarrow (x < y \vee y < x)$

Proof: immediate.

The problems we met in connection with the introduction of the inverse are inherent to intuitionistic mathematics, the theory of fields, etc.; one cannot avoid partially defined operations.

By means of counterexamples analogous to the ones given above we see that there is no evidence for $x \leqslant y$ and $\neg\, y > x$ to be equivalent. So we introduce the relations $\nless$ and $\ngtr$.

1.3.6.

Definition: $x \nless y := \neg x < y$

$x \ngtr y := \neg x > y$

One easily proves the following facts about the ordering of the reals.

1.3.7.

Theorem: $x \nless y \wedge y \nless x \rightarrow x = y$

$x < y \wedge y < z \rightarrow x < z$

$x \nless y \wedge y > z \rightarrow x > z$

$x > y \wedge y \nless z \rightarrow x > z$

$x \ngtr y \wedge y \ngtr z \rightarrow x \ngtr z$

(for proofs see [Heyting, 56], 2.2.6, [Troelstra, 69A], p. 26).

1.3.8.

Definition: $\max(\langle r_n\rangle, \langle s_n\rangle) = \langle \max(r_n, s_n)\rangle$

$\min(\langle r_n\rangle, \langle s_b\rangle) = \langle \min(r_n, s_n)\rangle$

$|\langle r_n\rangle| = \langle |r_n|\rangle$

The properties of the maximum, minimum and absolute value are laid down in:

1.3.9.

Theorem: $\max(x,y) + \min(x.y) = x+y$

$|x+y| \ngtr |x|+|y|$

$|x-y| \nless |x|-|y|$

$|x \cdot y| = |x| \cdot |y|$, $|-x| = |x|$

$\min(x,y) \ngtr \max(x,y)$

In defining a segment one has to take into account that not for every pair x,y it is decidable whether $x \nless y$ or $x \ngtr y$.

1.3.10.

Definition: $[x,y] = \{z \mid \neg(z > x \wedge z > y) \wedge \neg(z < x \wedge z < y)\}$

1.3.11.

Theorem: $[x,y] = [\min(x,y), \max(x,y)]$

$z \ngtr y \rightarrow [x,y] = \{z \mid y \nless z \nless x\}$

For proofs see [Heyting, 56], 2.2, 3.3.

We will now examine some elementary problems of analysis. In order not to get involved in sophisticated arguments concerning notions of sequence we will restrict ourselves to lawlike sequences. Thus we are dealing with lawlike intuitionistic analysis.

As one would guess, the applicability of a notion of (choice-) sequence in analysis strongly depends on the closure properties under certain operations. The closure of the reals, for example, under addition, requires a simple closure property of the underlying notion of sequence. So, once one knows that a certain notion of choicesequence satisfies some quite natural closure conditions, one can build a fair amount of analysis on it. For the above reasons many results in lawlike analysis generalize to analysis based on other notions of choice-sequence.

We will first show that the reals are lawlike complete with respect to their natural metric. Since we are dealing with lawlike reals we suppose a sequence of reals to be given by a lawlike "double sequence" i.e. $\langle x_n \rangle$ is given by a lawlike sequence $\langle\langle r_{m,n}\rangle_m\rangle_n$, such that $\langle r_{m,n}\rangle_m \in x_n$ for every n.

1.3.12.

Definition: (i) $\langle x_n \rangle$ is a Cauchy-sequence if

$\forall k \exists n \forall m (|x_{n+m} - x_n| < 2^{-k})$

(ii) $\langle x_n \rangle$ converges to x if $\forall k \exists n \forall m (|x - x_{n+m}| < 2^{-k})$.

We say that $\langle x_n \rangle$ is convergent if there exists an x such that $\langle x_n \rangle$

converges to x.

1.3.13.

Theorem ([Troelstra, 69A]). Every lawlike Cauchy-sequence converges to a lawlike real.

Proof: let $\langle\langle r_{m,n}\rangle_m\rangle_n$ be the lawlike sequence of rationals belonging to $\langle x_n\rangle$. For each x_n we have the Cauchy condition, so

$$\forall k\ \forall n\ \exists m\ \forall p\ (|r_{m+p,n}-r_{m,n}| < 2^{-k}).$$

Apply the selection principle, then

$$\exists a\ \forall k\ \forall n\ \forall p\ (|r_{a\{n,k\}+p,n} - r_{a\{n,k\},n}| < 2^{-k}).$$

We choose the "diagonal" sequence

$$s_n = r_{a\{n,n\}n}.$$

Claim: $\langle s_n\rangle$ is a Cauchy-sequence.
From the definition of a it follows that
$|x_m - r_{a\{m,m\}m}| \not> 2^{-m}$, or $|x_m - s_m| \not> 2^{-m}$. Now let n be such that $\forall m(|x_{n+m} - x_n| < 2^{-k})$ and $n \geq k$. Then $|s_{n+p} - s_n| \not> |s_{n+p} - x_{n+p}| +$ $+ |x_{n+p} - x_n| + |x_n - s_n| < 2^{-n-p} + 2^{-k} + 2^{-n} < 3\cdot 2^{-k}$.
So $\langle s_n\rangle$ determines a lawlike real x with
$|x - x_{n+p}| \not> |x - s_{n+p}| + |s_{n+p} - x_{n+p}| \not> 3\cdot 2^{-k} + 2^{-k} = 2^{-k+2}$ for $n \geq k$.
So $\langle x_n\rangle$ converges to x.
Note that this is the standard argument. The extra power comes from the selection principle.

Quite simple theorems of classical analysis turn out to fail here. An example is the statement

"Every bounded monotone sequence converges".

Example: define $x_n = \begin{cases} 0 \text{ if } \neg\exists m \leq n\ \Pi m \\ 1 \text{ if } \exists m \leq n\ \Pi m \end{cases}$

Suppose $\langle x_n \rangle$ converges to x, then

$$\forall k\ \exists n\ \forall m\ (|x_{n+m} - x| < 2^{-k}).$$

We know that $x \# 0$ or $x \# 1$. Let $x \# 0$, then $\exists n(x_n = 1)$ or $\exists n\ \Pi n$. Likewise $x \# 1$ implies $\neg \exists n\ \Pi n$. This reduction shows that there is no evidence for the convergence of $\langle x_n \rangle$.

Another well-known classisal theorem that does not hold intuitionistically is the so-called "intermediate value theorem": If $a < b$ and $f(a) < 0$, $f(b) > 0$ and f continuous on $[a,b]$, then $\exists x \in [a,b](f(x) = 0)$.

We will sketch the proof, the details can readily be supplied by the reader.

$$\text{Define}\quad a_n = \begin{cases} (-2)^{-n} & \text{if } \neg\exists m \leq n\ \Pi m \\ (-2)^{-m} & \text{if } \Pi m \wedge m \leq n \wedge (p < m \rightarrow \neg \Pi p). \end{cases}$$

Let $\langle a_n \rangle$ determine a.

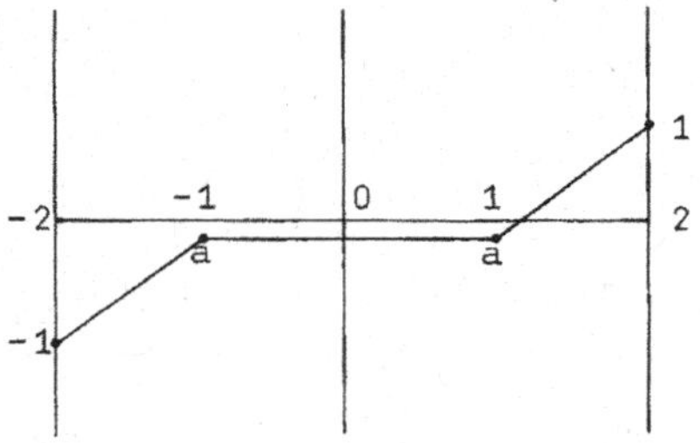

Determine the piecewise linear function f through the points $(-2,-1)$, $(-1,a)$, $(1,a)$, $(2,1)$ (e.g. by first defining f on the rationals). f is continuous on $[-2,2]$ and satisfies the premiss of the theorem.

However as above, one convinces oneself easily that the existence of a zero of f implies $\exists n\ \Pi n \vee \neg \exists n\ \Pi n$.

By a trick of Kreisel an analogy with methods of recursive analysis can be established (cf. [Troelstra, 69A], 8.3.2):

Let a and b be lawlike sequences such that

$$\neg(\exists n(an = 0) \wedge \exists n(bn = 0)).$$

Define $a_n = \begin{cases} (-2)^{-n} & \text{if } \neg\exists m \leq n\ (am = 0 \vee bm = 0) \\ 2^{-m} & \text{if } am = 0 \wedge \forall m' < m\ (am \neq 0) \wedge m \leq n \\ -2^{-m} & \text{if } bm = 0 \wedge \forall m' < m\ (bm \neq 0) \wedge m \leq n. \end{cases}$

Define f as before. Let $f(x_0) = 0$, then $x_0 \# -1$, of $x_0 \# 1$. In the first case we have $\forall n(an \# 0)$, in the second $\forall n(bn \# 0)$.

So $\forall a \forall b[\neg(\exists n(an = 0) \wedge \exists n(bn = 0)) \rightarrow \neg\exists n(an = 0) \vee \neg\exists n(bn = 0)]$ is implied by the intermediate value theorem.

Now

(1) $\forall n \neg(\exists m(a\{n,m\} = 0) \wedge \exists m\ (b\{n,m\} = 0)) \rightarrow$

$\forall n \exists n'[(n' = 0 \rightarrow \neg\exists n(a\{n,m\} = 0)) \wedge (n \neq 0 \rightarrow \neg\exists n(b\{n,m\} = 0)]$

(2) $(n' = 0 \rightarrow \neg\exists n(a\{n,m\} = 0)) \rightarrow (\exists n(a\{n,m\} = 0) \rightarrow n' \neq 0)$

(3) $(n' \neq 0 \rightarrow \neg\exists n(b\{n,m\} = 0)) \rightarrow (\exists n(b\{n,m\} = 0) \rightarrow n' = 0)$

So by the selection principle we get

(4) $\forall a \forall b(\forall n \neg(\exists m(a\{n,m\} = 0) \wedge \exists m(b\{n,m\} = 0) \rightarrow$

$\exists c \forall n[(\exists n(a\{n,m\} = 0) \rightarrow cn \neq 0) \wedge (\exists n(b\{n,m\} = 0) \rightarrow cn = 0)])$.

Paraphrasing the language of recursion theory (4) reads: Every pair of disjoint lawlike enumerable species is separated by a lawlike species.

A positive result is the following:

<u>Theorem</u>: If f is uniformly continuous on [0,1], then it possesses a least upper bound on [0,1].

Proof: see [Troelstra, 69A], p. 30.

Elementary intuitionistic analysis has been extensively studied by M.J. Belinfante, J.G. Dijkman; for references see the bibliography in [Heyting, 56].

2. LOGIC

2.1. The place of logic in intuitionism has been controversial. Brouwer pointed out, in his Ph.D.thesis, that mathematics precedes logic. That is, in constructive mathematics one performs (mental) constructions and the result is laid down in mathematical statements. These statements are proved by the mentioned constructions and not by applying certain logical laws or rules that were given beforehand. It is in this perspective that we will consider logic. In particular we will consider the logic of constructive reasoning. As pointed out in section 1, the evidence an intuitionist has for a mathematical statement consists of a proof, and a proof is understood to be a construction. Departing from proofs rather than from validity in the (idealised, if you want) external world, we have to reinterprete the logical connectives.

We make the following fundamental assumption about proofs: 'a proves A' is decidable for proofs a and statements A. Indeed, from the solipsist point of view, this is evident; one knows when a proves A, if there is doubt, then clearly a does not prove A (for the solipsist!).

Note that in e.g. arithmetic statements without 'logical structure' i.e. atoms (or generally speaking quantifier free statements, which are decidable and hence do not require a specific intuitionistic analysis of the meaning of the connectives), are decidable. In general we deal with mathematical objects, i.e. concrete objects and constructions, and with decidable properties of these objects, called <u>notions</u> by Kreisel([Kreisel, 65], p.123). For notions N(x) we state the principle that " 'a proves N(x) for all x' is a notion", i.e. it is decidable for a notion N(x) and a construction a whether 'a proves N(x) for all x' .

In formal systems proofs are represented in a specific symbolic way (note that formal proofs do not necessarily render the intended meaning of inferences, they can be employed because they are valid). In formal systems the above principle evidently holds, one can even mechanically check whether 'a proves N(x) for all x'. Moreover, the principle holds not only for notions but for statements in general. In our description of the meaning of the logical connectives, below, the general principle holds once it holds for atoms.

We will now give an informal description of the meaning of the logical connectives in terms of proofs:

(i) A proof of $A \wedge B$ consists of a proof of A and a proof of B,

(ii) A proof of $A \vee B$ consists of a proof of A or a proof of B,

(iii) A proof of $A \rightarrow B$ consists of a construction which associates to each proof of A a proof of B and a proof of this fact,

(iv) A proof of $\neg A$ is a proof of $A \rightarrow \bot$, where $\bot$ is some false statement like $0 \neq 1$.

(v) A proof of $\exists x\, Ax$ consists of a construction of an object c and a proof of Ac.

(vi) A proof of $\forall x\, Ax$ consists of a construction which associates to each object c a proof of Ac, and a proof of this fact.

For a more detailed description see [Goodman, 68,70], [Kreisel,65], [Troelstra, 69].

It may be thought that in (iii) and (vi) we introduce a vicious circle by requiring 'a proof of this fact'. Note, however, that we require a proof for "p is a proof of A $\Rightarrow$ c(p) is a proof of B" (in (iii)). Since both parts of this implication are decidable, this universal statement has (intuitionistically speaking) a trivial logical structure, i.e. can be handled as an atom.

It was left open how proofs of atomic statements (i.e. statements not involving logical connectives) are obtained. In the case of applied calculi, such as arithmetic, the proofs of atomic statements can actually be indicated. However, in general predicate calculus there is a problem. As Kreisel points out ([Kreisel, 71], p.159, footnote 17), interpretations in all possible domains must be considered. A 'logical' proof of a statement relative to the validity of the atoms must be uniform in the interpretation, but a 'non-logical' proof may use specific insights into closure properties of species (cf [Kreisel, 71], p.145).

The proof-interpretation goes back to [Heyting, 30,31]; it is related to Kolmogoroff's problem-interpretation [Kolmogoroff, 32].

A systematic development of a theory of constructions was initiated by Kreisel [Kreisel, 62A, 65].
Goodman presented in [Goodman, 68,70] a detailed description of a theory of constructions. Goodman's theory diverges from the one proposed by Kreisel in the respect that Goodman uses partial functions for his constructions, whereas Kreisel stresses the importance of total functions. E.g. in case of partial functions, equality between terms is not decidable, in particular $\rho(p,A) = 0$ would not be decidable for the characteristic function ρ of the proof predicate.
The same remark applies to extensional functions. If one takes functions (constructions) to be laws, then decidability of the equality relations can be expected.

The proof-interpretation of logic finds a concrete illustration in Gödel's Dialectica interpretation, at least if one uses functionals of finite types with intensional equality relation. The equality between terms is decidable at all types.

Also Kleene's realizability-interpretation has points in common with the proof-interpretation (see Kleene, this volume). A foundationally significant departure from the proof-interpretation is that 'e realizes A' is not decidable, furthermore Kleene employs partial functions (thus involving the extra condition of 'definedness'). The surprising feature of the realizability-interpretation actually is that the logical laws hold even if the reference to proofs is suppressed and if the notion of construction is suitably modified (see also [Kreisel, 71], p.160).

We will now give a few informal examples of the proof-interpretation. The 'together with a proof of this fact' stipulation in the cases of the implication and universal quantification is not taken into account. The verification is left to the reader.

1. $[A \rightarrow (B \rightarrow C)] \leftrightarrow (A \wedge B \rightarrow C)$.

How to transform a proof of $A \rightarrow (B \rightarrow C)$ into a proof of $A \wedge B \rightarrow C$? A proof of $A \rightarrow (B \rightarrow C)$ is a construction c that converts a proof of A into a proof of $B \rightarrow C$. That is, c applied to a proof a of A gives us a construction c', that converts any proof b of B into a proof of C. Now we have to indicate a construction c" that converts a pair of proofs $\langle a,b\rangle$ into a proof of C. Solution: apply c to a and the outcome to b. Clearly we have found the required conversion. Suppose now we have a proof d of $A \wedge B \rightarrow C$, i.e. a construction d that converts the pair of proofs $\langle a,b\rangle$ into a proof of C. In informal notations $d(a,b) = c$.

Now define $(c(a))(b) = d(a,b)$, for all a,b i.e.

$c = \lambda x(\lambda y\ d(x,y))$. This shows how to convert d into c, a proof of $A \rightarrow (B \rightarrow C)$.

From the above it follows that we have a proof of $[A \rightarrow (B \rightarrow C)] \leftrightarrow (A \wedge B \rightarrow C)$.

2. $A \to \neg\neg A$ or $A \to ((A \to \;) \to)$.

By 1. we may consider instead $A \wedge (A \to \bot) \to \bot$. Required is a construction c that converts the pair $\langle a,d \rangle$, where d is a construction converting a into f, into f (where f is a proof of $\bot$). Solution: define $c(a,d) = d(a)$, or $c = \lambda xy \cdot y(x)$.

3. $\bot \to A$ where $\bot$ is the 'falsumsymbol' .

It stands to reason that $\bot$ has no proof, so the justification of $\bot \to A$ requires some care. Required is a construction c such that for a proof f of $\bot$ c(f) is a proof of A. Since the 'proof relation' is decidable we can just as well ask for a construction c such that for all d 'd is not a proof of $\bot$ or c(d) is a proof of A' .

This statement is clearly correct under the intended interpretation of $\bot$, independent of the choice of c.

[Laüchli, 70] allows $\bot$ to possess proofs, but enough is required to show the provability of $\bot \to A$.

4. $(\exists x\, Ax \to B) \to \forall x(Ax \to B)$.

Let d be a proof of $\exists x\, Ax \to B$, then d(c) is a proof of B for every proof c of $\exists x\, Ax$. Now c is a proof of $\exists x\, Ax$ if c0 is a proof of Ac1. e is a proof of $\forall x(Ax \to B)$ if for each object p e(p) is a proof of $Ap \to B$, so (d(p))(g) is a proof of B for each proof g of Ap.

Now define c0 = g and c1 = p, then e is determined by (e(p))(g) = d(c). This establishes the effective transformation of d into e. (One can see that $e = \lambda p(\lambda g(d(c)))$ and $\lambda d(\lambda p(\lambda g(d(c))))$ is a proof of 4.)

A number of formalizations of intuitionistic logic can be found in the literature, see for example [Kleene, 52], [Prawitz, 65], [Schütte, 68],[Fitting, 69].

2.2. Arithmetic. In intuitionistic mathematics the concept of a natural number is a central one. Natural numbers are the outcome of a (mental) construction process which consists of the repeated conjoining of units (cf.[Brouwer, 48], [Heyting, 56]). The properties of the natural numbers are the outcome of reflection on the construction process. To quote Heyting ([Heyting, 56]): "The notion of natural number does not come to us as a bare notion, but from the beginning it is clothed in properties which I can detect by simple examination". The principle of complete induction, for instance is justified by the insight that our proving activity runs parallel to the construction activity.

Goodman has worked out the theory of intuitionistic arithmetic in terms of constructions (proofs), [Goodman, 68], [Goodman, 70]. The formalization of intuitionistic arithmetic was presented by Heyting [Heyting, 30A]. For an axiom system see [Kleene, 52], [Spector, 62]. We will call the formal theory of intuitionistic arithmetic Heyting's Arithmetic (HA).
HA is a proper subsystem of classical arithmetic P, but, as Gödel observed there is a translation of P into HA, preserving provability. The translation is inductively defined:

$$P^- = \neg\neg P \qquad \text{for atomic } P$$
$$(P \to Q)^- = P^- \to Q^-$$
$$(P \wedge Q)^- = P^- \wedge Q^-$$
$$(P \vee Q)^- = \neg(\neg P^- \wedge \neg Q^-)$$
$$(\forall x P)^- = \forall x P^-$$
$$(\exists x P)^- = \neg \forall x \neg P^-$$

Theorem: $\Gamma \vdash_C A \Leftrightarrow \Gamma^- \vdash_I A^-$, where C and I are the classical and intuitionistic versions of propositional logic, predicate logic, or arithmetic.
For details the reader is refered to [Kleene, 52], §81,

[Prawitz, Malmnäs, 68], Gentzen in [Szabo, 69]. The above result clearly shows that consistency for H̰Ḁ is equivalent to consistency for P̰.

The negative fragment of H̰Ḁ is the set of formulae built from atoms with the help of the connectives ∀,∧,¬ only. From the above we conclude that P̰ is a conservative extension of H̰Ḁ with respect to the negative fragment.

2.3. Semantics.

Since the formalizations of intuitionistic logic a number of systematic interpretations have been put forward, we list some of these here.

(1) Jaśkowski's truth tables   [Jaśkowski,36], [Rose, 53]

(2) The topological interpretation [Rasiowa, Sikorski, 63]

(3) The lattice interpretation   [Rasiowa, Sikorski, 63]

(4) Beth models   [Beth, 52], [Schütte, 68]

(5) Kripke models   [Schütte, 68], [Fitting, 69], [Kripke, 65],
                [Gregorczyk, 64].

All of these interpretations are related, for equivalence proofs see [Rasiowa, Sikorski, 63], [Schütte, 68], [Fitting, 69], [De Jongh-Troelstra,   ]. In this section we will consider the interpretation of intuitionistic logic via Kripke models.
The motivation for Kripke models is given as follows: consider an individual pursuing (let us say) mathematical research. This research proceeds by stages and at each stage a certain amount of evident facts is acquired. The possible stages are in general not linearly ordered, as there may be many ways to acquire new evidence. Let us assume that the new facts are atomic, then composite statements are treated by reference to their components, e.g. A ∧ B holds at stage α if both A and B hold at stage α; A → B

holds at stage α if, whenever A holds in any later stage β, also B holds in β.

We systematize this in the following definition.

2.3.1. Definition. A Kripke model is a triple $\mathcal{K} = \langle A,D,I\rangle$ where A is a non-empty partially ordered set, D is a mapping from A into a collection of non-empty sets and I is a mapping defined on pairs $\langle\alpha,P\rangle$, $(\langle\alpha,c\rangle)$ of elements of A and predicate symbols (resp. pairs of elements of A and constants) such that

(i) $\beta \leq \alpha \Rightarrow D(\alpha) \subseteq D(\beta)$

(ii) $I(\alpha,P) \subseteq (D(\alpha))^k$ for k-ary P

$I(\alpha,c) \in D(\alpha)$

(iii) $\beta \leq \alpha \Rightarrow I(\alpha,P) \subseteq I(\beta,P)$

$\beta \leq \alpha \rightarrow I(\alpha,c) = I(\beta,c)$

We also allow 0-ary predicate symbols, we then define $I(\alpha,P) \in \{t,f\}$ and require $\beta \leq \alpha \Rightarrow (I(\alpha,P) = t \Rightarrow I(\beta,P) = t)$. Here t and f are two suitably chosen, distinct elements.

D(I) is called the domainfunction (interpretationfunction). The interpretation of a first order language by a Kripke model is inductively defined. We suppose that the language contains constants for all elements of $\bigcup_{\alpha\in A} D(\alpha)$. Let a be denoted by the constant $\bar{a}$. In the definition below only closed formulae are considered.

2.3.2. Definition.

(i) $\alpha \models P(\bar{a} \ldots \bar{a}_k)$ if $\langle a_1,\ldots,a_k\rangle \in I(\alpha,P)$

$\alpha \models P(\text{0-ary})$ if $I(\alpha,P) = t$

(ii) $\alpha \models A \wedge B$ if $\alpha \models A$ and $\alpha \models B$

(iii) $\alpha \models A \vee B$ if $\alpha \models A$ or $\alpha \models B$

(iv) $\alpha \models A \rightarrow B$ if for all $\beta \leq \alpha$ $\beta \models A \Rightarrow \beta \models B$

(v) for no α $\alpha \models \bot$

(vi) $\alpha \models \forall x\, Ax$ if for all $\beta \leqslant \alpha$ and for all $b \in D(\beta)$ $\beta \models A\bar{b}$

(vii) $\alpha \models \exists x\, Ax$ if there exists an $a \in D(\alpha)$ such that $\alpha \models A\bar{a}$

From the definition of $\neg A$ it follows that

(viii) $\alpha \models \neg A$ if for all $\beta \leqslant \alpha$ $\beta \not\models A$.

From the definition one easily proves

2.3.3. <u>Lemma</u>. $\beta \leqslant \alpha$ and $\alpha \models A \Rightarrow \beta \models A$.

An interpretation of propositional logic is obtained by considering 0-ary predicates only (and by defining $D(\alpha) = \{\emptyset\}$, or by 'forgetting' D).

2.3.4. <u>Definition</u>. $\alpha \models B$ holds for arbitrary B if $\alpha \models B^*$, for the universal closure B^* of B

$\mathcal{A} \models B$ if for all $\alpha \in A$ $\alpha \models B$

$\models B$ if for all $\mathcal{A}$ $\mathcal{A} \models B$

$\Gamma \models B$ if for all $\mathcal{A}$ and all $\alpha \in A$ $\alpha \models C$ for all $C \in \Gamma \Rightarrow \alpha \models A$

It has been shown by several authors that intuitionistic predicate logic is strongly complete for the Kripke semantics, [Kripke, 65], [Aczel, 68], [Thomason, 68].

2.3.5. <u>Theorem</u> (strong completeness): $\Gamma \vdash A \Leftrightarrow \Gamma \models A$ (for closed Γ and A).

The proof of the completeness theorem essentially uses non-intuitionistic means.
On the whole the Kripke semantics belongs rather to (classical) model theory than to intuitionism. In a number of instances, however, a closer analysis shows that results obtained by way of Kripke models are intuitionistically acceptable.

We will give some applications of Kripke models here.

1. $\nvdash (\neg\neg P \rightarrow P) \rightarrow (P \vee \neg P)$

Proof: consider a Kripke model $\mathcal{A}$ with $A = \{\alpha,\beta,\gamma\}$ and $\beta,\gamma < \alpha$, and neither $\beta \leqslant \gamma$ nor $\gamma \leqslant \beta$. Let $I(\alpha,P) = I(\beta,P) = f$, $I(\gamma,P) = t$. $\alpha \models \neg\neg P \rightarrow P$ holds since only $\gamma \models \neg\neg P$. But clearly $\alpha \not\models P \vee \neg P$.

2. $\nvdash \neg\neg P \vee \neg P$.

Proof: consider the same $\mathcal{A}$ as in 1.

$\alpha \models \neg\neg P \Leftrightarrow \bigwedge\beta \leqslant \alpha \ \bigvee\gamma \leqslant \beta \ \ (\gamma \models P)$ (we use $\bigwedge,\bigvee$ as meta-quantifiers). Because $\beta \not\models P$ we have $\alpha \not\models \neg\neg P$, so $\alpha \not\models \neg\neg P \vee \neg P$.

3. $\nvdash \neg(P \wedge Q) \rightarrow \neg P \vee \neg Q$.

Proof: consider $\mathcal{A}$ with $A = \{\alpha,\beta,\gamma\}$, as in 1.

$I(\alpha,P) = I(\alpha,Q) = I(\beta,Q) = I(\gamma,P) = f$,

$I(\beta,P) = I(\gamma,Q) = t$.

4. $\nvdash (P \rightarrow \exists x\, Qx) \rightarrow \exists x(P \rightarrow Qx)$.

Proof: Let P and Q be 0-ary and 1-ary predicate symbols. Consider $\mathcal{A}$ with $A = \{\alpha,\beta\}$, $\beta < \alpha$.

Let $D(\alpha) = \{0\}$, $D(\beta) = \{0,1\}$,

$I(\alpha,P) = f$, $I(\beta,P) = t$, $I(\alpha,Q) = \emptyset$, $I(\beta,Q) = \{1\}$.

Then $\alpha \models P \rightarrow \exists x\, Qx$, but $\alpha \not\models \exists x(P \rightarrow Qx)$ so $\alpha \not\models (P \rightarrow \exists x\, Qx) \rightarrow$ $\rightarrow \exists x(P \rightarrow Qx)$.

Note that intuitively speaking $P \rightarrow \exists x\, Qx$ is weaker than $\exists x(P \rightarrow Qx)$. In the former the information about the proof of P can be used to construct the desired x, in the latter we must construct x independent of the proof of P.

2.3.6. <u>Functionsymbols and terms</u> (cf [Osswald, 70]. So far we dealt with a first order language with predicate letters only. At no extra cost we can introduce function symbols as well. Let us assume that, unless stated otherwise, we are dealing with predicate logic with identity, i.e. we require

$$x = x$$

$$x = y \wedge x = z \rightarrow y = z$$

$$x = y \rightarrow (Px \rightarrow Py)$$

$$x = y \rightarrow tx = ty \quad \text{for any term } t$$

to hold.

To extend the Kripke semantics to a language with function symbols, we extend I to function symbols $I(\alpha,F) : D(\alpha))^k \rightarrow D(\alpha)$ if F is a k-ary function symbol, such that $\beta \leqslant \alpha \Rightarrow$ $I(\alpha,F)(a_1,\ldots,a_k) = b \Rightarrow I(\beta,F)(a_1,\ldots,a_k) = b$.

In order to satisfy the laws of predicate logic we have to require the domains $D(\alpha)$ to be closed under the interpretations of the function symbols.

5. If $I(\alpha,=)$ is the real identity relation on $D(\alpha)$ for each α then $\mathcal{A} \models \forall x \; \forall y \quad (x = y \vee x \neq y)$.

Proof: choose any α and $a,b \in D(\alpha)$. If $a \neq b$ then $\bigwedge\beta \leqslant \alpha \quad (\beta \not\models \bar{a} = \bar{b})$, so $\alpha \models \bar{a} \neq \bar{b}$, if $a = b$, then $\alpha \models \bar{a} = \bar{b}$, so $\alpha \models \bar{a} = \bar{b} \vee \bar{a} \neq \bar{b}$.

This shows that we cannot restrict ourselves to "normal" models (i.e. with the natural equality interpretation) as in classical model theory. But one can show that any model of $\forall xy \quad (x = y \vee x \neq y)$ can be contracted (by the usual process of identification) to an elementary equivalent normal model. It can be advantageous to escape this complication by modifying the definition of a Kripke model such that injecting the structure belonging to α into a structure belonging to β with $\beta \leqslant \alpha$ is replaced by homomorphically embedding. For applications of this notion of Kripke model see [Osswald A], [Osswald, B].

6. $\underset{\sim}{HA} \vdash \exists x\, Px \Rightarrow \underset{\sim}{HA} \vdash P\bar{n}$ for some n.

Proof: Suppose that for each n $\underset{\sim}{HA} \nvdash P\bar{n}$, then there are Kripke models $\mathcal{A}_n$ of $\underset{\sim}{HA}$ such that $\mathcal{A}_n \not\models P\bar{n}$. Now construct a new Kripke model $\mathcal{A}$ with partially ordered set $A = \bigcup_{n \in w} A'_n \cup \{\alpha_0\}$, where $\alpha_0 \notin \bigcup_{n \in w} A'_n$.

The A'_n are pairwise disjoint isomorphic copies of the A_n's. The partial order extends all partial orders on the A'_n's and α_0 is a maximal element.

The structure at α_0 is the (classical) standard model of arithmetic. Clearly $\mathcal{A}$ is a Kripke model of $\underset{\sim}{HA}$, but by the construction we have $\mathcal{A} \not\models \exists x\, Px$, contradiction. So $\underset{\sim}{HA} \models P\bar{n}$ must hold for at least one n, and hence $\underset{\sim}{HA} \vdash P\bar{n}$ for some n.

6a. Cor. $\underset{\sim}{HA} \vdash P \vee Q \Rightarrow \underset{\sim}{HA} \vdash P$ or $\underset{\sim}{HA} \vdash Q$.

(apply $P \vee Q \leftrightarrow \exists x[(x = 0 \to P) \wedge (x \neq 0 \to Q)]$).

As $\underset{\sim}{HA}$ has the disjunctive and the existential properties (i.e. 6 and 6a hold) we say that $\underset{\sim}{HA}$ is a <u>saturated theory</u>. [Aczel, 68] gave necessary and sufficient conditions for a theory to be saturated. Compare also [Prawitz, 65], [Kleene, 62], [Harrop, 56].

7. The non derivability of IP_0 from Heyting's arithmetic plus Markov's principle [Smorynski, 70].

IP_0 is the scheme $[\forall x(Ax \vee \neg Ax) \wedge (\forall x\, Ax \to \exists y\, By)] \to \exists y(\forall x\, Ax \to By)$.

(independence of premiss for the existential quantifier)

Markov's principle is the scheme M:

$$[\forall x(Ax \vee \neg Ax) \wedge \neg\neg \exists x\, Ax] \to \exists x\, Ax.$$

<u>Theorem</u>: $\underset{\sim}{HA} + M \nvdash IP_0$

Proof: (1) Let a family F of Kripke models of $\underset{\sim}{HA}$ + M be given. Construct a new Kripke model as follows: take the disjoint union of the Kripke models of $F \cup \{\mathfrak{N}\}$ (i.e. the family F plus the

Kripke model consisting of the standard model of the system $\underset{\sim}{P}$ of Peano arithmetic). Next add to the resulting partially ordered set a vertex α_0 consisting of the set of standard natural numbers..

To be precise: $A = \bigcup_{\mathcal{A}_i \in F} A_i \cup \{\alpha_1, \alpha_0\}$

for all β $\quad \beta \leqslant \alpha_0$

and $\beta \leqslant \gamma$ if $\beta, \gamma \in A_i$ and $\beta \leqslant_i \gamma$.

$\beta \models P$ if $\beta \models_i P$ if $\beta \neq \alpha_0, \alpha_1$

$\alpha_j \models P$ if $\mathfrak{N} \models P$ (classically), $j = 0,1$

for atomic P.

Claim : if each $\mathcal{A}_i$ of F is a model of $\underset{\sim}{HA}$ + M, so is $\mathcal{A}$.

Suppose $\alpha_0 \models \forall x(Ax \vee \neg Ax) \wedge \neg\neg\exists x\, Ax$, then in particular $\alpha_1 \models \neg\neg\exists x\, Ax$, so $\alpha_1 \models \exists x\, Ax$.

Hence $\alpha_1 \models An$ for some $n \in N$.

Now we conclude from $\alpha_0 \models An \vee \neg An$ that $\alpha_0 \models An$ holds. Therefore we have $\alpha_0 \models \exists x\, Ax$.

The case $\alpha_0 \not\models \forall x(Ax \vee \neg Ax) \wedge \neg\neg\exists x\, Ax$ is trivial.

(2) Let $\forall x\, Ax$ be the Gödel sentence that asserts its own unprovability, for $\underset{\sim}{P}$ (which is true in the standard model $\mathfrak{N}$) and let $\exists z\, Cz$ be the negation of the Gödel sentence for $\underset{\sim}{P} + \forall x\, Ax$. ($\mathfrak{N} \not\models \exists z\, Cz$).

Let $\mathfrak{N}_1$ be a model of $\forall x\, Ax \wedge \exists z\, Cz$ and let $\mathfrak{N}_2$ be a model of $\neg\forall x\, Ax$.

Consider the Kripke model with partially ordered set $\{\alpha_0, \alpha_1, \alpha_2, \alpha_3\}$ and $\alpha_0 \geqslant \alpha_i$ $(i = 0,1,2,3)$. To $\alpha_1, \alpha_2, \alpha_3$ we assign the structures $\mathfrak{N}_1, \mathfrak{N}_2, \mathfrak{N}_3$ and $D(\alpha_0) = N$. In α_0 the true atomic statements are valid. In $\alpha_1, \alpha_2, \alpha_3$ the classically true statements hold.

Define $By =: (y = 0 \wedge \exists z\, Cz) \vee (y = 1 \wedge \neg\exists z\, Cz)$.

Clearly $\alpha_0 \models \forall x\, Ax \rightarrow \exists y\, By$, but

not $\quad \alpha_0 \models \exists y(\forall x\, Ax \rightarrow By)$, as

$\alpha_1 \models B\bar{0}$ and $\alpha_3 \models B\bar{1}$.

Therefore IP_0 is not valid in the model.

By (1) the above model is a model of HA + M.

Hence HA + M $\nvdash IP_0$.

8. We will end this section with an elegant application of Kripke models, that was presented by C.Smorynski in [Smorynski, 70]. He proved quite straight forwardly the completeness of HA for intuitionistic propositional logic as laid down in the following

Theorem ([De Jongh, A]).

If the proposition A is not derivable in intuitionistic propositional logic, then there is an arithmetical instance of A which is not derivable in HA.

In order to prove the theorem we first derive some lemmas.

Lemma. For each n there are n disjoint pairwise effectively inseparable sets.

Proof: For $i,j \leq n$ define

$$x \in A_i := \exists y(\bigwedge_{j \neq i} \exists z \leq yT((x)_j,x,z) \wedge \neg \exists z \leq yT((x)_i,x,z))$$

(T is Kleene's predicate, [Kleene, 52], §57).

Remark: the exact nature of the coding of finite sequences is not relevant here. Take e.g. the coding from 3.2 and let $(x)_k$ be a suitably defined inverse.

Clearly $A_i \cap A_j = \emptyset$ for $i \neq j$.

Now suppose $A_i \subseteq W_e$, $A_j \subseteq W_f$, $W_e \cap W_f = \emptyset$. Let x be a sequence number such that lth $x = n$ $(x)_i = e$, $(x)_j = f$, $(x)_k = t$ such that $W_t = N$ for $k \neq i,j$.

Then $x \in W_e \Rightarrow \exists y\ T((x)_i,x,y)$ and because $W_e \cap W_f = \emptyset$ we have $\neg \exists y\ T((x)_j,x,y)$, finally $\exists y\ T((x)_k,x,y)$ for $k \neq i,j$.

Hence $\exists y(\exists z \leq y\ T((x)_i,x,z) \wedge \neg \exists z \leq y\ T((x)_j,x,z) \wedge$

$$\wedge \bigwedge_{k \neq i,j} \exists z \leq y\ T((x)_k,x,z)).$$

Therefore $x \in A_j$ and thus $x \in W_f$. Contradiction.
Conclusion: $x \notin W_e$, likewise $x \notin W_f$. Put $\Psi(e,f) = x$, then the effective inseparability follows.

Lemma. For each n there are sentences $S_0,\dots,S_{n-1}$ such that for each i $\underset{\sim}{P} + S_i \wedge \bigwedge_{j\neq i} \neg S_j$ is consistent.

Proof: Let T^* represent the predicate T, then $n \in A_i$ iff there exists an m such that

$$\vdash_{\underset{\sim}{P}} \bigwedge_{j\neq i} \exists z \leq \bar{m}\, T^*((\bar{n})_j,\bar{n},z) \wedge \neg \exists z \leq \bar{m}\, T^*((\bar{n})_i,\bar{n},z).$$

Call that formula $R_i(\bar{n},\bar{m})$, one easily checks that

$\vdash_{\underset{\sim}{P}} \forall x \; \neg(\exists y\, R_i(x,y) \wedge \exists y\, R_j(x,y))$ for $i \neq j$.

Now let $n \in B_0 := \vdash_{\underset{\sim}{P}} \exists y\, R_0(\bar{n},y)$

$n \in B_0' := \vdash_{\underset{\sim}{P}} \neg\exists y\, R_0(\bar{n},y)$,

then $A_0 \subseteq B_0$, $A_1 \subseteq B_0'$, B_0 and B_0' are r.e., so by the effective inseparability there is a number n such that $n \notin B_0 \cup B_0'$.

Let $T_0 = \exists y\, R_0(\bar{n},y)$, then $\underset{\sim}{P} + T_0$ and $\underset{\sim}{P} + \neg T_0$ are consistent.

Now let $T_0,\dots,T_k$ be defined, such that $\underset{\sim}{P} + \neg T_0 + \dots + \neg T_k$ is consistent, put

$$\Delta = \underset{\sim}{P} + \neg T_0 + \dots + \neg T_k$$

and

$$n \in B_{k+1} := \vdash_{\Delta} \exists y\, R_{k+1}(\bar{n},y)$$

$$n \in B_{k+1}' := \vdash_{\Delta} \neg \exists y\, R_{k+1}(\bar{n},y).$$

As before we find that $\underset{\sim}{P} + \neg T_0 + \dots + \neg T_k + T_{k+1}$ and $\underset{\sim}{P} + \neg T_0 + \dots + \neg T_k + \neg T_{k+1}$ are consistent.

By the inductive procedure we have found a sequence of formulae $T_0,\dots,T_{n-1}$ such that $\underset{\sim}{P} + T_0$, $\underset{\sim}{P} + \neg T_0 + T_1, \dots, \underset{\sim}{P} + \neg T_0 + \dots + \neg T_{n-2} + T_{n-1}$ are consistent.

Now let $S_k = T_k \wedge \bigwedge_{i<k} \neg T_i$.

We see that $S_i \rightarrow \neg S_j$ (for $i \neq j$). As $\underset{\sim}{P} + S_i$ is consistent, now also $\underset{\sim}{P} + S_i \wedge \bigwedge_{j\neq i} \neg S_j$ is consistent.

Note that the sentences S_i are Δ_2^0.

Next we need some facts about Kripke models.

Lemma. If the proposition A is not derivable in propositional logic, then there is a finite Kripke model $\mathcal{A}$ such that $\mathcal{A} \not\models A$.
Proof: see [Fitting, 69] , [Schütte, 68].

Note that $\mathcal{A}$ can effectively be determined from A. The property stated in the above lemma is called the 'finite model property'.

Lemma. For each Kripke model $\mathcal{A}$ there is a model $\mathcal{A}'$ such that each non terminal node in $\mathcal{A}'$ has at least two immediate successors and such that $\mathcal{A} \models A \Leftrightarrow \mathcal{A}' \models A$ for all A.
Proof: $\mathcal{A}$ is transformed into $\mathcal{A}'$ as follows:
if α has just one immediate successor β add a copy of the subtree dominated by β so that its top node is an immediate successor of α. Work your way from the terminal nodes upward to the top node.
One easily checks that the resulting model has the required properties.

Proof of De Jongh's theorem: Let $\nvdash$ A, determine $\mathcal{A}'$ as in the above lemma. Say $\mathcal{A}'$ has n terminal nodes. Consider classical models $\mathfrak{N}_0, \ldots, \mathfrak{N}_{n-1}$ for $\underset{\sim}{P} + S_i \wedge \bigwedge_{j \neq i} \neg S_j$ $(i=0,\ldots,n-1)$.

The definition of the (quantificational) Kripke model $\mathcal{A}^*$ proceeds as follows:
for all non terminal nodes α set $D(\alpha) = N$
and for the terminal nodes α_i set $D(\alpha_i) = N_i$
$(i=0,\ldots,n-1$, N_i is the universe of $\mathfrak{N}_i$).
Now for each atomic formula P and node α define $\alpha \models P$ if P holds in the corresponding classical model. One easily checks that $\mathcal{A}^* \models \underset{\sim}{HA}$.

Finally we have to find an instance of A that fails in $\mathcal{A}^*$.

Let A_i be an atom of A, consider $V_i = \{\beta | \beta \models A_i\}$. If $V_i = \emptyset$, then we substitute $0 = 1$ for A_i. If $V_i \neq \emptyset$, then $V_i = \{\gamma | \gamma \leqslant \beta_i\} \cup \ldots \cup \{\gamma | \gamma \leqslant \beta_{i_k}\}$, i.e. the β's are the uppermost nodes in which A_i holds. As A is a proposition we have only to take care of substitution instances at these β's.
For each of these β's let $U_\beta = \{\alpha_i | \alpha_i \leqslant \beta\}$ (the set of terminals dominated by β).
Define $A_\beta = \bigwedge_{i \notin U_\beta} \neg S_i$, then $\gamma \leqslant \beta \Leftrightarrow \gamma \models A_\beta$, unless β is the top node. In that case put $A_\beta := 0 = 0$.

(i) $\gamma \leqslant \beta$ then $\gamma \models \neg S_i$ for $i \notin U_\beta$, so $\gamma \models \bigwedge_{i \notin U_\beta} \neg S_i$;

(ii) conversely let $\gamma \models \bigwedge_{i \notin U_\beta} \neg S_i$ and suppose $\gamma \not\leqslant \beta$, then $\alpha_j \leqslant \gamma$ for some $j \notin U_{\beta_j}$ (because every non terminal has at least two immediate successors), so $\gamma \not\models \neg S_j$. Contradiction. Therefore $\gamma \leqslant \beta$.
The formula A_β therefore holds exactly on the subtree determined by β. Now put $A_i^* = A_{\beta_i} \vee \ldots \vee A_{\beta_{i_k}}$, then $\gamma \models A_i \Leftrightarrow \gamma \models A_i^*$.

If the formula A^* is obtained by substituting A_i^* for A_i in A, then $\mathcal{A}^* \models A^*$. Hence $\not\vdash_{\underset{\sim}{HA}} A^*$.
Note that A^* can be effectively determined.

The strongest possible form of De Jongh's theorem has been proved by Myhill, who showed that there exists a sequence R(0),R(1),R(2), ... of Σ_1^0 sentences, which is completely independent over Peano arithmetic. That is - every Boolean combination of R(i)'s is consistent with $\underset{\sim}{P}$. By the technique, used above, one then obtains for an unprovable proposition an unprovable arithmetical instance A* obtained from A by substituting disjunctions of Π^0_1 formulas for the atoms.

One cannot expect that Π^0_1 formulas will do the job. E.g. $\neg\neg A \rightarrow A$ does not have a counterexample in HA with a substituti of Π^0_1 formulas (remember that $\vdash \neg\neg \forall x\, P(x) \rightarrow \forall x\, \neg\neg P(x)$).

3. CHOICE SEQUENCES

3.1.

Among the objects that are created by the intuitionist are the "infinitely proceeding sequences whose terms are chosen more or less freely from mathematical entities previously acquired; in such a way that the freedom existing at the first choice may be irrevocably subjected, again and again to progressive restrictions at subsequent choices, while all these restricting interventions, as well as the choices themselves, may, at any stage, be made to depend on possible future mathematical experiences of the creating subject" ([Brouwer, 54], p. 2). These memorable objects, which played a role in the development of intuitionism ever since Brouwer's fundamental expositions [Brouwer, 18], [Brouwer, 24], are called choice sequences. Since the notion is so central in the non-discrete parts of intuitionistic mathematics we will give ample attention to it.

Here we will consider choice sequences of natural numbers although this by no means exhausts the possible notions.

3.2. Lawless sequences

Suppose one generates a sequence of natural numbers such that after the construction of any initial segment one is allowed to choose freely any natural number to extend the segment. Here we mean that no restriction on next choices is imposed at any time. The sequences obtained in this way are called lawless; they can be compared to those sequences one obtains by successive throws of a die. This notion does not allow "interdependence" of sequences, e.g. in the form of $\alpha x = 2\beta x$. Although at any time only initial segments of α and β are known, one is not free to extend these segments ad libitum.
Kreisel in [Kreisel, 68] analysed the notion of lawless sequence.

Further use of lawless sequences was made in [Troelstra, 69B, 70] and [van Dalen-Troelstra, 70].

We will use the following tools, familiar from recursion theory.

Pairing function $\{x,y\}$

with inverses j_1, j_2

codings for finite sequences

$$\nu_0(x_0) = x_0$$
$$\nu_1(x_0,x_1) = \{x_0,x_1\}$$
$$\nu_k(x_0\ldots x_k) = \{x_0,\nu_{k-1}(x_1\ldots x_k)\}$$

coding for all finite sequence (onto N)

$$\mathrm{lth}\langle x_0\ldots x_k\rangle = k+1 \qquad \langle x_0,\ldots,x_k\rangle = \{k,\nu_k(x_0\ldots x_k)\} + 1$$
$$\langle\rangle = 0$$

where $\langle\rangle$ denotes the empty sequence.

Instead of $\langle x\rangle$ we will also write $\hat{x}$.

concatenation function: $*$

$$\langle x_0,\ldots,x_k\rangle * \langle x_{k+1},\ldots,x_n\rangle = \langle x_0,\ldots,x_n\rangle$$

concatenation of a sequence and a function:

$$\langle x_0,\ldots,x_k\rangle * \chi = \chi' \text{ iff}$$
$$\forall i \leq k(\chi' i = x_i) \qquad (\forall i)(\chi'(k+i) = \chi i)$$

$\overline{\chi}x = \langle \chi 0,\ldots,\chi(x-1)\rangle$

$\chi \in n$ iff $\overline{\chi}(\mathrm{lth}\ n) = n$.

Now we list the axioms for lawless sequences. The variables ranging over lawless sequences are denoted by α, β, $\gamma,\ldots$. We stipulate that in formulas all lawless variables will be shown.

$$\text{LS1 } \forall x\ \exists\alpha\ (\alpha \in x)$$

LS1 states that any initial segment can be extended to a lawless sequence, which is quite evident on the intended interpretation.

$$\text{LS2 } \alpha \equiv \beta \vee \alpha \not\equiv \beta$$

Here the decidability of the intensional equality relation is postulated. The justification of this statement lies in the consideration that a mathematician knows whether he is creating two identical lawless sequences or not. Note that two lawless sequences α and β are intensionally equal if they are given by the same generation process (e.g. arbitrary choices or casts of a die). As we at any moment have only initial segments available either we know beforehand that α and β denote the same object or it is impossible that we should know that they denote the same object.

Note that LS1 is not evident for sequences of numbers produced by a die. Unless one cheats it is hard to imagine that one produces a prescribed finite sequence of throws. It is better to think of prolonging a given finite sequence by use of a die.

Let $\not\equiv(\alpha,\alpha_0,\ldots,\alpha_k)$ denote $\alpha \not\equiv \alpha_0 \wedge \ldots \wedge \alpha \not\equiv \alpha_k$.

$$\text{LS3} \quad \not\equiv(\alpha,\alpha_0,\ldots,\alpha_k) \wedge A(\alpha,\alpha_0,\ldots,\alpha_k) \rightarrow$$
$$\exists n(\alpha \in n \wedge \forall\beta \in n\,(\not\equiv(\beta,\alpha_0,\ldots,\alpha_n) \rightarrow A(\beta,\alpha_0,\ldots,\alpha_k))$$

A special case is

$$A(\alpha) \rightarrow \exists n(\alpha \in n \wedge \forall\beta \in n\, A(\beta)).$$

The validity of the special case is argued as follows:

if $A(\alpha)$ holds and α is the only lawless parameter then we must be able to conclude $A(\alpha)$ on the basis of our knowledge of just an initial segment of α, but then it is irrelevant how we continue that segment.

In the general case one reasons as follows:

suppose $\not\equiv(\alpha,\alpha_0,\ldots,\alpha_k)$ and $A(\alpha,\alpha_0,\ldots,\alpha_k)$. α is generated freely in the sense that no information on any of the α_i's plays a role in its generation. As α develops completely individually, we must be able to conclude the validity of $A(\alpha,\alpha_0,\ldots,\alpha_k)$ on the strength of a finite initial segment of α, say n. Clearly now for any β which develops freely, regardless of $\alpha_0,\ldots,\alpha_k$ and which has the same initial segment n $A(\alpha,\alpha_0,\ldots,\alpha_k)$ must hold.

The importance of the condition $\not\equiv(\alpha,\alpha_0,\ldots,\alpha_k)$ is illustrated by the following example.

Apply LS3 without condition to the formula $\alpha \equiv \gamma$

$$\alpha \equiv \gamma \rightarrow \exists n(\alpha \in n \wedge \forall \beta \in n(\beta \equiv \gamma)).$$

Now replace γ by α, then

$$\alpha \equiv \alpha \rightarrow \exists n(\alpha \in n \wedge \forall \beta \in n(\beta \equiv \alpha)),$$

or
$$\exists n(\alpha \in n \wedge \forall \beta \in n(\beta \equiv \alpha))$$

which clearly is contradictory.

LS3 clearly shows that it is not sufficient to suppose that a lawless sequence as an individual is "completely free", i.e. that at any time the only available information consists of an initial sequence. Consider e.g. the sequences α, β and $\gamma = \lambda x \cdot \langle \alpha x, \beta x \rangle$.
Clearly $\not\equiv(\gamma, \alpha, \beta)$, so by LS3 we would have
$\exists n(\gamma \in n \wedge \forall \delta \in n(\not\equiv(\delta, \alpha, \beta) \rightarrow \forall x(\delta x = \langle \alpha x, \beta x \rangle))$, which is contradictory. This contradiction is caused by our introduction of the illegal sequence γ. So the sequences we consider here are not only individually lawless, but also lawless with respect to each other.

Now we can show that for lawless sequences intensional and extensional equality coincide.
We show $\alpha \equiv \beta \leftrightarrow \forall x(\alpha x = \beta x)$.
From left to right is immediate. From right to left we reason by contradiction.
Suppose $\alpha \not\equiv \beta$ and $\forall x(\alpha x = \beta x)$, then by LS3
$\exists n(\alpha \in n \wedge \forall \gamma \in n(\gamma \not\equiv \beta \rightarrow \forall x(\gamma x = \beta x)).$
By LS1 there is a γ_1 such that $\gamma_1 \in n * \langle \beta(\text{lth } n) + 1 \rangle$.
As $\gamma_1 \not\equiv \beta$ and $\gamma_1 \in n$, we have $\forall x(\gamma_1 x = \beta x)$, which is contradictory.
Hence $\neg\, \alpha \not\equiv \beta$, so by LS2 $\alpha \equiv \beta$.
Thus we showed $\forall x(\alpha x = \beta x) \rightarrow \alpha \equiv \beta$
As usual we write $\alpha = \beta$ for $\forall x(\alpha x = \beta x)$. As a corollary we have the

somewhat surprising result $\alpha = \beta \vee \alpha \neq \beta$.

From LS3 we can derive the $\forall\exists$ (weak) continuity properties:

$$\text{WC-N: } \forall\alpha\,\exists x\, A(\alpha,x) \rightarrow \forall\alpha\,\exists y\,\exists x\,\forall\beta(\overline{\beta}y = \overline{\alpha}y \rightarrow A(\beta,x))$$

and

$$\text{WC-F: } \forall\alpha\,\exists a\, A(\alpha,a) \rightarrow \forall\alpha\,\exists y\,\exists a\,\forall\beta(\overline{\beta}y = \overline{\alpha}y \rightarrow A(\beta,a))$$

where a is lawlike variable.

Consider the following selection principle:
$\forall\alpha\,\exists x\, A(\alpha,x) \rightarrow \exists\Psi\,\forall\alpha\, A(\alpha,\Psi(\alpha))$, where Ψ is a functional of type $((0)0)0$. It turns out that Ψ is continuous. By making special assumptions about the nature of the continuous functionals we may adopt stronger continuity principles.

Some properties of lawless sequences:

1. $\forall\alpha\,\neg\neg\exists x\ (\alpha x = 0)$.

   Suppose $\neg\exists x\ (\alpha x = 0)$, then by LS3
   $\exists n(\alpha \in n \wedge \forall\beta \in n\ (\neg\exists x(\beta x = 0)))$, now choose $\beta \in n * \langle 0\rangle$ (by LS1) then we obtain a contradiction.
   So $\neg\neg\exists x\ (\alpha x = 0)$, and hence $\forall\alpha\,\neg\exists x\ (\alpha x = 0)$.

2. $\forall\alpha\,\neg\exists\beta\,\forall x\ (\beta x = \alpha x+1)$.

   Suppose $\beta x = \alpha x+1$, then $\alpha \neq \beta$, so by LS3
   $\exists n(\alpha \in n \wedge \forall\gamma \in n\ (\gamma \neq \beta \rightarrow \forall x\ (\beta x = \gamma x+1)))$.
   Now choose $\gamma \in n * \langle\beta(\text{lth } n)+1\rangle$, then $\gamma \neq \beta$ but not $\forall x(\beta x = \gamma x+1)$.
   From this contradiction we conclude $\forall\alpha\,\forall\beta\,\neg\forall x(\beta x = \alpha x+1)$ or
   $\forall\alpha\,\neg\exists\beta\,\forall x(\beta x = \alpha x+1)$.

3. $\neg\exists\alpha\ \exists a\ (\alpha = a)$ (a is lawlike).

   Suppose $\exists a\,\forall x(\alpha x = ax)$. Apply LS3 $\exists n(\alpha \in n \wedge \forall\beta \in n\,\forall x(\beta x = ax))$.
   Choose $\beta \in n * a\,(\text{lth } n)+1$; then we obtain a contradiction. Hence
   $\forall\alpha\,\neg\exists a\ (\alpha = a)$.

Number 2 shows us that lawless sequences cannot enjoy even the most innocent relations (e.g. they are not closed under very simple continuous relations); for this reason the notion of lawless sequence is called anti-social.

Number 3 shows that lawless sequences are provably non-lawlike. Therefore their direct usefulness is small (cf. 3.8.), they are useful however in combination with other notions. Troelstra used lawless sequences in a construction of a model for choice sequences, [Troelstra, 70].

3.3. Continuity and Brouwer operations

We will go somewhat deeper into the properties of the class of continuous functionals. As this section bears on a wide class of sequences and functions, we will not restrict ourselves to lawless sequences. Henceforth ξ, η, ζ will denote arbitrary (unspecified) sequences of type (0)0.

The class of representing functions of continuous functionals (or moduli of continuity) plays an important role in intuitionistic mathematics. This class can be introduced without direct reference to continuity by an inductive definition (following [Kreisel, 63, 68]). Intuitively one thinks of a continuous functional $\Phi: N^N \rightarrow N$ as an operator that determines the value for a certain (function-) argument on the basis of a finite initial segment of the argument. That is, one can replace Φ by a function ϕ that operates on finite sequences of natural numbers. Stated picturesquely: one feeds ϕ (think of a black box) the consecutive values of a function ξ, for a time the black box asks for more information (say it prints zeros) until it has at its disposal a sufficiently long segment of ξ after which it calculates and prints the answer (or rather its successor to distinguish a possible zero output from the earlier zeros). Because of the continuity the box then keeps printing the same number.

This leads us to the definition of a class of functions which are zero on a well-founded subtree of the universal tree (universal spread, Baire space) and which are constant on the remaining part.

Consider a lawlike function a defined on sequence numbers, let $a_x(n) = a(\hat{x} * n)$.
Now we define K as the class of functions that contains a with $a0 = 0$ if it contains all a_x for all x.

More precisely we introduce a one place predicate letter for function arguments with the properties

$$\text{K1} \quad a = \lambda n \cdot Sx \rightarrow Ka$$

$$\text{K2} \quad a0 = 0 \wedge \forall x\, K(\lambda n \cdot a(\hat{x} * n)) \rightarrow Ka$$

To complete the inductive definition we have to stipulate that K is "the least class such that ..." .
Define $A_K(Q,a) := \exists y(a = \lambda x \cdot Sy) \vee (a0 = 0 \wedge \forall x\, Q(\lambda n \cdot a(\hat{x} * n))$.
Then the closure conditions expressed by K1 and K2 may be summarized for the class Q by

$$A_K(Q,a) \rightarrow Qa.$$

The minimality of K is then expressed by

$$\text{K3} \quad \forall a(A_K(Q,a) \rightarrow Qa) \rightarrow \forall a(Ka \rightarrow Qa).$$

For convenience we will denote the elements of K by e, f with or without subscripts.

If one wants to show that a lawlike function a belongs to K then one starts to check whether a is constant, if not, one checks all the functions $\lambda n \cdot a(\hat{x} * n)$ $(x = 0,1,2,\ldots)$, for those which are not constant one checks all the functions $\lambda n \cdot a(\langle x,y \rangle * n)$ etc. In this way one obtains a tree, which is well-founded if $a \in K$. The end points of the (proof-) tree correspond to conclusions $\lambda n \cdot a(m * n)$ is constant. The other nodes correspond to inferences of the type

$$\forall x(\lambda n\cdot a(\hat{x} * m * n) \in K) \rightarrow \lambda n\cdot a(m * n) \in K$$

and the top node represents the conclusion $a \in K$. One can say that a function a in K thus carries (in its definition) the natural proof that it belongs to K.

Compare this to the situation with the natural numbers where a natural number (in its definition or construction) carries the proof that it is a natural number.

The elements of K can be used to represent continuous functionals, this is a consequence of the following

3.3.1.

Theorem (i) $\forall e\, \forall \xi\, \exists x\, (e\bar{\xi}x \neq 0)$

(ii) $\forall e\, \forall m\, \forall n\, (em \neq 0 \rightarrow e(m * n) = em)$

Proof: by induction on K.

3.3.2.

Definition: A functional $\Phi: N^N \rightarrow N$ is continuous ($\Phi \in$ CONT) if $\forall \xi\, \exists x\, \forall \eta[\bar{\xi}x = \bar{\eta}x \rightarrow \Phi\xi = \Phi\eta]$.

The following lemma allows us to associate to each element e of K a functional Φ_e.

3.3.3.

Lemma. Let $R(\xi,x) := \exists y(e\bar{\xi}\, y = x+1)$, then $\forall \xi\, \exists! x\, R(\xi,x)$.

Proof: apply theorem 3.3.1.

3.3.4.

Definition. $\Phi_e(\xi) = x := \exists y(e\bar{\xi}y = x+1)$.

Notation: when convenient we will write $e(\xi)$ for $\Phi_e(\xi)$. The class of functionals Φ_e is denoted by K^*.

The inductive definition of K leads us directly to the following:

3.3.5.

Lemma. (i) $\lambda\xi\cdot x \in K^*$

(ii) let for e of type (00)0 e_x be defined by $e_x = \lambda\xi\cdot e(\hat{x} * \xi)$, then $\forall x(e_x \in K^*) \rightarrow e \in K^*$

(iii) K^* is the least class satisfying (i) and (ii).

One now easily shows

3.3.6.

Theorem. $K^* \subseteq$ CONT.

By classical means one can show K^* = CONT as follows: suppose for $\Phi \in$ CONT we have $\Phi \notin K^*$. By $\Phi \in$ CONT we have $\forall x(\Phi_{\hat{x}} \in \text{CONT})$, where $\Phi_n = \lambda\xi\cdot\Phi(n * \xi)$.

By 3.3.5. there is an x_0 such that $\Phi_{\hat{x}_0} \notin K^*$. Similarly we find for each n with $\Phi_n \notin K^*$ an x such that $\Phi_{n*\hat{x}} \notin K^*$, so an application of the axiom of dependent choices provides us with a ξ such that

$\forall x(\Phi_{\bar{\xi}x} \notin K^*)$. (†)

By $\Phi \in$ CONT we have $\exists x\, \forall\eta \in \bar{\xi}x(\Phi\xi = \Phi\eta)$, in particular for such an x $\Phi_{\bar{\xi}x}$ is constant and hence by 3.3.5. in K^*, contradicting (†). Therefore $\Phi \in K^*$.

We have seen that K enables us to define a sensible subclass of CONT, likewise we can use K to represent continuous functionals of type ((0)0)(0)0. The representation is brought about by considering initial segments of the argument and the values of the image.

3.3.7.

Lemma. Let $R(\xi,\eta) := [\eta x = y \leftrightarrow \exists z\, (e(\hat{x} * \bar{\xi}z) = y+1)]$, then $\forall\xi\, \exists!\eta\, R(\xi,\eta)$.

Proof: induction on K.

3.3.8.

Definition. $\Psi_e\xi = \eta := [\eta x = y \leftrightarrow \exists z\, (e(\hat{x} * \bar{\xi}z) = y+1)]$.
(cf. Kleene's functionrealizability [Kleene-Vesley, 65], § 8).

Notation: when convenient we write $e|\xi$ for $\Psi_e\xi$. The class of all functionals Ψ_e is denoted by K**.

3.3.9.

Definition. A functional $\Psi: N^N \to N^N$ is continuous ($\Psi \in$ CONT') if

$$\forall\xi\, \forall x\, \exists y\, \forall\eta \in \bar{\xi}y\, [\Psi\eta \in \overline{\Psi\xi}x].$$

3.3.10.

Theorem. $K^{**} \subseteq$ CONT'.

Proof: Consider Ψ_e and choose ξ and x. Let $\Psi_e\xi = \zeta$, $e(\hat{0} * \bar{\xi}k_0) =$
$= \zeta_0+1, \ldots, e(\langle x-1\rangle * \bar{\xi}k_{x-1}) = \zeta(x-1)+1$.
Put $k = \max\{k_0, \ldots, k_{x-1}\}$.
Then for $\eta \in \bar{\xi}k$ $\quad e(\hat{z} * \bar{\eta}k) = e(\hat{z} * \bar{\xi}k) = \zeta z+1 \quad (0 \leqslant z < x)$.
Hence for all $\eta \in \bar{\xi}k$ $\quad \Psi_e\eta \in \overline{\Psi\xi}x$.

The elements of K are called Brouwer operations after Brouwer who used essentially these operations in establishing that extensional functionals on choice sequences have continuity moduli in K. ([Brouwer, 24], also see [Van Heyenoort, 67], p. 447 and Parson's excellent introduction). Actually Brouwer used well-ordered species (i.e. intuitionistic well-orderings), but these can be defined in terms of K and vice versa, see [Troelstra, 69A].

Clearly the elements of K define well-founded subtrees of the universal tree (and hence of any tree), therefore it is not surprising that the following induction principle holds:

3.3.11.

$\forall n(en \neq 0 \rightarrow Qn) \wedge \forall n\, \forall y\, (Q(\hat{y} * n) \rightarrow Q(n)) \rightarrow Q0$

(induction over unsecured sequences).

Proof: by induction on K.

A species B of sequents is called a <u>bar</u> if $\forall \xi\, \exists x\, (\bar{\xi}x \in B)$. From Theorem 3.3.1. it appears that each element of K represents a bar.

3.4. Lawless sequences and barcontinuity

We have already seen that the class K represents continuous functionals (3.3.6.), in the sense that each element of K is a modulus of continuity. The next step is to postulate that K represents <u>all</u> continuous functions. We will do this via a strengthening of the weak continuity property WC-N. The Brouwer operations allow us to make explicit the class of functionals, involved in $\forall\alpha\, \exists x$-principles.

Let $\#(\alpha_0,\ldots,\alpha_p)$ denote $\bigwedge\limits_{i\neq j} \alpha_i \neq \alpha_j$.

LS4 $\quad \forall\alpha_0\ldots\forall\alpha_p[\#(\alpha_0,\ldots,\alpha_p) \rightarrow \exists x\, A(\alpha_0,\ldots,\alpha_p,x)] \rightarrow$
$\quad\quad \rightarrow \exists e\, \forall\alpha_0\ldots\forall\alpha_p[\#(\alpha_0,\ldots,\alpha_p) \rightarrow A(\alpha_0,\ldots,\alpha_p, e(\langle\alpha_0,\ldots,\alpha_p\rangle))]$,

where $\langle\alpha_0,\ldots,\alpha_p\rangle = \lambda x \cdot \langle\alpha_0 x,\ldots,\alpha_p x\rangle$.

In a special case we have

LS'4 $\quad \forall\alpha\, \exists x\, A(\alpha,x) \rightarrow \exists e\, \forall\alpha\, A(\alpha,e(\alpha))$.

LS4 is called the principle of barcontinuity. Applying LS4 to the formula $\Phi(\xi) = x$ we find $\exists e\, \forall\xi\, [\Phi(\xi) = e(\xi)]$, that is, Φ has a modulus of continuity in K.

Some more properties of lawless sequences:

3.4.2. $\neg \forall\alpha \exists x (\alpha x = 0)$.

Proof. Suppose $\forall\alpha \exists x (\alpha x = 0)$, then by LS4 $\exists e \forall\alpha (\alpha(e(\alpha)) = 0)$, or by definition of $e(\alpha)$

$$\forall n(en \neq 0 \rightarrow \forall\alpha \in n(\alpha(en \dot{-} 1) = 0)) \text{ for some } e.$$

Now consider $a = \lambda x \cdot 1$, by Theorem 3.3.1., which holds for arbitrary sequences so in particular for a, we have

$$\exists x \; e\bar{a}x \neq 0.$$

Let $\bar{a}x = n$. Choose m such that $a \in m$, $\exists n'(m = n * n')$ and lth $m > en \dot{-} 1$, then $\forall\alpha \in m(\alpha(en \dot{-} 1) = 0)$. But by the choice of m $\alpha(en \dot{-} 1) = 1$. Contradiction, hence

$$\neg \forall\alpha \exists x (\alpha x = 0).$$

3.4.3.

$$\neg \forall\alpha (\exists x(\alpha x = 0) \vee \neg \exists x(\alpha x = 0)).$$

3.4.4. $\neg \forall\alpha (\neg\neg\exists x(\alpha x = 0) \rightarrow \exists x(\alpha x = 0))$.

In the system for lawless sequences that includes LS4 we can prove the so-called Bar Theorem:

3.4.5.

$$\forall\alpha \exists x A(\bar{\alpha}x) \wedge \forall n(An \rightarrow Bn) \wedge \forall n \forall m(An \rightarrow An * m) \wedge \forall n(\forall x B(n * \hat{x}) \rightarrow Bn) \rightarrow \\ \rightarrow \forall n \, Bn.$$

Remark: the conclusion may be changed into B0.

The bar theorem can be viewed as a transfinite induction principle for well-founded trees. A determines a well-founded tree, such that at the end points A (and hence B) holds. The clause $\forall n(\forall x \, Bn * \hat{x} \rightarrow Bn)$ carries the property B upwards in the tree. For an extensive exposition see [Kleene-Vesley, 65], 6.5.

For a proof of the bar theorem from LS4 see [Troelstra, 69A], 9.8.

For reasons explained above the schema 3.4.5. is also called "the principle of bar induction". We will denote it by BI. It is noteworthy that, when K is defined by

$Ka := \forall\alpha\, \exists x\, (a\bar{\alpha}x \neq 0) \wedge \forall n\, \forall m\, (an \neq 0 \rightarrow a(n * m) = an)$, the axioms K1, K2, K3 are provable from the bar theorem.

3.4.6.

Theorem. $BI \Rightarrow Ka \leftrightarrow [\forall\alpha\, \exists x\, (a\bar{\alpha}x \neq 0) \wedge \forall n\, \forall m\, (an \neq 0 \rightarrow a(n * m) = an)]$

Proof: Let K_0a be the right hand side of the equivalence. The properties K1 and K2 are evident, for K3 assume we have $\forall a\, (A_K(P,a) \rightarrow Pa)$ and K_0a. We will show Pa.

Define $Cn := an \neq 0$

$Dn := P(\lambda m \cdot a(n * m))$

and apply the bar theorem.

Hyp 1 $\forall\alpha\, \exists x\, C(\bar{\alpha}x)$

$\forall\alpha\, \exists x\, C(\bar{\alpha}x) \leftrightarrow \forall\alpha\, \exists x\, (a\bar{\alpha}x \neq 0)$

the right hand side holds by K_0a.

Hyp 2 $\forall n\, \forall m\, (Cn \rightarrow C(n * m))$

this also holds by K_0a.

Hyp 3 $\forall n\, (Cn \rightarrow Dn)$

$\forall n\, (an \neq 0 \rightarrow P(\lambda m \cdot a(n * m)))$

The function $\lambda m \cdot a(n * m)$ has the property $a(n * \langle\rangle) \neq 0$, so it belongs to P by the closure properties.

Hyp 4 $\forall n\, (\forall x\, D(n * \hat{x}) \rightarrow Dn)$

$\forall n\, (\forall x\, P(\lambda m \cdot a(n * \hat{x} * m)) \rightarrow P(\lambda m \cdot a(n * m))$

Consider $b = \lambda m \cdot a(n * m)$.

If $b0 \neq 0$, then b is a constant positive function so Pb holds. If, on the other hand, $b0 = 0$, then $\forall x\ (P(\lambda m \cdot b(\hat{x} * m)) \rightarrow Pb)$ holds because of the closure properties. Therefore Hyp 4 holds.

Now BI allows us to conclude $D(\langle\rangle)$, i.e. $P(a)$. Hence we have shown $Ka \leftrightarrow K_0 a$.

Note that the fact that α ranged over lawless sequences was not used in the proof (cf. [Kreisel-Troelstra, 70], 5.6.2.).

BI can also be used to show that each continuous functional has a modulus of continuity in K.

The following theorem, although formulated for lawless sequences almost immediately extends to a wider class of choice sequences.

3.4.7.

Theorem. WC-N + AC-NF + BI $\Rightarrow$ LS'4.

Proof: We work in a theory of sequences with the usual properties (e.g. ELC of [Kreisel-Troelstra, 70]).

WC-N reads $\forall\alpha\ \exists x\ A(\alpha,x) \rightarrow \forall\alpha\ \exists y\ \ \exists x\ \forall\beta\ (\bar{\alpha}y = \bar{\beta}y \Rightarrow A(\beta,x))$

AC-NF reads $\forall x\ \exists a\ A(x,a) \rightarrow \exists b\ \forall x\ A(x,(b)_x)$, where $(b)_x = \lambda y \cdot b\{x,y\}$.

a, b, c stand for lawlike functions.

We will apply BI to the predicates

$P(n) := \exists x\ \forall\alpha \in n\ R(\alpha,x)$ and

$Q(n) := \exists a\ (K_0 a \wedge \forall\alpha \in n\ \forall m(\alpha \in m \wedge am \neq 0 \rightarrow R(\alpha, am \dot{-} 1)))$

to obtain LS'4 for the formula $R(\alpha,x)$, for which $\forall\alpha\ \exists x\ R(\alpha,x)$ is given.

We will first check the hypotheses.

Hyp 1 $\forall\alpha\ \exists x\ P(\bar{\alpha}x)$

or $\forall\alpha\ \exists x\ \ \exists y\ \forall\beta \in \bar{\alpha}x\ R(\beta,y)$. This is an instance of WC-N.

Hyp 2 $P(n) \rightarrow P(n * m)$

or $\exists x\ \forall \alpha \in n\ R(\alpha,x) \rightarrow \exists x\ \forall \alpha \in n * m\ R(\alpha,x)$.

This is evident because $\alpha \in n * m \rightarrow \alpha \in n$.

Hyp 3 $P(n) \rightarrow Q(n)$

or $\exists x\ \forall \alpha \in n\ R(\alpha,x) \rightarrow \exists a\ [K_0 a \wedge \forall \alpha \in n\ \forall m(\alpha \in m \wedge am \neq 0 \rightarrow$

$\rightarrow R(\alpha, am \dot{-} 1))]$.

Let for a choosen n an x be given such that $\forall \alpha \in n\ R(\alpha,x)$ holds. Define $am = x+1$ for all n, then a satisfies $K_0 a \wedge \forall \alpha \in n\ \forall m\ (\alpha \in m \wedge am \neq 0 \rightarrow R(\alpha, am \dot{-} 1))$.

Hyp 4 $\forall u\ Q(n * \hat{u}) \rightarrow Q(n)$.

Let $\forall n\ \exists a\ [K_0 a \wedge \forall \alpha \in n * \hat{u}\ \forall m(\alpha \in m \wedge am \neq 0 \rightarrow R(\alpha, am \dot{-} 1))]$ hold.

Apply AC-NF:

$\exists b\ \forall n [K_0 (b)_x \wedge \forall \alpha \in n * \hat{u}\ \forall m(\alpha \in m \wedge (b)_x m \neq 0 \rightarrow R(\alpha, (b)_x m \dot{-} 1))]$.

From b we define c primitive recursively:

$$cm = \begin{cases} 0 \text{ if } \mathrm{lth}(m) \leq \mathrm{lth}\ n \\ (b)_{\mathrm{lth}(n)}(n) \quad \text{otherwise} \end{cases}$$

Now we have

$$K_0 c \wedge \forall \alpha \in n\ \forall m(\alpha \in m \wedge cm \neq 0 \rightarrow R(\alpha, cm \dot{-} 1)),$$

that is Hyp 4 holds.

Applying BI we find that Q0 holds, so

$$\exists a [K_0 a \wedge \forall \alpha\ \forall m(\alpha \in m \wedge am \neq 0 \rightarrow R(\alpha, am \dot{-} 1))].$$

As we have shown K and K_0 can be identified under the assumption of BI. Furthermore from the definition of $a(\alpha)$ it follows that $\forall \alpha\ R(\alpha, a(\alpha))$ holds for the particular $a \in K$.

Recalling our assumption $\forall \alpha\ \exists x\ R(\alpha,x)$, we now conclude

$$\forall \alpha\ \exists x\ R(\alpha,x) \rightarrow \exists e \in K\ \forall \alpha\ R(\alpha, e(\alpha)).$$

3.5. Choice sequences

Between lawless sequences and lawlike sequences many intermediate notions of choice sequence can be found. We will give here an informal exposition of the general notion. For further reading one is referred to [Troelstra, 69A], [Kleene-Vesley, 65], [Kreisel-Troelstra, 70], [Troelstra, 69]. We will restrict ourselves to choice sequences with numerical values, a large part of what will be said can quite straightforwardly be generalized to other kinds of choice sequences.

The idea behind choice sequences is that one successively chooses natural numbers and at the same time restricts future choices. So the process can be described by a sequence of pairs $\langle x,R\rangle$ of natural numbers and conditions. As the conditions restrict the future numerical values, we suppose them to be extensional, i.e. $\xi = \eta \rightarrow (R\xi \rightarrow R\eta)$ (for $\xi,\eta \in (N)N$). We consider a class $\underset{\sim}{R}$ of conditions, to be used in the successive choices.

The conditions may become more and more restrictive each time (cf. [Brouwer, 54], p. 2), we will express that by a relation $\sqsubset$ between conditions (which need not be extensional).

Notations: $\xi \in (N)(N \times \underset{\sim}{R})$

let $\xi = \langle\langle x_n,R_n\rangle\rangle_n$, then

$\pi_0\xi := \langle x_n\rangle_n$

$\pi_1\xi := \langle R_n\rangle_n$

Property of choice sequences:

$\forall x\ (\pi_1\xi(x+1) \sqsubset \pi_1\xi x)$

(A) $\forall x\ (\pi_1\xi x(\pi_0\xi))$

$\forall x\ (\pi_1\xi x \in \underset{\sim}{R}_x)$, where $\underset{\sim}{R} = \bigcup \underset{\sim}{R}_x$, i.e. $\underset{\sim}{R}$ has some structure with respect to the possible conditions at different moments. Each of the classes $\underset{\sim}{R}_x$ is supposed to be inhabited (i.e. $\forall x\ \exists R(R \in \underset{\sim}{R}_x)$. It is fundamental to know what choices can actually be made, so we add the following condition of "local verifiability"

(B) to every $R \in \underset{\sim}{R}$ there exists an R^* such that

$\forall R \in \underset{\sim}{R}\ \forall x(R^*x \vee \neg R^*x)$ and

$\forall R \in \underset{\sim}{R}\ \forall \xi \in (N)N\ \ (R\xi \leftrightarrow \forall x\ R^*\overline{\xi}x)$.

Moreover the associated R^*'s have to satisfy a monotonicity condition:

(C) $R \in \underset{\sim}{R}_x \wedge R' \in \underset{\sim}{R}_{x+1} \wedge R' \sqsubset R \rightarrow \forall a\ \forall y(y > x+1 \rightarrow (R'^*\overline{a}y \rightarrow R^*\overline{a}y))$.

In some cases even a stronger condition is satisfied:

$$R' \sqsubset R \rightarrow \forall x(R'^*x \rightarrow R^*x).$$

The following conditions exclude some trivial cases:

(D) $\forall R \in \underset{\sim}{R}\ (R^*0)$, $\forall R \in \underset{\sim}{R}_0\ \exists x\ R^*\hat{x}$.

(E) $\forall a\ \forall x\ \forall R \in \underset{\sim}{R}_x\ [\forall y \leqslant x+1\ (R^*\overline{a}y) \rightarrow$

$\rightarrow \exists R' \in \underset{\sim}{R}_{x+1}\ \exists z(R' \sqsubset R \wedge \forall y \leqslant x+1\ R'^*\overline{a}y \wedge R'^*(\overline{a}(x+1) * \hat{z}))]$.

Condition E guarantees the possibility of extending a given initial segment.

The process of the creation of a choice sequence can now be described as follows:

step 0 - choose $R_0 \in \underset{\sim}{R}_0$ and choose x_0 such that $R_0 x_0$ (by D)

step 1 - choose $R_1 \in \underset{\sim}{R}_1$ such that $R_1 \sqsubset R_0$ and choose x_1 such that $R_1\hat{x}_0$, $R_1\langle x_0, x_1\rangle$ (by E)

step k+1 - choose $R_{k+1} \in \underset{\sim}{R}_{k+1}$ such that $R_{k+1} \sqsubset R_k$ and choose x_{k+1} such that $R_{k+1}\hat{x}_0, \ldots, R_{k+1}\langle x_0, \ldots, x_{k+1}\rangle$.

One easily checks that the notions of lawless and lawlike sequences fit into the framework presented above.
Also Myhill's elucidation of Brouwer's notion of choice sequence ruled by a spreadlaw can be described by the above systematization (see [van Dalen-Troelstra, 70], p. 176, [Troelstra, 69], p. 42).

The notions of choice sequence considered here are supposed to be anti-social, in the sense that the conditions of R are supposed not to contain choice parameters.

Another type of choice sequence was introduced by Troelstra in connection with the formal system CS (see [Troelstra, 69A], [Troelstra, 68], [Troelstra, 69]). He considered choice sequences generated by continuous operations.

3.6. Intensional and extensional continuity

When dealing with a certain notion of choice sequence, some finite sequences can actually serve as initial segments of choice sequences, we call these admissable, to be precise $\sigma = \langle\langle x_0, R_0\rangle, \ldots, \langle x_n, R_n\rangle\rangle$ is admissable if $\forall m < n(R_{m+1} \sqsubset R_m) \wedge \forall v \forall w (w \leqslant v \leqslant n \rightarrow R^*_v\langle x_0, \ldots, x_w\rangle)$.
Notation: $\sigma \in A_{adm}$.
Put $\langle\rangle \in A_{adm}$.

The following principle of intensional continuity can be justified on the same grounds as LS3 for lawless sequences:

$$A\xi \rightarrow \exists\sigma \in A_{adm} [\xi \in \sigma \wedge \forall\eta \in \sigma A\eta].$$

As a matter of fact one considers the sequence ξ as being lawless, although the general conditions must be obeyed. The principle is also justified for Troelstra's notion of continuous generated sequences.

The principle of extensional continuity is what we called WC-N:

$$\forall\xi \exists x A(\xi,x) \rightarrow \forall\xi \exists y \exists x \forall\eta(\overline{\pi_0\xi}y = \overline{\pi_0\eta}y \rightarrow A(\eta,x)).$$

For various notions of choice sequences this principle has been justified on the basis of the principle of intensional continuity. Inspired by Myhill [Myhill, 67] Troelstra introduced an abstraction operator for the purpose of getting rid of (abstracting from) the conditionpart of the choice sequences.

However, especially in the case of the "social" notions, for a "definite" treatment a still more formal and detailed analysis (which doubtless will involve a good deal of technical work) is wanting, so for the time being we will accept WC-N as an intuitionistic plausible principle.

Another approach is to be found in [Troelstra, 69B, 70] where first a "model" of the theory of lawless sequences is built on the following procedure: consider a single lawless sequence α and construct for each finite sequence n a sequence α^n as follows:

$$\alpha^n x = \begin{cases} n_x & \text{if } x < \text{lth } n \\ \alpha\{n, \text{lth } n - x\} & \text{for } x > \text{lth } n \end{cases}$$

The species $\{\alpha^n \mid n \in N\}$ turns out to satisfy the axioms of the theory of lawless sequences. Secondly one considers a model of the theory of lawless sequences and applies continuous operations to n-tuples of lawless sequences.

The resulting notion of sequence satisfies

$$\forall x \, \exists y \, A(x,y,\xi) \rightarrow \exists \eta \, \forall x \, A(x,\eta x,\xi)$$
$$\forall \xi \, \exists! x A(\xi,x) \rightarrow \exists e \, \forall \xi \, A(\xi,e(\xi))$$
$$\forall \xi \, \exists x \, A(\bar{\xi}x) \rightarrow \exists e \, \forall \xi \, A(\bar{\xi}(e(\xi))).$$

The weaker WC-N! is thus also satisfied. A combination of the two results provides a model derived from a single lawless sequence in which a number of important principles are valid.

3.7. More on bar induction

Various forms of BI have been considered in the literature. The form we presented is BI_M, i.e. BI with the monotonicity condition: $An \rightarrow An * \hat{x}$. Another version is BI_D, i.e. BI with monotonicity replaced by decidability: $\forall n(An \vee \neg An)$.

3.7.1.

Theorem. $BI_M \Rightarrow BI_D$

Proof: Let A and B satisfy the hypotheses of BI_D, we define A' and B' such that BI_M can be applied.

$$A'n := \exists m_1 \neq 0\, \exists m_2 \neq 0\, (n = m_1 * m_2 \wedge Am\,)$$
$$B'n := A'n \vee Bn$$

Hyp 1, Hyp 2, Hyp 3 are trivially satisfied. We check Hyp 4. Assume $\forall x\, B'(n * \hat{x})$, i.e. $\forall x(A'(n * \hat{x}) \vee B(n * \hat{x}))$. Note that A'n is also decidable, so either A'n or $\neg A'n$. In the latter case we have An and hence Bn, or $\neg An$ and hence $\neg A'(n * \hat{x})$ for all x, so $B(n * \hat{x})$ for all x. By Hyp 4 for BI_D this implies Bn. In all cases we have B'n. Apply BI_M: B'0 or A'0 v B0. As A'0 is impossible, we have B0.

One might wonder whether BI_M could be strengthened by omitting the monotonicity conditions. The following example of Kleene [Kleene-Vesley, 65], p. 87, shows that the resulting principle is inconsistent with WC-N.

$$\text{Let } An := (\text{lth } n = 0 \wedge \neg\forall x(\delta x = 0)) \vee (\text{lth } n = 1 \wedge \delta n_0 = 0)$$
$$Bn := An \vee \forall y\, A(n * \hat{y}).$$

Hyp 1 $\forall \alpha\, \exists x\, A\bar{\alpha}x$.
Determine $\alpha 0$, if $\delta(\alpha 0) = 0$ then $A\bar{\alpha}1$ holds, if $\delta(\alpha 0) \neq 0$ then $A\bar{\alpha}0$ holds, so there exists an x such that $A\bar{\alpha}x$ holds.

Hyp 3 is evident.

Hyp 4, suppose $\forall x\, (A(n * \hat{x}) \vee \forall y\, A(n * \hat{x} * \hat{y}))$.
As $\text{lth}(n * \hat{x} * \hat{y}) \geq 2$, we have $\forall x\, A(n * \hat{x})$, hence Bn holds.

Now apply the strengthened BI: B0, or A0 v $\forall y\, A\hat{y}$, this is equivalent to $\neg\forall x(\delta x = 0) \vee \forall x(\delta x = 0)$, which in turn is equivalent to

$\exists y[(y = 0 \wedge \forall x(\delta x = 0)) \vee (y = 1 \wedge \neg \forall x(\delta x = 0))]$, generalization gives

$$\forall \delta \exists y[(y = 0 \wedge \forall x(\delta x = 0)) \vee (y = 1 \wedge \neg \forall x(\delta x = 0))].$$

Apply WC-N

$$\forall \delta \exists z \exists y \forall \delta'(\bar{\delta}z = \bar{\delta}'z \rightarrow [(y = 0 \wedge \forall x(\delta x = 0)) \vee (y = 1 \wedge \neg \forall x(\delta x = 0))]).$$

Let $\delta = \lambda x.0$ and determine the z_0 and y_0 such that

$$\forall \delta'(\delta' \in \bar{\delta}z_0 \rightarrow [(y_0 = 0 \wedge \forall x(\delta' x = 0)) \vee (y_0 = 1 \wedge \neg \forall x(\delta' x = 0))]).$$

If $y_0 = 0$ choose $\delta' \in \bar{\delta}z_0 \langle 1 \rangle$,
if $y_0 = 1$ choose $\delta' = \delta$. In both cases a contradiction is obtained.

Kreisel and Howard in [Howard-Kreisel, 66] studied relations between BI and other principles of intuitionistic analysis, among which the principle of transfinite induction. We will consider one case here. A partial ordering $\prec$ is <u>strongly well-founded</u> if $\forall \alpha \exists x \neg \alpha(x+1) \prec \alpha x$. (SWF($\prec$)).
$\prec$ is <u>weakly well-founded</u> if $\forall \alpha \forall x \neg \forall y \leq x(\alpha(y+1) \prec \alpha y)$. (WWF($\prec$)).
The induction property for a predicate A w.r.t. $\prec$ is $I(\prec,A) := \forall x[\forall y \prec x\, Ay \rightarrow Ax] \rightarrow \forall x\, Ax$. Now we formulate the principle of transfinite induction

$$\text{TI} \quad \text{WF}(\prec) \rightarrow \text{TI}\ (\prec,A)$$

where WF is any of the predicates SWF or WWF. In [Howard-Kreisel, 66], § 5 it is shown that the principles of transfinite induction thus obtained are equivalent.

3.7.2.
<u>Theorem</u> $BI_M \Rightarrow TI$.
Proof: let $\forall \alpha \exists x \neg \forall y \leq x(\alpha(y+1) \prec \alpha y)$ and $\forall x[\forall y \prec x\, Ay \rightarrow Ax]$ hold.

Define $Bn := (\forall x < \text{lth } n \dot{-} 1)(n_{x+1} \prec n_x)$

$Pn := \neg Bn$

$Qn := (n = 0 \wedge \forall y\, Ay) \vee (n \neq 0 \wedge (Bn \rightarrow (\forall y \prec n_{\text{lth } n-1})Ay))$

One easily chechs that Hyp 1,...,Hyp 4 are satisfied, so by BI_M Q0 holds, i.e. $\forall y\, Ay$.

Corollary: $BI_{QF} \Rightarrow TI_{QF}$ and $BI_D \Rightarrow TI_D$ where BI_{QF} stands for BI_D with quantifierfree An. $TI_D(TI_{QF})$ stands for TI with decidable $\prec$ ($\prec$ expressed by a quantifierfree formula).

Proof: If $\prec$ is quantifierfree, then Pn above can be shown to be equivalent to a quantifierfree formula by making use of primitive recursion. Likewise Pn is decidable if $\prec$ is.

3.7.3.

<u>Theorem</u>. $TI_{QF} \Rightarrow BI_{QF}$ and $TI_{DPO} \Rightarrow BI_D$ where TI_{DPO} stands for TI_D with $\prec$ a partial ordering.

Proof: Let A and B satisfy Hyp 1, Hyp 2, Hyp 2_D, Hyp 3, Hyp 4. Define $Rn := \exists m_1 \exists m_2 \neq 0\ (n = m_1 * m_2 \wedge Am_1)$, geometrically speaking R consists of the nodes beyond the bar determined by A (the nodes which are past secured). Put $m \prec n := \exists m' \neq 0\ (m = n * m' \wedge \neg R(m))$. Clearly if A is quantifierfree (decidable) then $\prec$ is quantifierfree (a decidable partial ordering). We will apply transfinite induction with respect to Q, which is defined by $Qn := \neg Rn \rightarrow Bn$. First we show that $\prec$ is strongly well founded. Consider α and define a corresponding β by $\beta 0 = (\alpha 0)_0$.

$$\beta(n+1) = \begin{cases} (\alpha(n+1))_{n+1} & \text{if } \alpha(n+1) \text{ is a proper extension of } \beta(n) \\ \beta n * \hat{0} & \text{otherwise} \end{cases}$$

By Hyp 1 we have $\exists x\, A\bar{\beta}x$. Now check whether $\alpha \in \bar{\beta}x$, if so then $\neg\alpha(x-1) \prec \alpha x$, if not then for some $y < x$ we have $\neg\alpha(y-1) \prec \alpha y$. One can think of the function β following the sequences $\alpha 0, \alpha 1, \ldots$ as long as they are properly extending. As β must hit the bar, we can draw the desired conclusions about α.

Now we prove $\forall n\,(\forall m \prec n\ Qm \rightarrow Qn)$.

Let $\forall m \prec n\ Qm$, or $\forall m \prec n\ (\neg Rm \rightarrow Bm)$; by the definition of $\prec$ this is equivalent to $\forall m \prec n\ Bm$. Now assume $\neg Rn$. We will show Bn. Because A is decidable we can distinguish two cases:

1) An holds, then we have Bn (by Hyp 3)

2) $\neg$An holds. Because $\neg$Rn holds we have $\neg R(n * \hat{x})$ for every x. So by $\forall m \prec n\ (\neg Rm \rightarrow Bm)$ we have $\forall x\ B(n * \hat{x})$, and hence Bn (by Hyp 4).

We now apply TI: Q0 holds and since $\neg R0$, we conclude B0.

Remark: Howard and Kreisel derive these and other results in a basic formal system of analysis $\underset{\sim}{H}$. It is not hard to see that the above proof can be formalized.

3.8. The fan theorem

A well-known theorem in intuitionistic analysis is the so-called fan theorem, originally obtained by Brouwer as a corollary to the bar theorem [Brouwer, 24], p. 191. We need some preparation for the formulation.

3.8.1.

Definition: A lawlike function a is a spreadlaw (Spr(a)) if

(i) $a0 \neq 0$

(ii) $\forall m\,\forall n\,(a(n * m) \neq 0 \rightarrow an \neq 0)$

(iii) $\forall n\,\exists x\,(an \neq 0 \rightarrow a(n * \hat{x}) \neq 0)$

a is called finitary (fan-law) if moreover

(iv) $\forall n\,\exists z\,\forall x\,(a(n * \hat{x}) \neq 0 \rightarrow x \leqslant z)$

The connection between spreads and trees is familiar.

3.8.2.

<u>Definition</u>: $\alpha \in a \ := \ \forall n\, (a\bar{\alpha}n \neq 0)$ for a spreadlaw a .

3.8.3.

<u>Definition</u>: let a be a fan law. The top of e in a is defined by

$T_e^a \ = \ \{n \mid an \neq 0 \wedge en = 0\} \quad (e \in K)$.

(T_e^a is the species of unsecured nodes which are not past secured.)

3.8.4.

<u>Lemma</u>: If a is fan law and $e \in K$, then T_e^a is finite.

Proof: use induction over K.

(i) e is constant, then $T_e^a = \emptyset$

(ii) $e0 = 0$, then $T_e^a \ = \ \bigcup\limits_{a\hat{y} \neq 0} \{\hat{y} * n \mid n \in T^a_{\lambda m \cdot e(\hat{y} * m)}\}$.

By induction hypothesis this union consists of finite species, hence T_e^a is finite.

3.8.5.

<u>Theorem</u> (fan theorem). Let a be a fan law, then

$$\forall \alpha \in a\, \exists x\, A(\alpha,x) \rightarrow \exists z\, \forall \alpha \in a\, \exists x\, \forall \beta \in a(\bar{\alpha}z = \bar{\beta}z \rightarrow A(\beta,x))$$

Proof: Apply bar continuity to $\forall \alpha \in a\, \exists x\, A(\alpha,x)$, then we obtain $e(n) \neq 0 \rightarrow \exists x\, \forall \alpha \in a(\alpha \in n \rightarrow A(\alpha,x))$ for some $e \in K$ and for all n.

Let $k \ = \ \max\limits_{n \in T_e^a} \mathrm{lth}(n)+1$, then

$$\mathrm{lth}(n) \ = \ k \rightarrow \exists x\, \forall \alpha \in a(a \in n \rightarrow A(\alpha,x)), \text{ or equivalently}$$
$$\forall \alpha \in a\, \exists x\, \forall \beta \in a(\bar{\alpha}k = \bar{\beta}k \rightarrow A(\beta,x)).$$

Hence $\exists z\, \forall \alpha \in a\, \exists x\, \forall \beta \in a(\bar{\alpha}z = \bar{\beta}z \rightarrow A(\beta,x))$.

Cor. $\forall \alpha \in a\, \exists x\, A(\bar{\alpha}x) \rightarrow \exists z\, \forall \alpha \in a\, \exists x \leqslant z\, A(\bar{\alpha}x)$

Cor. Let Γ be an operation of type ((0)0)0, then

$$\exists z\ \forall\alpha\ \forall\beta[\bar{\alpha}z = \bar{\beta}z \rightarrow \Gamma\alpha = \Gamma\beta].$$

Proof: consider $\forall\alpha\ \exists x\ (\Gamma\alpha = x)$.

Remark: Bar continuity has boldly been generalized to arbitrary spreads (and forms). As long as the notion of choice sequence is not changed this is certainly justified. For a more careful treatment see [Troelstra, 69A], p. 52.

The fan theorem (and hence the bar theorem) does not hold irrespective of the class of sequences (functions) under consideration. The following example given by Kleene [Kleene-Vesley, 65], p. 112, shows that it fails when α ranges over recursive function.

Consider the sets $A_0 = \{x \mid \{x\}x = 0\}$ and
$A_1 = \{x \mid \{x\}x = 1\}$.

It is well known that A_0 and A_1 are recursively inseparable. Now for each sequence n and function f define

$V_{n0} = \{x \mid (n)_x = 0\}$, $V_{f0} = \{x \mid fx = 0\}$

$V_{n1} = \{x \mid (n)_x = 1\}$, $V_{f1} = \{x \mid fx = 1\}$

Define $Rn := \exists x[x \in (V_{n0} \cap A_0) \cup (V_{n1} \cap A_1)$.
We consider only functions f with $\forall x(fx \leqslant 1)$, therefore $V_{f0} = \overline{V}_{f1}$.

As A_0 and A_1 are recursively inseparable and because both V_{f0} and V_{f1} are recursive we have $x \in (V_{f0} \cap A_0) \cup (V_{f1} \cap A_1)$ for some x. Then by definition $x \in (V_{\bar{f}(x+1),0} \cap A_0) \cup (V_{\bar{f}(x+1),1} \cap A_1)$. Therefore we conclude $R(\bar{f}(x+1))$, so

$$\forall f \text{ rec } \exists x\ R(\bar{f}x). \tag{1}$$

However we can show

$$\forall z\ \exists f \text{ rec } \forall x \leqslant z\ \neg R(\bar{f}x) \tag{2}$$

Let z be given, $n_i = \begin{cases} 0 \text{ if } i \in A_1 \text{ for } i < z \\ 1 \text{ else} \end{cases}$

then $\neg R\langle n_0,\ldots,n_{x-1}\rangle$ for $x \leq z$, and hence $\neg R\ \bar{f}x$ for a recursive $f \in \langle n_0,\ldots,n_{z-1}\rangle$.

From (1) we conclude with the fan theorem $\exists z\ \forall f$ rec $\exists x \leq z\ R(\bar{f}x)$, which contradicts (2).

As a consequence also the principle of transfinite induction fails when the function variables are taken to range over recursive functions.

3.9. Spreads and applications to real numbers

At the first page of his series of fundamental papers on intuitionistic mathematics Brouwer introduced the concept of spread (german: Menge), which was to serve as a constructive substitute for the classical notion of set [Brouwer, 18], constructive in the sense that it regulated the generation of choice sequences. This notion has turned out to be extremely fruitful for the treament of such concrete things as topological spaces, the continuum, functionspaces.

We use the notion of spreadlaw as defined in 3.8. For all practical applications it is sufficient to consider lawlike spreadlaws, one can however extend the notion. The idea behind the concept of spread is that one generates certain sequences of objects, e.g. Cauchy sequences, or sequences of nested species (like intervals).

3.9.1.

Definition. ξ is a complementary mapping of a spreadlaw a (or a tree) if ξ associates elements of a species S to the elements of a (i.e. to these n with $an \neq 0 \wedge n \neq 0$).

Again ξ may be taken lawlike (as in most applications) or not. A pair $\langle a,\xi\rangle$ of a spreadlaw and a complementary law is called a dressed spread. A so-called undressed spread is obtained by taking $\xi\langle x_1,\ldots,x_k\rangle = x_k$. For convenience we will understand by the spread a the undressed spread associated to a .

The elements of $\langle a,\xi\rangle$ are the sequences of elements of S with the property that all initial segments are in the range of ξ. To be precise $\eta \in \langle a,\xi\rangle$ if $\exists \eta' \in (N)N[\forall x(a(\overline{\eta'}x) \neq 0) \wedge \forall x\ (\eta x = \xi(\overline{\eta'}(x+1))]$.

Example. Let $\langle r_n\rangle$ be a fixed lawlike enumeration of the rationals. We will now consider the dressed spread of all Cauchy sequences with a specified rate of convergence. The spreadlaw will be the universal one, i.e. $a_0 = \lambda x\cdot 1$. The rate of convergence will be 2^{-k+1} so we have to present a complementary law that picks out a suitably converging sequence of rationals.
We can find an enumeration $\langle r_{s(n,k,m)}\rangle_m$ of the species $\{r \mid |r_n - r| < 2^{-k}\}$, moreover we can take the enumeration such that no repetitions take place.
Now we define $\xi_0\langle k\rangle = r_k$

$$\xi_0\langle k_0,\ldots,k_{m-1},k\rangle = r_{s(n,m,k)} \text{ if } \xi_0\langle k_0,\ldots,k_{m-1}\rangle = r_n.$$

For $\langle s_n\rangle \in \langle a_0,\xi_0\rangle$ we have

$$|s_{n+m} - s_n| \ngtr |s_{n+m} - s_{n+m-1}| +\ldots+ |s_{n+1} - s_n| <$$
$$< 2^{-n-m-1} +\ldots+ 2^{-n} < 2^{-n+1},$$

so $\langle s_n\rangle$ is a Cauchy sequence.
We put $S_0 = \langle a_0,\xi_0\rangle$.
The example shows how to produce actual spreads. In order to obtain the reals we have to consider equivalence classes under the usual relation $\sim$ (see 1.3.1.).

We say that a species X is represented by a spread $\langle x,\xi\rangle$ with an equivalence relation $\sim$ (notation $X = \langle a,\xi,\sim\rangle$) if X is mapped bi-uniquely onto the species of equivalence classes $\langle a,\xi\rangle/\sim$.

In the example above $\mathbb{R} = \langle a_0, \xi_0, \sim\rangle$ as one easily checks.

Let us have a closer look at the possible notions of real number. Evidently the species of real numbers depends on the underlying species of sequences that is employed in the definition of Cauchy sequence. We denote the species of Cauchy sequences $\langle r_{\xi n}\rangle$ with $\xi \in X$ by Rng(X) and the species of reals belonging to Rng(X) by $\mathbb{R}(X)$. The species X is supposed to be extensional, i.e. $\forall x(\xi x = \eta x) \wedge \xi \in X \rightarrow \eta \in X$. Possible notions are recursive reals, lawlike reals, lawless reals etc.

The enumeration of the rationals upon which the definition of Rng(X) is based is irrelevant in the cases where X is closed under composition with lawlike functions, i.e. if $\forall \xi\ \forall a(\xi \in X \rightarrow a \circ \xi \in X)$ ($\circ$ denotes the composition of mappings).
Namely let r and r' be lawlike bi-unique enumerations of Q, then $\forall i\ \exists j\ (r_i = r'_j)$. Therefore $\exists a\ \forall i\ (r_i = r'_{ai})$, i.e. $r = r' \circ a$.
$r \circ \xi$ is the sequence $\langle r_{\xi n}\rangle$. By the hypothesis $r \circ \xi \in \mathrm{Rng}(X) \Rightarrow$
$\Rightarrow r' \circ a \circ \xi \in \mathrm{Rng}(X)$. So $\mathrm{Rng}^{r}(X) \subseteq \mathrm{Rng}^{r'}(X)$. Likewise $\mathrm{Rng}^{r'}(X) \subseteq \mathrm{Rng}^{r}(X)$.

If we consider a fixed enumeration r, the condition for a sequence $r \circ \xi$ to be a Cauchy sequence is $\forall k\ \exists n\ \forall m(|r\xi n - r\xi(n+m)| < 2^{-k})$. In the case of lawless sequences this condition cannot be fulfilled as follows from LS3. Thus $\mathbb{R}(LS) = \phi$. This fact is explained by the inability of lawless sequences to obey a condition on all future choices. As the lawless sequence so to speak has to pick out the consecutive members of the Cauchy sequence it is not allowed to pick the next member completely freely and this is impossible.

In the representation of Cauchy sequences by means of a spread, as in the example, the situation is different, the complementary law takes care of the convergence, so there is no restriction on the choices in

the underlying spread. The resulting notion of real number is however most unsatisfactory, as for example 0 is not represented.
Even worse, the equivalence relation among cauchy-sequences turns out to be the identity relation.
This has an amazing consequence for the collection of real-valued functions. A real valued function is represented by a relation $F(\alpha,\beta)$ such that $\forall\alpha\ \exists!\beta\ F(\alpha,\beta)$.
One derives however easily from LS3 that real valued function on the lawless reals is the identity function. The exemple of lawless reals makes it clear that the choice of the notion of sequences is worth attention. It also shows that the spread-approach and the spreadfree-approach are not simply equivalent.
Let us denote by Rng(S,X) ($\mathbb{R}$(S,X) the species of cauchy-sequences (reals) represented by the spread S with respect to the species X of sequences.
For lawlike sequences and suitable notions of choice sequence one can show Rng(S,X) = Rng(X) and $\mathbb{R}$(S,X) = $\mathbb{R}$(X) (see [Troelstra, 69A], p.61).

The use of fans in real analysis is demonstrated by

3.9.2.
Theorem: Each closed interval a,b of reals can be represented by a fan ([Brouwer, 19], cf. [Heyting, 56], p.42).

This theorem allows the derivation of a number of well-known theorems.

To avoid technical details we will consider a special fan that represents the interval [-1,1]. All conclusions are however valid for intervals and fans in general.

Example. Define
$$\begin{cases} a_1 0 = 1 \\ a_1\langle x_0 \ldots x_k\rangle = a_1\langle x_0,\ldots,x_{k-1}\rangle \cdot sg(3 \dot{-} x_k) \end{cases}$$
$$\begin{cases} \xi_1\langle x_0\rangle = (x_0-1)2^{-1} \\ \xi_1\langle x_0,\ldots,x_k\rangle = \xi_1\langle x_0,\ldots,x_{k-1}\rangle + (x_k-1)2^{-k-1} \end{cases}$$

The dressed spread defined by a_1 and ξ_1 is $S_1 = \langle a_1, \xi_1\rangle$.

The elements of S_1 are sequences of the form

$\langle i_1 2^{-1}, i_1 2^{-1} + i_2 2^{-2}, \ldots, i_1 2^{-1} + \ldots + i_k 2^{-k}, \ldots\rangle$ where $i_j \in \{-1,0,1\}$.

It is clear that S_1 represents $[-1,1]$.

Using the fan S_1 we will prove

3.9.3.

Theorem: A real valued function f on $[-1,1]$ is uniformly continuous ([Brouwer, 24], see [Heyting, 56]).

Proof: let α be an element of a_1, then for a given p

$\exists m\ (|f(\xi_1\alpha) - m\cdot 2^{-p}| < 2^{-p})$.

By the fan theorem we have

$$\exists n\ \forall\alpha \in a_1\ \exists m\ \forall\beta \in a_1(\bar{\alpha}n = \bar{\beta}n \to |f(\xi_1\beta) - m\cdot 2^{-p}| < 2^{-p}).$$

Let n_0 be such an n. For $x,y \in [-1,1]$ with $|x-y| < 2^{-n_0-1}$ it holds that there exists α and β with $\xi_1\alpha = x$, $\xi_1\beta = y$ and $\bar{\alpha}n_0 = \bar{\beta}n_0$ (this can be seen from the properties of the fan). So now we have

$$|x-y| < 2^{-n_0-1} \to |fx-fy| = |f\xi\alpha - f\xi\beta| \ngtr |f\xi\alpha - m\cdot 2^{-p}| + $$
$$+ |f\xi\beta + m\cdot 2^{-p}| < 2^{-p+1}.$$

This establishes the uniform continuity.

3.9.4.

Cor. Every real valued function on $\mathbb{R}$ is continuous.

3.9.5.

Cor. The only detachable subspecies of $[-1,1]$ are $[-1,1]$ and $\emptyset$ (or: $[-1,1]$ allows no proper partition (Zerlegung)).

An application of the fan theorem shows us that the apartness relation on the reals is definable from the inequality relation.

3.9.6.

Theorem: $x \# y \leftrightarrow \forall z \ (z \neq x \vee z \neq y)$ for $x,y \in [-1,1]$.

Proof. Let x and y be given, $x = \xi_1\alpha_0$, $y = \xi_1\alpha_1$.

Suppose $\forall z \ (z \neq x \vee z \neq y)$.

Then $\forall\alpha \exists n [(n = 0 \wedge \neg(\xi_1\alpha_0 \sim \xi_1\alpha)) \vee (n \neq 0 \wedge \neg(\xi_1\alpha_1 \sim \xi_1\alpha))]$.

Apply the fan theorem:

$\exists m \forall\alpha \exists n \forall\beta(\bar{\alpha}m = \bar{\beta}m \rightarrow [(n = 0 \wedge \neg(\xi_1\alpha_0 \sim \xi_1\beta)) \vee (n \neq 0 \wedge$
$\wedge \neg(\xi_1\alpha_1 \sim \xi_1\beta))])$.

Determine m_0 such that $\forall\alpha \ \exists n \forall\beta(\ldots)$ holds. Choose z such that $|z-x| < 2^{-m_0-1}$, then z is represented by a $\xi\alpha$ with $\bar{\alpha}m_0 = \bar{\alpha}_0 m_0$. Now determine n_0 (which depends on α).

Suppose $n_0 = 0$ then $\forall\beta \in \bar{\alpha}_0 m_0(\xi_1\alpha_0 \sim \xi_1\beta)$. By choosing $\beta = \alpha_0$ we obtain a contradiction, so $n_0 \neq 0$. Hence $\forall\beta \in \bar{\alpha}_0 m_0(\neg\xi_1\alpha_1 \sim \xi_1\beta)$, i.e. $\forall z \ (|z-x| < 2^{-m_0-1} \rightarrow z \neq y)$, or $|x-y| > 2^{-m_0-2}$.

From this one concludes $x \# y$.

The implication from left to right is trivial.

Remark: the theorem holds for spreads in general, one then applies WC-N.

More applications of the theory of spreads, the fan theorem, etc. can be found in [Heyting, 56], [Troelstra, 69A, 66], [Freudenthal, 36], [Ashwinikumar, 66, 69].

3.10. Formalizations of intuitionistic analysis.

At present there are three formal systems for the theory of choice sequences, those of Kleene-Vesley, of Kreisel-Troelstra and of Myhill. As the last one is less developed than the others, we will restrict our attention to the first two.

i) Kleene and Vesley presented in their monograph [Kleene-Vesley, 65] a system for analysis formulated in a language with two sorts of variables: numerical and function variables.

ii) Kreisel and Troelstra have developed a system with three sorts of variables; numerical variables, lawlike function variables and choice variables. A comprehensive treatment appeared in [Kreisel-Troelstra, 70].

The system of Kleene and Vesley contains apart from the logical axioms and the axioms for arithmetic axioms for the existence of certain primitive recursive functions, the axiom of bar induction and an axiom concerning the dependence of β on α under the assumption $\forall\alpha \ \exists\beta \ A(\alpha,\beta)$.
This last axiom is called "Brouwer's principle for functions", it reads

(27.1) $\forall\alpha \ \exists\beta \ A(\alpha,\beta) \rightarrow \exists\tau \ \forall\alpha\{\forall x \ \exists! y \ \tau(\hat{x} * \bar{\alpha}y) > 0 \ \wedge$
$\wedge \ \forall\beta[\forall x \ \exists y \ \tau(\hat{x} * \bar{\alpha}y) = \beta(x) + 1 \rightarrow A(\alpha,\beta)]$

From it the following 'Brouwer's principle for numbers' can be derived:

(27.2) $\forall\alpha \ \exists x \ A(\alpha,x) \rightarrow \exists\tau \ \forall\alpha \ \exists y \ \{\tau(\bar{\alpha}y) > 0 \ \wedge$
$\wedge \ \forall z[\tau(\bar{\alpha}z) > 0 \rightarrow y = z] \ \wedge \ A(\alpha,\tau(\bar{\alpha}y) \dot{-} 1)\}.$

(the numbers refer to the monograph [Kleene-Vesley, 65]).

We shall call the above system FIM.

Kreisel and Troelstra have put forward a formal system CS ([Kreisel-Troelstra, 70], §6). The formal aparatus contains variables for lawlike functions and for choice sequences. The system contains the axioms for K (actually a subsystem IDB dealing with lawlike analysis and K is treated separately (§3)). The axioms for choice sequences are

(i) (A) $A\alpha \rightarrow \exists e \ (\exists\gamma \ (\alpha = e|\gamma) \wedge \forall\beta \ A(e|\beta))$,

the so-called 'principle of analytic data'.

(A) states that if α has the property A, then α is contained in an analytic set (in the Baire space or the universal spread) such that A holds on that set. In the case of lawless sequences we have the corresponding principle LS3 (the principle of open data).

(ii) (BC-C) $\forall\alpha\ \exists\beta\ \ A(\alpha,\beta) \rightarrow \exists e\ \forall\alpha\ A(\alpha,e|\dot{\alpha})$.

(iii) (BC-F!) $\forall\alpha\ \exists! a\ A(\alpha,a) \rightarrow \exists a\ \forall b\ \ \forall\alpha\ A(\alpha(b)_{e(\alpha)})$.

The principle of analytic data becomes evident when seen in the light of Troelstra's GC-sequences (i.e. choice sequences generated by continuous operations, cf [Troelstra, 69A], 10.2).

The principle of analytic data allows us often to generalize theorems to cases with extra parameters.
We will sketch an example here [Troelstra, 69A], 10.5).

3.10.1.
Lemma: $\forall\alpha[A\alpha \rightarrow B\alpha] \leftrightarrow \forall e[\forall\alpha\ A(e|\alpha) \rightarrow \forall\alpha\ B(e|\alpha)]$.

Proof: from left to right is trivial. Consider $A\alpha$ given and let $\forall e(\forall\alpha\ A(e|\alpha) \rightarrow \forall\alpha\ B(e|\alpha))$ hold. Then $\alpha = f|\beta$ and $\forall\gamma\ A(f|\gamma)$ holds, therefore $\forall\gamma\ B(f|\gamma)$ holds, in particular $B\alpha$.

3.10.2.
Theorem: WC-N can be extended to formulae with choice parameters.

Proof: Consider $A(\alpha,x,\beta)$ (for more extra parameters an analogous reasoning suffices).
In order to show

$$\forall\beta[\forall\alpha\ \exists x\ A(\alpha,x,\beta) \rightarrow \forall\alpha\ \exists x\ \exists y\ \forall\gamma(\bar{\alpha}y = \bar{\gamma}y \rightarrow A(\gamma,x,\beta)] \qquad (1)$$

it is sufficient to show

$$\forall e[\forall\beta\ \forall\alpha\ \exists x\ A(\alpha,x,e|\beta) \rightarrow \forall\beta\ \forall\alpha\ \exists x\ \exists y\ \forall\gamma(\bar{\alpha}y = \bar{\gamma}y \rightarrow A(\gamma,x,e|\beta)] \qquad (2)$$

Let

$\forall\beta\ \forall\alpha\ \exists x\ A(\alpha,x,e|\beta)$ (3) be given.

(3) is equivalent to

$\forall\delta\ \exists x\ A(j_1\delta,x,e|j_2\delta)$ (4)

on the basis of closure of the universe of choice sequences under continuous operations (j_1,j_2 are inverses of the pairing operation).

WC-N applied to (4) gives

$$\forall\delta\,\exists y\,\exists x\,\forall\delta'(\bar{\delta}y = \overline{\delta'}y \rightarrow A(j_1\delta',x,e|j_2\delta'))$$

or $\forall\beta\,\forall\alpha\,\exists y\,\exists x\,\forall\alpha'\,\forall\beta'(\overline{\langle\alpha',\beta'\rangle}y \rightarrow A(\alpha',x,e|\beta'))$ (5)

Now choose in (5) $\beta' = \beta$ then

$$\forall\beta\,\forall\alpha\,\exists y\,\exists x\,\forall\alpha'(\bar{\alpha}y = \overline{\alpha'}y \rightarrow A(\alpha',x,e|\beta)) \quad (6)$$

Hence (3) → (6) and thus (1) holds.

For further results see [Kreisel-Troelstra, 70], 5.7.

A remarkable feature of CS is the eliminability of choice variables. The following theorems are proved:

1. First elimination theorem [Kreisel-Troelstra, 70], 7.2.
   There is a translation τ of formulas in the language of CS into formulas in the language of IDB, such that if A does not contain free choice variables then $\vdash_{CS} A \leftrightarrow \tau(A)$.

2. Second elimination theorem [Kreisel-Troelstra, 70], 7.3.8.
   Let A be a closed formula of CS, then we can prove finitistically
   $\vdash_{CS} A \leftrightarrow \vdash_{IDB_1} \tau(A)$
   (IDB_1 is a formal system for Brouwer operations defined in [Kreisel-Troelstra, 70], 3.6.2.)

Both theorems are extended to the theory of choice sequences CSS i.e. an extension of CS to a theory with species variables.

Recently Troelstra has found a striking application of the elimination theorems. He established that CS is a conservative extension of FIM, thus determining the relation between the two prominent systems of intuitionistic analysis [Troelstra, 71B].

4. COMPLETENESS OF INTUITIONISTIC LOGIC

The completeness results we discussed in 2.3 were deduced by non-intuitionistic means. It is natural to ask whether the intuitionistic predicate calculus (HPC) can also be proved to be complete by means of intuitionistic principles.

For propositional logic the answer (not surprisingly) turns out to be affirmative (cf.[Scott, 60],[Kreisel, 58]).

Heyting has expressed his opinion on the status of formalizations of intuitionistic theories as follows ([Heyting, 56], p.102):

"It must be remembered that no formal system can be proved to represent adequately an intuitionistic theory. There always remains a residue of ambiguity in the interpretation of the signs, and it can never be proved with mathematical rigour that the system of axioms really embraces every valid method of proof."

Notwithstanding the fundamental correctness of Heyting's enunciation (which by the way is very liberal compared to Brouwer's views, cf. [Brouwer, 08]) it is worthwhile to examine the possibility of intuitionistic correct completeness proofs. In order to attach foundational importance to the considerations one has to examine the state of affairs with respect to the 'intended' interpretation. Following [Kreisel, 62] we will consider the following notion of validity as the intended one:

A formula A with atomic subformulas $P_1,\ldots,P_k$ (denoted by $A(P_1,\ldots,P_k)$, if convenient) is valid, if for all species D^* and all relations $P_1^*,\ldots,P_k^*$ on D^*, the interpretation A^* of A holds intuitionistically.

We consider a language without function symbols.

Note that no restrictions have been placed on the species D^* or on the relations P_i^*, they may be discrete like N or incomplete, e.g.

depending on choice parameters.

Implicit in the above convention is the intuitionistic interpretation of the logical constants. As for the present purpose the actual interpretation of constants is of secondary interest we will leave this matter and just say that the evaluations of the interpreted statements A* is according to the principles of ordinary intuitionistic mathematics.

We will write Val(A) for A is valid and $\vdash$ stands for 'derivable in HPC'.

Note that validity à la Heyting is based on the notion of proof. The above notion of validity is more along the lines of Gödel's validity in structures.

4.1.

Definition:

(i) HPC is complete for A if $\text{Val}(A) \Rightarrow\ \vdash A$.

(ii) HPC is weakly complete for A if $\nvdash A \Rightarrow$ not Val(A).

If one wants to capture the definitions in a theory of species (containing arithmetic) then the definition could be formalized

4.2.

(i) $(\forall D^*)(\forall P_1^*)\ldots(\forall P_k^*)A^*(P_1^*,\ldots,P_k^*) \rightarrow \exists x\ \text{Prov}(x,\ulcorner A \urcorner)$,

(ii) $\neg\exists x\ \text{Prov}(x,\ulcorner A \urcorner) \rightarrow \neg(\forall D^*)\ldots(\forall P_k^*)A^*(P_1^*,\ldots,P_k^*)$,

where Prov(x,y) is the canonical proof-predicate and where $A^*(P_1^*,\ldots,P_k^*)$ is obtained from A by relativization to D^* and substitution of P_i^* (i=1...k).

The notion of completeness may be restricted by requiring all basic species D^* and relations P_i^* to be completely defined (lawlike, recursive etc.).

So far a number of positive results have been obtained:

(a) HPC is complete for prenex formulae [Kreisel, 58] (for these formulae derivability is even decidable).

(b) HPC is weakly complete for negative formulae (i.e. defined without $\vee$, $\exists$ and in which every atom is negated).

For theories with a primitive recursive proof predicate (e.g. HPC) completeness and weak completeness are equivalent if for each primitive recursive formula $A(x)$ $\neg\forall x\, A(x) \to \exists x\, \neg A(x)$ holds. For then $\neg\neg\exists x\, Prov(x, \ulcorner A \urcorner) \to \exists x\, Prov(x, \ulcorner A \urcorner)$ holds.

The connection between completeness and Markov's principle for primitive recursive formulae was noted by Gödel, he showed that for each primitive recursive formula $A(x)$ there is a (negative) formula B such that completeness of HPC for $\neg B$ implies $\neg\neg\exists x\, A(x) \to \exists x\, A(x)$.

Kreisel has extended these results in [Kreisel, 62]. Here we will see that the completeness of HPC (i.e. the completeness of HPC for all A contradicts Church's Thesis.
The arguments are drawn from [Kreisel, 62, 70] and unpublished material of C.Jockusch.

First we will show that for any primitive recursive relation $A(n,\alpha)$ there exists a formula A such that A 'expresses' $\forall\alpha$ n $A(n,\alpha)$. To obtain the result we mimick the definition of $A(n,\alpha)$ (for which only a fragment of number theory is required) in HPC.

4.3.

Definition: Consider the binary predicates =, S and the unary predicate Z. Let Suc be the conjunction of the closures of the following formulae:

$$x = x$$
$$x = y \to (x = z \to y = z)$$
$$x = y \to (Z(x) \to Z(y))$$
$$Z(x) \wedge Z(y) \to x = y$$
$$S(x,y) \wedge S(x,z) \to y = z$$
$$S(x,z) \wedge S(y,z) \to x = y$$

$x = y \rightarrow (S(x,z) \rightarrow S(y,z))$

$x = y \rightarrow (S(z,x) \rightarrow S(z,y))$

$Z(x) \rightarrow \neg S(y,x)$

Suc axiomatizes the theory of the successor relation.

In the 'computation' of values of a primitive recursive function we will need initial segments of the natural number sequence, therefore we put

$$G := \exists x\, Z(x) \wedge \forall x\, \exists y\, S(x,y)$$

$$G_p := \exists x\, \exists x_1 \ldots \exists x_p\, [Z(x) \wedge S(x,x_1) \wedge \ldots \wedge S(X_{p-1},x_p)]$$

Now $\mathrm{NUM} := \mathrm{Suc} \wedge G$

$\mathrm{NUM}_p := \mathrm{Suc} \wedge G_p$

NUM (NUM_p) says there is a sequence 0,1,2,... (0,1,2,...,p).

Now let $A(n,\alpha)$ be primitive recursive, that is $A(n,\alpha)$ has a primitive recursive characteristic function $f(n,\alpha)$.

The primitive recursive definition of $f(n,\alpha)$ consists of finitely many equations of the form

(1) $f_0(x) = \alpha(x)$

(2) $f_q(x) = 0$

(3) $f_q(x) = x'$

(4) $f_q(x_1,\ldots,x_r) = x_i \qquad (1 \leq i \leq r)$

(5) $f_q(x_1,\ldots,x_r) = f_s(f_{s_1}(x_1,\ldots,x_r),\ldots,f_{s_t}(x_1,\ldots,x_r))$

for $0 \leq s,s_1,\ldots,s_t < q$

(6) $f_q(0,x) = f_s(x) \qquad s < q$

$f_q(y',x) = f_t(y,x,f_q(y,x)) \qquad t < q \quad \text{for } q \leq k$

and $f(n,\alpha) = f_k(n)$.

We convert the defining equations into axioms in the language of predicate logic.

In the following think of $Q(x)$ as the predicate of which α is the characteristic function, and of $P_q(x,y)$ as the relation representing $f_q(x) = y$.

Let A_1 be the conjunction of the closure of the following formulae:

(1) $Q(x) \wedge Z(y) \to P_0(x,y)$

$\neg Q(x) \wedge Z(y) \wedge S(y,z) \to P_0(x,z)$

and further according to the defining equations above

(2) $Z(y) \to P_q(x,y)$

(3) $S(x,y) \to P_q(x,y)$

(4) $P_q(x_1,\ldots,x_r,x_i)$

(5) $P_{s_1}(x_1,\ldots,x_r,y_1) \wedge \ldots \wedge P_{s_t}(x_1,\ldots,x_r,y_t) \wedge P_s(y_1,\ldots,y_t,z) \to$

$\to P_q(x_1,\ldots,x_r,z)$

(6) $\begin{cases} P_s(x,y) \wedge Z(z) \to P_q(z,x,y) \\ P_t(y,x,z,w) \wedge P_q(y,x,z) \wedge S(y,y_0) \to P_q(y_0,x,w) \end{cases}$

(7) $P_i(x_1,\ldots,x_{j_i},y) \wedge P_i(x_1,\ldots,x_{j_i},z) \to y = z$ for $i=0,\ldots,k$.

(8) the identity axioms for the predicates $P_0,\ldots,P_k$.

Now let $U := NUM \wedge A_1 \wedge \forall x(Q(x) \vee \neg Q(x))$

and $H := \exists x\, \exists y\, (Z(y) \wedge P_k(x,y))$

H states that there exists an element n such that $A(n,\alpha)$.

We claim that the following holds:

4.4.

Lemma: $\forall\alpha\, \exists n\, A(n,\alpha) \leftrightarrow Val(U \to H)$

Proof: Let $\underset{\sim}{D}$ be a realization of the proper language. The interpretations of the predicates are denoted by asterisks.

(i) Suppose $\forall\alpha\, \exists n\, A(n,\alpha)$.

For the (arbitrary) realization $\underset{\sim}{D}$ we show that H holds in $\underset{\sim}{D}$ if U holds in $\underset{\sim}{D}$.

So let U hold in $\underset{\sim}{D}$.

Then, by G, there is an element 0^* in the species D^* of $\underset{\sim}{D}$, also for each $d \in D^*$ there is a $d' \in D^*$ such that $S^*(d,d')$ holds. Applying DC (the axiom of dependent choices) we conclude the existence of a sequence $0^*, 1^*, 2^*, \ldots,$ such that $S^*(i^*,(i+1)^*)$ holds. By the validity of NUM this sequence is, with respect to the successor relation, isomorphic to N.

From $Q^*(d) \vee \neg Q^*(d)$ for all $d \in D^*$, it follows that

$\exists a \in D^*[(a = 0^* \wedge Q^*(d)) \vee (a = 1^* \wedge \neg Q^*(d))]$ holds for all $d \in D^*$;

by the axiom of choice ($AC! - D^*D^*$) for elements of D^* we have:

there exists a function α^* such that $\alpha^*(d) = \begin{cases} 0^* \text{ if } Q^*(d) \\ 1^* \text{ if } \neg Q^*(d) \end{cases}$

The function α^* is thus the characteristic function of the decidable species Q^*. Choose α corresponding to α^*.

Now let f_q be one of the functions occurring in the definition of $f(n,\alpha)$. One proves by induction that $f_q(n_1,\ldots,n_r) = 0$ iff $P_q^*(n_1^*,\ldots,n_r^*,0^*)$. So $f(n,\alpha) = 0$ iff $P_k^*(n^*,0^*)$.

As we know that $\exists n\, f(n,\alpha) = 0$, we know that $\exists x\, \exists y(Z(y) \wedge P_k(x,y))$ holds in $\underset{\sim}{D}$.

Therefore we conclude Val($U \to H$).

(ii) Suppose Val($U \to H$). Choose any α and show $\exists n\, A(n,\alpha)$. As $U \to H$ is valid it is certainly valid in the natural realization of the natural numbers. Take Q to be the representing predicate of α, then from H we conclude $\exists n\, A(n,\alpha)$. As this holds for all α, we have $\forall\alpha\, \exists n\, A(n,\alpha)$.

4.5.

Cor.: $\forall\alpha \neg\neg\exists n\, A(n,\alpha) \leftrightarrow$ Val($\neg\neg(U \to H)$).

Proof: we only consider the implication from left to right.

$\neg\neg(U \to H) \leftrightarrow \neg(U \wedge \neg H)$ (cf.[Kleene, 52], §27, *609-i).

Suppose $U \wedge \neg H$ holds in $\underset{\sim}{D}$.

As in lemma 4.4. we obtain $0^*,1^*,2^*,\ldots,\alpha^*$. By $\neg H$, we have $\forall d \in D^*\ \forall d' \in D^*\ \neg(Z^*(d) \wedge P_k^*(d,d'))$, in particular $\neg P_k^*(n^*,0^*)$ for all n^*.

Hence $f_k(n,\alpha) \neq 0$ for all n and for α corresponding to α^*, so $\exists\alpha\, \forall n\, \neg A(n,\alpha)$ holds, this contradicts $\forall\alpha \neg\neg\exists n\, A(n,\alpha)$. Therefore we conclude that $\neg(U \wedge \neg H)$ holds in $\underset{\sim}{D}$. As the choice of the realization $\underset{\sim}{D}$ was arbitrary, we have $\mathrm{Val}(\neg\neg(U \to H))$.

Note that when we choose the sequences α to be constructive, lemma 4.4 holds with 'validity' replaced by 'constructive validity'.

Note that U can be transformed into prenex form and likewise H. Therefore $\neg\neg(U \to H)$ is equivalent to a formula T, which is the negation of a prenex formula.

4.6.

Lemma: For the formula T introduced above $\vdash T \Rightarrow \forall\alpha\exists n\, A(n,\alpha)$.

Proof: We can apply Herbrand's theorem to negations of prenex formulae (cf. [Kreisel, 58A], Thm.8), so from the proof of T in $\underset{\sim}{HPC}$ we can effectively find q and terms a_i, b_i, t_i $(1 \leq i \leq q)$ such that

$$\neg[T_1' \wedge Z(b_1) \wedge S(t_1,a_1) \wedge \ldots \wedge T_q' \wedge Z(b_q) \wedge S(t_q,a_q))] \qquad (*)$$

Here T_i' are substitution instances of T', which is a formula such that $T \leftrightarrow T' \wedge \exists x\, Z(x) \wedge \forall x\, \exists y\, S(x,y)$. The only term in T' is a_0; t_1 denotes a_0 and t_j denotes a term a_i or b_i with $i < j$.

We now interpret the terms in the obvious way:

$a_0, b_1, \ldots, b_q$ denote 0,

if t_i denotes $\bar{t}_i$, then a_i denotes $\bar{t}_i+1$,

$Z(x)$ is interpreted as $\{0\}$,

$S(x,y)$ is interpreted as the successor relation,

$Q(x)$ is interpreted as $\alpha x = 0$.

The predicates $\overline{P}_q$ are interpreted by the corresponding functions.

Now (*) is reduced to

$$\neg[\neg(P_k(x^1,y^1) \wedge Z(y^1)) \wedge \ldots \wedge \neg(P_k(x^q,y^q) \wedge Z(y^q))],$$

since all the remaining conjuncts are true on the natural numbers. As the predicates P_i are decidable, the above formula is equivalent to

$(P_k(x^1,y^1) \wedge Z(y^1)) \vee \ldots \vee (P_k(x^q,y^q) \wedge Z(y^q))$.

This disjunction must therefore hold on the natural numbers.

The interpretation reads

$f_k(x^1,\alpha) = 0 \vee \ldots \vee f_k(x^q,\alpha) = 0$.

We now effectively check the finitely many terms $x^1,\ldots,x^q$ and find an n such that $f_k(n,\alpha) = 0$.

So we proved $\forall\alpha\, \exists n\, A(n,\alpha)$.

4.7.

Theorem ([Kreisel, 62], Thm.1): If HPC is complete then for each primitive recursive predicate $A(n,\alpha)$ the following holds

$\forall\alpha\, \neg\neg\exists n\, A(n,\alpha) \rightarrow \forall\alpha\exists n\, A(n,\alpha)$.

Proof: by lemma 4.5 and cor.4.6.

4.8.

Cor.: Weak completeness of HPC implies

$\forall\alpha\, \neg\neg\exists n\, A(n,\alpha) \rightarrow \neg\neg\forall\alpha\, \exists n\, A(n,\alpha)$

for each primitive recursive predicate $A(n,\alpha)$.

Proof: immediate.

We will now, following [Kreisel, 70] p.133, specialize the predicate $A(n,\alpha)$. Consider a primitive recursive tree, i.e. a primitive recursive predicate Tn, such that T0 and $T(n*m) \rightarrow Tn$. We say that Tn determines the tree T.

Define $A(n,\alpha) := \neg T\bar{\alpha}n$.

Definition: T is well-founded if $\forall\alpha\, \exists n\, A(n,\alpha)$ (each path through T is finite).

T is weakly well-founded if $\forall\alpha\, \neg\neg\exists n\, A(n,\alpha)$ (each path through T is not infinite).

The sentence $U \rightarrow H$ we considered in lemma 4.4, associated to the tree T will be denoted by F_T. We will make use of constructive validity and constructive sequences α.

By lemma 4.4. and cor.4.5. we have:

Each constructive path in T is finite $\Leftrightarrow Val_c(F_T)$.

(Val_c stands for 'constructively valid').

Each constructive path in T is not infinite $\Leftrightarrow Val_c(\neg\neg F_T)$.

These facts will be used to show that the completeness of HPC is inconsistent with Church's Thesis: the species of valid sentences of HPC is not recursively enumerable.

The proof is based on the following unpublished work of C.Jockusch (December 1970).

Definition: W_e is the r.e. set with index e,

T_e is the primitive recursive tree with index e and which is a subtree of the binary tree (i.e. only sequences of zeroes and ones are allowed).

4.9.

Lemma: There is a primitive recursive function h such that for all e_0, e_1 $h(e_0,e_1)$ is the index of primitive recursive tree $T_{h(e_0,e_1)}$ with the property that α is an infinite path in $T_{h(e_0,e_1)}$ iff α is the characteristic function of a set C that separates W_{e_0} and W_{e_1} (i.e. $W_{e_0} \subseteq C \wedge W_{e_1} \subseteq C = \emptyset$).

Proof: Put $Q(e_0,e_1,\alpha_c) := W_{e_0} \subseteq C \wedge W_{e_1} \cap C = \emptyset$ where α_c is the characteristic function of C.

Q is Π^0_1, so $Q(e_0,e_1,\alpha_c) \leftrightarrow \forall x\, R(\overline{\alpha_c}x, e_0, e_1)$ for primitive recursive R [(Shoenfield, 67], p.163). This shows that $R(n,e_0,e_1)$ determines a tree.

Define $f(e_0,e_1,n) = \begin{cases} 0 & \text{if } R(n,e_0,e_1) \\ 1 & \text{else} \end{cases}$

An application of the s-m-n-theorem gives $f(e_0,e_1,n) = \{h(e_0,e_1)\}(n)$.

Clearly h is the required primitive recursive function.

4.10.

<u>Lemma</u>: There is a primitive recursive function k such that for all e

W_e finite $\Rightarrow W_{k(e)}$ finite.

W_e infinite $\Rightarrow W_{k(e)} = N$.

Proof: immediate.

4.11.

<u>Lemma</u>: There is a primitive recursive function f such that for all e

W_e is finite $\Rightarrow f(e) \in I$

W_e is infinite $\Rightarrow f(e) \in F$,

where $I = \{e | T_e$ has an infinite primitive recursive path$\}$

$F = \{e | T_e$ has no infinite recursive path$\}$

Proof: Choose two disjoint r.e. sets A and B which are recursively inseparable.

Let h and k be the functions considered in lemma 4.9. and lemma 4.10.

Let g_0 and g_1 satisfy $W_{g_0(e)} = A \cap W_{k(e)}$

$W_{g_1(e)} = B \cap W_{k(e)}$

and put $f(e) = h(g_0(e), g_1(e))$.

Now W_e is finite $\Rightarrow W_{k(e)}$ is finite $\Rightarrow W_{g_0(e)}, W_{g_1(e)}$ are finite and disjoint.

So $W_{g_0(e)}$ and $W_{g_1(e)}$ can be separated by a primitive recursive set, hence $T_{h(g_0(e),g_1(e))}$ contains an infinite primitive recursive path, i.e. $f(e) \in I$.

Next let W_e be infinite, then $W_{k(e)} = N$ and $W_{g_0(e)} = A$, $W_{g_1(e)} = B$. A and B are recursively inseparable, so $T_{h(g_0(e),g_1(e))}$ has no infinite recursive path, i.e. $f(e) \in F$.

In the following we use the following fact

4.12.

<u>Lemma</u>: $Inf = \{e | W_e$ is infinite$\}$ is productive and hence completely productive.

Proof: see [Rogers, 67] p.84, Example 2 and p.87, example following Cor.VI. Furthermore §11.3 Thm.VI.

4.13.

Lemma: There exists a recursive function p such that $p(e) \in (W_e \cap I) \cup (F - W_e)$.

Proof: By an application of s-m-n-theorem we find a primitive recursive q such that

$$W_{q(e)} = \{x \mid fx \in W_e\} = \overset{-1}{f}(W_e),$$ where f is the function from lemma 4.11.

Let ψ be the production function of the completely productive set Inf, then

$$\psi q(e) \in (W_{q(e)} - \mathrm{Inf}) \cup (\mathrm{Inf} - W_{q(e)}).$$

Hence $f\psi q(e) \in (W_e - F) \cup (F - W_e)$. It is no restriction to suppose that $f(n) \in I \cup F$ for all n. Therefore $f\psi q(e) \in (W_e \cap I) \cup (F - W_e)$. Now put $p(e) = f\psi q(e)$.

The above proofs can be formalized in classical arithmetic (cf. [Kreisel, 70], p.133).

From lemma 4.13. we conclude

$$p(e) \in W_e \to p(e) \in I \qquad (0)$$

$$p(e) \notin W_e \to p(e) \notin F \qquad (1)$$

$p(e) \in F$ is expressed in arithmetic as follows

$$\forall u\,[(\forall x\, \exists z\, T(u,x,z) \wedge Uz \leq 1) \to \exists n(\forall y \leq \mathrm{lth}(n)\ \forall z(T(u,y,z) \to U(z) = (n)_y) \wedge \neg T_{p(e)}(n))] \qquad (2)$$

(intuitively speaking: each recursive path has a node not in $T_{p(e)}$, or each recursive path in $T_{p(e)}$ is finite).

Applying the Gödeltranslation (2.2). to it we obtain

$$\forall u\,[\ \forall x\ \neg\neg\exists z\ T(u,x,z) \wedge Uz \leq 1) \to \to \neg\neg\exists n(\forall y < \mathrm{lth}(n)\ \forall z(T(u,y,z) \to Uz = (n)_y) \wedge \neg T_{p(e)}(n))] \qquad (3)$$

(intuitively speaking: every weakly recursive path in $T_{p(e)}$ is not infinite).

The Gödeltranslation leaves $p(e) \notin W_e$ (i.e. $\forall n \neg T(e,p(e),n)$) invariant, so from (1) we obtain the intuitionistically provable statement

$p(e) \notin W_e \rightarrow$ every recursive path is not infinite (4)

We will also show the converse.

Suppose 'every recursive path in $T_{p(e)}$ is not infinite' (5)

and assume $p(e) \in I$, then

$\exists u \forall x\, T_{p(e)}(\bar{g}_u(x))$, where g_u is an enumeration of the primitive recursive functions.

Clearly $p(e) \in I$ contradicts (5), so $p(e) \notin I$.

By the contraposition of (0) we find

$p(e) \notin I \rightarrow p(e) \notin W_e$, so the converse of (4) holds.

Therefore $p(e) \notin W_e \leftrightarrow$ every recursive path in $T_{p(e)}$ is not infinite.

Assuming Church's Thesis the following holds:

'All recursive paths through T_e are not infinite $\leftrightarrow Val_c(\neg\neg F_{T_e})$.

Now assume that the species of all constructively valid formulae is r.e., then

$\{\neg\neg F_{T_e} \mid Val_c(\neg\neg F_{T_e})\}$ is r.e., therefore

$\{e \mid$ All recursive paths through T_e are not infinite$\} = W_t$ for some t.

Now we obtain a contradiction as follows: $p(t) \notin W_t \leftrightarrow$ every recursive path in $T_{p(t)}$ is not infinite (by (6)), but by the definition of W_t $p(t) \in W_t \leftrightarrow$ every recursive path in $T_{p(t)}$ is not infinite.

Hereby we have established our claim that the species of valid sentences is not recursively enumerable, so HPC is incomplete, if Church's Thesis is assumed.

The consistency of Church's Thesis and a number of systems has been shown by various realizability methods. For example $IDBS_1$ + CA (analysis with variables for lawlike functions, Brouwer-operations and species and with full comprehension) is consistent with Church's thesis ([Kreisel-Troelstra, 70]). This shows that the completeness of HPC is a rather dubious commodity (as was already apparent from Theorem 4.7).

Let us consider completeness with respect to a universe of choice sequences satisfying the laws of CS. By [Kreisel-Troelstra, 70] §6, p.349, we have that Church's Thesis is refuted (although $\forall\alpha \neg\neg\exists a(\alpha=a)$ holds, i.e. a negative version of Church's Thesis is consistent with CS). So there could be a possibility for the completeness of HPC. However (weak) completeness implies

$$\forall\alpha \neg\neg\exists n\, A(n,\alpha) \rightarrow \neg\neg\forall\alpha\, \exists n\, A(n,\alpha).$$

Because in CS $\forall\alpha \neg\neg\exists n\, A(n,\alpha) \leftrightarrow \forall a \neg\neg\exists n\, A(n,\alpha)$ holds, we have

$$\forall a \neg\neg\exists n\, A(n,a) \rightarrow \neg\neg\forall\alpha\, \exists n\, A(n,\alpha) .$$

Now consider Kleene's counterexample for the fan theorem on recursive sequences ([Kleene-Vesley, 65], p.112, lemma 9.8, these lectures 3.7). Under (the consistent) assumption of Church's Thesis for the lawlike sequences both $\forall a \neg\neg\exists n\, A(n,a)$ and $\neg\forall\alpha\, \exists n\, A(n,\alpha)$ hold. Hence weak completeness is refuted in CS.

If one considers the universe of lawless sequences then Markov's principle is straightforwardly refuted (see 3.2 and 3.4.2), so weak completeness fails.

R E F E R E N C E S

P.H.G.Aczel

1968 Satured intuitionistic theories, in Contributions to mathematical logic, ed. Schmidt, Schütte, Thiele, Amsterdam, p. 1-11.

Ashwinikumar

1966 Hilbert spaces in intuitionism. Ph.D.Thesis, Amsterdam.

1969 Über katalogisierte Räume, Comp.Math. 21, p.431-456.

E.W.Beth

1952 The Foundations of mathematics. Amsterdam.

L.E.J.Brouwer

1908 De onbetrouwbaarheid der logische principes (The untrustworthyness of the logical principles), Tijdschrift voor Wijsbegeerte, p.152-158.

1918 Begründung der Mengenlehre unabhängig vom logischen Satz vom ausgeschlossenen Dritten, I. Proc.Akad.Amsterdam 12, p.3-43.

1919 Begründung der Mengenlehre unabhängig vom logischen Satz vom ausgeschlossenen Dritten, II. Verhandelingen Akad. Amsterdam 12, p.3-33.

1920 Besitzt jede reelle Zahl eine Dezimalbruchentwicklung? Proc.Akad.Amsterdam 23, p.955-965.

1924 Zur Begründung der intuitionistischen Mathematik, I. Math.Ann. 93, p.244-258.

1924 Beweiss dasz jede volle Funktion gleichmässig stetig ist. Proc.Akad.Amsterdam 27, p.89-194.

1948 Consciousness philosophy and mathematics, Proc.Xth intern. Congress Phil. Amsterdam, p.1235-1249.

1954 Points and Spaces, Can.J.Math.6, p.1-17.

D.van Dalen, C.E.Gordon

1971 Independence problems in subsystems of intuitionistic arithmetic. Indag.Math. 33, p.448-456.

D. van Dalen, A.S. Troelstra,

1970 Projections of lawless sequences, in Intuitionism and Proof Theory, ed. Kino, Myhill, Vesley. Amsterdam, p.163-186.

J.E. Fenstad

1971 (ed). Proceedings of the Second Scandinavian Logic Symposium, Oslo. Amsterdam.

M.C. Fitting

1969 Intuitionistic Logic Model Theory and Forcing. Amsterdam.

H. Freudenthal

1936 Zum intuitionistischen Raumbegriff. Comp.Math. 4, p.82-111.

A.O. Gelfond

Transcendental and Algebraic numbers. Dover Publications.

N.D. Goodman

1968 Intuitionistic arithmetic as a theory of constructions. Ph.D. Thesis Stanford.

1970 A theory of constructions equivalent to arithmetic, in Intuitionism and Proof Theory, ed. Kino, Myhill, Vesley, Amsterdam, p.101-120.

A. Grzegorczyk

1964 A philosophically plausible formal interpretation of intuitionistic logic, Indag.Math. 26, p.596-601

1964A Recursive objects in all finite types. Fund.Math. 54, p.73-93.

R. Harrop

1956 On disjunctions and existential statements in intuitionistic systems of logic. Math.Ann. 132, p.347--361.

J. van Heyenoort

1967 From Frege to Gödel, Cambridge, Mass.

A.Heyting

1930 Die formale Regeln der intuitionistischen Logik. Sitzungsber.preuss.Akad.Wiss. Berlin, p.42-56.

1930A Die formale Regeln der intuitionistischen Mathematik. Sitzungsber.preuss.Akad.Wiss. Berlin, p.57-71.

1931 Die intuitionistische Grundlegung der Mathematik, Erkenntnis, 2. p.106-115.

1953 Inleiding tot de Intuitionistische Wiskunde (mimeographed), with cooperation of J.J.de Iongh, Amsterdam.

1955 Les fondements des mathématiques. Intuitionisme. Théorie de la démonstration, Paris, Louvain.

1956 Intuitionism, An introduction. Amsterdam.

W.A.Howard, G.Kreisel

1966 Transfinite induction and bar induction of types zero and one, and the role of continuity in intuitionistic analysis. J.S.L. 31, p.325-358.

S.Jaśkowski

1936 Recherches sur le système de la logique intuitioniste. Actes du Congrès Intern Phil.Sc.VI Phil math. Paris,p.58-61.

D.H.J.de Jongh

A A characterization of the intuitionistic propositional calculus, to appear.

S.C.Kleene

1952 Introduction to Metamathematics, Amsterdam-New York.

1959 Countable Functionals, in Constructivity in Mathematics (ed.A.Heyting), p.81-100.

1959A Recursive Functionals and Quantifiers of Finite Types,I, Trans.Am.Math.Soc. 91, p.1-52.

1962 Disjunction and existence under implication in elementary intuitionistic formalisms. J.S.L., 27, p.11-18.

S.C.Kleene, R.E.Vesley

1965 The foundations of intuitionistic mathematics. Amsterdam.

A.Kolmogoroff

1932 Zur Deutung der intuitionistischen Logik. Math.Zeitschrift 35, p.58-65.

G.Kreisel

1958 A remark on free choice sequences and the topological completeness proofs. J.S.L.23, p.369-388.

1958A Elementary completeness properties of intuitionistic logic with a note on negations of prenex formulae. J.S.L.23

1959 Interpretation of Analysis by means of constructive functionals of finite type, in Constructivity in Mathematics (ed.A.Heyting) Amsterdam, p.101-128.

1962 On weak completeness of intuitionistic predicate logic. J.S.L. 27, p.139-158.

1962A Foundations of intuitionistic logic, in Logic,methodology and philosophy of science. Stanford, p.198-210.

1963 Stanford report on the foundations of mathematics (mimeographed), Stanford.

1965 Mathematical Logic, in Lectures on modern mathematics, p.95-195, ed. T.L.Saaty.

1968 Lawless sequences of natural numbers, Comp.Math.20, p.222-248.

1970 Church's thesis: a kind of reducibility axiom of constructive mathematics, in Intuitionism and Proof Theory ed. Kino, Myhill, Vesley. Amsterdam, p.121-150.

1971 A survey of proof theory II in [Fenstad, 71].

G.Kreisel, M.H.A.Newman

1970 L.E.J.Brouwer, Biographical Memoir, Royal Society

G.Kreisel, A.S.Troelstra

1970 Formal systems for some branches of intuitionistic analysis. Annals of math.logic 1, p.229-387.

S.Kripke

1965 Semantic analysis of intuitionistic logic I, in Formal systems and recursive functions. ed.Crossley and Dummett. Amsterdam, p.92-130.

H.Läuchli

1970 An abstract notion of realizability for which intuitionistic predicate calculus is complete, in Intuitionism and Proof Theory, ed. Kino, Myhill, Vesley, Amsterdam, p.227-234.

P.Martin-Löf

1971 Hauptsatz for the intuitionistic theory of iterated inductive definitions, in [Fenstad, 71].

1971A Hauptsatz for the intuitionistic theory of species, in [Fenstad, 71].

J.Myhill

1967 Notes toward a formalization of intuitionistic analysis. Logique et Analyse 35, p.280-297.

H.Osswald

A Homomorphie-invariante Formeln in der Kripke-Semantik (forthcoming).

B Unterstruktur-invariante Formeln in der Kripke-Semantik (forthcoming).

D.Prawitz

1965 Natural Deduction. Stockholm.

1971 Ideas and results in proof theory, in [Fenstad, 71].

D.Prawitz, P.E.Malmnäs

1968 A survey of some connections between classical, intuitionistic and minimal logic, in Contributions to mathematical logic, ed.Schmidt,Schütte,Thiele. Amsterdam, p.215-228.

H.Rasiowa, R.Sikorski

1963 The mathematics of metamathematics, Warsaw.

H.Rogers, Jr.

1967 Theory of recursive functions and effective computability. New York.

G.F.Rose

1953 Propositional calculus and realizability. Trans.Am.Math.Soc. 75, p.1-19.

K. Schütte

1968 Vollständige Systeme modaler und intuitionistischer Logik. Berlin.

D. Scott

1960 Completeness proofs for the intuitionistic sentential calculus, in Summaries of talks presented at the Summer Institute of Symbolic Logic in 1957 at Cornell University Princeton, N.J..

J.R. Shoenfield

1967 Mathematic Logic. Reading, Mass.

C.S. Smorynski

1970 Three papers on Intuitionistic Arithmetic. Stanford Studies in Logic, no.2

C. Spector

1962 Provably recursive functionals of analysis; a consistency proof of analysis by an extension of principles formulated in current intuitionistic mathematics, in Recursive function theory. Proc.Symp.Pure Mathematics, p.1-27.

W.P. van Stigt

1971 Brouwer's Intuitionism; a reappraisal of Brouwer's contribution to the study of the foundations of mathematics. Ph.D. Thesis. London.

M.E. Szabo

1969 The collected Papers of Gerhard Gentzen, Amsterdam.

W.W. Tait

1967 Intensional interpretations of functionals of finite type I. J.S.L. 32, p.198-212.

R.H. Thomason

1968 On the strong semantical completeness of the intuitionistic predicate calculus. J.S.L. 33, p.1-7.

A.S. Troelstra

1966 Intuitionistic general topology, Ph.D.Thesis, Amsterdam.

1967 Finite and infinite in intuitionistic mathematics. Comp. Math. 18, p.94-116.

A.S.Troelstra

1968 The theory of choice sequences, in Logic, Methodology and Phil. Science III, ed. van Rootselaar, Staal. Amsterdam, p.201-226.

1969 Informal theory of choice sequences. Studia logica 15, p.31-52.

1969A Principles of Intuitionism. Lecture Notes in Mathematics 95, Berlin.

1969B Notes on the intuitionistic theory of sequences (I). Indag.Math. 31, p.430-440.

1970 Ibid. III. Indag.Math. 32, p.245-252

1971 Notions of realizability for intuitionistic arithmetic in all finite types, in [Fenstad, 71].

1971A Computability of terms and notions of realizability for intuitionstic analysis. Report 71-02, Dept. of Math. University of Amsterdam.

1971B An addendum. Annals of Math.Logic, 3.

A.S.Yessenin-Volpin

1959 Le programme ultra-intuitioniste des fondements des mathématiques, in Infinistic Methods, Warsaw, p.201-223.

1970 The ultra-intuitionistic criticism and the anti-traditional program for foundations of mathematics, in Intuitionism and Proof theory, p.3-46 ed.Kino, Myhill, Vesley.

II. REALIZABILITY: A RETROSPECTIVE SURVEY

S. C. KLEENE
(The University of Wisconsin, Madison)

In Tables 1 and 2 (pp. 2 and 3) I show five definitions of realizability (Definition 2, using Definition 1) in Columns A, B, C, D, E. These definitions are not stated exactly as they originally appeared (in my papers 1945, 1945, 1962a, 1957, 1967a),[1] but in the versions that fit best in the present survey. In Table 1, "⊢" means provable in intuitionistic (Heyting) arithmetic HA; in Table 2, in intuitionistic analysis (as formalized in Kleene and Vesley "The Foundations of Intuitionistic Mathematics" 1965) FIM. The reader may postpone studying each definition until the point where the discussion has led up to it.

PART I. HEYTING ARITHMETIC HA (NUMBER REALIZABILITY)

In 1940, intuitionism had been on the scene (in Brouwer's proposals, and at successive stages of development) for a bit over thirty years, and the theory of general recursive functions (or λ-definable functions, or Turing computable functions) for a bit over five. Both theories claimed to deal, in different arenas, with effective or constructive processes. So far as I knew, no one had looked for any precise connection between the two theories. I had been working in the theory of general recursive functions, and had begun to read some of Brouwer's papers. It struck me the situation would be very anomalous if there were not some precise connection.

An intuitionist only claims to have proved an existential statement "(Ey)A(y)" (with a natural number variable "y") if he has found, or at least knows a method sufficient in principle (i.e. apart from practical limitations on the number of steps he can carry out) for finding, a natural number y such that A(y). What, then, can an intuitionist mean by asserting "(x)(Ey)A(x, y)"? Compatibly with his meaning of "(Ey)", what can he mean but that he has found a general method about which he knows in advance that, whatever natural number x may be chosen, the method will apply and enable him (in principle) to find a y such that A(x, y) for that x? In other words, before he has the right to assert "(x)(Ey)A(x, y)", does he not need to have

[1] An underlined date, such as "1945", is a reference to the bibliography at the end of the paper.

Table 1. Heyting Arithmetic HA (Number Realizability)

Column	A	B	C
Definition 1. For E a closed formula of HA,			
if E is	then $\underline{e}$ realizes E iff	then $\underline{e}$ realizes-($\vdash$) E iff	then $\mid$ E iff
prime	E is true.	E is true.	E is true.
$A \& B$	$(\underline{e})_0$ realizes A and $(\underline{e})_1$ realizes B.	$(\underline{e})_0$ realizes-($\vdash$) A and $(\underline{e})_1$ realizes-($\vdash$) B.	$\mid$ A and $\mid$ B.
$A \vee B$	$(\underline{e})_0 = 0$ and $(\underline{e})_1$ realizes A, or $(\underline{e})_0 \neq 0$ and $(\underline{e})_1$ realizes B.	$(\underline{e})_0 = 0$ and $(\underline{e})_1$ realizes-($\vdash$) A and $\vdash$ A, or $(\underline{e})_0 \neq 0$ and $(\underline{e})_1$ realizes-($\vdash$) B and $\vdash$ B.	$\mid$ A and $\vdash$ A, or $\mid$ B and $\vdash$ B.
$A \supset B$	for each $\underline{a}$, if $\underline{a}$ realizes A, then $\{\underline{e}\}(\underline{a})$ is defined and realizes B.	for each $\underline{a}$, if $\underline{a}$ realizes-($\vdash$) A and $\vdash$ A, then $\{\underline{e}\}(\underline{a})$ is defined and realizes-($\vdash$) B.	if $\mid$ A and $\vdash$ A, then $\mid$ B.
$\neg A$	for each $\underline{a}$, not $\underline{a}$ realizes A.	for each $\underline{a}$, not ($\underline{a}$ realizes-($\vdash$) A and $\vdash$ A).	not ($\mid$ A and $\vdash$ A).
$\forall x A(x)$	for each $\underline{x}$, $\{\underline{e}\}(\underline{x})$ is defined and realizes $A(\underline{x})$.	for each $\underline{x}$, $\{\underline{e}\}(\underline{x})$ is defined and realizes-($\vdash$) $A(\underline{x})$.	for each $\underline{x}$, $\mid A(\underline{x})$.
$\exists x A(x)$	for $\underline{x} = (\underline{e})_0$, $(\underline{e})_1$ realizes $A(\underline{x})$.	for $\underline{x} = (\underline{e})_0$, $(\underline{e})_1$ realizes-($\vdash$) $A(\underline{x})$ and $\vdash A(\underline{x})$.	for some $\underline{x}$, $\mid A(\underline{x})$ and $\vdash A(\underline{x})$.
Definition 2. For E a formula $E(y_1, \ldots, y_m)$ containing free only $y_1, \ldots, y_m$ $(m \geq 0)$,			
	E is realizable iff there is a general recursive function φ such that, for each $\underline{y}_1, \ldots, \underline{y}_m$, $\varphi(\underline{y}_1, \ldots, \underline{y}_m)$ realizes $E(\underline{y}_1, \ldots, \underline{y}_m)$.	E is realizable-($\vdash$) iff there is a general recursive function φ such that, for each $\underline{y}_1, \ldots, \underline{y}_m$, $\varphi(\underline{y}_1, \ldots, \underline{y}_m)$ realizes-($\vdash$) $E(\underline{y}_1, \ldots, \underline{y}_m)$.	$\Vert$ E iff, for each $\underline{y}_1, \ldots, \underline{y}_m$, $\mid E(\underline{y}_1, \ldots, \underline{y}_m)$.
Theorem.			
	If $\vdash$ E, then E is realizable.	If $\vdash$ E, then E is realizable-($\vdash$).	If $\vdash$ E, then $\Vert$ E.
Consequences, for closed formulas $A \vee B$, $\exists x A(x)$, $\forall x \exists y A(x,y)$.			
	(i) If $\vdash A \vee B$, then A is realizable or B is realizable. (ii) If $\vdash \exists x A(x)$, then, for some $\underline{x}$, $A(\underline{x})$ is realizable. (iii) If $\vdash \forall x \exists y A(x,y)$, then there is a general recursive function φ such that, for each $\underline{x}$, $A(\underline{x}, \underline{y})$ is realizable for $\underline{y} = \varphi(\underline{x})$.	(i) If $\vdash A \vee B$, then $\vdash$ A or $\vdash$ B. (ii) If $\vdash \exists x A(x)$, then, for some $\underline{x}$, $\vdash A(\underline{x})$. (iii) If $\vdash \forall x \exists y A(x,y)$, then there is a general recursive function φ such that, for each $\underline{x}$, $\vdash A(\underline{x}, \underline{y})$, and hence $A(\underline{x}, \underline{y})$ is true, for $\underline{y} = \varphi(\underline{x})$.	(i) as in Column B, much more simply. (ii) as in Column B, much more simply.

Table 2. Intuitionistic Analysis FIM (Function Realizability)

Column	D	E
Definition 1. For E a formula of FIM containing free only Ψ,		
if E is	then ϵ realizes-Ψ E iff	then $\epsilon \textcircled{q}$ E is
prime	E is true-Ψ.	E.
A & B	$(\epsilon)_0$ realizes-Ψ A and $(\epsilon)_1$ realizes-Ψ B.	$(\epsilon)_0 \textcircled{q} A \;\&\; (\epsilon)_1 \textcircled{q} B$.
A ∨ B	$(\epsilon(0))_0 = 0$ and $(\epsilon)_1$ realizes-Ψ A, or $(\epsilon(0))_0 \neq 0$ and $(\epsilon)_1$ realizes-Ψ B.	$[(\epsilon(0))_0 = 0 \;\&\; (\epsilon)_1 \textcircled{q} A \;\&\; A] \vee [(\epsilon(0))_0 \neq 0 \;\&\; (\epsilon)_1 \textcircled{q} B \;\&\; B]$.
A ⊃ B	for each α, if α realizes-Ψ A, then $\{\epsilon\}[\alpha]$ is defined and realizes-Ψ B.	$\forall\alpha(\alpha \textcircled{q} A \;\&\; A \supset\, !\{\epsilon\}[\alpha] \;\&\; [\{\epsilon\}[\alpha] \textcircled{q} B])$.
¬A	for each α, not α realizes-Ψ A.	$\forall\alpha \neg(\alpha \textcircled{q} A \;\&\; A)$.
∀x A(x)	for each x, $\{\epsilon\}[x]$ is defined and realizes-Ψ, x A(x).	$\forall x(!\{\epsilon\}[x] \;\&\; [\{\epsilon\}[x] \textcircled{q} A(x)])$.
∃x A(x)	$(\epsilon)_1$ realizes-Ψ, $(\epsilon(0))_0$ A(x).	$(\epsilon)_1 \textcircled{q} A((\epsilon(0))_0) \;\&\; A((\epsilon(0))_0)$.
∀α A(α)	for each α, $\{\epsilon\}[\alpha]$ is defined and realizes-Ψ, α A(α).	$\forall\alpha(!\{\epsilon\}[\alpha] \;\&\; [\{\epsilon\}[\alpha] \textcircled{q} A(\alpha)])$.
∃α A(α)	$\{(\epsilon)_0\}$ is defined and $(\epsilon)_1$ realizes-Ψ, $\{(\epsilon)_0\}$ A(α).	$!\{(\epsilon)_0\} \;\&\; [(\epsilon)_1 \textcircled{q} A(\{(\epsilon)_0\}) \;\&\; A(\{(\epsilon)_0\})]$.
Definition 2. For E as above,		
	E is realizable iff there is a general recursive function φ such that, for each Ψ, $\varphi[\Psi]$ realizes-Ψ E.	$\textcircled{q}$ E is $\exists f \forall\Psi \exists\epsilon(\lambda x\{f\}(x,\Psi) \simeq \epsilon \;\&\; \epsilon \textcircled{q} E)$.
Theorem.		
	If ⊢ E, then E is realizable.	If ⊢ E (for E as above), then ⊢ $\textcircled{q}$ E (and ⊢ $!u \;\&\; [u \textcircled{q} E]$ for some p-functor u, containing free only Ψ, and expressing a general recursive function φ such that, for each Ψ, $\varphi[\Psi]$ realizes-Ψ E).
Consequences, for closed formulas $A \vee B$, $\exists x A(x)$, $\forall x \exists y A(x,y)$, $\exists\alpha A(\alpha)$.		
	FIM is consistent. (i)-(iii) as in Column A, but now for FIM.	(i)-(iii) as in Column B, but now for FIM. (iv) If ⊢ $\exists\alpha A(\alpha)$, then ⊢ $\exists\alpha_{GR(\alpha)} A(\alpha)$ (and, for a suitable e, ⊢ $\exists\alpha\{\forall x \exists y[T(e,x,y) \;\&\; U(y)=\alpha(x)] \;\&\; A(\alpha)\}$).

found an "effectively calculable" function φ such that $(\underline{x})\underline{A}(\underline{x}, \underline{\varphi}(\underline{x}))$? But then by Church's thesis 1936, which I believed then and still do,[2] φ must be general recursive.

Does the intuitionists' actual practice conform to the aforesaid interpretation? Hopefully, it should be possible to tackle the question for such a stock of intuitionistic methods of proof as the intuitionists, by incorporating them into an intuitionistic formal system, have acknowledged to be intuitionistic.

Accordingly (early in 1940), I conjectured that, for an intuitionistic formal system embracing at least number theory, if a closed formula $\forall x \exists y A(x, y)$ is provable in the system (in symbols, if $\vdash \forall x \exists y A(x, y)$), then there is a general recursive function φ such that, for each x, $A(\underset{\sim}{x}, \underset{\sim}{y})$ is true for $y = \varphi(x)$, where $\underset{\sim}{x}$ and $\underset{\sim}{y}$ are the numerals for x and y.[3] Trying to settle the conjecture for a specific intuitionistic formalism would give me a concrete hold, I thought, on the problem of whether there are precise connections between intuitionism and general recursive functions.[4]

To start out as simply as possible, I proposed to try to resolve the conjecture for "intuitionistic number theory", or "Heyting arithmetic" as it is now often called (though it does not occur as a subsystem readily separated out from Heyting's full system of intuitionistic mathematics, 1930a with 1930). Heyting arithmetic is the system obtained by using the intuitionistic predicate calculus (originally introduced by Heyting) with equality, Peano's axioms and the recursion equations for addition and multiplication. Specifically, I had in mind the intuitionistic version of the number-theoretic formalism of my IM (1952b), which I had been using in courses at Wisconsin since the fall of 1936 (cf. IM p. 101).

That this plan was not altogether obvious in 1940 is illustrated by the reaction of a prominent logician to whom I explained it at a chance meeting early in 1940. He explained to me reasons why, in his view, the plan could not be expected to succeed. I did not succeed in understanding his reasons.

---

[2]Subsequent presentations of Church's thesis are in my "Introduction to Metamathematics" 1952b (cited as "IM") § 62 with p. 352 and § 70, and in my "Mathematical Logic" 1967 § 41.

[3]I first published this conjecture in 1943 p. 69.

[4]The reader should be alert to the distinction between logical and other symbols used informally, such as

$$(\underline{x}),\ (\underline{Ey}),\ \underline{A}(\underline{x}, \underline{y}),\ \underline{A} \rightarrow \underline{B},\ \underline{\alpha},\ \underline{\Psi},$$

and such symbols used in reference to a formal system, such as

$$\forall x,\ \exists y,\ A(x, y),\ A \supset B,\ \alpha,\ \Psi .$$

I was busy at the time obtaining my first results on the arithmetical hierarchy 1943,[5] and did not take up my plan until early in 1941.

How could one undertake to establish the conjecture? The conjecture attributes a non-classical feature to formulas of the one particular shape $\forall x \exists y A(x, y)$. But formulas of all sorts of shapes may occur in a proof of a formula $\forall x \exists y A(x, y)$. So evidently I needed to find a property or interpretation, defined for all shapes of formulas, which one could establish to be possessed successively by each formula in any proof in HA, and the possession of which by a formula of the shape $\forall x \exists y A(x, y)$ would entail the conclusion of my conjecture.

The clues of which I was conscious initially were two. First, the intuitionistic meaning of an existential statement, as explained above, can be expressed in the following way. For an intuitionist, the statement "(Ex)A(x)" is an incomplete communication of a fuller statement giving an x such that A(x). This concept of "(Ex)A(x)" as an incomplete communication appears explicitly in Hilbert-Bernays 1934 p. 32 in explanation of the meaning of existence under the "finitary standpoint" (cf. IM p. 63). Can we generalize this idea to think of all (except, trivially, the simplest) intuitionistic statements as incomplete communications? After giving an x such that A(x), the "A(x)" may again be an incomplete communication, requiring information to be supplied to complete it. Altogether then, "(Ex)A(x)" is an incomplete communication, to complete which one must supply an x such that A(x) together with such further information as may be required to complete "A(x)" for that x.

Similarly, "A∨B" is an incomplete communication, to complete which one must, either indicate that "A" is to be completed and supply such information as may be needed to complete "A", or indicate that "B" is to be completed and supply such information as may be needed to complete "B".

Now consider a generality statement "(x)A(x)". For each x, in general information will be required to complete "A(x)" for that x. The intuitionist, it would seem, would only be justified in asserting "(x)A(x)", if he knew in advance a procedure by which, upon demand for any particular x, he could (in principle) obtain information to complete "A(x)" for that x. Now let us try out the idea that all the lots of information that may be required to complete intuitionistically formulas of any shapes can be codified by natural numbers. Then it comes readily to mind to propose that "(x)A(x)" is to be completed by giving an effective process, and thence

---

[5] Presented to the American Mathematical Society September 11, 1940, with abstract in Bull. Amer. Math. Soc. vol. 46 (1940) p. 885 (received August 7, 1940).

by Church's thesis a general recursive function φ, such that, for each $\underline{x}$, the corresponding function value $\varphi(\underline{x})$ is the code for information completing "$\underline{A}(\underline{x})$" for that $\underline{x}$. Replacing φ by a Godel number $\underline{e}$ for it (so $\varphi(\underline{x})$ can be written "$\{\underline{e}\}(\underline{x})$", IM p. 340), we meet the condition that the information to complete "$(\underline{x})\underline{A}(\underline{x})$" itself be codified by a number, namely this $\underline{e}$.

Now we are well on our way toward interpreting all shapes of number-theoretic statements (except the simplest) as incomplete communications, to be completed by supplying information, codified by natural numbers $\underline{e}$. And furthermore, general recursive functions have been brought in, as clearly was necessary for my plan.

Thinking along these lines, I proposed to interpret each closed formula E of HA as an incomplete communication which can be completed by giving a number $\underline{e}$, said to "realize E", which is a code for information completing E. I proposed to say "E is realizable" if a number $\underline{e}$ can be found such that $\underline{e}$ realizes E. Open formulas could then be defined to be "realizable" if their closures are realizable (but an alternative treatment will be used in Table 1). My hope was to find a definition of when "$\underline{e}$ realizes E" that would meet the two criteria: (a) it can be shown that every provable formula of HA is realizable, (b) the realizability of a closed formula $\forall x \exists y A(x, y)$ entails the conclusion of my conjecture.

I still haven't discussed the interpretation of an implication "$\underline{A} \rightarrow \underline{B}$". This proved to be crucial. The second clue which I consciously tried to use in 1941 was Heyting's "proof-interpretation" as it is rendered in 1934 p. 14. Under this, an implication "$\underline{A} \rightarrow \underline{B}$" is to be established by giving a construction which from any proof of $\underline{A}$ leads to a proof of $\underline{B}$. In my first trial of a definition of realizability, I handled implication in a way based on this idea. But the resulting definition didn't "work" when the attempt was made to establish that every formula provable in HA is realizable. Thus Heyting's proof-interpretation failed to help me to my goal.[6]

In my second trial a few weeks later of a definition of realizability, I thought of an implication "$\underline{A} \rightarrow \underline{B}$" in terms of an effective process for converting information $\underline{a}$ to complete "$\underline{A}$" into information $\underline{b}$ to complete "$\underline{B}$". This idea "worked" when I took into account that the effective process need only lead to a number $\underline{b}$ when

---

[6] Hence, with all due respect to Heyting's proof-interpretation, I do not agree with the statement of van Dalen in his lectures at this Logic Summer School that the proof-interpretation of Heyting is at the back of my realizability notion. Also Kolmogoroff's problem-interpretation 1932, which I knew from Heyting 1934 pp. 14, 17, failed to help me in any way of which I am conscious.

applied to a number <u>a</u> which does complete "<u>A</u>". So the form of Church's thesis which it is appropriate to use is not the original form (IM p. 300, after Church <u>1936</u>) but my form calling for a partial recursive function not necessarily defined except for such <u>a</u>'s (IM p. 332 Thesis I† (b)). [7]

Now I have described fully how I was led (early in 1941) to the definition of "E is realizable" for number-theoretic formulas which I published in <u>1945</u>. [8]

The resulting definitions of "<u>e</u> realizes E" (essentially) and of "E is realizable" (with the improvement to be noted) are shown in Table 1 Column A. The first of these definitions (Definition 1) is by induction (or "recursion") on the number of logical symbols in E, with seven cases. For example, "<u>e</u> realizes A&B" (the second case) is defined presupposing the meanings of "$(\underline{e})_0$ realizes A" and of "$(\underline{e})_1$ realizes B"; [9] but the inductive assumption justifies this presupposition, since A and B each have at least one fewer logical symbols than A&B. As I announced in <u>1948</u>, the proof of the theorem is simplified by using in Definition 2 the "realization function" φ instead of the closure of E. The theorem of Column A is proved on this basis in IM § 82.

At the bottom of Column A are three consequences of the theorem. The third (iii) is an approximation to the conjecture with which I started the program of research, while (i) and (ii) are similarly related to the interpretations of A∨B and

---

[7] I had introduced the theory of partial recursive functions briefly in <u>1938</u>, so in 1941 it had little currency. (By IM p. 324 Example 1 or <u>1938</u> Footnote 3, Church's potentially recursive functions <u>1936</u> will not serve equivalently.) I subsequently promoted the theory of partial recursive functions in <u>1943</u> § 10 and in IM Chapter XII, as well as through the realizability application.

On an occasion during the academic year 1939-40 I drew Gödel's attention to partial recursive functions; and in the summer of 1941 he told me he had used partial recursive functions somewhat differently and about simultaneously to get one of the same results mentioned below (the unprovability in the intuitionistic predicate calculus of $\neg\neg\forall x(\mathcal{a}(x) \vee \neg\mathcal{a}(x))$). [11]

[8] Cf. <u>1945</u>, the last paragraph of Footnote 1 (where "spring of 1941" means the spring term with classes from February 6 to June 7, 1941, rather than the spring season April 21 to June 22, 1941). I withheld the (detailed) publication until after David Nelson had confirmed for me each of a list of six propositions (I) - (VI) which in 1941 and 1942 I gave him to establish, the first (established in the summer of 1941) being the theorem of Table 1 Column A. However, I presented a preliminary report to the Amer. Math. Soc. and the Assoc. for Symbolic Logic on December 31, 1941, with abstract published in Bull. Amer. Math. Soc. vol. 48 (1942) p. 51 (received November 21, 1941) and in Jour. Symbolic Logic vol. 7 (1942) p. 47.

[9] Here $(\underline{e})_i$ is the exponent of the $i+1$-st prime p_i in the representation of $\underline{e}$ as a product of powers of distinct primes; so, as $\underline{e}$ ranges over all natural numbers, $((\underline{e})_0, (\underline{e})_1)$ ranges (with repetitions) over all pairs of natural numbers (IM p. 230 #19).

$\exists x A(x)$. In (iii), the conclusion has "$A(\underset{\sim}{x}, \underset{\sim}{y})$ is realizable" instead of "$A(\underset{\sim}{x}, \underset{\sim}{y})$ is true". The logical connectives in Definitions 1 and 2 can be read either classically or intuitionistically. In 1945 p. 110, I declared that the notion of realizability is to be regarded as providing only a partial analysis of the intuitionistic meanings, since it takes over without analysis, or leaves unanalyzed, the component of evidence (which, under Heyting's proof-interpretation, the proofs should supply). But I suggested that, when the connectives in Definitions 1 and 2 are read intuitionistically, this component is supplied in the reading of the connectives, so that "E is realizable intuitionistically" should be necessary and sufficient for "E is true intuitionistically". If this suggestion is accepted, (iii), which is established intuitionistically, would imply the conclusion of the conjecture as originally stated.

However, I also pointed out (1945 p. 115) how to establish the conjecture without leaning on such an identification of realizability, intuitionistically established, and intuitionistic truth. And indeed, in retrospect, there are grounds for refraining from that identification; in particular, as we shall see, there are other realizability properties possessed by all the formulas of HA.

To obtain the conjecture unalloyed I introduced (in 1945 p. 115) another definition of realizability, called in IM "E is realizable-($\vdash$)" (Column B Definition 2), obtained by modifying the treatment of the cases for $\vee$, $\rightarrow$, $\neg$ and $\exists$ in Definition 1. Take for example $\vee$. Under the intuitionistic meaning of $\vee$, a formula $A \vee B$ should be provable only when A is provable (and hence true) or B is provable (and hence true). But under the definition in Column A, we might have $(\underline{e})_0=0$ (so $(\underline{e})_1$ realizes A) when in fact B but not A is provable in HA, or $(\underline{e})_0 \neq 0$ (so $(\underline{e})_1$ realizes B) when A but not B is provable. The modification in the case for $A \vee B$ in Column B prevents this. In Column B, Consequence (iii) includes the conjecture as originally stated. Likewise, Consequence (i) and (ii) in Column B confirm similar conjectures about the provability in HA of disjunctions $A \vee B$ and existence statements $\exists x A(x)$.

The theorem in Column A has as other consequences the unprovability in HA of some formulas provable in classical arithmetic (e.g. of a suitable instance of the least-number principle, IM p. 511), the existence of consistent extensions of intuitionistic arithmetic contradictory to classical arithmetic (1945, IM p. 514),[10] and the unprovability in the intuitionistic predicate calculus of some formulas provable

[10] The consistency is established initially by the realizability interpretation. However, by Nelson's work in 1947 formalizing the realizability notion, the consistency can be concluded metamathematically relative to that of HA, as is argued in my 1945 § 14.

in the classical predicate calculus (e.g. of $\neg\neg\forall x(\mathcal{A}(x) \vee \neg\mathcal{A}(x))$).[11] For these applications, the greater simplicity of the definition in Column A (than in Column B) is advantageous.

Around 1960, I became aware of work of Harrop in 1956 giving new proofs of (i) and (ii) in Column B and in 1960 discussing similarly closed formulas of the forms $C \supset A \vee B$ and $C \supset \exists x A(x)$. It then struck me that the definitions in Column B are a mixture of two elements: recursion-theoretic (as in Column A) and proof-theoretic (added in Column B). If starting from Column B, we leave out the recursion-theoretic elements, we are led to the definitions in Column C. Following the terminology in my 1962a, I am writing the resulting notions "$|$ E" and "$\|$ E" (instead of "realizes" and "realizable"). Indeed, we get from the cases for Definition 1 in Column B to those in Column C, almost literally, by simply replacing such phrases as "$(\underline{e})_0$ realizes-($\vdash$)" by "$|$". This gives much the simplest proofs of the (i) and (ii) of Column B that I know.

The definitions in Columns B and C were given with an unspecified list Γ (or $C_1, \ldots, C_\ell$) of assumption formulas (which I took to be closed for Column C, and could have for Column B) before the symbol "$\vdash$" (where "$\Gamma \vdash$" means "is deductible from Γ").[12] The theorems of Columns A, B and C then generalize from proofs in HA to deductions in HA from Γ with the assumption formulas Γ each meeting the condition claimed for E in the conclusion.

Taking Γ to be a single formula C, and using the deduction theorem, the (generalized) theorem of Column C has the following consequences, for closed formulas $C \supset A \vee B$ and $C \supset \exists x A(x)$: (i*) If $C | C$ and $\vdash C \supset A \vee B$, then $\vdash C \supset A$ or $\vdash C \supset B$. (ii*) If $C | C$ and $\vdash C \supset \exists x A(x)$, then, for some x, $\vdash C \supset A(\underset{\sim}{x})$. (Harrop had the like with a different condition than $C | C$. Harrop's condition is decidable;

---

[11]For this purpose there are other methods; cf. Heyting 1946 and my 1948, IM pp. 487-492, and FIM 7.9-7.10 pp. 81-85. (For some history of the particular example $\neg\neg\forall x(\mathcal{A}(x) \vee \neg\mathcal{A}(x))$, cf. my 1945 p. 117 including Footnote 13 and end p. 123, Heyting 1946 p. 121, my 1948 pp. 186-187, IM Theorem 58 (a) p. 487, and FIM *27.17 p. 84.) The method of realizability is not complete for this purpose, even for the propositional calculus, by Rose 1953.

[12]Then in Column C, the entry "E is true" for the case E is prime is to be replaced by "$\Gamma \vdash$ E", and "not (Γ $|$ A and $\Gamma \vdash$ A)" for the case E is $\neg$A by "Γ $|$ $E \supset 1=0$". For Γ empty, these are equivalent via the consistency of HA to the entries in Column C, which I elected for this survey in order to allow Column C to come from Column B in the literal fashion described. For the same reason, in Column A I elected not to use an alternative entry which avoids "or" (namely, "If $(\underline{e})_0=0$ then $(\underline{e})_1$ realizes A, and if $(\underline{e})_0 \neq 0$ then $(\underline{e})_1$ realizes B"), which I introduced (in the formalized version) in 1960 (Theorem 1 p. 158) in response to Shanin 1958a.

mine is not. Mine is necessary and sufficient for the conjunction of the results; Harrop's is only sufficient.)

The concepts in Column C, without or with Γ, were also applied in my 1962a (with the 1963 addendum) to the intuitionistic predicate and propositional calculi, with similar results.

This led me to the conjecture that Heyting's propositional calculus can be characterized as the consistent propositional calculus T with the greatest class of provable formulas whose class of provable formulas is closed under modus ponens and substitution and includes all the provable formulas of Heyting's propositional calculus and which has the property (i*) (with "$\vdash$" meaning provable in T). de Jongh 1970 established the conjecture.[13]

PART II. INTUITIONISTIC ANALYSIS FIM (FUNCTION REALIZABILITY)

During 1950 I first addressed myself seriously to the problem of obtaining similar results for intuitionistic analysis. Some of my thoughts on the problem are in my paper 1950a. There I proposed to express intuitionistic analysis in a language which I subsequently developed (via 1957) into the formal system presented in Chapter I of the monograph "The Foundations of Intuitionistic Mathematics" by Kleene and Vesley 1965. I am using "FIM" to designate either the formal system or the monograph, according to context.

I found it surprisingly difficult to formulate a definition of realizability that would "work" for this language, the immediate test of "working" being that I should be able to establish the realizability of every formula provable (or deducible from realizable formulas) using the two-sorted intuitionistic predicate calculus, with number and (one-place number-theoretic) function variables, and such other postulates for intuitionistic analysis as I then had in mind.

In 1950, just as in 1941, I was proposing to codify the information to be supplied to complete communications by natural numbers. So I was seeking a definition of "e realizes E relative to given values $\underline{\Psi}$ of Ψ", or briefly of "e realizes-$\underline{\Psi}$ E", where Ψ are the free function variables of E or sometimes for convenience a larger list of function variables, and E contains free no number variables. For, while the free number variables can be replaced in all possible ways by numerals, in a constructive formal system one cannot replace the function variables by constant

[13] The like with (i) instead of (i*) had been conjectured by Łukasiewicz in 1952, and disproved by Kreisel and Putman in 1957.

functors expressing all number-theoretic functions, since $2^{\aleph_0}$ functors would be needed. Hence I had to allow free function variables to remain in E but to interpret them.

However, doing this for the function variables, there is not very much point to replacing the free number variables by numerals. So in FIM and here in Table 2, realizations are relative to values of all of a list of variables Ψ including all the variables (both number and function) occurring free in the formula E to be realized. Up through 1957, I still substituted numerals for the number variables.

For several weeks in October 1950, I spent all my research time trying out, laboriously, one after another way of extending, or of modifying and extending, the realizability definition of Column A to intuitionistic analysis. All reasonably direct or obvious extension-modifications which I thought of failed to "work" on one or another of the postulates of the two-sorted intuitionistic predicate calculus. I then laid the work aside with one more extension-modification, hardly direct or obvious, in my notes but not yet tested. Thirteen months later (on November 27, 1951), I returned to this; it didn't "work" as it stood, but I found and tested (on November 30-December 2, 1951) a further modification which did. The resulting definition of realizability, the laborious proof that it "works", and various results about it, were incorporated into a binder of typed notes which accumulated by November 1955 to 82 pages. I published the definition, and a minimum of remarks about it, in 1957.

Since I subsequently transformed this definition into a totally new format (shown in Table II Column D), I will here only note the decisive idea. This entered (characteristically) in the treatment of an implication $A \supset B$. Previously, to complete the communication $A \supset B$, I asked for an effective process, codified by the index of a partial recursive function, that would convert information, (codified by) $\underline{a}$, completing A into information $\underline{b}$ to complete B. Here $\underline{a}$ was to complete A in an absolute sense, constructively. Thus, all implications $A \supset B$ with A not constructively completable are realizable "vacuously". The modification (discovered on November 30, 1951) in contrast demands that the effective process convert any completion $\underline{a}$ of A, whether constructive or not, into a completion $\underline{b}$ of B of an equal or lesser degree of non-constructiveness than $\underline{a}$. This was implemented by allowing $\underline{e}$ to operate not just on $\underline{a}$ but on a function $\underline{\theta}$ which can exemplify any degree of non-constructiveness (the case of constructiveness being represented by a general recursive $\underline{\theta}$), as well as on the functions $\underline{\Psi}$ used as values of the function variables Ψ. Specifically, where Ψ_A are those of the function variables Ψ which actually occur free in A, $\underline{e}$ realizes-$\underline{\Psi}$

$A \supset B$ iff, for each $\underline{a}$ and $\underline{\theta}$, if $\underline{a}$ realizes-$\underline{\Psi}_A, \underline{\theta}$ A then $\{\underline{e}\}(\underline{\Psi}, \underline{\theta}, \underline{a})$ realizes B.

In the case $A \supset B$ is a number-theoretic formula (taking Ψ to be empty), this entails an alteration of the previous definition of realizability (Column A), to which I was led through my failure to find any "workable" extension of the unaltered realizability notion to the formulas of analysis. Having been thus coerced into it, I did come to find an appeal in the less stark form of constructivism which it represents. For "contrary-to-fact" conditionals $A \supset B$ to be true under this interpretation, there does have to be a relation between A and B — they are not all simply true "vacuously", just on the ground that A does not hold constructively. This interpretation makes a place in intuitionistic mathematics for the theory of relative recursiveness.

The technical difficulties of working with this definition of realizability (found in 1951 and published in 1957) were very great. In April 1959, it occurred to me that I might be able to put the definition in another form by using as the realizing objects functions $\underline{\epsilon}$ instead of numbers $\underline{e}$. To realize a closed formula E, $\underline{\epsilon}$ would be a recursive function. To realize-$\underline{\Psi}$ a formula E with the variables Ψ (including all free in E) interpreted by functions and numbers $\underline{\Psi}$, $\underline{\epsilon}$ would be recursive in the functions among $\underline{\Psi}$. In the case of $A \supset B$, an $\underline{\alpha}$ realizing A could be recursive in the functions $\underline{\Psi}_A$ evaluating the function variables Ψ_A free in A and in a function $\underline{\theta}$ incorporating a degree of non-constructiveness as indicated above. What had required of me a "shock" or discontinuity in my thinking to discover in 1950 and 1951 might thereby take care of itself almost automatically.

The outcome of this was that (in the next weeks) I devised a definition of realizability, with functions $\underline{\epsilon}$ as the realizing objects, which reads almost literally as the result of translating numbers into functions in the definition of Column A. In particular, I chose a meaning of "$\{\underline{\epsilon}\}[\underline{\alpha}]$" whereby a function $\underline{\epsilon}$ operates effectively on a function $\underline{\alpha}$ to produce a function $\{\underline{\epsilon}\}[\underline{\alpha}]$, analogously to "$\{\underline{e}\}(\underline{x})$" with numbers. Similarly, $\{\underline{\epsilon}\}[\underline{x}]$ is a function produced by the effective operation of a function $\underline{\epsilon}$ on a number $\underline{x}$, and $\{\underline{\epsilon}\}$ is simply a function produced effectively by $\underline{\epsilon}$. The problem was to pick the definitions of "$\{\underline{\epsilon}\}[\underline{\alpha}]$", "$\{\underline{\epsilon}\}[\underline{x}]$", "$\{\underline{\epsilon}\}$" appropriately. The definitions I chose are given in FIM pp. 91-92;[14] the concept underlying them was provided by my theory of countable functionals 1959a (found during the academic year

[14] Briefly, for each $\underline{t}$, $\underline{\epsilon}$ operates on (a code for) a number-tuple $(\underline{t}, \underline{\alpha}(0), \ldots, \underline{\alpha}(\underline{y}-1))$, for a sufficiently big $\underline{y}$, to produce the function value $\{\underline{\epsilon}\}[\underline{\alpha}](\underline{t})$ (and, when $\underline{y}$ is not suitable, to so indicate). Then $\{\underline{\epsilon}\}[\underline{x}] = \{\underline{\epsilon}\}[\lambda \underline{t}\, \underline{x}]$ and $\{\underline{\epsilon}\} = \{\underline{\epsilon}\}[0]$.

1956-7).[15] Thus I arrived at the definition of realizability that was published in FIM (1965) § 8, and which (essentially) is given in Table 2 Column D. In Definition 2 there, the "$\varphi[\Psi]$" with "Ψ" in square brackets indicates that the value of φ for the arguments Ψ is a (one-place number-theoretic) function in contrast to the "$\varphi(\Psi)$" in Column A, where the value is a natural number.[16]

Now the proof of the theorem is very much easier than with the 1957 definition (to which the Column D definition is equivalent). The treatment of the postulates of the predicate calculus (now two-sorted) parallels the corresponding treatment for Column A as given in IM § 82, and similarly with the postulates of number theory that are taken over in analysis. Of course, there are also postulates that enter only in analysis. The treatment, and indeed the formulation, of these postulates was somewhat delicate. In fact, it was in the process of attempting (in the summer of 1957) to establish the realizability of the simple classical form of the "bar theorem" (FIM *26.1 p. 53) that I discovered a counterexample (later reworked as FIM *27.23 p. 87) showing me the necessity of a restriction for the intuitionistic version (variously chosen in FIM ^{x}26.3a-^{x}26.3d pp. 54-55).

The consistency of FIM (listed among the consequences of the theorem in Column D) might not be taken for granted by a classical mathematician, since FIM has a postulate ("Brouwer's principle" or "continuity", FIM ^{x}27.1 p. 73) contradicting classical mathematics (cf. FIM p. 90).

Variants of the notion of realizability in Column D are obtained by restricting the function variables in Definitions 1 and 2 to range over a class $\underline{C}$ closed under general recursiveness and/or allowing the φ in Definition 2 to be recursive in the functions of a class of list $\underline{T}$ (where $\underline{T} \subset \underline{C}$); thus we obtain a notion "E is $\underline{C}$/realizable/$\underline{T}$", with $\underline{C}$ acting as a ceiling on all the functions considered, and $\underline{T}$ as a threshold below which constructivity is not demanded (FIM pp. 96, 111). In another variant, called "special realizability" or "$_{s}$realizability" (FIM § 10) the functions used in realizing a formula E are restricted to be of a certain sort or "order" e determined by the shape of E; this is done in such a way that, in the case of A⊃B, the value of the analog $^{b}\{^{e}\underline{\epsilon}\}[^{a}\underline{\alpha}]$ of $\{\underline{\epsilon}\}[\underline{\alpha}]$ (where a, b, e are the orders of A, B, A⊃B) is completely defined for each $\underline{\alpha}$ of order a even though $\underline{\alpha}$ may not realize

---

[15] "Functionals" in the present context refers to functions whose values are one-place number-theoretic functions (rather than natural numbers); but in this survey I have usually just called them "functions".

[16] In Column D Definition 2, "φ is general recursive" means that $\varphi[\Psi] = \lambda t\, \varphi(\Psi, t)$ where $\lambda\Psi t\, \varphi(\Psi, t)$ is general recursive. Also, $(\underline{\epsilon})_{i} = \lambda t\, (\underline{\epsilon}(t))_{i}$.

A. The /C and $_s$ variants are used in establishing the independence of the bar theorem (FIM pp. 112-113) and of Markov's principle (FIM pp. 131-132), respectively. Different realizability notions for intuitionistic analysis FIM include different notions for intuitionistic arithmetic HA, which in general will be non-equivalent.

At this point in the development of the subject, it is questionable whether there is profit in trying to ask whether one or another notion is intuitively the "right one" for the interpretation of intuitionism. Realizability, in its various forms, has proved a flexible tool for the investigation of intuitionistic systems.

In conclusion, I turn to the question whether the conjecture with which we began, and which was established as (iii) in Column B for HA, can be established for FIM; and similarly for (i) and (ii).

An obstacle in Table 2 to following the route we took in Table 1 from Column A to Columns B and C is that we do not have functors to express all the values of our function variables. Thus defeats using the Column C treatment in a direct manner in Table 2. As for the Column B treatment, we might circumvent the obstacle if $\{(\underline{\epsilon})_0\}$ in the case of Column D Definition 1 for $\exists\alpha A(\alpha)$ could always be a general recursive function. But $\exists\alpha A(\alpha)$ may contain free function variables, and $\{(\underline{\epsilon})_0\}$ will then be recursive in functions interpreting those variables but not necessarily recursive.

Joan Rand Moschovakis in <u>1967</u> obtained results (i) and (ii) as stated in Column B for subsystems of FIM lacking Brouwer's principle, by entertaining some non-constructive systems in which $2^{\aleph_0}$ symbols are adjoined to provide constant functors evaluating the functions (so the Column C method is used indirectly).[17]

In Column E the definition in Column D is modified as Column B modified that in Column A but with two differences, the first of which makes it feasible to entertain the second.

First, the definition is formalized as a formula in a suitable intuitionistic formalism, which in fact is FIM with some extension of the symbolism and postulates to provide additional primitive recursive functions. Since we explicitly left open the possibility of such an extension (FIM p. 19), and there is no intuitionistic advantage in being parsimonious in the use of primitive recursive functions, we shall not distinguish between the formal system FIM as first published in <u>1965</u> and the extended version published in <u>1969</u>. (We could have stuck with the <u>1965</u> version at the cost of some inconvenience.)

---

[17] Later she extended her methods to cover subsystems with a weakened form of Brouwer's principle (cf. my <u>1969</u> Footnote 41 p. 104).

Second, in terms of the interpretation, we have used truth instead of provability. For example, consider the case of A∨B in Definition 1. In Column B, we added informally "and ⊢ A" to the alternative with $(\underline{e})_0=0$, and "and ⊢ B" to the alternative with $(\underline{e})_0\neq 0$. Now in Column E, but in the formal symbolism, we add respectively "& A" and "& B", expressing respectively the truth of A and the truth of B.

I published the resulting definition and plan for investigation in 1967a.[18] I carried the work out in 1969.

The reader can understand the notations in Column E as suitable formalizations of the corresponding informal notations in Column D. Full explanations of these notations are in Part I of 1969 (75 pages long). A p-functor is an expression, in an extension of the symbolism of FIM, for a partial recursive function (perhaps general recursive); and, via rules given in 1969, the expressions "u ≃ v" for any p-functors u and v, and "!u & [u ⓠ E]" for any p-functor u (e.g. {ε}[α]), abbreviate formulas in the symbolism of FIM without this extension (but with the aforesaid additional symbolism for primitive recursive functions).

Furthermore, in Part I, there is a formal development of all the theory of recursive functionals which informally was used to prove the theorem in Column D,[15] and which suffices (in Part II, of 28 pages) for the proof, not only of the theorem formalized,[19] but also of the new theorem stated in Column E.

Among the consequences (in Part II of 1969) of the new theorem are the author's conjecture of 1940 established now for the intuitionistic formal system of

---

[18] In 1967a I also gave a direct formalization ε ⓡ E of the "ε realizes-$\underline{\Psi}$ E" of Column D. Thus in the case for A∨B, "& A" and "& B" are not added. This served some aims formulated in 1964a and in FIM pp. 90, 110. For example, using ε ⓡ E, work done in 1969 converts the model-theoretic consistency proof for FIM (Column D) into a metamathematical consistency proof relative to the "basic system" common to the intuitionistic and classical systems of analysis (as did Nelson 1947 for extensions of HA diverging from classical arithmetic).[10] Also, using ε ⓡ E, 1969 with 1964a establishes the 1964a conjecture that any prenex formula provable in FIM is provable in the basic system.

[19] Formalized, it reads as the theorem of Column E but with ⓡ instead of ⓠ and with the "⊢" in the conclusion referring to the basic system.[18]

analysis FIM (included in (iii) in Column E),[20] and other results such as (iv) where "$\exists \alpha_{GR(\alpha)}$" expresses "there exists a general recursive function $\underline{\alpha}$ such that".

From publications, reports and remarks in 1970 and 1971 (including at this Logic Summer School), by Läuchli, Troelstra, Friedman and others,[21] it is evident that realizability investigations are continuing to be actively pursued.

BIBLIOGRAPHY

CHURCH, ALONZO

1936. An unsolvable problem of elementary number theory. Amer. jour. of math., vol. 58, pp. 345-363.

HARROP, RONALD

1956. On disjunctions and existential statements in intuitionistic systems of logic. Math. Ann., vol. 132, pp. 347-361.

1960. Concerning formulas of the types $A \rightarrow B \vee C$, $A \rightarrow (Ex)B(x)$ in intuitionistic formal systems. Jour. symbolic logic, vol. 25, pp. 27-32.

HEYTING, AREND

1930. Die formalen Regeln der intuitionistischen Logik. Sitzungsberichte der Preussischen Akademie der Wissenschaften, Physikalisch-mathematische Klasse, 1930, pp. 42-56.

1930a. Die formalen Regeln der intuitionistischen Mathematik. Ibid., pp. 57-71, 158-169.

1934. Mathematische Grundlagenforschung. Intuitionismus. Beweistheorie. Ergebnisse der Mathematik und ihrer Grenzgebiete, vol. 3, no. 4, Berlin (Springer), pp. iv+73.

1946. On weakened quantification. Jour. symbolic logic, vol. 11, pp. 119-121.

HILBERT, DAVID and BERNAYS, PAUL

1934. Grundlagen der Mathematik, vol. 1. Berlin (Springer), xii+471 pp.

JONGH, DICK H. J. de

1970. A characterization of the intuitionistic propositional calculus. Intuitionism and proof theory, proceedings of the summer conference at Buffalo, N. Y., 1968, ed. A. Kino, J. Myhill, R. E. Vesley, Amsterdam and London (North-Holland Pub. Co.) 1970, pp. 211-217.

KLEENE, STEPHEN COLE

1938. On notation for ordinal numbers. Jour. symbolic logic, vol. 3, pp. 150-155.

1943. Recursive predicates and quantifiers. Trans. Amer. Math. Soc., vol. 53, pp. 41-73.

---

[20] In (iii) for Column E, "$A(\underset{\sim}{x}, \underset{\sim}{y})$ is true" has to mean true intuitionistically, accepting Brouwer's principle (which is false classically), as is seen for example by taking $\forall x \exists y A(x, y)$ to be the formula $\forall x \exists y(x=y) \& B)$ where B is Brouwer's principle. Only for an intuitionistic system, e.g. the basic system, not diverging from classical mathematics could we have (iii) holding with "true" in its classical sense.

[21] Cf. Items 64, 12 and 13 in the bibliography of the notes of van Dalen, and the notes of Friedman, distributed at this Logic Summer School.

1945. On the interpretation of intuitionistic number theory. Jour. symbolic logic, vol. 10, pp. 109-124.

1948. On the intuitionistic logic. Proceedings of the Tenth International Congress of Philosophy (Amsterdam, Aug. 11-18, 1948), Amsterdam (North-Holland Pub. Co.) 1949, pp. 741-743 (fasc. 2).

1950a. Recursive functions and intuitionistic mathematics. Proceedings of the International Congress of Mathematicians (Cambridge, Mass., U.S.A., Aug. 30-Sept. 6, 1950), 1952, vol. 1, pp. 679-685.

1952b IM. Introduction to metamathematics. Amsterdam (North-Holland Pub. Co.), Groningen (Noordhoff), New York and Toronto (Van Nostrand), X+550 pp. Sixth reprint, Amsterdam (North-Holland Pub. Co.), Groningen (Wolters-Noordhoff Pub.), New York (American Elsevier Pub. Co.) 1971.

1957. Realizability. Summaries of talks presented at the Summer Institute of Symbolic Logic in 1957 at Cornell University, vol. 1, pp. 100-104. Reprinted in Constructivity in mathematics, Amsterdam (North-Holland Pub. Co.) 1959, pp. 285-289.

1959a. Countable functionals. Constructivity in mathematics, Amsterdam (North-Holland Pub. Co.), pp. 81-100.

1960. Realizability and Shanin's algorithm for the constructive deciphering of mathematical sentences. Logique et analyse, 3e Année, Oct. 1960, 11-12, pp. 154-165.

1962a. Disjunction and existence under implication in elementary intuitionistic formalisms. Jour. symbolic logic, vol. 27, pp. 11-18. An addendum, ibid., vol. 28 (1963), pp. 154-156.

1964a. Classical extensions of intuitionistic mathematics. Logic, methodology and philosophy of science, proceedings of the 1964 international congress (held at Jerusalem Aug. 26-Sept. 2), ed. Yehoshua Bar-Hillel, Amsterdam (North-Holland Pub. Co.) 1965, pp. 31-44.

1967. Mathematical logic. New York, London, Sydney (John Wiley & Sons), xiii+398 pp.

1967a. Constructive functions in "The foundations of intuitionistic mathematics". Logic, methodology and philosophy of science III, proceedings of the Third International Congress for L., M. and P. of S., Amsterdam (Aug. 25-Sept. 2) 1967, ed. B. van Rootselaar and J. F. Staal, Amsterdam (North-Holland Pub. Co.) 1968, pp. 137-144.

1969. Formalized recursive functionals and formalized realizability. Memoirs Amer. Math. Soc., no. 89, 106 pp.

KLEENE, STEPHEN COLE and VESLEY, RICHARD EUGENE

1965 FIM. The foundations of intuitionistic mathematics, expecially in relation to recursive functions. Amsterdam (North-Holland Pub. Co.), VIII+206 pp.

KOLMOGOROFF, A. N. (KOLMOGOROV, A. N.)

1932. Zur Deutung der intuitionistischen Logik. Math. Zeitschr., vol. 35, pp. 58-65.

KREISEL, GEORG and PUTNAM, HILARY

1957. Eine Unableitbarkeitsbeweismethode für den intuitionistischen Aussagenkalkül. Archiv für mathematische Logik und Gründlagenforschung, vol. 3, nos. 3-4, pp. 74-78.

ŁUKASIEWICZ, JAN

1952. On the intuitionistic theory of deduction. Koninklijke Nederlandsche Akademie van Wetenschappen (Amsterdam), Proceedings, series A, vol. 55 (or Indagationes mathematicae, vol. 14), pp. 202-212.

MOSCHOVAKIS, JOAN RAND

1967. Disjunction and existence in formalized intuitionistic analysis. Sets, models and recursion theory, Proceedings of the Summer School in Mathematical Logic and Tenth Logic Colloquium, Leicester, Aug. - Sept. 1965, ed. John R. Crossley, Amsterdam (North-Holland Pub. Co.) 1967, pp. 309-331.

NELSON, DAVID

1947. Recursive functions and intuitionistic number theory. Trans. Amer. Math. Soc., vol. 61, pp. 307-368.

ROSE, GENE F.

1953. Propositional calculus and realizability. Trans. Amer. Math. Soc., vol. 75, pp. 1-19.

SHANIN, N. A. (ŠANIN, N. A.)

1958a. Ob algorifme konstruktivnoĭ rasšifrovki matematičeskih suždeniĭ (Über einen Algorithmus zur konstruktiven Dechiffrierung mathematischer Urteile). Zeitschr. für math. Logik und Grundlagen der Math., vol. 4, pp. 293-303.

III. SOME APPLICATIONS OF KLEENE'S METHODS FOR INTUITIONISTIC SYSTEMS

Harvey Friedman[1]
Department of Philosophy, Stanford University, U.S.A.

INTRODUCTION. Much space would be needed to properly discuss the foundational and philosophical interest of intuitionism and intuitionistic systems, and to this day such matters are the source of heated debate. However, independently of such unresolved matters, much work has gone into the study of formal properties of intuitionistic systems (many of which crop up naturally even from several diverging philosophical points of view). The aim of this paper is twofold: to push the elegant, intelligible, attractive, powerful, and concise methods of S. C. Kleene to obtain both new proofs of known results, and new results; also to give a self-contained, uniform exposition of several of the principal results about intuitionistic systems, suitable even for those who have never studied them before.

The principal matters dealt with here are a) a syntactic characterization of the intuitionistic propositional calculus among so called intermediate calculi, b) the so called disjunction and existence properties in propositional, many-sorted predicate, arithmetical, 2nd-order, and (finite) typed calculi, c) propositionally independent sequences of sentences in the calculi above, d) the consistency of Church's Thesis with a certain 2-sorted set theory, numbers and sets (which properly contains type theory), e) a consistency proof of a 1-sorted set theory in which bounded quantification is provably decidable, but not unbounded quantification.

Our a) is crucial to our c). The results of b) (and presumably c)) extend nicely to certain set theories. Such extensions will be reported by other researchers elsewhere.

Appendix 1 is devoted to problems, and Appendix 2 to other methods and research.

Two very different methods of Kleene (which were also meshed by him for some of his results) are used here. The one used in a)-c), which is Sections 1-5, uses $\vdash$,

[1] Research partially supported by NSF GP 29254.

and there are no "realizing objects," and a formula is either "realized" or "not realized" in a given context. Thus some have suggested that it not be called "realizability" at all; Kleene uses $|$ in Kleene [7]. The second method, for d), e), in Section 6, is ultimately based on Kleene [5], and was suggested by looking at Kreisel-Troelstra [8]. In this realizability, no use is made of $\vdash$, but heavy use is made of recursion theory; indices of recursive functions realize formulae, now. For e), we use indices of meta-recursive functions.

In Section 1, propositional calculus only is considered, the "realizability" notion defined, and soundness proved. An unusual measure of complexity is used for the proof of Theorem 1.3, which is our a).

Theorem 1.3 was suggested by experience. People who are familiar with intuitionistic propositional calculus, when actually confronted with the problem of whether $T \vdash \varphi$, will first break the problem down into an equivalent set of problems. Then, if they get stuck, they will try to simplify the problem by going after implicative antecedents (like proving $T \vdash (\alpha \to \beta)$ for some $((\alpha \to \beta) \to \gamma) \in T$ and replacing $T \vdash \varphi$ by $T - ((\alpha \to \beta) \to \gamma) + \gamma \vdash \varphi$). This has the drawback of increasing the number of problems, but what else is the poor soul going to do (until he can see that there is no point in continuing at all). This whole scene suggests <u>backwards proofs</u>, and a whole <u>backward proof theory</u>. This is beyond the scope of this paper.

The Kreisel-Putnam scheme is written $(\sim A \to (B \vee C)) \to (\sim A \to B \vee \sim A \to C)$. We conclude Section 1 by proving a general theorem which has as a consequence that the scheme above introduces no new provable negative formulae, and that every subset has the disjunction property.

In Section 2, the "realizability" is introduced and soundness is proved for many-sorted predicate logic. The disjunction and existence properties, relative to a relational type (A,B), are introduced. A necessary and sufficient condition, in terms of the "realizability" is given for a theory to have both the existence and disjunction properties (relative to (A,B)). The rest of the section consists of three theorems which respectively show the $\exists, \vee$-properties, $\vee$-property, and a weak form of the $\exists$-property for 3 general classes of theories.

In Section 3, Heyting arithmetic (HA), and extensions by transfinite induction on well-orderings with index x (TI(x)) are defined, and proved to be covered by the three theorems of Section 2. A general theorem is then proved about the existence of sequences of propositionally independent Π_2^0 sentences in theories based on arithmetic. Theorem 1.3 is used in a crucial way. A consequence is that any sequence of Π_2^0 sentences classically independent over classical arithmetic + true Π_1^0 sentences is automatically independent over intuitionistic arithmetic.

In Section 4 we consider 2nd-order theories based on the comprehension axioms. We do not introduce abstraction constants or terms (except in the auxiliary theories). Consequently, we state the disjunction property proved in Theorem 4.1 to allow for 2nd-order parameters; and we also distinguish the existence property for existential 1st-order quantifiers and existential 2nd-order quantifiers. In the latter case, since we do not use closed 2nd-order terms, we must use formulae instead, in the obvious way. The rest of Section 4 deals with independent sequences of Π_2^0 sentences; e.g., allowing us to replace the results, in this connection, of Section 3 about arithmetic by 2nd-order arithmetic.

In Section 5, we first push the results of Section 4 as far as they can go, for type theory. All the results lift except the formulation of the existence property for existential higher-order quantifiers. This can be seen to be outright false. So we then add rested abstraction constants, on top of free constants. For our purposes, this is better than (though equivalent to) using, instead, nested abstraction terms; the free constants take the place of parameters in the term formulation.

In Section 6, we first introduce a set theory on top of number theory, which at least we know to be capable of formulating the cumulative hierarchy up to each stage below Church-Kleene ω_1. We give a recursion-theoretic realizability interpretation, using indices of partial recursive functions, whose novelty is in the atomic formula clause. We prove soundness, and prove that the set theory is consistent with, among other things, Church's Thesis.

In the second part of Section 6 we consider a set theory with bounded quantification decidable. We must therefore take separation only for decidable formulae, not for all formulae; otherwise the theory would be classical. (More philosophically, the idea is that not every unbounded quantification is well-determined; but sets are to have a well-determined membership relation.) We use a recursion-theoretic realizability similar to that of Kleene [5] for arithmetic. Instead, here, meta-recursive functions and indices are used - in particular, Σ_2 functions on L(K) for K a limit cardinal of cofinality $> \omega$ with L(K) satisfying Σ_2-replacement.[2/] We prove soundness, and prove that this second set theory is consistent, with the consistency proof formalizable in classical ZFC.

Throughout Section 6 we use indices of recursive and meta-recursive functionals, but not in an essential way. We could have used, instead, indices of recursive and meta-recursive functions of several arguments, but this is notationally more cumbersome.

[2/] This is the same as L(K) being admissible with respect to the power set operation; and in this case, Σ_2 functions on L(K) are L(K)-recursive w.r.t. power set.

SECTION 1. PROPOSITIONAL CALCULUS

In this section we concern ourselves only with the Heyting propositional calculus, which we describe presently.

DEFINITION 1.1. The atoms are symbols p_i, $i \in \omega$. The formulae are given by i) each atom is a formula, ii) $\perp$ (for absurdity) is a formula,, iii) if φ, ψ are formulae, then so are $(\varphi \& \psi)$, $(\varphi \vee \psi)$, and $(\varphi \rightarrow \psi)$. A theory T is a set of formulae.

DEFINITION 1.2. (Spector.) Let T be a theory. We define $T \vdash \varphi$ to be the least predicate on formulae φ such that for all formulae φ, ψ, ρ, the following holds: 1. if $\varphi \in T$, then $T \vdash \varphi$ 2. $T \vdash (\varphi \rightarrow \varphi)$ 3. if $T \vdash \varphi$, $T \vdash (\varphi \rightarrow \psi)$, then $T \vdash \psi$ 4. if $T \vdash (\varphi \rightarrow \psi)$, $T \vdash (\psi \rightarrow \rho)$, then $T \vdash (\varphi \rightarrow \rho)$ 5. $T \vdash ((\varphi \& \psi) \rightarrow \varphi)$, $T \vdash ((\varphi \& \psi) \rightarrow \psi)$, $T \vdash (\varphi \rightarrow (\varphi \vee \psi))$, $T \vdash (\psi \rightarrow (\varphi \vee \psi))$ 6. if $T \vdash (\varphi \rightarrow \rho)$, $T \vdash (\psi \rightarrow \rho)$, then $T \vdash ((\varphi \vee \psi) \rightarrow \rho)$ 7. if $T \vdash (\rho \rightarrow \varphi)$, $T \vdash (\rho \rightarrow \psi)$, then $T \vdash (\rho \rightarrow (\varphi \& \psi))$ 8. if $T \vdash ((\varphi \& \psi) \rightarrow \rho)$, then $T \vdash (\varphi \rightarrow (\psi \rightarrow \rho))$ 9. if $T \vdash (\varphi \rightarrow (\psi \rightarrow \rho))$, then $T \vdash ((\varphi \& \psi) \rightarrow \rho)$ 10. $T \vdash (\perp \rightarrow \varphi)$. We write $T \vdash T_1$ iff $(\forall \varphi \in T_1)(T \vdash \varphi)$. We write $\vdash \varphi$, $\vdash T_1$ for $\emptyset \vdash \varphi$, $\emptyset \vdash T_1$.

DEFINITION 1.3. Let T be a theory, φ a formula, S a set of atoms. We define $R(S,T,\varphi)$ by induction on φ: a) $R(S,T,p_i)$ iff $p_i \in S$ b) not $R(S,T,\perp)$ c) $R(S,T,(\varphi \& \psi))$ iff $R(S,T,\varphi)$ and $R(S,T,\psi)$ d) $R(S,T,(\varphi \vee \psi))$ iff $(R(S,T,\varphi), T \vdash \varphi)$ or $(R(S,T,\psi), T \vdash \psi)$ e) $R(S,T,(\varphi \rightarrow \psi))$ iff if $R(S,T,\varphi)$, $T \vdash \varphi$, then $R(S,T,\psi)$. Take $R(S,T,T_1)$ iff $(\forall \varphi \in T_1)(R(S,T,\varphi))$.

DEFINITION 1.4. A theory T is called disjunctive just in case $T \vdash (\varphi \vee \psi)$ implies $T \vdash \varphi$ or $T \vdash \psi$. A theory T is called strongly disjunctive iff every subset of T is disjunctive.

Note that T is strongly disjunctive if and only if every finite subset of T is disjunctive. Also $\{\varphi\}$ is strongly disjunctive iff it is disjunctive.

THEOREM 1.1. (Soundness.) If $T \vdash T_1$, $T_1 \vdash T_2$, $R(S,T,T_1)$, then $R(S,T,T_2)$.

Proof: Let $T \vdash T_1$, $T_1 \vdash T_2$, $R(S,T,T_1)$, $\varphi \in T_2$, S a set of atoms. We prove $R(S,T,\varphi)$ by induction on the proof of φ from T_1. In fact we show that the predicate on φ, "$T_1 \vdash \varphi$ and $R(S,T,\varphi)$" satisfies conditions 1-10 of Definition 1.2: 1. Let $\varphi \in T_1$. Then $T_1 \vdash \varphi$ and $R(S,T,\varphi)$ 2. Clearly $T_1 \vdash (\varphi \to \varphi)$, $R(S,T,(\varphi \to \varphi))$ 3. Assume $T_1 \vdash \varphi$, $T_1 \vdash (\varphi \to \psi)$, $R(S,T,\varphi)$, $R(S,T,(\varphi \to \psi))$. Then $T \vdash \varphi$, and hence $R(S,T,\psi)$, $T_1 \vdash \psi$ 4. Assume $T_1 \vdash (\varphi \to \psi)$, $T_1 \vdash (\psi \to \rho)$, $R(S,T,(\varphi \to \psi))$, $R(S,T,(\psi \to \rho))$. Then $T_1 \vdash (\varphi \to \rho)$. Assume $T \vdash \varphi$, $R(S,T,\varphi)$. Then $R(S,T,\psi)$, $T \vdash \psi$. Then $R(S,T,\rho)$. Hence $R(S,T,(\varphi \to \rho))$ 5. Clearly $T_1 \vdash ((\varphi \& \psi) \to \varphi)$. Assume $R(S,T,(\varphi \& \psi))$, $T \vdash (\varphi \& \psi)$. Then $R(S,T,\varphi)$. Clearly $T_1 \vdash ((\varphi \& \psi) \to \psi)$. Assume $R(S,T,(\varphi \& \psi))$, $T \vdash (\varphi \& \psi)$. Then $R(S,T,\psi)$. Clearly $T_1 \vdash (\varphi \to (\varphi \vee \psi))$. Assume $R(S,T,\varphi)$, $T \vdash \varphi$. Then $R(S,T,(\varphi \vee \psi))$. Clearly $T_1 \vdash (\psi \to (\varphi \vee \psi))$. Assume $R(S,T,\psi)$, $T \vdash \psi$. Then $R(S,T,(\varphi \vee \psi))$ 6. Assume $T_1 \vdash (\varphi \to \rho)$, $T_1 \vdash (\psi \to \rho)$, $R(S,T,(\varphi \to \rho))$, $R(S,T,(\psi \to \rho))$. Clearly $T_1 \vdash ((\varphi \vee \psi) \to \rho)$. Assume $R(S,T,(\varphi \vee \psi))$, $T \vdash (\varphi \vee \psi)$. Assume $R(S,T,\varphi)$, $T \vdash \varphi$, the other case being symmetric. Then $R(S,T,\rho)$ 7. Assume $T_1 \vdash (\rho \to \varphi)$, $T_1 \vdash (\rho \to \psi)$, $R(S,T,(\rho \to \varphi))$, $R(S,T,(\rho \to \psi))$. Clearly $T_1 \vdash (\rho \to (\varphi \& \psi))$. Assume $R(S,T,\rho)$, $T \vdash \rho$. Then $R(S,T,\varphi)$, $R(S,T,\psi)$. Hence $R(S,T,(\varphi \& \psi))$ 8. Assume $T_1 \vdash ((\varphi \& \psi) \to \rho)$, $R(S,T,((\varphi \& \psi) \to \rho))$, $T \vdash \varphi$, $R(S,T,\varphi)$, $T \vdash \psi$, $R(S,T,\psi)$. Then $T \vdash (\varphi \& \psi)$, $R(S,T,(\varphi \& \psi))$. Hence $R(S,T,\rho)$ 9. Assume $T_1 \vdash (\varphi \to (\psi \to \rho))$, $R(S,T,(\varphi \to (\psi \to \rho)))$, $T \vdash (\varphi \& \psi)$, $R(S,T,(\varphi \& \psi))$. Then $T \vdash \varphi$, $T \vdash \psi$, $R(S,T,\varphi)$, $R(S,T,\psi)$. Hence $R(S,T,(\psi \to \rho))$. Hence $R(S,T,\rho)$ 10. Assume $R(S,T,\bot)$. Then $R(S,T,\varphi)$ vacuously.

THEOREM 1.2. Let T be a theory. The following are equivalent: i) T is disjunctive and not $T \vdash \bot$ ii) $R(S,T,T)$ for some S iii) $R(S,T,T)$ for $S = \{p_i : T \vdash p_i\}$ iv) $R(S,T,T)$ if $\{p_i : T \vdash p_i\} \subset S$. Hence the problem of whether a finite theory is disjunctive is recursively decidable.

Proof: i) $\to$ iv). We prove $R(S,T,\varphi)$ for all $T \vdash \varphi$ by induction on the complexity of φ: a) if $T \vdash p_i$, then $R(S,T,p_i)$ b) not $T \vdash \bot$ c) if $T \vdash (\varphi \& \psi)$, then $T \vdash \varphi$, $T \vdash \psi$, and so $R(S,T,\varphi)$, $R(S,T,\psi)$, and hence $R(S,T,(\varphi \& \psi))$ d) if $T \vdash (\varphi \vee \psi)$, then $T \vdash \varphi$ or $T \vdash \psi$, and so $T \vdash \varphi$,

$R(S,T,\varphi)$ or $T \vdash \psi$, $R(S,T,\psi)$, and hence $R(S,T,(\varphi \vee \psi))$ e) if $T \vdash (\varphi \to \psi)$, $T \vdash \varphi$, $R(S,T,\varphi)$, then $T \vdash \psi$, and hence $R(S,T,\psi)$. ii) $\to$ i). Let $T \vdash (\varphi \vee \psi)$. Note $T \vdash T$, $T \vdash (\varphi \vee \psi)$, $R(S,T,T)$. Hence by Theorem 1.1, $R(S,T,(\varphi \vee \psi))$. Hence $T \vdash \varphi$ or $T \vdash \psi$. Also, not $T \vdash \bot$, since if $T \vdash \bot$, then by Theorem 1.1, $R(S,T,\bot)$. The recursive decidability of disjunctivity follows from the well-known decidability of $\vdash$, which can also be obtained from the proof of Theorem 1.3.

DEFINITION 1.5. T is called reduced iff for all $\varphi \in T$ either i) φ is an atom ii) φ is $(p_i \to \gamma)$ for some $p_i \notin T$, formula γ iii) φ is $((\alpha \to \beta) \to \gamma)$ for some α, β, γ.

LEMMA 1.3.1. If T is reduced and not $T \vdash (\alpha \to \beta)$ for all $((\alpha \to \beta) \to \gamma) \in T$, then T is disjunctive. If T is reduced and not $T \vdash (\alpha \to \beta)$ for all $((\alpha \to \beta) \to \gamma) \in T$, then the only atoms provable from T are elements of T.

Proof: Assume T is reduced, not $T \vdash (\alpha \to \beta)$ for all $((\alpha \to \beta) \to \gamma) \in T$. Let S be the set of atoms in T. We claim $R(S,T,T)$. To see this, let $\varphi \in T$. Case 1: φ is an atom. Then $R(S,T,\varphi)$. Case 2: φ is $(p_i \to \gamma)$, $p_i \notin T$. Then $p_i \notin S$, and hence $R(S,T,\varphi)$. Case 3: φ is $((\alpha \to \beta) \to \gamma)$. Then not $T \vdash (\alpha \to \beta)$, and so $R(S,T,\varphi)$. Hence by Theorem 1.2, T is disjunctive. For the 2nd part, let $T \vdash p_i$. Then $R(S,T,p_i)$, and so $p_i \in S$. Hence $p_i \in T$.

LEMMA 1.3.2. To every φ there is a finite T which is conjunctionless, such that $T \vdash \varphi$ and $\{\varphi\} \vdash T$.

Proof: By induction on the complexity of φ: i) if φ is $\bot$, take $T = \{\bot\}$; if φ is p_i, take $T = \{p_i\}$ ii) if φ is $(\psi \,\&\, \rho)$ and $T_1 \vdash \psi$, $T_2 \vdash \rho$, $\{\psi\} \vdash T_1$, $\{\rho\} \vdash T_2$, take $T = T_1 \cup T_2$ iii) if φ is $(\psi \vee \rho)$, $T_1 \vdash \psi$, $T_2 \vdash \rho$, $\{\psi\} \vdash T_1$, $\{\rho\} \vdash T_2$, take $T = \{(\alpha \vee \beta): \alpha \in T_1, \beta \in T_2\}$ iv) if φ is $(\psi \to \rho)$, $T_1 \vdash \varphi$, $T_2 \vdash \rho$, $\{\psi\} \vdash T_1$, $\{\rho\} \vdash T_2$, then we let $T_1 = \{\alpha_1,\ldots,\alpha_k\}$. Take $T = \{(\alpha_1 \to (\alpha_2 \to (\cdots(\alpha_k \to \beta)\cdots))): \beta \in T_2\}$.

DEFINITION 1.6. Let $f \in 2^{\omega}$. We define $f \models \varphi$ by induction on the complexity of φ: i) not $f \models \bot$, $f \models p_i$ iff $f(i) = 0$ ii) $f \models (\varphi \,\&\, \psi)$ iff $f \models \varphi$ and

$f \models \psi$ iii) $f \models (\varphi \vee \psi)$ iff $f \models \varphi$ or $f \models \psi$ iv) $f \models (\varphi \to \psi)$ iff (if $f \models \varphi$, then $f \models \psi$). We write $f \models T$ iff $(\forall\varphi \in T)(f \models \varphi)$, for theories T.

LEMMA 1.3.3. If no $f \models T$, then $T \vdash \bot$.

Proof: It suffices to prove this for finite T, since if no $f \models T$ then no $f \models T_o$, for some finite $T_o \subset T$. Hence it suffices to prove this for T of cardinality 1. Assume no $f \models \varphi$. We want $\{\varphi\} \vdash \bot$. Let us assume that this is true for all φ whose atoms are among $p_1,\ldots,p_k$. Let φ have atoms among $p_1,\ldots,p_k$, p_{k+1}. Let ψ_1 be the result of replacing $\bot$ for every occurrence of p_{k+1} in φ, and let ψ_2 be the result of replacing $(\bot \to \bot)$ for every occurrence of p_{k+1} in φ. Then no $f \models \psi_1$, no $f \models \psi_2$. Hence $\{\psi_1\} \vdash \bot$, $\{\psi_2\} \vdash \bot$. Now $\{\varphi,(p_k \to \bot), (\bot \to p_k)\} \vdash \psi_1$, $\{\varphi,(p_k \to (\bot \to \bot)), ((\bot \to \bot) \to p_k)\} \vdash \psi_2$. So $\{\varphi\} \vdash ((p_k \to \bot) \to \bot)$, $\{\varphi\} \vdash (p_k \to \bot)$. Hence $\{\varphi\} \vdash \bot$.

DEFINITION 1.7. Define a measure of complexity, μ, on formulae by $\mu(p_i) = \mu(\bot) = 3$, $\mu((\varphi \,\&\, \psi)) = \mu(\varphi) + \mu(\psi) + 1$, $\mu((\varphi \vee \psi)) = \mu(\varphi) + \mu(\psi)$, $\mu((\varphi \to \psi)) = \mu(\varphi)^{\mu(\psi)}$, where the latter indicates arithmetic exponentiation. For finite Theories T, $\mu(T) = \sum_{\varphi\in T} \mu(\varphi)$.

LEMMA 1.3.4. $\mu((\alpha \to \gamma)) + \mu((\beta \to \gamma)) < \mu(((\alpha \vee \beta) \to \gamma))$, $\mu(\alpha) + \mu(\beta) + \mu((\beta \to \gamma)) < \mu(((\alpha \to \beta) \to \gamma))$.

Proof: We have $a^c + b^c < (a + b)^c$ for $a, b, c \geq 3$ by the binomial theorem. We have $a + b + b^c < (a^b)^c$ for $a, b, c \geq 3$ by i) $6b < 3^b$ for $b \geq 3$ ii) $2b < 3^{b-1}$ for $b \geq 3$ iii) $2b < a^{b-1}$ for $a, b \geq 3$ iv) $2ab < a^b$ for $a, b \geq 3$ v) $2a^cb^c < (a^b)^c$ for $a, b, c \geq 3$ vi) $a + b < a^cb^c$ for $a, b, c \geq 3$ vii) $b^c < a^cb^c$ for $a, b, c \geq 3$ viii) $a + b + b^c < 2a^cb^c < (a^b)^c$ for $a, b, c \geq 3$.

THEOREM 1.3. The only binary relation $\vdash_*$ between theories satisfying the following conditions is $\vdash_* = \vdash$: 1) $T \vdash_* T_1$ iff $(\forall\varphi \in T_1)(\exists \text{ finite } T_2 \subset T)(T_2 \vdash_* \varphi)$ 2) if $T \vdash T_1$ then $T \vdash_* T_1$ 3) if $T \vdash_* T_1$, $T_1 \vdash_* T_2$ then $T \vdash_* T_2$ 4) if $T \vdash_* T_1$ then every $f \models T$ has $f \models T_1$ 5) if $T \vdash_* \{(\alpha \to \beta)\}$ then $T \cup \{\alpha\} \vdash_* \beta$ 6) if T is reduced, $T \vdash_* \{p_i\}$, then $p_i \in T$ or $T \vdash_* \{(\alpha \to \beta)\}$ for some

$((\alpha \to \beta) \to \gamma) \in T$ 7) if T is reduced, $T \vdash_* \{(\psi \vee \rho)\}$, then $T \vdash_* \{\psi\}$ or $T \vdash_* \{\rho\}$ or $T \vdash_* \{(\alpha \to \beta)\}$ for some $((\alpha \to \beta) \to \gamma) \in T$.

Proof: By Lemma 1.3.1, $\vdash$ satisfies the conditions.

Assume $\vdash_*$ satisfies the conditions. It suffices to prove that, for T finite, $T \vdash_* \{\varphi\}$ iff $T \vdash \varphi$, by conditions 1) and 2).

It suffices to prove that, for finite conjunctionless T, φ, $T \vdash_* \{\varphi\}$ iff $T \vdash \varphi$: Suppose this is true, $T \vdash_* \varphi$, T finite, T, φ possibly with conjunctions. By Lemma 1.3.2, choose conjunctionless T_1, φ_1 such that $T \vdash T_1$, $T_1 \vdash T$, $\{\varphi\} \vdash \varphi_1$, $\{\varphi_1\} \vdash \varphi$. Then $T_1 \vdash_* T$, $\{\varphi\} \vdash_* \{\varphi_1\}$, by condition 2). Hence $T_1 \vdash_* T \vdash_* \{\varphi\} \vdash_* \{\varphi_1\}$ by condition 3). So $T_1 \vdash \varphi_1$. Hence $T \vdash \varphi$. So it does suffice.

So henceforth we will assume T is finite, and T, φ are conjunctionless.

Let us further assume that for all finite, conjunctionless T_1, conjunctionless φ_1 with $\mu(T_1) + \mu(\varphi_1) < n$, we have $T_1 \vdash_* \varphi_1$ iff $T_1 \vdash \varphi_1$. Assume $T \vdash_* \varphi$, $\mu(T) + \mu(\varphi) \leq n$. Then we will be done after we demonstrate that $T \vdash \varphi$: I. T is not reduced. IA. $\bot \in T$. Then $T \vdash \varphi$. IB. $(\alpha \vee \beta) \in T$. Let $T' = T - \{(\alpha \vee \beta)\}$. Now $T' \cup \{\alpha\} \vdash T$, $T' \cup \{\beta\} \vdash T$. Hence $T' \cup \{\alpha\} \vdash_* T \vdash_* \{\varphi\}$, $T' \cup \{\beta\} \vdash_* T \vdash_* \{\varphi\}$, and hence $T' \cup \{\alpha\} \vdash \varphi$, $T' \cup \{\beta\} \vdash \varphi$, and so $T \vdash \varphi$. IC. $(\bot \to \gamma) \in T$. Let $T' = T - \{(\bot \to \gamma)\}$. Then $T' \vdash T$, and hence $T' \vdash_* T \vdash_* \{\varphi\}$. So $T' \vdash \varphi$, $T \vdash \varphi$. ID. $((\alpha \vee \beta) \to \gamma) \in T$. Let $T' = T - \{((\alpha \vee \beta) \to \gamma)\} \cup \{(\alpha \to \gamma), (\beta \to \gamma)\}$. Then by Lemma 1.3.4, $\mu(T') < \mu(T)$. Now $T' \vdash T$. Hence $T' \vdash_* T \vdash_* \{\varphi\}$. So $T' \vdash \varphi$, $T \vdash \varphi$. IE. $(p_i \to \gamma), p_i \in T$. Let $T' = T - \{(p_i \to \gamma)\} \cup \{\gamma\}$. Then $T' \vdash T$. So $T' \vdash_* T \vdash_* \{\varphi\}$. Hence $T' \vdash \varphi$, $T \vdash \varphi$. II. T is reduced. IIA. φ is $\bot$. Then $T \vdash_* \{\bot\}$, and hence by condition 4), no $f \models T$. Hence by Lemma 1.3.3, we have $T \vdash \bot$. IIB. φ is $(\alpha \to \beta)$. Then $T \vdash_* \{(\alpha \to \beta)\}$, $T \cup \{\alpha\} \vdash_* \{\beta\}$ by condition 5). So $T \cup \{\alpha\} \vdash \beta$, $T \vdash (\alpha \to \beta)$. IIC. φ is p_i. If $p_i \in T$ then $T \vdash p_i$. By condition 6), assume $T \vdash_* \{(\alpha \to \beta)\}$, $((\alpha \to \beta) \to \gamma) \in T$, $T' = T - \{((\alpha \to \beta) \to \gamma)\}$. So $T' \cup \{((\alpha \to \beta) \to \gamma), \alpha\} \vdash_* \{\beta\}$. Now $T' \cup \{(\beta \to \gamma), \alpha\} \vdash T' \cup \{((\alpha \to \beta) \to \gamma), \alpha\}$. Hence $T' \cup \{(\beta \to \gamma), \alpha\} \vdash_* \{\beta\}$. Now $\mu(T) = \mu(T') + \mu(((\alpha \to \beta) \to \gamma))$, $\mu(T' \cup \{(\beta \to \gamma), \alpha\}) + \mu(\beta) \leq \mu(T') + \mu((\beta \to \gamma)) + \mu(\alpha) + \mu(\beta) < \mu(T)$, by Lemma 1.3.4. Hence $T' \cup \{(\beta \to \gamma), \alpha\} \vdash \beta$. So $T \cup \{\alpha\}$

$\vdash \beta$, $T \vdash (\alpha \to \beta)$. So $T \vdash \gamma$. Now $T' \cup \{\gamma\} \vdash_* \{p_i\}$, since $T' \cup \{\gamma\} \vdash T$. Hence $T' \cup \{\gamma\} \vdash p_i$. So $T \vdash p_i$. IID. φ is $(\psi \vee \rho)$. If $T \vdash_* \psi$ or $T \vdash_* \rho$ then we are done. So we may assume by condition 7) that $T \vdash_* \{(\alpha \to \beta)\}$ for some $((\alpha \to \beta) \to \gamma) \in T$. Let $T' = T - \{((\alpha \to \beta) \to \gamma)\}$. So $T' \cup \{((\alpha \to \beta) \to \gamma),\alpha\} \vdash_* \{\beta\}$. Now $T' \cup \{(\beta \to \gamma),\alpha\} \vdash T' \cup \{((\alpha \to \beta) \to \gamma),\alpha\}$. Hence $T' \cup \{(\beta \to \gamma),\alpha\} \vdash_* \{\beta\}$. Now $\mu(T) = \mu(T') + \mu(((\alpha \to \beta) \to \gamma))$, $\mu(T' \cup \{(\beta \to \gamma),\alpha\}) + \mu(\beta) \leq \mu(T') + \mu((\beta \to \gamma)) + \mu(\alpha) + \mu(\beta) < \mu(T)$, by Lemma 1.3.4. Hence $T' \cup \{(\beta \to \gamma),\alpha\} \vdash \beta$. So $T \cup \{\alpha\} \vdash \beta$, $T \vdash (\alpha \to \beta)$. So $T \vdash \gamma$. Now $T' \cup \{\gamma\} \vdash_* \{(\psi \vee \rho)\}$, since $T' \cup \{\gamma\} \vdash T$. Hence $T' \cup \{\gamma\} \vdash (\psi \vee \rho)$. So $T \vdash (\psi \vee \rho)$.

DEFINITION 1.8. Let K be the least class A of formulae satisfying 1) $\bot \in A$, and each atom is in A 2) $(\varphi \,\&\, \psi) \in A$ if $\varphi,\psi \in A$ 3) $(\varphi \to \psi) \in A$ if $\psi \in A$ 4) $((\varphi \to (\psi \vee \rho)) \to ((\varphi \to \psi) \vee (\varphi \to \rho))) \in A$ if $\varphi \in A$. Let L_o be the least class A satisfying 1), 2), 3). Let L_1 be $\{((\varphi \to (\psi \vee \rho)) \to ((\varphi \to \psi) \vee (\varphi \to \rho))): \varphi \in L_o\}$. Let $L = L_o \cup L_1$. Obviously $L \subset K$, $L_o \cap L_1 = \emptyset$.

LEMMA 1.4.1. Let $S = \{p_i\colon i \in \omega\}$, $T \subset K$, T finite, not $T \vdash \bot$. Then $R(S,T,T)$. So K is strongly disjunctive.

Proof: By induction on $\mu(T)$. Assume that $R(S,T_o,T_o)$ for all $T_o \subset K$ with $\mu(T_o) < k$, not $T \vdash \bot$. Let $\mu(T) = k$, $T \subset K$, not $T \vdash \bot$. We want to show $R(S,T,\varphi)$ for all $\varphi \in T$. If φ is an atom, then $R(S,T,\varphi)$. If φ is $(\psi \,\&\, \rho)$, φ, $\psi \in K$, then let $T_o = T - \{\varphi\} \cup \{\varphi,\psi\}$. Then $T_o \vdash T$, $T \vdash T_o$, $\mu(T_o) < k$, $T_o \subset K$. So $R(S,T_o,T_o)$, $R(S,T,T)$. If φ is $(\psi \to \rho)$, $\rho \in K$, then assume $T \vdash \psi$, $R(S,T,\psi)$. Then let $T_o = T - \{\varphi\} \cup \{\rho\}$. Then $T_o \vdash T$, $T \vdash T_o$, $\mu(T_o) < k$, $T_o \subset K$. So $R(S,T_o,T_o)$, $R(S,T,T)$. If φ is $((\varphi_1 \to (\psi \vee \rho)) \to ((\varphi_1 \to \psi) \vee (\varphi_1 \to \rho)))$, $\varphi_1 \in K$, then assume $T \vdash (\varphi_1 \to (\psi \vee \rho))$, $R(S,T,(\varphi_1 \to (\psi \vee \rho)))$. Then note $\{\varphi_1\} \vdash \varphi$. Let $T_o = T - \{\varphi\} \cup \{\varphi_1\}$. Then $T_o \vdash (\psi \vee \rho)$. Note $\mu(T_o) < k$, $T_o \subset K$. If not $T_o \vdash \bot$ then $R(S,T_o,T_o)$. In any event, T_o is disjunctive by Theorem 1.2. Hence $T_o \vdash \psi$ or $T_o \vdash \rho$. So $T - \{\varphi\} \vdash ((\varphi_1 \to \psi) \vee (\varphi_1 \to \rho))$. Hence $T - \{\varphi\} \vdash \varphi$. But $\mu(T - \{\varphi\}) < k$, $T - \{\varphi\} \subset K$, not $T - \{\varphi\} \vdash \bot$. Hence $R(S, T - \{\varphi\}, T - \{\varphi\})$, $R(S,T,T)$. By Theorem 1.2, K is strongly disjunctive.

LEMMA 1.4.2. Let $T \subset L$, T finite. Then $R(S,T,T \cap L_1)$, (where S is arbitrary).

Proof: By induction on the complexity $\mu(T)$. Assume that $R(S,T_o,T_o \cap L_1)$ for all $T_o \subset L$ with $\mu(T_o) < k$. Let $T \subset L$ have $\mu(T) = k$. We want to show $R(S,T,\varphi)$ for all $\varphi \in T \cap L_1$. Let φ be $((\varphi_1 \to (\psi \vee \rho)) \to ((\varphi_1 \to \psi) \vee (\varphi_1 \to \rho)))$, $\varphi_1 \in L_o$. Suppose $R(S,T,(\varphi_1 \to (\psi \vee \rho)))$, $T \vdash (\varphi_1 \to (\psi \vee \rho))$. Our goal is $R(S,T,((\varphi_1 \to \psi) \vee (\varphi_1 \to \rho)))$. Then let $T_o = T - \{\varphi\} \cup \{\varphi_1\}$. Note that $\{\varphi_1\} \vdash \varphi$. So $T_o \vdash T$. Hence $T_o \vdash (\varphi_1 \to (\psi \vee \rho))$, and so $T_o \vdash (\psi \vee \rho)$. By Lemma 1.4.1, $T_o \vdash \psi$ or $T_o \vdash \rho$. Hence $T - \{\varphi\} \vdash (\varphi_1 \to \psi)$ or $T - \{\varphi\} \vdash (\varphi_1 \to \rho)$. We will only make use of $T \vdash (\varphi_1 \to \psi)$ or $T \vdash (\varphi_1 \to \rho)$. Assume $T \vdash (\varphi_1 \to \psi)$, the other case being symmetric. We must prove now that $R(S,T,((\varphi_1 \to \psi) \vee (\varphi_1 \to \rho)))$. We are done if we can show $R(S,T,(\varphi_1 \to \psi))$. We can assume that not $R(S,T,(\varphi_1 \to \psi))$. Then $T \vdash \varphi_1$, $R(S,T,\varphi_1)$, not $R(S,T,\psi)$. By $R(S,T,(\varphi_1 \to (\psi \vee \rho)))$, we have $R(S,T,(\psi \vee \rho))$. Hence $R(S,T,\rho)$, $T \vdash \rho$. Hence $R(S,T,(\varphi_1 \to \rho))$, $T \vdash (\varphi_1 \to \rho)$, and so $R(S,T,((\varphi_1 \to \psi) \vee (\varphi_1 \to \rho)))$.

THEOREM 1.4. K is strongly disjunctive. If φ has no occurrence of $\vee$ and $L_1 \vdash \varphi$ then $\vdash \varphi$.

Proof: It suffices to prove that for all finite T, φ, if T, φ have no occurrence of $\vee$ or $\&$, and $L_1 \cup T \vdash \varphi$ then $T \vdash \varphi$. We prove this by induction on $\mu(T) + \mu(\varphi)$. Thus we assume that for all finite T_o, φ_o, if T_o, φ_o have no occurrence of $\vee$ or $\&$, $L_1 \cup T_o \vdash \varphi_o$, $\mu(T_o) + \mu(\varphi_o) < k$, then $T_o \vdash \varphi_o$. Assume also that T, φ have no occurrences of $\vee$ or $\&$, $L_1 \cup T \vdash \varphi$, $\mu(T) + \mu(\varphi) = k$. We wish to prove $T \vdash \varphi$. This splits into cases. I. T is not reduced. IA. $\bot \in T$. Then $T \vdash \varphi$. IB. $(\bot \to \gamma) \in T$. Let $T' = T - \{(\bot \to \gamma)\}$. Then $L_1 \cup T' \vdash \varphi$. So $T' \vdash \varphi$, $T \vdash \varphi$. IC. $(p_i \to \gamma)$, $p_i \in T$. Let $T' = T - \{(p_i \to \gamma)\} \cup \{\gamma\}$. Then $L_1 \cup T' \vdash \varphi$. Hence $T' \vdash \varphi$, $T \vdash \varphi$. II. T is reduced. IIA. φ is $\bot$. Then no $f \models L_1 \cup T$. Now every $f \models L_1$. Hence no $f \models T$. So $T \vdash \bot$. IIB. φ is $(\alpha \to \beta)$. Then $L_1 \cup T \cup \{\alpha\} \vdash \beta$. So $T \cup \{\alpha\} \vdash \beta$, $T \vdash (\alpha \to \beta)$. IIC. φ is p_i, $p_i \in T$. Then $T \vdash \varphi$. IID. φ is p_i, $p_i \notin T$. Let $T_1 \subset L_1$, T_1 finite, $T_1 \cup T \vdash \varphi$. Take $S = \{p_k\colon p_k \in T\}$. By Lemma 1.4.2, $R(S,T_1 \cup T,T_1)$. By Theorem 1.1, if $R(S,T_1 \cup T,T)$ then $R(S,T_1 \cup T,p_i)$. Hence not $R(S,T_1 \cup T,T)$. Let $\psi \in T$

have not $R(S, T_1 \cup T, \psi)$. IIDa. ψ is p_k. Then $p_k \in S$, and so $R(S, T_1 \cup T, \psi)$. IIDb. ψ is $(p_k \rightarrow \gamma)$. Then $p_k \notin S$, and so $R(S, T_1 \cup T, \psi)$. IIDc. ψ is $((\alpha \rightarrow \beta) \rightarrow \gamma)$. Then $T_1 \cup T \vdash (\alpha \rightarrow \beta)$. Let $T' = T - \{((\alpha \rightarrow \beta) \rightarrow \gamma)\}$. Now $T_1 \cup T \cup \{\alpha\} \vdash \beta$. Hence $T_1 \cup T' \cup \{\alpha, (\beta \rightarrow \gamma)\} \vdash \beta$. So by Lemma 1.3.4, $T' \cup \{\alpha, (\beta \rightarrow \gamma)\} \vdash \beta$. Hence $T' \cup \{(\beta \rightarrow \gamma)\} \vdash (\alpha \rightarrow \beta)$. Hence $T' \cup \{((\alpha \rightarrow \beta) \rightarrow \gamma)\} \vdash (\alpha \rightarrow \beta)$. Now $L_1 \cup T' \cup \{\gamma\} \vdash p_i$. Hence $T' \cup \{\gamma\} \vdash p_i$. So $T' \cup \{((\alpha \rightarrow \beta) \rightarrow \gamma)\} \vdash p_i$. I.e., $T \vdash p_i$.

SECTION 2. MANY-SORTED PREDICATE LOGIC

We first introduce the many-sorted predicate logic, where each set represents a different sort.

DEFINITION 2.1. The function symbols are given by $F_x^{y_1-y_k,z}$, where $0 \leq k$, x, $y_1,\ldots,y_k$, z are sets. The relation symbols are given by $R_x^{y_1-y_k,z}$, where $0 \leq k$, x, $y_1,\ldots,y_k$, z are sets. $F_x^{y_1-y_k,z}$ is the xth function symbol which is k-ary and maps sorts $y_1,\ldots,y_k$ into sort z. $R_x^{y_1-y_k}$ is the xth relation symbol which is k-ary and defined on sorts $y_1,\ldots,y_k$. A constant symbol is just a 0-ary function symbol.

DEFINITION 2.2. The parameters are written a_n^y, where $0 \leq n$, y is a set; a_n^y is the nth parameter of sort y. The variables are written v_n^y, $0 \leq n$, y a set; v_n^y is the nth variable of sort y.

DEFINITION 2.3. The terms and their sorts are given by i) each parameter a_n^x is a term of sort x ii) $F_x^{y_1-y_k,z}(t_1,\ldots,t_k)$ is a term of sort z if each t_i is a term of sort y_i. A closed term is a term without parameters.

DEFINITION 2.4. The atomic formulae are written $R_x^{y_1-y_k}(t_1,\ldots,t_k)$, where $t_1,\ldots,t_k$ are terms of sorts $y_1,\ldots,y_k$. In addition, we also include the special atomic formula $\perp$ (absurdity). An atomic sentence is an atomic formula which mentions no parameters.

DEFINITION 2.5. Let λ be a string of symbols, let $x_1,\ldots,x_k$ be distinct symbols, and let $\theta_1,\ldots,\theta_k$ be strings of symbols. By $\lambda_{\theta_1-\theta_k}^{x_1-x_k}$ we mean the result of replacing, simultaneously, each and every occurrence in λ of x_i by θ_i. Let f be any function whose domain is a set of symbols, and whose range is a set of strings of symbols. By $\lambda(f)$ we mean the result of, simultaneously, replacing each and every occurrence of x in λ by $f(x)$, for each $x \in \text{Dom}(f)$.

DEFINITION 2.6. The formulae are given by i) the atomic formulae are formulae ii) if φ,ψ are formulae, then $(\varphi \,\&\, \psi)$, $(\varphi \vee \psi)$, $(\varphi \rightarrow \psi)$ are formulae iii) if a is a parameter of sort x, v a variable of sort x, v does not occur in the formula φ, then $(\forall v)(\varphi^{a}_{v})$ is a formula iv) $(\exists v)(\varphi^{a}_{v})$ is a formula under same conditions as iii). A sentence is a formula without parameters. Often we will let $(\varphi \longleftrightarrow \psi)$ abbreviate $((\varphi \rightarrow \psi)\&(\psi \rightarrow \varphi))$.

DEFINITION 2.7. A theory is a set of formulae. A closed theory is a set of sentences.

DEFINITION 2.8. (Spector.) Let T be a theory. We take $T \vdash \varphi$ to be the least predicate on formulae such that for all formulae φ, ψ, ρ, parameters a, variables v, terms t, all three of the same sort, v not in φ, the following conditions are satisfied: 1. if $\varphi \in T$, then $T \vdash \varphi$ 2. $T \vdash (\varphi \rightarrow \varphi)$ 3. if $T \vdash \varphi$, $T \vdash (\varphi \rightarrow \psi)$, then $T \vdash \psi$ 4. if $T \vdash (\varphi \rightarrow \psi)$, $T \vdash (\psi \rightarrow \rho)$, then $T \vdash (\varphi \rightarrow \rho)$ 5. $T \vdash ((\varphi \& \psi) \rightarrow \varphi)$, $T \vdash ((\varphi \& \psi) \rightarrow \psi)$, $T \vdash (\varphi \rightarrow (\varphi \vee \psi))$, $T \vdash (\psi \rightarrow (\varphi \vee \psi))$ 6. if $T \vdash (\varphi \rightarrow \rho)$, $T \vdash (\psi \rightarrow \rho)$, then $T \vdash ((\varphi \vee \psi) \rightarrow \rho)$ 7. if $T \vdash (\rho \rightarrow \varphi)$, $T \vdash (\rho \rightarrow \psi)$, then $T \vdash (\rho \rightarrow (\varphi \& \psi))$ 8. if $T \vdash ((\varphi \& \psi) \rightarrow \rho)$, then $T \vdash (\varphi \rightarrow (\psi \rightarrow \rho))$ 9. if $T \vdash (\varphi \rightarrow (\psi \rightarrow \rho))$, then $T \vdash ((\varphi \& \psi) \rightarrow \rho)$ 10. $T \vdash (\bot \rightarrow \varphi)$ 11. if a does not occur in ρ, $T \vdash (\rho \rightarrow \varphi)$, then $T \vdash (\rho \rightarrow (\forall v)(\varphi^{a}_{v}))$ 12. $T \vdash ((\forall v)(\varphi^{a}_{v}) \rightarrow \varphi^{a}_{t})$ 13. $T \vdash (\varphi^{a}_{t} \rightarrow (\exists v)(\varphi^{a}_{v}))$ 14. if a does not occur in ρ, $T \vdash (\varphi \rightarrow \rho)$, then $T \vdash ((\exists v)(\varphi^{a}_{v}) \rightarrow \rho)$.

DEFINITION 2.9. A relational type is a pair (A,B) such that A is a set of function and relation symbols, B is the set of all individual superscripts used in elements of A, and A contains a constant symbol of each sort in B. (Thus B is determined from A.)

DEFINITION 2.10. The (A,B)-function symbols, (A,B)-relation symbols, (A,B)-parameters, (A,B)-variables, (A,B)-terms, (A,B)-atomic formulae, (A,B)-formulae, (A,B)-theories are respectively the function symbols, relation symbols, parameters, variables, terms, atomic formulae, formulae, theories such that the only superscripts used are in B and the only function and relation symbols used are in A. It is customary to use "atomic (A,B)-formula" instead of "(A,B)-atomic formula."

PROPOSITION. Let φ be an (A,B)-formula, T an (A,B)-theory, $T \vdash \varphi$. Then $T \vdash \varphi$ would follow from Definition 1.8 if we used there only (A,B)-formulae,

(A,B)-parameters, and (A,B)-variables.

Proof: By a suitable transformation on formulae.

DEFINITION 2.11. An (A,B)-structure M is just a set of atomic (A,B)-sentences not including $\perp$.

DEFINITION 2.12. Define $R(M,A,B,T,\varphi)$ by i) M is an (A,B)-structure ii) φ is an (A,B)-sentence iii) $R(M,A,B,T,\varphi)$ iff $\varphi \in M$ for φ atomic iv) $R(M,A,B,T,(\varphi \& \psi))$ iff $R(M,A,B,T,\varphi)$ and $R(M,A,B,T,\psi)$ v) $R(M,A,B,T,(\varphi \vee \psi))$ iff $(R(M,A,B,T,\varphi), T \vdash \varphi)$ or $(R(M,A,B,T,\psi), T \vdash \psi)$ vi) $R(M,A,B,T,(\varphi \rightarrow \psi))$ iff (if $R(M,A,B,T,\varphi), T \vdash \varphi$, then $R(M,A,B,T,\psi)$) vii) $R(M,A,B,T,(\forall v)(\varphi^a_v))$ iff for all closed (A,B)-terms t, $R(M,A,B,T,\varphi^a_t)$ viii) $R(M,A,B,T,(\exists v)(\varphi^a_v))$ iff for some closed (A,B)-term t, $R(M,A,B,T,\varphi^a_t)$, $T \vdash \varphi^a_t$. We write $R(M,A,B,T,T_1)$ for $(\forall\varphi \in T_1)(R(M,A,B,T,\varphi))$.

DEFINITION 2.13. An (A,B)-substitution, α, is a partial function on the set of (A,B)-parameters, such that $\alpha(a)$ is a closed (A,B)-term of the same sort as a. For (A,B)-parameters a, closed (A,B)-terms t, take α^a to be given by $\alpha^a(x) = \alpha(x)$ if $x \neq a$; $\alpha^a(a)$ undefined, and take α^a_t to be given by $\alpha^a_t(x) = \alpha(x)$ if $x \neq a$; $\alpha^a_t(a) = t$.

LEMMA 2.1.1. Let φ be an (A,B)-formula, α a total (A,B)-substitution. If $T \vdash \varphi$, then $T \vdash \varphi(\alpha)$.

LEMMA 2.1.2. Let φ be an (A,B)-formula, a, v respectively an (A,B)-parameter, (A,B)-variable not in φ, of the same sorts. Then $\varphi^a_v(\alpha) = \varphi(\alpha^a)^a_v$, $\varphi^a_t(\alpha) = \varphi(\alpha^a)^a_{t(\alpha)}$.

LEMMA 2.1.3. Let T be a theory, T_1 a closed (A,B)-theory, $R(M,A,B,T,T_1)$, $T \vdash T_1$. Then the predicate on (A,B)-formulae, φ, "$T \vdash \varphi$ and $R(M,A,B,T,\varphi(\alpha))$ for all (A,B)-substitutions α" satisfies clauses 1)-10) when T is replaced by T_1 and "formulae φ, ψ, ρ" is replaced by "(A,B)-formulae φ, ψ, ρ."

Proof: The proof is essentially the same as that of Theorem 1.1.

LEMMA 2.1.4. Let T be a theory, φ an (A,B)-formula, a, v, t respectively an (A,B)-parameter, (A,B)-variable not in φ, (A,B)-term, of the same sorts, α a total (A,B)-substitution. Then $R(M,A,B,T,((\forall v)(\varphi^a_v) \to \varphi^a_t)(\alpha))$.

Proof: Assume hypotheses. We want $R(M,A,B,T,((\forall v)(\varphi^a_v)(\alpha) \to \varphi^a_t(\alpha)))$. So we want $R(M,A,B,T,((\forall v)(\varphi(\alpha^a)^a_v) \to \varphi(\alpha^a)^a_{t(\alpha)}))$. Assume $R(M,A,B,T,(\forall v)(\varphi(\alpha^a)^a_v))$. Then $R(M,A,B,T,\varphi(\alpha^a)^a_{t(\alpha)})$.

LEMMA 2.1.5. Let T be a theory, φ an (A,B)-formula, a, v, t respectively an (A,B)-parameter, (A,B)-variable not in φ, (A,B)-term, of the same sorts, α a total (A,B)-substitution. Then $R(M,A,B,T,(\varphi^a_t \to (\exists v)(\varphi^a_v))(\alpha))$.

Proof: Assume hypotheses. We want $R(M,A,B,T,(\varphi^a_t(\alpha) \to (\exists v)(\varphi^a_v)(\alpha)))$. I.e., $R(M,A,B,T,(\varphi(\alpha^a)^a_{t(\alpha)} \to (\exists v)(\varphi(\alpha^a)^a_v)))$. Assume $R(M,A,B,T,\varphi(\alpha^a)^a_{t(\alpha)})$, $T \vdash \varphi(\alpha^a)^a_{t(\alpha)}$. Then $R(M,A,B,T,(\exists v)(\varphi(\alpha^a)^a_v))$.

LEMMA 2.1.6. Let T be a theory, φ, ρ (A,B)-formulae, a, v, t respectively an (A,B)-parameter not in ρ, (A,B)-variable not in φ, (A,B)-term, of the same sorts, $R(M,A,B,T,(\rho \to \varphi)(\alpha))$ for all total (A,B)-substitutions α. Then $R(M,A,B,T,(\rho \to (\forall v)(\varphi^a_v))(\alpha))$ for all total (A,B)-substitutions α.

Proof: Assume hypotheses, and let β be a total (A,B)-substitution. We want $R(M,A,B,T,(\rho(\beta) \to (\forall v)(\varphi(\beta^a)^a_v)))$. Assume $R(M,A,B,T,\rho(\beta))$, $T \vdash \rho(\beta)$. We want $R(M,A,B,T,(\forall v)(\varphi(\beta^a)^a_v))$. Let t be a closed (A,B)-term. We want $R(M,A,B,T,\varphi(\beta^a)^a_t)$; i.e., $R(M,A,B,T,\varphi(\beta^a_t))$. We have $R(M,A,B,T,(\rho(\beta^a_t) \to \varphi(\beta^a_t)))$, and since $\rho(\beta^a_t) = \rho(\beta)$, we are done.

LEMMA 2.1.7. Let T be a theory, φ, ρ (A,B)-formulae, a, v, t respectively an (A,B)-parameter not in ρ, (A,B)-variable not in ρ, (A,B)-term, of the same sorts, $R(M,A,B,T,(\varphi \to \rho)(\alpha))$ for all total (A,B)-substitutions α. Then $R(M,A,B,T,((\exists v)(\varphi^a_v) \to \rho)(\alpha))$ for all total (A,B)-substitutions α.

Proof: Assume hypotheses, and let β be a total (A,B)-substitution. We want $R(M,A,B,T,((\exists v)(\varphi^a_v)(\beta) \to \rho(\beta)))$. Assume $R(M,A,B,T,(\exists v)(\varphi^a_v)(\beta))$. Then $R(M,A,B,T,(\exists v)(\varphi(\beta^a)^a_v))$. Let t be a closed (A,B)-term with $R(M,A,B,T,\varphi(\beta^a)^a_t)$, $T \vdash \varphi(\beta^a)^a_t$.

Then $R(M,A,B,T,\varphi(\beta_t^a)),T \vdash \varphi(\beta_t^a)$. We want $R(M,A,B,T,\rho(\beta))$. We have $R(M,A,B,T,(\varphi(\beta_t^a) \rightarrow \rho(\beta_t^a)))$, and since $\rho(\beta_t^a) = \rho(\beta)$, we are done.

DEFINITION 2.14. Let <u>$U(\varphi)$</u>, (the universal closure), be any sentence obtained from applying clause iii) of Definition 2.6, from φ. Let $U(T)$ be $\{U(\varphi): \varphi \in T\}$.

THEOREM 2.1. Let T be a theory, T_1 an (A,B)-theory, T_2 an (A,B)-theory, $R(M,A,B,T,U(T_1))$, $T \vdash T_1$, $T_1 \vdash T_2$. Then $R(M,A,B,T,U(T_2))$.

<u>Proof</u>: It suffices to prove the above if T_2 is replaced by the (A,B)-formula φ. From the Lemmas, we see that $R(M,A,B,T,\psi(\alpha))$ for all (A,B)-formulae ψ with $T_1 \vdash \psi$, and all total (A,B)-substitutions α. Hence $R(M,A,B,T,U(\psi))$, since there exist total (A,B)-substitutions.

DEFINITION 2.15. We say that an (A,B)-structure M is <u>T-complete</u> iff every atomic (A,B)-sentence φ with $T \vdash \varphi$ has $\varphi \in M$. We write $R(A,B,T_1)$ just in case for all T-complete M with $T \vdash T_1$, we have $R(M,A,B,T,U(T_1))$.

DEFINITION 2.16. A theory T is <u>(A,B)-disjunctive</u> just in case $T \vdash (\varphi \vee \psi)$ iff $T \vdash \varphi$ or $T \vdash \psi$, for (A,B)-sentences φ,ψ. T is <u>(A,B)-existential</u> just in case for all (A,B)-sentences $(\exists v)(\varphi_v^a)$ with $T \vdash (\exists v)(\varphi_v^a)$ we have $T \vdash \varphi_t^a$ for some closed (A,B)-term t. T is <u>(A,B)-semi-existential</u> iff for all (A,B)-sentences $(\exists v)(\varphi_v^a)$ with $T \vdash (\exists v)(\varphi_v^a)$ we have $T \vdash (\varphi_{t_1}^a \vee \cdots \vee \varphi_{t_n}^a)$ for some closed (A,B)-terms $t_1,\ldots,t_n$.

We now list some closure conditions on a class K of formulae: I. $\varphi \in K$ if φ is atomic II. $(\varphi \mathbin{\&} \psi) \in K$ if $\varphi,\psi \in K$ III. $(\varphi \vee \psi) \in K$ if $\varphi,\psi \in K$ IV. $(\varphi \rightarrow \psi) \in K$ if $\psi \in K$ V. $(\forall v)(\varphi_v^a) \in K$ if a is a parameter, v a variable of the same sort, the latter not occurring in $\varphi, \varphi \in K$ VI. $(\exists v)(\varphi_v^a) \in K$ under same conditions as V.

DEFINITION 2.17. The class of <u>formulae without strictly positive $\vee$</u> is the least class K satisfying conditions I, II, IV, V, VI. The class of <u>formulae without strictly positive $\exists$</u> is the least class K satisfying conditions I, II, III, IV, V. The class of <u>formulae without strictly positive $\exists,\vee$</u> is the least class K

satisfying conditions I, II, IV, V.

THEOREM 2.2. Let T be an (A,B)-theory with not $T \vdash \bot$. Then the following are equivalent: i) T is both (A,B)-disjunctive and (A,B)-existential ii) R(M,A,B,T,U(T)) for some M iii) R(M,A,B,T,U(T)) if M is T-complete.

Proof: i) $\rightarrow$ iii). Let M be T-complete. We prove by induction on (A,B)-sentences φ that if $T \vdash \varphi$, then $R(M,A,B,T,\varphi)$: a) φ is atomic, $T \vdash \varphi$. Then $R(M,A,B,T,\varphi)$ since M is T-complete b) φ is $(\psi \mathbin{\&} \rho)$, $T \vdash \varphi$. Then $T \vdash \psi$, $T \vdash \rho$. Hence $R(M,A,B,T,(\psi \mathbin{\&} \rho))$ c) φ is $(\psi \vee \rho)$, $T \vdash (\psi \vee \rho)$. Then $T \vdash \psi$ or $T \vdash \rho$, and in either case, $R(M,A,B,T,(\psi \vee \rho))$ d) φ is $(\psi \rightarrow \rho)$, $T \vdash (\psi \rightarrow \rho)$. Assume $T \vdash \psi$. Then $T \vdash \rho$, and so $R(M,A,B,T,\rho)$. Hence $R(M,A,B,T,(\psi \rightarrow \rho))$ e) φ is $(\forall v)(\psi^a_v)$, $T \vdash (\forall v)(\psi^a_v)$. Then $T \vdash \psi^a_t$ for all closed (A,B)-terms t. Hence $R(M,A,B,T,\psi^a_t)$ for all closed (A,B)-terms t. So $R(M,A,B,T,(\forall v)(\psi^a_v))$ f) φ is $(\exists v)(\psi^a_v)$, $T \vdash (\exists v)(\psi^a_v)$. Then let t be a closed (A,B)-term with $T \vdash \psi^a_t$. Then $R(M,A,B,T,\psi^a_t)$, and so $R(M,A,B,T,(\exists v)(\psi^a_v))$. ii) $\rightarrow$ i). Let R(M,A,B,T,U(T)), $T \vdash (\varphi \vee \psi)$, φ,ψ (A,B)-sentences. By Theorem 2.1, we have $R(M,A,B,T,(\varphi \vee \psi))$. Hence $T \vdash \varphi$ or $T \vdash \psi$. Let $T \vdash (\exists v)(\rho^a_v)$, $(\exists v)(\rho^a_v)$ an (A,B)-sentence. Then $R(M,A,B,T,(\exists v)(\rho^a_v))$, and hence $T \vdash \rho^a_t$ for some closed (A,B)-term t.

LEMMA 2.3.1. Let φ be an (A,B)-formula such that φ is without strictly positive $\exists$,v. Then $R(A,B,\{\varphi\})$.

Proof: Let M be T-complete. We wish to prove by induction on φ that if φ is an (A,B)-sentence without strictly positive $\exists$,v and $T \vdash \varphi$, then $R(M,A,B,T,\varphi)$: a) φ is atomic, $T \vdash \varphi$. Then $R(M,A,B,T,\varphi)$ since M is T-complete b) φ is $(\psi \mathbin{\&} \rho)$, $T \vdash (\psi \mathbin{\&} \rho)$. Then $R(M,A,B,T,(\psi \mathbin{\&} \rho))$ c) φ is $(\psi \rightarrow \rho)$, $T \vdash (\psi \rightarrow \rho)$. Assume $T \vdash \psi$. Then $T \vdash \rho$, and so $R(M,A,B,T,\rho)$. Hence $R(M,A,B,T,(\psi \rightarrow \rho))$ d) φ is $(\forall v)(\psi^a_v)$, $T \vdash \varphi$. Then $T \vdash \psi^a_t$ for all closed (A,B)-terms t. Hence $R(M,A,B,T,\psi^a_t)$ for all closed (A,B)-terms t. So $R(M,A,B,T,(\forall v)(\psi^a_v))$.

LEMMA 2.3.2. Suppose $R(A,B,T)$, $(\forall v)\varphi^a_v)$ is an (A,B)-formula, $T \vdash \varphi^a_t$ for all closed (A,B)-terms t. Then $R(A,B,T \cup \{\varphi\})$.

<u>Proof</u>: Let $T' \vdash T \cup \{\varphi\}$, M be T'-complete. Then $R(M,A,B,T',U(T))$. Now by Theorem 2.1, we have $R(M,A,B,T',U(\varphi^a_t))$ for all closed (A,B)-terms t. Hence $R(M,A,B,T',U(T \cup \{\varphi\}))$, and so $R(A,B,T \cup \{\varphi\})$.

LEMMA 2.3.3. Let S be a set of (A,B)-theories such that R(A,B,T) for all $T \in S$. Then $R(A,B, \bigcup S)$.

THEOREM 2.3. Let K be the least class of (A,B)-theories satisfying 1) K is closed under arbitrary unions 2) $\{\varphi\} \in K$ if φ is without strictly positive $\exists, \vee$ 3) $T \cup \{\varphi\} \in K$ if $T \in K$ and $T \vdash \varphi^a_t$ for all closed (A,B)-terms t (of the same sort as a). Then every element of K is both (A,B)-disjunctive and (A,B)-existential.

<u>Proof</u>: It is clear from Lemmas 2.3.1, 2.3.2, and 2.3.3 that every $T \in K$ has R(A,B,T). Let $T \in K$. We can assume that not $T \vdash \bot$. Let $M = \{\varphi\colon T \vdash \varphi$ and φ is an atomic (A,B)-sentence$\}$. Then M is T-complete. Hence R(M,A,B,T,U(T)). By Theorem 2.2, T is both (A,B)-disjunctive and (A,B)-existential.

DEFINITION 2.18. We write $R(\exists,A,B,T)$ just in case for all C with $A \subset C$, all T' with $T' \vdash T$ and the (C,B)-existence property, all (C,B)-structures M which are T'-complete, we have R(M,C,B,T',T).

LEMMA 2.4.1. Let φ be an (A,B)-formula such that φ is without strictly positive $\vee$. Then $R(\exists,A,B,\{\varphi\})$.

<u>Proof</u>: Let $A \subset C$, T have the (C,B)-existence property, M a (C,B)-structure which is T-complete. We wish to prove by induction on φ that if φ is a (C,B)-sentence without strictly positive $\vee$, and $T \vdash \varphi$, then $R(M,C,B,T,\varphi)$: a) φ is atomic, $T \vdash \varphi$. Then $R(M,C,B,T,\varphi)$ since M is T-complete b) φ is $(\psi \,\&\, \rho)$, $T \vdash (\psi \,\&\, \rho)$. Then $R(M,C,B,T,\varphi)$ c) φ is $(\psi \to \rho)$, $T \vdash (\psi \to \rho)$. Assume $T \vdash \psi$. Then $T \vdash \rho$, and so $R(M,C,B,T,\rho)$. So $R(M,C,B,T,(\psi \to \rho))$ d) φ is $(\forall v)(\psi^a_v)$, $T \vdash (\forall v)(\psi^a_v)$. Then $T \vdash \psi^a_t$ for all closed (C,B)-terms t, and so $R(M,C,A,B,T,\psi^a_t)$ for all closed (C,B)-terms t. So $R(M,C,B,T,(\forall v)(\psi^a_v))$ e) φ is

$(\exists v)(\psi_v^a)$, $T \vdash (\exists v)(\psi_v^a)$. Then $T \vdash \psi_t^a$ for some closed (C,B)-term t. Hence $R(M,C,B,T,(\exists v)(\psi_v^a))$.

LEMMA 2.4.2. Suppose $R(\exists,A,B,T)$, φ is an (A,B)-formula, $T \vdash \varphi_t^a$ for all closed (A,B)-terms t. Then $R(\exists,A,B,T \cup \{\varphi\})$.

Proof: Let $T' \vdash T \cup \{\varphi\}$, $A \subset C$, T' be (C,B)-existential, M a (C,B)-structure which is T'-complete. Then $R(M,C,B,T',U(T))$. Now by Theorem 2.1, we have $R(M,C,B,T',U(\varphi_t^a))$ for all closed (C,B)-terms t. Hence $R(M,C,B,T',U(T \cup \{\varphi\}))$, and so $R(\exists,A,B,T \cup \{\varphi\})$.

LEMMA 2.4.3. Let S be a set of (A,B)-theories such that $R(\exists,A,B,T)$ for all $T \in S$. Then $R(\exists,C,B,\bigcup S)$.

LEMMA 2.4.4. Let α be a one-one substitution taking only constants as values. Suppose $T \vdash \varphi(\alpha)$ and no $\alpha(a)$ occurs in T for any parameter a. Then $T \vdash \varphi$.

LEMMA 2.4.5. Let T be an (A,B)-theory. Then there is a C, T' with $A \subset C$, T' a (C,B)-theory, T' is (C,B)-existential, and $T' \vdash \varphi$ iff $T \vdash \varphi$ for (A,B)-formulae φ.

Proof: For each n we define C_n, T'_n, where T'_n is a (C_n,B)-theory conservative over T, and take $C = \bigcup_n C_n$, $T' = \bigcup_n T'_n$: $T'_o = T$, $C_o = C$. Take C_{n+1} to be C_n together with a new constant c_φ of sort x for each (C_n,B)-sentence $(\exists v)(\varphi_v^a)$ such that a occurs in φ, $T'_n \vdash (\exists v)(\varphi_v^a)$. Take $T'_{n+1} = \{\varphi_{c_\varphi}^a : c_\varphi \in C_{n+1}\} \cup T'_n$. It is clear that T' is (C,B)-existential. To see that T' is conservative over T it suffices to prove that T'_{n+1} is conservative over T under the assumption that T'_n is. Fix an (A,B)-sentence ψ. It suffices to prove by induction on the cardinality k of $Y \subset C_{n+1}$ that $T'_n \cup \{\varphi_{c_\varphi}^a : c_\varphi \in Y\} \vdash \psi$ implies $T \vdash \psi$. This is true for $k = 0$ by hypothesis. Suppose $T'_n \cup \{\varphi_{c_\varphi}^a : c_\varphi \in Y\} \vdash (\rho_{c_\rho}^b \to \psi)$, $c_\rho \in C_{n+1}$, $c_\rho \notin Y$, Y has cardinality k, ψ an (A,B)-sentence. It suffices to prove $T'_n \cup \{\varphi_{c_\varphi}^a : c_\varphi \in Y\} \vdash \psi$. Let α be the substitution $\alpha(b) = c_\rho$. Since ψ is closed and c_ρ does not occur in $T'_n \cup \{\varphi_{c_\varphi}^a : c_\varphi \in Y\}$, we have by Lemma 2.4.4, that $T'_n \cup \{\varphi_{c_\varphi}^a : c_\varphi \in Y\} \vdash (\rho \to \psi)$. Hence $T'_n \cup \{\varphi_{c_\varphi}^a :$

$c_\varphi \in Y\} \vdash ((\exists v)(\rho^b_v) \rightarrow \psi)$. But $T'_n \vdash (\exists v)(\rho^b_v)$. Hence $T'_n \cup \{\varphi^a_{c_\varphi}:\ c_\varphi \in Y\} \vdash \psi$.

THEOREM 2.4. Let K be the least class of (A,B)-theories satisfying 1) K is closed under arbitrary unions 2) $\{\varphi\} \in K$ if φ is without strictly positive v 3) $T \cup \{\varphi\} \in K$ if $T \in K$ and $T \vdash \varphi^a_t$ for all closed (A,B)-terms t (of the same sort as a). Then every element of K is (A,B)-disjunctive.

__Proof:__ It is clear from Lemmas 2.4.1, 2.4.2, and 2.4.3 that every $T \in K$ has $R(\exists,A,B,T)$. Let $T \in K$. We can assume that not $T \vdash \bot$. By Lemma 2.4.5, let $A \subset C$, T' a (C,B)-theory, T' be (C,B)-existential, and $T' \vdash \varphi$ iff $T \vdash \varphi$ for (A,B)-formulae φ. Let $M = \{\psi:\ T' \vdash \psi$ and ψ is an atomic (C,B)-sentence$\}$. Then M is T'-complete. Hence $R(M,A,B,T',T)$. Let $T \vdash (\varphi \vee \psi)$, φ, ψ (A,B)-sentences. Since $T' \vdash T$ we have $R(M,A,B,T',(\varphi \vee \psi))$ by Theorem 2.1. Hence $T' \vdash \varphi$ or $T' \vdash \psi$. By the way T' was chosen, $T \vdash \varphi$ or $T \vdash \psi$.

DEFINITION 2.19. We write __$R(v,A,B,T)$__ just in case for all (A,B)-disjunctive T' with $T' \vdash T$, (A,B)-structures M, we have $R(M,A,B,T',T)$.

LEMMA 2.5.1. Let φ be an (A,B)-formula without strictly positive $\exists$. Then $R(v,A,B,\{\varphi\})$.

__Proof:__ Let T be (A,B)-disjunctive, M an (A,B)-structure, M be T-complete. We wish to prove by induction on φ that if $T \vdash \varphi$, φ is an (A,B)-sentence without strictly positive $\exists$, then $R(M,A,B,T,\varphi)$: a) φ is atomic, $T \vdash \varphi$. Then $R(M,A,B,T,\varphi)$ since M is T-complete b) φ is $(\psi \,\&\, \rho)$, $T \vdash (\psi \,\&\, \rho)$. Then $T \vdash \psi$, $T \vdash \rho$, and so $R(M,A,B,T,(\psi \,\&\, \rho))$ c) φ is $(\psi \vee \rho)$, $T \vdash (\psi \vee \rho)$. Then $T \vdash \psi$ or $T \vdash \rho$. In either case, $R(M,A,B,T,(\psi \vee \rho))$ d) φ is $(\psi \rightarrow \rho)$, $T \vdash (\psi \rightarrow \rho)$. Assume $T \vdash \psi$. Then $T \vdash \rho$, and so $R(M,A,B,T,\rho)$. Hence $R(M,A,B,T,(\psi \rightarrow \rho))$ e) φ is $(\forall v)(\psi^a_v)$, $T \vdash \varphi$. Then $T \vdash \psi^a_t$ for all closed (A,B)-terms t. Hence $R(M,A,B,T,\psi^a_t)$ for all closed (A,B)-terms t. So $R(M,A,B,T,(\forall v)(\psi^a_v))$.

LEMMA 2.5.2. Let T be an (A,B)-theory, $R(v,A,B,T)$, $T \vdash \varphi^a_t$ for all closed (A,B)-terms t. Then $R(v,A,B,T \cup \{\varphi\})$.

Proof: Let $T' \vdash T \cup \{\varphi\}$, T' be (A,B)-disjunctive, M be a T-complete (A,B)-structure. Then R(M,A,B,T',T). Now by Theorem 2.1, $R(M,A,B,T',U(\varphi^a_t))$. Hence $R(M,A,B,T',U(T \cup \{\varphi\}))$. So $R(v,A,B,T \cup \{\varphi\})$.

LEMMA 2.5.3. Let S be a set of (A,B)-theories such that R(v,A,B,T) for all $T \in S$. Then $R(v,A,B,\bigcup S)$.

LEMMA 2.5.4. Assume T is an (A,B)-theory, $(\exists v)(\varphi^a_v)$ is an (A,B)-sentence, $T \vdash (\exists v)(\varphi^a_v)$, and for no sequence of closed (A,B)-terms $t_1,\ldots,t_k$ do we have $T \vdash (\varphi^a_{t_1} \vee \cdots \vee \varphi^a_{t_k})$. Then there is a T' which is (A,B)-disjunctive, $T' \vdash T$, and for no sequence of closed (A,B)-terms $t_1,\ldots,t_k$ do we have $T' \vdash (\varphi^a_{t_1} \vee \cdots \vee \varphi^a_{t_k})$.

Proof: Let *(T') mean that for no sequence of closed (A,B)-terms $t_1,\ldots,t_k$ do we have $T' \vdash (\varphi^a_{t_1} \vee \cdots \vee \varphi^a_{t_k})$. It suffices to note that if *(T'), $T' \vdash (\psi \vee \rho)$, ψ, ρ (A,B)-sentences, then either $*(T' \cup \{\psi\})$ or $*(T' \cup \{\rho\})$.

THEOREM 2.5. Let K be the least class of (A,B)-theories satisfying 1) K is closed under arbitrary unions 2) $\{\varphi\} \in K$ if φ is without strictly positive $\exists$ 3) $T \cup \{\varphi\} \in K$ if $T \in K$ and $T \vdash \varphi^a_t$ for all closed (A,B)-terms t (of the same sort as a). Then every element of K is (A,B)-semi-existential.

Proof: It is clear from Lemmas 2.5.1, 2.5.2, and 2.5.3 that any $T \in K$ has R(v,A,B,T). Let $T \in K$, and assume T is not (A,B)-semi-existential. Let $(\exists v)(\varphi^a_v)$ be an (A,B)-sentence, $T \vdash (\exists v)(\varphi^a_v)$, and for no sequence of closed (A,B)-terms $t_1,\ldots,t_k$ do we have $T \vdash (\varphi^a_{t_1} \vee \cdots \vee \varphi^a_{t_k})$. Choose T' according to Lemma 2.5.4. Let $M = \{\psi: \psi$ is an atomic (A,B)-sentence with $T' \vdash \psi\}$. Then M is T'-complete. Hence R(M,A,B,T',T). By Theorem 2.1, we have $R(M,A,B,T',(\exists v)(\varphi^a_v))$. Hence for some closed (A,B)-term t we have $T' \vdash \varphi^a_t$, which is a contradiction.

SECTION 3. THEORIES BASED ON ARITHMETIC

DEFINITION 3.1. We let $(\varphi \longleftrightarrow \psi)$ abbreviate $((\varphi \to \psi) \& (\psi \to \varphi))$. We define $E(A,B,C)$ (the (A,B)-theory of equality based on C), by the axioms $R_o^x(a,a)$, $(R_o^{x,x}(a,b) \to (\varphi \longleftrightarrow \varphi_b^a))$, where $x \in C$, φ is an atomic (A,B)-formula, a,b of sort in B.

LEMMA 3.1.1. We have $E(A,B,C) \vdash (R_o^{x,x}(a,b) \to (\varphi \longleftrightarrow \varphi_b^a))$ for all (A,B)-formulae φ, $x \in C$.

<u>Proof</u>: By induction on the complexity of φ: a) φ is atomic. b) φ is $(\psi \& \rho)$. c) φ is $(\psi \vee \rho)$. d) φ is $(\psi \to \rho)$. e) φ is $(\forall v)(\psi_v^c)$. We may assume that $c \neq a,b$. We have $E(A,B,C) \vdash (R_o^{x,x}(a,b) \to (\forall v)(\psi_v^c \longleftrightarrow \psi_{bv}^{ac}))$. So $E(A,B,C) \vdash (R_o^{x,x}(a,b) \to ((\forall v)(\psi_v^c) \longleftrightarrow (\forall v)(\psi_v^c)_b^a))$. f) φ is $(\exists v)(\psi_v^c)$. Similar.

DEFINITION 3.2. We let $(x_1,\ldots,x_k)$ be the ordered k-tuple consisting of $x_1,\ldots,x_k$. The <u>n-ary primitive recursion indices</u> are given by i) 0 is a 0-ary p.r. index ii) 1 is a 1-ary p.r. index iii) $(2,n,m)$, $1 \leq m \leq n$, is an n-ary p.r. index iv) $(3,x,y_1,\ldots,y_k)$ is an n-ary p.r. index if $y_1,\ldots,y_k$ are n-ary p.r. indices and x is a k-ary p.r. index, $0 \leq n$, $1 \leq k$ v) $(4,x,y)$ is an $n+1$-ary p.r. index if x is an n-ary p.r. index and y is an $n+2$-ary p.r. index, $0 \leq n$.

DEFINITION 3.3. Take PR to be the class of function symbols $F_x^{o,o,\ldots,o}$, where x is an n-ary p.r. index and there are $n+1$ 0's as superscripts, together with the relation symbol $R_o^{o,o}$. We may henceforth leave off superscripts on function symbols in PR. We also replace $R_o^{o,o}(s,t)$ by $s = t$. We let 0 stand for F_o, S stand for F_1.

DEFINITION 3.4. The theory <u>PRE</u> (primitive recursion equations) is given by $F_{(2,n,m)}(a_1,\ldots,a_n) = a_m$, $F_{(3,x,y_1,\ldots,y_k)}(a_1,\ldots,a_n) = F_x(F_{y_1}(a_1,\ldots,a_n),\ldots,F_{y_k}(a_1,\ldots,a_n))$, $F_{(4,x,y)}(a_1,\ldots,a_n,0) = F_x(a_1,\ldots,a_n)$, $F_{(4,x,y)}(a_1,\ldots,a_n,S(a_{n+1})) = F_y(a_1,\ldots,a_n,a_{n+1},F_{(4,x,y)}(a_1,\ldots,a_n,a_{n+1}))$, where $(2,n,m)$, $(3,x,y_1,\ldots,y_k)$ and $(4,x,y)$ are p.r. indices which are respectively n-ary, n-ary, and $n+1$-ary. (We have left off the superscript 0 on the parameters.)

DEFINITION 3.5. The theory SA (successor axioms) is given by $(S(a_o) = 0 \to \bot)$, $(S(a_o) = S(a_1) \to a_o = a_1)$.

DEFINITION 3.6. $I(\varphi)$ is given by $((\varphi^a_o \;\&\; (\forall v)((\varphi^a_v \to \varphi^a_{S(v)}))) \to \varphi)$, a a parameter of sort 0. Take $I(A,B) = \{I(\varphi)\colon\ \varphi$ is an (A,B)-formula$\}$. Take $HA(A,B) = E(A,B,\{0\}) + PRE + SA + I(A,B)$. Take $HA = HA(PR,\{0\})$.

DEFINITION 3.7. We define $|x|$, for primitive recursion indices x, as follows: $|0|$ is 0, $|1|$ is the successor function, $|(2,n,m)|$ is the function given by $|(2,n,m)|\,(x_1,\ldots,x_n) = x_m$, $|(3,x,y_1,\ldots,y_k)|$ is the function given by $|(3,x,y_1,\ldots,y_k)|(x_1,\ldots,x_n) = |x|(|y_1|(x_1,\ldots,x_n),\ldots,|y_k|(x_1,\ldots,x_n))$, $|(4,x,y)|$ is the function given by $|(4,x,y)|(x_1,\ldots,x_n,0) = |x|(x_1,\ldots,x_n)$, $|(4,x,y)|(x_1,\ldots,x_n,z+1) = |y|(x_1,\ldots,x_n,z,|(4,x,y)|(x_1,\ldots,x_n,z))$.

DEFINITION 3.8. Define $|t|$, for closed $(PR,\{0\})$-terms t, by $|0| = 0$, $|F_x(t_1,\ldots,t_k)| = |x|(|t_1|,\ldots,|t_k|)$. Let $\bar{n}$ be the term given by $\bar{0}$ is 0, $\overline{k+1}$ is $S(\bar{k})$. For closed $(PR,\{0\})$-terms t, we let $\bar{t}$ be $\overline{|t|}$.

LEMMA 3.1.2. For each closed $(PR,\{0\})$-term t we have $E(PR,\{0\},\{0\}) + PRE \vdash t = \bar{t}$.

Proof: By induction on the construction of the term t. We will not carry this out.

DEFINITION 3.9. An arithmetical type is a relational type (A,B) with $PR \subset A$, and every closed (A,B)-term is a $(PR,\{0\})$-term.

THEOREM 3.1. Let (A,B) be an arithmetical type, and let $E(A,B,\{0\}) + PRE \subset T \subset HA(A,B)$. Then T is in the class K of Theorem 2.3. Consequently, T is both (A,B)-disjunctive and (A,B)-existential.

Proof: It suffices to prove that $E(A,B,\{0\}) + PRE + \rho$ is in K, for every $\rho \in I(A,B)$. It suffices to prove that $E(A,B,\{0\}) + PRE \vdash \rho^a_t$ for each closed (A,B)-term t, where a (and t) have sort 0. Thus fix t as a closed

(PR,{0})-term. By Lemma 3.1.2, we have $E(A,B,\{0\}) + PRE \vdash t = \bar{t}$. By Lemma 3.1.1, it suffices to prove $E(A,B,\{0\}) + PRE \vdash \rho^a_{\bar{n}}$ for each n. Let ρ be $((\varphi^a_0 \,\&\, (\forall v)((\varphi^a_v \to \varphi^a_{S(v)}))) \to \varphi)$. We prove $\vdash \rho^a_{\bar{n}}$ by induction on n: clearly $\vdash \rho^a_{\bar{0}}$. Assume $\vdash \rho^a_{\bar{k}}$. Then $\vdash ((\varphi^a_0 \,\&\, (\forall v)((\varphi^a_v \to \varphi^a_{S(v)}))) \to \varphi^a_{\bar{k}})$. Now $\vdash ((\varphi^a_0 \,\&\, (\forall v)((\varphi^a_v \to \varphi^a_{S(v)}))) \to (\varphi^a_{\bar{k}} \to \varphi^a_{\overline{k+1}}))$. Hence $\vdash ((\varphi^a_0 \,\&\, (\forall v)((\varphi^a_v \to \varphi^a_{S(v)}))) \to \varphi^a_{\overline{k+1}})$.

DEFINITION 3.10. Let x be a 2-ary primitive recursion index. Then x is a <u>primitive recursive well-ordering index</u> if and only if i) $|x|(n,m)$ is 0 or 1, all n,m ii) $|x|(n,n)$ is 1, all n iii) $|x|(n,m)$ is 0 iff $|x|(m,n)$ is 1, provided $n \neq m$ iv) if $|x|(n,m)$ and $|x|(m,r)$ are 0 then $|x|(n,r)$ is 0 v) if Y is a nonempty set of natural numbers then for some $n \in Y$ we have $(\forall m \in Y)(|x|(m,n)$ is 1).

DEFINITION 3.11. Let x be a p.r. well-ordering index. We take <u>Prog(x,φ)</u> to be the formula $((\forall v)((F^{0,0}_x(v,a) = 0 \to \varphi^a_v)) \to \varphi)$. We take $TI(x,\varphi)$ to be the formula $((\forall w)((Prog(x,\varphi))^a_w \to \varphi)$. We take $TI(A,B)$ to be $\{TI(x,\varphi)\colon \varphi\}$ is an (A,B)-formula and x is a p.r. well-ordering index. We let $\tilde{x}$ be the ordinal of the well-ordering $(R(n,m)$ iff $|x|(n,m)$ is 0). We let $\tilde{x}(k)$ be the ordinal of the ordering $(R(n,m)$ iff $|x|(n,m) = 0)$ restricted to $\{j\colon R(j,k)\}$.

THEOREM 3.2. Let (A,B) be an arithmetical type, and let $E(A,B,\{0\}) + PRE + SA \subset T \subset HA(A,B) + TI(A,B)$. Then T is provably equivalent to some element of the K of Theorem 2.3. (I.e., $T \vdash T'$, $T' \vdash T$, for some $T' \in K$.) Consequently T is (A,B)-disjunctive and (A,B)-existential.

<u>Proof</u>: Let x be a p.r. well-ordering index, φ an (A,B)-formula. From Theorem 3.1, it suffices to prove that $E(A,B,\{0\}) + PRE + SA + TI(x,\varphi)$ is provably equivalent to some element of K. We define a hierarchy of theories T_α for all $\alpha \leq \tilde{x}$ as follows: $T_\alpha = E(A,B,\{0\}) + PRE + SA + \{(F_x(a,\bar{k}) = 0 \to TI(x,\varphi))\colon \tilde{x}(k) < \alpha\}$. As a preliminary remark, note that $T_\alpha \vdash (TI(x,\varphi))^a_{\bar{k}}$ if $\tilde{x}(k) < \alpha$. We first prove by induction on α that each T_α is in the K of Theorem 2.3. Suppose this is true for all $\beta < \alpha \leq \tilde{x}$. <u>Case 1</u>. $\alpha = 0$. Then T_α is in the K

of Theorem 2.3. Case 2. α is a limit ordinal. Then $T_\alpha = \bigcup_{\beta < \alpha} T_\beta$, and so T_β is in the K of Theorem 2.3. Case 3. $\alpha = \beta + 1$. Then $T_\alpha = T_\beta \cup \{(F_x(a,\bar{n}) = 0 \rightarrow TI(x,\varphi))\}$, where $\tilde{x}(n) = \beta$. Note that $T_\beta \vdash (TI(x,\varphi))^a_{\bar{k}}$ if $\tilde{x}(k) < \beta$. Note for each k, $(T_\beta \vdash F_x(\bar{k},\bar{n}) = 0)$ iff $\tilde{x}(k) < \beta$, and that $(T_\beta \vdash (F_x(\bar{k},\bar{n}) = 0 \rightarrow \bot))$ iff $\tilde{x}(k) \geq \beta$, by Lemma 3.1.2. Hence for each k, $T_\beta \vdash (F_x(\bar{k},\bar{n}) = 0 \rightarrow (TI(x,\varphi))^a_{\bar{k}})$. By Lemmas 3.1.1 and 3.1.2 we have $T_\beta \vdash (F_x(a,\bar{n}) = 0 \rightarrow TI(x,\varphi))^a_t$ for all closed (A,B)-terms t of sort 0. Hence T_α is in the K of Theorem 2.3, all $\alpha \leq \tilde{x}$.

In particular, $T_{\tilde{x}}$ is in the K of Theorem 2.3. Now $T_{\tilde{x}} \vdash (TI(x,\varphi))^a_{\bar{k}}$ for all k, and hence $T_{\tilde{x}} \vdash (TI(x,\varphi))^a_t$ for all closed (A,B)-terms of sort 0. Hence $T_{\tilde{x}} + TI(x,\varphi)$ is in the K of Theorem 2.3. It is easily seen that $T_{\tilde{x}} + TI(x,\varphi)$ is provably equivalent to $E(A,B,\{0\}) + PRE + SA + TI(x,\varphi)$, and so we are done.

DEFINITION 3.13. We inductively define the $\underline{\Pi^0_k \text{ formulae}}$ and the $\underline{\Sigma^0_k \text{ formulae}}$: i) the Π^0_0 formulae as well as the Σ^0_0 formulae are just the atomic (PR,{0})-formulae ii) if φ is a Π^0_k formula, then $(\exists v)(\varphi^a_v)$ is a Σ^0_{k+1} formula, provided $\underline{a}$ is of sort 0 iii) if φ is a Σ^0_k formula, then $(\forall v)(\varphi^a_v)$ is a Π^0_{k+1} formula, provided $\underline{a}$ is of sort 0.

DEFINITION 3.14. Let $\underline{Tr(n)}$ be the set of all true Π^0_n sentences. Let $\underline{C}$ be $\{(\varphi \vee (\varphi \rightarrow \bot)) : \varphi$ is a (PR,{0})-formula$\}$.

DEFINITION 3.15. An $\underline{\text{(A,B)-metasubstitution}}$ β is a function from all the propositional atoms into (A,B)-formulae. We define $\varphi(\beta)$ for propositional formulae φ by $p_n(\beta) = \beta(p_n)$, $(\varphi \,\&\, \psi)(\beta) = (\varphi(\beta) \,\&\, \psi(\beta))$, $(\varphi \vee \psi)(\beta) = (\varphi(\beta) \vee \psi(\beta))$, $(\varphi \rightarrow \psi)(\beta) = (\varphi(\beta) \rightarrow \psi(\beta))$, $\bot(\beta) = \bot$. An (A,B)-metasubstitution β is called $\underline{\text{independent over } T}$ iff for all propositional formulae φ, we have $T \vdash \varphi(\beta)$ iff $\vdash \varphi$. Also β is called $\underline{\text{classically independent over } T}$ iff for all propositional formulae φ, we have $T \vdash \varphi(\beta)$ iff $f \models \varphi$ for all $f \in 2^\omega$.

LEMMA 3.3.1. Suppose T is a recursive (PR,{0})-theory such that $T \cup Tr(1) \cup HA \cup C$ is consistent. Then there exists a (PR,{0})-metasubstitution, β, which is classically independent over $T \cup Tr(1) \cup HA \cup C$ and such that β takes on only Π^o_2 sentences as values.

<u>Proof</u>: By Gödel self-reference, and we merely sketch the proof. By self-reference, choose a (PR,{0})-formula φ with only the parameter a, in which, in $HA \cup C$, is provably equivalent to the following (PR,{0})-formula φ^* with only the parameter a: for each proof of a disjunction of sentences $\pm\varphi^a_o \vee\cdots\vee \pm \varphi^a_{\underline{a-1}} \vee \varphi^a_{\underline{a}}$ (where $\pm\varphi^a_{\underline{i}}$ is either $\varphi^a_{\underline{i}}$ or $(\varphi^a_{\underline{i}} \to \bot)$) from the theory $T \cup Tr(1) \cup HA \cup C$, there is a smaller proof of some other disjunction $\pm\varphi^a_o \vee\cdots\vee \pm\varphi^a_{\underline{a-1}} \vee (\varphi^a_{\underline{a}} \to \bot)$ from the theory $T \cup Tr(1) \cup HA \cup C$. Note that φ^* is of the form $(\forall n)(An \to (\exists m < n)(Bm))$, where A,B are Π^o_1. Hence φ^* is Π^o_2 (in the theory $HA \cup C$).

We wish to prove by induction on n that not $T \cup Tr(1) \cup HA \cup C \vdash \pm\varphi^a_o \vee\cdots\vee \pm\varphi^a_{\underline{n}}$. This is enough since every propositional formula is classically equivalent to a conjunction of disjunctions of atoms and negated atoms, and we can define $\beta(j) = \varphi^a_{\underline{j}}$.

We first do the basis case n = 0. Suppose $T \cup Tr(1) \cup HA \cup C \vdash \varphi^a_o$. If φ^a_o is true, then $T \cup Tr(1) \cup HA \cup C \vdash (\varphi^a_o \to \bot)$, which is impossible since $T \cup Tr(1) \cup HA \cup C$ is consistent. If φ^a_o is false, then, since φ^a_o is Π^o_2, $Tr(1) \cup HA \cup C \vdash (\varphi^a_o \to \bot)$, which is again a contradiction. Suppose $T \cup Tr(1) \cup HA \cup C \vdash (\varphi^a_o \to \bot)$, with a smallest proof of size k. If φ^a_o is true, then $Tr(1) \cup HA \cup C \vdash \varphi^a_o$, by enumerating the proofs of size $\leq k$, which is a contradiction. If φ^a_o is false, then $T \cup Tr(1) \cup HA \cup C \vdash \varphi^a_o$, which is a contradiction.

Assume that $n > 0$ and for all $m < n$ we have not $T \cup Tr(1) \cup HA \cup C \vdash \pm\varphi^a_o \vee\cdots\vee \pm\varphi^a_{\underline{m}}$. Suppose some $T \cup Tr(1) \cup HA \cup C \vdash \pm\varphi^a_o \vee\cdots\vee \pm\varphi^a_{\underline{n-1}} \vee \pm\varphi^a_{\underline{n}}$. Then it is easily seen that $Tr(1) \cup HA \cup C \vdash \pm\varphi^a_{\underline{n}}$. Suppose $Tr(1) \cup HA \cup C \vdash \varphi^a_{\underline{n}}$. Then $\varphi^a_{\underline{n}}$ is true, and hence some $T \cup Tr(1) \cup HA \cup C \vdash \pm\varphi^a_o \vee\cdots\vee \pm\varphi^a_{\underline{n-1}} \vee (\varphi^a_{\underline{n}} \to \bot)$. So some $T \cup Tr(1) \cup HA \cup C \vdash \pm\varphi^a_o \vee\cdots\vee \pm\varphi^a_{\underline{n-1}}$, contradicting induction hypothesis. Suppose $Tr(1) \cup HA \cup C \vdash (\varphi^a_{\underline{n}} \to \bot)$. Then $\varphi^a_{\underline{n}}$ is false, and hence some $T \cup Tr(1) \cup HA \cup C \vdash \pm\varphi^a_o \vee\cdots\vee \pm\varphi^a_{\underline{n-1}} \vee \varphi^a_{\underline{n}}$. So some $T \cup Tr(1) \cup HA \cup C \vdash \pm\varphi^a_o \vee\cdots\vee \pm\varphi^a_{\underline{n-1}}$, which again contradicts induction hypothesis.

DEFINITION 3.16. Let T be a recursive (PR,{0})-theory, which is recursive with index e. Let T_1 be a finite, closed (PR,{0})-theory. Whenever R(Tr(0),PR,{0}, T,T_1) occurs either to the left or to the right of $\vdash$, we mean the (PR,{0})-sentence R(Tr(0),PR,{0},T,T_1) formalized in the natural way using the index e. We will use this formalization only in the presence of C.

LEMMA 3.3.2. Let φ be a Π^0_2 sentence, T any recursive (PR,{0})-theory. Then HA + C $\vdash$ (R(Tr(0),PR,{0}[T $\cup$ HA,φ) $\longleftrightarrow$ φ).

LEMMA 3.3.3. Let T, T_1, T_2 be (PR,{0})-theories, T_1, T_2 finite, T recursive. Then HA + C $\vdash$ ((T $\vdash T_1$ & $T_1 \vdash T_2$ & R(Tr(0),PR,{0},T,U(T_1))) $\rightarrow$ R(Tr(0),PR,{0},T, U(T_2))).

LEMMA 3.3.4. If φ is a Σ_0 sentence which is false, then E(PR,{0},{0}) $\cup$ PRE $\cup$ SA $\vdash$ ($\varphi \rightarrow \bot$). Hence if T $\vdash$ E(PR,{0},{0}) $\cup$ PRE $\cup$ SA then either T $\vdash \bot$ or Tr(0) is a T-complete (PR,{0})-structure.

DEFINITION 3.17. Let <u>R(M,A,B,T)</u> mean that for all T' $\vdash$ T, not T' $\vdash \bot$, we have R(M,A,B,T',U(T)). Let T^0 = {φ: T $\vdash \varphi$ and φ is a (PR,{0})-formula}.

LEMMA 3.3.5. Let (A,B) be an arithmetical type, T an (A,B)-theory with R(A,B,T), and E(PR,{0},{0}) + PRE + SA $\subset$ T. Then R(Tr(0),PR,{0},T^0).

<u>Proof</u>: Let T' $\vdash T^0$, not T' $\vdash \bot$. Then not $(T')^0 \cup T \vdash \bot$. By Lemma 3.3.4, choose an (A,B)-structure M such that $(T')^0 \cup T$ is M-complete and {φ: φ is a PR,{0})-formula, $\varphi \in$ M} = Tr(0). Hence R(M,A,B,$(T')^0 \cup$ T,U(T)). Hence R(M,A,B, $(T')^0 \cup$ T,U(T^0)). Since $(T')^0 \cup T$ and T' have the same provable (PR,{0})-formulae, and every closed (A,B)-term of sort 0 is a (PR,{0})-term, we have R(Tr(0),A,B,T',U(T^0)).

THEOREM 3.3. Suppose T is a (PR,{0})-theory, T is recursive, and each element of T is (universally) true. Suppose further that R(Tr(0),PR,{0},T). Then there is a (PR,{0})-metasubstitution β which is independent over T, and such that β takes on only Π^0_2 sentences as values.

<u>Proof</u>: Fix T as in hypotheses. We may assume without loss of generality

that $T \vdash E(PR,\{0\},\{0\}) + PRE + SA$. Let T^* be the $(PR,\{0\})$-theory consisting of $T \cup Tr(1) \cup HA \cup C \cup \{(R(Tr(0),PR,\{0\},T \cup T_1,U(\varphi)) \vee T \cup T_1 \vdash \bot): T_1$ is any finite $(PR,\{0\})$-theory and $\varphi \in T\}$. Then T^* is consistent since every element of T^* is true.

According to Lemma 3.3.1, let β be any $(PR,\{0\})$-metasubstitution which is classically independent over T^*, and such that β takes on only Π_2^0 sentences as values. We will prove that β is independent over T.

We fix $\vdash_*$ to be the following binary relation between propositional theories S_1, S_2: $S_1 \vdash_* S_2$ iff $T \cup S_1(\beta) \vdash S_2(\beta)$. It is enough to prove that $\vdash_*$ satisfies the conditions laid down in Theorem 1.3. Conditions 1)-5) are easily verified, condition 4) using the classical independence of β over T. We firstly concentrate on condition 7).

Let S be a reduced finite propositional theory, A, B propositional formulae, and $T \cup S(\beta) \vdash (A(\beta) \vee B(\beta))$. Let $T' \subset T$, T' finite, $T' \cup S(\beta) \vdash (A(\beta) \vee B(\beta))$. We wish to prove $T' \cup S(\beta) \vdash A(\beta)$ or $T' \cup S(\beta) \vdash B(\beta)$ or $(T' \cup S(\beta) \vdash (C(\beta) \rightarrow D(\beta))$ for some $((C \rightarrow D) \rightarrow E) \in S)$.

Let $T_o = \{p_n(\beta): p_n \in S\} \cup \{(p_n(\beta) \rightarrow \bot): p_n \notin S\}$. Note that since β is classically independent over T^* we have that $T_o \cup T^*$ is consistent. Hence any Σ_1^0 sentence provable from $T_o \cup T^*$ is true. So it suffices to prove in $T_o \cup T^*$ that $T' \cup S(\beta) \vdash A(\beta)$ or $T' \cup S(\beta) \vdash B(\beta)$ or $(T' \cup S(\beta) \vdash (C(\beta) \rightarrow D(\beta))$ for some $((C \rightarrow D) \rightarrow E) \in S)$. We argue within $T_o \cup T^*$ as follows: we have $R(Tr(0),$ $PR,\{0\},T' \cup S(\beta),p_n(\beta))$ for all $p_n \in S$, by Lemma 3.3.2. We have $R(Tr(0),PR,\{0\},$ $T' \cup S(\beta),(p_k \rightarrow C)(\beta))$ for all $(p_k \rightarrow C) \in S$, by Lemma 3.3.2. We have $R(Tr(0),$ $PR,\{0\},T' \cup S(\beta),U(T'))$ or $T' \cup S(\beta) \vdash \bot$. Hence we have $R(Tr(0),PR,\{0\},$ $T' \cup S(\beta),U(T' \cup S(\beta)))$ or $T' \cup S(\beta) \vdash \bot$ or (not $R(Tr(0),PR,\{0\},T' \cup S(\beta),$ $((C(\beta) \rightarrow D(\beta)) \rightarrow E(\beta)))$ for some $((C \rightarrow D) \rightarrow E) \in S)$. Hence we have $R(Tr(0),$ $PR,\{0\},T' \cup S(\beta),U(T' \cup S(\beta))$ or $(T' \cup S(\beta) \vdash (C(\beta) \rightarrow D(\beta))$ for some $((C \rightarrow D) \rightarrow E) \in S$. By Lemma 3.3.3, since $T' \cup S(\beta) \vdash (A(\beta) \vee B(\beta))$, we have $R(Tr(0),PR,\{0\},T' \cup S(\beta),(A(\beta) \vee B(\beta)))$ or $(T' \cup S(\beta) \vdash (C(\beta) \rightarrow D(\beta))$ for some $((C \rightarrow D) \rightarrow E) \in S)$. Hence $T' \cup S(\beta) \vdash A(\beta)$ or $T' \cup S(\beta) \vdash B(\beta)$ or $(T' \cup S(\beta)$ $\vdash (C(\beta) \rightarrow D(\beta))$ for some $((C \rightarrow D) \rightarrow E) \in S)$.

Next, condition 6). Let S be a reduced finite propositional theory, p a propositional atom, $T \cup S(\beta) \vdash p(\beta)$, and assume $p \notin S$. Let $T' \subset T$, T' finite, $T' \cup S(\beta) \vdash p(\beta)$. We wish to prove $T' \cup S(\beta) \vdash (C(\beta) \to D(\beta))$ for some $((C \to D) \to E) \in S$.

Let T_o be as before, and note as before that it suffices to prove, in $T_o \cup T^*$, that $T' \cup S(\beta) \vdash (C(\beta) \to D(\beta))$ for some $((C \to D) \to E) \in S$. We argue within $T_o \cup T^*$ as follows: just as before, we have $R(Tr(0), PR, \{0\}, T' \cup S(\beta), U(T' \cup S(\beta)))$ or $(T' \cup S(\beta) \vdash (C(\beta) \to D(\beta))$ for some $((C \to D) \to E) \in S)$. By Lemma 3.3.3, since $T' \cup S(\beta) \vdash p(\beta)$, we have $R(Tr(0), PR, \{0\}, T' \cup S(\beta), p_n(\beta))$ or $(T' \cup S(\beta) \vdash (C(\beta) \to D(\beta))$ for some $((C \to D) \to E) \in S)$. By Lemma 3.3.2, the first disjunct is false, and so $T' \cup S(\beta) \vdash (C(\beta) \to D(\beta))$ for some $((C \to D) \to E) \in S$.

COROLLARY 3.3.1. Let (A,B) be an arithmetical type, T an (A,B)-theory in the class K of Theorem 2.3, T^o is recursively enumerable and consists only of (universally) true formulae. Then there is a $(PR, \{0\})$-metasubstitution β which is independent over T, and such that β takes on only Π^o_2 sentences as values.

Proof: By the proof of Theorem 2.3, we have $R(A,B,T)$. By Lemma 3.3.5, we have $R(Tr(0), PR, \{0\}, T^o)$. Hence we are done by Theorem 3.3, since every r.e. theory has a recursive axiomatization.

COROLLARY 3.3.2. Let Y be any set of primitive recursive well-ordering indices. Let $T = HA \cup \{TI(x,\varphi)\colon x \in Y \text{ and } \varphi \text{ is a } (PR,\{0\})\text{-formula}\}$. Then any $(PR,\{0\})$-metasubstitution classically independent over $T \cup Tr(1) \cup C$ is independent over T.

Proof: Upon examination of the proof of Theorem 3.3. What was shown there is that any $(PR,\{0\})$-metasubstitution classically independent over T^* there, taking Π^o_2 sentences as values, is independent over T. Redefine T^* for the purposes of this Corollary as $T \cup Tr(1) \cup C \cup \{R(Tr(0), PR, \{0\}, T' \cup T_1, U(\varphi)\colon T' \text{ is a finite subset of } T,\ T_1 \text{ is a finite } (PR,\{0\})\text{-theory},\ \varphi \in T\}$. Then the same will hold for the new T^*, and it can be seen that this new T^* is equivalent to $T \cup Tr(1) \cup C$.

SECTION 4. SECOND-ORDER LOGIC AND ARITHMETIC

DEFINITION 4.1. An <u>integer relational type</u> is a relational type (A,B) such that $B \subset \omega$, and all subscripts of elements of A are in ω.

Throughout Section 4 we fix (A,B) as an integer relational type, and it will be convenient to assume that no R_o^x is in A.

DEFINITION 4.2. We define $\underline{B^*} = \{(k_o,\dots,k_n)\colon\ k_o,\dots,k_n \in \omega\} \cup B$. The elements of $B^* - B$ are called the <u>2nd-order sorts</u>.

DEFINITION 4.3. The <u>2nd-order free constants</u> are given by F^x_α, where x is a 2nd-order sort and $\alpha < \omega$. The <u>2nd-order abstraction constants</u> are given by $F^x_{(\alpha,y)}$, where x is a 2nd-order sort, $\omega \leq \alpha < \omega \times 2$, $x = (k_o,\dots,k_n)$, and y is a set of n+1-tuples $t_o,\dots,t_n$ such that each t_i is a closed (A,B)-term of sort k_i. The <u>2nd-order constants</u> are the free 2nd-order constants together with the abstraction constants.

DEFINITION 4.4. Let $\underline{A^*} = A \cup \{R_o^{(k_o-k_n),k_o-k_n}\colon\ k_o,\dots,k_n \in B\} \cup$ the set of all 2nd-order constants. Let $\underline{A^-} = A^*$ without any 2nd-order constants. Let $\underline{A^{*-}} = A^*$ without 2nd-order abstraction constants.

The theorems of this section involve only (A^-,B^*)-theories. We use the 2nd-order constants only as necessary auxiliary constructions.

DEFINITION 4.5. Let φ be an (A^*,B^*)-formula, $b,b_o,\dots,b_n$ distinct parameters, b not occurring in φ, b of sort $(k_o,\dots,k_n)$, $k_o,\dots,k_n \in B$, $b_o,\dots,b_n$ of respective sorts $k_o,\dots,k_n$. We write <u>$Comp(\varphi,b,b_o,\dots,b_n)$</u> for $(\forall v_o^{k_o})\cdots(\forall v_n^{k_n})$ $(R_o(b,v_o,\dots,v_n) \longleftrightarrow \varphi^{b_o-b_n}_{v_o-v_n})$. Whenever we write $Comp(\varphi,b,b_o,\dots,b_n)$ we assume that the above conditions on parameters are met.

DEFINITION 4.6. A theory T is called <u>2nd-order abstraction complete</u> iff if b is of some 2nd-order sort x, φ has only parameters among $b_o,\dots,b_n$, and $T \vdash (\exists v)$ $(Comp(\varphi,b,b_o,\dots,b_n)^b_v)$, then for some $\omega \leq \alpha \leq \omega \times 2$ we have $T \vdash$ $Comp(\varphi,b,b_o,\dots,b_n)^b_F$ for all 2nd-order constants $F = F^x_{(\alpha,y)}$.

DEFINITION 4.7. A <u>T-special structure</u> is an (A^*,B^*)-structure M such that i) if φ is an atomic (A,B)-sentence and $T \vdash \varphi$, then $\varphi \in M$ ii) we have $\varphi = R_o(F_{(\beta,y)},t_o,\ldots,t_n) \in M$ iff φ is an (A^*,B^*)-sentence, and $(t_o,\ldots,t_n) \in y$.

DEFINITION 4.8. Let T be an (A^*,B^*)-theory. We write <u>$R^*(T)$</u> iff for all 2nd-order abstraction-complete $T' \vdash T$, all T'-special structures M, we have $R(M,A^*,B^*,T',U(T))$.

DEFINITION 4.9. We let Z^* be the least class of (A^*,B^*)-formulae satisfying i) all atomic (A,B)-formulae are in Z^* ii) if $\varphi,\psi \in Z^*$, then $(\varphi \,\&\, \psi) \in Z^*$ iii) if $\varphi \in Z^*$, ψ is an (A^*,B^*)-formula, then $(\psi \to \varphi) \in Z^*$ iv) if $\varphi \in Z^*$, then $(\forall v)(\varphi^a_v) \in Z^*$ (provided <u>a</u> is an (A^*,B^*)-parameter) v) each $(\exists v)(\mathrm{Comp}(\varphi,b,b_o,\ldots,b_n)^b_v) \in Z^*$.

LEMMA 4.1.1. We have $R^*(\{(\exists v)(\mathrm{Comp}(\varphi,b,b_o,\ldots,b_n)^b_v)\})$.

<u>Proof</u>: Let T' be 2nd-order abstraction-complete, $T' \vdash (\exists v)(\mathrm{Comp}(\varphi,b,b_o,\ldots,b_n)^b_v)$, M be T'-special. We may assume φ has only the parameters $b_o,\ldots,b_n$ and prove $R(M,A^*,B^*,T',(\exists v)(\mathrm{Comp}(\varphi,b,b_o,\ldots,b_n)^b_v))$. Let b have the 2nd-order sort x, $\omega \leq \alpha < \omega \times 2$, and $T' \vdash \mathrm{Comp}(\varphi,b,b_o,\ldots,b_n)^b_F$ for all 2nd-order abstraction constants $F = F^x_{(\alpha,y)}$, by 2nd-order abstraction-completeness. Fix F as $F^x_{(\alpha,y)}$, where $y = \{(t_o,\ldots,t_n)\colon t_o,\ldots,t_n$ are closed (A,B)-terms with $R(M,A^*,B^*,T',\varphi^{b_o-b_n}_{t_o-t_n})\}$. We must verify that $R(M,A^*,B^*,T',(\forall v_o)\cdots(\forall v_n)(R_o(F,v_o,\ldots,v_n) \longleftrightarrow \varphi^{b_o-b_n}_{t_o-t_n}))$. Let $t_o,\ldots,t_n$ be closed (A,B)-terms of the same sorts as $v_o,\ldots,v_n$. We need to prove $R(M,A^*,B^*,T',R_o(F,t_o,\ldots,t_n) \longleftrightarrow \varphi^{b_o-b_n}_{t_o-t_n})$. But this is clear by the choice of F and the specialty of M.

LEMMA 4.1.2. For each $\varphi \in Z^*$ we have $R^*(\{\varphi\})$.

<u>Proof</u>: Fix a 2nd-order abstraction-complete theory, T, M a T-special structure. It suffices to prove, by induction, that for all closed $\psi \in Z^*$, if $T \vdash \psi$, then $R(M,A^*,B^*,T,\psi)$: i) ψ is an atomic (A,B)-sentence, $T \vdash \psi$. Then $R(M,A^*,B^*,T,\psi)$ by specialty ii) ψ is $(\varphi_1 \,\&\, \varphi_2)$, $T \vdash \psi$. Then $T \vdash \varphi_1$,

$T \vdash \varphi_2$, and so $R(M,A^*,B^*,T,\psi)$ iii) ψ is $(\varphi_1 \rightarrow \varphi_2)$, $T \vdash \psi$. Then assume $T \vdash \varphi_1$. Hence $T \vdash \varphi_2$, and so $R(M,A^*,B^*,T,\psi)$ iv) ψ is $(\forall v)(\rho^a_v)$, $T \vdash \psi$. Then $T \vdash \rho^a_t$ for all appropriate t. Hence $R(M,A^*,B^*,T,\rho^a_t)$ for all appropriate t, and so $R(M,A^*,B^*,T,\psi)$ v) ψ is $(\exists v)(\mathrm{Comp}(\varphi,b,b_o,\ldots,b_n)^b_v)$, $T \vdash \psi$. Then $R(M,A^*,B^*,T,\psi)$ by Lemma 4.1.1.

LEMMA 4.1.3. Suppose T is an (A^*,B^*)-theory with $R^*(T)$. Suppose φ is an (A^*,B^*)-formula, $\underline{a}$ an (A^*,B^*)-parameter with $T \vdash \varphi^a_t$ for all closed (A^*,B^*)-terms t of the same sort as $\underline{a}$. Then $R^*(T \cup \{\varphi\})$.

<u>Proof</u>: Let $T' \vdash T \cup \{\varphi\}$, T' be 2nd-order abstraction-complete. Then $R(M,A^*,B^*,T',U(T))$. Hence $R(M,A^*,B^*,T',U(\varphi^a_t))$ for all appropriate terms t. Hence $R(M,A^*,B^*,T',U(\varphi))$.

DEFINITION 4.10. Let T be a theory. Then T_1 is <u>abstraction-minimal over T</u> iff i) $T_1 \vdash T$ ii) T_1 and T have the same provable (A^{*-},B^*)-formulae iii) to each 2nd-order constant F there is an (A^{*-},B^*)-formula φ such that $T_1 \vdash \mathrm{Comp}(\varphi,b,b_o,\ldots,b_n)^b_F$.

LEMMA 4.1.4. To each T there is a T^* which is 2nd-order abstraction-complete and abstraction-minimal over T.

<u>Proof</u>: Let $X = \{(\varphi,b,b_o,\ldots,b_n)\colon\ T \vdash (\exists v)(\mathrm{Comp}(\varphi,b,b_o,\ldots,b_n)^b_v)$, and φ is an (A^{*-},B^*)-formula whose parameters are among $b_o,\ldots,b_n\}$. Then X is countably infinite and we let $|\ |$ map X one-one onto ω. We take $T^* = T \cup \{\mathrm{Comp}(\varphi,b,b_o,\ldots,b_n)^b_F\colon (\varphi,b,b_o,\ldots,b_n) \in X$ and F is a 2nd-order abstraction constant whose left subscript is $|(\varphi,b,b_o,\ldots,b_n)|$. We leave it to the reader to make the necessary "elimination of abstraction" argument to verify ii) of Definition 4.10, and abstraction-completeness. Note that an extensionality axiom would be needed to carry out this elimination argument in the event that 3rd-order and higher-order theories were involved.

Below, we will use the notation T^*.

LEMMA 4.1.5. Let φ be an (A^{*-},B^*)-sentence, ψ be the result of replacing the second-order constants appearing in φ by distinct parameters of the appropriate sorts, the same constants being replaced by the same parameter. Then $\vdash \varphi$ iff $\vdash \psi$.

Proof: See Lemma 2.4.4.

THEOREM 4.1. (Relative to (A,B).) Let K_1 be the least class of (A^-,B^*)-formulae satisfying i) all atomic (A,B)-formulae are in K_1 ii) if $\varphi,\psi \in K_1$, then $(\varphi \,\&\, \psi) \in K_1$ iii) if $\varphi \in K_1$, ψ an (A^-,B^*)-formula, then $(\psi \to \varphi) \in K_1$ iv) if $\varphi \in K_1$, then $(\forall v)(\varphi^a_v) \in K_1$ (provided $\underline{a}$ is an (A^-,B^*)-parameter) v) each $(\exists v)(\mathrm{Comp}(\varphi,b,b_o,\ldots,b_n)^b_v) \in K_1$, provided φ is an (A^-,B^*)-formula. Let K be the least class of (A^-,B^*)-theories satisfying i) K is closed under arbitrary unions ii) $\{\varphi\} \in K$ if $\varphi \in K_1$ iii) if b is of sort x in B, $T \vdash \varphi^b_t$ for all closed (A,B)-terms t of sort x, and $T \in K$, then $T \cup \{\varphi\} \in K$. Then every $T \in K$ has the following properties: I. For all (A^-,B^*)-formulae $(\varphi \vee \psi)$ with no parameters of sort in B, we have $T \vdash (\varphi \vee \psi)$ iff $T \vdash \varphi$ or $T \vdash \psi$ II. For all (A^-,B^*)-formulae $(\exists v)(\varphi^b_v)$ with no parameters of sort in B, b of sort $x \in B$, we have $T \vdash (\exists v)(\varphi^b_v)$ iff $T \vdash \varphi^b_t$ for some closed (A,B)-term t of sort x III. Let b be of sort in $B^* - B$, $(\exists v)(\varphi^b_v)$ an (A^-,B^*)-formula with no parameters of sort in B. Then there are ψ, $b_o,\ldots,b_n$ such that ψ is an (A^-,B^*)-formula, and $T \vdash (\exists v)(\varphi^b_v)$ iff $T \vdash (\exists v)(\mathrm{Comp}(\psi,b,b_o,\ldots,b_n)^b_v \,\&\, \varphi^b_v)$.

Proof: Let $T \in K$. By lemmas 4.1.2, 4.1.3 we have $R^*(T)$. By Lemma 4.1.4, T^* is 2nd-order abstraction-complete, and abstraction-minimal over T. We may assume that not $T \vdash \bot$. By minimality, not $T^* \vdash \bot$. Hence there is some T^*-special structure M which we fix. Hence $R(M,A^*,B^*,T^*,U(T))$. I. Let φ,ψ be as in hypotheses. Let φ_1,ψ_1 be the result of simultaneously replacing the 2nd-order parameters in φ_1,ψ_1 by distinct free constants. Then $T \vdash (\varphi_1 \vee \psi_1)$. Hence $R(M,A^*,B^*,T^*,(\varphi_1 \vee \psi_1))$. So $T^* \vdash \varphi_1$ or $T^* \vdash \psi_1$. By minimality over T, we have $T \vdash \varphi_1$ or $T \vdash \psi_1$. Hence by Lemma 4.1.5, we have $T \vdash \varphi$ or $T \vdash \psi$ II. Let $(\exists v)(\varphi^b_v)$, x be as in hypotheses. Let $(\exists v)(\psi^b_v)$ be the result of simultaneously replacing the 2nd-order parameters in $(\exists v)(\varphi^b_v)$ by distinct free constants. Then $T \vdash (\exists v)(\psi^b_v)$. Hence $R(M,A^*,B^*,T^*,(\exists v)(\psi^b_v))$. So for some closed

(A,B)-term t of sort x we have $T \vdash \psi^b_t$. Hence by Lemma 4.1.5, we have $T \vdash \varphi^b_t$ III. Let $b, k_o, \ldots, k_n, (\exists v)(\varphi^b_v)$ be as in hypotheses. Let $(\exists v)(\rho^b_v)$ be the result of simultaneously replacing the 2nd-order parameters in $(\exists v)(\varphi^b_v)$ by distinct free constants. Then $T \vdash (\exists v)(\rho^b_v)$. Hence $R(M, A^*, B^*, T^*, (\exists v)(\rho^b_v))$. There is a 2nd-order constant F such that $T^* \vdash \rho^b_F$. By minimality over T, choose an (A^{*-}, B^*)-formula ψ_1, and $b_o, \ldots, b_n$, with $T^* \vdash Comp(\psi_1, b, b_o, \ldots, b_n)^b_F$, where ψ_1 has only parameters among $b_o, \ldots, b_n$. Then $T^* \vdash (\exists v)(Comp(\psi_1, b, b_o, \ldots, b_n)^b_v \& \rho^b_v)$. So by minimality over T, we have $T \vdash (\exists v)(Comp(\psi_1, b, b_o, \ldots, b_n)^b_v \& \rho^b_v)$. Finally let ψ be the result of replacing the new constants by the old parameters, and any additional free constants by parameters. Then $T \vdash (\exists v)(Comp(\psi, b, b_o, \ldots, b_n) \& \varphi^b_v)$, by Lemma 4.1.5.

LEMMA 4.2.1. Let T be in the K of Theorem 4.1. Let $T_o = \{\varphi: \varphi$ is an (A,B)-formula with $T \vdash \varphi\}$. Then $R(A, B, T_o)$.

Proof: Let $T \vdash T_o$, T an (A,B)-theory, M_o be a T-complete (A,B)-structure. Since T^* is a conservative extension of T for (A,B)-formulae, we have that there is a T^*-special structure M which has the same (A,B)-sentences as M_o. Since $R^*(T_o)$, we have $R(M, A^*, B^*, T^*, U(T))$. Hence $R(M, A^*, B^*, T^*, U(T_o))$. Since T^* is a conservative extension of T for (A,B)-formulae, we have $R(M_o, A, B, T, U(T_o))$.

THEOREM 4.2. Suppose T is in the K of Theorem 4.1 with (A,B) an arithmetical type. Furthermore, assume T^o is recursively enumerable and consists only of (universally) true formulae. Then there is a (PR,{0})-metasubstitution β which is independent over T, and such that β takes on only Π^o_2 sentences as values.

Proof: By Lemma 4.2.1, we have $R(A, B, T_o)$. Without loss of generality, we can assume that $E(PR, \{0\}, \{0\}) + PRE + SA \subset T$. Since (A,B) is an arithmetical type, we have $R(Tr(0), PR, \{0\}, (T_o)^o)$, by Lemma 3.3.5. Since $(T_o)^o = T^o$, we have $R(Tr(0), PR, \{0\}, T^o)$. Hence by Theorem 3.3, there is a (PR,{0})-metasubstitution β which is independent over T^o, and such that β takes on only Π^o_2 sentences. Clearly β is also independent over T. (Every r.e. theory has a recursive axiomatization).

DEFINITION 4.11. (Using notation of Section 3.) We take $\underline{Z_2}$, (2nd-order arithmetic), as $HA(PR^-,\{0\}*) + \{(\exists v)(Comp(\varphi,b,b_o,\ldots,b_n)^b_v): \varphi$ is a $(PR^-,\{0\}*)$-formula$\}$. Let $\underline{C(*)} = \{(\varphi \vee (\varphi \to \bot)): \varphi$ is a $(PR^-,\{0\}*)$-formula$\}$.

COROLLARY 4.2.1. Let Y be any set of primitive recursive well-ordering indices. Let $T = Z_2 \cup \{TI(x,\varphi): x \in Y$ and φ is a $(PR^-,\{0\}*)$-formula$\}$. Then T is provably equivalent to some element of the K of Theorem 4.1, and so satisfies I, II, III there. Also any $(PR,\{0\})$-metasubstitution classically independent over $T \cup Tr(1) \cup C(*)$ is independent over T.

<u>Proof</u>: For the first part see the proof of Theorem 3.2. The rest is related to Theorem 4.2 as Corollary 3.3.2 is related to Corollary 3.3.1.

SECTION 5. TYPE LOGIC AND TYPE THEORY

Throughout Section 5 we fix (A,B) as an integer relational type, and it will be convenient to assume that no R_o^x is in A.

DEFINITION 5.1. We define $\underline{B_o^+} = B$, $\underline{B_{k+1}^+} = \{(x_o,\ldots,x_n)\colon\ x_o,\ldots,x_n \in B_k^+\}$, $\underline{B^+} = \bigcup_k B_k^+$. The elements of $B^+ - B$ are called <u>higher-order sorts</u>.

DEFINITION 5.2. The <u>higher-order free constants</u> are given by F_α^x, where x is a higher-order sort and $\alpha < \omega$.

DEFINITION 5.3. We define $\underline{A_x^+}$, for $x \in B^+$, inductively as follows: $A_x^+ = \{F\colon F \in A,\ F$ has sort $x\}$, for $x \in B$. $A^+_{(x_o,\ldots,x_n)} = \{F\colon F$ is a higher-order free constant of sort $(x_o,\ldots,x_n)\} \cup \{F_{(\alpha,y)}^{(x_o-x_n)}\colon\ \omega \leq \alpha < \omega \times 2$, y is a set of $n+1$-tuples $t_o,\ldots,t_n$ such that for each higher-order x_i we have $t_i \in A^+_{x_i}$, and for each $x_i \in B$ we have t_i is a closed (A,B)-term of sort $x_i\}$. Take $A^+ = (\bigcup_{x\in B^+} A_x^+) \cup \{R_o^{(x_o-x_n),x_o-x_n}\colon\ x_o,\ldots,x_n \in B^+\}$.

DEFINITION 5.4. The <u>higher-order constants</u> are the elements of the A_x^+ for x a higher-order sort. The <u>higher-order abstraction constants</u> are the higher-order constants that are not free. Let $\underline{A^\sim} = A^+$ without higher-order constants. Let $\underline{A^{+\sim}} = A^+$ without higher-order abstraction constants.

DEFINITION 5.5. Let φ be an (A^+,B^+)-formula, $b,b_o,\ldots,b_n$ distinct parameters, b not occurring in φ, b of sort $(x_o,\ldots,x_n) \in B^+$, $b_o,\ldots,b_n$ of respective sorts $x_o,\ldots,x_n$. We write <u>$\mathrm{Comp}(\varphi,b,b_o,\ldots,b_n)$</u> for $(\forall v_o^{x_o})\cdots(\forall v_n^{x_n})(R_o(b,v_o,\ldots,v_n) \longleftrightarrow \varphi^{b_o-b_n}_{v_o-v_n})$. Whenever we write $\mathrm{Comp}(\varphi,b,b_o,\ldots,b_n)$ we assume that the above conditions on parameters are met.

DEFINITION 5.6. A theory T is called <u>higher-order abstraction-complete</u> iff if b is of some higher-order sort x, φ has only parameters among $b_o,\ldots,b_n$, and $T \vdash (\exists v)(\mathrm{Comp}(\varphi,b,b_o,\ldots,b_n)^b_v)$, then for some $\omega \leq \alpha < \omega \times 2$, we have $T \vdash \mathrm{Comp}(\varphi,b,b_o,\ldots,b_n)^b_F$ for all higher-order constants $F = F^x_{(\alpha,y)}$.

DEFINITION 5.7. A <u>T-adequate structure</u> is an (A^+, B^+)-structure M such that
i) if φ is an atomic (A,B)-sentence and $T \vdash \varphi$, then $\varphi \in M$ ii) we have
$\varphi = R_0(F_{(\beta,y)}, t_0, \ldots, t_n) \in M$ iff φ is an (A^+, B^+)-sentence, and $(t_0, \ldots, t_n) \in y$.

DEFINITION 5.8. Let T be an (A^+, B^+)-theory. We write $\underline{R^+(T)}$ iff for all higher-order abstraction-complete $T' \vdash T$, all T'-adequate structures M, we have
$R(M, A^+, B^+, T', U(T))$.

DEFINITION 5.9. We let Z^+ be the least class of (A^+, B^+)-formulae satisfying
i) all atomic (A,B)-formulae are in Z^+ ii) if $\varphi, \psi \in Z^+$, then $(\varphi \& \psi) \in Z^+$
iii) if $\varphi \in Z^+$, ψ is an (A^+, B^+)-formula, then $(\psi \to \varphi) \in Z^+$ iv) if $\varphi \in Z^+$,
then $(\forall v)(\varphi_v^a) \in Z^+$, (provided $\underline{a}$ is an (A^+, B^+)-parameter) v) each $(\exists v)(Comp(\varphi, b, b_0, \ldots, b_n)_v^b) \in Z^+$.

LEMMA 5.1.1. We have $R^+(\{(\exists v)(Comp(\varphi, b, b_0, \ldots, b_n)_v^b)\})$.

Proof: See the proof of Lemma 4.1.1.

LEMMA 5.1.2. For each $\varphi \in Z^+$ we have $R^+(\{\varphi\})$.

Proof: See the proof of Lemma 4.1.2.

LEMMA 5.1.3. Suppose T is an (A^+, B^+)-theory with $R^+(T)$. Suppose φ is an (A^+, B^+)-formula, $\underline{a}$ an (A^+, B^+)-parameter with $T \vdash \varphi_t^a$ for all closed (A^+, B^+)-terms t of the same sort as $\underline{a}$. Then $R^+(T \cup \{\varphi\})$.

Proof: See the proof of Lemma 4.1.3.

DEFINITION 5.10. Let T be a theory. Then T_1 is said to be <u>conservative over T</u>
iff $T_1 \vdash T$ and for all $(A^{+\sim}, B^+)$-formula φ, we have $T_1 \vdash \varphi$ iff $T \vdash \varphi$.

LEMMA 5.1.4. To each T there is a T' which is conservative over T and higher-order abstraction-complete.

Proof: See the proof of Lemma 2.4.5.

THEOREM 5.1. (Relative to (A,B).) Let Q_1 be the least class of $(A^\sim, B^+)$-formulae satisfying i) all atomic (A,B)-formulae are in Q_1 ii) if $\varphi, \psi \in Q_1$, then $(\varphi \,\&\, \psi) \in Q_1$ iii) if $\varphi \in Q_1$, ψ an $(A^{\sim}, B^{+})$-formula, then $(\psi \rightarrow \varphi) \in Q_1$ iv) if $\varphi \in Q_1$, then $(\forall v)(\varphi^{a}_{v}) \in Q_1$ (provided $\underline{a}$ is an $(A^{\sim}, B^{+})$-parameter) v) each $(\exists v)(\mathrm{Comp}(\varphi, b, b_o, \ldots, b_n)^{b}_{v}) \in Q_1$, provided φ is an $(A^{\sim}, B^{+})$-formula. Let Q be the least class of $(A^{\sim}, B^{+})$-theories satisfying i) Q is closed under arbitrary unions ii) $\{\varphi\} \in Q$ if $\varphi \in Q_1$ iii) if b is of sort x in B, $T \vdash \varphi^{a}_{t}$ for all closed (A,B)-terms t of sort x, and $T \in Q$, then $T \cup \{\varphi\} \in Q$. Then every $T \in Q$ has the following properties: I. For all $(A^{\sim}, B^{+})$-formulae $(\varphi \vee \psi)$ with no parameters of sort in B, we have $T \vdash (\varphi \vee \psi)$ iff $T \vdash \varphi$ or $T \vdash \psi$. II. For all $(A^{\sim}, B^{+})$-formulae $(\exists v)(\varphi^{b}_{v})$ with no parameters of sort in B, b of sort $x \in B$, we have $T \vdash (\exists v)(\varphi^{b}_{v})$ iff $T \vdash \varphi^{b}_{t}$ for some closed (A,B)-term t of sort x.

Proof: As in I, II of Theorem 4.1. Whenever 2nd-order abstraction minimality of T^* over T is used there, here we use the conservation of T' over T (see Lemma 5.1.4).

LEMMA 5.2.1. Let T be in the Q of Theorem 5.1. Let $T_o = \{\varphi: \varphi$ is an (A,B)-formula with $T \vdash \varphi\}$. Then $R(A,B,T_o)$.

Proof: See the proof of Lemma 4.2.1. Use T' instead of T^*.

THEOREM 5.2. Suppose T is in the Q of Theorem 5.1 with (A,B) an arithmetical type. Furthermore assume T^o is recursively enumerable and consists only of (universally) true formulae. Then there is a $(PR, \{0\})$-metasubstitution β which is independent over T, and such that β takes on only Π^{o}_{2} sentences as values.

Proof: See the proof of Theorem 4.2.

DEFINITION 5.11. (Using the notation of Section 3.) We take $\underline{Z_{\omega}}$, (type theory), as $HA(PR^{\sim}, \{0\}^{+}) + \{(\exists v)(\mathrm{Comp}(\varphi, b, b_o, \ldots, b_n)^{b}_{v}): \varphi$ is a $(PR^{\sim}, \{0\}^{+})$-formula$\}$. Let $C(+) = \{(\varphi \vee (\varphi \rightarrow \bot)): \varphi$ is a $(PR^{\sim}, \{0\}^{+})$-formula$\}$.

COROLLARY 5.2.1. Let Y be any set of primitive recursive well-ordering indices. Let $T = Z_{\omega} \cup \{TI(x,\varphi): x \in Y$ and φ is a $(PR^{\sim}, \{0\}^{+})$-formula$\}$. Then T is provably equivalent to some element of the Q of Theorem 5.1, and so satisfies I, II there. Also any $(PR,\{0\})$-metasubstitution classically independent over $T \cup Tr(1) \cup C(+)$ is independent over T.

Proof: See the proof of Corollary 4.2.1.

We now wish to obtain an analogue, Theorem 5.3, of III of Theorem 4.1. We fix T as an element of the Q of Theorem 5.1, until the end of the proof of Theorem 5.3.

DEFINITION 5.12. We inductively define $\underline{T_n^+}$ and $\underline{L_n^+}$ as follows: $L_0^+ = A^{+\sim}$, $T_0^+ = T$, $L_{n+1}^+ = \{F^x_{(\varphi,b,b_0,\dots,b_n)}: T_n^+ \vdash (\exists v)(Comp(\varphi,b,b_0,\dots,b_n)^b_v)$, φ has only parameters among $b_0,\dots,b_n$, b of sort x, and φ is an (L_n^+,B^+)-formula$\}$, $T_{n+1}^+ = T \cup \{Comp(\varphi,b,b_0,\dots,b_n)^b_F: F \in L_{n+1}^+$ and $F = F^x_{(\varphi,b,b_0,\dots,b_n)}\}$. Take $\underline{T^+} = \bigcup_n T_n^+$, $\underline{L^+} = \bigcup_n L_n^+$.

DEFINITION 5.13. Since for each higher-order sort x, $L^+ - A^{+\sim}$ contains countably infinitely many elements of sort x, we let $\underline{|x,\alpha|}$, for each higher-order sort x, be a one-one map from $\omega \times 2 - \omega$ onto the set of subscripts of elements of $L^+ - A^{+\sim}$. We extend $|\ |$ to all higher-order abstraction constants $F = F^x_{(\alpha,y)}$ by $|F| = F^x_{|x,\alpha|}$. We also extend $|\ |$ to all (A^+,B^+)-formulae by $|\varphi|$ = the (L^+,B^+)-formula that results when each and every higher-order abstraction constant, F, in φ is replaced by $|F|$. For (A^+,B^+)-theories T, let $|T| = \{|\varphi|: \varphi \in T\}$.

DEFINITION 5.14. Let $T^{++} = \{\varphi: \varphi$ is an (A^+,B^+)-formula with $T^+ \vdash |\varphi|\}$.

LEMMA 5.3.1. For each (A^+,B^+)-formula φ, we have $T^{++} \vdash \varphi$ iff $T^+ \vdash |\varphi|$.

Proof: Show that $|\ |$ preserves derivations; details left to the reader.

LEMMA 5.3.2. T^{++} is higher-order abstraction-complete, and T^{++} is conservative over T.

Proof: The first part is immediate from Lemma 5.3.1. The second is proven straightforwardly: first show that T^+ is conservative over T for $(A^{+\sim},B^+)$-formulae by showing by induction that each T_n^+ is. Next, apply Lemma 5.3.1.

THEOREM 5.3. Let T be in the Q of Theorem 5.1, and let T^+ be given as in Definition 5.12. Then T^+ is both (L^+,B^+)-disjunctive and (L^+,B^+)-existential.

Proof: By Lemmas 5.1.2 and 5.1.3, we have $R^+(T^{++})$. By Lemma 5.3.1, we may as well assume that not $T^{++} \vdash \perp$. Hence let M be T^{++}-special. Then $R(M,A^+,B^+,T^{++},U(T))$. Assume firstly that $T^+ \vdash (\varphi \vee \psi)$, where φ,ψ are (L^+,B^+)-sentences. Choose $V \subset T^+ - T$, V finite, such that $T \cup V \vdash (\varphi \vee \psi)$. Let V^-, φ^-, ψ^- be the result of simultaneously replacing the constants in $L^+ - A^{+\sim}$ in V, φ, ψ by distinct parameters. Then $T \vdash ((\bigwedge V^-) \to (\varphi^- \vee \psi^-))$ since T is an $(A^{+\sim},B^+)$-theory. Hence $R(M,A^+,B^+,T^{++},U((\bigwedge V^- \to (\varphi^- \vee \psi^-))))$. From the proof of Lemma 5.1.1, we see that there is an (A^+,B^+)-substitution β such that $T^{++} \vdash (\bigwedge V^-)(\beta)$, $R(M,A^+,B^+,T^{++},(\bigwedge V^-)(\beta))$, and $|V^-(\beta)| = V$, $|\varphi^-(\beta)| = \varphi$, $|\psi^-(\beta)| = \psi$. Hence $R(M,A^+,B^+,T^{++},(\varphi^-(\beta) \vee \psi^-(\beta)))$, and so $T^{++} \vdash \varphi^-(\beta)$ or $T^{++} \vdash \psi^-(\beta)$. By Lemma 5.3.1, we have $T^+ \vdash \varphi$ or $T^+ \vdash \psi$.

Secondly, assume that $T^+ \vdash (\exists v)(\rho^b_v)$, where ρ has only the parameter b, ρ an (L^+,B^+)-formula. Just as above, define V, V^-, $(\exists v)(\rho^b_v)^-$, and β. We then have $R(M,A^+,B^+,T^{++},(\exists v)(\rho^b_v)^-(\beta))$. Hence $R(M,A^+,B^+,T^{++},(\exists v)(\rho^-(\beta)^b_v))$. So for some closed (A^+,B^+)-term s we have $T^{++} \vdash \rho^{-b}_s$. Since $|(\exists v)(\rho^b_v)^-(\beta)| = (\exists v)(\rho^b_v)$, we have $|\rho^{-b}_s| = \rho^b_t$, for some (L^+,B^+)-term t. Hence by Lemma 5.3.1, we have $T^+ \vdash \rho^b_t$.

COROLLARY 5.3.1. If $T^+ \vdash (\varphi \vee \psi)$, φ, ψ are (L^+,B^+)-formulae with no parameters of sort in B, then $T^+ \vdash \varphi$ or $T^+ \vdash \psi$.

Proof: This follows from T^+ being (L^+,B^+)-disjunctive, using higher-order free constants.

SECTION 6. RECURSION-THEORETIC REALIZABILITY IN SET THEORY

Until the end of the proof of Theorem 6.2, we work entirely within the sorts 0, 1. Sort 0 for natural numbers, as in Section 3, and sort 1 for sets.

DEFINITION 6.1. We let (A,B) be the following relational type: $B = \{0,1\}$, $A = PR \cup \{R_0^{0,1}, R_1^{1,1}, R_0^{1,1}\}$. Here $R_0^{0,1}$ denotes the ϵ-relation between natural numbers and sets (of transfinite type); $R_1^{1,1}$ denotes the ϵ-relation between sets; $R_0^{1,1}$ denotes the equality relation between sets. We use ε for $R_1^{1,1}$, $=^1$ for $R_0^{1,1}$, and ε^0 for $R_0^{0,1}$.

DEFINITION 6.2. We let <u>ZFTI</u> be the following theory: 1) $HA(A,B) + TI(x,\varphi)$, for all (A,B)-formulae φ, primitive recursive well-ordering indices x 2) $E(A,B,\{1\})$ 3) pairing. $(\exists v)(a^0 \varepsilon v \;\&\; b^0 \varepsilon v \;\&\; c^1 \varepsilon v \;\&\; d^1 \varepsilon v)$ 4) union. $(\exists v)(\forall w)(\forall z^1)(\forall z^0)(((w \varepsilon a \;\&\; z^1 \varepsilon w) \rightarrow z^1 \varepsilon v) \;\&\; ((w \varepsilon a \;\&\; z^0 \varepsilon w) \rightarrow z^0 \varepsilon v))$ 5) power set. $(\exists v)(\forall w)((\forall z^1)(\forall z^0)((z^1 \varepsilon w \rightarrow z^1 \varepsilon a) \;\&\; (z^0 \varepsilon w \rightarrow z^0 \varepsilon a)) \rightarrow w \varepsilon v)$ 6) infinity. $(\exists v)(\forall z^0)(z^0 \varepsilon v)$ 7) separation. $(\exists v)(\forall z^1)(\forall z^0)((z^1 \varepsilon v \longleftrightarrow (\varphi^b_{z^1} \;\&\; z^1 \varepsilon c)) \;\&\; (z^0 \varepsilon v \longleftrightarrow (\psi^a_{z^0} \;\&\; z^0 \varepsilon c)))$ 8) replacement. $(\forall z^1)(\forall z^0)((z^1 \varepsilon a \rightarrow (\exists y^1)(\varphi^{bc}_{z^1 y^1})) \;\&\; (z^0 \varepsilon a \rightarrow (\exists y^1)(\psi^{dc}_{z^0 y^1}))) \rightarrow (\exists x)(\forall z^0)(\forall z^1)((z^1 \varepsilon a \rightarrow (\exists y^1)(\varphi^{bc}_{z^1 y^1} \;\&\; y^1 \varepsilon x)) \;\&\; (z^0 \varepsilon a \rightarrow (\exists y^1)(\psi^{dc}_{z^0 y^1} \;\&\; y^1 \varepsilon x)))$.

DEFINITION 6.3. We let <u>$\{e\}(n)$</u> be the partial recursive function of one argument n, with gödel number e. We let <u>$\{e\}(f)$</u> be the partial recursive functional from $f \in \omega^\omega$ into ω, with gödel number e. For $f \in \omega^\omega$, we let $\overline{f}(n) = \overline{f(n)}$.

DEFINITION 6.4. Let f be any function, $y \in Dom(f)$. We let <u>f^y_z</u> be the function given by $f^y_z(x) = f(x)$ if $x \neq y$; $f^y_z(y) = z$. We use $=$ when both sides are defined and are equal. We use $\simeq$ when either both sides are undefined or both sides are defined and equal.

DEFINITION 6.5. Define $V(0) = \emptyset$, $V(\alpha+1) = \mathbb{P}(V(\alpha))$, <u>$V(\lambda)$</u> $= \bigcup_{\alpha < \lambda} V(\alpha)$, <u>$V$</u> $= \bigcup_\alpha V(\alpha)$. Define <u>$rk(x)$</u> $= (\mu\alpha)(x \in V(\alpha))$. Let <u>$(\;)$</u> be a map from $\omega \times V$ one-one onto $V - V(\omega)$, such that, for each $\alpha > \omega$, $(\;)$ maps $\omega \times V(\alpha)$ one-one onto $V(\alpha) - V(\omega)$. Define <u>$(x)_1$</u>, <u>$(x)_2$</u> so that $x = ((x)_1,(x)_2)$ for all sets $x \in V - V(\omega)$. Take

$\underline{[n,m]}$ to be a recursive function from $\omega \times \omega$ one-one onto ω, and let $[k]_1$, $[k]_2$ be so that $[[k]_1,[k]_2] = k$, all $k \in \omega$.

DEFINITION 6.6. We define $\underline{R(e,\varphi,f,g)}$, for $e \in \omega$, φ an (A,B)-formula, $f \in \omega^\omega$, $g \in V^\omega$ as follows: i) $R(e,s = t,f,g)$ iff $|s(\bar{f})| = |t(\bar{f})|$, for $(PR,\{0\})$-terms s,t ii) $R(e,s\,\varepsilon^0\,a^1_n,f,g)$ iff $[e,|s(\bar{f})|] \in g(n)$, for $(PR,\{0\})$-terms s iii) $R(e,a^1_n = a^1_m,f,g)$ iff $g(n) = g(m)$ iv) $R(e,a^1_n\,\varepsilon\,a^1_m,f,g)$ iff $(e,g(n)) \in g(m)$ v) not $R(e,\bot,f,g)$ vi) $R(e,(\varphi\ \&\ \psi),f,g)$ iff $R([e]_1,\varphi,f,g)$ and $R([e]_2,\varphi,f,g)$ vii) $R(e,(\varphi \vee \psi),f,g)$ iff either $[e]_1 = 1$ and $R([e]_1,\varphi,f,g)$, or $[e]_1 = 2$ and $R([e]_2,\psi,f,g)$ viii) $R(e,(\varphi \to \psi),f,g)$ iff for all n with $R(n,\varphi,f,g)$, we have that $R(\{e\}(n),\psi,f,g)$ ix) $R(e,(\forall v^0)(\varphi\genfrac{}{}{0pt}{}{a^0_n}{v^0_0}),f,g)$ iff for all $m \in \omega$, $R(\{e\}(n),\varphi,f^n_m,g)$ x) $R(e,(\exists v^0)(\varphi\genfrac{}{}{0pt}{}{a^0_n}{v^0_0}),f,g)$ iff $R([e]_2,\varphi,f^n_{[e]_1},g)$ xi) $R(e,(\forall v^1)(\varphi\genfrac{}{}{0pt}{}{a^1_n}{v^1_1}),f,g)$ iff for all sets x, $R(e,\varphi,f,g^n_x)$ xii) $R(e,(\exists v^1)(\varphi\genfrac{}{}{0pt}{}{a^1_n}{v^1_1}),f,g)$ iff for some set x, $R(e,\varphi,f,g^n_x)$.

DEFINITION 6.7. We write $R(e,\varphi)$ iff for all f,g we have $\{e\}(f) = \{e\}(f)$ and $R(\{e\}(f),\varphi,f,g)$.

In lemmas below, we will be choosing various e with $R(e,\varphi)$. Consequently, we will always insist that the e be chosen so that $(\forall f)(\{e\}(f) = \{e\}(f))$, in each case.

LEMMA 6.1.1. The predicate on (A,B)-formulae, φ, "$(\exists e)(R(e,\varphi))$," satisfies clauses 1)-10) of Definition 2.8 when T is replaced by $\emptyset$ and "formulae φ, ψ, ρ" is replaced by "(A,B)-formulae φ, ψ, ρ."

<u>Proof</u>: 1. not $\varphi \in \emptyset$ 2. let $(\forall f)(\forall k)(\{\{e\}(f)\}(k) = k)$ 3. let $R(n,\varphi)$, $R(m,(\varphi \to \psi))$. Let $(\forall f)(\forall k)(\{e\}(f) = \{\{m\}(f)\}(\{n\}(f)))$. Then $R(e,\psi)$ 4. let $R(n,(\varphi \to \psi))$, $R(m,(\psi \to \rho))$. Let $(\forall f)(\forall k)(\{\{e\}(f)\}(k) \simeq \{\{m\}(f)\}(\{\{n\}(f)\}(k)))$. Then $R(e,(\varphi \to \rho))$ 5. a) let $(\forall f)(\forall k)(\{\{e\}(f)\}(k) = [k]_1)$. Then $R(e,((\varphi\ \&\ \psi) \to \varphi))$ b) let $(\forall f)(\forall k)(\{\{e\}(f)\}(k) = [k]_2)$. Then $R(e,((\varphi\ \&\ \psi) \to \psi))$ c) let $(\forall f)(\forall k)(\{\{e\}(f)\}(k) = [1,k])$. Then $R(e,(\varphi \to (\varphi \vee \psi)))$ d) let $(\forall f)(\forall k)(\{\{e\}(f)\}(k) = [2,k])$. Then $R(e,(\psi \to (\varphi \vee \psi)))$ 6. let $R(n,(\varphi \to \rho))$, $R(m,(\psi \to \rho))$. Let $(\forall f)(\forall k)(\{\{e\}(f)\}(k) \simeq \{\{n\}(f)\}([k]_2)$ iff $[k]_1 = 1$; $\{\{m\}(f)\}([k]_2)$ if $[k]_1 = 2)$. Then $R(e,((\varphi \vee \psi) \to \rho))$ 7. let $R(n,(\rho \to \varphi))$, $R(m,(\rho \to \psi))$. Let

$(\forall f)(\forall k)(\{\{e\}(f)\}(k) \simeq [\{\{n\}(f)\}(k),\{\{m\}(f)\}(k)])$. Then $R(e,(\rho \to (\varphi \& \psi)))$ 8. let $R(n,((\varphi \& \psi) \to \rho))$. Let $(\forall f)(\forall p)(\forall q)(\{\{\{e\}(f)\}(p)\}(q) \simeq \{\{n\}(f)\}([p,q]))$. Then $R(e,(\varphi \to (\psi \to \rho)))$ 9. let $R(e,(\varphi \to (\psi \to \rho)))$. Let $(\forall f)(\forall k)(\{\{e\}(f)\}(k) \simeq \{\{\{n\}(f)\}([k]_1)\}([k]_2))$. Then $R(e,((\varphi \& \psi) \to \rho))$ 10. let $(\forall f)(\{e\}(f) = \{e\}(f))$. Then $R(e,(\bot \to \varphi))$.

LEMMA 6.1.2. Let f, g, φ be given, v, a_n, t of sort 0. Then $R(e,\varphi_t^{a_n},f,g)$ iff $R(e,\varphi,f^n_{|t(\bar{f})|},g)$.

Proof: By straightforward induction on φ.

LEMMA 6.1.3. We have, for all f, g, φ, that $R(e, \varphi, f, g^m_x) \longleftrightarrow R(e,\varphi,f,g)$, provided a^1_m does not occur in φ.

Proof: Straightforward induction on φ.

LEMMA 6.1.4. Let φ be an (A,B)-formula, a, v, t respectively a parameter of sort 0, a variable of sort 0 not in φ, and a (PR,{0})-term. Then for some e, $R(e,((\forall v)(\varphi^a_v) \to \varphi^a_t))$.

Proof: Let $(\forall f)(\forall k)(\{\{e\}(f)\}(k) \simeq \{k\}(|t(\bar{f})|))$. Then $R(e,((\forall v)(\varphi^a_v) \to \varphi^a_t))$, by Lemma 6.1.2.

LEMMA 6.1.5. Let φ be an (A,B)-formula, a, v respectively a parameter of sort 1 and a variable of sort 1 not in φ. Then for some e, $R(e,((\forall v)(\varphi^a_v) \to \varphi))$.

Proof: Let $(\forall f)(\forall k)(\{\{e\}(f)\}(k) = k)$. Then $R(e,((\forall v)(\varphi^a_v) \to \varphi))$. To see this, we must check that $(\forall f)(\forall g)(\forall k)($if $R(k,(\forall v)(\varphi^a_v),f,g)$ then $R(k,\varphi,f,g))$. If $R(k,(\forall v)(\varphi^a_v),f,g)$ then for all sets x, $R(k,\varphi,f,g^n_x)$, where $a = a^1_n$. In particular, this is true for $x = g(n)$.

LEMMA 6.1.6. Let φ be an (A,B)-formula, a, v, t respectively a parameter of sort 0, a variable of sort 0 not in φ, and a (PR,{0})-term. Then for some e, $R(e,(\varphi^a_t \to (\exists v)(\varphi^a_v)))$.

Proof: Let $(\forall f)(\forall k)(\{\{e\}(f)\}(k) \simeq [|t(\bar{f})|,k]$. Then $R(e,(\varphi^a_t \to (\exists v)(\varphi^a_v)))$, by Lemma 6.1.2.

LEMMA 6.1.7. Let φ be an (A,B)-formula, a, v respectively a parameter of sort 1, a variable of sort 1 not in φ. Then for some e, $R(e,(\varphi \rightarrow (\exists v)(\varphi^a_v)))$.

Proof: Let $(\forall f)(\forall k)(\{\{e\}(f)\}(k) = k)$. Then $R(e,(\varphi \rightarrow (\exists v)(\varphi^a_v)))$.

LEMMA 6.1.8. Let φ, ρ be (A,B)-formulae, a, v respectively a parameter of sort 0 not in ρ, and a variable of sort 0 not in φ. Suppose $R(n,(\rho \rightarrow \varphi))$. Then for some e, $R(e,(\rho \rightarrow (\forall v)(\varphi^a_v)))$.

Proof: Let $(\forall f)(\forall p)(\forall q)(\{\{\{e\}(f)\}(p)\}(q) \simeq \{\{n\}(f^m_q)\}(p))$, where $a = a^1_m$. Then $R(e,(\rho \rightarrow (\forall v)(\varphi^a_v)))$, by Lemma 6.1.2. To see this, we must prove that $(\forall f)(\forall g)(R(\{e\}(f),(\rho \rightarrow (\forall v)(\varphi^a_v)),f,g))$. Let $R(p,\rho,f,g)$. We must check $R(\{\{e\}(f)\}(p),(\forall v(\varphi^a_v),f,g)$. We must show, for all q, $R(\{\{\{e\}(f)\}(p)\}(q),\varphi,f^m_q,g)$. Note that $R(p,\rho^a_{\underline{q}},f^m_q,g)$, by Lemma 6.1.2. Now $\rho^a_{\underline{q}} = \rho$. Hence $R(p,\rho,f^m_q,g)$. Hence $\{\{\{e\}(f)\}(p)\}(q) = \{\{n\}(f^m_q)\}(p)$, and since $R(\{\{n\}(f^m_q)\}(p),\varphi,f^m_q,g)$, we have $R(\{\{\{e\}(f)\}(p)\}(q),\varphi,f^m_q,g)$.

LEMMA 6.1.9. Let φ, ρ be (A,B)-formulae, a, v respectively a parameter of sort 1 not in ρ, and a variable of sort 1 not in φ. Suppose $R(n,\rho \rightarrow \varphi))$. Then for some e, $R(e,(\rho \rightarrow (\forall v)(\varphi^a_v)))$.

Proof: We claim $R(n,(\rho \rightarrow (\forall v)(\varphi^a_v)))$. We must prove $(\forall f)(\forall g)(R(\{n\}(f),(\rho \rightarrow (\forall v)(\varphi^a_v)))$. Let $R(k,\rho,f,g)$. We must prove $R(\{\{n\}(f)\}(k),\varphi,f,g^m_x)$ for all sets x, (where $a = a^1_m$). Fix x. We have $R(k,\rho,f,g^m_x)$, by Lemma 6.1.3. Hence $R(\{\{n\}(f)\}(k),g^m_x)$.

LEMMA 6.1.10. Let φ, ρ be (A,B)-formulae, a, v respectively a parameter of sort 0 not in ρ, a variable of sort 0 not in φ. Assume $R(n,(\varphi \rightarrow \rho))$. Then for some e, $R(e,((\exists v)\varphi^a_v) \rightarrow \rho))$.

Proof: Let $(\forall f)(\forall k)(\{\{e\}(f)\}(k) \simeq \{\{n\}(f^m_{[k]_1})\}([k]_2))$, where $a = a^1_m$. Then $R(e,((\exists v)(\varphi^a_v) \rightarrow \rho))$, by Lemma 6.1.2. To see this, we must prove that $(\forall f)(\forall g)(R(\{e\}(f),((\exists v)(\varphi^a_v) \rightarrow \rho),f,g))$. Let $R(k,(\exists v(\varphi^a_v),f,g)$. We must prove that

$R(\{\{e\}(f)\}(k),\rho,f,g)$. We have $R(\{n\}(f^m_{[k]_2}),(\varphi \to \rho),f^m_{[k]_1},g)$, (where $a = a^1_m$). Also $R([k]_1,\varphi,f^m_{[k]_2},g)$. Hence $R(\{\{n\}(f^m_{[k]_2})\}([k]_1),\rho,f^m_{[k]_2},g)$. By Lemma 6.1.2, $R(\{\{e\}(f)\}(k),\rho^a_{[k]_2},f,g)$, and so $R(\{\{e\}(f)\}(k),\rho,f,g)$.

LEMMA 6.1.11. Let φ, ρ be (A,B)-formulae, a, v respectively a parameter of sort 1 not in ρ and a variable of sort 1 not in φ. Assume $R(n,(\varphi \to \rho))$. Then for some e, $R(e,((\exists v)(\varphi^a_v) \to \rho))$.

<u>Proof</u>: We claim $R(n,((\exists v)(\varphi^a_v) \to \rho))$. We prove that $(\forall f)(\forall g)(R(\{n\}(f), ((\exists v)(\varphi^a_v) \to \rho)))$. Let $R(k,(\exists v)(\varphi^a_v),f,g)$. Let x be a set with $R(k,\varphi,f,g^m_x)$, (where $a = a^1_m$). We must prove $R(\{\{n\}(f)\}(k),\rho,f,g)$. We have $R(\{\{n\}(f)\}(k),\rho, f,g^m_x)$, by $R(n,(\varphi \to \rho))$. Hence by Lemma 6.1.3, $R(\{\{n\}(f)\}(k),\rho,f,g)$.

THEOREM 6.1. If φ is an (A,B)-formula with $\vdash \varphi$, then for some e, $R(e,\varphi)$.

<u>Proof</u>: By Lemmas 6.1.1 and 6.1.4-6.1.11.

LEMMA 6.2.1. If φ is an (A,B)-formula then for some e, $R(e,I(\varphi))$.

<u>Proof</u>: $I(\varphi)$ is $((\varphi^{a}_{0}{}^{o} \,\&\, (\forall v^o)((\varphi^{a}_{v^o}{}^{o} \to \varphi^{a}_{s(v^o)}{}^{o}))) \to \varphi)$. Let H be given by $H(p,0) = [p]_1$, $H(p,n+1) \simeq \{\{[p]_2\}(n)\}(H(p,n))$. Let $(\forall f)(\forall p)(\{\{e\}(f)\}(p) \simeq H(p,f(0)))$. Then $R(e,I(\varphi))$.

LEMMA 6.2.2. Let x be a primitive recursive well-ordering index, φ an (A,B)-formula. Then for some e, $R(e,TI(x,\varphi))$.

<u>Proof</u>: $TI(x,\varphi)$ is given by $(\forall w)((\forall v)(F_x(v,w) = 0 \to \varphi^{a}_{v}{}^{o}) \to \varphi^{a}_{w}{}^{o}) \to \varphi$. By the recursion theorem, choose a partial recursive J with $(\forall p)(\forall n)(J(p,n) \simeq \{\{p\}(n)\}(\#(\lambda v \lambda k(J(p,v)))))$, where $\#(\lambda v \lambda k(J(p,v)))$ denotes that godel number u obtained through the S^m_n-theorem such that $(\forall v)(\forall k)(\{\{u\}(v)\}(k) \simeq J(p,v))$. Then by transfinite induction on $\tilde{x}$, it can be seen that $(\forall f)(\forall g)(\forall p)(\forall n)(R(p,(\forall w)((\forall v)(F_x(v,w)\ 0 \to \varphi^{a}_{v}{}^{o}) \to \varphi^{a}_{w}{}^{o}),f,g)$ implies that $R(J(p,n),\varphi,f^o_n,g))$. Let $(\forall f)(\forall p)(\{\{e\}(f)\}(p) \simeq J(p,f(0)))$. Then $R(e,TI(x,\varphi))$.

LEMMA 6.2.3. For each ρ in $HA(A,B) + TI(x,\varphi)$, x a p.r. well-ordering index, φ an (A,B)-formula, there is an e such that $R(e,\rho)$.

<u>Proof</u>: Use Lemmas 6.2.1, 6.2.2.

LEMMA 6.2.4. For some e, $R(e,(\exists v)(a_o^o \in v \;\&\; a_1^o \in v \;\&\; a_o^1 \in v \;\&\; a_1^1 \in v))$.

<u>Proof</u>: We must find e such that $(\forall f)(\forall g)(\exists x)(R(\{e\}(f),(a_o^o \in a_2^1 \;\&\; a_1^o \in a_2^1 \;\& a_o^1 \in a_2^1 \;\&\; a_1^1 \in a_2^1),f,g_x^2))$. Let $(\forall f)(\{e\}(f) = [[[0,0],0],0])$. Then let f, g be given. Choose α such that $f, g \in V(\alpha)$. Then set $x = V(\alpha)$.

LEMMA 6.2.5. For some e, $R(e,\varphi)$, where $\varphi \in E(A,B,\{1\})$.

<u>Proof</u>: Left to the reader.

LEMMA 6.2.6. For some e, $R(e,(\exists v)(\forall w)(\forall z^1)(\forall z^o)(((w \in a_o^1 \;\&\; z^1 \in w) \to z^1 \in v) \;\& ((w \in a_o^1 \;\&\; z^o \in w) \to z^o \in v)))$.

<u>Proof</u>: We must find e such that $(\forall f)(\exists v)(\forall w)(\forall z)(\forall n)(R(\{\{e\}(f)\}(n), (((a_2^1 \in a_o^1 \;\&\; a_3^1 \in a_2^1) \to a_3^1 \in a_1^1) \;\&\; ((a_2^1 \in a_o^1 \;\&\; a_o^o \in a_2^1) \to a_o^o \in a_1^1)),f_n^o,g_{vwz}^{123}))$.

Let f be given. Let $(\forall f)(\forall k)([\{\{e\}(f)\}(n)]_1(k) = 0 \;\&\; [\{\{e\}(f)\}(n)]_2(k) = 0)$. Then e will work if $v = V(\alpha + \omega^2)$ is used, where $g(0) \in V(\alpha)$.

LEMMA 6.2.7. For some e, $R(e,(\exists v)(\forall w)((\forall z^1)(\forall z^o)((z^1 \in w \to z^1 \in a_o^1) \;\&\; (z^o \in w \to z^o \in a_o^1)) \to w \in v))$.

<u>Proof</u>: We must find e such that for all f, g, $(\exists v)(\forall w)(R(\{e\}(f),(\forall z^1)(\forall z^o) ((z^1 \in a_2^1 \to z^1 \in a_o^1) \;\&\; (z^o \in a_2^1 \to z^o \in a_o^1)) \to a_2^1 \in a_1^1),f,g_{vw}^{12}))$.

Let $(\forall f)(\forall k)(\{\{e\}(f)\}(k) = 0)$. Choose $v = V(\alpha + \omega^3)$, where $g(0) \in V(\alpha)$. It suffices to prove that for all w, f, g if $(\exists k)(R(k,(\forall z^1)(\forall z^o)((z^1 \in a_2^1 \to z^1 \in a_o^1) \;\&\; (z^o \in a_2^1 \to z^o \in a_o^1)),f,g_w^2))$ then $w \in V(\alpha + \omega^2)$. Let w, f, g, k be as above. Choose k' such that $R(k',(\forall z^1)(z^1 \in a_2^1 \to z^1 \in a_o^1),f,g_w^2)$. Hence $(\forall z)(R(k',(a_3^1 \in a_2^1 \to a_3^1 \in a_o^1),f,g_{wz}^{23})$. Suppose $w \notin V(\alpha + \omega^2)$. Let $x \in w$ with $rk(x) > \alpha + \omega$. Then $R((x)_1,a_3^1 \in a_2^1,f,g_{w(x)_2}^{2\,3})$. Hence for some p, $R(p,a_3^1 \in a_o^1, f,g_{w(x)_2}^{2\,3})$, and hence for some p, $(p,(x)_2) \in g(0)$. Hence $(x)_2 \in V(\alpha + \omega)$. But this contradicts $rk(x) > \alpha + \omega$.

LEMMA 6.2.8. For some e, $R(e,(\exists v)(\forall z^o)(z^o \in v))$.

<u>Proof</u>: We must find e such that for all f, g, $(\exists v)(\forall n)(R(\{e\}(f), a^o_o \,\varepsilon\, a^1_o, f^o_n, g^o_v))$. Let $(\forall f)(\{e\}(f) = 0)$, and use $v = V(\omega)$.

LEMMA 6.2.9. Let φ, ψ be (A,B)-formulae without variables v, z^o, z^1, φ without a^1_1, ψ without a^o_o. Then for some e, $R(e, (\exists v)(\forall z^1)(\forall z^o)((z^1 \,\varepsilon\, v \longleftrightarrow (\psi{}^{a^1_1}_{z^1} \;\&\; z^1 \,\varepsilon\, a^1_o))$ $\&\; (z^o \,\varepsilon\, v \longleftrightarrow \varphi{}^{a^o_o}_{z^o})))$.

<u>Proof</u>: We must find e such that for all f, $(\exists v)(\forall z)(R(\{e\}(f), (\forall z^o)$ $((a^1_2 \,\varepsilon\, a^1_1 \longleftrightarrow (\psi \;\&\; a^1_2 \,\varepsilon\, a^1_o)) \;\&\; (z^o \,\varepsilon\, a^1_1 \longleftrightarrow \varphi{}^{a^o_o}_{z^o})), f, g^{12}_{vz}))$. So we must find e such that for all f, g, $(\exists v)(\forall z)(\forall n)(R(\{\{e\}(f)\}(n), (a^1_2 \,\varepsilon\, a^1_1 \longleftrightarrow (\psi \;\&\; a^1_2 \,\varepsilon\, a^1_o)) \;\&$ $(a^o_o \,\varepsilon\, a^1_1 \longleftrightarrow \varphi), f^o_n, g^{12}_{vz}))$. Let $(\forall f)(\forall k)(\{[[\{\{e\}(f)\}(n)]_1]_1\}(k) = k \;\&\; \{[[\{\{e\}(f)\}(n)]_1]_2$ $(k) = k \;\&\; \{[[\{\{e\}(f)\}(n)]_2]_1\}(k) = k \;\&\; \{[[\{\{e\}(f)\}(n)]_2]_2\}(k) = k)$. Choose $v = \{(k,z)\colon\; R(k, \psi \;\&\; a^1_2 \,\varepsilon\, a^1_o, f, g^2_z)\} \cup \{[k,n]\colon\; R(k, \varphi \;\&\; a^o_o \,\varepsilon\, a^1_o, f^o_n, g)\}$. (The reader can easily verify that v is a set.)

LEMMA 6.2.10. Let φ, ψ be (A,B)-formulae without the variables z^o, z^1, y^1, x, φ without a^1_2, a^1_3, ψ without a^o_o, a^1_3. Then for some e, $R(e, (\forall z^o)(\forall z^1)((z^o \,\varepsilon\, a^1_o \rightarrow$ $(\exists y^1)(\varphi{}^{a^o_o\,a^1_1}_{z^o\,y^1})) \;\&\; (z^1 \,\varepsilon\, a^1_o \rightarrow (\exists y^1)(\psi{}^{a^1_2\,a^1_1}_{z^1\,y^1}))) \rightarrow (\exists x)(\forall z^o)(\forall z^1)((z^o \,\varepsilon\, a^1_o \rightarrow (\exists y^1)$ $(\varphi{}^{a^o_o\,a^1_1}_{z^o\,y^1} \;\&\; y^1 \,\varepsilon\, x)) \;\&\; (z^1 \,\varepsilon\, a^1_o \rightarrow (\exists y^1)(\psi{}^{a^1_2\,a^1_1}_{z^1\,y^1} \;\&\; y^1 \,\varepsilon\, x))))$.

<u>Proof</u>: Let $(\forall f)(\forall k)(\forall k_1)(\forall k_2)(\{[\{\{\{e\}(f)\}(k)\}(k_1)]_1\}(k_2) \simeq [\{[\{k\}(k_1)]_1\}(k_2), 0]$ $\&\; \{[\{\{\{e\}(f)\}(k)\}(k_1)]_2\}(k_2) \simeq [\{[\{k\}(k_1)]_2\}(k_2), 0])$. We claim e works. Fix f, g. Let β such that $(\forall n)(\forall k)(\forall y^1)(R(k, \varphi, f^o_n, g^1_{y^1})$ implies $\exists y \in V(\beta)$ with $R(k, \varphi, f^o_n, g^1_y))$. Let $g(0) \in V(\alpha)$. Let γ be such that $(\forall z^1 \in V(\alpha + \omega^2))(\forall k)(\forall y^1)(R(k, \psi,$ $f, g^{2\,1}_{z^1 y^1})$ implies $\exists y \in V(\gamma)$ with $R(k, \psi, f, g^{2\,1}_{z^1 y}))$. Let $\delta = \max(\beta, \gamma) + \omega^2$. Then, in the verification that e works, always use $x = V(\delta)$.

THEOREM 6.2. If φ is an (A,B)-formula such that $ZFTI \vdash \varphi$, then for some e, $R(e, \varphi)$.

<u>Proof</u>: Let $\psi_1, \ldots, \psi_k$ $ZFTI$, $\vdash U(\bigwedge_i \psi_i) \rightarrow \varphi$. By Lemmas 6.2.1-6.2.9, let

$R(n, U(\bigwedge_i \psi_i))$. By Theorem 6.1.1, let $R(m, U(\bigvee_i \psi_i) \to \varphi)$. Let $(\forall f)(\{e\}(f) = \{\{m\}(f)\}(\{n\}(f)))$. Then $R(e, \varphi)$.

DEFINITION 6.8. A <u>{ }-formula</u> is a Σ^0_1 formula φ with exactly the parameters a^0_0, a^0_1, a^0_2 such that $(\forall n)(\forall m)(\forall r)(\{n\}(m) = r$ iff $\varphi^{a_0 a_1 a_2}_{\bar{n}\,\bar{m}\,\bar{r}}$ is true).

COROLLARY 6.2.1. ZFTI is consistent with the simultaneous addition of the following (A,B)-theories: i) MP (Markov's Principle). $(\forall v^0)(\varphi^{a^0_0}_{v^0} \vee (\varphi^{a^0_0}_{v^0} \to \bot)) \to (((\forall v^0)(\varphi^{a^0_0}_{v^0}) \to \bot) \to (\exists v^0)(\varphi^{a^0_0}_{v^0} \to \bot))$ ii) $(\forall v^1)(\exists v^0)(\varphi^{a^1_0 a^0_0}_{v^1 v^0}) \to (\exists v^0)(\forall v^1)(\varphi^{a^1_0 a^0_0}_{v^1 v^0})$ iii) CT (Church's Thesis). $(\forall v^0)(\exists w^0)(\varphi^{a_1 a_2}_{v^0 w^0}) \to (\exists z^0)(\forall v^0)(\exists w^0)(\psi^{a_0 a_1 a_2}_{z^0 v^0 w^0} \;\&\; \varphi^{a_1 a_2}_{v^0 w^0})$, where ψ is a { }-formula iv) TR(2).

<u>Proof</u>: By Theorem 6.2, it suffices to prove that for each ρ in i), ii), iii) or iv), there is an e with $R(e, \rho)$. i) Let $(\forall f)(\forall p)(\forall q)(\{\{\{e\}(f)\}(p)\}(q) \simeq \mu([m,n])([\{p\}(m)]_1 = 2 \;\&\; [\{p\}(m)]_2 = n))$. Then $R(e, \rho)$ for all ρ in i). ii) Let $(\forall f)(\forall k)(\{\{e\}(f)\}(k) = k)$. Then $R(e, \rho)$ for all ρ in ii). To see this, let $R(k, (\forall v^1)(\exists v^0)(\varphi^{a^1_0 a^0_0}_{v^1 v^0}), f, g)$. Then $(\forall v^1)(R([k]_1, \varphi, f^0_{[k]_1}, g^0_{v^1}))$. Hence $R([k]_2, (\forall v^1)(\varphi^{a^1_0}_{v^1}), f^0_{[k]_1}, g)$, and so $R(k, (\exists v^0)(\forall v^1)(\varphi^{a^1_0 a^0_0}_{v^1 v^0}), f, g)$. iii) Let ψ be given. Fix h as a 2-ary partial recursive function such that $(\forall n)(\forall m)(\forall r)(\{n\}(m) = r \to (\forall f)(\forall g)(R(h(n,m), \psi, f^{012}_{nmr}, g)))$. Let $(\forall f)(\forall p)(\forall q)(\{[\{\{e\}(f)\}(p)]_1\}(q) \simeq [\{p\}(q)]_1 \;\&\; \{[\{e\}(f)\}(p)]_2\}(q) \simeq [h(p,q), [\{p\}(q)]_2])$. Then $R(e, \rho)$ for every ρ in iii) based on ψ. iv) left to the reader.

We now define the theory $ZFC^{1/2}$. Until the end of Section 6, we work entirely within the sort 1 (for sets). We use ε for $R^{1,1}_1$, and $=$ for $R^{1,1}_0$. We work entirely within the relational type $(\{\varepsilon, =\}, \{1\})$.

DEFINITION 6.9. Let φ be a formula. We write $WF(\varphi)$ for $(\forall v)(\varphi^{a_0 a_1}_{v} \to \bot) \;\&\; (\forall v)(\forall w)(\forall z)((\varphi^{a_0 a_1}_{v\,w} \;\&\; \varphi^{a_0 a_1}_{w\,z}) \to \varphi^{a_0 a_1}_{v\,z}) \;\&\; (\forall v)(\forall w)(w \,\varepsilon\, v \to (\exists z)(z \,\varepsilon\, v \;\&\; (\forall y)(y \,\varepsilon\, v \to (\varphi^{a_0 a_1}_{y\,z} \to \bot)))) \;\&\; (\forall v)(\forall w)(\varphi^{a_0 a_1}_{v\,w} \vee (\varphi^{a_0 a_1}_{v\,w} \to \bot))$.

DEFINITION 6.10. We write <u>$TI(\varphi, \psi)$</u> for $(\forall w)((\forall v)(\varphi^{a_0 a_1}_{v\,w} \to \psi^{a_0}_{v}) \to \psi^{a_0}_{w}) \to \psi$.

DEFINITION 6.11. <u>$ZFC^{1/2}$</u> is the following theory: 1) $E(\{\varepsilon,=\},\{1\},\{1\})$ 2) extensionality. $(\forall v)(v \varepsilon a_o \longleftrightarrow v \varepsilon a_1) \to a_o = a_1$ 3) pairing. $(\exists v)(a_o \varepsilon v \& a_1 \varepsilon v)$ 4) union. $(\exists v)(\forall w)(\forall z)((w \varepsilon a_o \& z \varepsilon w) \to z \varepsilon v)$ 5) power set. $(\exists v)(\forall w)((\forall z)(z \varepsilon w \to z \varepsilon a_o) \to w \varepsilon v)$ 6) infinity. $(\exists v)((\exists w)(w \varepsilon v \& (\forall z)(z \varepsilon w \to \bot)) \& (\forall w)(w \varepsilon v \to (\exists z)(z \varepsilon v \& (\forall y)(y \varepsilon z \longleftrightarrow (y \varepsilon w \vee y = w)))))$ 7) decidability of bounded quantification. $(a_o \varepsilon a_1 \vee (a_o \varepsilon a_1 \to \bot))$, $(a_o = a_1 \vee (a_o = a_1 \to \bot))$, $((\forall v)(\varphi_v^{a_1} \vee (\varphi_v^{a_1} \to \bot)) \to ((\forall v)(v \varepsilon a_o \to \varphi_v^{a_1}) \vee (\exists v)(v \varepsilon a_o \& (\varphi_v^{a_1} \to \bot))))$ 8) separation. $(\forall v)(\varphi_v^{a_1} \vee (\varphi_v^{a_1} \to \bot)) \to (\exists w)(\forall v)(v \varepsilon w \longleftrightarrow (\varphi_v^{a_1} \& v \varepsilon a_o))$ 9) foundation. $(\forall x)(\forall y)(y \varepsilon x \to (\exists z)(z \varepsilon x \& (\forall y)(y \varepsilon x \to (z \varepsilon x \to \bot))))$, and $WF(\varphi) \to TI(\varphi,\psi)$ 10) replacement. $(\forall v)(v \varepsilon a_o \to (\exists w)(\varphi_{v\,w}^{a_1 a_2})) \to (\exists z)(\forall v)(v \varepsilon a_o \to (\exists w)(\varphi_{v\,w}^{a_1 a_2} \& w \varepsilon z))$ 11) choice. $((\forall v)(v \varepsilon a_o \to (\exists w)(w \varepsilon v)) \& (\forall v_1)(\forall v_2)((v_1 \varepsilon a_o \& v_2 \varepsilon a_o) \to (\forall v_3)((v_3 \varepsilon v_1 \& v_3 \varepsilon v_2) \to \bot))) \to (\exists z)(\forall v)(v \varepsilon a_o \to ((\exists w)(w \varepsilon v \& w \varepsilon z) \& (\forall w_1)(\forall w_2)((w_1 \varepsilon z \& w_2 \varepsilon z \& w_1 \varepsilon v \& w_2 \varepsilon v) \to w_1 = w_2)))$.

We now fix k as a limit of infinite cardinals, with cofinality of k greater than ω, and $L(\kappa) \models \Sigma_2$-replacement.

DEFINITION 6.12. The <u>Δ_o-formulae</u> are the least class K of formulae satisfying i) $a_n \varepsilon a_m \in K$, $a_n = a_m \in K$, $\bot \in K$ ii) $(\varphi \& \psi)$, $(\varphi \to \psi)$, $(\varphi \vee \psi) \in K$ if $\varphi,\psi \in K$ iii) $(\exists v)(v \varepsilon a_n \& \varphi_v^{a_m})$, $(\forall v)(v \varepsilon a_n \to \varphi_v^{a_m}) \in K$ if $\varphi \in K$.

DEFINITION 6.13. The <u>Σ_2-formulae</u> are given by $(\exists v)(\forall w)(\varphi_{v\,w}^{a_o a_1})$, where φ is a Δ_o-formula.

DEFINITION 6.14. We let <u>$L(\kappa)$</u> be the constructible sets of level $< k$. For $g \in L(k)^{\omega}$ we let <u>$Tr(\varphi,g)$</u> mean that φ is (classically) true over $(L(\kappa),\in)$ at the assignment to parameters, g. A relation, $R(x_1,\ldots,x_n)$, on $L(\kappa)$ is said to be <u>Σ_2 on $L(\kappa)$</u> if and only if for some Σ_2-formula φ with parameters exactly $a_1,\ldots,a_n,\ldots,a_{n+m}$, some $x_1,\ldots,x_m \in L(\kappa)$, we have $(\forall g)((\forall i \le m)(g(n+i) = x_i) \to (Tr(\varphi,g) \longleftrightarrow R(g(1),\ldots,g(n))))$. We let <u>$\{x\}(y)$</u> be any binary partial function on $L(\kappa)$ (whose graph is) Σ_2 on $L(\kappa)$, and such that for every monadic such h, there is an $x \in L(\kappa)$ such that $(\forall y)(\{x\}(y) \simeq h(y))$. We let <u>$\{x\}(g)$</u> be any binary partial continuous function on $L(\kappa) \times L(\kappa)^{\omega}$ which is Σ_2 as operating on finite sequences of elements of $L(\kappa)$, and such that for every monadic such H, there is an $x \in L(\kappa)$ such that $(\forall g)(\{x\}(g) \simeq H(g))$. We let $<$ be a well-ordering of $L(\kappa)$, of type k, which is Σ_1 on $L(\kappa)$, and we use μ in reference to $<$. ($<$ is used to verify that $\{\ \}$ exists.)

DEFINITION 6.15. We let $(\ )$ be any binary Σ_2 function from $L(\kappa)$ one-one onto $L(\kappa)$. Let $((x)_1,(x)_2) = x$. Do not confuse this with Definition 6.5.

DEFINITION 6.16. We define $R^{1/2}(x,\varphi,g)$, for $x \in L(\kappa)$, φ a formula, $g \in L(\kappa)^{\omega}$, as follows: i) $R^{1/2}(x,a_n \mathcal{E}\, a_m,g)$ iff $g(n) \in g(m)$ ii) $R^{1/2}(x,a_n = a_m,g)$ iff $g(n) = g(m)$ iii) $R^{1/2}(x,(\varphi \,\&\, \psi),g)$ iff $R^{1/2}((x)_1,\varphi,g)$ and $R^{1/2}((x)_2,\varphi,g)$ iv) $R^{1/2}(x,(\varphi \vee \psi),g)$ iff either $(x)_1 = 1$ and $R^{1/2}((x)_2,\varphi,g)$, or $(x)_1 = 2$ and $R^{1/2}((x)_2,\psi,g)$ v) $R^{1/2}(x,(\varphi \rightarrow \psi),g)$ iff for all y with $R^{1/2}(y,\varphi,g)$, we have $R^{1/2}(\{x\}(y),\psi,g)$ vi) $R^{1/2}(x,(\forall v)(\varphi_v^{a_n}),g)$ iff for all v, $R^{1/2}(\{x\}(v),\varphi,g_v^n)$ vii) $R^{1/2}(x,(\exists v)(\varphi_v^{a_n}),g)$ iff $R^{1/2}((x)_2,\varphi,g^n_{(x)_1})$ viii) not $R^{1/2}(x,\perp,g)$. We write $R^{1/2}(x,\varphi)$ iff for all g, $R^{1/2}(\{x\}(g),\varphi,g)$, (and $(\forall g)(\{x\}(g) = \{x\}(g))$.

THEOREM 6.3. If $\vdash \varphi$, then for some x, $R^{1/2}(x,\varphi)$.

<u>Proof</u>: $R^{1/2}(x,\varphi)$ is similar to $R(e,\varphi)$, for formulae φ with no parameters or variables of sort 1. Consequently, we refer the reader to the proof of Theorem 6.1.

LEMMA 6.4.1. For each Δ_0-formula φ there is an x such that $(\forall g)(R^{1/2}(\{x\}(g),\varphi,g)$ iff $Tr(\varphi,g))$, and $\{x\}(g) = \{x\}(g)$ for all g.

<u>Proof</u>: We prove more: for $\Delta_0\varphi$, we prove there is an x such that $(\forall g)$ $(R^{1/2}(\{x\}(g),\varphi,g)$ iff $Tr(\varphi,g)$ iff $(\exists t)(R^{1/2}(t,\varphi,g)))$, by induction on φ. i) φ is $a_n \mathcal{E}\, a_m$. Take x total. ii) φ is $a_n = a_m$. Take x total. iii) φ is $(\psi \,\&\, \rho)$, $(\forall g)(R^{1/2}(\{y\}(g),\psi,g)$ iff $Tr(\psi,g)$ iff $(\exists t)(R^{1/2}(t,\psi,g)))$, and $(\forall g)$ $(R^{1/2}(\{z\}(g),\rho,g)$ iff $Tr(\rho,g)$ iff $(\exists t)(R^{1/2}(t,\rho,g)))$. Let $(\forall g)(\{x\}(g) = (\{y\}(g),\{z\}(g)))$. iv) φ is $(\psi \vee \rho)$, $(\forall g)(R^{1/2}(\{y\}(g),\psi,g)$ iff $Tr(\psi,g)$ iff $(\exists t)(R^{1/2}(t,\psi,g)))$ and $(\forall g)(R^{1/2}(\{z\}(g),\rho,g)$ iff $Tr(\rho,g)$ iff $(\exists t)(R^{1/2}(t,\rho,g)))$. Let $(\forall g)(\{x\}(g) = (1,\{y\}(g))$ if $Tr(\psi,g)$; $(2,\{z\}(g))$o.w. v) φ is $(\psi \rightarrow \rho)$,

$(\forall g)(R^{1/2}(\{y\}(g),\psi,g)$ iff $Tr(\psi,g)$ iff $(\exists t)(R^{1/2}(t,\psi,g))), (\forall g)(R^{1/2}(\{z\}(g),\rho,g)$ iff $Tr(\rho,g)$ iff $(\exists t)(R^{1/2}(t,\rho,g)))$. Let $(\forall g)(\forall w)(\{\{x\}(g)\}(w) = \{z\}(g))$. vi) φ is $(\exists v)(v \in a_n \& \psi_v^{a_m}), (\forall g)(R^{1/2}(\{y\}(g),\psi,g)$ iff $Tr(\psi,g)$ iff $(\exists t)(R^{1/2}(t,\psi,g)))$. Let $(\forall g)((\{x\}(g))_1 = \mu v \in g(n)(Tr(\psi,g_v^m))$; 0 otherwise, & $(\{x\}(g))_1 \simeq (0,\{y\}(g^m_{(\{x\}(g))_1})))$. vii) φ is $(\forall v)(v \in a_n \to \psi_v^{a_m})$, $(\forall g)(R^{1/2}(\{y\}(g),\psi,g)$ iff $Tr(\psi,g)$ iff $(\exists t)(R^{1/2}(t,\psi,g)))$. Let $(\forall g)(\forall v)(\forall w)(\{\{\{x\}(g)\}(v)\}(w) = \{y\}(g_v^m))$. viii) if $\varphi = \bot$, let x be total. The reader can see that, in order to take care of case v), it is necessary to prove this stronger statement.

LEMMA 6.4.2. Let φ be $(\exists v)(\forall w)(\psi_{v\,w}^{a_2 a_3})$, where ψ is Δ_0. Suppose that $(\forall g)(Tr((\forall w)(\psi_w^{a_3}), g^2_{\{x\}(g)}))$. Then for some y, $R^{1/2}(y,\varphi)$.

<u>Proof</u>: Let $(\forall g)(R^{1/2}(z,\psi,g)$ iff $Tr(\{z\}(g),\psi))$, by Lemma 6.4.1. Let $(\forall g)(\forall w)((\{y\}(g))_1 = \{x\}(g)$ and $\{(\{y\}(g))_2\}(w) = \{z\}(g^{2\;\;3}_{\{x\}(g),w}))$.

LEMMA 6.4.3. If φ is either in $E(\{\in,=\},\{1\},\{1\})$, or is extensionality, pairing, union, power set, infinity, or is $(a_0 \in a_1 \vee (a_0 \in a_1 \to \bot))$, $(a_0 = a_1 \vee (a_0 = a_1 \to \bot))$, $(\forall x)(\forall y)(y \in x \to (\exists z)(z \in x \& (\forall y)(y \in x \to (z \in x \to \bot))))$, then for some x, $R^{1/2}(x,\varphi)$.

<u>Proof</u>: Directly from Lemma 6.4.2. For power set, we use the fact that power set holds in $L(k)$.

LEMMA 6.4.4. For each φ there is some x such that $R^{1/2}(x,((\forall v)(\varphi_v^{a_1} \vee (\varphi_v^{a_1} \to \bot)) \to ((\forall v)(v \in a_0 \to \varphi_v^{a_1}) \vee (\exists v)(v \in a_0 \& (\varphi_v^{a_1} \to \bot)))))$.

<u>Proof</u>: Let $(\forall g)(\forall y)((\{\{x\}(g)\}(y))_1 = 1$ and $(\forall v)(\forall w)(\{\{(\{\{x\}(g)(y))_2\}(v)\}(w) \simeq ((\{y\}(v))_2)$ if $(\forall v)(v \in g(0) \to (\{y\}(v))_1 = 1)$; $(\{\{x\}(g)\}(y))_1 = 2$ and $(\{\{x\}(g)\}(y))_2 \simeq ((\emptyset,(\{y\}(v))_2),v)$, where v is μ with $(v \in g(0)$ and $(\{y\}(v))_1 = 2)$, if $(\exists v)(v \in g(0) \& (\{y\}(v))_1 = 2))$.

LEMMA 6.4.5. For each φ there is an x such that $R^{1/2}(x,(\forall v)(\varphi_v^{a_1} \vee (\varphi_v^{a_1} \to \bot)) \to (\exists w)(\forall v)(v \in w \longleftrightarrow (\varphi_v^{a_1} \& v \in a_0)))$.

<u>Proof</u>: Similar to argument for Lemma 6.4.4. We merely remark that we use

$w = \{v: v \in g(0) \& (\{y\}(v))_1 = 1\}$. Use the fact that $R^{1/2}(z,\varphi_v^{a_1} \to \bot, g)$ iff $(\forall z)$ (not $R^{1/2}(z,\varphi,g)$).

LEMMA 6.4.6. For each φ,ψ there is an x such that $R^{1/2}(x,WF(\varphi) \to TI(\varphi,\psi))$.

Proof: Once again, we merely sketch the details. From a realizer of $WF(\varphi)$ we can produce the obvious Δ_2 partial ordering which is well-founded with respect to sets in $L(\kappa)$. But since $cf(k) > \omega$, this partial ordering must be well-founded. Hence we can apply the proof of Lemma 6.2.2 to get a realizer of $TI(\varphi,\psi)$.

LEMMA 6.4.7. For each φ there is an x such that $R^{1/2}(x,(\forall v)(v \mathcal{E}\ a_o \to (\exists w)(\varphi_{v\ w}^{a_1 a_2})) \to (\exists z)(\forall v)(v\ \mathcal{E}\ a_o \to (\exists w)(\varphi_{v\ w}^{a_1 a_2} \& w \mathcal{E}\ z)))$.

Proof: This is a bounding argument. Use the Σ_2-replacement axiom in $L(\kappa)$ to produce the bound, z.

LEMMA 6.4.8. There is an x with $R^{1/2}(x,((\ v)(v\,\mathcal{E}\ a_o \to (\exists w)(w\,\mathcal{E}\ v)) \& (\forall v_1)(\forall v_2)((v_1\,\mathcal{E}\ a_o \& v_2\,\mathcal{E}\ a_o) \to (\forall v_3)((v_3\,\mathcal{E}\ v_1 \& v_3\,\mathcal{E}\ v_2) \to \bot))) \to (\exists z)(\forall v)(v\,\mathcal{E}\ a_o \to ((\exists w)(w\,\mathcal{E}\ v \& w\,\mathcal{E}\ z) \& (\forall w_1)(\forall w_2)((w_1\,\mathcal{E}\ z \& w_2\,\mathcal{E}\ z \& w_1\,\mathcal{E}\ v \& w_2\,\mathcal{E}\ v) \to w_1 = w_2))))$.

Proof: This formula is of the form $\varphi \to (\exists z)(\psi_z^{a_1})$, where φ,ψ are Δ_o. Assume $(\forall g)(R^{1/2}(\{y\}(g),\varphi,g)$ iff $Tr(\varphi,g)),(\forall g)(R^{1/2}(\{z\}(g),\psi,g))$ iff $Tr(\psi,g))$. Let $(\forall g)(\forall w)((\{\{x\}(g)\}(w))_1 \simeq \mu z(Tr(\psi,g_z^1))$ and $(\{\{x\}(g)\}(w))_2 \simeq \{z\}(g^1_{(\{\{x\}(g)\}(w))_1}))$. Then by Lemma 6.4.1, together with the fact that $\varphi \to (\exists z)(\psi_z^{a_1})$ is universally true, we have $R^{1/2}(x,\varphi \to (\exists z)(\psi_z^{a_1}))$.

THEOREM 6.4. If $ZFC^{1/2} \vdash \varphi$ then for some x, $R^{1/2}(x,\varphi)$.

Proof: From Theorem 6.3 and Lemmas 6.4.3-6.4.8.

COROLLARY 6.4.1. A consistency proof of $ZFC^{1/2}$ can be given within classical ZFC.

Proof: This follows from Theorem 6.4, since within classical ZFC we can prove the existence of a limit of cardinals, k, of cofinality $> \omega$, such that $L(\kappa)$ satisfies Σ_2 replacement.

APPENDIX I. PROBLEMS

1. Propositional Calculus. We say that α is an assignment just in case α maps ω into propositional formulae. We let $\varphi(\alpha)$ be the result of replacing each atom p_n in φ by $\alpha(n)$. A valid rule for T, where T is a propositional theory, is a pair (φ,ψ) such that for all α, if $T \vdash \varphi(\alpha)$, then $T \vdash \psi(\alpha)$. Is there a consistent, disjunctive theory T such that, for all φ,ψ, we have (φ,ψ) is a valid rule for T iff $T \vdash (\varphi \rightarrow \psi)$?

Is the set of all valid rules for $\emptyset$ decidable?

Is $\{\varphi(\alpha)\colon \alpha$ is an assignment$\}$ a decidable theory for every φ?

Can Theorem 1.3 be extended to cover either predicate logic or infinitary logic?

2. Predicate Calculus. Find the proper analogue to Theorem 1.4 in the context of the predicate calculus.

3. Theories Based on Arithmetic. Let β be any $(PR,\{0\})$-metasubstitution classically independent over $HA \cup C$, and taking only Π^0_2 sentences as values (see Definitions 3.14 and 3.15). Then is β independent over HA? Alternatively, it is not known if there is a metasubstitution independent over HA which takes on only Σ^0_1 sentences as values. The first implies the second, since there are metasubstitutions β classically independent over $HA \cup C$ which takes on only Σ^0_1 sentences as values.

4. Second-Order Logic and Type Theory. In Theorems 4.1 and 5.1, the comprehension axioms (written with Comp) play a special role in the formulation. In Section 2, the formulations only involve purely syntactic conditions. Can the Theorems 4.1 and 5.1 be generalized so as to involve only purely syntactic conditions?

5. Set Theory. Can a consistency proof of ZFTI (formulated without $TI(x,\varphi)$, but just with HA), be given in classical ZFC? Can Corollary 6.2.1 be improved so as to include set theories with various forms of **foundation?** For what classes of formulae can classical ZFC prove the reflection principle for $ZFC^{1/2}$? Use modifications of the realizabilities used in Section 6 to obtain further consistency and independence results from ZFTI (and $ZFC^{1/2}$).

APPENDIX 2. SOME RELATED RESEARCH

The $R(S,T,\varphi)$ used in Section 1 and the $R(M,A,B,T,\varphi)$ used in Section 2 are respective generalizations of the $|$ of Kleene [7] applied to propositional and predicate logic. There he proves what amounts to a form of our Soundness Theorems 1.1, 2.1, and a form of our Theorems 1.2, 2.2. The second part of Theorem 2.2 is proved in Kleene [7], and a proof of the decidability of Heyting propositional calculus can be found in Kleene [6].

A special case of the second part of Theorem 1.4 is that the Kreisel-Putnam scheme for propositional calculus is a conservative extension of Heyting propositional calculus for formulae without v. This consequence has since been proved by different methods using Gabbay [1], [2] by Dov Gabbay (oral communication).

In connection with the decidability questions in Appendix 1, we mention the interesting decidability results of Gabbay [1] and McKay [10].

The results 2.13 of Kleene [7], Corollaries 6, 7 of Pravitz [12], p. 55-56, are special cases of Theorems 2.3, 2.4, and 2.5 respectively. The idea of Section 2 was to find general formulations in which these earlier results, as well as for various systems of number theory, are all special cases. (There are errors in the formulation of Corollary 7 which have since been corrected.)

The existence of intuitionistic propositionally independent sequences of sentences over HA, as well as a formulation using intuitionistic predicate calculus independence, was established in de Jongh [4]. Our theorems on independence throughout the paper relate classical propositional independence with intuitionistic propositional independence. Improvements of de Jongh [4] along different lines will be found in Smorynski [13] when completed.

A proof theory has been developed for second-order logic which will yield the existence and disjunction properties for second-order logic, and for certain systems based on second-order logic (see Troelstra [14]). There is also a proof theory for type theory (see Girand [3] and a forthcoming paper announced in 12 of Martin-Löf [9]). We expect that future publications by proof theorists will clarify for what classes of systems these proof-theoretical methods yield existence and disjunction properties.

The methods of this paper extend nicely to obtain existence and disjunction properties for certain intuitionistic set theories - including set theories which have the power set and replacement axioms; e.g., forms of ZF. See Myhill [11].

The realizability interpretation used in the first part of Section 6 was inspired by Troelstra's realizability interpretation of second-order arithmetic, which he used to obtain similar results for that subtheory of ZFTI (see Kreisel, Troelstra [8]).

The realizability used in the second part of Section 6 is essentially the same as that used in L.H. Tharp, A Quasi-Intutionistic Set Theory, JSL, vol. 36 No. 3, September, 1971, a paper which has just come to our attention after completing this manuscript. Tharp gives a consistency proof of his IZF in classical ZF^1. The method used is essentially the same as ours, but the relationship between his IZF and our $ZFC^{1/2}$ is unclear. (IZF can be interpreted in $ZFC^{1/2}$). To give a consistency proof of IZF in the style of Section 6, we would not need the cofinality condition on k.

Tharp claims that IZF can be interpreted in a classical set theory ZF^1. He identifies ZF^1 with "classical ZF with the replacement axiom limited to Σ_1 formulas." However the correctness of his claim depends sensitively on just what particular axioms one takes for "classical ZF."

The issue centers around what form of separation, axiom of choice, and foundation are taken to be in ZF^1. We let C be the base theory with full law of excluded middle, consisting of 1), 2), 3), 4), 5), 6), first half of 9) in Definition 6.11 above, and replacement (formulated as in Cohen, Set Theory and the Continuum Hypothesis, 1966, p. 52), but where A_n is required to be (the expansion of) a Σ_1-formula in the sense of Tharp.

Thus C has no specific form of separation, no form of the axiom of choice, and a weak form of foundation. But C will prove separation of Tharp's bounded formulas.

Let Sep be 8) in Definition 6.11 above.

We distinguish two forms of the axiom of choice: AxC_o = 11) of Definition 6.11 above; AxC_1 = 6 of Tharp.

We distinguish three additional forms of foundation: TI(ε); $WF(\phi) \rightarrow TI(\phi,\psi)$; $F = (WF(\phi) \ \& \ (\exists w)(\forall v_o)(\forall v_1)(\phi^{a_o a_1}_{v_o v_1} \rightarrow (v_o \ \varepsilon \ w \ \& \ v_1 \ \varepsilon \ w))) \rightarrow TI(\phi,\psi)$.

We now have the following organization, where each group consists of equiconsistent theories, and each entry proves the consistency of individual entries in earlier groups.

I. C, C + AxC_o + "every ordinal is less than $\omega+\omega$," ordinary theory of types.

II. C+TI(ε), C+TI(ε) + AxC_o + "every ordinal is recursive"

III. C + AxC_1

IV. C + AxC_1 + TI(ε), C + F, C + AxC_1 + F

V. C + Sep, C + Sep + AxC_1 + F.

We have not established the status of, e.g., C + $CF(\phi) \rightarrow TI(\phi,\psi)$. Tharp's interpretation of his IZF can be formalized in IV. We conjecture that C + F and IZF are equiconsistent. We know that IZF proves the consistency of each entry in I, II.

The interpretation of our $ZFC^{1/2}$ can be formalized in C + $WF(\phi) \rightarrow TI(\phi,\psi)$. We conjecture that C + $WF(\phi) \rightarrow TI(\phi,\psi)$ and $ZFC^{1/2}$ are equiconsistent.

IZF-AxC_1 + AxC_o, $ZFC^{1/2}$ - (AxC_o + $WF(\phi) \rightarrow TI(\phi,\psi)$), C, ordinary type theory, are equiconsistent.

I wish to thank R. Wolf for calling some major points of Tharp's paper to my attention.

[1] D. Gabbay, The Dedidability of the Kreisel-Putnam System, JSL, 35 (1970), 431-437.

[2] D. Gabbay, Applications of Trees to Intermediate Logics I, JSL, 36 (1971).

[3] J-Y. Girard, Une Extension De L'Interpretation De Gödel A L'Analyse, Et Son Application A L'Elimination Des Coupures Dans L'Analyse Et La Theorie Des Types, in: J. E. Fenstad (editor), Proceedings of the Second Scandinavian Logic Symposium, North-Holland, 1971, 63-92.

[4] D. deJongh, The Maximality of the Intuitionistic Predicate Calculus with Respect to Heyting's Arithmetic, presented at the Second Scandinavian Logic Symposium at Oslo, June 1970.

[5] S. C. Kleene, On the Interpretation of Intuitionistic Number Theory, JSL, 10 (1945), 109-124.

[6] S. C. Kleene, Introduction to Metamathematics, Von Nostrand, 1950.

[7] S. C. Kleene, Disjunction and Existence under Implication in Elementary Intuitionistic Formalisms, JSL, 27 (1962), 11-18.

[8] G. Kreisel and A. S. Troelstra, Formal Systems for Some Branches of Intuitionistic Analysis, Annals of Mathematical Logic, 1 (1970), 229-387.

[9] P. Martin-Löf, Hauptsatz for the Theory of Species, in: J. E. Fenstad (editor), Proceedings of the Second Scandinavian Logic Symposium, North-Holland, 1971, 217-233.

[10] C. McKay, Decidability of Certain Intermediate Logics, JSL, 33 (1968), 258-265.

[11] J. Myhill, Some Properties of Intuitionistic Zermelo-Frankel Set Theory, this volume.

[12] D. Prawitz, Natural Deduction, A Proof-Theoretical Study, Stockholm, Göteborg, Uppsala (Almqvist and Wiksell), 1965.

[13] C. Smorynski, Doctoral Dissertation, University of Illinois at Chicago Circle.

[14] A. S. Troelstra, Notes on Intuitionistic Second Order Arithmetic, this volume.

IV. NOTES ON INTUITIONISTIC SECOND ORDER ARITHMETIC

A. S. Troelstra.[1]

1. Introduction

In the present paper, the term "second order arithmetic" is reserved for theories with variables for natural numbers and species (sets or n-ary relations, $n \geq 0$) of natural numbers, with full impredicative comprehension. Theories with function variables instead of species variables require a separate proof-theoretical treatment (see e.g. Scarpellini 1970, 1971).

In this paper we discuss some results and proof-theoretic methods, and their possible significance for intuitionistic second order arithmetic. We presuppose acquaintance with Troelstra 1971, Prawitz 1965, 1971. The discussion does not attempt to be exhaustive; further results are reserved for future publications.

1.1. Intuitionistic second order arithmetic, denoted by HAS (from: Heyting's arithmetic with species variables) is supposed to be based on two-sorted intuitionistic logic, with variables for numbers (denoted by x, y, z, u, v, w) and for species (i.e. variables for n-ary relations, $n \geq 0$, denoted by X^n, Y^n, Z^n; superscripts are often omitted). For natural-deduction techniques it is convenient to distinguish bound variables (X^n, Y^n, Z^n, x, y, z etc.) and parameters (free variables: P^n, Q^n, a, b, c, as in Prawitz 1965).
There is one individual constant 0 (zero), one function constant s (successor), and a binary predicate constant = (equality).
$\neg$ is everywhere treated as a defined constant, i.e. $\neg A \equiv_{def} A \rightarrow \curlywedge$.
The absurdity constant may be treated as a primitive, obeying the intuition-

1) The author is indebted to G. Kreisel and D. Prawitz for stimulating discussions and helpful criticisms.

istic absurdity rule λ_I, or be identified with $0 = 1$, or considered to be an abbreviation for $\forall X^0 . X^0$; it depends on the context which convention is the most convenient.

The axioms consist of

1°) the usual axioms for equality[2] and successor: $\forall x(x = x)$, $\forall xy(x = y \rightarrow y = x)$, $\forall xyz(x = y \,\&\, y = z \rightarrow x = z)$, $\forall xy(x = y \rightarrow sx = sy)$, $\forall xy(sx = sy \rightarrow x = y)$, $\forall x(\neg\, 0 = sx)$; their conjunction is abbreviated as ES.

2°) the induction axiom $\forall X^1[X0 \,\&\, \forall y(Xy \rightarrow Xsy) \rightarrow \forall x Xx]$, and

3°) the schema for full comprehension (A not containing X free):

$$\text{CA} \quad \exists X^n \forall x_1 \ldots x_n[A(x_1 \ldots x_n) \longleftrightarrow Xx_1 \ldots x_n].$$

1.2. Instead of using CA, we may use the λ-notation and introduce second order terms by λ-abstraction, similar to the use of λ-terms in the formulation of second order logic (second version) in chapter V of Prawitz 1965.

1.3. There are various ways of interpreting HAS in extensions of intuitionistic second order logic (= minimal second order logic):

a) Let the basic system S (in the sense of Prawitz 1971, § 1.5) consist of the language 0, s, =, and no rules, and let A^N denote the relativization of an arbitrary formula A of intuitionistic second order logic $I_2(S)$ w.r.t. $N \equiv \lambda x . \forall X[X0 \,\&\, \forall y(Xy \rightarrow Xsy) \rightarrow Xx]$. (The relativization replaces $\forall x(\ldots)$ by $\forall x(Nx \rightarrow \ldots)$, $\exists x(\ldots)$ by $\exists x(Nx \,\&\, \ldots)$.) Then, if the parameters of A are among $a_1, \ldots, a_n$, we have

$$\text{HAS} \vdash A(a_1 \ldots a_n) \iff I_2(S) + ES + Na_1 + \ldots + Na_n \vdash A^N(a_1 \ldots a_n).$$

(Cf. Prawitz 1965, Chapter V.)

b) Let S' be the basic system with language 0, s, =, λ and Post-rules

$$t = t, \quad \frac{t_1 = t_2}{t_2 = t_1}, \quad \frac{t_1 = t_2 \quad t_2 = t_3}{t_1 = t_3}, \quad \frac{t_1 = t_2}{st_1 = st_2}, \quad \frac{st_1 = st_2}{t_1 = t_2}, \quad \frac{0 = st}{\lambda}.$$

Then as before

$$\text{HAS} \vdash A(a_1 \ldots a_n) \iff I_2(S') + Na_1 + \ldots + Na_n \vdash A^N(a_1 \ldots a_n).$$

2) In stating certain reduction rules (example II in 2.5) we found the use of "symmetry" and "transitivity" slightly more convenient than the use of $\forall xyz(x = y \,\&\, z = y \rightarrow x = z)$.

c) Addition of an induction rule, to $\underset{\sim}{I}_2(S')$

$$\text{IND} \quad \frac{A0 \qquad \begin{matrix}[Aa] \\ A(sa)\end{matrix}}{At} .$$

The resulting system is then simply a natural-deduction version of $\underset{\sim}{HAS}$.

1.4. Extensionality. In the formalisms $\underset{\sim}{HAS}$, $\underset{\sim}{I}_2(S)$ as described above, we did not assume "extensionality"[3]; to include this we may add

$$\text{EXT}' \quad \forall X \forall xy[Xx \,\&\, x = y \rightarrow Xy]$$

or the equivalent rule

$$\text{EXT} \quad \frac{At \quad t = s}{As} .$$

Let us assume we are working in the formulation with parameters and (bound) variables. $\underset{\sim}{HAS}$ can be mapped into $\underset{\sim}{HAS}$ + EXT in two different ways:

(i) Replace each pseudo-formula $X^n t_1 \dots t_n$ by

$$\exists x_1 \dots x_n (t_1 = x_1 \,\&\, \dots \,\&\, t_n = x_n \,\&\, X^n x_1 \dots x_n) \qquad (1).$$

This induces a mapping φ such that for closed A

$$\underset{\sim}{HAS} + \text{EXT} \vdash A \iff \underset{\sim}{HAS} \vdash \varphi(A) \qquad (2).$$

For A containing parameters $P_1, \dots, P_n$ we either must also replace $P_i^n t_1 \dots t_n$ by

$$\exists x_1 \dots x_n (t_1 = x_1 \,\&\, \dots \,\&\, t_n = x_n \,\&\, P_i^n x_1 \dots x_n)$$

or (2) must be extended to

$$\underset{\sim}{HAS} + \text{EXT} \vdash A \iff \underset{\sim}{HAS} + \text{Ext}(P_1) + \dots + \text{Ext}(P_n) \vdash \varphi(A) \qquad (3)$$

where

$$\text{Ext}(P_i^m) \equiv_{\text{def}} \forall x_1 \dots x_m y_1 \dots y_m (P_i^m x_1 \dots x_m \,\& \\ \& \, x_1 = y_1 \,\&\, \dots \,\&\, x_m = y_m \rightarrow P_i^m y_1 \dots y_m).$$

(ii) Another mapping, ψ, which achieves the same, is obtained by relativizing second order quantifiers to extensional species, i.e. $\psi[\forall X^n A] \equiv \forall X^n(\text{Ext}(X^n) \rightarrow \psi[A])$, $\psi[\exists X^n A] \equiv \exists X^n(\text{Ext}(X^n) \,\&\, \psi[A])$. Then (3) holds with φ replaced by ψ.

3) "extensionality" is not a particularly appropriate term in the case of natural numbers, but we have kept the terminology because of the analogy with other contexts where the same rule indeed expresses extensionality.

Remark. Note that 1°) the same procedures apply to $\underset{\sim}{I}_2(S)$, $\underset{\sim}{I}_2(S')$ mentioned in 1.2, 2°) the prescriptions are easily adapted to the variants which do not distinguish between variables and parameters, and 3°) N might have been made "extensional" by defining it as

$$\lambda x \, . \, \forall X^1[\text{Ext}(X) \, \& \, (X0 \, \& \, \forall y(Xy \rightarrow Xsy) \rightarrow Xx)] \, .$$

1.5. Relation to classical second order logic HAS^c.

HAS^c is obtained from HAS by adding the principle of the excluded third, or the rule of indirect proof $\wedge_c$ as in Prawitz 1965.

As has been remarked in Kreisel 1968, § 5, Gödel's well-known translation of classical arithmetic and predicate logic extends to second order logic and arithmetic. To be specific, let $^-$ denote the translation[4], then $^-$ is given inductively by $(A)^- \equiv \neg\neg A$ for prime (quasi-)formulae A, $(A \, \& \, B)^- \equiv A^- \, \& \, B^-$, $(A \rightarrow B)^- \equiv A^- \rightarrow B^-$, $(\forall x A)^- \equiv \forall x A^-$, $(\forall X A)^- \equiv \forall X A^-$, $(A \vee B)^- \equiv \neg(\neg A^- \, \& \, \neg B^-)$, $(\exists x A)^- \equiv \neg \forall x \neg A^-$, $(\exists X A)^- \equiv \neg \forall X \neg A^-$. Then we have

$$\text{HAS}^c \vdash F \iff \text{HAS} \vdash F^-$$

and similarly for $\underset{\sim}{I}_2(S)$, $\underset{\sim}{I}_2(S')$.

1.6. Results and methods.

As a first approximation, the investigations of HAS can be grouped together according to results and methods as follows.

Results : (a) Conservative extension and (hence) consistency results,
        (b) closure conditions on the set of theorems (this includes the so-called "derived rules") of HAS itself and certain extensions,
        (c) structural results concerning formal proofs, closure conditions on the set of formal proofs of given formal systems which represent HAS or an extension of HAS, which cannot be trivially rephrased as results of type (b).

Methods : (d) Study of proof structure in natural deduction systems or calculus of sequents,

---

4) The variant chosen is Gentzen's translation (Gentzen 1933), which is logically equivalent to Gödel's definition.

(e) realizability and functional interpretations,

(f) extensions of Kleene's $\Gamma | C$ - relation (Kleene 1962).

The methods under (d) generally aim at normal form or normalization theorems, thereby obtaining structural results (c), and as corollaries also results of type (b). Realizability interpretations primarily yield relative consistency and conservative extension results; the Dialectica interpretation yields also results of type (a) (cf. Troelstra 1971 for the first order case), but occasionally also results of type (b) (closure under Markov's rule, see Girard 1971, Girard A).

The methods under (f) yield results of type (b); they are explored in Friedman A. The present paper explores normalization theorems and realizability methods.

A type of result (coming under the heading (c)), which up till now has scarcely been investigated, concerns the combinatorial equivalence between different formulations (formal systems) with the same set of theorems. To give a (rather trivial) instance of such a result, compare the formulations $\underset{\sim}{I}_2(S)$ and $\underset{\sim}{I}_2(S')$ mentioned in 1.2. There exist mappings with φ, φ' which map proofs of $\underset{\sim}{I}_2(S)$ into proofs of $\underset{\sim}{I}_2(S')$ and conversely such that $\varphi\varphi'\Pi$ and Π, and Π' and $\varphi'\varphi\Pi'$ respectively have the same normal form for each Π, Π' (disregarding Prawitz's immediate simplifications, which might disturb the uniqueness of normal forms).

For example, an application of $\forall xyz(x = y \;\&\; y = z \rightarrow x = z)$ in a proof of $\underset{\sim}{I}_2(S)$ is to be replaced by

$$\dfrac{\dfrac{\dfrac{\dfrac{\dfrac{a = b \;\&\; b = c}{a = b} \qquad \dfrac{a = b \;\&\; b = c}{b = c}}{a = c}}{a = b \;\&\; b = c \rightarrow a = c}}{\cdots}}{\forall xyz(x = y \;\&\; z = y \rightarrow x = z)} \quad \text{three times } \forall_1 I$$

and conversely,

$$\dfrac{t_1 = t_2 \qquad t_3 = t_2}{t_1 = t_3}$$

is under the mapping φ' replaced by

$$\dfrac{\dfrac{t_1 = t_2 \quad t_2 = t_3}{t_1 = t_2 \,\&\, t_2 = t_3} \qquad \text{3 times } \forall_1 E \; \dfrac{\dfrac{\forall xyz(x = y \,\&\, y = z \to x = z)}{\cdots}}{t_1 = t_2 \,\&\, t_2 = t_3 \to t_1 = t_3}}{t_1 = t_3 \,.}$$

(The remainder of the proof may be easily supplied by the reader.)

1.7. Outline of contents. We first discuss (in the remainder of this section) reasons for studying HAS and the possible significance of the results obtained.

In section 2 normalization theorems and their corollaries for intuitionistic second order logic and some extensions are discussed at length; applications to second order arithmetic are given. This section relies heavily on Prawitz's work (Prawitz 1965, 1970, 1971).

Section 3 discusses realizability, and section 4 modified realizability for HAS + EXT. (For comparison with the first order case we refer to Troelstra 1971; acquaintance with this paper will be presupposed.) (Most of the results in § 3,4 appeared earlier in the mimeographed Troelstra 1971 A.)

An extension HRO^2 of HRO (hereditarily recursive operations, see Troelstra 1971, 2.7), which is a model of (a slight variant of) Girard's theory of functionals as described in Girard 1971, is also discussed in § 4; the relation to modified realizability and the analogy to the situation in the first order case are pointed out.

1.8. Reasons for studying HAS; significance of results obtained.

Often, when mathematicians describe a proof of an existential statement as "constructive", what they mean is that an explicit realization of the statement is provided. This suggests the following problem: find formal systems such that the rule of existential instantiation holds:

$$\vdash \exists x\, Ax \implies \vdash At \quad \text{for a suitable term } t \quad (A \text{ closed}).$$

Kreisel noted (Kreisel 1971, § 4 (c)) that the usual formal systems based on intuitionistic logic provide a solution for this problem. Obviously, the stronger the system the better: classical mathematicians with "constructivist" leanings (in the sense described above) would prefer subsystems of classical formal systems with the mathematical content and the logical principles as strong as possible[5]. The best system for this purpose offered here is HAS + EXT + IP, where IP denotes the schema

$$\text{IP} \quad (\neg A \rightarrow \exists x B) \rightarrow \exists x(\neg A \rightarrow B).$$

From the same point of view, HAS + EXT + IP + UP, with

$$\text{UP} \quad \forall X \exists x A(X,x) \rightarrow \exists x \forall X A(X,x)$$

is not satisfactory, since UP conflicts with classical logic.

From the viewpoint of constructive justification however, most constructivists (including the author) will have misgivings about full comprehension for species. (In fact, the analysis of what would constitute such a justification seems to me to present a principal problem.) The cheapest way out of the difficulty, i.c. rejecting full comprehension, conflicts with accepting the standard ("intended") interpretation of intuitionistic implication. This interpretation, when applied to iterated implications, has the same degree of impredicativity as full comprehension itself in the sense that the property of being a proof of such an implication is defined by a formula containing quantifiers over all proofs of (arbitrary) logical complexity (cf. Kreisel 1968, pp. 154 - 155).

It is true that the laws of first order intuitionistic logic and arithmetic are also satisfied by much more elementary ("harmless") interpretations, in the context of a restricted concept of proof, e.g. the proofs in a given formal system (cf. e.g. Gödel's Dialectica interpretation for HA in Gödel 1958, and Prawitz 1971, Appendix A, page 288); but this is not a sufficient reason for rejecting the impredicative interpretation[6]. In this context, the question about the difference in character between the impredicative and the more elementary interpretations naturally presents itself, and suggests to us another approach: to study the use of impredicative principles in their own right, i.e. to investigate how they function in mathematical proofs, without immediately asking whether they are constructively justified. As we shall see, most of the results obtained so

5) After reading a preliminary draft of this paper, G. Kreisel remarked that he considered $\vee$, $\exists$ as "classically unemployed" operations (i.e. conceptually different from the classically defined $\vee$, $\exists$); then it is not necessary to insist on subsystems of a classical system, it suffices that the principles involving $\vee$, $\exists$ do not affect the classical fragment of the system.

6) On the other hand, of course, we do not suggest that such elementary interpretations cannot be interesting in their own right.

far do not shed light on this question: it is difficult to find marked differences between HA and HAS. It would even constitute progress if we could find mathematically interesting assertions whose proof would require in an essential way the comprehension schema for formulae of high complexity (w.r.t. second order quantification).

As will be clear from § 1.5, the problems concerning impredicative reasoning are basically the same for classical and intuitionistic second order arithmetic.

Let us now discuss the significance of some specific formal rules and schemata which play a role in the technical results of the sequel.

The principle of existential definability, as mentioned above, and its corollary (in HAS, not in I_2), the disjunction property

$$\vdash A \vee B \Longrightarrow \vdash A \text{ or } \vdash B \quad (A, B \text{ closed})$$

have sometimes been presented as tests for the constructive (intuitionistic) character of a theory. This point of view has been attacked (quite rightly, I think) in Kreisel 1970 and Kreisel 1971, § 4c. In Kreisel 1971, the discussion concerns pure logic, but here we shall restrict our attention to applied calculi, i.e. we think of HA, HAS as systems describing to some extent the intuitionistic properties of the natural numbers.

As a criterion for the intuitionistic character (in the sense of the intended interpretation) the existential definability property is neither necessary, nor sufficient. As pointed out in Kreisel 1970, p. 125, it is perfectly well possible that we can add to an intuitionistic system, e.g. HA, a property $\exists x Ax$ which is intuitionistically true, but such that $\exists x Ax \rightarrow A\bar{n}$ is not provable in HA for any numeral $\bar{n}$. The following simple example was communicated to the author[7] by G. Kreisel: Let "Prov" denote the canonical proof-predicate for HA, and let

$$Ax \equiv_{def} \mathrm{Prov}(x, \ulcorner 0 = 1 \urcorner) \vee \forall y \neg \mathrm{Prov}(y, \ulcorner 0 = 1 \urcorner).$$

Since HA is intuitionistically consistent (on the intended interpretation), $\forall y \neg \mathrm{Prov}(y, \ulcorner 0 = 1 \urcorner)$ is intuitionistically true, hence also $\exists x Ax$. Further note that because of $\forall y \neg \mathrm{Prov}(y, \ulcorner 0 = 1 \urcorner)$, we must have $\vdash \neg \mathrm{Prov}(\bar{n}, \ulcorner 0 = 1 \urcorner)$ for any numeral $\bar{n}$. Therefore

$$\vdash A\bar{n} \longleftrightarrow \forall y \neg \mathrm{Prov}(y, \ulcorner 0 = 1 \urcorner), \quad \text{for any numeral } \bar{n}.$$

7) In a letter dated October 5th, 1971.

Also

$$\vdash \exists x\, Ax \longleftrightarrow [\exists y\, Prov(y, \ulcorner 0 = 1 \urcorner) \vee \forall y \neg Prov(y, \ulcorner 0 = 1 \urcorner)] .$$

Now assume $\vdash \exists x\, Ax \rightarrow A\bar{n}$, then it would follow that
$\vdash [\exists y\, Prov(y, \ulcorner 0 = 1 \urcorner) \vee \forall y \neg Prov(y, \ulcorner 0 = 1 \urcorner)] \rightarrow \forall y \neg Prov(y, \ulcorner 0 = 1 \urcorner)$,
hence $\vdash \exists y\, Prov(y, \ulcorner 0 = 1 \urcorner) \rightarrow \forall y \neg Prov(y, \ulcorner 0 = 1 \urcorner)$, and thus
$\vdash \forall y \neg Prov(y, \ulcorner 0 = 1 \urcorner)$: contradiction (with Gödel's theorem). Hence the property of existential definability is not necessary ; the example given above applies to any intuitionistic formal system to which Gödel's theorem applies, e.g. to HAS.

That it is also not sufficient, can be shown as follows. There exist divergent extensions HA', HA" of intuitionistic arithmetic (i.e. HA' ∪ HA" is inconsistent) such that HA', HA" both satisfy the existential definability property : take for HA' : HA + M + CT_0, and for HA" : $HA^{\omega} + IP^{\omega}$. Then by Troelstra 1971, 7.4 (i) HA' possesses the existential definability property, and by Troelstra 1971, 7.4 (ii) HA" possesses the existential definability property. HA' ∪ HA" is inconsistent by Troelstra 1971, 3.20. This counter-example may also be transposed to the second order case. The divergence shows that the meaning of the logical operators is not uniquely determined.

Nevertheless, properties like the existential definability and disjunction property are combinatorially interesting. As shown by Friedman A, they are obtained rather easily for a variety of systems by extending Kleene's $\Gamma \mid C$ - relation. In Troelstra 1971, § 7 it is shown (for the first order case) how q - and mq - realizability may also be utilized for this purpose.

From the viewpoint of the intended interpretation, we are actually interested in a stronger property : how to obtain from a proof of $A \vee B$ either a proof of A or a proof of B. For under the assumption that the given formal system studied, is intuitionistically acceptable under the intended interpretation, which implies that the formal proofs may be viewed as representing certain intuitive proofs, it is natural to ask whether the set of formal proofs of a given system possesses the same closure property which is possessed by the universe of all intuitive proofs, namely that an intuitive proof of $A \vee B$ contains an intuitive (sub)proof of A or a (sub)proof of B.

From the validity of the statement $\vdash A \vee B \Longrightarrow \vdash A$ or $\vdash B$ we may indeed infer the existence of an effective method for constructing a proof of A or a proof of B whenever a proof of $A \vee B$ is given : list all proof of the formal system, and wait till a proof of A or a proof of B turns up. However, this procedure is not canonical, the proof of A or B found will in general depend on the enumeration of proofs chosen, and we wish a proof of A or a proof of B which is actually constructed from the proof of $A \vee B$, i.e. the method should depend on the form of the proof of $A \vee B$ only.

Now there is a rather trivial way of making the construction canonical, by asking, not for the first derivation of either A or B in a given enumeration, but for the shortest derivation which proves either A or B. (Cf. Kreisel's instructive example in Kreisel 1970, § 1, page 124 and top of page 125.) This shows that the procedure being "canonical" does not guarantee that the proof of A or the proof of B found is in any sense a subproof of the given proof of $A \vee B$. Kreisel's example shows that the procedures of taking the "subproof" and the "shortest proof" are distinct.

A detailed inspection of the proof by induction on the length of derivations of the soundness theorem for $\underset{\sim}{q}$ - realizability for $\underset{\sim}{HA}$:

$$\vdash A \Longrightarrow \vdash \bar{n}\,\underset{\sim}{q}\,A \quad \text{for some numeral } \bar{n} \quad (A \text{ closed})$$

shows that we actually construct in the proof recursive functions ψ_1, ψ_2 which satisfy

$$\mathrm{Prov}(x, \ulcorner A \urcorner) \Longrightarrow \mathrm{Prov}(\psi_1 x, \ulcorner \overline{\psi_2 x}\, \underset{\sim}{q}\, A \urcorner),$$

and this construction is canonical in the sense described above. This implies indeed the existence of a canonical recursive operation such that

$$\mathrm{Prov}(x, \ulcorner \exists y\, Ay \urcorner) \Longrightarrow \mathrm{Prov}(j_2 \xi(x), \ulcorner A(\overline{j_1 \xi(x)}) \urcorner)$$

($\bar{t}$ denoting the value (a numeral) corresponding to t on the intended interpretation, for each closed term t).

But it is not clear in which sense the proof with number $j_2 \xi(x)$ can be said to be a subproof of the proof with number x, although we may feel that the relationship between the two proofs is not likely to be of the trivial kind obtained by reference to "shortest derivations" as in the example described above. But the precise character of this relationship can be established only after a much more detailed analysis.

However, modulo the hypothesis that the reduction rules for systems of natural deduction preserve the intuitive proof-idea associated to the formal proof, we have in this case, for the obvious definition of subproof, that each derivation of $A \vee B$ contains (modulo some reduction steps) a sub-derivation of A or a subderivation of B. Hence to date the study of

natural deduction techniques provides us with the most satisfactory (from the point of view described above) result.
(The comments in the preceding paragraph concern the disjunction property, but equally apply, mutatis mutandis, to the property of existential definability.)

Another example of a result which has to be interpreted with care is the consistency of UP relative HAS, and the closure of HAS under the corresponding rule UR (2.5, example III):

$$\vdash \forall X \exists x A(X,x) \implies \vdash \exists x \forall X A(X,x) .$$

Are these results to be interpreted as expressing the "diffuseness" or "vagueness" of species? I.e. can we read UP as stating that species are so "non-discrete" that the only way to prove an assertion of the form $\forall X \exists x A(X,x)$ is to find a number x such that $\forall X A(X,x)$? Species are more "diffuse" than functions in the following sense: we can determine arbitrarily many values of a function, so we can decide for any two functions whether they are different or equal up to a certain degree of accuracy, whereas on the other hand for two species it may be impossible to show that they differ w.r.t. at least one element or that they have at least one element in common.

The following simple example seems to lend support to this way of viewing UP. Assume X to run through all species X_α, where

$$X_\alpha \equiv_{def} \lambda x . \forall y \exists z > y(\alpha z = x) ,$$

α an arbitrary choice sequence. Note that $\forall n(X_{n*\alpha} = X_\alpha)$ (here $n*\alpha$ denotes the function such that $(n*\alpha)(x) = g(n,x)$ for $x < \ell th(n)$, $\alpha(x \dot{-} \ell th(n))$ for $x \geq \ell th(n)$ in the notation of Kreisel and Troelstra 1970). For assume $\forall X_\alpha \exists x A(X_\alpha,x)$, i.e. $\forall \alpha \exists x A(X_\alpha,x)$, then $\forall n(en \neq 0 \rightarrow \forall \alpha \in n A(X_\alpha, en \dot{-} 1))$ for some $e \in K$. Take $m = (\overline{\lambda x . 0})y$ such that $em \neq 0$, then $\forall \alpha \in m A(X_\alpha, em \dot{-} 1)$, so $\exists y \forall \alpha A(X_\alpha,y)$, i.e. $\exists y \forall X_\alpha A(X_\alpha,y)$.

Here UP is validated for the universe of the X_α, because we have made the X_α very non-discrete by throwing in choice sequences.

Nevertheless, looking at the matter from another angle, it is easy to pick holes in the preceding explanation. In fact, the explanation is not at all obvious. For the "non-discreteness" of species holds with respect to their extensions only, and not with respect to their definitions. From the intuitionistic point of view, the mathematical objects are primarily constructions as they are given to us as mental objects; and a basic idealization is that we know whether we are thinking of the same object or not. So if the assignment of a natural number to each species would be given by reference to their definition, it is perfectly conceivable that different species would receive different numbers.

A more plausible interpretation of the validity of UR and the consistency of UP seems to be that HAS simply does not contain (sufficient) means of talking about definitions of species, i.e. in HAS we can talk about their extensions only. On this interpretation, the validity of UR would point to a defect (from a foundational, not necessarily a mathematical point of view) of HAS, namely its limited power of expression. This suggests a line of research : the search for systems which can refer to some extent to the definitions of species (such systems would not necessarily be mathematically interesting, but one expects them to be relevant from a logical point of view).

There is yet another aspect to the study of HAS, connected with the preceding ones. The study of HAS may serve as a testing ground for certain metamathematical methods. This is especially worthwhile, if the methods have independent interest , as is the case with natural deduction techniques.

In section 2 we exploit some of Prawitz's natural deduction techniques.

The fact that the results given there can be obtained by such techniques, using the insight gained in the structure of the proofs in the formal systems, is in itself satisfactory.

§ 2. Normalization theorems for intuitionistic second order logic

2.1. Let $\underset{\sim}{M}_2(S) = \underset{\sim}{I}_2(S)$ denote the system of intuitionistic (= minimal) second order logic in a language with constants $=$, 0, s, over a basic Post system S; for our applications to second order arithmetic, S is either empty or consists of the Post rules for equality and successor (i.e. S or S' of 1.3 (a), (b)).

In $\underset{\sim}{I}_2(S)$ $\curlywedge$, $\exists_1$, $\exists_2$, $\vee$, & are definable (see Prawitz 1965, Chapter V: $\curlywedge \equiv \forall X^0 . X^0$, $\exists x A \equiv \forall X^0(\forall x(A \to X^0) \to X^0)$, $\exists Y . A \equiv \forall X^0(\forall Y(A \to X^0) \to X^0)$, $A \& B \equiv \forall X^0[(A \to ((B \to X) \to X)]$, $A \vee B \equiv \forall X^0((A \to X^0) \to ((B \to X^0) \to X^0))$).

Because of this definability result, we may restrict ourselves to the $\to\forall$ - language, if we are interested in results (such as derived rules) which depend on the set of theorems only. For this reason, and because of the simplicity of the proofs w.r.t. this restricted language, we shall deal below with the $\to\forall$ - fragment before considering the full language. The $\to\forall$ - fragment is also sufficient for some basic applications to $\underset{\sim}{HAS}$ (see 2.9). But if we wish to consider certain extensions of $\underset{\sim}{HAS}$, by adding certain rules with corresponding reduction rules, it may happen (as we shall see below) that a natural formulation of the rules plus reduction rules requires an extension of the language.

2.2. Definition (Prawitz 1970). The strictly positive parts (s.p.p.'s) of a formula A (in the $\to\forall$ or $\to\forall\curlywedge$ - language) are defined inductively as follows:

(1) A is a strictly positive part of A.

(2) If $C \to B$, $\forall x B$, $\forall X^n B$ is a strictly positive part of A, then so is B.

(3) If $\{\lambda x_1 \dots x_n . B(x_1, \dots, x_n)\} t_1 \dots t_n$ is a strictly positive part of A, then so is $B t_1 \dots t_n$.

Note that each formula in the $\to\forall\curlywedge$ - language contains a uniquely determined atomic s.p.p.

2.3. Theorem. Let $C_1, \dots, C_n$ be formulae (in the $\forall\to$ - fragment) such that each C_i is either negated (i.e. of the form $D \to \forall X^0 . X^0$) or does not contain a predicate variable in an atomic strictly positive part. Then

(i) $\vdash C_1 \& \dots \& C_n \to A \vee B \implies \vdash C_1 \& \dots \& C_n \to A$ or $\vdash C_1 \& \dots \& C_n \to B$.

(ii) $\vdash C_1 \& \dots \& C_n \to \exists x A x \implies \vdash C_1 \& \dots \& C_n \to A t$ for some term t.

(iii) $\vdash C_1 \& \ldots \& C_n \rightarrow \exists X^n A(X^n) \implies \vdash C_1 \& \ldots \& C_n \rightarrow A(T^n)$
for a suitable second order term T^n.

<u>Remark</u>. A slightly weaker version (i.e. without admitting negated C_i) of the present theorem is stated without proof in <u>Prawitz</u> 1970. Prawitz gives a proof for $n = 0$ (i.e. empty premiss). We include a proof here.

<u>Proof</u>. As an example we prove (ii): in the other cases the proofs are similar. Assume $\vdash C_1 \& \ldots \& C_n \rightarrow \exists x A(x)$; for simplicity we conceive $\exists x Ax$ as an abbreviation of a formula in the $\forall\rightarrow$-fragment. So then there is a <u>normal</u> proof Π in the $\forall\rightarrow$-fragment of $\forall X(\forall x(Ax \rightarrow X) \rightarrow X)$ from $C_1 \& \ldots \& C_n$. (Here we combine either the normal form theorem for the calculus of sequents in <u>Prawitz</u> 1970 with the equivalence proof for calculi of sequents and systems of natural deduction in <u>Prawitz</u> 1965, Appendix A, or alternatively, we use the normalization theorem in <u>Prawitz</u> 1971, Appendix B, restricted to the $\rightarrow\forall$-language.) We may assume (inversion principle) that the proof ends with $\rightarrow I$, $\forall_2 I$, and hence is of the following form

$$\dfrac{\dfrac{\dfrac{\begin{array}{c}[C_1, \ldots, C_n, \ \forall x(Ax \rightarrow P)]\\ \Sigma\end{array}}{P}}{\forall x(Ax \rightarrow P) \rightarrow P}}{\forall X(\forall x(Ax \rightarrow X) \rightarrow X)}$$

(P does not occur in $C_1, \ldots, C_n, A$). Obviously, P is a minimal formula in the main branch. The top formula of the main branch in Π is either $\forall x(Ax \rightarrow P)$ or a C_i of the form $D \rightarrow \forall X.X$. For assume C_i to be a formula which does not contain a predicate variable in an atomic s.p.p. The atomic s.p.p. of C_i is of the form $Tt_1 \ldots t_n$, T a predicate constant or parameter. If $C_i = D_o, D_1, \ldots, D_k = P$ is the elimination part of the main branch, then each D_m contains an atomic s.p.p. of the form $Tt'_1 \ldots t'_n$ (as is seen by induction on m). But then T would be identical with P, which is false.

Assume now that $C_i \equiv D \rightarrow \forall X.X$ is the top formula of the main branch. Then Π has the following form:

$$\frac{\dfrac{\begin{matrix}[C_1, \ldots, C_n, \forall x(Ax \to P)] \\ \Sigma'\end{matrix}}{D} \qquad D \to \forall X . X}{(\forall X . X)}$$
$$\Pi' .$$

If we substitute $\exists x\, Ax$ for P, we obtain a proof of $\forall X . X$ from $C_1, \ldots, C_n$, i.e. $\vdash C_1 \& \ldots \& C_n \to \Lambda$. But then also $\vdash C_1 \& \ldots \& C_n \to At$ for arbitrary t.

Suppose finally $\forall x(Ax \to P)$ to be the top formula of the main branch. Then Π has the form:

$$\frac{\dfrac{\begin{matrix}[C_1, \ldots, C_n, \forall x(Ax \to P)] \\ \Sigma'\end{matrix} \qquad \forall x(Ax \to P)}{At \qquad\qquad At \to P}}{(P)}$$
$$\Pi^* .$$

Substituting $\exists x\, Ax$ for P, we obtain $\vdash C_1 \& \ldots \& C_n \to At$.

2.4. Normalization theorems in general.

In Prawitz 1971 a strong normalization theorem is proved (i.e. each sequence of successive reduction steps terminates) for second order logic w.r.t. the full language; this proof is much more flexible. We wish to include here some general observations.

Below we shall suppose the reduction rules to have "local" character, i.e. the conditions under which the rule is applicable refer to a finite set of nodes of the proof tree, the structure (i.e. number of nodes, partial ordering and the rules applied) of which is entirely determined by the type of the reduction to be applied.

In all examples considered the set S of nodes involved is also closed w.r.t. the ordering ($\prec$) of the proof tree, i.e.

$\nu \prec \nu' \prec \nu'' \ \& \ \nu \in S \ \& \ \nu'' \in S \to \nu' \in S$.

The usual reduction rules ($\&, \forall, \exists, \vee, \to$-reduction and $\vee E$, $\exists E$-reduction) all have local character. But a rule for constructing expanded normal forms, which is stated as follows: "apply immediate expansions to a tree which is normal w.r.t. the other reduction rules" clearly refers to reduction rules of non-local character.

A first point of interest which should be noted, is that strong normalization theorems, asserting the termination of <u>every</u> reduction sequence starting from a certain proof, possess certain monotonicity properties :

(i) If $\mathfrak{R}$, $\mathfrak{R}^*$ are sets of (local) reduction rules such that $\mathfrak{R} \subset \mathfrak{R}^*$, then a strong normalization theorem relative to $\mathfrak{R}^*$ implies the strong normalization theorem relative to $\mathfrak{R}$. For consider, for any given derivation Π, the class $\mathcal{C}$ of reduction sequences starting from Π which are such that rules of $\mathfrak{R}^* - \mathfrak{R}$ are not applied as long as a rule of $\mathfrak{R}$ is applicable. Since all reduction sequences relative to $\mathfrak{R}^*$ terminate, a fortiori all sequences in $\mathcal{C}$ terminate.

If for each $\sigma \in \mathcal{C}$ we take the maximal initial segment σ' of σ which does not contain applications of reductions in $\mathfrak{R}^* - \mathfrak{R}$, we obtain the class $\mathcal{C}'$ consisting of all reduction sequences relative to $\mathfrak{R}$ starting from Π.

(ii) Suppose we extend our calculus of natural deduction $\mathfrak{S}$ to a calculus $\mathfrak{S}^*$ by adding certain rules and correspondingly extend the reduction rules $\mathfrak{R}$ to $\mathfrak{R}^*$, where the rules $\mathfrak{R}^* - \mathfrak{R}$ involve derivations containing applications of rules in $\mathfrak{S}^* - \mathfrak{S}$. The strong normalization theorem relative to $\mathfrak{R}^*, \mathfrak{S}^*$ implies the strong normalization theorem relative to $\mathfrak{R}, \mathfrak{S}$. For if Π is a derivation in $\mathfrak{S}$, then any reduction sequence relative to $\mathfrak{R}^*$ starting from Π is a reduction sequence relative to $\mathfrak{R}$, since Π does not contain applications of rules in $\mathfrak{S}^* - \mathfrak{S}$; and applications of such rules are not introduced by application of reduction procedures in $\mathfrak{R}$.

In view of these monotonicity properties, there is a certain interest in finding "large" $\mathfrak{R}^*, \mathfrak{S}^*$. If we could find unique "maximal" $\mathfrak{R}^*, \mathfrak{S}^*$, the corresponding normalization theorem would imply normalization theorems for all extensions of $\underset{\sim}{I}_2(S)$ for which such theorems are valid ; however, the existence of such maximal $\mathfrak{R}^*, \mathfrak{S}^*$ is not particularly plausible. But finding "large" $\mathfrak{R}^*, \mathfrak{S}^*$ with a simple description, covering the known cases would already constitute progress. This direction of research is pursued in <u>Prawitz</u> A, where a general condition on reduction processes is formulated, which ensures the provability of a normalization theorem.

Another general remark seems to be in order. In connection with the hypothesis that intuitively equivalent proofs correspond to proofs with the same normal form, it is to be noted that (i) a proof in normal form in $\underset{\sim}{I}_2(S)$ still may contain redundancies in an intuitive sense ; e.g. one should like to simplify

$$\dfrac{\dfrac{P}{P \to P} \qquad \dfrac{\forall X . X}{(P \to P) \to Q}}{Q} \quad \text{to} \quad \dfrac{\forall X . X}{Q}$$

but the left hand derivation is in normal form; (ii) more than one possible reduction at a given inference may spoil the uniqueness of the normal form obtained for a given derivation (cf. Jervell 1971, bottom of page 107).

2.5. Examples of additional reduction rules.

Example I. Addition of $\curlywedge$ as a primitive, together with the intuitionistic absurdity rule $\curlywedge_I : \frac{\curlywedge}{A}$, and reduction rules which break down the complexity of A in applications of $\curlywedge_I$ (cf. Prawitz 1971, Appendix B3), such as ("contr." indicating "contracts to"):

$$\dfrac{\curlywedge}{(A \,\&\, B)} \atop \Pi \quad \text{contr.} \quad \dfrac{\dfrac{\curlywedge}{A} \quad \dfrac{\curlywedge}{B}}{(A \,\&\, B)} \atop \Pi , \qquad \dfrac{\curlywedge}{(A \to B)} \atop \Pi \quad \text{contr.} \quad \dfrac{\dfrac{\curlywedge}{B}}{(A \to B)} \atop \Pi ,$$

$$\dfrac{\curlywedge}{(\forall x\, Ax)} \atop \Pi \quad \text{contr.} \quad \dfrac{\dfrac{\curlywedge}{A(a)}}{(\forall x\, Ax)} \atop \Pi \quad \text{etc., etc.}$$

The method of Prawitz 1971, Appendix B cannot be straightforwardly extended to cover these additional reduction rules; a separate argument is needed. R. Statman (unpublished) gave a treatment by a method which applies not only to this example, but also to example II (extensionality) and example VI (induction), reducing e.g. normalization for $\underset{\sim}{I}_2(S) + \curlywedge_I$ to normalization for (a variant of) $\underset{\sim}{I}_2(S)$ by means of a suitable homomorphism (w.r.t. the reducibility relation). If we are interested in normalization only, we can get by with Prawitz's method with a slight supplementary argument.

Example II. Addition of EXT

$$\frac{At \qquad t = s}{As}$$

to the system $\underset{\sim}{I}_2(S')$ of 1.3 (b). We add to the reduction rules reduction procedures which reduce the applications of EXT to atomic A , for example

$$\begin{array}{c}\dfrac{\dfrac{\Sigma}{At \to Bt} \quad t = s}{(As \to Bs)} \\ \Pi\end{array} \quad \text{contr.} \quad \begin{array}{c}\dfrac{\dfrac{At \to Bt \qquad \dfrac{As \qquad \dfrac{t = s}{s = t}}{At}}{Bt} \quad t = s}{\dfrac{Bs}{(As \to Bs)}} \\ \Pi\end{array}$$

and

$$\begin{array}{c}\dfrac{\dfrac{\Sigma}{\forall X\, A(X,t)} \quad t = s}{(\forall X\, A(X,s))} \\ \Pi\end{array} \quad \text{contr.} \quad \begin{array}{c}\dfrac{\dfrac{\dfrac{\dfrac{\Sigma}{\forall X\, A(X,t)}}{A(P,t)} \quad t = s}{A(P,s)}}{(\forall X\, A(x,s))} \\ \Pi\end{array}$$

and

$$\begin{array}{c}\dfrac{\dfrac{\Sigma}{\exists X\, A(X,t)} \quad t = s}{(\exists X\, A(X,s))} \\ \Pi\end{array} \quad \text{contr.} \quad \begin{array}{c}\dfrac{\dfrac{\Sigma}{\exists X\, A(X,t)} \qquad \dfrac{\dfrac{A(P,t) \quad t = s}{A(P,s)}}{\exists X\, A(X,s)}}{(\exists X\, A(X,s))} \\ \Pi\end{array}$$

etc., etc.

See the remarks at the end of Example I.

Example III. Addition of the axiom of choice.
This example is discussed in Appendix B3 of Prawitz 1971.

Example IV. Addition of the uniformity principle.
We add to $I_2(S)$ a rule

$$\text{UP}\quad \frac{\forall X\, \exists x\, A(X,x)}{\exists x\, \forall X\, A(X,x)}$$

and a corresponding UP - reduction :

$$\begin{array}{l} \exists_1 I \\ \forall_2 I \\ \text{UP} \end{array} \quad \cfrac{\cfrac{\cfrac{\cfrac{\Sigma}{A(P,t)}}{\exists x\, A(P,x)}}{\forall X\, \exists x\, A(X,x)}}{\begin{array}{c}(\exists x \forall X\, A(X,x))\\ \Pi\end{array}} \qquad \text{contr.} \qquad \cfrac{\cfrac{\cfrac{\Sigma}{A(P,t)}}{\forall X\, A(X,t)}}{\begin{array}{c}(\exists x \forall X\, A(X,x))\\ \Pi\end{array}} \quad \begin{array}{l} \forall_2 I \\ \exists_1 I \,. \end{array}$$

UP is to be counted as an elimination rule. We first show that UP preserves derivability from null - assumptions (in other words, UR is a derived rule in $\underset{\sim}{I}_2(S)$).

Proof. Assume we have a derivation Π in $\underset{\sim}{I}_2(S)$ of $\forall X\, \exists x\, A(X,x)$. Then there is also a derivation Π' of $\exists x\, A(P,x)$; by the rule of existential definability for second order logic (2.3) we have a derivation Π'' of $A(P,t)$ for a suitable term t , hence by $\forall_2 I$, $\exists_1 I$ a derivation of $\exists x \forall X\, A(X,x)$.

In extending the proof of the strong normalization theorem, the only point which needs some care is the auxiliary lemma. Condition (ii) in the statement of the auxiliary lemma has to be extended with the case where UP is the last rule applied. Assume that a derivation

$$\cfrac{\cfrac{\cfrac{\cfrac{\Sigma\ (P)}{A(P,t)}}{\exists x\, A(P,x)}}{\forall X\, \exists x\, A(X,x)}}{\exists x\, \forall X\, A(X,x)}$$

satisfies (i), (ii), (iii) of the auxiliary lemma. Then it follows, by condition (ii) and the definition of strong validity relative $\mathcal{N}$, that $\frac{\Sigma\ (T)}{A(T,t)}$ is strongly valid relative $\mathcal{N} + \binom{T}{N}$ for each regular N and each suitable second order term T . Hence, the reduced derivation $\Sigma(P) \,/\, A(P,t) \,/\, \forall X\, A(X,t) \,/\, \exists x\, \forall X\, A(X,x)$ is strongly valid relative $\mathcal{N}$.

Example V. Addition of the schema IP.
Now we add the rule

$$\mathrm{IP}\quad \frac{\neg A \to \exists x B}{\exists x(\neg A \to B)}$$

which is to be treated as an elimination rule.
The corresponding reduction rule has the form

$$\text{IP-reduction}\qquad \begin{array}{c} [\neg A] \\ \Sigma \\ \hline \mathrm{Bt} \\ \hline \exists x \mathrm{Bx} \\ \hline \neg A \to \exists x \mathrm{Bx} \\ \hline (\exists x(\neg A \to \mathrm{Bx})) \\ \Pi \end{array} \begin{array}{l} \\ \\ \\ \exists_1 I \\ \to I \\ \mathrm{IP} \\ \\ \end{array} \qquad \text{contr.} \qquad \begin{array}{c} [\neg A] \\ \Sigma \\ \hline \mathrm{Bt} \\ \hline \neg A \to \mathrm{Bt} \\ \hline (\exists x(\neg A \to \mathrm{Bx})) \\ \Pi \end{array} \begin{array}{l} \\ \\ \\ \to I \\ \exists_1 I\,. \\ \\ \end{array}$$

It follows from 2.3 that IP preserves derivability from null - assumptions. The normalizability theorem is proved in the same manner as for example III.

Example VI [8]. Addition of the rule of induction, together with reduction schemata :

$$\frac{\dfrac{\Sigma}{\mathrm{AO}} \qquad \dfrac{\begin{array}{c}[\mathrm{Aa}]\\ \Sigma'(a)\end{array}}{\mathrm{Asa}}}{\begin{array}{c}(\mathrm{AO}) \\ \Pi\end{array}} \qquad \text{contr.} \qquad \frac{\Sigma}{\begin{array}{c}(\mathrm{AO}) \\ \Pi\end{array}}$$

and

8) The author's attention was drawn to this example by D. Prawitz.

$$\dfrac{\dfrac{\Sigma}{AO} \qquad \dfrac{\begin{array}{c}[Aa]\\ \Sigma'(a)\end{array}}{Asa}}{\begin{array}{c}(Ast)\\ \Pi\end{array}} \qquad \text{contr.} \qquad \dfrac{\dfrac{\Sigma}{AO} \qquad \dfrac{\begin{array}{c}[Aa]\\ \Sigma'(a)\end{array}}{Asa}}{\dfrac{\begin{array}{c}[At]\\ \Sigma'(t)\end{array}}{\begin{array}{c}(Ast)\\ \Pi\end{array}}} .$$

IND is to be counted as a non-introduction rule, and the proof and statement of the first lemma in Prawitz 1971, App. B, proof of 2.2.1 (the analogue of Appendix A, 3.4.5) have to be adapted accordingly. The induction value must now be made to depend also on the complexity of induction terms if the last inference used is IND.

Examples III - V show a common pattern: in each case, a certain rule R, to be counted as an elimination rule has been added; the corresponding reduction rule to deductions where an application of R is preceded by a string of applications of I-rules; after the reduction, the application of R has been removed, and the I-rules used have been permuted. In such cases the proof of the normalization theorem seems to extend almost automatically. In fact, Prawitz A gives general conditions on reduction processes which permit one to prove a normalization theorem. All our examples satisfy his conditions.

2.6. Extension of theorems on existential definability.

In Prawitz 1971, Appendix B3 it is already shown that the addition of AC preserves existential definability. This result is not very useful for applications to HAS, since AC does not obviously imply AC-NS:

$$\forall x \in N\, \exists X^n A(x,X) \rightarrow \exists Y^{n+1} \forall x \in N\, A(x, [1/x]Y)$$

in the notation of Kreisel and Troelstra 1970.
The situation is different for UP, IP.

2.7. Theorem. Let H be a system obtained by adding to $I_2(S)$ or $I_2(S')$ (1.3) none, one or more of the principles EXT, IP, UP, and let $C_1, \ldots, C_n$ be formulae in the $\rightarrow\forall\wedge$-fragment which do not contain a predicate variable in an atomic s.p.p. Then

(i) $H \vdash C_1 \& \ldots \& C_n \rightarrow A \vee B \Longrightarrow H \vdash C_1 \& \ldots \& C_n \rightarrow A$ or
$H \vdash C_1 \& \ldots \& C_n \rightarrow B$.

(ii) $\underset{\sim}{H} \vdash C_1 \& \ldots \& C_n \to \exists x Ax \Longrightarrow \underset{\sim}{H} \vdash C_1 \& \ldots \& C_n \to At$
for a suitable term t.

(iii) $\underset{\sim}{H} \vdash C_1 \& \ldots \& C_n \to \exists X A(X) \Longrightarrow \underset{\sim}{H} \vdash C_1 \& \ldots \& C_n \to A(T)$
for a suitable second order term T.

Proof. For $\underset{\sim}{H} \equiv \underset{\sim}{I}_2(S)$ we already gave a proof in using the $\forall \to$ - language. But for the purpose of readily extending the proof to other systems, it is advantageous to give a proof using the normalization theorem with respect to the full language. In this case we can for the greater part copy the proofs of Corollaries 6, 7 in Chapter IV of Prawitz 1965.

$\underset{\sim}{I}_2(S)$ - (i). We adapt the proof of Corollary 6 in Chapter IV of Prawitz 1965. Let Π be a normal deduction (normal also w.r.t. the reduction rules in Example I) of $A \vee B$ from $\Gamma = \{C_1, \ldots, C_n\}$.
As in Prawitz 1965, we show that there is exactly one end segment σ in Π. For assume there were two end segments, then they would contain a minor premiss of an application of $\vee E$; let $C \vee D$ be the corresponding major premiss. The first formula occurrence F in a path π through $C \vee D$ cannot be discharged in Π, since $C \vee D$ belongs to the E - part of π, and there are no applications of $\to I$ below $C \vee D$. Hence $F \in \Gamma$, i.e. $F \equiv C_i$ for some i.

Let $C_i \equiv D_0, D_1, \ldots, D_k$ be the E - part of π up till $C \vee D$, i.e. $D_k \equiv C \vee D$. For each j, either D_j, D_{j+1} belong to the same segment, or at the step from D_j to D_{j+1} the main logical operator of D_j is eliminated; occurrences of logical operators in D_j deriving from occurrences of the same operators in C_i are eliminated before any operator occurrence introduced by $\forall_2 E$ is eliminated. Each D_j $(0 \leq j \leq k)$ contains the same predicate constant or parameter T in its atomic s.p.p. So there must be a D_j $(0 \leq j < k)$ (since $\vee$ in $C \vee D$ is necessarily introduced by $\forall_2 E$) which is atomic and of the form $Tt_1 \ldots t_n$. However, to this D_j no E - rules apply and thus D_k would have to belong to the I - part, which is contradictory.

We have not explained in detail the expression "operator occurrence deriving from", assuming its meaning to be intuitively clear. A more precise description is obtained as follows.
We start constructing a sequence $D'_0, D'_1, \ldots$ parallel to $D_0, D_1, \ldots$ (the new sequence may have length $< k$) such that each D'_i has the same main logical operator as D_i. The definition is given by induction: $D'_0 \equiv D_0$;
a) if D_i, D_{i+1} belong to the same segment, then $D'_{i+1} = D'_i$;
b) if D_i, D_{i+1} belong to different segments, then, if $D'_i \equiv A' \to B'$, we put $D'_{i+1} \equiv B'$; if $D_i \equiv \forall x Ax$, $D'_i \equiv \forall x A'x$, $D_{i+1} \equiv At$, we put

$D'_{i+1} \equiv A't$; if $D'_i \equiv \{\lambda x_1 \ldots x_n B'(x_1,\ldots,x_n)\}t_1 \ldots t_n$, we put $D'_{i+1} \equiv B'(t_1,\ldots,t_n)$; if $D_i \equiv \forall X\, A(X)$, $D'_i \equiv \forall X\, A'(X)$, $D_{i+1} \equiv A(T)$, then we put $D'_{i+1} \equiv A'(P_{i+1})$. (The parameters P_j are assumed not to occur in Π and to be pairwise different.) Logical operators in D'_i derive from operators in $D_o \equiv D'_o$; for each i for which D_i, D'_i are defined, they contain the same predicate parameter or constant in their atomic s.p.p. The sequence $D'_o, D'_1, D'_2, \ldots$ breaks off if an atomic formula has been reached.

We further note that σ must be a consequence of an application α of an I-rule (it cannot be the consequence of an application of the $\wedge_I$-rule or EXT, since we have added reduction rules so as to obtain only atomic applications of $\wedge_I$, EXT), otherwise it would be a minimum segment of any path π' to which it belongs, and then as above we would obtain that the first formula F' of π' would belong to Γ, yielding a contradiction. The premiss of α is therefore either A or B, and then as in Prawitz 1965 we are able to construct a proof of either A or B. (The final part of the proof according to Prawitz 1965 may be simplified slightly, if we treat $\exists x, \exists X$ as defined; then the end segment has length 1.)

$\underset{\sim}{I}_2(S)$ - (iii). We discuss (iii): (ii) is proved in a similar way. (iii) is proved according to the argument used to establish (ii) of Corollary 7 in Chapter IV[9] of Prawitz 1965. If we treat $\vee$ as defined, a slight simplification occurs in the final part of the argument.

Extensions. Let us consider for example $\underset{\sim}{I}_2(S)$ + UP. A path in a normal derivation w.r.t. $\underset{\sim}{I}_2(S)$ + UP has the following form: it consists of a sequence of segments $\sigma_1, \ldots, \sigma_n$, divided into an E-part $\sigma_1, \ldots, \sigma_{i-1}$ and an I-part $\sigma_i, \ldots, \sigma_n$; σ_i is the minimum segment. In the I-part σ_{k+1} is obtained from σ_k $(k \geq i)$ by an application of an I-rule; σ_i may have been obtained from $\sigma_{i-1} = \wedge$ by an atomic instance of $\wedge_I$. In the E-part, there are applications of E-rules, UP and $\forall_2 I$; two or more consecutive applications of I-rules do not occur. It is possible to adapt the proof of (i) for $\underset{\sim}{I}_2(S)$ as follows.

Let $D_o, \ldots, D_i$ be the E-part of a path through a top formula $D_o \in \Gamma$. Assume D_k, $k \leq i$, to be the first formula which is obtained as a conclusion of an UP-application. If, on the other hand, we construct $D'_o, D'_1, \ldots$ parallel to $D_o, D_1, \ldots$ as before, the construction stops at an atomic formula D'_j; no $\exists$-operator occurs in any of the $D'_o, D'_1, \ldots, D'_j$.

9) D. Prawitz indicated an error and its correction in the statement and proof of Corollary 7; the part of the argument which we need, is not affected by the error, however.

But then $D'_j \equiv D_j$, and no eliminations nor UP-rules are applicable. Hence such a D_k does not exist, and the E-part $D_o, \ldots, D_i$ has the same structure as E-parts in the case of $\underset{\sim}{I}_2(S)$.
The other systems can be dealt with analogously.

2.8. Remark. From the proof of 2.7 it will be clear that we have actually established that for the systems considered, we obtain, after normalizing, from a proof of $A \vee B$, $\exists x Ax$, $\exists X A(X)$ respectively, a "sub-proof" of A or a "sub-proof" of B, a "sub-proof" of At for suitable t, a "sub-proof" of $A(T)$ for suitable T respectively.

2.9. The results in the preceding subsections may be applied to obtain results about HAS and certain extensions of HAS, using the embeddings of HAS in extensions of second order logic, described in 1.3.
As a first corollary to 2.7 we obtain

2.10. Theorem. Let $\underset{\sim}{H}$ be any one of the systems obtained by adding to HAS none, one or more of the rules EXT, UP, IP. Let C denote a conjunction of closed formulae $C_1 \& \ldots \& C_n$, each C_i belonging to the $\rightarrow\forall\wedge$-fragment, and not containing a predicate variable in an atomic s.p.p.

(i) $\underset{\sim}{H} \vdash C \rightarrow A \vee B \Longrightarrow \underset{\sim}{H} \vdash C \rightarrow A$ or $\underset{\sim}{H} \vdash C \rightarrow B$ (A, B closed)

(ii) $\underset{\sim}{H} \vdash C \rightarrow \exists x D(x) \Longrightarrow \underset{\sim}{H} \vdash C \rightarrow D(t)$ for a suitable term t ($\exists x D(x)$ closed)

(iii) $\underset{\sim}{H} \vdash C \rightarrow \exists X D(X) \Longrightarrow \underset{\sim}{H} \vdash C \rightarrow D(T)$ for a suitable second order term T ($\exists X D(X)$ closed)

(iv) $\underset{\sim}{H}$ satisfies UR w.r.t. closed formulae
$\underset{\sim}{H} \vdash \forall X \exists x A(X,x) \Longrightarrow \underset{\sim}{H} \vdash \exists x \forall X A(X,x)$

(v) (Corollary to (ii), (iii).) $\underset{\sim}{H}$ satisfies IR in two forms:
$\underset{\sim}{H} \vdash C \rightarrow \exists x D(x) \Longrightarrow \underset{\sim}{H} \vdash \exists x (C \rightarrow D(x))$ ($\exists x D(x)$ closed)
$\underset{\sim}{H} \vdash C \rightarrow \exists X D(X) \Longrightarrow \underset{\sim}{H} \vdash \exists X (C \rightarrow D(X))$ ($\exists X D(X)$ closed).

Proof. (i)-(iii) are consequences of Theorem 2.7; to see this, we use the representation described in 1.3 (a,b), and note that the equality and successor axioms ES satisfy our conditions on the C_i in 2.7.
Note that for the C_i satisfying these conditions, C_i^N satisfies the same condition; and also ψC^N after replacing & by its definition in the $\forall\rightarrow$-fragment (here ψ is the mapping of 1.4 (ii)).
(iv) follows from our remarks under Example IV in 2.5.

2.11. Remark. As will be apparent from te preceding argument, the remark of 2.8 carries over to HAS and its extensions when regarded as extensions of $I_2(S)$ in the sense of 1.3 (a). A closer analysis, which we shall not carry out here, yields similar results when we add the rule IND instead of relativizing to N.

A derivation of $\forall X \exists x\, A(X,x)$ is after normalization transformed in a similar way into a derivation of $\exists x \forall X\, A(X,x)$.

2.12. The applications of Theorem 2.7 in 2.10 have been restricted to closed formulae. A method of getting rid of this restriction is indicated in Kreisel 1970, Technical note II (page 135): combination of a reflection principle for formulae of bounded[10] complexity, together with a formalization in HAS of the proof of the normal form theorem for subsystems of $I_2(S)$ of bounded complexity. Kreisel uses this method to obtain closure under Markov's rule (for prime formulae with parameters) and under Church's rule for HAS

$$\vdash \forall x \exists y\, A(x,y) \implies \vdash \exists y \forall x \exists z\, [T(y,x,z)\ \&\ A(x, Uz)]\,.$$

The long round-about way via Spector's result, sketched in Kreisel 1970, can now be avoided: cf. Kreisel 1971, § 2a (ii) (an example of the superiority of normalization theorems over normal form theorems). Another proof of closure under HAS under Markov's rule has been given by Girard (Girard A) using formalization of a normalization theorem for the terms in his Dialectica-style interpretation (Girard 1971) together with a reflection principle for finitely axiomatized subsystems.

The combination of a reflection principle for finitely axiomatized subsystems of HAS together with a formalization of the normalization theorem for subsystems of bounded complexity would enable us to multiply our applications of Prawitz's proof, getting rid of restrictions to closed formulae in many cases[11].

10) Compare the first order case in Kreisel 1959.

11) We hope to give a more detailed treatment of this method on a future occasion. We only note on this occasion that it is convenient, in formalizing the proof of normalization in Prawitz 1971, for the full language (i.e. including $\exists$, $\vee$) to use a slightly stronger definition of strongly valid: if the derivation ends with an application of $\exists_1 E$, $\exists_2 E$, $\vee E$, the reduction tree of the derivation of the major premiss is assumed to be finite.

§ 3. Realizability for HAS + EXT

3.1. Let a fixed bi-unique assignment of (n+1) - ary species variables to n - ary species variables be given, for each $n \geq 0$, and let us denote the (n+1) - place variable assigned to an n - place variable X by X^*.
We obtain a notion of realizability which is a straightforward extension of Kleene's numerical realizability (Kleene 1945, 1952, § 82) by adding to the definition in Troelstra 1971, 3.2 the following clauses

$$\begin{array}{lll} \text{r (viii)} & x \, r \, Xy_1 \ldots y_n & \equiv X^*xy_1 \ldots y_n \\ \text{r (ix)} & x \, r \, \forall X \, A(X) & \equiv \forall X^*(x \, r \, A(X)) \\ \text{r (x)} & x \, r \, \exists X \, A(X) & \equiv \exists X^*(x \, r \, A(X)) . \end{array}$$

3.2. Definition. A formula A is called almost negative, if A does not contain $\vee$, and contains $\exists_1$ only in subformulae of the form $\exists x(t = s)$, and each prime formula of the form $Xt_1 \ldots t_n$ occurs negated. A is called negative if A is almost negative and does not contain $\exists$.

3.3. Lemma. For almost negative formulae containing variables of a non-empty sequence $\underline{a}$ free, there exists a partial recursive function $\varphi_A(\underline{a})$ such that $u \, r \, A(\underline{a}) \to \, ! \varphi_A(\underline{a}) \; \& \; \varphi_A(\underline{a}) \, r \, A(\underline{a})$.

Proof. As in Troelstra 1971, 3.8 but with one additional clause in the definition of φ: for subformulae of the form $\neg Xt_1 \ldots t_n$ we take $\varphi_{\neg Xt_1 \ldots t_n}(\underline{a}) = \Lambda \underline{a} . 0$.

3.4. Definition. ECT_0 is supposed to be defined as in the first order case (Troelstra 1971, 3.12), but now with respect to our extended language and definition of almost negative.

3.5. Theorem (Soundness theorem for r - realizability).
For any closed formula A, and for $H \equiv$ HAS + EXT, HAS + EXT + M (M denotes Markov's schema, see Troelstra 1971, 3.15), HAS + EXT + AC - NS or HAS + EXT + AC-NS + M :

$$H + ECT_0 + UP \vdash A \Longrightarrow H \vdash \bar{n} \; r \; A \quad \text{for a suitable numeral } \bar{n}.$$

Proof. Routine; the reader may compare section 3.7 in Kreisel and Troelstra 1970.

3.6. Corollary. (i) For $\underset{\sim}{H} \equiv \underset{\sim}{HAS}$, $\underset{\sim}{HAS}$ + AC - NS :
$\underset{\sim}{H}$ + M + UP + ECT_0 + EXT is consistent relative to $\underset{\sim}{H}$;

(ii) IP is not derivable, in fact refutable in $\underset{\sim}{HAS}$ + M + UP + ECT_0 + EXT (cf. Troelstra 1971, 3.20);

(iii) As an example we mention one conservative extension result:
$\underset{\sim}{HAS}$ + EXT + UP + ECT_0 is conservative over $\underset{\sim}{HA}$ w.r.t. the formulae of Γ_0 as described in Troelstra 1971, 6.2.
More examples can easily be constructed along the lines of § 6 of Troelstra 1971.

3.7. A variant of $\underset{\approx}{r}$ - realizability. Let us briefly discuss a variant of $\underset{\approx}{r}$ - realizability, to be called $\underset{\approx}{r}'$ - realizability, which is described as follows: the clauses $\underset{\approx}{r}'$ (i) - $\underset{\approx}{r}'$ (vii) are similar to $\underset{\approx}{r}$ (i) - $\underset{\approx}{r}$ (vii), and further we put

$$\begin{array}{lll} \underset{\approx}{r}' \text{(viii)} & x \,\underset{\approx}{r}\, Xy_1 \ldots y_n & \equiv \neg\neg\, Xy_1 \ldots y_n \\ \underset{\approx}{r}' \text{(ix)} & x \,\underset{\approx}{r}\, \forall X\, A(X) & \equiv \forall X(x \,\underset{\approx}{r}\, A(X)) \\ \underset{\approx}{r}' \text{(x)} & x \,\underset{\approx}{r}\, \exists X\, A(X) & \equiv \exists X(x \,\underset{\approx}{r}\, A(X)) . \end{array}$$

Note that now $x \,\underset{\approx}{r}\, A$ is always almost negative.
We then may prove a soundness theorem for a system $\underset{\sim}{HAS}^-$ + M, where $\underset{\sim}{HAS}^-$ is similar to $\underset{\sim}{HAS}$ + EXT, but with the schema for comprehension restricted to negative formulae A.

The only difficulty arises in verifying the $\underset{\approx}{r}'$ - realizability of the weakened comprehension schema. $y \,\underset{\approx}{r}\, Ax$ is an almost negative formula: in the presence of Markov's schema almost negative formulae are equivalent to negative formulae, so $\neg\neg\, y \,\underset{\approx}{r}\, Ax \longleftrightarrow y \,\underset{\approx}{r}\, Ax$ in $\underset{\sim}{HAS}^-$ + M. Choosing X to be $\neg\neg\, Ax$, and noting that $\exists y(y \,\underset{\approx}{r}\, Ax) \longleftrightarrow \forall y(y \,\underset{\approx}{r}\, Ax)$ for negative A, we easily see the instance $\exists X \forall x[Ax \longleftrightarrow Xx]$ of the weakened comprehension schema to be realizable.

By altering the clauses for $\vee$, $\rightarrow$, $\exists_1$ we may define a notion $\underset{\approx}{q}'$ - realizability, similar to $\underset{\approx}{q}$ - realizability (Troelstra 1971). By means of this notion we can obtain existential definability theorems etc. for $\underset{\sim}{HAS}^-$ + M and certain extensions. (Cf. Troelstra 1971, § 7.)

§ 4. Modified realizability for HAS + EXT

4.1. We suppose the reader to be acquainted with modified realizability in the first order case ; see Troelstra 1971, § 4.
HRO - mr - realizability is rather similar to r - realizability, but the types, in HRO - mr - realizability represented by species V_σ of natural numbers, have the effect of assigning to each formula A a domain D_A, depending on the logical structure of A only, such that realizing objects (partial recursive functions) of $A \to B$ are defined on D_A.

If we attempt to extend HRO - mr - realizability to HAS, it is natural to represent $D_{Xt_1 \dots t_n}$ (in short D_X) by a variable ranging over unary species. But if we try to verify the HRO - mr - realizability of $1 = 0 \to P$, we see that this is not enough, since $D_{1=0} = N$, the set of natural numbers (in analogy with the first order case ; in any case, $D_{1=0}$ should be non-empty, since $D_{x=x}$ must be non - empty) ; so we must be able to find at least one element in D_P for arbitrary P. In order to be able to do this in a uniform way, we would like to have a fixed number, say 0, belonging to each D_P. This can be arranged if we select a gödelnumbering and a pairing function j such that

$$j(0,0) = 0$$
$$\{0\}(x) \simeq 0 \text{ for all } x$$

(we use Kleene brackets {} w.r.t. this numbering also).
Such a gödelnumbering is easily constructed : Let T denote some standard T - predicate, satisfying

$$Txyz \,\&\, Txyz' \to z = z',$$

and let x_0 be a gödelnumber of $\lambda x.0$ in the enumeration.
We construct a new numbering by defining a predicate T' :

$$T'(x,y,z) \equiv_{def} (x = x_0 \,\&\, T(0,y,z)) \vee (x = 0 \,\&\, T(x_0,y,z)) \vee$$
$$\vee (x \neq 0 \,\&\, x \neq x_0 \,\&\, T(x,y,z)),$$

and define $\{x\}(y)$

$$\{x\}(y) \simeq U \min_z T'(x,y,z).$$

All the usual results of elementary recursion theory : s - m - n - theorem, recursion theorem remain available for this new numbering.

4.2. Definition. To each formula A of $\underset{\sim}{HAS}$, we associate a unary predicate D_A, depending on the logical structure of A only.
D_A is inductively defined as follows :

(i) If $A \equiv [t=s]$, then $D_A x \equiv [x=x]$.

(ii) If $A \equiv Xt_1 \ldots t_n$, then $D_{Xt_1 \ldots t_n}$ (abbreviated D_A, and depend- on X only) is regarded as a variable for species of one argument containing 0. (Adding such variables to $\underset{\sim}{HAS}$ is obviously a conservative extension.)
To distinct X we assign distinct variables D_X.

(iii) $D_{B \& C}(x) \equiv D_B(j_1 x) \& D_C(j_2 x)$.

(iv) $D_{B \vee C}(x) \equiv (j_1 x = 0 \to D_B(j_2 x)) \& (j_1 x \neq 0 \to D_C(j_2 x))$.

(v) $D_{B \to C}(x) \equiv \forall y (D_B(y) \to {!}\{x\}(y) \& D_C(\{x\}(y)))$.

(vi) $D_{\neg B}(x) \equiv D_{B \to 1=0}(x)$.

(vii) $D_{\forall y\, By}(x) \equiv \forall y\, ({!}\{x\}(y) \& D_{By}(\{x\}(y)))$.

(viii) $D_{\exists y\, By}(x) \equiv D_{B(j_1 x)}(j_2 x)$.

(ix) $D_{\forall X\, B(X)}(x) \equiv \forall D_X\, (D_{B(X)}(x))$.

(x) $D_{\exists X\, B(X)}(x) \equiv \exists D_X\, (D_{B(X)}(x))$.

By the induction on the complexity of formulae we easily verify for each A, $D_A(0)$.
There is an analogy with Girard's (completely independent) treatment of variable types in Girard 1971 ; see 4.8.

4.3. Definition. Now we associate to each formula A of $\underset{\sim}{HAS}$ a modified-realizability predicate $x \;\underset{\approx}{mr}\; A$. Let Γ, Δ be bi-unique mappings of species variables into species of species variables $\underset{=}{D}$, $\underset{=}{E}$ respectively. $\underset{=}{D}$ consists of unary species variables ranging over species containing zero, Γ corresponds to $X \to D_X$ in 4.2. ΔX is denoted by X^*; if X is n-ary, X^* is $(n+1)$-ary. $\underset{=}{D} \cap \underset{=}{E} = \emptyset$.
$x \;\underset{\approx}{mr}\; A$ is now defined inductively :

$\underset{\approx}{mr}$ (i) $x \;\underset{\approx}{mr}\; [t=s] \equiv [t=s]$.

$\underset{\approx}{mr}$ (ii) $x \;\underset{\approx}{mr}\; Xt_1 \ldots t_n \equiv D_X(x) \& X^* x t_1 \ldots t_n$.

$\underset{\approx}{mr}$ (iii) $x \;\underset{\approx}{mr}\; B \& C \equiv j_1 x \;\underset{\approx}{mr}\; B \& j_2 x \;\underset{\approx}{mr}\; C$.

$\underset{\approx}{mr}$ (iv) $x \;\underset{\approx}{mr}\; B \vee C \equiv ((j_1 x = 0 \to j_2 x \;\underset{\approx}{mr}\; B) \& (j_1 x \neq 0 \to j_2 x \;\underset{\approx}{mr}\; C))$.

$\underset{\approx}{mr}$ (v) $x \;\underset{\approx}{mr}\; B \to C \equiv D_{B \to C}(x) \& \forall u (u \;\underset{\approx}{mr}\; B \to {!}\{x\}(u) \& \{x\}(u) \;\underset{\approx}{mr}\; C)$.

mr (vi) $\quad x \ \underline{mr} \neg B \quad \equiv x \ \underline{mr} \ (B \to 1 = 0)$.

mr (vii) $\quad x \ \underline{mr} \ \forall y \, By \quad \equiv \forall y \, (!\{x\}(y) \ \& \ \{x\}(y) \ \underline{mr} \ By)$.

mr (viii) $\quad x \ \underline{mr} \ \exists y \, By \quad \equiv j_2 x \ \underline{mr} \ B(j_1 x)$.

mr (ix) $\quad x \ \underline{mr} \ \forall X \, B(X) \equiv \forall X^* \ \forall D_X (x \ \underline{mr} \ B(X))$.

mr (x) $\quad x \ \underline{mr} \ \exists X \, B(X) \equiv \exists X^* \ \exists D_X (x \ \underline{mr} \ B(X))$.

Note that $x \ \underline{mr} \ A$ implies $D_A(x)$, and that if we omit the clauses (ii), (ix), (x) we obtain HRO - mr - realizability for HA.

4.4. Lemma. Let $A(x_1, \ldots, x_n)$ be any negative formula of HA.
Then $0 \ \underline{mr} \ A(x_1, \ldots, x_n) \longleftrightarrow A(x_1, \ldots, x_n)$.

Proof. Routine, using $\{0\}(x) \simeq 0$ and $j(0,0) = 0$.

4.5. Lemma. The following instance of M_{PR} (Markov's schema for primitive recursive predicates, cf. Troelstra 1971) is refutable in HA + CT_0 + IP :

$$\forall x(\neg \neg \exists y \, T(x, x, y) \to \exists y \, T(x, x, y)) .$$

Proof. See Troelstra 1971, 3.20.

4.6. Theorem.

(i) For closed A, and H ≡ HAS + EXT, HAS + EXT + AC-NS :

$$H + UP + IP + CT_0 \vdash A \Longrightarrow H \vdash \bar{n} \ \underline{mr} \ A$$

for a suitable numeral $\bar{n}$.

(ii) (Corollary to (i).) H + UP + IP + CT_0 is consistent relative to H for H = HAS + EXT, HAS + AC-NS + EXT.

(iii) (Corollary to (i), lemma 4.5.) M_{PR} is not derivable in HAS + EXT + AC-NS.

Proof. The proof is completely routine, by induction on the length of derivations.

4.7. Remark. From the developments in § 6 of Troelstra 1971 it will be clear how to refine (ii) of the preceding theorem to a conservative extension result for a suitable subclass of (first order) arithmetical formulae; e.g. HAS + EXT + IP + CT_0 is conservative over HA w.r.t. formulae of Γ_1.

4.8. <u>Comparison with the type structure and functionals of Girard.</u>

The type structure $\underset{\sim}{T}$ of <u>Girard</u> 1971 can be described as follows :

(a) $0 \in \underset{\sim}{T}$.

(b) Variable types (denoted by α, α', α'', β, β', β'') belong to $\underset{\sim}{T}$.

(c) $\sigma, \tau \in \underset{\sim}{T} \Longrightarrow (\sigma)\tau,\ \sigma \times \tau \in \underset{\sim}{T}$.

(d) $\sigma[\alpha] \in \underset{\sim}{T} \Longrightarrow \forall\alpha.\sigma[\alpha],\ \exists\alpha.\sigma[\alpha] \in \underset{\sim}{T}$.

We now construct HRO^2, hereditarily recursive operations for second order arithmetic, by defining $V_\sigma \subseteq N$ for each $\sigma \in \underset{\sim}{T}$.

(A) $V_0 = N$.

(B) V_α ranges over sets of natural numbers containing 0.

(C) $V_{(\sigma)\tau}$ is defined as for HRO, $V_{\sigma\times\tau} = \{x \mid j_1x \in V_\sigma \;\&\; j_2x \in V_\tau\}$.

(D) $V_{\forall\alpha.\sigma[\alpha]} = \{x \mid \forall V_\alpha (x \in V_{\sigma[\alpha]})\}$,

$V_{\exists\alpha.\sigma[\alpha]} = \{x \mid \exists V_\alpha (x \in V_{\sigma[\alpha]})\}$.

The objects of type σ in HRO^2 consist of pairs (x, σ) with $x \in V_\sigma$. Application is interpreted by Kleene brackets.
We can find numbers representing Successor, $\Pi_{\sigma,\tau}$, $\Sigma_{\rho,\sigma,\tau}$, R_σ as for HRO ; $D_{\sigma,\tau}$, $D'_{\sigma,\tau}$, $D''_{\sigma,\tau}$ are represented by $(\Lambda x \Lambda y\,.\,j(x,y),\ (\sigma)(\tau)\sigma\times\tau)$, $(\Lambda x\,.\,j_1x,\ (\sigma\times\tau)\sigma)$, $(\Lambda x\,.\,j_2x,\ (\sigma\times\tau)\tau)$ respectively.
We have a gödelnumbering such that $j(0,0) = 0$ and $0\,(x) \simeq 0$, there is an element 0_σ, represented by $(0, \sigma)$ for each type σ.
There are two injection functionals : $I_{\sigma,\tau}$ and $I'_{\sigma,\tau}$. $I_{\sigma,\tau}$ is represented by

$$(\Lambda x\,.\,x,\ (\forall\alpha.\sigma[\alpha])(\tau)\sigma[\tau])\,.$$

$I'_{\sigma,\tau}$ is represented by

$$(\Lambda x\,.\,x,\ (\sigma[\tau])\,\exists\alpha.\sigma[\alpha])\,.$$

There is also an equality functional E_σ represented in the model.
If we now replace in Girard's system of functionals the λ-abstraction by Π, Σ, we see from the preceding remarks that HRO^2 is closed under this modification of Girard's schemata for introducing functionals. (Note that Girard's DT, ST do not correspond to operations with a type, and as such they are not represented in HRO^2 ; they correspond to schemata for introducing elements in $V_{\forall\alpha.\sigma[\alpha]}$, $V_{\exists\alpha.\sigma[\alpha]}$.)

Now assume variable types to be assigned to species variables. If to A is assigned σ, to B type τ, we assign to A & B type $\sigma\times\tau$, to $A \rightarrow B$ type $(\sigma)\tau$, to $\forall x A$ $0(\sigma)$, to $\exists x A$ $0 \times \sigma$, to $\forall X A$

$\forall\alpha.\sigma$ (if α is assigned to X), to $\exists X A$ we assign $\exists\alpha.\sigma$ ($A \vee B$ does not fit in, but we may translate $A \vee B$ as $\exists x((x=0 \rightarrow A) \& (x \neq 0 \rightarrow B))$; then we see that for each formula A, D_A corresponds to V_σ, if σ is the type assigned to A.
Our mr-realizability is HRO^2-modified realizability; our version of Girard's functionals is embedded in HRO^2 in the same way as the terms of $I\text{-}HA^\omega$ are embedded in HRO (Troelstra 1971).

4.9. Extension to theories with choice sequences.

If we consider theories with variables for choice sequences and species variables with choice sequences as arguments, it is possible to define a modified-realizability notion similar to the one discussed above, but based on the continuous application of functions instead:

$$\alpha|\beta \simeq \gamma \equiv_{def} \forall x \exists y(\alpha(\langle x\rangle * \bar{\beta}y) = \gamma x + 1) . \quad (1)$$

(Cf. Kleene's special realizability in Kleene and Vesley 1965, which is (almost) modified realizability w.r.t. this new application operation.)
Instead of using directly "$|$" we have to use in the presence of species variables a slightly different application operation $\|$, given by

$$\alpha\|\beta \simeq \gamma \equiv_{def} \Gamma\alpha|\beta \simeq \gamma$$

where Γ is given by

$$(\Gamma\alpha)x = (1 \dot{-} \alpha x) + sg(\alpha x \dot{-} 1) . \alpha x .$$

Then

$$\lambda x.0 \| \alpha \simeq \lambda x.0, \; j(\lambda x.0, \lambda x.0) = \lambda x.j(0,0) = \lambda x.0$$

for suitable j.

<u>References</u>

H. Friedman

A Some applications of Kleene's methods for intuitionistic systems. This volume.

G. Gentzen

1933 Über das Verhältnis zwischen intuitionistischer und klassischer Arithmetik. First published in English translation in: M.E. Szabo (editor), The collected papers of Gerhard Gentzen, Amsterdam - London (North-Holland), 1969, pp. 53 - 67.

J.Y. Girard

1971 Une extension de l'interprétation de Gödel à l'analyse et son application à l'élimination des coupures dans l'analyse et la théorie des types, in: J.E. Fenstad (editor), Proceedings of the second Scandinavian Logic Symposium, Amsterdam - London (North-Holland).

A Quelques résultats sur les interprétations fonctionnelles. Lecture Notes in Mathematics, vol. 337 (this volume).

K. Gödel

1958 Über eine bisher noch nicht benutzte Erweiterung des finiten Standpunktes. Dialectica <u>12</u> (1958), pp. 280 - 287.

H.R. Jervell

1971 A normal form in first order arithmetic, in: J.E. Fenstad (editor), Proceedings of the second Standinavian Logic Symposium, Amsterdam - London (North-Holland).

S.C. Kleene

1945 On the interpretation of intuitionistic number theory. J. Symb. Logic <u>10</u> (1945), pp. 109 - 124.

1952 Introduction to metamathematics. Amsterdam, Groningen (North-Holland, Noordhoff) and New York (Van Nostrand).

1962 Disjunction and existence under implication in elementary intuitionistic formalisms. J. Symb. Logic <u>27</u> (1962), pp. 11 - 18.

S. C. Kleene and R. E. Vesley

1965 Foundations of intuitionistic mathematics, especially in relation to recursive function. Amsterdam (North-Holland).

G. Kreisel

1959 Reflection principle for subsystems of Heyting's arithmetic (abstract). J. Symb. Logic 24 (1959), p. 322.

1968 Functions, ordinals, species, in : J.F. Staal, and B. van Rootselaar (editors), Logic, Methodology and Philosophy of Sciences III, Amsterdam (North-Holland).

1970 Church's thesis : a kind of reducibility axiom for constructive mathematics, in : A. Kino, J. Myhill, R.E. Vesley (editors), Intuitionism and Proof Theory, Amsterdam (North-Holland).

1971 A survey of proof theory II, in : J.E. Fenstad (editor), Proceedings of the second Scandinavian Logic Symposium, Amsterdam - London (North-Holland).

G. Kreisel and A.S. Troelstra

1970 Formal systems for some branches of intuitionistic analysis, Annals of Mathematical Logic 1 (1970), pp. 229 - 387.

D. Prawitz

1965 Natural deduction, a proof-theoretical study, Stockholm, Göteborg, Uppsala (Almqvist and Wiksell).

1970 Some results for intuitionistic logic with second order quantification rules, in : A. Kino, J. Myhill, R.E. Vesley (editors), Intuitionism and Proof Theory, Amsterdam (North-Holland).

1971 Ideas and results in proof theory, in : J.E. Fenstad (editor), Proceedings of the second Scandinavian Logic Symposium, Amsterdam - London (North-Holland).

A Towards a foundation of general proof theory. To appear in : Logic, Methodology and Philosophy of Science IV.

B. Scarpellini

1970 On cut elimination in intuitionistic systems of analysis, in : A. Kino, J. Myhill, R.E. Vesley (editors), Intuitionism and Proof Theory, Amsterdam (North-Holland).

1971 Proof theory and intuitionistic systems. Berlin, Heidelberg, New York (Springer-Verlag).

A. S. Troelstra

1971 Notions of realizability for intuitionistic arithmetic and intuitionistic arithmetic in all finite types, in : J.E. Fenstad (editor), Proceedings of the second Scandinavian Logic Symposium, Amsterdam, London (North-Holland).

<u>Some corrections</u> : page 374, line -8, replace last "σ" by " $(\sigma)\rho$ " ; page 380, line -8, replace "$\{u\}(\psi_A(x))$" by "$\{\{u\}(x)\}(\psi_A(x))$" ; page 393, line 11, replace "$\Longrightarrow$" by "$\Longrightarrow \vdash$" ; same page, line -13, delete "$(\underset{\sim}{HA}^{\omega} + A)$" and "$(\underset{\sim}{HA}^{\omega})$" ; line -12, add "closed" before "Σ_1^0" ; page 396, line -3, replace "$Y_1, \ldots, Y_n$" by "$y_1, \ldots, y_n$" ; page 397, line -3, replace "7.2" by "8.2".

1971 A Computability of terms and notions of realizability for intuitionistic analysis, Report 71 - 02 of the Department of Mathematics, University of Amsterdam (mimeographed).

V. SOME PROPERTIES OF INTUITIONISTIC ZERMELO-FRANKEL SET THEORY [1]

JOHN MYHILL

§1 The systems Z, $Z_c^{\rightarrow}$, $Z_c^{\rightarrow}$

In his paper [1], Harvey Friedman introduced a modification of Kleene's 'realizability' and used it to prove the existence and disjunction properties for a variety of systems based on intuitonistic predicate logic, in particular systems of type-theory and higher-order arithmetic. The present paper extends these results to an intuitionist version of Zermelo-Frankel set-theory. Except for the last two sections it represents joint work of Friedman and the author.

The basic system to which these results apply is a one-sorted first-order theory called Z, having one two-place predicate letter and no function symbols or individual constants. The underlying logic is Heyting's predicate calculus. The non-logical axioms and schemata are the customary ones (including replacement), except that choice is omitted and regularity is omitted and regularity is replaced by the following schema, called TI($\in$) (transfinite induction on $\in$):

$$(\forall \underline{x})[(\forall \underline{y})(y \in \underline{x} \rightarrow \phi\frac{a}{\underline{y}}) \rightarrow \phi\frac{a}{\underline{x}}] \rightarrow (\forall \underline{x})\phi\frac{a}{\underline{x}} .$$

The reason for this change is that full regularity, i.e.

$$\underline{a} \in \underline{b} \rightarrow (\exists \underline{x})(\underline{x} \in \underline{b} \wedge (\forall \underline{y})\neg(\underline{y} \in \underline{b} \wedge \underline{y} \in \underline{x})) \qquad (1)$$

would yield the classical system. This is seen as follows: Let ϕ be any formula not containing the parameter a. Let $0 \equiv \{\underline{x}|\underline{x} \neq \underline{x}\}$, $1 \equiv \{\underline{x}|\underline{x} = 0\}$, $\underline{A} \equiv \{\underline{x}|\underline{x} = 1 \vee (\underline{x} = 0 \wedge \phi)\}$.

---

1 This paper was supported by NSF grant GP21189

Clearly $1 \in \underline{A}$. By (1)

$$(\exists \underline{x})(\underline{x} \in \underline{A} \wedge (\forall \underline{y}) \neg (\underline{y} \in \underline{A} \wedge \underline{y} \in \underline{x})). \tag{2}$$

Now assume the hypothesis

$$\underline{a} \in \underline{A} \wedge (\forall \underline{y}) \neg (\underline{y} \in \underline{A} \wedge \underline{y} \in \underline{a}). \tag{3}$$

Then

$$(\underline{a} = 1 \wedge (\forall \underline{y}) \neg (\underline{y} \in \underline{A} \wedge \underline{y} \in 1))$$
$$\vee (\underline{a} = 0 \wedge \phi)$$

whence

$$\phi \vee \neg (0 \in \underline{A} \wedge 0 \in 1)$$
$$\phi \vee \; 0 \notin A$$
$$\phi \vee \neg (0 = 1 \vee (0 = 0 \wedge \phi))$$
$$\phi \vee \neg \phi \tag{4}$$

Since the hypothesis (3) yields (4) in which $\underline{a}$ does not occur, and since (2) is a theorem, so is (4). By change of parameters $\phi \vee \neg \phi$ is a theorem for all ϕ and so the system is classical. (For legibility we have carried out the above proof using comprehension-terms which are not in the primitive notation of Z; however, it is clear how to eliminate these).

For technical purposes we shall need certain modifications of Z. The first of these, denoted by $Z_{c_1, \ldots c_{\underline{n}}}$ or for brevity $Z_{\vec{c}}$, is an inessential extension of Z by the constants $c_1, \ldots c_{\underline{n}}$. The second, called $Z'_{\vec{c}}$, is formed by adding comprehension-terms to $Z_{\vec{c}}$, i.e. by adjoining the rule:

If $Z_{\vec{c}} \vdash (\exists \underline{x})(\forall \underline{y})(\underline{y} \in \underline{x} \leftrightarrow \phi\frac{\underline{a}}{\underline{y}})$, where ϕ has no parameters except $\underline{a}$, add a new constant c_ϕ with the defining axiom

$$\underline{a} \in c_\phi \leftrightarrow \phi.$$

(c_ϕ is what is normally written $\{x | \phi\frac{a}{x}\}$). In particular, if $\vec{c}$ is empty, $Z'_{\vec{c}}$ is simply Z with comprehension terms: henceforth we call this Z'.

The contents c_ϕ count as atomic: other constants appearing in ϕ do not count as parts of c_ϕ.

§2 Realizability - basic properties

Consider a first-order intuitionistic theory T* with variables $\underline{x}$, $\underline{y}$, $\underline{z}$ etc., parameters $\underline{a}$, $\underline{a}_0$, $\underline{a}_1$ ..., relation symbols $\underline{R}^n_{\underline{i}}$ for relations with n argument-terms ($\underline{n} \geq 0$; if $\underline{n} = 0$ they are proposition symbols), and function symbols $\underline{f}^n_{\underline{i}}$ for functions of $\underline{n}$ arguments (again $\underline{n} \geq 0$; if $\underline{n} = 0$ they are constants). Let $\underset{\sim}{M}$ be a model for T*, given by its positive diagram; i.e. we identify a model with the set of atomic sentences which hold in that model. Thus $\underset{\sim}{M}$ can be any set of atomic sentences of T*; in particular they need not be consistent with T*. Let finally ϕ be a sentence (= closed formula) of T*; we define the notion

$$\underset{\sim}{R}(\underset{\sim}{M}, T^*, \phi)$$

(T* realizes ϕ relative to the model $\underset{\sim}{M}$) inductively as follows:

$\underset{\sim}{R}(\underset{\sim}{M}, T^*, \phi) \equiv \phi \in \underline{M}$ if ϕ is atomic

$\neg \underset{\sim}{R}(\underset{\sim}{M}, T^*, \bot)$

$\underset{\sim}{R}(\underset{\sim}{M}, T^*, \phi \wedge \psi) \equiv \underset{\sim}{R}(\underset{\sim}{M}, T^*, \phi) \wedge \underset{\sim}{R}(\underset{\sim}{M}, T^*, \psi)$

$\underset{\sim}{R}(\underset{\sim}{M}, T^*, \phi \vee \psi) \equiv [\underset{\sim}{R}(\underset{\sim}{M}, T^*, \phi) \wedge \vdash^* \phi] \vee [\underset{\sim}{R}(\underset{\sim}{M}, T^*, \psi) \wedge \vdash^* \psi]$

$\underset{\sim}{R}(\underset{\sim}{M}, T^*, \phi \to \psi) \equiv [\underset{\sim}{R}(\underset{\sim}{M}, T^*, \phi) \wedge \vdash^* \phi] \to \underset{\sim}{R}[\underset{\approx}{M}, T^*, \psi]$

$\underset{\sim}{R}(\underset{\sim}{M}, T^*, (\forall \underline{x})\phi\frac{a}{\underline{x}}) \equiv (\forall ⓣ)\underset{\sim}{R}(\underset{\sim}{M}, T^*, \phi\frac{a}{ⓣ})$

$\underset{\sim}{R}(\underset{\sim}{M}, T^*, (\exists \underline{x})\phi\frac{a}{\underline{x}}) \equiv (\exists ⓣ)\left[\underset{\sim}{R}(\underset{\sim}{M}, T^*, \phi\frac{a}{ⓣ}) \wedge \vdash^* \phi\frac{a}{ⓣ}\right]$

[Notations (in the metalanguage). $\equiv$ means as before definitional identity or equivalence; later it will also mean syntactical identity, but the context will always make clear which is meant. $\vdash^*$ means provability in T^*. ⓣ is a metavariable ranging over closed terms of T^*; likewise later $ⓣ_1$, $ⓣ_2$ etc. $\phi\frac{a}{ⓣ}$ is the result of replacing every occurrence of $\underline{a}$ in ϕ by ⓣ; likewise $\phi\frac{a}{\underline{x}}$, $\phi\frac{a}{ⓣ_1}, \frac{a_1}{\underline{x}}, \frac{a_2}{ⓣ_2}$ etc. Where several replacements are indicated they are to be performed simultaneously.]

The fundamental property of $\underset{\sim}{R}$ is given by the following

SOUNDNESS THEOREM ([1], Theorem 2.1). Say that $\underset{\sim}{R}(\underline{M}, T^*, T)$ if $\underset{\sim}{R}(\underline{M}, T^*, \phi)$ for every closed ϕ provable in T. (Here T is a theory whose formulas are in the notation of T^*.) Likewise say that $T^* \vdash T$ if all formulas in T are deducible from formulas belonging to T^*, etc. Then

$$\underset{\sim}{R}(\underline{M}, T^*, T),\ T^* \vdash T \vdash T_o \quad \text{imply } \underset{\sim}{R}(\underline{M}, T^*, T_o).$$

Let $\phi \vee \psi$ be a closed theorem of a theory T. If $\underset{\sim}{R}(\underline{M}, T^*, T)$ and T^* is an extension of T then we see that (taking $T_o \equiv \{\phi \vee \psi\}$) $\underset{\sim}{R}(\underline{M}, T^*, \phi \vee \psi)$, whence, by the definition of $\underset{\sim}{R}$, $\vdash^*\phi$ or $\vdash^*\psi$. If T^* is furthermore a conservative extension of T, we get that either ϕ or ψ is a theorem of T, so that T possesses the disjunction-property.

Suppose again that $(\exists \underline{x})\phi\frac{a}{\underline{x}}$ (closed) is provable in T, and that $\underset{\sim}{R}(\underline{M}, T^*, T)$ where T^* is a conservative extension of T. The soundness theorem yields (with $T_o \equiv (\exists \underline{x})\phi\frac{a}{\underline{x}}$) $\underset{\sim}{R}(\underline{M}, T^*, (\exists \underline{x})\phi\frac{a}{\underline{x}})$. Again using the definition of $\underset{\sim}{R}$, there is a closed term ⓣ of T^* for which $\vdash^* \phi\frac{a}{ⓣ}$. Since T^* is conservative, $\phi\frac{a}{ⓣ}$ is provable in T and T has the existence-property provided that ⓣ is a term of T.

In order to prove the disjunction - and existence-properties for a theory T, then, it suffices to pick cleverly a conservative extension $T^* \supset T$ and a model $\underline{M}$ for T^*, such that $\underset{\sim}{R}(\underline{M}, T^*, T)$. (For the existence-property, it is required also that T^* has the same terms as T; but we can sometimes avoid this by a trick; see the proof of Corollary 3 below.)

We shall make tacit use of the following lemma, which is an immediate consequence of the soundness theorem:
If $R(M, T^*, \phi)$ for ϕ (the closure of) each non-logical axiom of a theory T, then $R(M, T^*, T)$.

The rest of this paper is divided into sections as follows. In §3 we state our main result, that $Z_{\vec{c}}$ is realizable by a suitable $Z^*_{\vec{c}}$ and M having a certain additional property, and show that this implies the disjunction properties for Z and Z' and the existence property for Z', as well as certain generalizations of these properties. In §4 we set up the system $Z^*_{\vec{c}}$ and define our M. In §§5-7 we show that all axioms of $Z_{\vec{c}}$ are realizable (this is the heart of the paper and constitutes the proof of our main result). §8 is by way of an appendix and shows that every term in Z' which provably denotes an integer is provably equal to some numeral, and that every provable real function of Z' is continuous. It also lists some open questions.

§3 Consequences of realizability

The bulk of this paper is devoted to proving the following

MAIN THEOREM. There is a conservative extension $Z^*_{\vec{c}}$ of $Z_{\vec{c}}$ and a model $M_{\vec{c}}$ such that (a) $R(M_{\vec{c}}, Z^*_{\vec{c}}, Z_{\vec{c}})$ and (b) there is a mapping $Ⓣ \to Ⓣ^-$ of terms of $Z^*_{\vec{c}}$ onto terms of $Z'_{\vec{c}}$ such that if ϕ is a formula of $Z_{\vec{c}}$ with the single free variable a, then

$$\vdash^* \phi^{a}_{Ⓣ} \text{ iff } \vdash' \phi^{a}_{Ⓣ^-}$$

(where we write $\vdash, \vdash', \vdash^*$ for $Z_{\vec{c}}$, $Z'_{\vec{c}}$, $Z^*_{\vec{c}}$ respectively).

This theorem has the following corollaries.

COROLLARY 1 If ϕ and ψ are closed and $Z \vdash \phi \vee \psi$, then $Z \vdash \phi$ or $Z \vdash \psi$; likewise for Z'.

Proof. For Z take $\vec{c}$ empty in the main theorem and apply the remarks following the soundness theorem. For Z' assume that $Z' \vdash \phi \vee \psi$ and let ϕ', ψ' be closed formulas of Z such that $Z' \vdash (\phi \leftrightarrow \phi') \wedge (\psi \leftrightarrow \psi')$. Then $Z' \vdash \phi' \vee \psi'$, $Z \vdash \phi' \vee \psi'$, $Z \vdash \phi'$ or $Z \vdash \psi'$, $Z \vdash \phi'$ or $Z' \vdash \psi'$ and $Z' \vdash \phi$ or $Z' \vdash \psi$.

COROLLARY 2 If ϕ and ψ are formulas (not necessarily closed) and $Z \vdash \phi \vee \psi$, then $Z \vdash \phi$ or $Z \vdash \psi$. (Thus if $Z \vdash (\forall \underline{x})(\phi\frac{a}{\underline{x}} \vee \psi\frac{a}{\underline{x}})$, either $Z \vdash (\forall \underline{x})(\phi\frac{a}{\underline{x}})$ or $Z \vdash (\forall \underline{x})(\psi\frac{a}{\underline{x}})$). Likewise for Z'.

Proof. Assume $Z \vdash \phi \vee \psi$ and let $\underline{a}_1, \ldots, \underline{a}_n$ be a list of all parameters occuring in ϕ and ψ. Then $Z_{\vec{c}} \vdash \phi\frac{\vec{a}}{\vec{c}} \vee \psi\frac{\vec{a}}{\vec{c}}$,

where now $\phi\frac{\vec{a}}{\vec{c}}$ and $\psi\frac{\vec{a}}{\vec{c}}$ are closed. By the main theorem and the remarks following the soundness theorem, $Z_{\vec{c}} \vdash \phi\frac{\vec{a}}{\vec{c}}$ or $Z_{\vec{c}} \vdash \psi\frac{\vec{a}}{\vec{c}}$. So $Z \vdash \phi$ or $Z \vdash \psi$. The result for Z' follows as in Corollary 1.

COROLLARY 3 If $(\exists \underline{x})\phi\frac{a}{\underline{x}}$ is closed and $Z' \vdash (\exists \underline{x})\phi\frac{a}{\underline{x}}$ (a fortiori if $Z \vdash (\exists \underline{x})\phi\frac{a}{\underline{x}}$) there is a term ⓒ of Z' such that $Z' \vdash \phi\frac{a}{ⓒ}$.

Proof. Let ϕ' be a formula of Z such that $Z' \vdash \phi \leftrightarrow \phi'$; let it have no parameters except $\underline{a}$. Then $Z \vdash (\exists \underline{x})\phi'\frac{a}{\underline{x}}$. By the main theorem and the remarks following the soundness theorem, there is a constant ⓣ of $Z^*_{\vec{c}}$ (with $\vec{c}$ empty) for which $\vdash^* \phi'\frac{a}{ⓣ}$. By the second part of the main theorem, $\vdash' \phi'\frac{a}{ⓣ}$, i.e. $Z' \vdash \phi'\frac{a}{ⓣ}$. Then $Z' \vdash \phi\frac{a}{ⓣ}$.

COROLLARY 4 If $(\forall \underline{x}_1, \ldots \underline{x}_n)(\exists \underline{y})\phi\frac{\vec{a}}{\vec{x}}, \frac{b}{\underline{y}}$ is a closed theorem of Z' (a fortiori of Z), there is a formula ψ of Z having $a_1, \ldots, a_n$, b as its only free variables, such that

$$Z' \vdash (\forall \underline{x}_1 \ldots x_n)(\exists \underline{y})(\phi\frac{\vec{a}}{\vec{x}}, \frac{b}{\underline{y}} \wedge (\forall \underline{z})(z \in \underline{y} \leftrightarrow \psi\frac{\vec{a}}{\vec{x}}, \frac{b}{\underline{z}})).$$

(Intuitively this means that

$$Z' \vdash (\forall \underline{x}_1, \ldots, \underline{x}_n) \phi^{\vec{\underline{a}}, b}_{\vec{\underline{x}}, \{z | \psi^{\vec{\underline{a}}, \underline{b}}_{\vec{x}, \underline{z}}\}}$$

but we have no comprehension-terms with free variables.)

Proof. Let ϕ' be as in Corollary 3. Then $Z' \vdash (\exists \underline{y}) \phi'^{\underline{b}}_{\underline{y}}$ and $Z_{c_1, \ldots, c_n} \vdash (\exists \underline{y}) \phi'^{\vec{\underline{a}}, \underline{b}}_{\vec{c}, \underline{y}}$. As in Corollary 3, there is a constant Ⓣ of $Z^*_{\vec{c}}$ for which $\vdash^* \phi'^{\vec{\underline{a}}, \underline{b}}_{\vec{\underline{c}}, Ⓣ}$. By the second part of the main theorem $\vdash' \phi'^{\vec{\underline{a}}, \underline{b}}_{\vec{c}, Ⓣ}$. Let Ⓣ be c_θ; then

$$\vdash' (\exists \underline{y}) (\phi'^{\vec{\underline{a}}, \underline{b}}_{\vec{\underline{c}}, \underline{y}} \wedge (\forall \underline{z}) (\underline{z} \in \underline{y} \leftrightarrow \theta^{\underline{a}}_{\underline{z}}).$$

Then

$$Z' \vdash (\exists \underline{y}) (\phi'^{\underline{b}}_{\underline{y}} \wedge (\forall \underline{z}) (\underline{z} \in \underline{y} \leftrightarrow \theta^{\vec{\underline{c}}, \underline{a}}_{\vec{\underline{a}}, \underline{z}}))$$

and

$$Z' \vdash (\forall \vec{\underline{x}}) (\exists \underline{y}) (\phi^{\vec{\underline{a}}, \underline{b}}_{\vec{\underline{x}}, \underline{y}} \wedge (\forall \underline{z}) (\underline{z} \in \underline{y} \leftrightarrow \theta^{\vec{\underline{c}}, \underline{a}}_{\vec{\underline{x}}, \underline{z}})$$

which proves the Corollary with $\psi \equiv \theta^{\vec{\underline{c}}, \underline{a}}_{\vec{\underline{a}}, \underline{b}}$.

§4 The system $Z^*_{\vec{c}}$ and the model $\underset{\sim}{M}$.

The system $Z^*_{\vec{c}}$ is obtained by splitting each term c_ϕ of $Z'_{\vec{c}}$ into many terms $c_{\phi,\underline{X}}$. Thus the terms of $Z^*_{\vec{c}}$ are the c_i and certain $c_{\phi,\underline{X}}$ where c_ϕ is a term of $Z'_{\vec{c}}$ and $\underline{X}$ is a set of terms of $Z^*_{\vec{c}}$. Roughly $c_{\phi,\underline{X}}$ denotes the set of $\{\underline{x} | \phi^{\underline{a}}_{\underline{x}}\}$ and $\underline{X}$ is the "reason" that we know $\phi^{\underline{a}}_{\underline{X}}$. Formally this means that we postulate

$$a \in c_{\phi^-, \underline{X}} \leftrightarrow \phi$$

where ϕ^- is obtained by suppressing the "reasons" from ϕ, and that an atomic formula Ⓣ $\in c_{\phi,\underline{X}}$ is realizable (i.e. (Ⓣ $\in c_{\phi,\underline{X}}) \in \underset{\sim}{M}$) iff Ⓣ $\in \underline{X}$.

We proceed to make all this more precise.

We first assign a <u>rank</u> $\rho\copyright$ to every constant $\copyright$ of $Z_{\vec{c}}$, in such a manner that

$$\vdash' \copyright_1 \in \copyright_2 \text{ implies } \rho\copyright_1 < \rho\copyright_2. \qquad (5)$$

To this end we define recursively for each term $\copyright$ (formula ϕ) of $Z_{\vec{c}}$ a term $\copyright^{\circ}$ (respectively formula ϕ°) as follows:

$$\underline{x}_{\underline{i}}^{\circ} \equiv \underline{x}_{\underline{i}},\quad \underline{a}_{\underline{i}}^{\circ} \equiv \underline{a}_{\underline{i}}$$

$$c_{\underline{i}}^{\circ} \equiv 0 \equiv c_{a\neq a}$$

$$(c_{\phi})^{\circ} \equiv c_{(\phi^{\circ})}$$

$$\bot^{\circ} \equiv \bot$$

$$(\copyright_1 \in \copyright_2)^{\circ} \equiv \copyright_1^{\circ} \in \copyright_2^{\circ}$$

$$(\phi \wedge \psi)^{\circ} \equiv \phi^{\circ} \wedge \psi^{\circ}, \text{ likewise } \vee, \rightarrow.$$

$$((\exists \underline{x})\phi)^{\circ} \equiv (\exists \underline{x})(\phi^{\circ}), \text{ likewise } \forall.$$

Then we define $\rho\copyright$ to be the rank of the set denoted by $\copyright^{\circ}$ in the standard interpretation. (Actually $\copyright^{\circ}$ is a term of $Z_{\vec{c}}$, but we do not need this). To verify (5) assume $\vdash' \copyright_1 \in \copyright_2$. Then $\copyright_1 \in \copyright_2$ expresses a truth of set-theory under <u>every</u> interpretation of $c_1, \ldots, c_{\underline{n}}$. In particular it is true if we set $c_1 = \ldots = c_{\underline{n}} = 0$. Thus $\copyright_1^{\circ} \in \copyright_2^{\circ}$ is <u>true</u>, and $\rho\copyright_1 = \rho\copyright_1^{\circ} < \rho\copyright_2^{\circ} = \rho\copyright_2$. Hence (5).[2]

Now we are ready to define the theory $Z_{\vec{c}}^{*}$. Its terms are the constants $c_1, \ldots, c_n$ and certain terms $c_{\phi,\underline{X}}$, where c_{ϕ} is a term of $Z'_{\vec{c}}$ and X is a set of terms of $Z_{\vec{c}}^{*}$. To facilitate the definition, set

$$(c_{\phi,\underline{X}})^{-} \equiv c_{\phi},\quad (c_{\phi,\underline{X}})^{+} \equiv \underline{X}$$

---

2 We can if we like replace this semantical argument by a purely finitary one to the effect that if $\vdash' \copyright_1 \in \copyright_2$, then $\vdash' \copyright_1^{\circ} \in \copyright_2^{\circ}$. The details are messy, and we forego them here.

$$(c_{\underline{i}})^- \equiv c_{\underline{i}}, \quad (c_{\underline{i}})^+ \equiv 0$$

and define

$$ⓣ_1 \sim ⓣ_2 \equiv \vdash' (\forall \underline{x})(x \in t_1^- \leftrightarrow x \in t_2^-)$$
$$\wedge \ t_1^+ = t_2^+$$

A set X of 'possible terms' is <u>extensional</u> if

$$ⓣ \sim ⓣ' \in \underline{X} \rightarrow ⓣ \in X.$$

Now the constants of Z_c^* are the elements of

$$\Delta \equiv \bigcup_{\alpha} \Delta_\alpha$$

where

$$\Delta_0 \equiv \{c_1, \ldots c_n\} \cup \{c_{\phi,0} | c_\phi \text{ is a term of } Z_c^{\pm} \text{ and } \rho\, c_\phi = 0\}$$

and for $\alpha > 0$

$$\Delta_\alpha \equiv \{c_{\phi,\underline{X}} | c_\phi \text{ is a term of } Z_c' \wedge$$
$$\rho c_\phi = \alpha \wedge$$
$$\underline{X} \subset \bigcup_{\beta<\alpha} \Delta_\beta \wedge$$
$$\underline{X} \text{ is extensional } \wedge$$
$$ⓣ \in \underline{X} \rightarrow \vdash' t^- \in c_\phi\}.$$

The axioms of $Z_c^{\pm *}$ are the axioms of $Z_c^{\pm}$ together with all formulas

$$\underline{a} \in c_{\phi^-,\underline{X}} \leftrightarrow \phi$$

where ϕ^- is obtained from ϕ by replacing each constant ⓣ by t^-, and where $c_{\phi^-,X}$ and all constants occuring in ϕ belong to Δ. Evidently $\vdash_c^* \phi$ iff $\vdash_c' \phi^-$, whence (b) of the main theorem. The next three sections are devoted to proving (a).

We define the model $\underline{M}$ by

$$(ⓣ \in c_i) \notin \underline{M}, \quad (ⓣ \in c_{\phi,\underline{X}}) \in \underline{M} \text{ iff } ⓣ \in \underline{X}.$$

§5 Realizability of the extensionality axiom

We must show

$$\underset{\sim}{R}(\underset{\sim}{M}, T^{*}_{EXT}, (\forall \underline{x})(\underline{x} \in \underline{a}_1 \leftrightarrow \underline{x} \in \underline{a}_2) \wedge \underline{a}_1 \in \underline{a}_3 \rightarrow \underline{a}_2 \in \underline{a}_3).$$

This amounts to showing that the hypotheses

$$\underset{\sim}{R}((\forall \underline{x})(\underline{x} \in \textcircled{t}_1 \leftrightarrow \underline{x} \in \textcircled{t}_2)) \qquad (6)$$

$$\underset{\sim}{R}\textcircled{t}_1 \in \textcircled{t}_3 \qquad (7)$$

$$\vdash^{*} (\forall \underline{x})(\underline{x} \in \textcircled{t}_1 \leftrightarrow \underline{x} \in \textcircled{t}_2) \qquad (8)$$

$$\vdash^{*} \textcircled{t}_1 \in \textcircled{t}_3$$

imply the conclusion

$$\underset{\sim}{R}(\textcircled{t}_2 \in \textcircled{t}_3).$$

By (7), $\textcircled{t}_1 \in \textcircled{t}_3^{+}$. $\textcircled{t}_3^{+}$ is extensional, i.e.

$$\vdash' (\forall x)(x \in \textcircled{t}^{-}_1 \leftrightarrow x \in \textcircled{t}^{-}_2) \wedge \textcircled{t}_1^{+} = \textcircled{t}_2^{+} \wedge \textcircled{t}_1 \in \textcircled{t}_3^{+} \rightarrow \textcircled{t}_2 \in \textcircled{t}_3^{+}.$$

But $\vdash' (\forall \underline{x})(\underline{x} \in \textcircled{t}^{-}_1 \leftrightarrow \underline{x} \in \textcircled{t}^{-}_2)$ by (8). So all we need is $\textcircled{t}_1^{+} = \textcircled{t}_2^{+}$.

We shall deduce this from (6). Since (6) is symmetric in $\textcircled{t}_1$ and $\textcircled{t}_2$, it is enough to show $\textcircled{t}_1^{+} \subset \textcircled{t}_2^{+}$). Let $\textcircled{t}_4 \in \textcircled{t}_1^{+}$. By (6) $\underset{\sim}{R}(\textcircled{t}_4 \in \textcircled{t}_1 \rightarrow \textcircled{t}_4 \in \textcircled{t}_2)$, i.e.

$$\underset{\sim}{R}(\textcircled{t}_4 \in \textcircled{t}_2) \wedge \vdash^{*} \textcircled{t}_4 \in \textcircled{t}_1 \rightarrow \underset{\sim}{R}(\textcircled{t}_4 \in \textcircled{t}_2). \qquad (9)$$

Since $\textcircled{t}_4 \in \textcircled{t}_1^{+}$, $\underset{\sim}{R}(\textcircled{t}_4 \in \textcircled{t}_1)$. Since $t_1 \in \Delta$, $\textcircled{t}_4 \in \textcircled{t}_1^{+} \rightarrow \vdash' \textcircled{t}_4 \in \textcircled{t}^{-}_1$.
Hence $\vdash^{*} \textcircled{t}_4 \in \textcircled{t}_1$.
Hence $\underset{\sim}{R}(\textcircled{t}_4 \in \textcircled{t}_2)$ by (9). Hence $\textcircled{t}_4 \in \textcircled{t}_2^{+}$.
This shows that $\textcircled{t}_1^{+} = \textcircled{t}_2^{+}$ and completes the proof of realizability of extensionality.

§6 Realizability of the existential axioms

Replacement has the form

$$\chi \rightarrow (\exists \underline{x})(\forall \underline{y})(\underline{y} \in \underline{x} \leftrightarrow \phi\frac{a}{\underline{y}}) \tag{10}$$

with parameters in χ and ϕ. All the other existential axioms except infinity have the simpler form

$$(\exists \underline{x})(\forall \underline{y})(\underline{y} \in \underline{x} \leftrightarrow \phi\frac{a}{\underline{y}})$$

with ϕ possibly having parameters: we can regard this as a special case of (10) with a trivial χ. We shall first show that every axiom of the form (10) is realizable: this enables our results to be extended to stronger systems than Z, e.g. systems with axioms asserting the existence of large cardinals. Then we shall consider the special case of the axiom of infinity.

To say that (10) is realizable is to say that

$$\underset{\sim}{R}(\chi\frac{\vec{a}}{\vec{\textcircled{t}}}) \wedge \vdash^{*} \chi\frac{\vec{a}}{\textcircled{t}} \rightarrow \underset{\sim}{R}(\exists \underline{x})(\forall \underline{y})(y \in x \leftrightarrow \phi\frac{a, \vec{a}}{\underline{y}, \vec{\textcircled{t}}}).$$

To show this we assume $\underset{\sim}{R}(\chi\frac{\vec{a}}{\vec{\textcircled{t}}})$ and $\vdash^{*} \chi\frac{\vec{a}}{\vec{\textcircled{t}}}$ and try to find $\textcircled{t}'$ for whic

$$\vdash^{*}(\forall \underline{y})(y \in \textcircled{t}' \leftrightarrow \phi\frac{a, \vec{a}}{\underline{y}, \vec{\textcircled{t}}}) \tag{11}$$

and

$$R((\forall \underline{y})(y \in \textcircled{t}' \leftrightarrow \phi\frac{a, \vec{a}}{\underline{y}, \vec{\textcircled{t}}}). \tag{12}$$

Evidently there is a constant c_{ψ} of $Z'_{\vec{c}}$ for which

$$\vdash' a \in c_{\psi} \leftrightarrow \phi\frac{\vec{a}}{\vec{\textcircled{t}}},$$

namely we take $\psi \equiv \phi\frac{\vec{a}}{\vec{\textcircled{t}}}$. Then (11) is satisfied by any $\textcircled{t}' \equiv c_{\psi,\underline{X}} \in \Delta$. So we need $\underline{X}$ for which

$$c_{\psi,\underline{X}} \in \Delta \tag{13}$$

and

$$\textcircled{t}' \equiv c_{\psi,\underline{X}} \text{ satisfies (12)} \tag{14}$$

Set

$$\underline{X} \equiv \{ (t) \mid \underset{\sim}{R}(\phi^{\vec{\underline{a}}\ \underline{a}}_{\vec{(t)}(t)''}) \wedge$$

$$\vdash^{*} \phi^{\vec{\underline{a}},\underline{a}}_{\vec{(t)},(t)''} \} .$$

Re (13). This amounts to showing firstly that

$$X \subset \bigcup_{\beta<\rho c_{\psi}} \Delta_{\beta} ; \tag{15}$$

secondly that $\underline{X}$ is extensional, i.e.

$$(t)_1 \in \underline{X} \wedge \vdash' (\forall \underline{x})(\underline{x} \in (t)_1^- \leftrightarrow \underline{x} \in (t)_2^-) \wedge (t)_1^+ = (t)_2^+ \rightarrow (t)_2 \in \underline{X} ; \tag{16}$$

and finally that

$$(t) \in \underline{X} \rightarrow \vdash' (t)^- \in c_{\psi} . \tag{17}$$

Re (15). If $(t)_1 \in \underline{X}$ then $\vdash^{*} \phi^{\vec{\underline{a}}\ \underline{a}}_{\vec{(t)},(t)_1}$; hence $\vdash' \phi^{\vec{\underline{a}}\ \underline{a}}_{\vec{(t)}(t)_1^-}$,

$\vdash' (t)_1' \in c_{\psi}$, $\rho\ (t)_1^- < \rho c_{\psi}$ and $(t)_1 \in \Delta_{\rho (t)_2^-}$.

Re (16). Assume the hypotheses

$$(t)_1 \in \underline{X} \tag{18}$$

$$\vdash' (t)_1^- = (t)_2^- \tag{19}$$

$$(t)_1^+ = (t)_2^+ . \tag{20}$$

By (18)

$$\vdash^{*} \phi^{\vec{\underline{a}}\ ,\underline{a}}_{\vec{(t)}\ ,(t)_1} .$$

By (19)

$$\vdash^{*} (t)_1 = (t)_2 .$$

By extensionality in $T^{*}_{\vec{c}}$

$$\vdash^{*} \phi^{\vec{\underline{a}}\ \underline{a}}_{\vec{(t)}(t)_2} .$$

So all we need show is $\underset{\sim}{R}(\phi\frac{\vec{a}}{\textcircled{t}}, \frac{a}{\textcircled{t}_2})$. This follows from (18) - (20) and the

<u>LEMMA</u> If $\vdash^* \textcircled{t}_1 = \textcircled{t}_2$, $\underset{\sim}{R}(\psi\frac{a}{\textcircled{t}_1})$ and $\textcircled{t}_1^+ = \textcircled{t}_2^+$, then $\underset{\sim}{R}(\psi\frac{a}{\textcircled{t}_2})$, where ψ may contain constants but no parameters besides $\underline{a}$.

The proof is by induction on ψ.

If ψ is atomic it can have the same four forms

$$\perp$$
$$\underline{a} \in \textcircled{t}$$
$$\textcircled{t} \in \underline{a}$$
$$\underline{a} \in \underline{a}$$

where $\textcircled{t}$, $\textcircled{t}_1$ and $\textcircled{t}_2$ can have the form $c_{\underline{i}}$ as well as the form $c_{\theta,\underline{A}}$.

Case I is trivial. Case II. $\underset{\sim}{R}(\textcircled{t}_1 \in \textcircled{t})$ implies $\textcircled{t}_1 \in \textcircled{t}^+$. Since $\textcircled{t}^+$ is extensional (vacuously if $\textcircled{t}$ is some $c_{\underline{i}}$) $\vdash_c' \textcircled{t}_1^- = \textcircled{t}_2^-$ and $\textcircled{t}_1^+ = \textcircled{t}_2^+$ imply $\textcircled{t}_2 \in \textcircled{t}^+$, whence $\underset{\sim}{R}\ \textcircled{t}_2 \in \textcircled{t}$. Case III. $\underset{\sim}{R}(\textcircled{t} \in \textcircled{t}_1) \rightarrow \textcircled{t} \in \textcircled{t}_1^+ = \textcircled{t}_2^+ \rightarrow \underset{\sim}{R}(\textcircled{t} \in \textcircled{t}_2)$. Case IV. Same, reading $\textcircled{t}_1$ for $\textcircled{t}$.

Inductive step.

Case I. ψ is $\psi_1 \wedge \psi_2$. Then

$$\begin{aligned}\underset{\sim}{R}(\psi\tfrac{a}{\textcircled{t}_1}) &\rightarrow \underset{\sim}{R}(\psi_1\tfrac{a}{\textcircled{t}_1}) \wedge \underset{\sim}{R}(\psi_2\tfrac{a}{\textcircled{t}_1}) \\ &\rightarrow \underset{\sim}{R}(\psi_1\tfrac{a}{\textcircled{t}_2}) \wedge \underset{\sim}{R}(\psi_2\tfrac{a}{\textcircled{t}_2}) \\ &\rightarrow \underset{\sim}{R}(\psi\tfrac{a}{\textcircled{t}_2}).\end{aligned}$$

Case II. ψ is $\psi_1 \wedge \psi_2$. Then

$$\underset{\sim}{R}(\psi\frac{a}{\textcircled{t}_1}) \rightarrow (\underset{\sim}{R}(\psi_1\frac{a}{\textcircled{t}_1}) \wedge \vdash^* \psi_1\frac{a}{\textcircled{t}_1})$$

$$\vee$$

$$(\underset{\sim}{R}(\psi_2\frac{a}{\textcircled{t}_1}) \wedge \vdash^* \psi_2\frac{a}{\textcircled{t}_2})$$

$$\rightarrow (\underset{\sim}{R}(\psi_1\frac{a}{\textcircled{t}_2}) \wedge \vdash^* \psi_1\frac{a}{\textcircled{t}_2})$$

$$(\underset{\sim}{R}(\psi_2\frac{a}{\textcircled{t}_2}) \wedge \vdash^* \psi_2\frac{a}{\textcircled{t}_2})$$

$$\rightarrow \underset{\sim}{R}(\psi\frac{a}{\textcircled{t}_2}).$$

Case III. ψ is $\psi_1 \rightarrow \psi_2$. Then

$$\underset{\sim}{R}(\psi\frac{a}{\textcircled{t}_1}) \rightarrow ((\underset{\sim}{R}(\psi_1\frac{a}{\textcircled{t}_1}) \wedge \vdash^* \psi_1\frac{a}{\textcircled{t}_1} \rightarrow \underset{\sim}{R}(\psi_2\frac{a}{\textcircled{t}_2}))$$

$$\rightarrow ((\underset{\sim}{R}(\psi_1\frac{a}{\textcircled{t}_2}) \wedge \vdash^* \psi_1\frac{a}{\textcircled{t}_2} \rightarrow \underset{\sim}{R}(\psi_2\frac{a}{\textcircled{t}_2}))$$

$$\rightarrow \underset{\sim}{R}(\psi\frac{a}{\textcircled{t}_2}).$$

Case IV. ψ is $(\exists \underline{x})\psi_o\frac{b}{\underline{x}}$. Then

$$\underset{\sim}{R}(\psi\frac{a}{\textcircled{t}}) \rightarrow (\exists \textcircled{t}_3)[\underset{\sim}{R}(\psi_o\frac{a}{\textcircled{t}_1},\frac{b}{\textcircled{t}_3}) \wedge \vdash^* \psi_o\frac{a}{\textcircled{t}_1},\frac{b}{\textcircled{t}_3}]$$

$$\rightarrow \underset{\sim}{R}(\psi_o\frac{a}{\textcircled{t}_2},\frac{b}{\textcircled{t}_3}) \wedge \vdash^* \psi_o\frac{a}{\textcircled{t}_2},\frac{b}{\textcircled{t}_3}$$

$$\rightarrow \underset{\sim}{R}(\psi\frac{a}{\textcircled{t}_2}).$$

Case V. ψ is $(\forall \underline{x})\psi_o\frac{b}{\underline{x}}$. Then

$$\underset{\sim}{R}(\psi\frac{a}{\textcircled{t}_1}) \rightarrow (\forall \textcircled{t}_3)\underset{\sim}{R}(\psi_o\frac{a}{\textcircled{t}_1},\frac{b}{\textcircled{t}_3})$$

$$\rightarrow (\forall \textcircled{t}_3)\underset{\sim}{R}(\psi_o\frac{a}{\textcircled{t}_2},\frac{b}{\textcircled{t}_3})$$

$$\rightarrow \underset{\sim}{R}(\psi\frac{a}{\textcircled{t}_2}).$$

This proves the Lemma and therewith (16). The proof of (17) is contained in that of (15). Thus (13) is proved and we need only prove (14) i.e.

$$\underset{\sim}{R}((\forall \underline{y})(\underline{y} \in \textcircled{t} \leftrightarrow \phi^{a,\vec{a}}_{\textcircled{y},\vec{t}}))$$

where

$$\textcircled{t}' \equiv c_{\psi,\underline{x}}, \quad \psi \equiv \phi^{a}_{\textcircled{t}}$$

and

$$\underline{x} \equiv \{ \textcircled{t}'' \mid \underset{\sim}{R}(\phi^{a,a}_{\textcircled{t},\textcircled{t}'}) \wedge \vdash^{*} \phi^{\vec{a},a}_{\textcircled{t},\textcircled{t}''} \} .$$

Fix $\textcircled{t}''$; we must show

$$\underset{\sim}{R}(\textcircled{t}'' \in \textcircled{t}') \wedge \vdash^{*} \textcircled{t}'' \in \textcircled{t}' \rightarrow \underset{\sim}{R}(\phi^{a,\vec{a}}_{\textcircled{t}'',\vec{\textcircled{t}}}) \tag{21}$$

and

$$\underset{\sim}{R}(\phi^{a,a}_{\textcircled{t}'',\textcircled{t}}) \wedge \vdash^{*} \phi^{a,\vec{a}}_{\textcircled{t}'',\vec{\textcircled{t}}} \rightarrow \underset{\sim}{R}(\textcircled{t}'' \in \textcircled{t}') . \tag{22}$$

Re (21). We have

$$\begin{aligned} \underset{\sim}{R}(\textcircled{t}'' \in \textcircled{t}') &\rightarrow \textcircled{t}'' \in \underline{x} \\ &\rightarrow \underset{\sim}{R}(\phi^{a,a}_{\textcircled{t}'',\textcircled{t}}) . \end{aligned}$$

Re (22). We have

$$\begin{aligned} \underset{\sim}{R}(\phi^{a,a}_{\textcircled{t}'',\textcircled{t}}) \wedge \vdash^{*} \phi^{a,\vec{a}}_{\textcircled{t}'',\vec{\textcircled{t}}} &\rightarrow \textcircled{t}'' \in \underline{x} \\ &\rightarrow \underset{\sim}{R}(\textcircled{t}'' \in \textcircled{t}) . \end{aligned}$$

This completes the proof of realizability of the existential axioms of the form (10), i.e. all of them except Infinity, to which we now turn.

The axiom of infinity in $Z^{\prime +}_{c}$ is

$$(\exists \underline{x})(0 \in \underline{x} \wedge (\forall \underline{y})(\underline{y} \in \underline{x} \rightarrow \underline{y} \cup \{\underline{y}\} \in \underline{x}))$$

(except that we do not have the abstraction term $y \cup \{y\}$). We introduce some definitions so that we can write this as a formula of $Z_c^{\rightarrow}$. The following are abbreviations in $Z_c^{\rightarrow}$, $Z_c^{\rightarrow\prime}$ and $Z_c^{\rightarrow *}$:

$$Z \equiv (\forall \underline{x})(\underline{x} \notin \underline{a})$$

$$\underline{a} = \underline{b} \equiv (\forall \underline{x})(\underline{x} \in \underline{a} \leftrightarrow \underline{x} \in \underline{b})$$

$$\text{Suc} \equiv (\forall \underline{x})(\underline{x} \in \underline{b} \leftrightarrow \underline{x} \in \underline{a} \vee \underline{x} = \underline{a})$$

$$\text{Cl} \equiv (\forall \underline{xyz})(Z\underline{x} \rightarrow \underline{x} \in \underline{a}) \wedge (\underline{y} \in \underline{a} \wedge \text{Suc}\,\frac{\underline{a}\;\underline{b}}{\underline{y},\underline{z}} \rightarrow \underline{z} \in \underline{a}$$

$$N \equiv (\forall \underline{x})(\text{Cl}\frac{\underline{a}}{\underline{x}} \rightarrow \underline{a} \in \underline{x}) .$$

The following are abbreviations in $Z_c^{\rightarrow\prime}$ only:

$$0 \equiv c_{\underline{a} \neq \underline{a}}$$

$$s(ⓣ) \equiv c_{\underline{a}} \in ⓣ \vee \underline{a} = ⓣ$$

$$\omega \equiv c_{N(a)} .$$

$Z\frac{\underline{a}}{\underline{x}}$, $\text{Suc}\frac{\underline{a},\underline{b}}{\underline{x},\underline{y}}$, $\text{Cl}\frac{\underline{a}}{\underline{x}}$, $N\frac{\underline{a}}{\underline{x}}$ say respectively that $\underline{x}$ is empty, that $\underline{y}$ is the successor of $\underline{x}$, that $\underline{x}$ contains 0 and is closed under successor, and that $\underline{x}$ is a natural number. 0 denotes the empty set, s(ⓣ) denotes the successor of the set denoted by ⓣ, and ω denotes the set of natural numbers. A <u>numeral</u> is one of the terms

$$0,\ s(0),\ ss(0),\ \ldots\ .$$

The axiom of infinity in $Z_c^{\rightarrow}$ is

$$(\exists \underline{x})(\forall \underline{yzw})[Z\frac{\underline{a}}{\underline{y}} \rightarrow \underline{y} \in \underline{x} \wedge (\underline{z} \in \underline{x} \wedge \text{Suc}\,\frac{\underline{a},\underline{b}}{\underline{z},\underline{w}} \rightarrow \underline{w} \in \underline{x})]. \tag{23}$$

We aim to show that (23) is realizable, i.e. to find a term

$$\overline{\omega} \equiv c_{N(a),\underline{x}}$$

of $Z_c^{\rightarrow *}$ such that for all terms ⓣ$_1$, ⓣ$_2$, ⓣ$_3$ of $Z_c^{\rightarrow *}$ we have

$$\underset{\sim}{R}(Z^{a}_{Ⓣ_1}) \wedge \vdash^{*} Z^{a}_{Ⓣ_1} \rightarrow \underset{\sim}{R}(Ⓣ_1 \in \bar{\omega}) \tag{24}$$

and

$$\underset{\sim}{R}(Ⓣ_2 \in \bar{\omega}) \wedge \vdash^{*} Ⓣ_2 \in \bar{\omega} \wedge \underset{\sim}{R}(Suc^{a,\ b}_{Ⓣ_2, Ⓣ_3})$$

$$\wedge \vdash^{*} Suc^{a\ b}_{Ⓣ_2, Ⓣ_3} \rightarrow \underset{\sim}{R}(Ⓣ_3 \in \bar{\omega}). \tag{25}$$

We set

$$\underline{X} \equiv \{Ⓣ_4 \in \Delta \mid \text{for some numeral } ⓝ, \vdash' Ⓣ_4 ⓝ\}.$$

Re (24). The hypothesis yields $\vdash' \bar{Ⓣ}_1 = 0$, $Ⓣ_1 \in \underline{X}$ and $\underset{\sim}{R}(Ⓣ_1\ \bar{\omega})$.

Re (25). By the first hypothesis $Ⓣ_2 \in \underline{X}$, i.e.

$$\vdash' Ⓣ_{\bar{2}} = ⓝ$$

for some numeral ⓝ .

By the fourth hypothesis $\vdash' Suc^{a\ ,\ b}_{Ⓣ_{\bar{2}}, Ⓣ_{\bar{3}}}$, i.e.

$$\vdash' (\forall \underline{x})(\underline{x} \in Ⓣ_{\bar{3}} \leftrightarrow \underline{x} \in Ⓣ_{\bar{2}} \vee \underline{x} = Ⓣ_{\bar{2}}).$$

Hence $\vdash' Ⓣ_{\bar{3}} = s(ⓝ)$, $Ⓣ_3 \in \underline{X}$ and $\underset{\sim}{R}(Ⓣ_3 \in \bar{\omega})$ q.e.d.

§7 Realizability of TI(∈)

The axiom schema TI(∈) (§1) is

$$(\forall \underline{x})[(\forall \underline{y})(\underline{y} \in \underline{x} \rightarrow \phi^{a}_{\underline{y}}) \rightarrow \phi^{a}_{\underline{x}}] \rightarrow (\forall \underline{x})\phi^{a}_{\underline{x}}.$$

We have to prove that this is realizable, i.e. that

$$\underset{\sim}{R}((\forall \underline{x})[(\forall \underline{y})(\underline{y} \in \underline{x} \rightarrow \phi^{a}_{\underline{y}}) \rightarrow \phi^{a}_{\underline{x}}]) \tag{26}$$

and

$$\vdash^{*} (\forall \underline{x})[(\forall \underline{y})(\underline{y} \in \underline{x} \rightarrow \phi^{a}_{\underline{y}}) \rightarrow \phi^{a}_{\underline{x}}] \tag{27}$$

imply $\underset{\sim}{R}(\phi^{a}_{Ⓣ})$ for every constant Ⓣ of T^{*}.

Assume then (26) - (27) and use induction on $\rho\,Ⓣ^-$. The inductive hypothesis is

$$\rho\,Ⓣ'^- < \rho\,Ⓣ^- \rightarrow \underset{\sim}{R}(\phi\frac{a}{Ⓣ'}),$$

and we aim to prove $\underset{\sim}{R}(\phi\frac{a}{Ⓣ})$. By (26)

$$\underset{\sim}{R}[(\forall \underline{y})(\underline{y} \in Ⓣ \rightarrow \phi\frac{a}{\underline{y}}) \rightarrow \phi\frac{a}{Ⓣ}]$$

i.e.

$$\underset{\sim}{R}((\forall \underline{y})(\underline{y} \in Ⓣ \rightarrow \phi\frac{a}{\underline{y}}) \wedge \vdash^* (\forall \underline{y})(\underline{y} \in Ⓣ \rightarrow \phi\frac{a}{\underline{y}}) \rightarrow \underset{\sim}{R}(\phi\frac{a}{Ⓣ}). \tag{28}$$

By (26) and $\vdash^* TI(\in)$, $\vdash^* (\forall \underline{x})\phi\frac{a}{\underline{x}}$ and a fortiori

$\vdash^* (\forall \underline{y})(\underline{y} \in Ⓣ \rightarrow \phi\frac{a}{\underline{y}})$. By (28)

$$\underset{\sim}{R}((\forall \underline{y})(\underline{y} \in Ⓣ \rightarrow \phi\frac{a}{\underline{y}})) \rightarrow \underset{\sim}{R}(\phi\frac{a}{t}). \tag{29}$$

Let $Ⓣ'$ be any constant satisfying

$$\vdash^* Ⓣ' \in Ⓣ.$$

Then $\vdash Ⓣ'^- \in Ⓣ^-$, whence $\rho\,Ⓣ'^- < \rho\,Ⓣ^-$ and by the inductive hypothesis $\underset{\sim}{R}(\phi\frac{a}{Ⓣ'})$. Thus

$$(\forall Ⓣ')(\vdash^* Ⓣ' \in Ⓣ \rightarrow \underset{\sim}{R}(\phi\frac{a}{Ⓣ'}))$$

and a fortiori

$$(\forall Ⓣ')(\underset{\sim}{R}\,Ⓣ' \in Ⓣ \wedge \vdash^* Ⓣ' \in Ⓣ \rightarrow \underset{\sim}{R}(\phi\frac{a}{Ⓣ'}))$$

i.e. $\underset{\sim}{R}(\forall \underline{y})(\underline{y} \in Ⓣ \rightarrow \phi\frac{a}{\underline{y}})$. By (29) $\underset{\sim}{R}(\phi\frac{a}{Ⓣ})$, q.e.d.

This proves the Main Theorem and therewith Corollaries 1-4 of §3.

§8 Consequences of realizability concluded

We prove two more corollaries. Firstly the so-called evaluation-property.

COROLLARY 5 If $Z' \vdash \textcircled{c} \in \omega$, then for some numeral $\textcircled{n}$, $Z' \vdash \textcircled{c} = \textcircled{n}$.

Proof. (We use throughout the notation of §6; also some of the results therein for $\vec{c}$ empty.) Let c be c_{ϕ}, and assume for the moment that ϕ is a formula of Z (i.e. no constants appear in ϕ). For any set $\underline{Y}$ for which $c_{\phi,\underline{Y}} \in \Delta$, we have

$$\vdash^* (\forall \underline{x})(\underline{x} \in c_{\phi,\underline{Y}} \leftrightarrow \phi\tfrac{a}{\underline{x}}) \tag{30}$$

and for suitable such $\underline{Y}$, by the argument of §6, we have

$$\underset{\sim}{R}(\forall \underline{x})(\underline{x} \in c_{\phi,\underline{Y}} \leftrightarrow \phi\tfrac{a}{\underline{x}}). \tag{31}$$

We also have

$$(\forall \underline{y})[(\forall \underline{x})(\underline{x} \in \underline{y} \leftrightarrow \phi\tfrac{a}{\underline{x}}) \to N\tfrac{a}{\underline{y}}].$$

Since $\underset{\sim}{R}(Z)$ this yields

$$\underset{\sim}{R}(\forall \underline{y})[(\forall \underline{x})(\underline{x} \in \underline{y} \leftrightarrow \phi\tfrac{a}{\underline{x}}) \to N\tfrac{a}{\underline{y}}];$$

i.e. for every $\textcircled{t} \in \Delta$

$$\{\underset{\sim}{R}(\forall \underline{x})(\underline{x} \in \textcircled{t} \leftrightarrow \phi\tfrac{a}{\underline{x}}) \wedge \vdash^* (\forall \underline{x})(\underline{x} \in \textcircled{t} \leftrightarrow \phi\tfrac{a}{\underline{x}})\}$$
$$\to \underset{\sim}{R}N\tfrac{a}{\textcircled{t}} . \tag{32}$$

(30) - (32) now yield $\underset{\sim}{R}(N\tfrac{a}{c\phi,\underline{Y}})$, i.e.

$$(\forall \textcircled{t}_1)\underset{\sim}{R}(Cl\tfrac{a}{\textcircled{t}_1} \to c_{\phi,Y} \in \textcircled{t}_1).$$

In particular

$$\underset{\sim}{R}(Cl\tfrac{a}{\overline{\omega}}) \wedge \vdash^* Cl\tfrac{a}{\overline{\omega}} \to \underset{\sim}{R}(c_{\phi,\underline{Y}} \in \overline{\omega}).$$

But evidently $\vdash' Cl\tfrac{a}{\overline{\omega}}$, so $\vdash^* Cl\tfrac{a}{\overline{\omega}}$. Also by the definition of $\overline{\omega}$ $\underset{\sim}{R}(c_{\phi,\underline{Y}} \in \overline{\omega})$ implies $\vdash' c_{\phi} = \textcircled{n}$ for a numeral $\textcircled{n}$. So all we need is $\underset{\sim}{R}Cl\tfrac{a}{\overline{\omega}}$, i.e. for all $\textcircled{t}_1$, $\textcircled{t}_2$, $\textcircled{t}_3 \in \Delta$ we must show that

$$\underset{\sim}{R}(Z^{\underline{a}}_{ⓣ_1}) \rightarrow ⓣ_1 \in \bar{\omega} \tag{33}$$

and that

$$\underset{\sim}{R}(ⓣ_2 \in \bar{\omega}) \wedge \vdash^* ⓣ_2 \in \bar{\omega} \wedge \underset{\sim}{R}(suc^{\underline{a}}_{ⓣ_2}{}^{,\underline{b}}_{ⓣ_3}) \wedge \vdash^* Suc^{\underline{a}}_{ⓣ_2}{}^{\underline{b}}_{ⓣ_3} \rightarrow \underset{\sim}{R}(ⓣ_3 \in \bar{\omega}). \tag{34}$$

(Warning! There is a temptation to infer $\underset{\sim}{R}(Cl^{\underline{a}}_{\underline{\omega}})$ directly from $\vdash^* Cl^{\underline{a}}_{\underline{\omega}}$, but this is invalid: $\vdash^*\psi$ implies $\underset{\sim}{R}(\psi)$ in general only if ψ contains no constants.)

Re (33). This amounts to

$$\underset{\sim}{R}(Z^{\underline{a}}_{ⓣ_1}) \wedge \vdash^* Z^{\underline{a}}_{ⓣ_1} \rightarrow \underset{\sim}{R}(ⓣ_2 \in \bar{\omega})$$

which is obvious since $(\vdash^* Z^{\underline{a}}_{ⓣ_1}) \rightarrow (\vdash' Z^{\underline{a}}_{ⓣ_{\bar{1}}}) \rightarrow (\vdash' ⓣ_{\bar{1}} = 0) \rightarrow$ $ⓣ_1 \in \underline{X} \rightarrow \underset{\sim}{R}(ⓣ_1 \in \bar{\omega})$, where $\underline{X}$ is as in $\S\underset{\sim}{6}$.

Re (34). The first hypothesis yields $ⓣ_2 \in \underline{X}$ and hence

$$\vdash' ⓣ_{\bar{2}} = ⓜ$$

for some numeral ⓜ. The fourth hypothesis yields

$$\vdash' Suc^{\underline{a}}_{ⓣ_2}{}^{,\underline{b}}_{ⓣ_{\bar{3}}}$$

i.e. $\vdash' ⓣ_{\bar{3}} = s(ⓣ_{\bar{2}}) = s(ⓜ)$. Hence $ⓣ_3 \in \underline{X}$ and $\underset{\sim}{R}(ⓣ'_3 \in \bar{\omega})$.

Thus (33) - (34) are proved: hence $\underset{\sim}{R}(CL^{\underline{a}}_{\omega})$ and we are through.

If $\emptyset$ contains constants, let $\emptyset'$ be a formula of Z such that $\vdash' \emptyset \leftrightarrow \emptyset'$. Then $\vdash' c_{\emptyset} = c_{\emptyset'} \in \omega$ and so $\vdash' c_{\emptyset} = ⓝ$ for some ⓝ.

<u>COROLLARY 6</u> If $\vdash' (\exists \underline{x} \in \omega)\emptyset^{\underline{a}}_{\underline{x}}$(closed), then $\vdash' \emptyset^{\underline{a}}_{ⓝ}$ for a numeral ⓝ.

<u>Proof</u> by Corr. 3 and 5.

Our final result is

<u>COROLLARY 7</u> If $Z' \vdash (\forall \underline{x}:\omega \to \omega)(\exists! \underline{y}:\omega \to \omega)\phi^{\underline{a},\underline{b}}_{\underline{x},\underline{y}}$, there is a continuous functional $\Phi:(\omega \to \omega) \to (\omega \to \omega)$ such that

$$(\forall \underline{x}:\omega \to \omega)\phi^{\underline{a},\underline{b}}_{\underline{x},\Phi(x)} \tag{35}$$

is true.

<u>Proof</u> (in outline). The idea is as follows: There evidently exists a functional satisfying (35). We require to show that, for each $\underline{n} \in \omega$, $(\Phi\chi)(\underline{n})$ depends on only a finite number of values of χ. We form a system $Z(\chi)$ by adjoining to Z the diagram of χ and certain other axioms sufficient to prove

$$\chi:\omega \to \omega.$$

We construct $Z'(\chi)$ and $Z^*(\chi)$ roughly as Z' and Z^* were constructed from Z and show that for cleverly-chosen $\underset{\sim}{M}_\chi$ we have

$$\underset{\sim}{R}(\underset{\sim}{M}_\chi, Z^*(\chi), Z(\chi)).$$

From this it follows that $Z(\chi)$ has the existence-and evaluation-properties. Hence for some numeral $\underset{\sim}{m}$ (henceforth we write $\underset{\sim}{m}$ for $s^{\underline{m}}(0)$) we have

$$Z'(\chi) \vdash (\exists \underline{g}:\omega\to\omega)[\phi^{\underline{a},\underline{b}}_{Ⓕ,\underline{g}} \wedge \underline{g}(\underset{\sim}{n}) = \underline{m}] \tag{36}$$

where

$$Ⓕ \equiv {}^{c}(\exists \underline{x}\underline{y})(\underline{a}=\langle \underline{x},\underline{y}\rangle \wedge \underline{F}(\underline{x},\underline{y})).$$

But this formula was deduced using the values of χ for only finitely many arguments, and hence any χ' which agrees with it on those arguments satisfies

$$(\Phi\chi')(\underline{n}) = \underline{m}$$

q.e.d.[3]

We proceed to give the details of the construction of $Z(\chi)$, $Z'(\chi)$ and $Z^*(\chi)$ and of the proof that $Z'(\chi)$ has the existence-and evaluation-properties. Some computations are left to the reader; after the preceding proofs he should have no difficulty filling them in.

$Z(\chi)$ is formed from Z by adding a binary relation-symbol $\underline{F}$ with the following axioms:

$$\text{F1.}\quad \underline{F}(\underset{\sim}{m},\underset{\sim}{n}),\ \text{if}\ \chi(\underline{m}) = \underline{n},$$

$$\text{F2.}\quad \underline{F}(\underline{a},\underline{b}) \to \underline{a} \in \omega \wedge \underline{b} \in \omega,$$

$$\text{F3.}\quad (\forall \underline{x} \in \omega)(\exists \underline{y} \in \omega)\underline{F}(\underline{x},\underline{y}),$$

$$\text{F4.}\quad \underline{F}(\underline{a},\underline{b}) \wedge \underline{F}(\underline{a},\underline{c}) \to \underline{b} = \underline{c}.$$

Of course the schemata of Z are now construed as applying to formulas in the extended vocabulary. Strictly speaking F1 - F3 should be written without comprehension-terms as follows:

$$\text{F1.}\quad (Z_{\underline{m}})\frac{a}{\underline{b}} \wedge (Z_{\underline{n}})\frac{a}{\underline{c}} \to \underline{F}(\underline{b},\underline{c}),\quad \text{if}\ \chi(\underline{m}) = \underline{n},$$

$$\text{F2.}\quad \underline{F}(\underline{b},\underline{c}) \to N\frac{a}{\underline{b}} \wedge N\frac{a}{\underline{c}},$$

$$\text{F3.}\quad (\forall \underline{x})(N\frac{a}{\underline{x}} \to (\exists \underline{y})(N\frac{a}{\underline{y}} \wedge \underline{F}(\underline{x},\underline{y})),$$

where N is defined as in §$\underset{\sim}{6}$, and where

$$Z_o \equiv Z,$$

$$Z_{\underline{n}+1} \equiv (\exists \underline{x})((Z_{\underline{n}})\frac{a}{\underline{x}} \wedge \mathrm{Suc}\frac{a,b}{\underline{x},\underline{a}}).$$

$Z'(\chi)$ and $Z^*(\chi)$ are formed from $Z(\chi)$ as Z' and Z^* respectively were formed from Z, except that $Z^*(\chi)$ has additional axioms

$$\underline{F}(ⓣ_1, ⓣ_2)$$

[3] We really need the evaluation- as well as the existence-property for $Z'(\chi)$; if we had a term ⓣ instead of a numeral m in (36) it could a priori have a different interpretation in $Z'(\chi)$ and $Z'(\chi')$.

wherever $\vdash'_\chi \underline{F}(\textcircled{t}_1^-, \textcircled{t}_2^-)$ (i.e. $Z'(\chi) \vdash \underline{F}(\textcircled{t}_1^-, \textcircled{t}_2^-)$).

Realizability of formulas of $Z^*(\chi)$ is defined as for Z^*, except that a formula $\underline{F}(\textcircled{t}_1, \textcircled{t}_2)$ is defined to be a realizable iff (a) there are numerals $\underset{\sim}{m}$, $\underset{\sim}{n}$ of $Z'(\chi)$ such that

$$\vdash'_\chi \textcircled{t}_1^- = \underset{\sim}{m}, \quad \vdash'_\chi \textcircled{t}_2^- = \underset{\sim}{n}, \ \chi(\underline{m}) = \underline{n}$$

and (b) $\textcircled{t}_1$ and $\textcircled{t}_2$ are <u>standard</u> <u>numerals</u> of $Z^*(\chi)$, i.e. $\textcircled{t}_1^+ = \underline{X}_{\underline{m}}$, $\textcircled{t}_2^+ = \underline{X}_{\underline{n}}$ where

$$\underline{X}_0 \equiv 0$$

$$\underline{X}_{n+1} \equiv \underline{X}_{\underline{n}} \cup \{\textcircled{t} \mid \textcircled{t}^+ = \underline{X}_{\underline{n}} \wedge \vdash'_\chi \textcircled{t}^- = \underline{n}\}.$$

Now we prove $\underset{\sim}{R}(\theta)$ for all axioms θ of $Z(\chi)$. The proof for the set-theoretic axioms is exactly the same as in §§$\underset{\sim}{5}$ - $\underset{\sim}{7}$, except that in the Lemma of §$\underset{\sim}{6}$ we have to consider three more cases, i.e. $\psi \equiv \underline{F}(\underline{a}, \textcircled{t})$, $\psi \equiv \underline{F}(\textcircled{t}, \underline{a})$, $\psi \equiv \underline{F}(\underline{a},\underline{a})$. We leave these to the reader. Realizability of F1 - F4 is easy after the

<u>LEMMA</u> ($\textcircled{t}$ is a standard numeral) $\leftrightarrow$ $(\underset{\sim}{R}(N^{\underline{a}}_{\textcircled{t}}))$ $\leftrightarrow (\exists \underline{n})(\underset{\sim}{R}((Z_{\underline{n}})^{\underline{a}}_{t}) \wedge \vdash^*_\chi (Z_{\underline{n}})^{\underline{a}}_{\textcircled{t}})$.

By the soundness theorem, $\underset{\sim}{R}(Z(\chi))$. The existence- and evaluation-properties for $Z'(\chi)$ follow from this by the arguments used to prove Corollaries 1 and 5 respectively above. Alternatively we can prove the evaluation-property as follows: If $\vdash'_\chi \textcircled{c} \in \omega$ where $\textcircled{c} \equiv c_\phi$, and where without loss of generality we can assume that ϕ contains no constants, then the formula

$$(\exists \underline{x})(N^{\underline{a}}_{\underline{x}} \wedge (\forall \underline{y})(\underline{y} \quad \underline{x} \leftrightarrow \phi^{\underline{a}}_{\underline{y}}))$$

is provable in $Z(\chi)$ and hence realizable. Thus for some term

Ⓣ of $Z^*(\chi)$, $N^{\underline{a}}_{Ⓣ}$ is <u>realizable</u> (whence by the Lemma $\vdash_\chi Ⓣ^- = \underset{\sim}{n}$ for some $\underline{n}$); and $(\forall \underline{y})(\underline{y} \in Ⓣ \leftrightarrow \phi^{\underline{a}}_{\underline{y}}$ is <u>provable</u> (whence $\vdash_\chi Ⓣ^- =$ Ⓒ). Hence $\vdash_\chi$ Ⓒ $= \underset{\sim}{n}$.[4] We now have everything we need to complete the proof of the Corollary.

---

4 Since the Lemma holds for Z as well as $Z(\chi)$, this gives another better proof of Corollary 5 above.

APPENDIX - PROPOSED RESEARCH PROBLEMS

I. Investigate the relation of our results to recursiveness and "Church's Rule". Specifically: if $Z' \vdash (\forall \underline{x} \in \omega)(\exists \underline{y} \in \omega)\phi\frac{a,b}{\underline{x},\underline{y}}$, there evidently exists a recursive function $\underline{f}$ such that $(\forall \underline{x} \in \omega)\phi\frac{a,b}{\underline{x},\underline{f}(\underline{x})}$ is true. Does this hold reading "provably recursive" for "recursive" and "provable" for "true"?

II. Investigate the relation of our results to continuity and "Brouwer's Rule". Specifically: if $Z' \vdash (\forall \underline{x}:\omega\to\omega)(\exists !\underline{y}:\omega\to\omega)\phi\frac{a,b}{\underline{x},\underline{y}}$, Corollary 7 says that there exists a continuous functional $\Phi:(\omega\to\omega) \to (\omega\to\omega)$ such that $(\forall \underline{x}:\omega\to\omega)\phi\frac{a,b}{\underline{x},\Phi(\underline{x})}$ is true. Does this hold reading "provably continuous" for "continuous" and "provable" for "true"? Does it hold reading $\exists$ for $\exists !$?

III. Investigate the proof-theoretic strength of intuitionistic Zermelo-Frankel. Specifically: it is clearly intermediate between classical Zermelo-Frankel and classical Zermelo. Is it equiconsistent with one of these, or does it lie strictly between them?

IV. Investigate the extent to which our results can be made constructive. Specifically: is there an effective method whereby given a proof of $(\exists \underline{x})\phi\frac{a}{\underline{x}}$ in Z' we can transform it into a proof of $\phi\frac{a}{\copyright}$ for some constant $\copyright$?

It is evident that the answer to the corresponding problem for formulas $(\exists \underline{x} \in \omega)\phi\frac{a}{\underline{x}}$ is negative. For let f be a recursive function such that if $\underline{n}$ is the Gödel-number of a proof of

$(\exists \underline{x} \in \omega)\phi\frac{a}{\underline{x}}$, then $Z' \vdash \phi\frac{a}{\underset{\sim}{m}}$, where $\underline{m} = f(\underline{n})$. In terms of f we can define a recursive function g which enumerates the provably recursive functions of Z'. If f were provably recursive g would be provably recursive too, contradiction. Thus the fact that the search for m terminates cannot be proved using all the resources of Z', and so there is no reasonable sense of 'construdence' in which the proof of Corollary 6 can be made constructive. On the other hand current work suggests that we can give a constructive, perhaps even a primitive recursive version of Corollary 1. If this turns out to be true the situation will be analogous to that with the Gödel Dialectica translation of Heyting arithmetic: if all we need, given a proof of $(\exists \underline{x})\phi\frac{a}{\underline{x}}$ in HA, is to find a term ⓣ satisfying $\phi\frac{a}{ⓣ}$, we can get one primitive recursively, while if we want a numeral we need ordinal recursive functions of order ϵ_0.

REFERENCE

(1) H. FRIEDMAN, Some applications of Kleene's methods to intuitionistic systems, these Proceedings,

Department of Mathematics
SUNY at Buffalo
BUFFALO, N.Y. / USA

VI. QUELQUES RESULTATS SUR LES INTERPRETATIONS FONCTIONNELLES

J.Y. Girard

Considérons les trois systèmes fonctionnels suivants, considérés comme des systèmes de calcul de termes clos :

le système T de Gödel (voir [Gd])

le système $\sum_4$ de Spector (voir [Sp])

le système F de l'auteur (voir [Gi])

En un certain sens, F et $\sum_4$ sont des"extensions"de T. Une grande partie de ce papier étant consacrée à l'étude d'autres "extensions", il convient de préciser cette notion.

Si on comprend un système fonctionnel comme la donnée d'un ensemble de termes clos et d'une égalité décidable entre ces termes, nous dirons que F est, par exemple une extension de T dans le sens que si a et b sont dans T, ils sont dans F, et que, si de plus ils sont égaux dans F, ils le sont dans T, et réciproquement. Avec une telle conception d'un système fonctionnel, une seule notion de validité est possible :

A(x) est valide ssi pour tout b clos, $A(b)=\bar{o}$

Il va de soit, que si A(x) est valide dans T, il peut ne pas l'être dans F, puisque l'ensemble des b sur lequel s'effectue la verification est en général plus grand. En ce sens, F n'est pas une extension de T.

La position plus généralement adoptée est la suivante : un système fonctionnel est un système logique formel dans lequel la validité est établie au moyen d'un certain nombre de règles. Bien entendu, la notion de validité précédente, que nous appellerons <u>validité calculatoire</u>, est un modèle de cette nouvelle notion, que nous appellerons <u>validité logique</u>. Par rapport a la validité logique, nous dirons de F et T (supposés munis d'axiomes et de règles) que la premier est une extension du second dans le sens que toutes les règles de T sont valides dans F. Bien que nous n'ayons jamais formulé explicitement de systèmes de règles pour F, il serait très facile de le faire, en faisant une liste des propriétés que nous utilisons en nous servant de F. Nous pourrions faire la même chose pour les nouveaux systèmes que nous allons introduire, et c'est donc en ce sens que nous devrons interpréter la notion d'extension.

Cependant, à notre avis, cette notion de validité logique, bien que supérieure à la validité **calculatoire** de nombreux points de vue, fait perdre une grande partie de son intérêt à la construction précise des fcl de T, par exemple : en effet, HRO (voir [Tr]) est un modèle de T considéré comme système de règles.

Nous montrerons dans un autre papier qu'il est possible de modifier les règles de calcul des systèmes de termes de manière à ce que :

1- La notion d'égalité s'étende aux termes ayant des variables libres.

2- La notion d'égalité tienne compte de certaines formes de validité **calculatoire** (Par exemple $(x \vee \neg x)=\bar{o}$)

3- L' **égalité est décidable.**

La notion de validité correspondante sera appelée validité intentionnelle faible. Par definition VIF $\models A(x)$ ssi $A(x)=\bar{o}$.

La validité intentionnelle faible est un modèle de tous les systèmes d'axiomes usuels correspondant au système fonctionnel sur lequel elle est définie, sauf ce qui a trait à l'induction.

Pour tenir compte de l'induction, on est amené à definir la validité intentionnelle. Par definition, A(x) est VI si on peut trouver B(z,y), z et y étant de type o, et t tel que :

$$\text{VIF} \models B(tx, \bar{o}) \rightarrow A(x)$$

$$\text{VIF} \models B(\bar{o},y)$$

$$\text{VIF} \models B(z, Sy) \rightarrow B(Sz,y)$$

La VI est récursivement énumérable non récursive, et satisfait à tous les axiomes de la validité logique. Elle permet de plus une réduction théoriquement utilisable du problème de la **prouvabilité** des formules Σ_1^0 de l'arithmétique.

Cette longue digression sur la validité était rendue nécessaire par nombre de critiques pertinentes de Kreisel sur le flou de la notion de système fonctionnel, de validité, d'extension, utilisées jusque là par l'auteur. Il n'est malheureusement pas possible de décrire ici ce qu'est exactement la VIF, car cette notion, bien que très simple, entraine de trop nombreux boulversements. Il importe cependant de souligner que tout ce qui suit peut être traité intégralemen sous l'angle de la validité logique, ou sous celui de la validité intentionnelle.

Nous montrons dans la première section que F peut être étendu de manière à interpreter la théorie des types par un systèms fonctionnel E. E est très similaire à F quant à son mode de fonctionnement.

Dans les sections 2 et 3, nous intéressons au principe de Markov et à la thèse de Church. Remarquons que ces deux principes ont en commun d'être en relation avec la complétude du calcul dés prédica de Heyting, à savoir que

1- La thèse de Church implique l'incomplétude

2- La négation du principe de Markov, pour A récursif primitif, implique l'incomplétude.

Dans la section 2, nous établissons la clôture de la théorie des types par rapport à la règle de Markov

$$\frac{\forall x^{0}(Ax \vee \neg Ax) \qquad \neg\neg\exists x^{0} Ax}{\exists x^{0} Ax} \quad \text{pour A quelconque.}$$

Bien entendu, le principe de Markov lui-même n'est pas démontrable, pas plus que la théorie des types n'est close par rapport à la règle

$$\frac{\forall x^{0}(Ax \vee \neg Ax) \rightarrow \neg\neg\exists x^{0} Ax}{\forall x^{0}(Ax \vee \neg Ax) \rightarrow \exists x^{0} Ax} \quad \text{(voir [Kr3], p. 122)}$$

Le résultat de clôture est obtenu combinaison de l'interprétation fonctionnelle E, de l'élimination des coupures, et du principe de réflexion pour les sous-systèmes finis de l'analyse.

Dans la section 3, nous étendons les systemes T,F,E, en des systèmes T^+,F^+,E^+, qui, en plus des axiomes déjà interprétés par T,F,E, interprètent la formulation mathématique de la thèse de Church : toute fonction est récursive, et même plus : toute fonction est représentable dans T^+,F^+,E^+. On en déduit que toute fonction récursive est dans E^+ et on pourrait meme montrer l'interprétation de "il est faux que tous les termes de E^+ soient calculables". En effet, on peut s'en assurer simplement comme suit : soit f de type (o,o → o); on veut trouver e tel que T'(e,e,fee) . Il suffit de poser e= GD λx S(U'(f(x,x))) .

Dans la section 4, nous étendons les systèmes T,F, et E en des systèmes TB,FB,EB, TB etant Σ_4. Ces systèmes sont des systèmes avec Bar-récursion à tous les types du système, y compris les types non clos. Ce qui montre la compatibilité totale de la Bar-récursion et des diverses formes de stratification et d'extraction. La bar-récursion permet d'interpréter la négation de la thèse de Church. Ceci montre que ce principe est indécidable par rapport à T,F,E, pour les notions de validité logique attachées à ces systèmes (et aussi pour la VI). Un argument de continuité nous montre d'ailleurs que , par exemple EB^+ est contradictoire...

Ce travail doit beaucoup aux critiques et suggestions du Professeur Kreisel. Notamment, les sections 2 et 3 sont des réponses à des questions formulées dans sa correspondance.

Je remercie également M.Reznikoff qui a vérifié de près les détails des démonstrations.

1. INTERPRETATION DE LA THEORIE DES TYPES

Nous allons construire des systèmes fonctionnels Z et E, qui sont très proches des systèmes Y et F; aussi n'indiquons-nous dans ce qui suit que les différences entre nos nouveaux systèmes et les systèmes correspondants de [Gi].

1.1. Types

Il est nécessaire d'introduire la notion d'opérateur ; pour chaque entier i, on définira des opérateurs de rang i; les opérateurs de rang o étant les types.

Les types et opérateurs sont donnés par la définiation inductive :

1+ Pour chaque i, il y a une infinité d'opérateurs indétermines de rang i. Dans ce qui suit on notera IO l'expression "opérateur indéterminé" et IO(i) l'expression "opérateur indéterminé de rang i".

2+ Si U_1,U_2,U_3 sont des opérateurs de rangs respectifs i+1,i et i, $U_1U_2U_3$ est un opérateur de rang o, c'est à dire un type.

3+ Si θ et θ' sont des IO(i), si V est un opérateur de rang o, $\lambda\theta\theta'V$ est un opérateur de rang i+1.

4+ Les schémas 1.3.4. de [Gi] p. 66. pour les types.

5+ Les schémas 5 et 6 de [Gi] p. 66. en remplaçant "indeterminée" par IO(i).

Un type sera dit normal s'il ne contient pas d'expression de la forme $\lambda\theta\theta'U\ VV'$; a l'aide de la règle de conversion $\lambda\theta\theta'U\ VV' \Rightarrow U[V,V']$ on peut remplacer tout type par un type normal, sa forme normale. Deux types seront égaux s'ils ont même forme normale, et nous n'aurons pas lieu de les distinguer l'un de l'autre. Remarquons que l'égalité entre types ainsi définie est décidable.

1.2. Système fonctionnel

Le système Z est défini exactement comme Y, dans [Gi] pp. 67-68. Il y a lieu cependant d'apporter les modifications suivantes aux schémas (vii) a (x) :

(vii) et (x) : remplacer "α une indéterminée" par θ une IO".

(viii) et (ix) : remplacer la lettre α par θ, et la condition "s est un type" par "U est un opérateur". Ainsi on forme a{U}, et la constante $I_{\sigma\theta r, U}$.

1.3. On définit E comme le sous-système de Z formé des termes sans variable ni indéterminée. On peut reprendre pour E et Z les définitions de la page 69.

1.4. Réduction

1.4.1. On peut recopier les axiomes de la réduction immédiate de la page 70, avec la modification évidente qui revient à remplacer les α par des θ, et les types dans les extractions et les seconds indices de I par des lettres désignant des opérateurs. Il y a cependant un schéma que l'on ne peut pas transcrire tout à fait mécaniquement : il s'agit de (1) : pour pouvoir écrire $ST\theta u(O_{\sigma\theta r}) \Rightarrow u(O_r)^{O}_{\theta}$, il faut avoir défini un o pour tous les rangs; pour le rang o, c'est chose faite, pour le rang i+1, on peut le prendre égal à $\lambda\theta\theta'o$.

1.4.2. Les définitions 3-7 des pages 70-71 se transcrivent sans difficulte pour E.

1.5. Théorème de reduction : Tout terme de E a une forme normale unique. Nous ne démontrerons pas ce résultat, étant donné que la démonstration est pratiquement faite dans [Gi]. Il suffit de remarquer que E est en fait le "croisement" de F et de $H^{\omega}F$, et qu'il n'y a qu'à juxtaposer les démonstrations de réductibilité pour ces deux systèmes. En particulier le théorème de réduction est (localement) démontrable dans la théorie des types.

** Voir la fin du Chapitre 1.

1.6. Interprétation de la théorie des types

Nous considérons le formalisme de la page 83, où les variables de type n sont notées X_n...

On définit par induction les types $\tau_n(\theta,\theta')$, on θ et θ' sont des IO de même rang n :

$\tau_o(\theta,\theta') = (\theta,\theta',o\to o)$

$\tau_{n+1}(\theta,\theta') = \delta\eta\zeta(\theta\eta\zeta, \theta\eta\zeta, \tau_n(\eta,\zeta)\to o)$

A la variable X_{n+1}, on attachera injectivement deux indéterminées distinctes θ et θ' de rang n, et une variable G de type $\tau_n(\theta,\theta')$.

A la variable X_o, on attachera une variable de type o.

Remarquons que ces définitions coïncident pour X_o et X_1 avec les définitions de [Gi] p. 84.

Les interprétations des enoncés ne renfermant pas de variable de type supérieur à 1, et des termes d'abstraction $\lambda X_o A$ se fait comme pp.84-85

<u>Interpretation des énconcés atomiques</u> :

Soit $T_{n+1} \in X_{n+2}$ un énoncé atomique; soit T* l'interprétation de T; alors T* est un terme de type $\tau_n(U,V)$, où U et V sont des opérateurs de rang n; si θ,θ',G sont associés à X_{n+2}, l'interprétation de l'énoncé est alors

$$\exists X_{\theta UV} \ \forall Y_{\theta' UV} \ G\{U,V\}(X,Y,T^*)$$

<u>Interprétation des termes d'abstraction</u>

Soit $A[X_{n+1}]$ un énoncé, et $\exists X_{r[\theta,\theta']} \ \forall Y_{s[\theta,\theta']} \ A^*[X,Y,G,\underline{Z}]$ son interprétation, où θ,θ',G sont associés à X_{n+1}. Alors $(\lambda X_{n+1} A\ X_{n+1})^*$ est défini comme $DT\theta\theta' \quad \lambda XYGA[X,Y,G,\underline{Z}]$. La stratification est légitime, puisque les variables $\underline{Z}$ ne renferment ni θ, ni θ'.

<u>Interprétation des quantifications</u>

Soit $A[X_{n+1}]$, un énoncé, $\exists X_{r[\theta,\theta']} \ \forall Y_{s[\theta,\theta']} \ A^*[X,Y,G,\underline{Z}]$ son interprétation. Les interprétations des quantifications universelle et existentielle par rapport à X_{n+1} sur A sont respectivement:

$$\exists X \quad \forall Y \; (ST\theta\theta'(\lambda z A^* [X\{\theta,\theta'\}(\pi^2 z),\ \pi^1 z,\ \pi^2 z, \underline{Z}\ ]\ ))(Y)$$

$$\exists X \quad \forall Y \; (ST\theta\theta'(\lambda z A^* \ \ \pi^1 z, Y\{\theta,\theta'\}(\pi^2 z),\ \pi^2 z, \underline{Z}\ \ ))\ (X)$$

La forme des interprétations étant, inchangée, ils s'ensuit que les schémas de la page 86 sont interprétables pour tous les types finis, pourvu que l'on puisse s'assurer de l'identité entre (A[T])* et A*[T*] .

Pour vérifier ceci, le plus simple est de supposer que dans la théorie des types, nous admettons des formules atomiques de la forme $T \in T'$, où T et T' sont deux termes quelconques, l'égalité entre deux enoncés étant définie par réduction à une même forme normale, c.a.d. un énoncé où les formules atomiques sont de la forme $T \in X_{n+1}$, au moyen du schéma ; $T \in \lambda X_n A\ X_n \Rightarrow A[T]$. Il faut, bien entendu, donner une interprétation pour tous les nouveaux énoncés intreduits; il suffit en fait de poser $(T \in T')^* = \exists X \forall Y\ T'^*\{U,V\}\ (X,Y,T^*)$, si T* est de type $\tau_n\ (U,V)$.

Si T' est $\lambda X_{n+1} B [X_{n+1}]$, il nous faut vérifier que (B [T']* et $(T \in T')^*$ sont identiques; mais le premier est de la forme $\exists X \quad \forall Y\ B^*[X,Y,T^*,\underline{Z}]$, et le second égal à $\exists X \quad \forall Y\ (DT\theta\theta'\lambda\ XYG\ B^*[X,Y,G,\underline{Z}])\ \{U,V\}\ (X,Y,T^*)$ c.a.d. qu'ils sont identiques à une réduction près.

1.7. En particulier, nous venons de montrer que les fonctionnelles de type 1 de E, et les récursives-prouvables de la théorie des types sont extensionnellement identiques. Remarquons que notre interprétation n'inclut pas l'axiome d'extensionalité, mais que ce fait agit pas sur la détermination des récursives-prouvables.

1.8. Du fait que la théorie des types avec extensionalité est interprètable dans la théorie sans extensionalité, au moyen de la transformation A $\Rightarrow$ A+, on peut interpréter cette première théorie dans E, au moyen de $(A^{+})^{*}$.

** Cette dernière assertion mériterait une dimonstration plus détaillée. Nous donnerons dans un autre article un theoréme de reductibilité général ainsi qu une méthode pour formaliser la dimonstartion pour les sous-systèmes finis.

2. LE PRINCIPE DE MARKOV

Le principe de Markov est le schéma:

$$(\forall z\ (Az \vee \neg Az) \wedge \neg\neg\exists zAz) \rightarrow \exists zAz \text{ avec } z \text{ de type o.}$$

Kreisel a montré [Kr 1] que cet énoncé est interprétable dans T. On remarque immédiatement que cette interprétation n'utilise aucune information particulière sur A* (ni meme que z soit de type o). Comme conséquence évidente, on a donc l'interprétation dans F de la quantification universelle de ce résultat par rapport à A :

$$\forall \theta((\forall z(\theta z \vee \neg \theta z) \wedge \neg\neg\exists z\theta z) \rightarrow \exists z\theta z)$$

Considérons un énoncé A de l'analyse . Supposons que nous puissions démontrer

$$Az \vee \neg Az \qquad\qquad \neg\neg\exists zAz$$

Peut-on alors démontrer $\exists zAz$?

Ce problème, posé par Kreisel, peut encore s'énoncer sous la forme : l'analyse est elle close par rapport à la règle de Markov ?

On sait (voir [Tr]) que ce résultat est vrai pour l'arithmétique, ou si A est récursif primitif dans le cas qui nous intéresse. ([Kr2])

Supposons avoir donné une réponse positive à la question ci-dessus quand A n'a qu'une variable libre par une démonstration formalisable dans l'analyse sans le tiers-exclu. Soit S_n un sous-système fini de l'analyse dans lequel on a démontré

$$\forall A \quad S_m \vdash \ulcorner \forall z(Az \vee \neg Az)\urcorner \wedge \quad S_m \vdash \ulcorner \neg\neg\exists zAz\urcorner \rightarrow \quad S_m \vdash \ulcorner \exists zAz\urcorner \quad \text{, (A clos)}$$

ou S_m est un sous-système de S donné à l'avance. Soit Byz un énoncé à deux variables libres tel que $S_m \vdash Byz \vee \neg Byz$ et $S_m \vdash \neg\neg\exists zByz$.

On peut démontrer dans S_n :

$$\forall y\ (S_m \vdash \ulcorner \forall z(B\bar{y}z \vee \neg B\bar{y}z)\urcorner \wedge \quad S_m \vdash \ulcorner \neg\neg\exists zB\bar{y}z\urcorner)$$

et donc nous avons démontré dans S_n : $\forall y \quad S_m \vdash \ulcorner \exists zB\bar{y}z\urcorner$, ce qui, combiné avec le principe de réflexion pour S_m (supposé démontrable dans S_n), on obtient $S_n \vdash \forall y\, \exists zByz$, c.a.d. que $\exists zByz$ est démontrable dans

l'analyse. Nous examinerons plus loin le cas où B contient des variables d'ordre superieur.
Nous venons de montrer que la solution du problème de Kreisel revient à donner une démonstration du résultat dans le cas où A n'a qu'une variable libre. Il faut bien entendu que le résultat soit formalisable dans S.

Nous considérons une transformation standard qui transforme tout énoncé A en un énoncé A+, et telle que, si D désigne la conjonction des axiomes pour l'égalité et le successeur, on ait $(S \vdash A) \leftrightarrow (D \to A+$ est démontrable dans $G^1LC)$

Supposons que $Az \vee \neg Az$ et $\neg\neg\exists zAz$ soient démontrables dans S, c'est à dire dans un sous-système fini S_m. Si on désigne par G_m le calcul des prédicats d'ordre deux correspondant à S_m (module la transformation +) on peut trouver un sous-système S_m de S dans lequel le théorème de normalisation pour G_m, et le théorème d'interprétation fonctionnelle pour S_m soient démontrables. Le raisonnement qui suit se place dans S_n :
1) On peut trouver un entier p, tel que $A^*(\bar{p})$ soit valide dans F^n. C'est une conséquence évidente du théorème d'interprétation fontionnelle et du résultat énoncé plus haut.
2) On peut démontrer dans S_m un des deux énoncés suivants ; $A(\bar{p})$,$\neg A(\bar{p})$. En effet, $D\to$ $A+(\bar{p}) \vee \neg A+(\bar{p})$ est démontrable dans S_m. Par le théorème de normalisation, on peut supposer l'existence d'une déduction normale de $A+(\bar{p}) \vee \neg A+(\bar{p})$ sous l'hypothèse D : si cette déduction admet une branche principale, vu que D ne contient ni v, ni $\exists$, ni $\forall$ d'ordre supérieur, la conclusion est alors sous-formule de D, ce qui est absurde La déduction se termine donc par une introduction, ce qui signifie que $D\to A+(\bar{p})$ ou $D\to\neg A+(\bar{p})$ est démontrable. En clair, $A(\bar{p})$ est démontrable, ou sa négation l'est.
3) $A(\bar{p})$ est démontrable dans S_m. Si $\neg A(\bar{p})$ était démontrable, il serait

interprétable dans F^n, ce qui est absurde d'après 1. Seule l'autre possibilité subsiste.
Enfin, $\exists zAz$ est démontrable dans S_m par $\exists$-introd.

REMARQUE

Nous avons en fait montré le résultat suivant : si A est un énoncé de l'analyse ne contenant que des variables de type o, et si A est décidable, A est équivalent à son interprétation fonctionnelle.
Remarquons de plus que tout le raisonnement utilisé plus haut s'applique parfaitement à la théorie des types, modulo l'interprétation E.

Pour être complet sur la clôture par rapport à Markov, il faut examiner le cas où A contient des variables libres d'ordre supérieur.

Nous nous plaçons ici pour plus de généralité dans la théorie des types. Supposons que A renferme, les variables Y et z. On peut démontrer, dans $G^{\omega}LC$, sous les hypothèses D et EXT(X) (voir [Gi] p. 84), l'énoncé $A+(X,\bar{p}) \vee \neg A+(X,\bar{p})$, donc aussi sans coupure. La déduction sans coupure ne peut avoir de branche principale, car l'hypothèse principale ne saurait être D, et pas d'avantage Ext(X), car une succession d'éliminations sur cet énoncé ne peut pas faire apparaître de disjonction. La déduction se termine donc par une introduction. Si $A+(X,\bar{p})$ est la prémisse de cette introduction, $\forall X\ A\ (X,\bar{p})$ est démontrable. Si au contraire c'est sa négation qui est la prémisse de l'introduction, $\forall X\ \neg A(X,\bar{p})$ est démontrable, et donc $\neg\forall XA(X,\bar{p})$ est démontrable. Dans les deux cas, on peut démontrer $B(\bar{p}) \vee \neg\ B(\bar{p})$, ou B(z) est l'énoncé : $\forall\ X\ A(X,z)$.
En applicant le principe de réflexion au sous-système fini dans lequel sont faites les démonstrations ci-dessous, on obtiant les théorèmes suivants de S : $\forall z(B(z) \vee \neg B(z))$, $\forall z(\forall X\ A(X,z) \vee \forall X \neg A(X,z))$
Si on considère le second théorème, on voit qu'il implique
$\exists zA(X,z) \rightarrow \exists zB(z)$. En effet si A(X,z), comme on ne peut avoir

simultanément $\forall X \neg A(X,z)$, on a $B(z)$. Finalement, si $\neg\neg\exists zA(X,z)$ est démontrable, $\neg\neg\exists zB(z)$ l'est par ce qui précède, donc $\exists zB(z)$ est démontrable, et, a fortiori $\exists zA(X,z)$ l'est.

3 LA THESE DE CHURCH

La Thèse de Church est l'énoncé :

$$\forall x\ \exists y\ A(x,y) \rightarrow \exists e\ \forall x\ \exists y\ (\ T_1(e,x,y) \wedge A(x,U(y))\ )$$

avec le T_1 et le U de Kleene. [Kr2]

Kreisel a posé le problème suivant : peut-on étendre le système F en un système F^+ de manière à ce que la thèse de Church soit interprétable dans F^+? [Kr3]

Nous allons donner une réponse positive à cette question, non seulement pour F, mais aussi pour T, et pour E... Nous verrons par contre que cette solution ne s'applique pas au système $\sum_4$, ce que nous savions déjà par ailleurs, l'interprétation de la thèse de Church étant réfutable dans $\sum_4$.

Dans la suite, nous nous placons dans F pour fixer les idées, mais tous les raisonnements faits sont valables pour T et pour E.

Le langage F^+

F^+ est obtenu à partir de F par adjonction de deux constantes, RED et GD, de types respectifs ((o→o),o→o) et (o→o)→o.

Numerotation de F^+

Il faut maintenant distribuer injectivement des numéros à tous les assemblages et à tous les types de F^+. Nous n'explicitons que la partie du numérotage strictement nécessaire à la démonstration.

$\ulcorner APuv \urcorner = 3^2.5^{\ulcorner u \urcorner}.7^{\ulcorner v \urcorner}$		$\ulcorner GD \urcorner = 4$	
$\ulcorner EXTur \urcorner = 3^4.5^{\ulcorner u \urcorner}.7^{\ulcorner r \urcorner}$		$\ulcorner S \urcorner = 6$	
$\ulcorner RED \urcorner = 2$		$\ulcorner \bar{o} \urcorner = 8$	etc...

On peut alors former les prédicats récursifs primitifs et les fonctions:

Term(a) : a est le numéro d'un terme de F^+

Type(a,b): Term(a), et b est le n° du type du terme désigné par a.

Num(a) : a est le n° d'un numéral : a=8 ; ou ($a=3^2.5^6.7^{a_3}$ et Num(a_3)).

$q(x)$: le numéro de $\bar{x}$: $q(o)=8$; $q(Sx)=3^2 .5^6 .7^{q(x)}$.

$r(x)$: le numéral représenté par z si Num(x), o sinon :

$r(x) = o$ si $x=8$ ou (non) Num(x) ; $r(x) = r(x_3)$ sinon .

Réduction dans F^+

Dans F^+, un terme est normal s'il ne contient aucun sous-terme de la forme de ceux énumérés en [Gi] p. 70, et aucun sous-terme de la forme GD(a) ou RED(a,b), (a et b clos). On peut donc construire un prédicat récursif primitif Norm(a) qui signifie que a est le n° d'un terme normal. Pour définir la notion de réduction dans F^+ , il suffit de donner la notion de réduction immédiate dans la cas de GD et RED.

$GD(a) \triangleright \ulcorner a \urcorner$ (a est alors normal).

$RED(a,\bar{n}) = p_o^{\ulcorner a_o \urcorner} . \ldots . p_m^{\ulcorner a_m \urcorner}$, ou $p_o \ldots p_m$ sont les (m+1) premiers nombres premiers, et ou $a_o, \ldots, a_m$ sont tels que :

1) $a_o = a(\bar{n})$ (bien entendu, a est normal ; ici, a_o est l'assemblage $a(\bar{n})$)

2) a_m est un numéral.

3) a_i se réduit atomiquement a a_{i+1} ($o \leq i < m$)

4) La réduction s'opère en réduisant toujours le sous-terme minimal le plus à gauche. (Réduction "à gauche")

La condition 4) est une condition d'unicité.

Une conséquence évidente de [Ttl] est le

LEMME 1

Toute réduction peut être remplacée mécaniquement par une réduction "a gauche".

Arithmétisation des réductions

Il faut montrer que la réduction atomique dans F^+ est récursive primitive. D'après le lemme 1, il suffit de travailler sur des réductions "à gauche". Au niveau des numéros, T(u,v) signifiera que u se réduit atomiquement (à gauche) en v.

Remarquons que les axiomes de la page 70 de [Gi] peuvent tous se mettre sous la forme $v=f(u)$, avec une fonction f récursive primitive, que nous ne pouvons expliciter, n'ayant pas complètement défini la numérotation. $T(u,v) \leftrightarrow$ Term(u) et Term(v) et soit (3.1) ou (3.2)

3.1. $u=3^2 . 5^{u_2} . 7^{u_3}$ et soit (3.1.1.) ou (3.1.2) ou (3.1.3)

3.1.1. (non) Norm(u_2) et $v=3^2 . 5^{v_2} . 7^{u_3}$ et $T(u_2,v_2)$

3.1.2. Norm(u_2) et (non) Norm(u_3) et $v=3^2 . 5^{u_2} . 7^{v_3}$ et $T(u_3,v_3)$

3.1.3. Norm(u_2) et Norm(u_3) et (3.1.3.1) ou (3.1.3.2) ou (3.1.3.3.)

3.1.3.1. $u_2=4$ et $v=q(u_3)$ (Axiome pour GD)

3.1.3.2. $u_2=3^2 . 5^2 . 7^{u_{23}}$ et Num(u_3) et Num(v) et $r(v) \neq o$ et

$(r(v))_o=3^2 . 5^{u_{23}} . 7^{u_3}$ et Num($(r(v))_{lh(r(v)) \dot- 1}$) et

$1<lh(r(v)) \dot- 1$ $T((r(v))_i,(r(v))_{i+1})$ (Axiome pour RED)

3.1.3.3. $v=f(u)$ sinon.

3.2. $u=3^4 . 5^{u_2} . 7^{u_3}$ et (3.2.1) ou (3.2.2.)

3.2.1. (non) Norm(u_2) et $v=3^4 . 5^{v_2} . 7^{u_3}$ et $T(u_2,v_2)$

3.2.2. Norm(u_2) et $v=f(u)$.

<u>LEMME 2</u>

T est récursif primitif.

Considérons T comme prédicat d'un argument : $T(u,v)=T_o(2^u . 3^v)$. Il faut montrer que les définitions précédentes sont bien fondées.
En 3.1.1., 3.1.2., 3.2.1., la valeur de T_o sur $2^u . 3^v$ est définie à partir de sa valeur sur $2^{u_i} . 3^{u_i}$ pour $i=2,3$. Il suffit de voir que $2^{u_i} . 3^{v_i} < 2^u . 3^v$. Ceci est évident puisque $a_i < a$.
En 3.1.3.2. il faut vérifier que $2^{r(v)_i} . 3^{r(v)_{i+1}} < 2^u . 3^v$. Mais $2^{r(v)_i} . 3^{r(v)_{i+1}}$ n'est pas plus grand que $p_i^{r(v)_i} . p_{i+1}^{r(v)_{i+1}}$, lui-même borné par r(v). Mais $r(v) \leq v < 2^u . 3^v$.

Il est immédiat, par induction sur $2^u . 3^v$, que, si on a $T(u,v)$ et $T(u,v')$ alors $v=v'$. En particulier, si une forme normale existe, elle est unique.

REMARQUE : Bien que T soit récursif primitif, la fonction qui à u associe le v qui vérifie T(u,v) quand u n'est pas normal , n'est pas récursive-prouvable dans l'analyse.

On peut maintenant construire un prédicat récursif primitif T'(e,x,y) qui signifie que e est le n d'un terme normal f de type o→o, et que y représente la suite des réductions "a gauche" de $f(\bar{x})$ en un numéral. $T'(e,x,y) \leftrightarrow Norm(e)$ et $Type(e, o \to o)$ et $y_o = 3^2 . 5^e . 7^{q(x}$ et $Num(y_{lh(y)\dot{-}1})$ et $\forall i < lh(y) \dot{-} 1 \; T(y_i, y_{i+1})$.

On définit de même $U'(y) = r(y_{lh(y)\dot{-}1})$

Théorème de réductibilité pour F^+

Pour démontrer ce théorème pour F^+ , remarquons que GD et RED sont de types sans indéterminée, et que donc, la réductibilité est la réductibilité habituelle, pour des types.

Réductibilité de GD : soit a réductible, en forme normle. Alors $\overline{\ulcorner a \urcorner}$ est un numéral, et puisque GD(a) R $\bar{a}$, GD(a) est réductible.

Réductibilité de RED : soient a et $\bar{n}$ réductibles et normaux; alors $a(\bar{n})$ est réductible, c'est à dire qu'il existe une réduction de $a(\bar{n})$ en un numéral, et on peut supposer cette réduction "à gauche". Si $a_o \ldots a_m$ dénotent les étapes de cette réduction, $RED(a,\bar{n})$ R $\overline{p_o^{\ulcorner a_o \urcorner} \ldots p_m^{\ulcorner a_m \urcorner}}$, qui est un numéral, et donc $RED(a,\bar{n})$ est réductible.

Interprétation fonctionelle de la thèse de Church

Remarquons que tous les prédicats construits sont récursifs primitifs, ils sont représentables dans F, et donc dans F^+ , par des fonctions que nous noterons de la même façon.

Remarquons que T'(GD(f),x,RED(f,x)) est valide dans Y^+. Il suffit de vérifier que $T'(GD(a),\bar{n},RED(a,\bar{n})) = \bar{o}$ dans F^+ pour tout n et tout a normal, c'est à dire vérifier que $T'(\ulcorner a \urcorner , n \,.\, p_o^{\ulcorner a_o \urcorner} \ldots p_m^{\ulcorner a_m \urcorner})$, où $a_o \ldots a_m$ est la réduction "à gauche" de $a(\bar{n})$. C'est exactoment ce qui est exprimé dans la définition de T'.

De même, $U'(RED(f,x))=f(x)$ est valide dans Y^{+}, car si l'on pose $x=\bar{n}$, $RED(f,\bar{n})_{lh(RED(f,\bar{n}))\dot{-}1}$ est égal à $\ulcorner\bar{p}\urcorner$, si $\bar{p}$ est la forme normale de $f(\bar{n})$, et donc $U'(RED(f,\bar{n}))$ est égal à $\bar{p}$.

Au lieu d'interpréter la thèse de Church sous la forme annoncée, nous allons en donner une interprétation qui utilise T' et U' :

$$\forall x\ \exists y\ A(x,y) \rightarrow \exists e\ \forall x\ \exists y\ (\ T'(e,x,y) \wedge A(x,U'(y))\ )$$

Cet énoncé est en fait strictement plus fort que la thèse de Church, puisqu'il affirme que toute fonction est non seulement récursive, mais est représentable dans F^{+}.

On sait qu'on peut adjoindre à l'interprétation celle des prédicats récursifs primitifs, en particulier T' et U'.

Dans la formule suivante,

$$A^{*}[x,f(x),X(x),Y,\underline{Z}] \rightarrow T'(e,x',y(x')) \wedge A^{*}[x',U'(y(x')),X'(x'),Y',\underline{Z}.]$$

il faut trouver e, y, X' en fonction de f et X, et x, Y en fonction de f, X, x', Y'.

On pose $x=x'$, $Y=Y'$, $X'=X$, $e=GD(f)$, $y=RED(f)$. La vérification est alors immédiate.

Comme nous l'avons dit, cette méthode s'applique tout aussi bien au système T ou au système E, ce qui permet d'étendre l'interprétation à la thèse de Church pour l'analyse, l'arithmétique, la théorie des types, et donne donc en particulier un théorème de non-contradiction relative.

Dans le cas du système de Spector, il est nécessaire voir [Kr3] que le raisonnement ci'dessus soit inapplicable. En fait, la démonstration de Tait [Tt2], untilise explicitement le fait que toutes les fonctionnelles sont continues, alors que GD et RED ne le sont pas.

Nous retrouvons ainsi l'opposition bien connue (voir[H,K]) entre la Thése de Church et la continuité.

4. COMPATIBLITE DE LA BAR-RECURSION ET DE LA STRAFIFICATION

Remarquons tout d'abord que les interprétations E et F supposaient que les sortes (species) sont à arguments de type o. En fait, on peut sans modifier l'esprit de l'interprétation, supposer que les sortes sont à arguments de type quelconque, y compris même les types apparaissant dans les systemes E et F, si l'on pense que celà a un quelconque sens. Per exemple l'interprétation de sortes de fonctions se fera au moyen de types de la forme $\theta, \theta', 1 \to o$.

Supposons donc que nous sommes en présence d'une interprétation de la théorie des sortes de fonctions. (Et éventuellement la théorie des types qui la prolonge). Nous pouvons très certainement interpréter l'axiome du choix : $\forall x \exists y\, A(x,y) \to \exists f \forall x\, A(x,f(x))$ (AC_{oo})
mais nous ne pouvons pas interpréter l'axiomes de compréhension $\exists f\ \forall x\ (\ A(x) \leftrightarrow f(x)=o\ )$, ni même sa double négation : en effet, outre une raison évidente qui tient au second théorème d'incomplétude de Gödel, il suffit de remarquer, que si A est récursivement énumérable non-récursif, $\neg\neg\exists f\ \forall x(A(x) \leftrightarrow f(x)=o)$ implique **la nègation de la** thèse de Church, **négation** qui ne saurait être **interprétée dans** E ou F. (voir § précédent)

Si on veut donc interpréter la double négation du schéma de compréhension et, par là-même hisser la force théorique de F du niveau 2 au niveau 3, (Bien entendu, la force théorique de E ("proof-theoretical strength") n'augmente pas, elle) il faut adjoindre un nouveau schéma à F, la Bar-récursion.

Spector a en effet montré que la Bar-récursion de type fini permet d'interpréter la schéma

$$\forall x\ \neg\neg p(x) \to \neg\neg\forall x p(x)$$

qui est suffisant pour le but cherché. (Voir [Sp])

Nous ne saurions nous contenter de la bar-récursion de type fini, car l'interprétation de p(x) peut utiliser tous les types de F (ou de E). Il nous faut donc une bar-récursion de type quelconque, et de plus stratifiable, pour pouvoir utiliser toutes les ressources de F (ou de E). Nous introduisons une certaine constante B_{rs} d'un type assez complexe $\mu(r,s)$, pour les types r et s, avec ou sans IO. Le type de B_{rs} est donné dans [Tt2] p.357. La notion de réduction immédiate pour B est donnée dans [Tt2,] p.363. La possibilité d'effectuer l'interprétation du schéma de Spector dans le modéle des termes clos est conséquence de ce que si $na\bar{n}=b\bar{n}$, alors T(a) = T(b) (T(a) de type o) (voir[Tt2]), ce qui est le seul cas de "Weak extensionality" réellement utilise par Spector.

La seule chose que nous ayons à vérifier pour montrer que les systèmes EB et FB ainsi obtenus sont réductibles, est que B_{rs} est réductible quand la réductibilité de type $\mu(r,s)$ est définie à partir de c.r. arbitraires de types r et s.

Examinons maintenant la démonstration de Tait [Tt2]; on remarque que les seules propriétés de la réductiblité de type r et s, sont celles que nous avons nommees "proprietés des candidats de réductibilité". (Voir [Gi] p. 72). La démonstration de Tait est donc toujours valable pour FB et EB.

Remarquons qu'un c.r. est un ensemble de termes; les termes de Tait sont codés, non par des entiers, mais par des fonctions, c.a.d. qu'un c.r. est un objet d'ordre 3, et non pas d'ordre 2, ce qui correspond bien au fait que dans ces interprétations, tous les ordres augmentent d'une unité.

On pout adjoindre aux théories que nous venons d'interpréter le principe de Markov. On a alors indentié extensionnelle entre les récursives-prouvables des deux théories et les fonctions de EB ou FB, suivant le cas.

Si E^+ et F^+ sont deux extensions de E et F qui vérifient la thèse de Church, EB et FB sont deux extensions qui la réfutent. Ce qui montre bien que la question est indécidable.

Reférences

[Gd] K.GÖDEL "Über eine bisher noch nicht benutzte Etweiterung des finiten Standpunktes." Dialectica 12 (1958)

[Gi] J.Y. GIRARD "Une extension de l'interprétation de Gödel à l'analys et son application à l'élimination des coupures dans l'analyes et la théorie des types." Proc.2nd Scand.Log.Symp. , ed Fenstad (North Holland Amsterdam)

[H,K] W.A.HOWARD & G.KREISEL "Transfinite induction and Bar induction of types zero and one, and the rôle of continuity in intuitionistic analysis." J.S.L. 31(1966).

[Kr1] G.KREISEL "Interpretation of classical analysis by means of constructive functionals of finite types." Con uctivity in Mathematics ed. Heyting (North Holland, Amsterdam, 1959)

[Kr2] G.KREISEL "Church's Thesis; a kind of reducibility axiom for constructive mathematics." Intuitionism and proof theory, ed. Myhill, kino, Vesley (North Holland, Amsterdam, 1970)

[Kr3] G.KREISEL "A survey of proof theory II." Proc.2nd.Scand.Log. Symp., ed. Fedstad (North Holland, Amsterdam 1971)

[Sp] C.SPECTOR "Provably recursive functionals of analysis : a consistency proof of analysis by an extension of principles formulated in current intuitionistic mathematics." Recursive function theory, Proc.Symp. Pure Math. 5, AMS Providence R.I. 1962.

[Tr] A.S.TROELSTRA "Notions of realizability", Proc.2nd.Scand.Log. Symp., ed. Fenstad (North Holland, Amsterdam, 1971)

[Tt1] W.W.TAIT "Intentional interpretation of functionals of finite type." J.S.L. 32 (1967)

[Tt2] W.W.TAIT "Normal form theorem for bar recursive functions of finite type." Proc.2nd.Scand.Log.Symp. ed. Fenstad (North Holland, Amsterdam 1971)

Monsieur J.Y. Girard
8, Rue du Moulin d'Amboile
94 Sucy en Brie
France

VII. COMBINATOR REALIZABILITY OF CONSTRUCTIVE FINITE TYPE ANALYSIS

John Staples[1]

Introduction In this paper proofs are given of some of the conjectures which I presented at the 1971 Cambridge Logic Institute. The proof method is a modification of Harvey Friedman's technique, presented at the same meeting, for realizing in a weaker sense axioms of comprehensions and extensionality.

The notion of a combinator-realization of a theory (relative to a given calculus of combinators) is defined; a formal theory of constructive finite type analysis is given, together with an assignment of combinatory terms to provable formulas; it is then proved that all assigned terms have normal forms and that the assignment of normal form terms to provable formulas so defined is a combinator-realization of the theory. It is shown that the argument can be extended to cover an extended theory which has an axiom of countable choice.

As with the notion of realizability used by Friedman, this method gives existence and disjunction properties for the theories concerned; e.g. if $A \vee B$ is a provable sentence then either A is provable or B is. The difference between the two methods (assuming a theory to which both apply) lies in what can be said about universal sentences. Consider for example a sentence of finite type analysis of the form $(\forall x)(\exists y)A(x,y)$, where the structure of A can be arbitrarily complex. If this is realizable in Friedman's sense then one knows that for each constant c of the type of x there is a constant d of the type of y such that $A(c,d)$ is realizable. If however this sentence is combinator-realizable by

[1] This work was supported by an S.R.C. research fellowship.

a term T in the sense to be defined below then one knows that for each constant c of the type of x the term $(T\bar{c})$ reduces to a normal form $\langle \bar{d}, U \rangle$, where d is a constant of the type of y, U combinator-realizes A(c,d) and $\bar{c}$, $\bar{d}$ are codings of c, d respectively by normal terms, in some fixed coding of all constants of the theory by normal terms. In particular, in the cases below when x, y are of the type of the natural numbers and numerals are coded in an obvious way, one concludes that the procedure for producing $\bar{d}$ from a given $\bar{c}$ is recursive. The point is that this property holds even though the proof of $(\forall x)(\exists y)A(x,y)$ may involve notions from higher order analysis; indeed the formula A(x,y) may contain constants, quantifiers etc. of higher type. Corresponding points can be made about universal sentences of other forms.

It is of course a matter of opinion whether the theories to be discussed are rightly called constructive, and whether combinator realizability is necessary or sufficient for calling a theory constructive; in any case this notion of realizability provides a more severe test of the constructivity of a theory than does Friedman's.

This paper is intended to be read either constructively or classically. Since it includes a proof of the syntactic consistency of a finite type theory of arithmetic, it cannot (assuming that result is true) be formalised within that theory. It is thus an example of a constructive result needing more than finite type analysis for its formalisation, and a counterexample to claims that constructive mathematics can be formalised in some fragment of such a finite type analysis.

I am indebted to Harvey Friedman for presenting and explaining his technique; also to Roger Hindley for discussions and notes on combinatory logic and to John Cleave for discussions on constructivity.

1. Sketch of the method of realization

1.1 Define an untyped calculus of combinators, with primitive constants S, K, O, s, R, D, D_1, D_2 and the following reduction rules; $SXYZ \geq XZ(YZ)$, $KXY \geq X$, $RXYO \geq X$, $RXY(sn) \geq Yn(RXYn)$, $D_1(DXY) \geq X$, $D_2(DXY) \geq Y$. Write $\langle X,Y \rangle$ for DXY and $\langle X_1, \ldots, X_n \rangle$ for $\langle X_1, \langle X_2, \ldots, X_n \rangle\rangle$; write I for SKK, and F for a term in normal form such that $FXY\langle Z,O \rangle \geq XZ$, $FXY\langle Z,sn \rangle \geq YZ$. For definiteness choose F to be $[x].[y].[w].(R(x(D_1w))([u].[v].(y(D_1w)))(D_2w))$. Define $\bar{A}$ to denote the normal form of A (undefined if A has no normal form). Define $D_{i,n}$ for $n \geq 1$, $1 \leq i \leq n$ by: $D_{1,1} = I$; for $1 \leq i \leq n-1$, $D_{i,n+1}$ is $D_{i,n}$; $D_{n,n+1}$ is $[x].D_1(D_{n,n}x)$; $D_{n+1,n+1}$ is $[x].D_2(D_{n,n}x)$.

1.2 Define an <u>algorithm</u> to be a combinatory term in normal form. It is well-known that a term has at most one normal form. An algorithm is regarded as having all other algorithms as possible inputs, and A has output C for input B if and only if C is the normal form of (AB). It is well-known (for the calculus given) that if algorithms are restricted to numeral inputs and regarded as having an output if and only if a normal form results which is a numeral, then the functions so defined are just the partial recursive functions.

1.3 Define a <u>combinator-realization</u> of a theory, by analogy with Kleene's notions of realizability, to be a coding of the constants of the theory by algorithms and an assignment of algorithms to theorems which together satisfy the following conditions:

(a) If $A(x_1, \ldots, x_n)$ is a predicate, if t_A is any algorithm assigned to it and if $c_1, \ldots, c_n$ code constants of the appropriate types then $t_A\langle c_1, \ldots, c_n \rangle \geq t_{A(c_1, \ldots, c_n)}$, where the latter is some algorithm assigned to $A(c_1, \ldots, c_n)$.

(b) For sentences, in the corresponding notation,

(i) any $t_{A\&B}$ has the form $\langle t_A, t_B \rangle$ for some t_A, t_B

(ii) any $t_{A\vee B}$ has the form $\langle t_A, 0 \rangle$ or $\langle t_B, 1 \rangle$ for some t_A or t_B respectively

(iii) any $t_{A\supset B}$ is such that for any t_A, $t_{A\supset B}t_A \geq t_B$ for some t_B

(iv) for any ~A which has algorithms assigned, A has none

(v) any $t_{(\exists x)A(x)}$ has the form $\langle c, t_{A(c)} \rangle$ for some constant c and some $t_{A(c)}$

(vi) any $t_{(\forall x)A(x)}$ is some $t_{A(x)}$.

1.4 For an appropriate theory and coding of its constants define an assignment of terms to formulas thus:

(i) assign appropriate algorithms to the axioms of the theory

(ii) use the "obvious" procedure (see later) for assigning a term to an inferred formula, given terms assigned to the hypotheses of the inference.

1.5 Prove that all assigned terms have normal forms, and that the assignment of algorithms to theorems defined by reducing assigned terms to normal form is a combinator-realization.

2. The axiom scheme $A \supset G(A)$

The notion of realization given has the property that negative formulas may be realized trivially. If one regards constructive analysis as a theory of computation then it is clearly desirable to make use of the corresponding classical theory where possible, e.g. to shorten proofs by contradiction (c.f.[1].) This can sometimes be done by translating the classical theory into a subtheory of the constructive theory, as Gödel and Gentzen did for first order arithmetic. A more powerful and convenient procedure is to adopt the above-mentioned axiom scheme, where the G-translation G(A) of a formula A is defined as follows, after Gentzen:

(i) for atomic formulas A, G(A) is $\sim\sim A$

(ii) for compound formulas,

$G(A \& B)$ is $G(A) \& G(B)$; $G(A \vee B)$ is $\sim(\sim G(A) \& \sim G(B))$

$G(A \supset B)$ is $G(A) \supset G(B)$; $G(\sim A)$ is $\sim G(A)$

$G((\forall x)A(x))$ is $(\forall x)G(A(x))$; $G((\exists x)A(x))$ is $\sim(\forall x)\sim G(A(x))$.

Since the intuitionist predicate calculus implies $\sim\sim G(A) \supset G(A)$, then $A \supset G(A)$ is equivalent to $\sim G(A) \supset \sim A$ and the classical theory is always available to prove negative theorems. It is technically convenient to assume it below in the equivalent form $A \supset \sim\sim G(A)$.

Note that since this axiom contradicts any classically false axiom it does not represent an intuitionistically valid statement. For example formulas of the scheme $(\forall x)\sim\sim A(x) \supset \sim\sim(\forall x)A(x)$ become provable; in fact this scheme is equivalent in Heyting's predicate calculus to the scheme $A \supset G(A)$, as may be proved by induction on the length of formulas.

In all of the following it is optional whether this scheme is included amongst the logical axioms; the modifications required to deal with it are noted in the appropriate places.

3. The theory and its assigned terms

For simplicity a theory F without countable choice is treated first; then the extension required to include it is indicated.

3.1 Here we give the vocabulary of F. The notion of finite type which is required is defined inductively by; O is a type, and if $t_1, \ldots, t_n$ are types then so is $(t_1, \ldots, t_n)$. To save argument we assume from the start that all symbols of the language are taken from (or coded within) the algorithms defined above, in a way which is natural enough to satisfy some weak restrictions to be given later. We may for brevity confuse a symbol with its code. A detailed discussion of coding is given in 3.4.

The following is an inductive definition of the vocabulary of F.

(i) 0 is a constant of type 0 (the algorithm 0 is assumed to code the constant 0)

(ii) successor, ', is a function symbol of type (0,0)

(iii) for each type t there is a sequence $x_t^0, x_t^1, \ldots$ of variables of that type

(iv) 0 is a term of type 0; any variable of type t is a term of type t; if z is a term of type 0 then so is z'; closed terms are called constants (if z' is a closed term of type 0 and z is coded by $\bar{z}$ then z' is coded by $s\bar{z}$)

(v) there is a relation symbol $=_0$ of type (0,0); for each type $t = (t_1, \ldots, t_n)$ there is an (n+1)-ary relation symbol $\in_t$ of type $(t, t_1, \ldots, t_n)$. We write $\langle r_1, \ldots, r_n \rangle \in_t r$ to denote $\in_t(r, r_1, \ldots, r_n)$ and may also omit the subscript t; similar conventions apply to the relation symbols for equality introduced above and later

(vi) connectives $\&$, $\vee$, $\supset$, $\sim$ and quantifiers $(\forall)$, $(\exists)$ are introduced as usual and the usual definitions of formulas and conventions for abbreviating them apply; in particular, $A \& (B \& C)$ may be abbreviated to $A \& B \& C$.

(vii) if $A(x_1, \ldots, x_n)$ is a predicate with $n \geq 1$ free variables occuring in the order displayed, x_i of type t_i, $i = 1, \ldots, n$, then $\{\langle x_1, \ldots, x_n \rangle : A(x_1, \ldots, x_n)\}$ is a constant of type $(t_1, \ldots, t_n)$.

3.2 Here the axioms of F are listed together with the algorithms which are initially assigned to them. For definiteness the list is given in detail. In schemes 10. and 11. where the term p occurs, p* denotes the following algorithm:

(i) if the axiom is a sentence then p* codes the constant which is the closed term p

(ii) if the axiom is a predicate with free variables $x_1, \ldots, x_n$ in the order displayed then for any constants $c_1, \ldots, c_n$ of the corresponding types, $p^*\langle c_1, \ldots, c_n\rangle$ has as normal form the algorithm coding the constant defined by the closed term obtained from p by substituting c_i for x_i wherever it occurs in p, $i = 1, \ldots, n$.

	Axiom	Sentence algorithm	Predicate algorithm
1.	$A \supset (B \supset A)$	K	[w].K
2.	$(A \supset (B \supset C)) \supset ((A \supset B) \supset (A \supset C)$	S	[w].S
3.	$A \supset (B \supset A \& B)$	[x].[y].Dxy	[w].[x].[y].Dxy
4a.	$A \& B \supset A$	D_1	$[w].D_1$
4b.	$A \& B \supset B$	D_2	$[w].D_2$
5a.	$A \supset A \vee B$	[x].DxO	[w].[x].DxO
5b.	$B \supset A \vee B$	[x].Dx(sO)	[w].[x].Dx(sO)
6.	$(A \supset C) \supset ((B \supset C) \supset (A \vee B \supset C))$	F	[w].F
7.	$(A \supset B) \supset ((A \supset {\sim}B) \supset {\sim}A)$	K	[w].K
8.	$A \supset ({\sim}A \supset B)$	K	[w].K
9.	$(\forall x)(C \supset A(x)) \supset (C \supset (\forall x)A(x))$	[x].[y].[z].xyz	[w].[x].[y].[z].xzy
10.	$(\forall x)A(x) \supset A(p)$	[x].(xp*)	[w].[x].(x(p*w))
11.	$A(p) \supset (\exists x)A(x)$	$[x].\langle p^*, x\rangle$	$[w].[x].\langle p^*w, x(p^*w)\rangle$
12.	$(\forall x)(A(x) \supset C) \supset ((\exists x)A(x) \supset C)$	$[x].[y].(x(D_1y))$	$[w].[x].[y].(x(D_1y))$
13.	$0 \in z \supset ((\forall x)(x \in z \supset x' \in z) \supset (\forall x)(x \in z)$		[w].R
14.	$x =_0 x$		[w].O
15.	$x =_0 y \supset y =_0 x$		[w].[x].O
16.	$x =_0 y \supset (y =_0 z \supset x =_0 z)$		[w].[x].[y].O
17.	$x =_0 y \equiv x' =_0 y'$		[w]. [x].O,[y].O
18.	${\sim}(0 =_0 y')$		[w].O

To give the remaining axioms we first define higher type extensional equality, $x =_t y$ for any type $t = (t_1, \ldots, t_n)$, to abbreviate

$$(\forall x_1) \ldots (\forall x_n)(\langle x_1, \ldots, x_n \rangle \in x \equiv \langle x_1, \ldots, x_n \rangle \in y).$$

Then for $n \geq 1$ and t as above the extensionality axiom is

19. $x_1 = y_1 \supset (\ldots \supset (x_n = y_n \supset (\langle x_1, \ldots, x_n \rangle \in_t z \supset \langle y_1, \ldots, y_n \rangle \in_t z)) \ldots).$

To motivate the algorithm assigned to this axiom it is necessary to describe the intended interpretation of a closed atomic formula, say $a \in \{x: B(x)\}$. For brevity call a combinatory term which is assigned to a formula a <u>test</u> of that formula. The idea is that the tests of $a \in \{x: B(x)\}$ should be just the tests of $(\exists x)(x = a \& B(x))$ (and of course tests of the latter should be as required by the definition of combinator-realisation). We ensure that two such formulas do have the same tests by introducing the auxillary axioms

$$a \in \{x: B(x)\} \equiv (\exists v)(v = a \& B(v)),$$

to which is assigned the term $\langle I, I \rangle$. Then to interpret extensionality the problem is to give a term which will construct a test with the desired properties for, say, $c \in \{x: B(x)\}$, given such tests for $a \in \{x: B(x)\}$ and $a = c$. In view of the supposed form of tests of $a \in \{x: B(x)\}$ the problem reduces to constructing a test for, say, $d = c$ from ones for $d = a$ and $a = c$. We assume (the interesting case) that the type of a, d and c is not 0; it is sufficient to give a test for $x = y \supset (y = z \supset x = z)$. Although this formula is provable from the axioms already given, the simplest way of finding a test of it which is easily managed is to introduce this formula as an auxilary axiom and assign to it an appropriate test; say $[w].T_t$, where T_t is

$$[u].[z].[p].\langle [x].(D_1(zp))(D_1(up)x), [x].(D_2(up))(D_2(zp)x) \rangle .$$

The axiom $x = y \supset (x \in z \supset y \in z)$ is now assigned $[u].W_t$, where W_t is

$$[v].[w].\langle D_{1,3}w,\ T_t(D_{2,3}w)v,\ D_{3,3}w\rangle .$$

Other instances of extensionality are treated correspondingly.

In view of the above discussion an appropriate test to assign to a sentence of the comprehension scheme,

$$20.\quad (\exists z)(\forall x_1)\ldots(\forall x_n)(\langle x_1,\ldots,x_n\rangle \in z \equiv (\exists y_1)\ldots(\exists y_n)(y_1=x_1 \& \ldots \& y_n=x_n \& B(y_1,\ldots,y_n)))$$

is $\langle\{\langle u_1,\ \ldots,u_n\rangle : B(u_1,\ \ldots,u_n)\},\ [v_1].\ \ \ldots\ [v_n].\langle I,I\rangle\rangle$.

This assumes that, in the case $n = 1$ for example, $[v_1].\langle I,I\rangle$ has been assigned to the sentence

$$(\forall x_1)(x_1 \in \{u_1 :\ B(u_1)\} \equiv (\exists y_1)(y_1 = x_1 \& B(y_1)));$$

we justify this assumption by adding the latter as an auxiliary axiom, to which is assigned the desired test.

To interpret predicates of the scheme appropriately we assume that the coding of constants as algorithms satisfies the following restriction. For each predicate $B(x_1,\ \ldots,x_n)$ with variables $x_1,\ \ldots,\ x_n$ as displayed and (perhaps) undisplayed variables $z_1,\ \ldots,\ z_m$, $m \geq 1$, there is an algorithm A_B such that for arbitrary constants $c_1,\ \ldots,\ c_m$, where c_i has the type of z_i, $i = 1,\ \ldots,\ m$, the term $A_B\langle c_1,\ \ldots,c_m\rangle$ reduces to the normal form $\{\langle x_1,\ \ldots,x_n\rangle : B^+(x_1,\ \ldots,x_n)\}$, where B^+ denotes the predicate obtained from B by substituting c_i for z_i, $i = 1,\ \ldots,\ m$. Since we allow a different A_B for each B and each selection of undisplayed variables of B this is a very weak requirement; see 3.4.

Then predicates of the comprehension scheme are assigned $[u].\langle A_B u,[v_1].\ \ \ldots\ [v_n].\langle I,I\rangle\rangle$. To complete the list of axioms and their assigned algorithms we add the auxiliary scheme $(\forall x_1)\ \ldots\ (\forall x_n)(A \equiv A)$; sentences of this scheme are assigned $[v_1].\ \ \ldots\ \ [v_n].\langle I,I\rangle$, and predicates $[w].[v_1].\ \ \ldots\ [v_n].\langle I,I\rangle$.

3.3 Here we complete the total assignment of terms to provable formulas by giving the rules of inference (generalization and modus ponens) and the "obvious" procedure for assigning a test to an inferred formula, given tests for the hypotheses. To indicate this procedure suppose that A, $A \supset B$ are sentences with tests t_A, $t_{A \supset B}$ respectively; in view of the definition of combinator-realization it is then natural to assign the term $t_{A \supset B} t_A$ to the inference B (in the combinator-realization which it is to be proved is defined by the present assignment of terms to theorems it would of course be the normal form of this term which would be assigned; but in the absence of a proof we do not know that such a normal form exists). For predicate instances of this rule of inference the necessary modifications are made; the details are as follows.

Define for each type t a constant o_t of that type, to be set aside for substituting for occurrences of variables of that type which occur in A but not in B. Say for definiteness that o_0 is 0 and $o_{(t_1, \ldots, t_n)}$ is $\{\langle x_1, \ldots, x_n \rangle : x_1 = o_{t_1} \& \ldots \& x_n = o_{t_n}\}$. Now if the free variables of $A \supset B$ are (in order of appearance from left to right) $x_1, \ldots, x_n$, where x_i has type t_i, $i = 1, \ldots, n$ and those of A are the subsequence $x_{a_1}, \ldots, x_{a_k}$ and those of B the subsequence $x_{b_1}, \ldots, x_{b_m}$, $k, m \geq 0$, then if r_A is assigned to A and $r_{A \supset B}$ to B, assign $Mr_{A \supset B} r_A$ to B, where M is

$$[a].[c].[d].(a\langle q_1, \ldots, q_n\rangle (c\langle r_1, \ldots, r_k\rangle)),$$

r_i denotes q_{a_i}, $i = 1, \ldots, k$ and q_i is defined as follows. If i is some b_j then q_i is $(D_{j,m} d)$; otherwise q_i is o_{t_i}. (If $k = 0$, $\langle r_1, \ldots, r_k\rangle$ is omitted; if $m = 0$, $[d]$. is omitted.)

Generalization is treated similarly. If the free variables of A are as above and $(\forall y)A$ is inferred where y is some x_j, and if r_A is assigned to A, then assign (Gr_A) to $(\forall x_j)A$, where

G is defined to be [a].[b].[c].a$\langle d_1, \ldots, d_p \rangle$ and where d_i is defined as follows. For $i < j$, d_i is $(D_{i,n-1}b)$; for $i = j$, d_i is c; for $i > j$, d_i is $(D_{i-1,n-1}b)$. (If $n = 0$, omit [b]. and $\langle d_1, \ldots, d_n \rangle$; if $n = 1$, omit [b].).

If y is not some x_j repeat the above except that for $n = 1$ retain [b]., and for $n \geq 1$ define $d_i = (D_{i,n}b)$.

3.4 Here is given an appropriate coding of constants of F (indeed of all formulas of F). These details are given only to emphasize that the coding problem is trivial. Since we are coding symbols of a formal language F into symbols of another formal language (the calculus of combinators) we can when this is done redefine F so that the symbols are identical with their codes, and so forget about the distinction. A reference to coding is then an explanatory device. Recall we may denote the term sO by 1, ssO by 2, etc.. We indicate the coding of B by *(B).

<u>Coding of types:</u> *(O) is $\langle 0,0 \rangle$, $*((t_1, \ldots, t_n))$ is $\langle 0, *(t_1), \ldots, *(t_n) \rangle$

<u>Coding of symbols of F:</u> *(O) is O, *(O') is sO, etc.,
*(') is $\langle 0,s \rangle$, $*(x^n_t)$ is $\langle 1,*(t),n \rangle$, $*(=_0)$ is $\langle 2,0 \rangle$, $*(\in_t)$ is $\langle 2,*(t) \rangle$,
*(&) is $\langle 2,2 \rangle$, $*(\vee)$ is $\langle 2,3 \rangle$, $*(\supset)$ is $\langle 2,4 \rangle$, $*(\sim)$ is $\langle 2,5 \rangle$,
*(() is $\langle 2,6 \rangle$, *( )) is $\langle 2,7 \rangle$, $*(\exists)$ is $\langle 2,8 \rangle$, $*(\forall)$ is $\langle 2,9 \rangle$,
*(:) is $\langle 2,10 \rangle$, *({) is $\langle 2,11 \rangle$, *(}) is $\langle 2,12 \rangle$, *(,) is $\langle 2,13 \rangle$,
$*(\langle)$ is $\langle 2,14 \rangle$, $*(\rangle)$ is $\langle 2,15 \rangle$.

<u>Coding of formulas of F:</u> If $s_1 \ldots s_n$ is a string of symbols of F (other than a type or a numeral O', O'' etc. for which codings have already been given) then $*(s_1 \ldots s_n)$ is $*(s_1) \ldots *(s_n)$. In particular this applies to the constants
$\{ \langle x_1, \ldots, x_n \rangle : B(x_1, \ldots, x_n) \}$.

<u>Definition of</u> p* <u>and</u> A_B: If p occurs in an open instance of schemes 10. or 11. in which the distinct free variables in order of appearance are, say, $x_1, \ldots, x_n$, then p is either a constant

(define p^* to be $[u].*(p)$), a variable x_j (define p^* to be $[u].(D_{j,n}u)$) or of the form $x_j{}'\cdots{}'$ (define p^* to be $[u].(s\ldots s(D_{j,n}u))$).

Similarly, to define A_B for a formula $B(x_1, \ldots, x_n)$ and the undisplayed variables $z_1, \ldots, z_m$, replaces in $*(B(x_1, \ldots, x_n))$ any occurrences of $*(z_i)$ not followed by $*(')$, by $(D_{i,n}u)$; other occurrences of $*(z_i)$ are as part of some $*(z_i{}'\cdots{}')$; the latter is replaced by $s\ldots s(D_{i,m}u)$. The resulting term is prefaced by $[u].$ to define the required term A_B.

4. Proof that a combinator-realization of F has been defined

First we define an auxillary language F^*, which is the same as F except for its constants. The following is an inductive definition of the set F_t of constants of F^* of type t, and for each element e of F_t the corresponding constant e^- of F:

(i) F_0 is the set of constants of F of type 0 and e^- is e

(ii) for $t = (t_1, \ldots, t_n)$, F_t is the set of all ordered pairs (g,S) where g is a constant of F of type t and S is a subset of $F_{t_1} \times \ldots \times F_{t_n} \times A$ which satisfies a certain condition C_g to be stated below (here A is the set of all algorithms). Define $(g,S)^-$ to be g.

Clearly we can now define for any formula f of F^* the corresponding formula f^- of F.

4.2 Now we define a notion of $*$-realizability for F^*, as follows. A combinatory term r $*$-realizes a formula f of F^* if and only if r is the normal form of a term assigned in the previous section to f^- and in the case that f has the form

a) a closed atomic formula,

(i) $\langle a_1, \ldots, a_n\rangle \in (\{\langle x_1, \ldots, x_n\rangle : B(x_1, \ldots, x_n)\}, S)$; then

$(a_1, \ldots, a_n, r) \in S$, (ii) $t_1 =_0 t_2$; then r is 0.

b) a closed formula of the form

(i) $A \& B$; then r has the form $\langle r_A, r_B \rangle$ where r_A, r_B *-realize A, B respectively

(ii) $A \vee B$; then r has the form $\langle C, 0 \rangle$ where C *-realizes A, or $\langle C, 1 \rangle$ where C *-realizes B

(iii) $A \supset B$; then r is such that for every r_A which *-realizes A, $\overline{rr_A}$ exists and *-realizes B

(iv) $\sim A$; then A is not *-realizable

(v) $(\exists x) A(x)$; then r has the form $\langle c^-, r_{A(c)} \rangle$ where c is a constant of F^* of the type of x and $r_{A(c)}$ *-realizes $A(c)$

(vi) $(\forall x) A(x)$; then r *-realizes $A(x)$

c) an open formula $A(x_1, \ldots, x_n)$ with just the free variables as displayed; then r is such that for each n-tuple of constants $c_1, \ldots, c_n$ of F^*, c_i of the type of x_i, $i = 1, \ldots, n$, $\overline{r \langle c_1^-, \ldots, c_n^- \rangle}$ exists and *-realizes $A(c_1, \ldots, c_n)$.

4.3 The definition of condition C_g in 4.1 can now be motivated by considering what is necessary in order to *-realize extensionality. Suppose for example that t is (σ) where $\sigma \neq 0$; then we will require that $x = y \supset (x \in_{(\sigma)} z \supset y \in_{(\sigma)} z)$ be *-realized by $[u].[v].[w].W_{(\sigma)}$. In other words for any constants a, b, c of the appropriate types, if R_1 *-realizes $(a, S_a) = (b, S_b)$ and R_2 *-realizes $(a, S_a) \in (g, S_g)$ then $(b, S_b) \in (g, S_g)$ is *-realized by $\overline{W_{(\sigma)} R_1 R_1}$. In view of the definition of *-realizability this amounts to a closure condition C_g on S_g which can (and must to avoid circularity in the definition of *-realizability) be stated without mentioning *-realizability. The case when σ is (ρ) is typical and relatively simple to discuss. In this case the condition C_g is:

For all (a, S_a), $(b, S_b) \in F_\sigma$, if

a) R_1 is the normal form of a term assigned to $a = b$ and is such that for every constant c of F^* of type ρ, $\overline{R_1 c^-}$ exists and has the form $\langle R', R'' \rangle$, where

(i) R' is such that if $(c^-, R_c) \in S_a$ and R_c is the normal form of a term assigned to $c^- \in a$, then $\overline{R' R_c}$ exists and is the normal form of a term assigned to $c^- \in b$ and $(c^-, \overline{R' R_c}) \in S_b$,

(ii) R'' is such that if $(c^-, R_c) \in S_b$ and R_c is the normal form of a term assigned to $c^- \in b$, then $\overline{R'' R_c}$ exists and is the normal form of a term assigned to $c^- \in a$ and $(c^-, \overline{R'' R_c}) \in S_a$,

(iii) $W_{(\sigma)} R$, is defined, and if

b) R_2 is the normal form of a term assigned to $a \in g$ and $(a, R_2) \in S_g$, then $\overline{W_{(\sigma)} R_1 R_2}$ exists and is the normal form of a term assigned to $b \in g$, and $(b, \overline{W_{(\sigma)} R_1 R_2}) \in S_g$.

For other types the definition of C_g is modified in the natural way so as to allow extensionality to be *-realized.

4.4 Next we associate to each constant c of F a constant c^* of F^* by the following inductive definition:

(i) if c is a numeral then c^* is c

(ii) if c is $\{\langle x_1, \ldots, x_n \rangle : B(x_1, \ldots, x_n)\}$, we may assume that all constants d of F occuring in B are such that d^* is defined, and so define c^* to be (c, S) where S is $\{(c_1, \ldots, c_n, r) : r$ *-realizes $(\exists y_1) \ldots (\exists y_n)(y_1 = c_1 \& \ldots \& y_n = c_n \& B^*(y_1, \ldots, y_n)\}$. In order that this define a constant of F^* we need to check that S satisfies condition C_c; this is straightforward from the definition of C_c.

Having defined c^* for all constants c of F we extend the definition in the natural way to define f^* for all formulas f of F. Then we define f to be <u>realized</u> by r if and only if r *-realizes f^*. We check that

(i) all provable formulas of F are realized in this sense

by all the terms assigned to them. (Proof by induction on the length of proof. The axioms are treated individually; the only interesting cases are extensionality and comprehension, and we have taken care that these can be dealt with. It is also easy to see that the property is preserved under deduction.)

(ii) this notion of realizability satisfies the definition of a combinator-realization. This is proved by induction on the complexity of formulas; the only interesting case is closed formulas of the form $(\exists x)A(x)$, realized by r say. Then r has the form $\langle c^{-}, r_{A^*(c)} \rangle$ where $r_{A^*(c)}$ *-realizes $A^*(c)$ and c is a constant of F^* of the type of x. We show that $r_{A^*(c)}$ also *-realizes $A^*((c^{-})^*)$, i.e. that it also realizes $A(c^{-})$.

Consider that by definition of *-realizability $r_{A^*(c)}$ is the normal form of a term assigned to $A(c^{-})$ by a proof in F. Each line of this proof has a unique term assigned to it (the term defined by the subproof which proves that line). Transform the proof by changing each line, say l, to l^*; then each l^* is *-realized by the normal form of the term assigned by the proof to l, as we proved above. In particular the last line is $A^*((c^{-})^*)$ and is *-realized by $r_{A^*(c)}$ as required.

The proof is now complete.

4.5 Observe that the proof did not assume the synactic consistency of F; this is proved since every theorem is realized but e.g. $0 = 1$ is not realized. A similar remark can be made for the later treatment of FC, the extension of F given below which has an axiom of countable choice; but if $A \supset G(A)$ is an axiom then the argument will explicitly assume the syntactic consistency of the theory in question.

5. Adding A G(A) to F

5.1 It is easy to see that the above argument can be repeated with the scheme O. $A \supset \sim\sim G(A)$ added to the logical axioms. Sentences of the scheme may be assigned I and predicates [w].I. To see this we have only to check for a sentence A of F* that if A is provable, $\sim G(A)$ is not; this is a consequence of the assumption of syntactic consistency of the theory.

In the same way we can add other negative (i.e. of the form $\sim A$) or conditionally negative (of the form $C \supset \sim A$) formulas as axioms without altering the above argument. For example a negative form of the full axiom of choice could usefully be assumed in conjunction with $A \supset G(A)$.

6. Adding countable choice

6.1 The theory FC for which we shall give a combinator-realization is not presented as a formal theory. The reason for this is that the theory has some rules of inference which are not presented as formal rules. It is however a corollary of the proof that these rules can be stated as formal rules and hence that the theory can be presented as a formal theory.

6.2 Here we define FC and the assignment of terms to theorems. The language of FC is defined by induction by adding to the clauses which define F the further clause (viii) below. The coding of constants of FC by algorithms is defined at the same time; we agree to code FC by an extension of the coding of F given in 3.4. In particular, code the additional primitive symbols [and] by $\langle 2,16\rangle$, $\langle 2,17\rangle$ respectively; then the additional clause is:

(viii) if r is an algorithm and c is a constant of type $t = (t_1, \ldots, t_n)$ and $1 \leq p < n$ then r,c,p is a constant. Considered as a string of symbols the code of this constant is defined by defining the code of r to be $\langle 3,r\rangle$.

The axioms of F are assigned the same terms as previously; modus ponens and generalization are also treated as before. Next we describe the terms to be assigned to formulas inferred by the auxiliary rules of inference. In the case of b) assign

$$[u_1].\ldots[u_k].\langle\langle m_1,\ldots,m_h,c_1,\ldots,c_k,[x_1].\langle I,I\rangle,\ldots,[x_{h+k}].\langle I,I\rangle,r\rangle,D_{k+1,k+1}(rm_1\ldots m_h),\ [v].\langle D_{h+k+1,2(h+k)+1}v,\ldots,D_{h+k+h,2(h+k)+1}v\rangle\rangle.$$

Call this term Q. In the case of a) assign

$$[q_1].\ldots[q_h].\langle c_1,\ldots,c_k,Q\rangle\ ,$$

where throughout c_i is regarded as an abbreviation for $D_{i,k+1}(rq_1\ldots q_h)$.

Call the above term Q'. Finally we describe the term which is assigned to the choice axiom. We write $E_{h,k}$ for an algorithm (definition omitted) such that for each constant c of FC, $\overline{E_{h,k}c}$ exists and for each algorithm r such that [r,c,h] is a constant of FC, $\overline{E_{h,k}cr}$ exists and is the code of [r,c,h]. Then the term assigned to the choice axiom is $[p].[q].\langle E_{h,k}pq,Q'\rangle$, where each instance of r in this term is regarded as denoting an instance of q.

6.3 We now prove that all terms assigned to theorems of FC have normal forms and that a combinator-realization of FC is thereby defined. The method is an elaboration of the previous argument. As before an auxiliary language is defined, differing only in its constants; the constants of FC* are defined by induction on the following clauses; the corresponding constant of FC is defined simultaneously.

(i) the set F_0 of constants of FC* of type 0 is the set of constants of FC of type 0, and for each $c \in F_0$ the corresponding constants c^- is c

(ii) if $t = (t_1, \ldots, t_n)$ then F_t is defined to be the set of ordered pairs (g,S) such that g is a constant of FC of type t and S is a subset of $F_{t_1} \times \ldots \times F_{t_n} \times A$ satisfying condition C'_g (see below); the corresponding constant of FC is g.

The next step is to define a notion of *-realizability for FC*. The definition is as before except that there is the following additional clause in the definition of *-realizability of closed atomic formulas;

$c_1, \ldots, c_n \in ([r,c,p],S)$ is *-realized by an algorithm r if and only if r is the normal form of a term assigned to $\langle c_1^-, \ldots, c_n^- \rangle \in [r,c,p]$, and $(c_1, \ldots, c_n, r) \in S$.

It is now clear how C'_g is to be defined; changes of notation in the previous definition of C_g suffices.

Next a map $c \to c^*$ from the constants of FC to constants of FC* is defined, by induction following the definition of the vocabulary of FC:

(i) for type O constants, c* is c

(ii) for a constant $\{\langle x_1, \ldots, x_n \rangle : B(x_1, \ldots, x_n)\}$, where all constants in B have been treated at previous stages, write B* for the formula of FC* obtained by changing each constant c of B to c*; then $\{\langle x_1, \ldots, x_n \rangle : B(x_1, \ldots, x_n)\}^*$ is to be $(\{\langle x_1, \ldots, x_n \rangle : B(x_1, \ldots, x_n), S)$ where S is $\{(c_1, \ldots, c_n, r): r \text{ *-realizes } (\exists y_1)\ldots(\exists y_n)(y_1 = c_1 \& \ldots \& y_n = c_n \& B^*(y_1, \ldots, y_n))\}$. As before one checks that S satisfies the appropriate condition.

(iii) for a constant [r,c,p] we may assume that c* has already been defined and we define [r,c,p]* to be ([r,c,p],S) where S is the set of $(c_1, \ldots, c_n, r')$ (where $t = (t_1, \ldots, t_n)$ is the type of c) such that r *-realizes $(\forall n_1)\ldots(\forall n_p)(\exists y_{p+1})\ldots(\exists y_n)(\langle n_1, \ldots, n_p, y_{p+1}, \ldots, y_n \rangle \in c^*$, $c_1, \ldots, c_p$ are constants of type O, r' has the form $\langle d_1, \ldots, d_n, r_1, \ldots, r_n, r \rangle$ where d_i is c_i, $i = 1, \ldots, p$, d_{p+i} is $\overline{D_{i,n-p+1}(rd_1 \ldots d_p)}$, $i = 1, \ldots, n-p$, d r_i *-realizes $c_i = d_i$ *, i = 1, ..., n.

Again, S satisfies the appropriate condition.

Now we must show for every formula f of FC and every term r ·signed to f that $\bar{r}$ exists and *-realizes f*. The argument is · transfinite induction following the definition of prooof in FC. ıe interesting cases are the axiom of countable choice and the related ıxiliary rules of inference. Consider the axiom first.

Assume in the notation used in stating the axiom that c is a ›nstant of FC* of the type of z and that

$\forall n_1) \dots (\forall n_h)(\exists y_1) \dots (\exists y_k)(\langle n_1, \dots, n_h, y_1, \dots, y_k \rangle \in c)$ is *-realized · r. Then $[r, c^-, h]$ is a constant of FC and for each choice of ımerals $m_1, \dots, m_h$ and each i, $1 \leq i \leq h$, $\overline{rm_1 \dots m_i}$ and $\overline{D_{i,k+1}(rm_1 \dots m_h)}$ ·e defined and the latter denotes a constant of FC, c_i^- say, of the ·pe of y_i such that for some c_i

$\overline{{}_{i,i}(rm_1 \dots m_h)}$ exists and *-realizes $(\exists y_i) \dots (\exists y_k)(\langle m_1, \dots, m_h, c_1, \dots,$ $_{i-1}, y_i, \dots, y_k \rangle \in c)$, i = 1, ..., k+1.

Use countable choice in the metatheory to select for each h-tuple $_1, \dots, m_h$ and each i a unique such c_i; denote it by $c_i(m_1, \dots, m_h)$. efine a set S to be the set of $(d_1, \dots, d_{n+k}, r')$ such that r' has ıe form

$\langle d_1, \dots, d_h, c_1(d_1, \dots, d_h), \dots, c_k(d_1, \dots, d_k), r_1, \dots, r_{h+k}, r \rangle$ here r_i *-realizes $c_i = d_i$ and d_i is a numeral, i = 1, ..., h, and r_i -realizes $d_i = c_{i-h}$, i = h+1, ..., k.

Clearly S satisfies the appropriate condition so $([r, c^-, h], S)$ is constant of FC*; denote it by f. It is now straightforward to heck that

$$\forall p_1) \dots (\forall p_h)(\exists z_1) \dots (\exists z_k)(\forall w_1) \dots (\forall w_k)(\langle p_1, \dots, p_h, z_1, \dots, z_k \rangle \in f \;\&$$
$$p_1, \dots, p_h, z_1, \dots, z_k \rangle \in c \;\&\; (\langle p_1, \dots, p_h, w_1, \dots, w_k \rangle \in f \supset z_1 = w_1 \;\& \dots \&\; z_k = w_k))$$

is *-realized by the normal form of the term Q' given in 6.2, and hence that countable choice is *-realized by the normal form of the term given at the end of 6.2.

Consider now the auxiliary rule of inference a). We can assume that $(\forall n_1)\dots(\forall n_h)(\exists y_1)\dots(\exists y_k)(\langle n_1,\dots,n_h,y_1,\dots,y_k\rangle \in c^*$ is *-realized by r and show that the normal form of the term Q' given in 6.2 *-realizes

$$(\forall p_1)\dots(\forall p_h)(\exists z_1)\dots(\exists z_k)(\forall w_1)\dots(\forall w_k)(\langle p_1,\dots,p_h,z_1,\dots,z_k\rangle \in [r,c,h]^* \,\&\, \langle p_1,\dots,p_h,z_1,\dots,z_k\rangle \in c^* \,\&\, (\langle p_1,\dots,p_h,w_1,\dots,w_k\rangle \in [r,c,h]^* \supset z_1=w_1 \,\&\dots\&\, z_k=w_k);$$

We show for any numerals $m_1, \dots, m_h$ that $\overline{D_{k+1,k+1}(rm_1 \dots m_h)}$ *-realizes $\langle m_1, \dots, m_h, c_1^*, \dots, c_k^*\rangle \in c^*$, where c_i is $\overline{D_{i,k+1}(rm_1 \dots m_h)}$ $i = 1, \dots, k$. Consider that by hypothesis $\langle m_1, \dots, m_h, c_1, \dots, c_k\rangle \in c$ has been proved already, by a proof which assigns to it a term whose normal form is $\overline{D_{k+1,k+1}(rm_1 \dots m_h)}$; therefore by the inductive hypothesis this algorithm *-realizes $\langle m_1, \dots, m_h, c_1^*, \dots, c_k^*\rangle \in c^*$ as required.

The argument for the rule b) is now clear and is omitted; we have thus proved that if f is provable with assigned term r then f* is *-realized by $\bar{r}$.

The remainder of the argument is as before: define a provable formula f of FC to be realized by an algorithm r if and only if r*-realizes f*, and show that this is a notion of combinator-realizability by induction on the complexity of f.

6.4 It was remarked at the beginning of this section that the auxiliary rules of inference could be presented as formal rules. This is because in view of the above result we can state e.g. a) as follows:

a) if $(\forall n_1 \dots (\forall n_h)(\exists y_1) \dots (\exists y_k)(\langle n_1, \dots, n_h, y_1, \dots, y_k\rangle \in c)$ has been proved by a proof which assigns a term r, then r has a normal form and the appropriate sentence is inferred, in which the constant denoting the choice function is $[\bar{r},c,h]$.

7. Further results

The other conjectures which I presented at Cambridge referred to analogous statements for a suitable version of Morse set theory. These can also be proved by modification of the above method, and will be discussed elsewhere.

References

1. Staples, J., On constructive fields. To appear, Proc. London Math. Soc.

2. Heyting, A., Infinitistic methods from a finite point of view, pp 185-192 of "Infinitistic methods", Proceedings of the symposium on foundations of mathematics, Warsaw, 1959.

School of Mathematics
BRISTOL, BS8 1TW / England

VIII. THE ARITHMETIC THEORY OF CONSTRUCTIONS

Nicolas D. Goodman[1]

The present paper is essentially an expansion of [1]. In that paper we described a certain theory of constructions which we asserted to be equivalent to intuitionistic arithmetic. This theory was based on a theory of combinators and discriminators which we studied in detail in [2]. The purpose of the present paper is to describe a certain simplification of the theory in [1], called the *arithmetic theory of constructions* (ATC), and to prove the equivalence of that theory to intuitionistic arithmetic for positive sentences. The proof of the full equivalence of the two theories requires techniques of a rather different character and will therefore be postponed to a later paper. We will assume throughout that the reader is familiar with the contents of [2].

1. *The primitive theory of constructions.* Suppose $\mathfrak{A}$ is a set of atomic constants. Then, as described in [2], the theory of discriminators $DL(\mathfrak{A})$ is obtained by adjoining the combinators $\underset{\sim}{K}$ and $\underset{\sim}{S}$ and the discriminators $\underset{\sim}{Q}, \underset{\sim}{\delta}_1$, and $\underset{\sim}{\delta}_2$ to $\mathfrak{A}$ and writing down the evident axioms about these rules. The theory $DL(\mathfrak{A})$, however, contains no minimality principle to guarantee that, as intended, every rule is built up by a finite number of steps of application from atomic rules. As is usual, we formulate this minimality principle as a rule of induction.

1.1. A *theory of induction over* $\mathfrak{A}$ is a theory of discriminators over $\mathfrak{A}$ containing the following additional rule of inference, where we continue the numbering of axioms and rules from our [2]:

XXV. Suppose x does not occur in Δ or in a. Suppose

[1] This work was supported in part by NSF Grant GP13019.

$$\Delta,\ \delta x \equiv \bot \vdash ax \equiv T$$

and

$$\Delta,\ \delta x \equiv T,\ a(\delta_1 x) \equiv T,\ a(\delta_2 x) \equiv T \vdash ax \equiv T.$$

Then $\Delta \vdash ax \equiv T$.

We let PTC($\mathfrak{C}$), the <u>primitive theory of constructions over</u> $\mathfrak{C}$, be the theory of induction over $\mathfrak{C}$ whose axioms and rules of inference are exactly the axioms and rules of inference of DL($\mathfrak{C}$) together with the above <u>rule of induction</u> XXV.

2. <u>The natural numbers</u>. Let $\mathfrak{C}$ be a set of atomic constants, and let Σ be a theory of induction over $\mathfrak{C}$. The first step toward the interpretation of arithmetic in Σ is to see that in Σ we can prove the decidability of the species N of natural numbers and justify induction over N. Recalling how N was defined in our [2], it is plausible that we could do this if we had a principle of induction on pairs -- that is, a principle asserting that everything either fails to be a pair or else is built up in a finite number of steps by D from objects which are not pairs. What we have instead is a principle asserting that everything either is atomic or else is built up in a finite number of steps by application from objects which are atomic. Since x and y are components, although not immediate components, of Dxy, the latter principle should imply the former. Indeed, it suffices to be able to justify a course-of-values induction on components. To do that, we must have a notion of finite quantification over the components of a given object. Thus, given a decidable predicate a of one argument, we wish to define a predicate a" in such a way that a"x has the value T iff ay has the value T for all components y of x, including x itself. We let

$$(") = (\lambda x.\ \Phi(\lambda yz.\ Q(xy)T(\delta y((z\circ\delta_1) \cap_1 (z\circ\delta_2))(KT))(K\bot)y)).$$

For any term a, we let a" = (")a. Using the decidability of Q and δ, we have

2.1. i) $x''y \equiv \underset{\sim}{T} \vdash_{\Sigma} xy \equiv \underset{\sim}{T}$.

ii) $x''y \equiv \underset{\sim}{T} \vdash_{\Sigma} x''(\underset{\sim}{\delta}_i y) \equiv \underset{\sim}{T}$ for $i = 1, 2$.

iii) $xy \equiv \underset{\sim}{T},\ x''(\underset{\sim}{\delta}_1 y) \equiv \underset{\sim}{T},\ x''(\underset{\sim}{\delta}_2 y) \equiv \underset{\sim}{T} \vdash_{\Sigma} x''y \equiv \underset{\sim}{T}$.

2.2 Proposition (Induction on pairs). The following rule is derivable in Σ: Suppose x does not occur in Δ or in a. Suppose

$$\Delta,\ \underset{\sim}{P}x \equiv \underset{\sim}{\bot} \vdash ax \equiv \underset{\sim}{T}$$

and

$$\Delta,\ \underset{\sim}{P}x \equiv \underset{\sim}{T},\ a(\underset{\sim}{D}_1 x) \equiv \underset{\sim}{T},\ a(\underset{\sim}{D}_2 x) \equiv \underset{\sim}{T} \vdash ax \equiv \underset{\sim}{T}.$$

Then $\Delta \vdash ax \equiv \underset{\sim}{T}$.

Proof: Assume the premises of the rule. Evidently we have

$$\Delta,\ \underset{\sim}{\delta}x \equiv \underset{\sim}{\bot} \vdash \underset{\sim}{P}x \equiv \underset{\sim}{\bot}.$$

Hence, using the first premise,

1) $\Delta,\ \underset{\sim}{\delta}x \equiv \underset{\sim}{\bot} \vdash a''x \equiv \underset{\sim}{T}$.

Now let Γ be

$$\Delta,\ \underset{\sim}{\delta}x \equiv \underset{\sim}{T},\ a''(\underset{\sim}{\delta}_1 x) \equiv \underset{\sim}{T},\ a''(\underset{\sim}{\delta}_2 x) \equiv \underset{\sim}{T}.$$

We have

$$\Gamma,\ \underset{\sim}{P}x \equiv \underset{\sim}{\bot} \vdash ax \equiv \underset{\sim}{T},$$

and so

2) $\Gamma,\ \underset{\sim}{P}x \equiv \underset{\sim}{\bot} \vdash a''x \equiv \underset{\sim}{T}$.

Now, we know from [2] that

$$\Gamma,\ \underset{\sim}{P}x \equiv \underset{\sim}{T} \vdash x \equiv \underset{\sim}{D}(\underset{\sim}{D}_1 x)(\underset{\sim}{D}_2 x).$$

Hence

$$\Delta,\ \underset{\sim}{\delta}x \equiv \underset{\sim}{\bot} \vdash ax \equiv \underset{\sim}{T}$$

and

$$\Delta,\ \underset{\sim}{\delta}x \equiv \underset{\sim}{T},\ a(\underset{\sim}{\delta}_1 x) \equiv \underset{\sim}{T},\ a(\underset{\sim}{\delta}_2 x) \equiv \underset{\sim}{T} \vdash ax \equiv \underset{\sim}{T}.$$

Then $\Delta \vdash ax \equiv \underset{\sim}{T}$.

We let PTC($\mathfrak{C}$), the primitive theory of constructions over $\mathfrak{C}$, be the theory of induction over $\mathfrak{C}$ whose axioms and rules of inference are exactly the axioms and rules of inference of DL($\mathfrak{C}$) together with the above rule of induction XXV.

2. The natural numbers. Let $\mathfrak{C}$ be a set of atomic constants, and let Σ be a theory of induction over $\mathfrak{C}$. The first step toward the interpretation of arithmetic in Σ is to see that in Σ we can prove the decidability of the species $\underset{\sim}{N}$ of natural numbers and justify induction over $\underset{\sim}{N}$. Recalling how $\underset{\sim}{N}$ was defined in our [2], it is plausible that we could do this if we had a principle of induction on pairs -- that is, a principle asserting that everything either fails to be a pair or else is built up in a finite number of steps by $\underset{\sim}{D}$ from objects which are not pairs. What we have instead is a principle asserting that everything either is atomic or else is built up in a finite number of steps by application from objects which are atomic. Since x and y are components, although not immediate components, of $\underset{\sim}{D}xy$, the latter principle should imply the former. Indeed, it suffices to be able to justify a course-of-values induction on components. To do that, we must have a notion of finite quantification over the components of a given object. Thus, given a decidable predicate a of one argument, we wish to define a predicate a" in such a way that a"x has the value $\underset{\sim}{T}$ iff ay has the value $\underset{\sim}{T}$ for all components y of x, including x itself. We let

$$('') = (\lambda x.\ \Phi(\lambda yz.\ \underset{\sim}{Q}(xy)\underset{\sim}{T}(\underset{\sim}{\delta}y((z\circ\underset{\sim}{\delta}_1) \cap_1 (z\circ\underset{\sim}{\delta}_2))(\underset{\sim}{K}\underset{\sim}{T}))(\underset{\sim}{K}\underset{\sim}{\bot})y)).$$

For any term a, we let a" = ('')a. Using the decidability of $\underset{\sim}{Q}$ and $\underset{\sim}{\delta}$, we have

2.1. i) $x''y \equiv \mathbf{T} \vdash_{\Sigma} xy \equiv \mathbf{T}$.

ii) $x''y \equiv \mathbf{T} \vdash_{\Sigma} x''(\boldsymbol{\delta}_i y) \equiv \mathbf{T}$ for $i = 1, 2$.

iii) $xy \equiv \mathbf{T},\ x''(\boldsymbol{\delta}_1 y) \equiv \mathbf{T},\ x''(\boldsymbol{\delta}_2 y) \equiv \mathbf{T} \vdash_{\Sigma} x''y \equiv \mathbf{T}$.

2.2 __Proposition__ (Induction on pairs). The following rule is derivable in Σ: Suppose x does not occur in Δ or in a. Suppose

$$\Delta,\ \mathbf{P}x \equiv \boldsymbol{\perp} \vdash ax \equiv \mathbf{T}$$

and

$$\Delta,\ \mathbf{P}x \equiv \mathbf{T},\ a(\mathbf{D}_1 x) \equiv \mathbf{T},\ a(\mathbf{D}_2 x) \equiv \mathbf{T} \vdash ax \equiv \mathbf{T}.$$

Then $\Delta \vdash ax \equiv \mathbf{T}$.

__Proof:__ Assume the premises of the rule. Evidently we have

$$\Delta,\ \boldsymbol{\delta}x \equiv \boldsymbol{\perp} \vdash \mathbf{P}x \equiv \boldsymbol{\perp}.$$

Hence, using the first premise,

1) $\Delta,\ \boldsymbol{\delta}x \equiv \boldsymbol{\perp} \vdash a''x \equiv \mathbf{T}$.

Now let Γ be

$$\Delta,\ \boldsymbol{\delta}x \equiv \mathbf{T},\ a''(\boldsymbol{\delta}_1 x) \equiv \mathbf{T},\ a''(\boldsymbol{\delta}_2 x) \equiv \mathbf{T}.$$

We have

$$\Gamma,\ \mathbf{P}x \equiv \boldsymbol{\perp} \vdash ax \equiv \mathbf{T},$$

and so

2) $\Gamma,\ \mathbf{P}x \equiv \boldsymbol{\perp} \vdash a''x \equiv \mathbf{T}$.

Now, we know from [2] that

$$\Gamma,\ \mathbf{P}x \equiv \mathbf{T} \vdash x \equiv \mathbf{D}(\mathbf{D}_1 x)(\mathbf{D}_2 x).$$

Hence

$$\Gamma,\ \underset{\sim}{P}x \equiv \underset{\sim}{T} \vdash \underset{\sim}{\delta}_2 x \equiv \underset{\sim}{K}(\underset{\sim}{D}_2 x),$$

and so

$$\Gamma,\ \underset{\sim}{P}x \equiv \underset{\sim}{T} \vdash a''(\underset{\sim}{K}(\underset{\sim}{D}_2 x)) \equiv \underset{\sim}{T}.$$

But then, by the same reasoning,

$$\Gamma,\ \underset{\sim}{P}x \equiv \underset{\sim}{T} \vdash a''(\underset{\sim}{D}_2 x) \equiv \underset{\sim}{T}.$$

Thus

3) $\Gamma,\ \underset{\sim}{P}x \equiv \underset{\sim}{T} \vdash a(\underset{\sim}{D}_2 x) \equiv \underset{\sim}{T}.$

A similar argument gives

$$\Gamma,\ \underset{\sim}{P}x \equiv \underset{\sim}{T} \vdash a(\underset{\sim}{D}_1 x) \equiv \underset{\sim}{T}.$$

Therefore, using 3) and the second premise,

$$\Gamma,\ \underset{\sim}{P}x \equiv \underset{\sim}{T} \vdash ax \equiv \underset{\sim}{T},$$

and so

$$\Gamma,\ \underset{\sim}{P}x \equiv \underset{\sim}{T} \vdash a''x \equiv \underset{\sim}{T}.$$

Combining this with 2) by the decidability of $\underset{\sim}{P}$, we have $\Gamma \vdash a''x \equiv \underset{\sim}{T}$. But then, from 1) by the rule of induction, $\Delta \vdash a''x \equiv \underset{\sim}{T}$, and so $\Delta \vdash ax \equiv \underset{\sim}{T}$, as was to be derived.

We can now turn to the treatment of the natural numbers.

2.3. Proposition. The term $\underset{\sim}{N}$ is a decidable predicate of one argument in Σ.

Proof: We let

$$a = (\lambda x.\ \underset{\sim}{Q}\underset{\sim}{T}(\underset{\sim}{N}x) \cup_0 \underset{\sim}{Q} \underset{\sim}{\perp} (\underset{\sim}{N}x)).$$

By Proposition 8.1 of our [2], it suffices to show that $\vdash_{\Sigma} ax \equiv \underset{\sim}{T}$. We do this by induction on pairs. We have $\underset{\sim}{Q}\underset{\sim}{O}x \equiv \underset{\sim}{T} \vdash \underset{\sim}{N}x \equiv \underset{\sim}{T}$, and so

4) $\underset{\sim}{Q}\underset{\sim}{O}x \equiv \underset{\sim}{T} \vdash ax \equiv \underset{\sim}{T}.$

Moreover,

$$QOx \equiv \perp \;,\; Px \equiv \perp \vdash ax \equiv T,$$

Combining this with 4) by the decidability of Q,

5) $Px \equiv \perp \vdash ax \equiv T.$

Now let Δ be

$$Px \equiv T, \quad a(D_1x) \equiv T, \quad a(D_2x) \equiv T.$$

Since, $PO \equiv \perp$, we have $\Delta \vdash QOx \equiv \perp$. Therefore

$$\Delta,\; QK(D_2x) \equiv \perp \vdash Nx \equiv \perp,$$

and so

6) $\Delta,\; QK(D_2x) \equiv \perp \vdash ax \equiv T.$

On the other hand, let Γ be

$$\Delta,\; QK(D_2x) \equiv T.$$

We have

$$\Gamma,\; QT(N(D_1x)) \equiv T \vdash Nx \equiv T,$$

and

$$\Gamma,\; Q\perp(N(D_1x)) \equiv T \vdash Nx \equiv \perp.$$

Thus we clearly have $\Gamma \vdash ax \equiv T$, and so, combining this with 6), $\Delta \vdash ax \equiv T$. Hence, by 5) and the rule of induction on pairs, $\vdash ax \equiv T$, as was to be shown.

2.4. Proposition (Induction on the natural numbers). The following rule is derivable in Σ: Suppose x does not occur in Δ or in a. Suppose $\Delta \vdash aO \equiv T$ and

$$\Delta,\; Nx \equiv T, \quad ax \equiv T \vdash a(DxK) \equiv T.$$

Then

$$\Delta, \underset{\sim}{N}x \equiv \underset{\sim}{T} \vdash ax \equiv \underset{\sim}{T}.$$

<u>Proof</u>: Assume the premises. Let $b = \underset{\sim}{N} \supset_1 a$. We have $\underset{\sim}{NO} \equiv \underset{\sim}{T}$ and $\Delta \vdash a\underset{\sim}{O} \equiv \underset{\sim}{T}$. Hence

7) $\Delta, \underset{\sim}{QO}x \equiv \underset{\sim}{T} \vdash bx \equiv \underset{\sim}{T}.$

However,

$$\Delta, \underset{\sim}{P}x \equiv \underset{\sim}{\perp}, \underset{\sim}{QO}x \equiv \underset{\sim}{\perp} \vdash \underset{\sim}{N}x \equiv \underset{\sim}{\perp},$$

and so

$$\Delta, \underset{\sim}{P}x \equiv \underset{\sim}{\perp}, \underset{\sim}{QO}x \equiv \underset{\sim}{\perp} \vdash bx \equiv \underset{\sim}{T}.$$

Combining this with 7) by the decidability of $\underset{\sim}{Q}$,

8) $\Delta, \underset{\sim}{P}x \equiv \underset{\sim}{\perp} \vdash bx \equiv \underset{\sim}{T}.$

Now let Γ be

$$\Delta, \underset{\sim}{P}x \equiv \underset{\sim}{T}, b(\underset{\sim}{D}_1 x) \equiv \underset{\sim}{T}, b(\underset{\sim}{D}_2 x) \equiv \underset{\sim}{T}.$$

Clearly $\Gamma \vdash \underset{\sim}{QO}x \equiv \underset{\sim}{\perp}$. Thus

$$\Gamma, \underset{\sim}{N}x \equiv \underset{\sim}{T} \vdash \underset{\sim}{QK}(\underset{\sim}{D}_2 x) \equiv \underset{\sim}{T},$$

and so

$$\Gamma, \underset{\sim}{N}x \equiv \underset{\sim}{T} \vdash \underset{\sim}{N}(\underset{\sim}{D}_1 x) \equiv \underset{\sim}{T}.$$

Therefore

$$\Gamma, \underset{\sim}{N}x \equiv \underset{\sim}{T} \vdash a(\underset{\sim}{D}_1 x) \equiv \underset{\sim}{T}.$$

Hence, by the second premise,

$$\Gamma, \underset{\sim}{N}x \equiv \underset{\sim}{T} \vdash ax \equiv \underset{\sim}{T}.$$

But then, by the decidability of $\underset{\sim}{N}$, we have $\Gamma \vdash bx \equiv \underset{\sim}{T}$. Therefore, from 8) by the rule of induction on pairs, $\Delta \vdash bx \equiv \underset{\sim}{T}$. That is,

$$\Delta,\ \underset{\sim}{N}x \equiv \underset{\sim}{T} \vdash ax \equiv \underset{\sim}{T},$$

as was to be derived.

3. The theory SA. We describe a variant, which we call SA for Skolem arithmetic, of the theory PRA^0 of Tait [3]. We suppose given a numerical constant $\overline{0}$ and function constants $\phi^m, \psi^m, \chi^m, \ldots$ of degree m for each $m > 0$. Then we define the terms of SA inductively. First, $\overline{0}$ is a term and each variable v_i is a term. If s is a term, so is s'. If $s_1, \ldots, s_m$ are terms, then $\phi^m(s_1, \ldots, s_m)$ is a term. These are all the terms of SA. As usual, we use x,y,z,u,v,w to denote variables v_i. We use r,s,t to denote terms of SA.

Next we define the formulas of SA inductively. If s and t are terms, then $s = t$ is a formula. If $\mathfrak{A}$ is a formula, so is $\sim\mathfrak{A}$. If $\mathfrak{A}$ and $\mathfrak{B}$ are formulas, then so is $\mathfrak{A} \vee \mathfrak{B}$. These are all the formulas. We use script letters $\mathfrak{A}, \mathfrak{B}, C, \ldots$ to denote formulas. We introduce the other propositional connectives as usual.

We list the axioms and rules of inference of SA:

3.1. i) All tautologies.

ii) If $\mathfrak{A}$ and $\mathfrak{A} \to \mathfrak{B}$, then $\mathfrak{B}$.

iii) If $\mathfrak{A}(x)$, then $\mathfrak{A}(t)$.

iv) $x = x$.

v) $x = y \to (\mathfrak{A}(x) \to \mathfrak{A}(y))$.

vi) $x' \neq 0$.

vii) $x' = y' \to x = y$.

viii) If $\mathfrak{A}(\overline{0})$ and $\mathfrak{A}(x) \to \mathfrak{A}(x')$, then $\mathfrak{A}(x)$.

ix) If all the variables occuring in t are among $v_0, \ldots, v_n$, then introduce ϕ^{n+1} with the axiom $\phi(v_0, \ldots, v_n) = t$.

x) If ψ^{n+1} and χ^{n+2} have been introduced, then introduce ϕ^{n+1} with the axioms

$$\phi(v_1, \dots, v_n, \overline{0}) = \psi(v_1, \dots, v_n, \overline{0})$$

$$\phi(v_1, \dots, v_n, v_0') = \chi(v_1, \dots, v_n, \phi(v_1, \dots, v_n, v_0), v_0).$$

We assume that every function constant of SA is introduced as in ix) or x).

We write SA $\vdash \mathfrak{A}$ to mean that $\mathfrak{A}$ is a theorem of SA.

4. Primitive recursive terms. Again let $\mathfrak{C}$ be a set of atomic constants and Σ a theory of induction over $\mathfrak{C}$. We assign a term $\|t\|$ of Σ to each term t of SA and a term $\|\phi\|$ of Σ to each function constant ϕ of SA recursively as follows:

4.1. i) $\|v_i\| = v_i$.

ii) $\|\overline{0}\| = \underset{\sim}{0}$.

iii) $\|s'\| = \underset{\sim}{D}\|s\|\underset{\sim}{K}$.

iv) $\|\phi(s_0, \dots, s_n)\| = \|\phi\| \, \|s_0\| \dots \|s_n\|$.

v) If ϕ is introduced by 3.1. ix), then $\|\phi\| = (\lambda v_1 \dots v_n . t)$.

vi) If ϕ is introduced by 3.1. x), then

$$\|\phi\| = (\lambda v_1 \dots v_n v_0 . \underset{\sim}{R}(\|\psi\| v_1 \dots v_n \underset{\sim}{0})(\|\chi\| v_1 \dots v_n) v_0).$$

4.2. If t is a term of SA and the variables occuring in t are exactly $x_0, \dots, x_{k-1}$, then we let $\Gamma(t)$ be the set

$$\underset{\sim}{N}x_0 \equiv \underset{\sim}{T}, \dots, \underset{\sim}{N}x_{k-1} \equiv \underset{\sim}{T}.$$

Intuitively, $\Gamma(t)$ asserts that all the variables in t denote natural numbers. A straightforward induction gives

4.3. Lemma. If t is a term of SA, then $\Gamma(t) \vdash_\Sigma \underset{\sim}{N}\|t\| \equiv \underset{\sim}{T}$. Moreover, if ϕ^{k+1} is a function constant of SA, then

$$\underset{\sim}{N}v_0 \equiv \underset{\sim}{T}, \dots, \underset{\sim}{N}v_k \equiv \underset{\sim}{T} \vdash_\Sigma \underset{\sim}{N}(\|\phi\| v_0 \dots v_k) \equiv \underset{\sim}{T}.$$

We also easily have

4.4. Lemma. Suppose $t(x)$, s_0, s_1 are terms of SA. Let $\Delta = \Gamma(t(s_0)) \cup \Gamma(t(s_1))$. Then

$$\Delta, \|s_0\| \equiv \|s_1\| \vdash_{\Sigma} \|t(s_0)\| \equiv \|t(s_1)\|.$$

5. The interpretation of SA. Next we assign a term $\|\mathfrak{A}\|$ of Σ to each formula $\mathfrak{A}$ of SA recursively as follows:

5.1. i) $\|s = t\| = \underset{\sim}{Q}\|s\|\ \|t\|.$

ii) $\|\sim \mathfrak{A}\| = \|\mathfrak{A}\| \underset{\sim}{\perp} \underset{\sim}{T}.$

iii) $\|\mathfrak{A} \vee \mathfrak{B}\| = \|\mathfrak{A}\| \cup_0 \|\mathfrak{B}\|.$

Intuitively, $\|\mathfrak{A}\|$ is $\underset{\sim}{T}$ or $\underset{\sim}{\perp}$ depending on the truth-value of the formula $\mathfrak{A}$.

We let $\Gamma(s = t)$ be $\Gamma(s) \cup \Gamma(t)$. Moreover, $\Gamma(\sim\mathfrak{A})$ is $\Gamma(\mathfrak{A})$, and $\Gamma(\mathfrak{A} \vee \mathfrak{B})$ is $\Gamma(\mathfrak{A}) \cup \Gamma(\mathfrak{B})$. Thus, for any $\mathfrak{A}$, the set $\Gamma(\mathfrak{A})$ asserts that all the variables occuring in $\mathfrak{A}$ denote natural numbers.

In order to show that this interpretation is correct, we need a few preliminary lemmas. All of them are proved by induction on the structure of the formulas involved.

5.2. Lemma. Suppose $\mathfrak{A}$ is a formula of SA. Then

$$\Gamma(\mathfrak{A}) \vdash_{\Sigma} \underset{\sim}{Q}\underset{\sim}{T}\|\mathfrak{A}\| \cup_0 \underset{\sim}{Q}\ \underset{\sim}{\perp}\|\mathfrak{A}\| \equiv \underset{\sim}{T}.$$

5.3. Lemma. Suppose $\mathfrak{A}$ is a formula of SA. Suppose the distinct equations occuring in $\mathfrak{A}$ are $\mathfrak{A}_0, \ldots, \mathfrak{A}_k$. Let ζ be an assignment of the truth-values $\underset{\sim}{T}$ and $\underset{\sim}{\perp}$ to the $\mathfrak{A}_i$, and let $\zeta(\mathfrak{A})$ be the corresponding truth-value of $\mathfrak{A}$. Let Δ_ζ consist of the equations $\|\mathfrak{A}_i\| \equiv \zeta(\mathfrak{A}_i)$ for $i = 0, \ldots, k$. Then

$$\Delta_\zeta \vdash_{\Sigma} \|\mathfrak{A}\| \equiv \zeta(\mathfrak{A}).$$

5.4. Lemma. Suppose $\mathfrak{A}(x)$ is a formula of SA and t and s are terms. Let $\Delta = \Gamma(\mathfrak{A}(t)) \cup \Gamma(\mathfrak{A}(s))$. Then

$$\Delta, \|t\| = \|s\| \vdash_{\Sigma} \|\mathfrak{A}(t)\| \equiv \|\mathfrak{A}(s)\|.$$

5.5. Lemma. Suppose $\mathfrak{A}$ and $\mathfrak{B}$ are formulas of SA and $\Gamma(\mathfrak{A}) \vdash_{\Sigma} \|\mathfrak{B}\| \equiv \underset{\sim}{T}$. Then $\Gamma(\mathfrak{B}) \vdash_{\Sigma} \|\mathfrak{B}\| \equiv \underset{\sim}{T}$.

Proof: Suppose the variables free in $\mathfrak{A}$ but not in $\mathfrak{B}$ are exactly $x_1, \ldots, x_k$. Then, by a weakening,

$$\Gamma(\mathfrak{B}), \underset{\sim}{N}x_1 \equiv \underset{\sim}{T}, \ldots, \underset{\sim}{N}x_k \equiv \underset{\sim}{T} \vdash_{\Sigma} \|\mathfrak{B}\| \equiv \underset{\sim}{T}.$$

But no x_i occurs in $\Gamma(\mathfrak{B})$ or in $\|\mathfrak{B}\|$. Hence, substituting,

$$\Gamma(\mathfrak{B}), \underset{\sim}{N}\underset{\sim}{O} \equiv \underset{\sim}{T}, \ldots, \underset{\sim}{N}\underset{\sim}{O} \equiv \underset{\sim}{T} \vdash_{\Sigma} \|\mathfrak{B}\| \equiv \underset{\sim}{T}.$$

Then the result follows by a cut.

We now come to the theorem which asserts the correctness of our interpretation of SA.

5.6. Theorem. Suppose $SA \vdash \mathfrak{A}$. Then

$$\Gamma(\mathfrak{A}) \vdash_{\Sigma} \|\mathfrak{A}\| \equiv \underset{\sim}{T}.$$

Proof: By induction on the length of the proof of $\mathfrak{A}$ in SA. All the axioms and rules are quite easily treated. As an illustration, we consider the rule of induction 3.1. viii), which is the most difficult. Suppose we have

$$\Gamma(\mathfrak{A}(\overline{0})) \vdash_{\Sigma} \|\mathfrak{A}(\overline{0})\| \equiv \underset{\sim}{T},$$

and

$$\Gamma(\mathfrak{A}(\overline{0})), \underset{\sim}{N}x \equiv \underset{\sim}{T} \vdash_{\Sigma} \|\mathfrak{A}(x) \rightarrow \mathfrak{A}(x')\| \equiv \underset{\sim}{T}.$$

Let $a = (\lambda x. \|\mathfrak{A}(x)\|)$. Then clearly

$$\Gamma(\mathfrak{A}(\overline{0})) \vdash_{\Sigma} a\underset{\sim}{0} \equiv \underset{\sim}{T}.$$

Moreover,

$$\Gamma(\mathfrak{A}(\overline{0})),\ \underset{\sim}{N}x \equiv \underset{\sim}{T},\ ax \equiv \underset{\sim}{T} \vdash_{\Sigma} a(\underset{\sim}{D}x\underset{\sim}{K}) \equiv \underset{\sim}{T}.$$

Hence, by Proposition 2.4,

$$\Gamma(\mathfrak{A}(\overline{0})),\ \underset{\sim}{N}x \equiv \underset{\sim}{T} \vdash_{\Sigma} ax \equiv \underset{\sim}{T}.$$

That is,

$$\Gamma(\mathfrak{A}(x)) \vdash_{\Sigma} \|\mathfrak{A}(x)\| \equiv \underset{\sim}{T},$$

as was to be shown.

6. The arithmetic theory of constructions. All of constructive mathematics is ultimately based on finitary computation and on reasoning about finitary computation. The purpose of the theory of discriminators in [2] was to describe these computations and to axiomatize their most immediately evident properties. By reflecting on the structure of the universe described in that theory, we arrived above at a theory PTC which contains among its theorems insights which are not immediately evident but require proof. These insights can always be put into the following form: a particular rule a, applied to any element of this basic universe, always gives the value $\underset{\sim}{T}$. Such an insight is evidently in AE form. It asserts that a certain rule is actually a function - specifically, the constant function whose value is always $\underset{\sim}{T}$. We can think of the insight as the visualization, or grasping, of the totality of the computations of the values of the function. The formal proof, which is a finite object, is not the insight but only a guide to aid us in the visualization of this infinite structure of computations.

We now have objects of two different kinds. On the one hand, there are the rules which are the subject matter of PTC. Let us say that they are of level zero. On the other hand, there are insights about the objects of level zero and rules built up from these insights and acting on these insights. Let us say that the latter are of level one. Given an object we can tell, simply by looking at that object and without reflecting on any infinite totality, whether or not it is of level zero. Thus we add an atomic constant $\underset{\sim}{L}_0$ with the intended interpretation that $\underset{\sim}{L}_0 x$ is $\underset{\sim}{T}$ or $\underset{\sim}{\perp}$ depending on whether or not x is an object of level zero. Evidently $\underset{\sim}{L}_0$ is itself of level zero. Observe that the level zero is <u>closed</u> in the sense that no rule of level zero can involve reflecting on the whole of level zero, and that therefore any such rule, if it is defined on an argument of level zero, gives a value of level zero.

Given a rule of level zero, it always makes sense to try to visualize the totality of the computations of a on arguments of level zero and to see that they always give the value $\underset{\sim}{T}$. If we succeed in doing this, we construct a new object, an insight, which is itself of level one. In general, of course, we will not succeed, since the desired insight may be false or since we may not be able, at the moment, to settle the question one way or the other. Thus we have a rule, acting on level zero, which is necessarily partial and which gives values in level one. We introduce an atomic constant $\underset{\sim}{\rho}_1$ to denote this rule. If $\underset{\sim}{\rho}_1 x$ is defined, then x is of level zero, $\underset{\sim}{\rho}_1 x$ is of level one, and $\underset{\sim}{\rho}_1 x$ is the visualization of the totality of computations of x on arguments of level zero together with the recognition that they always give the value $\underset{\sim}{T}$.

Let us suppose we have a rule a which is not of level zero. Suppose we apply a to an object b of level zero and obtain a result, ab, which is again of level zero. Then it might happen that the computation of the value ab involves the visualization of the whole of level zero or of some even more complicated totality. It seems to us, however, that that visualization cannot actually be essential to the computation. For, the visualization of such a totality of computations cannot make it possible for us to make any computation other than the ones being visualized. Moreover, since the result of the computation does not involve any infinite insight, the computation of ab can only be using, so to speak, a bounded part of the infinite visualization. In other words, we are asserting that if an infinite visualization is essential to a computation, then it must occur in the result of the computation. We may imagine, for example, that the computations are given in a normal form in which the act of visualization plays the role of an

ω-rule. Then, following Brouwer's proof of the bar theorem, we are saying that any operation which plays the role of a cut-rule can be eliminated, so that only visualizations which occur in the result are necessary for the computation.

It follows from this line of reasoning that if we are given a rule a which is not of level zero, then we can think of it as a rule c of level zero by simply ignoring any request for a visualization which cannot be carried out at level zero. This new rule c, on arguments of level zero, will give the same value as a whenever that value is itself of level zero. The rule which leads from a to c is itself of level zero, and we call it $\underset{\sim}{F}_0$. The rule $c \equiv \underset{\sim}{F}_0 a$ is essentially simpler than a, and we cannot hope to recapture a from it. Nevertheless, a is involved in its meaning, so that we can have $\underset{\sim}{F}_0 a$ *intensionally* equal to $\underset{\sim}{F}_0 b$ only if $a \equiv b$.

For convenience, let us think of our levels as cumulative, so that everything of level zero is automatically also of level one. Then we are in a position to write down a theory about level one and prove theorems about it. Thus we can repeat the entire construction above and consider insights which involve visualizing the whole of level one. These insights will be *of level two*. We introduce constants $\underset{\sim}{L}_1$, $\underset{\sim}{P}_2$, and $\underset{\sim}{F}_1$ with the obvious interpretations. Continuing in this way, we can construct the *nth* level for arbitrary n.

Several people have suggested to us that it ought to be possible to iterate this construction into the transfinite. Formally, of course, this should not be difficult. Nevertheless, we cannot bring ourselves to believe that such an iteration would make constructive sense. After all, the idea of the construction is just that if we are given a domain as a well-defined potential totality, then we can understand what it means to prove a universal assertion about that totality. The principle which generates the totality also explains the meaning of a universal quantifier over the totality. But this principle of generation changes, becoming more complex and less surveyable, as we go up the hierarchy of levels. The rule which leads from the *nth* level to the (n+1)*st* level is not a rule which we can understand. If it were, then we could understand the notion of proof in an absolute sense and could visualize the entire constructive universe. But that leads at once to self-reflexive paradoxes. Let us put the point another way. We understand computation, thinking about computation, and so on. But we do not understand thinking about thinking in an absolute sense. Thus it seems to us that we can never do constructive mathematics except within some level. Every proof should be interpretable by taking the variables as ranging only over objects of some fixed level (depending on the proof). Hence we can see no way to give a constructive interpretation to $\underset{\sim}{L}_{\omega}$.

We assume that for each $p = 0,1,2, \ldots$ we are given atomic constants $\underset{\sim}{L}_p$, $\underset{\sim}{F}_p$, and $\underset{\sim}{\wp}_{p+1}$. Suppose $\mathcal{G}$ is a set of atomic constants.

6.1. An <u>arithmetic theory of constructions over</u> $\mathcal{G}$ is a theory Σ of induction over $\mathcal{G}$ together with all the constants $\underset{\sim}{L}_p$, $\underset{\sim}{F}_p$, $\underset{\sim}{\wp}_{p+1}$ containing the following additional axioms and rules of inference (for further motivation, see [1]):

XXVI. $\underset{\sim}{L}_p$ is a decidable predicate of one argument.

XXVII. $\underset{\sim}{L}_p x \equiv \underset{\sim}{T} \vdash \underset{\sim}{L}_{p+1} x \equiv \underset{\sim}{T}$.

XXVIII. $\underset{\sim}{L}_p x \equiv \underset{\sim}{T},\ \underset{\sim}{L}_p y \equiv \underset{\sim}{T},\ yx \equiv yx \vdash \underset{\sim}{L}_p (yx) \equiv \underset{\sim}{T}$.

XXIX. If A is $\underset{\sim}{K}$, $\underset{\sim}{S}$, $\underset{\sim}{Q}$, $\underset{\sim}{\delta}_1$, $\underset{\sim}{\delta}_2$, then $\vdash \underset{\sim}{L}_0 A \equiv \underset{\sim}{T}$.

XXX. $\vdash \underset{\sim}{L}_p \underset{\sim}{L}_p \equiv \underset{\sim}{T}$.

XXXI. $\vdash \underset{\sim}{L}_p \underset{\sim}{L}_{p+1} \equiv \underset{\sim}{\perp}$.

XXXII. $\vdash \underset{\sim}{L}_p \underset{\sim}{\wp}_{p+1} \equiv \underset{\sim}{\perp}$.

XXXIII. $\vdash \underset{\sim}{L}_{p+1} \underset{\sim}{\wp}_{p+1} \equiv \underset{\sim}{T}$.

XXXIV. $\underset{\sim}{\wp}_{p+1} x \equiv \underset{\sim}{\wp}_{p+1} x \vdash \underset{\sim}{L}_p x \equiv \underset{\sim}{T}$.

XXXV. $\underset{\sim}{\wp}_{p+1} x \equiv \underset{\sim}{\wp}_{p+1} x \vdash \underset{\sim}{\delta}_1 (\underset{\sim}{\wp}_{p+1} x) \equiv \underset{\sim}{\wp}_{p+1}$.

XXXVI. $\underset{\sim}{\wp}_{p+1} x \equiv \underset{\sim}{\wp}_{p+1} x \vdash \underset{\sim}{\delta}_2 (\underset{\sim}{\wp}_{p+1} x) \equiv x$.

XXXVII. $\underset{\sim}{\wp}_{p+1} y \equiv \underset{\sim}{\wp}_{p+1} y,\ \underset{\sim}{L}_p x \equiv \underset{\sim}{T} \vdash yx \equiv \underset{\sim}{T}$.

XXXVIII. Suppose x does not occur in Δ or in a. Suppose $\Delta, \underset{\sim}{L}_p x \equiv \underset{\sim}{T} \vdash ax \equiv \underset{\sim}{T}$. Then $\Delta, \underset{\sim}{L}_p a \equiv \underset{\sim}{T} \vdash \underset{\sim}{\wp}_{p+1} a \equiv \underset{\sim}{\wp}_{p+1} a$.

XXXIX. $\vdash \underset{\sim}{L}_p \underset{\sim}{F}_p \equiv \underset{\sim}{T}$.

XL. $\vdash \underset{\sim}{L}_p \underset{\sim}{F}_{p+1} \equiv \underset{\sim}{\bot}$.

XLI. $\vdash \underset{\sim}{L}_p (\underset{\sim}{F}_p x) \equiv \underset{\sim}{T}$.

XLII. $\vdash \underset{\sim}{L}_p (\underset{\sim}{F}_{p+1} x) \equiv \underset{\sim}{\bot}$.

XLIII. $\vdash \underset{\sim}{\delta}_i (\underset{\sim}{F}_p x) \equiv \underset{\sim}{F}_p x$ for $i = 1, 2$.

XLIV. $\underset{\sim}{L}_p x \equiv \underset{\sim}{T}, \underset{\sim}{L}_p (yx) \equiv \underset{\sim}{T} \vdash \underset{\sim}{F}_p yx \equiv yx$.

XLV. If A is a constant, then $\vdash \underset{\sim}{Q} A (\underset{\sim}{F}_p x) \equiv \underset{\sim}{\bot}$.

XLVI. $\underset{\sim}{F}_p x \equiv \underset{\sim}{F}_p y \vdash x \equiv y$.

We let ATC($\mathfrak{C}$) be the arithmetic theory of constructions over $\mathfrak{C}$ whose only axioms and rules of inference are those required by the above definition.

7. The formalization of intuitionistic arithmetic. We give an inductive definition of the formulas of HA: If s and t are terms of SA, then $s = t$ is a formula. If $\mathfrak{A}$ and $\mathfrak{B}$ are formulas, then so are $\mathfrak{A} \vee \mathfrak{B}$, $\mathfrak{A} \wedge \mathfrak{B}$, and $\mathfrak{A} \rightarrow \mathfrak{B}$. If $\mathfrak{A}$ is a formula, then so are $\bigvee x \mathfrak{A}$ and $\bigwedge x \mathfrak{A}$. These are the only formulas.

It is convenient for our purposes to formulate HA as a sequent calculus. We use Δ, Γ, Λ to denote finite sets of formulas of HA. Then a sequent of HA is an expression $\Delta \Rightarrow \mathfrak{A}$, where Δ is a finite set of formulas of HA and $\mathfrak{A}$ is a formula of HA.

7.1. We list the axioms and rules of HA:

i) $\mathfrak{A} \Rightarrow \mathfrak{A}$.

ii) If $\Delta \Rightarrow \mathfrak{A}$, then $\Delta, \mathfrak{B} \Rightarrow \mathfrak{A}$.

iii) If $\Delta \Rightarrow \mathfrak{B}$ and $\Gamma, \mathfrak{B} \Rightarrow \mathfrak{A}$, then $\Delta, \Gamma \Rightarrow \mathfrak{A}$.

iv) If $\Delta \Rightarrow \mathfrak{A}$ and $\Delta \Rightarrow \mathfrak{B}$, then $\Delta \Rightarrow \mathfrak{A} \wedge \mathfrak{B}$.

v) If $\Delta, \mathfrak{A} \Rightarrow C$ or $\Delta, \mathfrak{B} \Rightarrow C$, then $\Delta, \mathfrak{A} \wedge \mathfrak{B} \Rightarrow C$.

vi) If $\Delta \Rightarrow \mathfrak{A}$ or $\Delta \Rightarrow \mathfrak{B}$, then $\Delta \Rightarrow \mathfrak{A} \vee \mathfrak{B}$.

vii) If $\Delta, \mathfrak{A} \Rightarrow C$ and $\Delta, \mathfrak{B} \Rightarrow C$, then $\Delta, \mathfrak{A} \vee \mathfrak{B} \Rightarrow C$.

viii) If $\Delta, \mathfrak{A} \Rightarrow \mathfrak{B}$, then $\Delta \Rightarrow \mathfrak{A} \rightarrow \mathfrak{B}$.

ix) If $\Delta \Rightarrow \mathfrak{A}$ and $\Delta, \mathfrak{B} \Rightarrow C$, then $\Delta, \mathfrak{A} \rightarrow \mathfrak{B} \Rightarrow C$.

x) If $\Delta \Rightarrow \mathfrak{A}(t)$, then $\Delta \Rightarrow \mathrm{V}x\mathfrak{A}(x)$.

xi) If x is not free in Δ or in $\mathfrak{A}$, and if $\Delta, \mathfrak{B}(x) \Rightarrow \mathfrak{A}$, then $\Delta, \mathrm{V}y\mathfrak{B}(y) \Rightarrow \mathfrak{A}$.

xii) If $\Delta, \mathfrak{B}(t) \Rightarrow \mathfrak{A}$, then $\Delta, \Lambda x\,\mathfrak{B}(x) \Rightarrow \mathfrak{A}$.

xiii) If x is not free in Δ, and if $\Delta \Rightarrow \mathfrak{A}(x)$, then $\Delta \Rightarrow \Lambda y\,\mathfrak{A}(y)$.

xiv) $\Rightarrow x = x$.

xv) $x = y, \mathfrak{A}(x) \Rightarrow \mathfrak{A}(y)$.

xvi) $0 = x' \Rightarrow 0 = 1$.

xvii) $x' = y' \Rightarrow x = y$.

xviii) If $\mathfrak{A}$ is a defining equation for one of the primitive recursive functions, then $\Rightarrow \mathfrak{A}$.

xix) Suppose x is not free in Δ. Suppose $\Delta \Rightarrow \mathfrak{A}(0)$ and $\Delta, \mathfrak{A}(x) \Rightarrow \mathfrak{A}(x')$. Then $\Delta \Rightarrow \mathfrak{A}(x)$.

8. The interpretation of HA. Throughout this section and the next, let Σ be a fixed arithmetic theory of constructions. Then, roughly as in our [1], we give an interpretation of HA in Σ. The motivation for the particular interpretation we give is exactly as in [1].

8.1. We assign a level, $\ell(\mathfrak{A})$, to each formula $\mathfrak{A}$ of HA:

i) $\ell(s = t) = 0$.

ii) $\ell(\mathfrak{A} \vee \mathfrak{B}) = \ell(\mathfrak{A} \wedge \mathfrak{B}) = \max\{\ell(\mathfrak{A}), \ell(\mathfrak{B})\}$.

iii) $\ell(\mathfrak{A} \rightarrow \mathfrak{B}) = 1 + \max\{\ell(\mathfrak{A}), \ell(\mathfrak{B})\}$.

iv) $\ell(\mathrm{V}x\mathfrak{A}) = \ell(\mathfrak{A})$.

v) $\ell(\Lambda x\,\mathfrak{A}) = 1 + \ell(\mathfrak{A})$.

Next we assign a term $|\mathfrak{A}|$ to each formula $\mathfrak{A}$. Intuitively, $|\mathfrak{A}|x$ is $\underset{\sim}{T}$ or $\underset{\sim}{\perp}$ depending on whether or not x is a proof of $\mathfrak{A}$.

8.2. The definition of $|\mathfrak{A}|$ is recursive:

i) $|s = t| = (\lambda x.\ Q\|s\|\ \|t\| \cap_0 QKx)$.

ii) $|\mathfrak{A} \vee \mathfrak{B}| = |\mathfrak{A}| \cup_1 |\mathfrak{B}|$.

iii) $|\mathfrak{A} \wedge \mathfrak{B}| = (\lambda x.\ |\mathfrak{A}|(D_1 x) \cap_0 |\mathfrak{B}|(D_2 x))$.

iv) Let $p = \max\{\ell(\mathfrak{A}), \ell(\mathfrak{B})\}$. Then

$$|\mathfrak{A} \to \mathfrak{B}| = (\lambda x.\ L_p(D_1 x) \cap_0 \delta(D_2 x) \cap_0 Q(\delta_1(D_2 x))P_{p+1} \cap_0 Q(\delta_2(D_2 x))(|\mathfrak{A}| \supset_1 |\mathfrak{B}| \circ (D_1 x))).$$

v) $|\vee y\mathfrak{A}| = (\lambda x.\ N(D_1 x)) \cap_0 (\lambda y.\ |\mathfrak{A}|)(D_1 x)(D_2 x))$.

vi) Let $p = \ell(\mathfrak{A})$. Then

$$|\wedge y\,\mathfrak{A}| = (\lambda x.\ L_p(D_1 x) \cap_0 \delta(D_2 x) \cap_0 Q(\delta_1(D_2 x))P_{p+1} \cap_0 Q(\delta_2(D_2 x))(N \supset_1 S(\lambda y.\ |\mathfrak{A}|)(D_1 x))).$$

Given any formula $\mathfrak{A}$, let $\Gamma(\mathfrak{A})$ be the set of all equations $Ny \equiv T$ for variables y free in $\mathfrak{A}$. For any term t of SA, we have

$$\Gamma(t) \vdash_\Sigma L_0\|t\| \equiv T,$$

since, using induction, we certainly have

$$Nx \equiv T \vdash_\Sigma L_0 x \equiv T.$$

Using this we see that, if $\ell(\mathfrak{A}) = p$, then

$$\Gamma(\mathfrak{A}) \vdash_\Sigma L_p|\mathfrak{A}| \equiv T.$$

Moreover, the term $|\mathfrak{A}|$ is a decidable predicate of one argument in Σ over $\Gamma(\mathfrak{A})$. If $\ell(\mathfrak{A}) = p$, then

$$\Gamma(\mathfrak{A}),\ |\mathfrak{A}|y \equiv T \vdash_\Sigma L_p y \equiv T.$$

Finally, we need a lemma to cope with substitution.

8.2. Lemma. Suppose $\mathfrak{A}(x)$ is a formula of HA and s and t are terms of SA. Then there is a term a of Σ such that

$$\Gamma(\mathfrak{A}(x)),\ \Gamma(s),\ \Gamma(t),\ \|s\| \equiv \|t\|,\ |\mathfrak{A}(s)|y \equiv T \vdash_\Sigma |\mathfrak{A}(t)|(ay) \equiv T.$$

9. <u>The correctness of the interpretation of HA</u>. Suppose that $\mathfrak{A}_1, \ldots, \mathfrak{A}_k, \mathfrak{A}$ are formulas of HA. Then we write

$$\mathfrak{A}_1, \ldots, \mathfrak{A}_k \Rightarrow_a \mathfrak{A}$$

to mean that

$$\Gamma(\mathfrak{A}_1), \ldots, \Gamma(\mathfrak{A}_k), |\mathfrak{A}_1|y_1 \equiv \underset{\sim}{T}, \ldots, |\mathfrak{A}_k|y_k \equiv \underset{\sim}{T} \vdash_\Sigma |\mathfrak{A}|a \equiv \underset{\sim}{T}.$$

If Δ is $\mathfrak{A}_1, \ldots, \mathfrak{A}_k$, we often write $\Gamma(\Delta)$ for $\Gamma(\mathfrak{A}_1), \ldots, \Gamma(\mathfrak{A}_k)$, and write $|\Delta|y$ for

$$|\mathfrak{A}_1|y_1 \equiv \underset{\sim}{T}, \ldots, |\mathfrak{A}_k|y_k \equiv \underset{\sim}{T}.$$

Then $\Delta \Rightarrow_a \mathfrak{A}$ means that

$$\Gamma(\Delta), \Gamma(\mathfrak{A}), |\Delta|y \vdash_\Sigma |\mathfrak{A}|a \equiv \underset{\sim}{T}.$$

9.1. <u>Theorem</u>. Suppose the sequent $\Delta \Rightarrow \mathfrak{A}$ is provable in HA. Then there is a <u>term</u> a of Σ such that $\Delta \Rightarrow_a \mathfrak{A}$.

<u>Proof</u>: By induction on the length of the given proof of the sequent $\Delta \Rightarrow \mathfrak{A}$. We must consider cases according to the form of the last step of the given proof. Most of these cases are either trivial or routine. We consider those few cases which present some interest, numbering them as in 7.1. The reader should note how closely our arguments resemble the usual intuitive justifications of these rules.

iii) Suppose $\Delta \Rightarrow_a \mathfrak{B}$ and $\Gamma, \mathfrak{B} \Rightarrow_a \mathfrak{A}$.
Suppose, as a matter of fact, that

$$\Gamma(\Gamma), \Gamma(\mathfrak{B}), |\Gamma|y, |\mathfrak{B}|z \equiv \underset{\sim}{T} \vdash_\Sigma |\mathfrak{A}|b \equiv \underset{\sim}{T}.$$

Then, taking $c = [a/z]b$, it is clear that $\Delta, \Gamma \Rightarrow_c \mathfrak{A}$.

vi) Suppose $\Delta, \mathfrak{A} \Rightarrow_a C$ and $\Delta, \mathfrak{B} \Rightarrow_b C$.

In order to verify $\Delta, \mathfrak{A} \vee \mathfrak{B} \Rightarrow C$, we must show how to get from a proof z of $\mathfrak{A} \vee \mathfrak{B}$ to a proof of C. For this it suffices to determine whether z is a proof of $\mathfrak{A}$ or not. For, if z is a proof of $\mathfrak{A}$ give a. If z is not a proof of $\mathfrak{A}$, it must be a proof of $\mathfrak{B}$ and so give b. Thus, suppose

$$\Gamma(\Delta),\ \Gamma(\mathfrak{A}),\ \Gamma(C),\ |\Delta|y,\ |\mathfrak{A}|z \equiv \underset{\sim}{T} \vdash_{\Sigma} |C|a \equiv \underset{\sim}{T}$$

and

$$\Gamma(\Delta),\ \Gamma(\mathfrak{B}),\ \Gamma(C),\ |\Delta|y,\ |\mathfrak{B}|z \equiv \underset{\sim}{T} \vdash_{\Sigma} |C|b \equiv \underset{\sim}{T}.$$

Let $c = |\mathfrak{A}|zab$. Then, using the decidability of $|\mathfrak{A}|$, we have

$$\Gamma(\Delta),\ \Gamma(\mathfrak{A} \vee \mathfrak{B}),\ |\Delta|y,\ |\mathfrak{A} \vee \mathfrak{B}|z \equiv \underset{\sim}{T} \vdash_{\Sigma} |C|c \equiv \underset{\sim}{T}.$$

viii) Suppose we have

$$\Gamma(\Delta),\ \Gamma(\mathfrak{A}),\ \Gamma(\mathfrak{B}),\ |\Delta|y,\ |\mathfrak{A}|z \equiv \underset{\sim}{T} \vdash_{\Sigma} |\mathfrak{B}|a \equiv \underset{\sim}{T}.$$

Then, in order to have a proof of $\mathfrak{A} \to \mathfrak{B}$, we must find a function from proofs of $\mathfrak{A}$ to proofs of $\mathfrak{B}$. But $(\lambda z.\ a)$ is just such a function. The only difficulty is that $(\lambda z.\ a)$ could be of arbitrarily high level. So let $p = \max\{\ell(\mathfrak{A}),\ \ell(\mathfrak{B})\}$, and let $b = \underset{\sim}{F}_p(\lambda z.\ a)$. Since any proof of $\mathfrak{A}$ or of $\mathfrak{B}$ is of level p, we have

$$\Gamma(\Delta), \Gamma(\mathfrak{A}),\ \Gamma(\mathfrak{B}),\ |\Delta|y,\ |\mathfrak{A}|z \equiv \underset{\sim}{T} \vdash_{\Sigma} |\mathfrak{B}|(bz) \equiv \underset{\sim}{T}.$$

Therefore, by the decidability of $|\mathfrak{A}|$,

$$\Gamma(\Delta), \Gamma(\mathfrak{A}),\ \Gamma(\mathfrak{B}),\ |\Delta|y \vdash_{\Sigma} (|\mathfrak{A}| \supset_1 |\mathfrak{B}|\circ b)z \equiv \underset{\sim}{T}.$$

Hence, if

$$c = \underset{\sim}{\rho}_{p+1}(|\mathfrak{A}| \supset_1 |\mathfrak{B}|\circ b),$$

then

$$\Gamma(\Delta), \Gamma(\mathfrak{A} \to \mathfrak{B}),\ |\Delta|y \vdash_{\Sigma} c \equiv c.$$

Thus, taking $d = \underset{\sim}{D}bc$, we have $\Delta \Rightarrow_d \mathfrak{A} \to \mathfrak{B}$.

ix) Suppose we have

$$\Gamma(\Delta), \Gamma(\mathfrak{A}),\ |\Delta|y \vdash_{\Sigma} |\mathfrak{A}|a \equiv \underset{\sim}{T}$$

and

$$\Gamma(\Delta), \Gamma(\mathfrak{B}),\ \Gamma(C),\ |\Delta|y,\ |\mathfrak{B}|z \equiv \underset{\sim}{T} \vdash_{\Sigma} |C|b \equiv \underset{\sim}{T}.$$

Then, taking $c = [\underset{\sim}{D}_1 wa/z]b$, we have

$$\Gamma(\Delta),\ \Gamma(\mathfrak{A} \to \mathfrak{B}),\ \Gamma(C),\ |\Delta|y,\ |\mathfrak{A} \to \mathfrak{B}|w \equiv \underset{\sim}{T} \vdash_{\Sigma} |C|c \equiv \underset{\sim}{T}.$$

xix) Suppose x is not free in Δ. Suppose

$$\Gamma(\Delta),\ \Gamma(\mathfrak{A}(0)),\ |\Delta|y \vdash_{\Sigma} |\mathfrak{A}(0)| \equiv \underset{\sim}{T}$$

and

$$\Gamma(\Delta), \Gamma(\mathfrak{A}(0)),\ \underset{\sim}{N}x \equiv \underset{\sim}{T},\ |\Delta|y, |\mathfrak{A}(x)|z \equiv \underset{\sim}{T} \vdash_{\Sigma} |\mathfrak{A}(x')|b \equiv \underset{\sim}{T}.$$

Then define a function f by primitive recursion so that

$$\Gamma(\Delta), \Gamma(\mathfrak{A}(0)),\ |\Delta|y \vdash_{\Sigma} f\underset{\sim}{0} \equiv a$$

and

$$\Gamma(\Delta), \Gamma(\mathfrak{A}(0)),\ |\Delta|y, \underset{\sim}{N}x \equiv \underset{\sim}{T},\ |\mathfrak{A}(x)|(fx) \equiv \underset{\sim}{T} \vdash_{\Sigma} f(\underset{\sim}{D}x\underset{\sim}{K}) \equiv [fx/z]b.$$

Then it is easy to prove by induction on x in Σ that

$$\Gamma(\Delta), \Gamma(\mathfrak{A}(x)),\ |\Delta|y \vdash_{\Sigma} |\mathfrak{A}(x)|(fx) \equiv \underset{\sim}{T}.$$

9.2. Corollary. Suppose $\mathfrak{A}$ is a sentence of HA and HA $\vdash \mathfrak{A}$. Then there is a closed term a of Σ such that $\vdash_{\Sigma} |\mathfrak{A}|a \equiv \underset{\sim}{T}$.

10. The consistency of ATC. Let ATC be ATC($\emptyset$). If we are intuitionists, and if we believe the informal considerations of [1] and above, then each axiom of ATC and each application of one of the rules of inference of ATC is simply evident. Accordingly, each proof in ATC is correct, and each sequent provable in ATC is true. Thus the question of the consistency of ATC does not arise for us. Looking at ATC from a classical standpoint, on the other hand, the theory ATC is not obviously consistent. What we shall do is to give a proof of the consistency of ATC relative to HA by describing in HA a semantics for the restriction of ATC to some level p. For the remainder of this section we are working informally in HA, and the reader should convince himself that our arguments can actually be formalized there.

The central problem in interpreting ATC is in making formal sense of the reducibility operators $\underset{\sim}{F}_p$. That is, we must show how, given a computation involving objects of high level, to replace it with a computation not essentially involving those objects. The idea is that whenever a complicated visualization is called for, we just pretend that we have carried it out and go on with the computation. If our argument above is correct, then the fact that we do not really have the insight we are pretending to have can make no difference.

In order to make this precise, we define a credulous interpretation, which accepts every claim as true and which allows $\underset{\sim}{F}_p xy$ to be defined without looking at the value. We simultaneously define the following: A set $\mathfrak{A}_*$ of atomic constants, a set $\mathfrak{B}_*$ of constants, and an assignment of a number $\ell(A)$, called the level of A, to each A in $\mathfrak{B}_*$.

10.1. The definition is inductive.

i) If $A = \underset{\sim}{K}, \underset{\sim}{S}, \underset{\sim}{Q}, \underset{\sim}{\delta}_1, \underset{\sim}{\delta}_2$, then $A \in \mathfrak{A}_*$ and $\ell(A) = 0$.

ii) If $p = 0, 1, 2, \ldots$, then $\underset{\sim}{L}_p \in \mathfrak{A}_*$, $\underset{\sim}{F}_p \in \mathfrak{A}_*$, $\underset{\sim}{\rho}_{p+1} \in \mathfrak{A}_*$, $\ell(\underset{\sim}{L}_p) = p$, $\ell(\underset{\sim}{F}_p) = p$, and $\ell(\underset{\sim}{\rho}_{p+1}) = p + 1$.

iii) If $A \in \mathfrak{A}_*$, then $A \in \mathfrak{B}_*$.

iv) If $A \in \mathfrak{B}_*$, and if $B = \underset{\sim}{K}, \underset{\sim}{S}, \underset{\sim}{Q}$, then $(B;A) \in \mathfrak{B}_*$ and $\ell((B;A)) = \ell(A)$.

v) If $A, B \in \mathfrak{B}_*$, then $((\underset{\sim}{S};A);B) \in \mathfrak{B}_*$, and $\ell(((\underset{\sim}{S};A);B)) = \max\{\ell(A), \ell(B)\}$.

vi) If $A \in \mathfrak{B}_*$, then $\underset{\sim}{F}_p^A \in \mathfrak{A}_*$ and $\ell(\underset{\sim}{F}_p^A) = p$.

vii) If $A \in \mathfrak{B}_*$ and $\ell(A) \leq p$, then $(\underset{\sim}{\rho}_{p+1};A) \in \mathfrak{B}_*$ and $\ell((\underset{\sim}{\rho}_{p+1};A)) = p+1$.

Now we can define the credulous interpretation μ_*.

10.2. We give an inductive definition of the interpretation μ_* of $\mathfrak{B}_*$:

i) If A is $\underset{\sim}{K}, \underset{\sim}{S}, \underset{\sim}{Q}$, then $\mu_*(A,B) = (A;B)$.

ii) $\mu_*((\underset{\sim}{K};A),B) = A$.

iii) $\mu_*((\underset{\sim}{S};A),B) = ((\underset{\sim}{S};A);B)$.

iv) $\mu_*(((\underset{\sim}{S};A);B),C) = AC(BC)$.

v) $\mu_*((\underset{\sim}{Q};A),B) = \underset{\sim}{T}$ if $A = B$, and $\mu_*((\underset{\sim}{Q};A),B) = \underset{\sim}{\perp}$ if $A \neq B$.

vi) $\mu_*(\underset{\sim}{\delta}_i, A) = A$ if $A \in \mathfrak{A}_*$, and $\mu_*(\underset{\sim}{\delta}_i, A) = A_i$ if $A = (A_1;A_2)$.

vii) $\mu_*(\underset{\sim}{L}_p, A) = T$ if $\ell(A) \leq p$, and $\mu_*(\underset{\sim}{L}_p, A) = \underset{\sim}{\perp}$ if $\ell(A) > p$.

viii) If $\ell(A) \leq p$, then $\mu_*(\underset{\sim}{\rho}_{p+1}, A) = (\underset{\sim}{\rho}_{p+1}, A)$.

ix) $\mu_*(\underset{\sim}{F}_p, A) = \underset{\sim}{F}_p^A$.

x) If $\ell(B) \leq p$, then $\mu_*(\underset{\sim}{F}_p^A, B) = AB$.

For each fixed p, we can now construct a model of the pth level.

10.3. Let $\mathfrak{A}_p$ be the set consisting of $\underset{\sim}{K}$, $\underset{\sim}{S}$, $\underset{\sim}{Q}$, $\underset{\sim}{\delta}_1$, $\underset{\sim}{\delta}_2$, all $\underset{\sim}{L}_q$ for $q \leq p$, all $\underset{\sim}{F}_q$ for $q \leq p$, all $\underset{\sim}{F}_q^A$ for $q \leq p$ and $A \in \mathfrak{B}_*$, and all $\underset{\sim}{\rho}_{q+1}$ for $q < p$.

10.4. We define $\mathfrak{B}_0$ inductively:

i) If $A \in \mathfrak{A}_0$, then $A \in \mathfrak{B}_0$.

ii) If $A \in \mathfrak{B}_0$ and $B = \underset{\sim}{K}$, $\underset{\sim}{S}$, $\underset{\sim}{Q}$, then $(B;A) \in \mathfrak{B}_0$.

iii) If $A, B \in \mathfrak{B}_0$, then $((\underset{\sim}{S};A);B) \in \mathfrak{B}_0$.

10.5. Suppose $\mathfrak{B}_p$ has been defined. Then we define the interpretation μ_p of $\mathfrak{B}_p$ inductively:

i) If $(A;B) \in \mathfrak{B}_p$, then $\mu_p(A,B) = (A;B)$.

ii) $\mu_p((\underset{\sim}{K};A),B) = A$.

iii) $\mu_p(((\underset{\sim}{S};A);B),C) = AC(BC)$.

iv) $\mu_p((\underset{\sim}{Q};A),B) = \underset{\sim}{T}$ if $A = B$, and $\mu_p((\underset{\sim}{Q};A),B) = \underset{\sim}{\perp}$ if $A \neq B$.

v) $\mu_p(\underset{\sim}{\delta}_i, A) = A$ if $A \in \mathfrak{A}_p$, and $\mu_p(\underset{\sim}{\delta}_i, A) = A_i$ if $A = (A_1;A_2)$.

vi) $\mu_p(\underset{\sim}{L}_q, A) = \underset{\sim}{T}$ if $\ell(A) \leq q$, and $\mu_p(\underset{\sim}{L}_q, A) = \underset{\sim}{\perp}$ if $\ell(A) > q$.

vii) $\mu_p(\underset{\sim}{F}_q, A) = \underset{\sim}{F}_q^A$.

viii) If $\ell(B) \leq q$, $\ell(C) \leq q$, $C \in \mathfrak{B}_p$, and $\hat{\mu}_*(AB) = C$, then $\mu_p(F_q^A, B) = C$.

10.6. Suppose $\mathfrak{B}_p$ and μ_p have been defined. Then we define $\mathfrak{B}_{p+1}$ inductively:

i) If $A \in \mathfrak{A}_{p+1}$ or $A \in \mathfrak{B}_p$, then $A \in \mathfrak{B}_{p+1}$.

ii) If $A \in \mathfrak{B}_{p+1}$ and $B = \underset{\sim}{K}$, $\underset{\sim}{S}$, $\underset{\sim}{Q}$, then $(B;A) \in \mathfrak{B}_{p+1}$.

iii) If $A, B \in \mathfrak{B}_{p+1}$, then $((\underset{\sim}{S};A);B) \in \mathfrak{B}_{p+1}$.

iv) Suppose $A \in \mathfrak{B}_p$. Suppose that for every $B \in \mathfrak{B}_p$ we have $\hat{\mu}_p(AB) = T$. Then $(\underset{\sim}{\rho}_{p+1};A) \in \mathfrak{B}_{p+1}$.

We need some basic facts about our models $< \mathfrak{B}_p, \mu_p >$, all of which are quite straightforward to check. First of all, we always have $\mathfrak{B}_p \subseteq \mathfrak{B}_{p+1}$ and $\mu_p \subseteq \mu_{p+1}$. If $\hat{\mu}_p(a) = A$, then $\hat{\mu}_*(a) = A$, as can be shown by induction on the length of the μ_p-computation of a. If $\ell(A) = q$ and $A \in \mathfrak{B}_p$, then $q \leq p$ and $A \in \mathfrak{B}_q$. Moreover, if $\ell(A) \leq q$ and $\ell(B) \leq q$ and $\hat{\mu}_p(AB) = C$, then we already have $\hat{\mu}_q(AB) = C$, and so $C \in \mathfrak{B}_q$. Finally, it is clear that $< \mathfrak{B}_p, \mu_p >$ is a system of discriminators.

10.7. <u>Theorems</u>: In HA we can prove that every axiom or rule of inference of ATC involving no constants of level $> p$ is μ_p-valid.

<u>Proof</u>: The validity of each of the axioms and rules is clear either from the results of [2], directly from the definitions of the models, or from the remarks above.

From this theorem, the consistency of ATC relative to HA is immediate.

11. <u>Positive formulas</u>. We say that a formula of HA is <u>positive</u> if it is built up without using implication. In particular, then, every prenex formula is positive.

11.1. <u>Lemma</u>: Suppose $\mathfrak{A}(x_1, \ldots, x_k)$ is a positive formula of HA with all of its free variables displayed. Suppose $\ell(\mathfrak{A}(x_1, \ldots, x_k)) \leq p$. Then we can prove in HA that, for any $n_1, \ldots, n_k$, if there is a closed term a of ATC such that $\hat{\mu}_p(|\mathfrak{A}(\bar{n}_1, \ldots, \bar{n}_k)|a) = \underset{\sim}{T}$, then $\mathfrak{A}(n_1, \ldots, n_k)$.

<u>Proof</u>: By induction on the structure of $\mathfrak{A}(x_1, \ldots, x_k)$. If it is an equation, the result is clear since the natural numbers and the primitive recursive functions are standard in all systems of discriminators.

Suppose $\mathfrak{A}(x_1, \ldots, x_k)$ is $\mathfrak{B}_1(x_1, \ldots, x_k) \vee \mathfrak{B}_2(x_1, \ldots, x_k)$. Suppose

$$\hat{\mu}_p(|\mathfrak{A}(\bar{n}_1, \ldots, \bar{n}_k)|) = \underset{\sim}{T}.$$

Then, for $i = 1$ or 2, we have

$$\hat{\mu}_p(|\mathfrak{B}_i(\bar{n}_1, \ldots, \bar{n}_k)|a) = \underset{\sim}{T}.$$

By the induction hypothesis, then, $\mathfrak{B}_i(n_1,\ldots,n_k)$, and so $\mathfrak{A}(n_1,\ldots,n_k)$.

Suppose $\mathfrak{A}(x_1,\ldots,x_k)$ is $\mathfrak{B}_1(x_1,\ldots,x_k) \wedge \mathfrak{B}_2(x_1,\ldots,x_k)$. Suppose

$$\hat{\mu}_p((\mathfrak{A}(\bar{n}_1,\ldots,\bar{n}_k)|a) = \underset{\sim}{T}.$$

Then, for $i = 1, 2$,

$$\hat{\mu}_p(|\mathfrak{B}_i(\bar{n}_1,\ldots,\bar{n}_k)|(\underset{\sim}{D}_i a)) = \underset{\sim}{T}.$$

Hence, by the induction hypothesis, $\mathfrak{B}_i(n_1,\ldots,n_k)$, and so $\mathfrak{A}(n_1,\ldots,n_k)$.

Suppose $\mathfrak{A}(x_1,\ldots,x_k)$ is $\vee y\mathfrak{B}(y, x_1,\ldots,x_k)$. Suppose

$$\hat{\mu}_p(|\vee y\mathfrak{B}(y,x_1,\ldots,x_k)|a) = \underset{\sim}{T}.$$

Then $\hat{\mu}_p(\underset{\sim}{N}(\underset{\sim}{D}_1 a)) = \underset{\sim}{T}$. But the natural numbers are standard. Therefore there is a number m such that $\hat{\mu}_p(\underset{\sim}{D}_1 a) = \hat{\mu}_p(\underset{\sim}{m})$. But then there is a closed term b such that

$$\hat{\mu}_p(|\mathfrak{B}(\bar{m},\bar{n}_1,\ldots,\bar{n}_k)|b) = \underset{\sim}{T}.$$

Hence, by the induction hypothesis, $\mathfrak{B}(m,n_1,\ldots,n_k)$, and so $\mathfrak{A}(n_1,\ldots,n_k)$.

Finally, suppose $\mathfrak{A}(x_1,\ldots,x_k)$ is $\wedge y\,\mathfrak{B}(y,x_1,\ldots,x_k)$. Suppose

$$\hat{\mu}_p(|\wedge y\,\mathfrak{B}(y,\bar{n}_1,\ldots,\bar{n}_k)|a) = \underset{\sim}{T}.$$

Consider any number m. Then $\hat{\mu}_{p-1}(\underset{\sim}{N}\underset{\sim}{m}) = \underset{\sim}{T}$. Thus there is a closed term b with

$$\hat{\mu}_{p-1}(|\mathfrak{B}(\bar{m},\bar{n}_1,\ldots,\bar{n}_k)|b) = \underset{\sim}{T}.$$

Hence, by the induction hypothesis, $\mathfrak{B}(m,n_1,\ldots,n_k)$. But m was arbitrary, and so $\mathfrak{A}(n_1,\ldots,n_k)$. That completes the proof.

11.2. Theorem: Suppose $\mathfrak{A}$ is a positive sentence of HA. Then HA $\vdash\mathfrak{A}$ if and only if there is a closed term a of ATC such that $\vdash_{ATC} |\mathfrak{A}|a \equiv \underset{\sim}{T}$.

Proof: One direction is just Corollary 9.2. For the converse, suppose we have a closed term a such that $\vdash_{ATC} |\mathfrak{A}|a \equiv T$. Using Theorem 10.7, there is a number $p \geq \ell(\mathfrak{A})$ such that, in HA, we can prove that $\hat{\mu}_p(|\mathfrak{A}|a) \equiv \underset{\sim}{T}$. But then, by Lemma 11.1, HA $\vdash\mathfrak{A}$.

References

1. N. D. Goodman, A theory of constructions equivalent to arithmetic, in: Intuitionism and Proof Theory, ed. A. Kino, J. Myhill, and R. E. Vesley (North-Holland, Amsterdam, 1970) 101-120.

2. N. D. Goodman, A simplification of combinatory logic, to appear in J. Symb. Log.

3. W. W. Tait, Functionals defined by transfinite recursion, J. Symb. Log., 30(1965) 155-174.

IX. THE PRIORITY METHOD FOR THE CONSTRUCTION OF RECURSIVELY ENUMERABLE SETS

A.H. LACHLAN

For each $i < \omega$ let $C(i) \subset P(\omega)$ the power set of ω. Suppose that an r.e. set $\underset{\sim}{A} \in \bigcap\{C(i) \mid i < \omega\}$ is found by a construction which appears to succeed because the need for $\underset{\sim}{A} \in C(0)$ is viewed as having first importance, the need for $\underset{\sim}{A} \in C(1)$ is viewed as having second importance, and so on, then one says that the construction uses the "priority method". Beyond this, apart from providing a long list of examples, it is difficult to be more precise about what the priority method is. Discussions about the general nature of the method can be found in [8], [4], and [1]. Historically, the term arose from the way in which Friedberg [2] and Mučnik [5] independently solved Post's problem.

In his review [7] of [6] Myhill remarked that Friedberg's method seemed to be an effective analogue of Baire's theorem from topology. Below we shall develop this analogy of Myhill's for three different constructions: (1) a way of solving Post's problem due to Sacks, (2) the construction of a minimal pair of r.e. degrees achieved independently by Yates and the author, and (3) the density theorem for r.e. degrees again due to Sacks. In each case we shall isolate the way in which $A \in C(i)$ is ensured for a particular $i < \omega$.

Most of our notation is standard. $\underset{\sim}{N}$ denotes the set of natural numbers, $P(S)$ denotes the power set of S, $P_{fin}(S)$ denotes the set of finite subsets of S. For any set $\underset{\sim}{A}$, $c_{\underset{\sim}{A}}$ denotes the characteristic function of $\underset{\sim}{A}$. $L(x)$ denotes $\{y \mid y \in \underset{\sim}{N} \ \& \ y < x\}$. If F is a function and $\underset{\sim}{R}$ a set $F|\underset{\sim}{R}$ denotes the restriction of F to $\underset{\sim}{R}$; if $\underset{\sim}{R}$ is $L(x)$ we write $F|x$ for $F|L(x)$. We use $\leq_T$ to denote the relation of Turing reducibility and $<_T$ for proper Turing reducibility.

1. Baire's theorem. For the purposes of the analogy that we wish to draw below we shall take Baire's theorem to be the one which states that in a compact Hausdorff space X, if $X - C(i)$ is nowhere dense for each $i < \omega$, then $\bigcap\{C(i) \mid i < \omega\} \neq \emptyset$. This can be broken down as follows. Call a closed set

0-closed if its interior is non-empty. We have

Lemma 1.1. Let X be compact. Let $\langle C(i) \mid i < \omega \rangle$ be a sequence in $P(X)$ such that for each $i < \omega$ and each 0-closed set V there exists a 0-closed set $V' \subset V \cap C(i)$. Then $\cap\{C(i) \mid i < \omega\} \neq \emptyset$.

Lemma 1.2. If X is compact and Hausdorff, $X - C$ is nowhere dense, and V is 0-closed there exists a 0-closed set $V' \subset V \cap C$.

Many applications of the priority method can be presented in such a way that they become effective analogues of Baire's theorem construed as consisting of the two lemmas we have just stated.

2. "Finite injury" applications of the priority method. Consider the space X consisting of all mappings $F : \underset{\sim}{N} \to P_{fin}(\underset{\sim}{N})$ such that $F(s) \subset F(s+1)$ for all $s < \omega$. The members of X are called enumerations. For $F \in X$ the set $\cup\{F(s) \mid s < \omega\}$ enumerated by F is denoted by $\underset{\sim}{F}$. Let $X^* = \{G \mid \exists F \exists x(F \in X \ \& \ G = F|x)\}$. For $G \in X^*$ let $U(G) = \{F \mid F \in X \ \& \ G \subset F\}$. Topologize X by letting $\{U(G) \mid G \in X^*\}$ be a basis for the open sets.

Definition 2.1. $V \subset X$ is called 1-closed if there are recursive functions $f : X^* \to \underset{\sim}{N}$, $G : X^* \to P_{fin}(\underset{\sim}{N})$ such that

(i) $V = \{F \mid F \in X \ \& \ \forall s(F(s) \supset G(F|s) \ \& \ F(s) - F(s-1) \geq f(F|s))\}$

(ii) if $F \in X^*$ and $(\forall s \in \text{Dom } F)(F(s) \supset G(F|s) \ \& \ F(s) - F(s-1) \geq f(F|s))$ then there exists an extension of F in V, and

(iii) if $F \in V$ and F is recursive then $\lim_x f(F|x)$ exists and $\lim_x G(F|x) = \emptyset$, i.e. $f(F|x)$ is fixed and $G(F|x) = \emptyset$ for all sufficiently large x.

Y is a recursively closed subset of X if there exists a recursive subset Y^* of X^* such that

$$Y = \{F \mid F \in X \ \& \ \forall s(F|s \in Y^*)\}$$

and

$$Y^* = \{G \mid \exists F \exists x(F \in Y \ \& \ G = F|x)\} .$$

The definition of 1-closed can be extended in a natural way to apply to subsets of a specified recursively closed subset of X - in Definition 2.1 we read Y, Y^* for X, X^* respectively. Y is <u>recursively compact</u> if it is recursively closed and further $\{G|G \in Y^* \& \text{Dom } G = L(s)\}$ is finite for each s, its cardinality being a recursive function of s.

The analogue of Lemma 1.1 which is relevant to the easiest applications of the priority method is

Lemma 2.1. Let Y be a recursively compact subset of X. Let $\langle C(i) \mid i < \omega\rangle$ be a sequence in $P(Y)$ such that given i and 1-closed $V \subset Y$ we can effectively find 1-closed $V' \subset V \cap C(i)$. Then $\cap\{C(i) \mid i < \omega\}$ has a recursive member.

We shall not prove this because it is an easy exercise. We apply Lemma 2.1 to the solution of Post's problem. Let $X^c = \{F|F \in X \& \forall s(F(s) \subset L(s))\}$ so that X^c is recursively compact. We must construct a recursive enumeration F such that $\underset{\sim}{F}$ is neither recursive nor complete. Let $\langle \underset{\sim}{W}_i\rangle$ be a standard enumeration of the r.e. sets, $\underset{\sim}{K}$ be a complete r.e. set, and $\{i\}^{\underset{\sim}{B}}$ denote the i-th partial function recursive in $\underset{\sim}{B}$. F must satisfy $\underset{\sim}{F} \neq \underset{\sim}{N} - \underset{\sim}{W}_i$ and $\{i\}^{\underset{\sim}{F}} \neq c_{\underset{\sim}{K}}$ for all i. Let $C(2i) = \{F|F \in X^c .\&. \underset{\sim}{F} \neq \underset{\sim}{N} - \underset{\sim}{W}_i \vee F \text{ not recursive}\}$ and $C(2i+1) = \{F|F \in X^c .\&. \{i\}^{\underset{\sim}{F}} \neq c_{\underset{\sim}{K}} \vee F \text{ not recursive}\}$. It clearly suffices to find recursive F in $\cap\{C(i) \mid i < \omega\}$. To be able to apply Lemma 2.1 we need:

Lemma 2.2. Given 1-closed $V \subset X^c$ and an r.e. set $\underset{\sim}{W}$ we can effectively find 1-closed $V' \subset V \cap \{F|F \in X^c .\&. \underset{\sim}{F} \neq \underset{\sim}{N} - \underset{\sim}{W} \vee F \text{ not recursive}\}$.

Lemma 2.3. Given 1-closed $V \subset X^c$ and $i < \omega$ we can effectively find 1-closed $V' \subset V \cap \{F|F \in X^c .\&. \{i\}^{\underset{\sim}{F}} \neq c_{\underset{\sim}{K}} \vee F \text{ not recursive}\}$.

Establishing these two lemmas will conclude the proof that there exists a non-recursive incomplete r.e. set. To prove the lemmas let V be specified by f, G as in the definition. We shall define f', G' which specify the appropriate 1-closed V'. Let $F \in (X^c)^*$ and $\text{Dom } F = L(s)$. Let $W(s)$ be the finite set of numbers enumerated in $\underset{\sim}{W}$ by stage s. For Lemma 2.2 let $f'(F) = f(F)$ and $G'(F) = G(F)$ if $G(F) \neq \emptyset$ or $W(s) \cap \underset{\sim}{F} \neq \emptyset$, where $\underset{\sim}{F} = \cup\{F(t) \mid t < s\}$. If

$G(F) = W(s) \cap \underset{\sim}{F} = \emptyset$ and there exists $m \in W(s)$, $f(s) \leq m < s$, then let $f'(F) = f(F)$ and $G'(F) = \{m\}$ taking the least such m. Otherwise $f'(F) = \mu x[x \geq f(F) \;\&\; x \notin F] + 1$ and $G'(F) = \emptyset$. We leave the reader to verify that f', G' specify the V' required for Lemma 2.2.

For the proof of Lemma 2.3, we need some more notation. Let $\{i\}^{\underset{\sim}{B},s}$ be the approximation to $\{i\}^{\underset{\sim}{B}}$ defined as follows: $\{i\}^{\underset{\sim}{B},s}(x) = \{i\}^{\underset{\sim}{B}}(x)$ if $x < s$ and the value on the right can be found in $< s$ steps while asking no questions "$n \in \underset{\sim}{B}$?" for $n \geq s$; $\{i\}^{\underset{\sim}{B},s}(x)$ is undefined otherwise. For $B \in X^c \cup (X^c)^*$ let $\{i\}^{B,s}(x) = \{i\}^{B(s-1),s}(x)$. Let K be a recursive enumeration of $\underset{\sim}{K}$. As before let $F \in (X^c)^*$ and $\mathrm{Dom}\, F = L(s)$. Let

$$k(F) = \mu x[c_{K(s)}(x) \neq \{i\}^{F,s}(x)],$$

$$\ell(F) = \mu y[(\forall x \leq k(F))(\{i\}^{\underset{\sim}{F},y}(x) = \{i\}^{F,s}(x))].$$

Let $f'(F)$ be $\max\{\ell(F), f(F)\}$ if $G(F) = \emptyset$ and be $f(F)$ if $G(F) \neq \emptyset$. Let $G'(F) = G(F)$.

Notice that f', G' specify unique $V' \subset V$ satisfying (i) and (ii) of Definition 2.1. Let F be a recursive member of V' we have to show that $\lim_x f'(F|x)$ exists and $\lim_x G'(F|x) = \emptyset$. Since V is 1-closed there exists s_0 such that $f(F|s) = f(F|s_0)$ and $G(F|s) = \emptyset$ for all $s \geq s_0$. Suppose that $k(F|x)$ is not non-decreasing for $x \geq s_0$, then for some $s > s_0$ we have $p < k(F|s)$, $p \in K(s) - K(s-1)$, and $k(F|s) = p$. Choosing the least such p it is easy to see that $k(F|x) = p$ and $\ell(F|x) = \ell(F|s)$ for all $x \geq s$, which is enough. Now suppose that $k(F|x)$ is non-decreasing for $x \geq s_0$, there are two cases. If $\lim_x k(F|x)$ exists the result again follows easily. If $\lim_x k(F|x)$ does not exist we can effectively compute $c_{\underset{\sim}{K}}(n)$ as follows. Choose $s \geq s_0$ such that $n < k(F|s)$ then $c_{K(x)}(n) = \{i\}^{F,s}(n)$ for all $x \geq s$ whence $c_{\underset{\sim}{K}}(n) = c_{K(s)}(n)$. Since $\underset{\sim}{K}$ is not recursive $\lim_x k(F|x)$ exists. This completes the proof that V' is 1-closed. Since $\lim_x k(F|x)$ exists we have $\{i\}^{\underset{\sim}{F}} \neq c_{\underset{\sim}{K}}$ which is the rest of the conclusion of the lemma.

The proof of Lemma 2.3 was invented by Sacks [8, Theorem 1].

3. The minimal pair construction. As a first example of more complicated priority methods we shall discuss the construction of non-zero r.e. degrees $\underset{\sim}{f}_0$, $\underset{\sim}{f}_1$ such that $\underset{\sim}{f}_0 \cap \underset{\sim}{f}_1 = \underset{\sim}{0}$. Here we need another kind of closed set. A recursive function is called <u>repeating</u> if there is some value it takes infinitely often. Let $Y \subset X$ be recursively closed.

Definition 3.1. $V \subset Y$ is called 2-<u>closed</u> if there are recursive functions f, G such that (i) and (ii) of Definition 2.1 hold and (iii) if $F \in V$ and F is recursive then $\lambda x[f(F|x)]$ is repeating and $\lim_x G(F|x) = \emptyset$.

Lemma 3.1. Let Y be a recursively compact subset of X. Let $\langle C(i)|i < \omega\rangle$ be a sequence in P(Y) such that given i and 2-closed $V \subset Y$ we can effectively find 2-closed $V' \subset V \cap C(i)$. Then $\cap\{C(i) \mid i < \omega\}$ has a recursive member.

Like Lemma 2.1 this is easy. For $F \in X \cup X^*$ let $F^0(s) = \{x|2x \in F(s)\}$ and $F^1(s) = \{x|2x+1 \in F(s)\}$. Let X^c be as in §2 but now with the added requirement that $F(s) - F(s-1)$ has cardinality at most 1. To obtain a minimal pair of r.e. degrees we must find a recursive F in X such that for all i and i′, $\underset{\sim}{F}^0 \neq \underset{\sim}{N} - \underset{\sim}{W}_i$, $\underset{\sim}{F}^1 \neq \underset{\sim}{N} - \underset{\sim}{W}_i$, and $(\{i\}^{\underset{\sim}{F}^0}$ and $\{i'\}^{\underset{\sim}{F}^1}$ are the same total function$) \rightarrow \{i\}^{\underset{\sim}{F}^0}$ recursive. The existence of such an F is obvious from Lemma 3.1 and the two lemmas which follow.

Lemma 3.2. Given 2-closed $V \subset X^c$ and an r.e. set W we can effectively find 2-closed $V' \subset V \cap \{F|F \in X^c .\&. \underset{\sim}{F}^0 \neq \underset{\sim}{N} - \underset{\sim}{W} \vee F$ not recursive$\}$. Similarly with $\underset{\sim}{F}^1$ for $\underset{\sim}{F}^0$.

Like Lemma 2.2 this is very easy so we omit the proof.

Lemma 3.3. Given 2-closed $V \subset X^c$ and $i, i' < \omega$ we can effectively find 2-closed $V' \subset V \cap \{F|F \in X^c .\&. F$ not recursive $\vee$ $(\{i\}^{\underset{\sim}{F}^0}$ and $\{i'\}^{\underset{\sim}{F}^1}$ are the same total function $\rightarrow \{i\}^{\underset{\sim}{F}^0}$ recursive$)\}$.

Proof. Let V be specified by f, G as before. We define f′, G′ specifying suitable V′. Let $F \in (X^c)^*$ and Dom F = L(s). Define

$$k(F) = \mu x[\{i\}^{F^0,s}(x) \text{ undefined, or } \{i'\}^{F^1,s}(x) \text{ undefined, or } \{i\}^{F^0,s}(x) \neq \{i'\}^{F^1,s}(x)].$$

$$p(F) = \mu x[x \le s \ \& \ f(x) = f(s) \ \& \ k(F|x) = \max\{k(F|y) \| y \le s \ \& \ f(y) = f(s)\}].$$

Let $G'(F) = G(F)$. If $G(F) \neq \emptyset$ let $f'(F) = f(F)$. If $G(F) = \emptyset$ and $p(F) = s$ let $f'(F)$ take the value

$$\mu x[(\forall y \le s)(f(F|y) < f(F) \to 2p(F|y) \le x)].$$

If $p(F) \neq s$ let $f'(F)$ take the same value but with $\le$ for $<$. There is a unique V' satisfying (i) and (ii) of Definition 3.1 with respect to f' and G'. Let F be a recursive member of V' then clearly $F \in V$, whence $\lambda x[f(F|x)]$ takes some value infinitely often. Let n be the least such value. Choose s_0 such that $f(F|s) \ge f(F|s_0) = n$ and $G(F|s) = \emptyset$ for all $s \ge s_0$. If s_j has been defined let s_{j+1} be the least $x > s_j$ such that $f(F|x) = n$ and $p(F|x) = x$.

Suppose that s_{j+1} is not defined, then $k(F|x)$ is bounded as x runs through the infinite set of s such that $f(F|s) = n$, whence $\{i\}^{F^0}_{\sim}$ and $\{i'\}^{F^1}_{\sim}$ can not be the same total function. Also for any $s > s_j$ such that $f(F|s) = n$ we have $f'(F|s) = f'(F|s_j)$ because $p(F|s) = p(F|s_j)$.

Now suppose that s_j is defined for all $j < \omega$. Since F is recursive s_j is a recursive function of j. Clearly $k(F|s_j) < k(F|s_{j+1})$ for each $j < \omega$. Let $y < k(F|s_j)$ then $\{i\}^{F^0,s_j}(y)$ and $\{i'\}^{F^1,s_j}(y)$ are both defined and equal. By choice of X^c, $F(s_j) - F(s_j-1)$ has at most one member. Without loss of generality $F^0(s_j-1) = F^0(s_j)$. If $s_j < s < s_{j+1}$ then either $f(F|s_j) < f(F|s)$, or $f(F|s_j) = f(F|s)$ and $p(F|s) \neq s$. It follows that $f'(F|s) \ge 2p(F|s_j)$ for $s_j < s < s_{j+1}$, whence if $j > 0$ $f'(F|s) \ge 2s_j$, and so $F^0(s) - F^0(s_j) \ge s_j$. Thus $\{i\}^{F^0,s_{j+1}}(y) = \{i\}^{F^0,s_j}(y)$, whence we have $\{i'\}^{F^1,s_{j+1}}(y) = \{i'\}^{F^1,s_{j+1}}(y)$. This shows that if $\{i\}^{F^0}$ is total then it is recursive. Further, for $j > 0$, $f'(F|s_j) = f'(F|s_1)$.

We have shown above that whether s_j is defined for all j or not F satisfies:

$$\{i\}^{F^0}_{\sim} \text{ and } \{i'\}^{F^1}_{\sim} \text{ are the same total function} \to \{i\}^{F^0}_{\sim} \text{ recursive}$$

and that $\lambda x[f(F|x)]$ is repeating. Since $\lim_x G'(F|x)$ is obviously $\emptyset$, the proof is complete.

The idea for the proof of Lemma 3.3 comes from Yates [12] and Lachlan [3].

4. The density theorem. In [10] Sacks showed that if C, A are r.e. sets such that $\underset{\sim}{C} <_T \underset{\sim}{A}$ then there exists an r.e. set B such that $\underset{\sim}{C} <_T \underset{\sim}{B} <_T \underset{\sim}{A}$. We shall now present essentially the same proof of the theorem that Sacks gave, but following the pattern established above we shall analyse completely the way in which the individual conditions are satisfied. Let X^c be as in §2 and let $C, A \in X^c$ be fixed recursive enumerations of $\underset{\sim}{C}, \underset{\sim}{A}$ respectively.

Let $\langle \underset{\sim}{R}_i \rangle$ be a recursive disjoint sequence of infinite recursive sets such that $\cup\{\underset{\sim}{R}_i | i \in \underset{\sim}{N}\} = \underset{\sim}{N}$. Let r_i, r_i' be indices of $\underset{\sim}{R}_i, \underset{\sim}{R}_0 \cup \ldots \cup \underset{\sim}{R}_i$ respectively as recursive sets. Without loss of generality assume that $\underset{\sim}{C} \subset \underset{\sim}{R}_0$.

Let g be a recursive function and D be a recursive enumeration of a set $\underset{\sim}{D}$. We say D <u>respects</u> g if for all s, $(D(s) - D(s-1)) \cap L(g(s)) = \emptyset$. We say g <u>respects</u> D if D respects g and

$$g(s) < g(t) \ \& \ t < s. \rightarrow (D(s) - D(t)) \cap L(g(t)) \neq \emptyset.$$

We say g is <u>repeating</u> if it takes some value infinitely often. Note that if g respects D then $g^* \leq_T \underset{\sim}{D}$ by an index which may be found effectively from indices of g and D, where $g^*(x) =_{dfn} \mu y[x \leq y \ \& \ (\forall z \geq y)(g(z) \geq g(y))]$. Note that g^* is recursive if g is recursive and repeating.

Below we shall construct a recursive sequence $\langle B_i \rangle \in X^c$ and a recursive sequence $\langle f_i \rangle$ of repeating functions such that $B = \cup\{B_i | i < \omega\}$ is an enumeration of the desired $\underset{\sim}{B}$. We set $B_0 = C$. For each $i < \omega$, B_{i+1} will enumerate a recursive set $\underset{\sim}{B}_{i+1} \subset \underset{\sim}{R}_{i+1}$ such that $\{i\}^{\underset{\sim}{C}} | \underset{\sim}{R}_{i+1} \neq c_{\underset{\sim}{B}} | \underset{\sim}{R}_{i+1}$. Also, we shall have $f_0 \leq f_1 \leq \ldots$ and for each $i < \omega$, B_{i+1} will respect f_i. We shall use f_i which may be regarded as a system of restraints on $B_{i+1} \cup B_{i+2} \cup \ldots$ to ensure that $\{i\}^{\underset{\sim}{B}} \neq c_{\underset{\sim}{A}}$. We now prove the requisite lemmas.

Lemma 4.1. There exists a recursive function g with the following property. Let $i \in N$, r be a recursive index of a recursive set $\underset{\sim}{R}$, and B_0 be a recursive enumeration enumerating $\underset{\sim}{B}_0 \subset \underset{\sim}{R}$. If B_1 is a recursive enumeration of $\underset{\sim}{B}_1 \subset \underset{\sim}{N} - \underset{\sim}{R}$ then $f = \lambda x g(i, r, B_0 | (x+1), B_1 | x)$ respects B_0. Further, if $\underset{\sim}{A} \not\leq_T \underset{\sim}{B}_0$ and B_1

respects f then $\{i\}^{\underset{\sim}{B}_0 \cup \underset{\sim}{B}_1} \neq c_{\underset{\sim}{A}}$ and f is repeating.

Proof. Define $G(i,r,B_0|s, B_1|s)$ to be the set of all triples $\langle t,\ell,k\rangle$ such that $t < s$, $k \le t$, $(\forall x < \ell)(\{i\}^{B_0(t) \cup B_1(t),k}(x) = c_{A(t)}(x))$, and k is the least such number. Define

$$g(i,r,B_0|(s+1), B_1|s) = \mu x[\forall t \forall \ell \forall k\ (\langle t,\ell,k\rangle \in G(i,r,B_0|s, B_1|s)\ \&\ B_0(s) - B_0(t) \ge k. \rightarrow k \le x)],$$

then f respects B_0 from the form of the definition of g.

Suppose that B_1 is recursive and respects f. For proof by contradiction suppose that $\{i\}^{\underset{\sim}{B}_0 \cup \underset{\sim}{B}_1} = c_{\underset{\sim}{A}}$. Let a $\underset{\sim}{B}_0$-oracle be given. For any m we can effectively find t, k such that $\langle t,m+1,k\rangle \in G(i,r,B_0|(t+1), B_1|(t+1))$ and $\underset{\sim}{B}_0 - B_0(t) \ge k$. It is easy to see by induction on s that $\langle t,m+1,k\rangle \in G(i,r,B_0|s, B_1|s)$ and $B_1(s) - B_1(t) \ge k$ for all $s > t$. Thus $\{i\}^{\underset{\sim}{B}_0 \cup \underset{\sim}{B}_1}(m) = \{i\}^{\underset{\sim}{B}_0 \cup \underset{\sim}{B}_1,k}(m) = \{i\}^{B_0(t) \cup B_1(t),k}(m) = c_{A(t)}(m)$. We conclude that $c_{\underset{\sim}{A}}(m) = c_{A(t)}(m)$. We have shown that $\underset{\sim}{A} \le_T \underset{\sim}{B}_0$, contradiction. Thus $\{i\}^{\underset{\sim}{B}_0 \cup \underset{\sim}{B}_1} \neq c_{\underset{\sim}{A}}$. Let m now be the least number such that $\{i\}^{\underset{\sim}{B}_0 \cup \underset{\sim}{B}_1}(m) \neq c_{\underset{\sim}{A}}(m)$. Let k_0 be the least number such that $(\forall x < m)(\{i\}^{\underset{\sim}{B}_0 \cup \underset{\sim}{B}_1,k_0}(x) = c_{\underset{\sim}{A}}(x))$. Choose t_0 such that $\underset{\sim}{A} - A(t_0) \ge m + 1$, $(\underset{\sim}{B}_0 \cup \underset{\sim}{B}_1) - (B_0(t_0) \cup B_1(t_0)) \ge k_0$, and $k_0 \le t_0$. Let t_1 be a stage $> t_0$ such that for any x in $\underset{\sim}{B}_0 - B_0(t_1)$ there exists $y < x$, $y \in B_0(t_1 - 1)$. We claim that $f(t_1) \le t_0$. If not there exists $\langle t,\ell,k\rangle \in G(i,r,B_0|t_1, B_1|t_1)$ such that $B_0(t_1) - B_0(t) \ge k$ and $t_0 < k$. Since $k > t_0 \ge k_0$ we must have $\ell > m$ whence $\{i\}^{B_0(t) \cup B_1(t),k}(m) = c_{A(t)}(m)$. Since $B_0(t_1) - B_0(t) \ge k$, by choice of t_1 it follows that $\underset{\sim}{B}_0 - B_0(t) \ge k$. As above we have $\{i\}^{\underset{\sim}{B}_0 \cup \underset{\sim}{B}_1}(m) = c_{A(t)}(m)$. By choice of t_0, $c_{A(t)}(m) = c_{\underset{\sim}{A}}(m)$. This contradicts the choice of m. Thus the claim that $f(t_1) \le t_0$ has been established. However, once t_0 has been chosen, there are infinitely many possibilities for t_1 whence f is repeating. This completes the lemma.

Lemma 4.2. There exists a recursive function H with the following property. Let $i < \omega$, r be a recursive index of an infinite recursive set $\underset{\sim}{R}$, and f be a recursive function then

(i) $B = \lambda x H(i,r,f|(x+1),x)$ is an enumeration respecting f of a set $\underset{\sim}{B} \subset \underset{\sim}{R}$

(ii) $\underset{\sim}{B}$ is recursive uniformly in f^* and $\underset{\sim}{A}$

(iii) if f is repeating then $\underset{\sim}{B}$ is recursive and $\{i\}^{\underset{\sim}{C}}|\underset{\sim}{R} \neq c_{\underset{\sim}{B}}|\underset{\sim}{R}$.

Proof. Since it makes hardly any difference we shall assume that $\underset{\sim}{R} = \underset{\sim}{N}$. Let $H(i,r,f|1,0) = \emptyset$. If $s > 0$ a number is in $H(i,r,f|(s+1),s)$ if either it is in $H(i,r,f|s,s-1)$ or has the form $\langle t,y,k\rangle$ where $f(s) \leq \langle t,y,k\rangle < s$, $t < s$, $y \in A(s)$, $(\forall x < y)(\{i\}^{C(t),k}(x) = c_{B(t)}(x))$, and $C(s) - C(t) \geq k$. It is immediate that B respects f. Since $\underset{\sim}{C} \leq_T \underset{\sim}{A}$ it is clear that $\underset{\sim}{B}$ is recursive in f^* and $\underset{\sim}{A}$; we say that this is "uniform" because an index of the reduction can be computed effectively from an index of f. Now suppose that f is repeating and that n is the least value which f takes infinitely often.

For proof by contradiction suppose that $\{i\}^{\underset{\sim}{C}} = c_{\underset{\sim}{B}}$. For any m with the help of a $\underset{\sim}{C}$-oracle we can effectively find t and k such that $n \leq \langle t,m,k\rangle$, $(\forall x < m)(\{i\}^{C(t),k}(x) = c_{B(t)}(x))$, and $\underset{\sim}{C} - C(t) \geq k$. Clearly $m \in \underset{\sim}{A}$ if and only if $\langle t,m,k\rangle \in B$, whence $\underset{\sim}{A} \leq_T \underset{\sim}{C}$ which is impossible. Thus $\{i\}^{\underset{\sim}{C}} \neq c_{\underset{\sim}{B}}$. Let m now be the least number such that $\{i\}^{\underset{\sim}{C}}(m) \neq c_{\underset{\sim}{B}}(m)$. Choose t_0 such that $c_{B(t_0)}(m) = c_{\underset{\sim}{B}}(m)$. Let $\langle t,y,k\rangle$ satisfy $t \geq t_0$, $y > m$, and $(\forall x < y)(\{i\}^{C(t),k}(x) = c_{B(t)}(x))$, then by choice of m and t_0 we have $C(s) - C(t) < k$ for some $s > t$. Thus for $t \geq t_0$ and $y > m$ we can effectively decide whether or not $\langle t,y,k\rangle \in B$. Let c be the least member of $\underset{\sim}{C} - C(t_0-1)$ and suppose that t_0 has been chosen such that $c \in C(t_0)$. If $t < t_0$ and $y > c$ then $\langle t,y,k\rangle \in \underset{\sim}{B}$ if and only if $\langle t,y,k\rangle \in B(t_0)$, because $\langle t,y,k\rangle \in \underset{\sim}{B}$ implies that $\{i\}^{C(t),k}(y-1)$ is defined which implies $y \leq k$ which implies $C(t_0) - C(t) < k$. Choose t_1 such that for all $y \leq \max\{c,m\}$, $y \in \underset{\sim}{A}$ implies $y \in A(t_1)$. If $y \leq \max\{c,m\}$, $\langle t,y,k\rangle \in \underset{\sim}{B}$ if and only if $\langle t,y,k\rangle \in B(f^*(\max\{t+1, t_1\}))$. Thus $\underset{\sim}{B}$ is recursive, which completes the lemma.

We shall now define the sequences $\langle B_i \rangle$ and $\langle F_i \rangle$ mentioned at the beginning of the section. For induction on s suppose that $B_i|s$ and $f_i|s$ have been defined for all i, let $B_0(s) = C(s)$

$$f_0(s) = g(0,r_0',B_0|(s+1),\ (B_1 \cup B_2 \cup \ldots)|s)$$

$$B_{i+1}(s) = H(i,r_{i+1},f_i|(s+1),s)$$

$$f_{i+1}(s) = \max\{f_i(s),\ g(i+1,r_{i+1},(B_0 \cup \ldots \cup B_{i+1})|s+1,$$

$$(B_{i+2} \cup B_{i+1} \cup \ldots)|s)\}.$$

These equations define $B_i(s)$ and $f_i(s)$ uniquely by simultaneous induction on i. Let $f_i' = \lambda x g(i,r_i,(B_0 \cup \ldots \cup B_i)|(x+1),\ (B_{i+1} \cup B_{i+2} \cup \ldots)|x)$. We can observe immediately that $f_{i+1} \geq f_i \geq f_i'$. Also B_{i+1} respects f_i by Lemma 4.2, and f_i' respects $B_0 \cup \ldots \cup B_i$ by Lemma 4.1. We claim that f_i respects $B_0 \cup \ldots \cup B_i$ for all i. If $i = 0$ this is immediate. Suppose it is true for $i = j$ then f_j respects $B_0 \cup \ldots \cup B_j$ and hence $B_0 \cup \ldots \cup B_{j+1}$. But f_{j+1}' also respects $B_0 \cup \ldots \cup B_{j+1}$, whence $f_{j+1} = \lambda x[\max\{f_j(x),\ f_{j+1}'(x)\}]$ also respects $B_0 \cup \ldots \cup B_{j+1}$ and the induction is complete.

Since $\underset{\sim}{B}_0 = \underset{\sim}{C}$, by hypothesis there is a reduction of $\underset{\sim}{B}_0$ to $\underset{\sim}{A}$ and $\underset{\sim}{A} \not\leq_T \underset{\sim}{B}_0$. Suppose for induction on i that $\underset{\sim}{A} \not\leq_T \underset{\sim}{B}_0 \cup \ldots \cup B_i$, that a reduction of $\underset{\sim}{B}_0 \cup \ldots \cup \underset{\sim}{B}_i$ to $\underset{\sim}{A}$ has been found and that f_{i-1} is repeating if $i > 0$. Since B_j respects f_i' for all $j > i$, by Lemma 4.1 $\{i\}^{\underset{\sim}{B}} \neq c_{\underset{\sim}{A}}$ and f_i' is repeating. If $i > 0$, we have f_{i-1} repeating whence $f_i = \lambda x[\max\{f_{i-1}(x),\ f_i'(x)\}]$ is also repeating, because as noted above f_{i-1} and f_i' both respect $\underset{\sim}{B}_0 \cup \ldots \cup \underset{\sim}{B}_i$. Thus f_i is repeating whether $i = 0$ or not. Since f_i respects $\underset{\sim}{B}_0 \cup \ldots \cup \underset{\sim}{B}_i$ we can effectively find a reduction of f_i^* to $\underset{\sim}{B}_0 \cup \ldots \cup \underset{\sim}{B}_i$ and hence to $\underset{\sim}{A}$. By Lemma 4.2 we can effectively find a reduction of $\underset{\sim}{B}_{i+1}$ to f_i^* and $\underset{\sim}{A}$, and hence to $\underset{\sim}{A}$. Further since f_i is repeating $\underset{\sim}{B}_{i+1}$ is recursive and $\{i\}^{\underset{\sim}{C}}|\underset{\sim}{R}_{i+1} \neq c_{\underset{\sim}{B}}|\underset{\sim}{R}_{i+1}$. Obviously $\underset{\sim}{A} \not\leq_T \underset{\sim}{B}_0 \cup \ldots \cup \underset{\sim}{B}_{i+1}$, so the induction is complete.

The truth of the density theorem is now apparent. We conclude this section by identifying the sort of closed subset of X^c which is appropriate to the density theorem. Let $\underset{\sim}{A}$ be a fixed non-recursive r.e. set.

Definition 4.1. $V \subset X^c$ is 3-closed if there are given a co-infinite recursive set R, and recursive mappings $B : (X^c)^* \to P_{fin}(\underset{\sim}{N})$ and $f : (X^c)^* \to \underset{\sim}{N}$ such that

(i) $V = \{F \mid F \in X^c \;\&\; \forall s(F(s) \cap \underset{\sim}{R} = B(F|s) \;\&\; f(F|s) \leq F(s) - F(s-1))\}$

(ii) for each $F \in V$, $\lambda x[f(F|x)]$ respects $\lambda x[B(F|x)]$ and for each s $B(F|s) \subset \underset{\sim}{R}$

(iii) if $F \in V$ is recursive then $B(F) = \cup\{B(F|x) \mid x < \omega\} <_T \underset{\sim}{A}$, $B(F)$ is recursive in $\underset{\sim}{A}$ uniformly, and $\lambda x[f(F|x)]$ is repeating.

Had we chosen to, we could have presented the proof of the density theorem as three lemmas following the pattern of §2 and §3. For example, the first lemma would say that if given $i < \omega$ and 3-closed V we can effectively find 3-closed $V' \subset V \cap C(i)$, then $\cap\{C(i) \mid i < \omega\}$ has a recursive member F with $\underset{\sim}{F} \leq_T \underset{\sim}{A}$.

5. Conclusion. Our aim above has been to throw some light on the priority method by analysing in complete detail how the individual conditions $C(i)$ are satisfied. There are many applications of the priority method other than the three mentioned here, some of them extremely difficult. It is our hope that the kind of analysis we have carried out above will make the proofs of theorems in recursion theory easier to understand.

References

1. D.A. Martin, The priority method of Sacks, mimeographed.

2. R.M. Friedberg, Two recursively enumerable sets of incomparable degrees of unsolvability, Proc. Nat. Acad. Sci. U.S.A. 43 (1957), 236-238.

3. A.H. Lachlan, Lower bounds for pairs of recursively enumerable degrees, Proc. London Math. Soc. (3) 16 (1966), 537-569.

4. __________, The priority method I, Zeitschr. f. math. Logik und Grundlagen Math. 13 (1967), 1-10.

5. A.A. Mučnik, Negative answer to the problem of reducibility of the theory of algorithms (Russian), Dokl. Akad. Nank SSSR 108 (1956), 194-197.

6. __________, Solution of Post's reduction problem and of certain other problems in the theory of algorithms (Russian), Trudy Moskov. Mat. Obsc. 7 (1958), 391-405.

7. J. Myhill, Review of [6], Math. Reviews #5570 22 (1967), 930.

8. G.E. Sacks, On the degrees less than $0'$, Ann. of Math. 77 (1963), 211-231.

9. __________, Degrees of Unsolvability, Ann. of Math. Studies 55, Princeton, 1963.

10. __________, The recursively enumerable degrees are dense, Ann. of Math. 80 (1964), 300-312.

11. C.E.M. Yates, A minimal pair of recursively enumerable degrees, J. Symb. Logic 31 (1966), 159-168.

Simon Fraser University
Department of Mathematics
BURNABY 2, B.C. / Canada

X. ADMISSIBLE ORDINALS AND PRIORITY ARGUMENTS

Manuel Lerman[1]

Yale University

Although the word "priority" has become very popular recently, especially in the political arena, there is a mystique surrounding the use of this word in Recursion Theory. If a Recursion Theory paper is written, where theorems are proved using a priority method construction, it is a foregone conclusion that very few people will read the paper. The situation has reached the point where some lecturers on the priority method set, as one of their goals (though not always seriously), preserving the mystique of priority arguments. Thus the lecture is never a total failure. We view this situation as somewhat disconcerting since we do not believe that the priority method in itself is difficult. Rather, it is many theorems proved using the priority method which are justifiably referred to as difficult, messy, and impossible to understand. The major difficulty in writing lies in trying to describe the clear picture the author has of a given construction. Words, symbols, and equations are usually inadequate, and the reader has to try to translate these words into a mental picture of his own.

We have two purposes in writing this paper. The first is to give the reader an idea of some of the obstacles one can expect to encounter in trying to generalize a theorem proved by the priority method, from Recursion Theory on ω to Recursion Theory on admissible ordinals; and to present methods which can be used to circumvent some of these obstacles. We take the construction of a minimal pair of α - r.e. sets due independently to Lachlan [5] and Yates [12], and discuss the proof of this theorem in both ordinary and admissible Recursion Theory. We consider this theorem particularly suitable for

(1) Research partially supported by NSF contract GP-29218.

our purposes, since its proof involves neither the most complicated, nor the easiest kind of priority argument.

Our second aim is to present a construction of a minimal pair so that the reader can get a picture of the construction without a struggle. To do so, we separate the combinatories of the construction from the recursion theory. We first describe a generalized pinball machine, complete with picture and rules for games played on it, with no reference to recursion theory. Since everybody is an expert on pinball machines, and since a typical game is not very complicated, we feel that this approach will yield a clear picture of the combinatorics of the construction. The construction will just be a specific game played on the machine. After presenting the construction, we show why this particular game yields the desired result: this is where the recursion theory will play its role. We will present the recursion theoretic facts needed, and it will not be difficult to obtain the desired lemmas from these facts.

Many people, perhaps justifiably, feel that a lucid presentation of a priority argument is impossible, and so this paper would be doomed to failure. We therefore reserve the right to say that we lied, and that the second aim of this paper is to preserve the mystique of priority arguments. We thus offer an alternate title for this paper, "How to Succeed with Priority Argument without Really Trying".

The paper is organized as follows:
Section 1 presents most of the definitions we will need. We will also try to describe the basic features of a priority argument there. In section 2, we discuss some obstacles one might encounter in trying to generalize a priority argument construction.

Section 3 is devoted to the construction of a minimal pair in ordinary recursion theory, and it is here that we devise our first pinball machine. We follow this in section 4 with a discussion of the problems encountered in generalizing this theorem, and devise a new pinball machine for the revised construction. We then construct a minimal pair for certain admissible ordinals. Section 5 concludes with an open question.

SECTION 1: PRELIMINARIES

We begin by describing what a priority argument is. Our description will not necessarily include all priority arguments, but will hopefully give the reader the general idea.

Assume that one is trying to construct a set or collection of sets with certain properties; and that these properties can be well-ordered into a sequence of requirements $\{R_i : i < \alpha\}$. One tries to satisfy requirements by placing certain ordinals in specified sets, and keeping other ordinals out of specified sets. Unfortunately, placing an ordinal in a set to satisfy one requirement may interfere with satisfying another requirement, or may force us into the situation where a requirement, previously satisfied, is no longer satisfied. In order to satisfy all requirements, we assign priorities to requirements: We say that R_i has higher priority than R_j if $i < j$. We always try to satisfy the requirement of highest priority which we are able to satisfy at a given stage of the construction, and never allow the satisfaction of requirement R_i to interfere with the satisfaction of requirements of higher priority. We then prove by induction that all requirements are satisfied, and so the sets constructed have the desired properties.

We now present some definitions which we will need.

Let α be an ordinal. Assume that we are given the language of the predicate calculus with two additional binary relation symbols $=$ and $\in$. Formulas are defined in the usual way. Let β be a transitive set. We say that a formula is <u>limited</u> over β if all its parameters are elements of β and all quantifiers are of the form

$(\exists x \in y)$ or $(\forall x \in y)$. A formula is Σ_0 and Π_0 over β if it is limited over β. A formula is $\Sigma_n(\Pi_n)$ over β if it is of the form $(\exists x)G$ $((\forall x)G)$ where G is $\Pi_{n-1}(\Sigma_{n-1})$ over β. Let Fodo (β) = the set of sets first order definable over $\beta = \{A : A$ is Σ_n definable over β or A is Π_n definable over β for some $n\}$. Gödel's hierarchy of constructible sets is now defined as follows:

$$L_0 = \phi$$
$$L_{\alpha+1} = \text{Fodo}\ (L_\alpha)$$
$$L_\lambda = \bigcup_{\delta < \lambda} L_\delta \quad \text{for } \lambda \text{ a limit ordinal.}$$

We say that a function f is <u>partial</u> α-<u>recursive</u> if its graph is Σ_1 definable over L_α. An α-<u>recursive</u> function is a partial α-recursive function whose domain is all the ordinals less than α. (Here we are considering functions $f : L_\alpha \longrightarrow L_\alpha$.) $A \subseteq \alpha$ is α-<u>recursively enumerable</u> (α - r.e.) if it is the domain of a partial α-recursive function. $A \subseteq \alpha$ is α-<u>recursive</u> if its characteristic function is α-recursive, or equivalently, if both A and $\bar{A} = \{x : x < \alpha$ and $x \notin A\}$ are α - r.e. (The characteristic function of A, also denoted by A, is defined by $A(x) = 0$ if $x \notin A$ and $A(x) = 1$ if $x \in A$.) $A \subseteq \alpha$ is α-<u>bounded</u> if there is a $\beta < \alpha$ such that $(\forall x)(x \in A \longrightarrow x < \beta)$. $A \subseteq \alpha$ is α-<u>finite</u> if A is α-recursive and α-bounded, or equivalently, if $A \in L_\alpha$.

An ordinal α is said to be <u>admissible</u> if L_α satisfies Σ_1 replacement, that is, given any partial α-recursive function f, and any $A \in L_\alpha$ such that A is a subset of the domain of f, then $f(A) \in L_\alpha$.

Cardinals of L_α play an important role in priority arguments. We say that β is an α-<u>cardinal</u> if β is an ordinal, $\beta < \alpha$, and there is no function $f \in L_\alpha$ such that f is one-one, has domain β, and image an ordinal $< \beta$. β is a <u>singular</u> α-<u>cardinal</u> if

$\beta = \bigcup_{i \in I} S_i$ where I has α-cardinality $< \beta$, for each $i \in I$ S_i has α-cardinality $< \beta$, and the function $f : I \longrightarrow \{S_i : i \in I\}$ defined by $f(i) = S_i$ is an element of L_α. Otherwise, β is a <u>regular α-cardinal</u>.

We now define two "reducibilities" on subsets of admissible ordinals. Other reducibilities exist, but we are only interested right now in generalizations of Turing reducibility which are relevant to priority arguments. Let $A \subseteq \alpha$, $B \subseteq \alpha$ where α is an admissible ordinal. We say that A <u>is weakly</u> α-<u>recursive in</u> B ($A \leq_{w\alpha} B$) if there are partial α-recursive functions Φ and Ψ such that

$$x \in A \iff (\exists M)(\exists N)(\Phi(x,M,N) = 0 \text{ and } M \subseteq B \text{ and } N \subseteq \bar{B}) \text{ , and}$$
$$x \in \bar{A} \iff (\exists M)(\exists N)(\Psi(x,M,N) = 0 \text{ and } M \subseteq B \text{ and } N \subseteq \bar{B}),$$

where M and N range over the α-finite sets. This straightforward generalization of Turing reducibility, however, is not transitive for all admissible ordinals α, as was discovered by Driscoll [1], and hence is not a real reducibility. We therefore define A <u>is</u> α-<u>recursive in</u> B ($A \leq_\alpha B$) if there are partial α-recursive functions Φ and Ψ such that for all α-finite sets K,

$$K \subseteq A \iff (\exists M)(\exists N)\ (\Phi(K,M,N) = 0 \text{ and } M \subseteq B \text{ and } N \subseteq \bar{B}), \text{ and}$$
$$K \subseteq \bar{A} \iff (\exists M)(\exists N)\ (\Psi(K,M,N) = 0 \text{ and } M \subseteq B \text{ and } N \subseteq \bar{B})$$

where M and N range over the α-finite sets. We note that $\leq_{w\omega} = \leq_\omega = \leq_T$ (Turing reducibility), and that $\leq_\alpha$ is transitive for all admissible ordinals α.

Various projectums of α are also important in priority arguments. We define α^*, the <u>projectum</u> of α, to be the least ordinal $\beta \leq \alpha$ such that there is a total one-one α-recursive function $f : \alpha \longrightarrow \beta$. We define $tp2(\alpha)$, the <u>tame</u> Σ_2-<u>projectum</u> of α, to be the least ordinal $\beta \leq \alpha$ such that there is a total one-one Σ_2 definable function f from α onto β, and such that if $\gamma < \beta$, then $f^{-1}(\{x : x < \gamma\})$ is α-finite.

SECTION 2: ON GENERALIZING PRIORITY ARGUMENTS

In this section, we discuss briefly certain problems which are encountered in trying to generalize specific constructions of r.e. sets to constructions of α - r.e. sets, where the sets are constructed via a priority argument. The problems arise from two sources; the construction itself and the priority method.

We say that $M \subseteq \omega$ is a maximal set if M is r.e., and given any r.e. set A, then either $A \cap \overline{M}$ is finite or $\overline{M} - A$ is finite. Friedberg [2] constructed a maximal set. In trying to generalize Friedberg's construction, one notes that it relies heavily on the fact that if $x < \omega$, then x, considered as an ordinal, has a greatest element. Hence under any possible definition of maximal α - r.e. set, one has trouble generalizing the construction to admissible $\alpha > \omega$, for ω itself does not have this property. Priorities can be rearranged so that maximal sets can be constructed for at least some countable admissible ordinals. However, Sacks [10] has shown that there is no maximal α - r.e. set if α is the $\aleph_1$ of $L = \bigcup_\gamma L_\gamma$.

In the construction of a minimal pair (Section 3), we encounter the difficulty that the construction uses the fact that $\alpha = \omega$ in the following way; if $x < \alpha$ and we reverse the order of the ordinals $< x$, then we still have a well-ordering. To prove the existence of a minimal pair for all α would thus have to involve modifying the construction for $\alpha = \omega$.

Friedberg's construction of two r.e. sets of incomparable Turing degree [3] is an example of a construction which does not have to be modified except with respect to the priority ordering of requirements. Sacks and Simpson [11] proved, for all admissible ordinals α, the existence of α - r.e. sets A and B such that $A \not\leq_\alpha B$ and $B \not\leq_\alpha A$. (A subsequent proof can be found in [7].) The source of all

difficulties here is the priority argument. The proof that all requirements are satisfied uses the fact that ω is a regular cardinal of L. But not all admissible ordinals are regular cardinals of L. And arguments counting the cardinality of sets are very important in the proof that the construction works. One way to circumvent this difficulty is to use the shortest indexing of requirements possible, which in many cases is a regular cardinal of L_α. This indexing seems to be of length $tp2(\alpha)$. For Σ_2 functions can be approximated to by a recursive sequence of α-finite functions, yielding priorities recursively at each stage of the construction. And if the function is tame, any proper initial segment of the priority ordering will be correct forever after some stage $\beta < \alpha$ (β will depend on the initial segment). The method for the case $\alpha^* < \alpha$ will be discussed in more detail in Section 4.

SECTION 3: A MINIMAL PAIR FOR ω

If $A \subseteq \omega$, $B \subseteq \omega$, we say that A and B are a minimal pair if $A \not\leq_T B$, $B \not\leq_T A$, and for all $C \subseteq \omega$, if $C \leq_T A$ and $C \leq_T B$, then C is recursive. Minimal pairs of r.e. sets were constructed by Lachlan [5] and Yates [12]. Subsequently, a slightly different construction was given by Jockusch and Soare [4].

The aim of this section is to make the construction of a minimal pair transparent. Our construction is very similar to that of Jockusch and Soare, but the wording is different, as is the organization of the proof.

Our motivation stems from the state the priority argument now finds itself in. The following situation is not uncommon. One has before him a paper, in which a theorem is proved by a priority argument, and finds that it is nearly impossible to understand or even verify the proof. Hence he proceeds as follows. First he tries to prove the theorem himself. If he fails, he skims through the paper getting some new ideas, and makes a second attempt at proving the theorem himself. This process is repeated finitely often until the reader comes up with a proof of his own, using some ideas of the paper and some of his own. The resulting construction and proof may be different from the proof in the paper, but at this stage the reader has enough of an idea of what is happening in the paper to verify the original proof with less difficulty. There may be some theorems for which this state if affairs is unavoidable. For other theorems, however, we view the style of writing as the major cause of difficulty. All the ideas used are intertwined in such a way, that the theorem must be viewed as a whole, and so seems very complicated. One makes finitely many approximations at proving the theorem in order to break the proof down into a few digestible pieces, and then assembles these pieces into a proof. We will attempt to present the proof of the existence of a minimal pair here, by breaking it down into digestible pieces.

A theorem proved by the priority method can usually be viewed as follows. One is trying to construct sets with given properties. Hence one gives a construction of the sets, and then proves that the sets constructed have the desired properties. One can view a typical construction as a dynamic system whose motion is described by recursion theoretic considerations. The proofs involve certain lemmas regarding the possible dynamics of the system and independent of recursion theoretic considerations, and other lemmas in which the recursion theory plays an important part. Attempts have been made at separating the recursion theory from the combinatorics. A very beautiful presentation of this form is due to Lachlan [6]. However, these methods of presentation still leave it to the reader to reconstruct a considerable portion of the dynamics of the construction for himself, and we feel that a clear picture of the dynamics is required to really understand the proof. Rogers' use of movable markers [9] is, in our opinion, the best presentation of the dynamics of a construction heretofore available. Even there, however, one still has to work to obtain the desired mental picture.

We present the proof of the existence of a minimal pair of r.e. sets as follows. We describe and draw a picture of a pinball machine, and give a set of rules for games on this machine. From this description and picture, we hope that the reader will get a mental picture of the dynamics of the construction, and a clearer picture of the proofs of the combinatorial lemmas. The construction will then be described as the play of a particular game on this pinball machine, and the proof of the theorem will follow from a few recursion theoretic facts.

Let us begin by describing the pinball machine (Figure 3.1).

The pinball machine consists of a _track_, _gates_, _holes_, _doors_, _pockets_, and _balls_. The _surface of the machine_ is that portion of the machine covered by arrows. Balls, may roll only on the surface of the machine, and must always roll in the direction of the arrows which we call _down_. Before any ball moves, each gate and door below the ball must be in the position of one of the dotted line segments intersecting it. We call a gate or door _closed_ if it is in the horizontal position; otherwise it is _open_. Holes are numbered 0,1,2, ... as are doors, and holes and doors are alternated on the surface of the machine. Balls may be _marked_ or _unmarked_. Let p and q be any two balls which have been marked before or at move n, and assume p originated from hole i and q from hole j. We say that p has _higher priority than_ q _at move n_ if $i < j$, or $i = j$ and p was marked before q was marked.

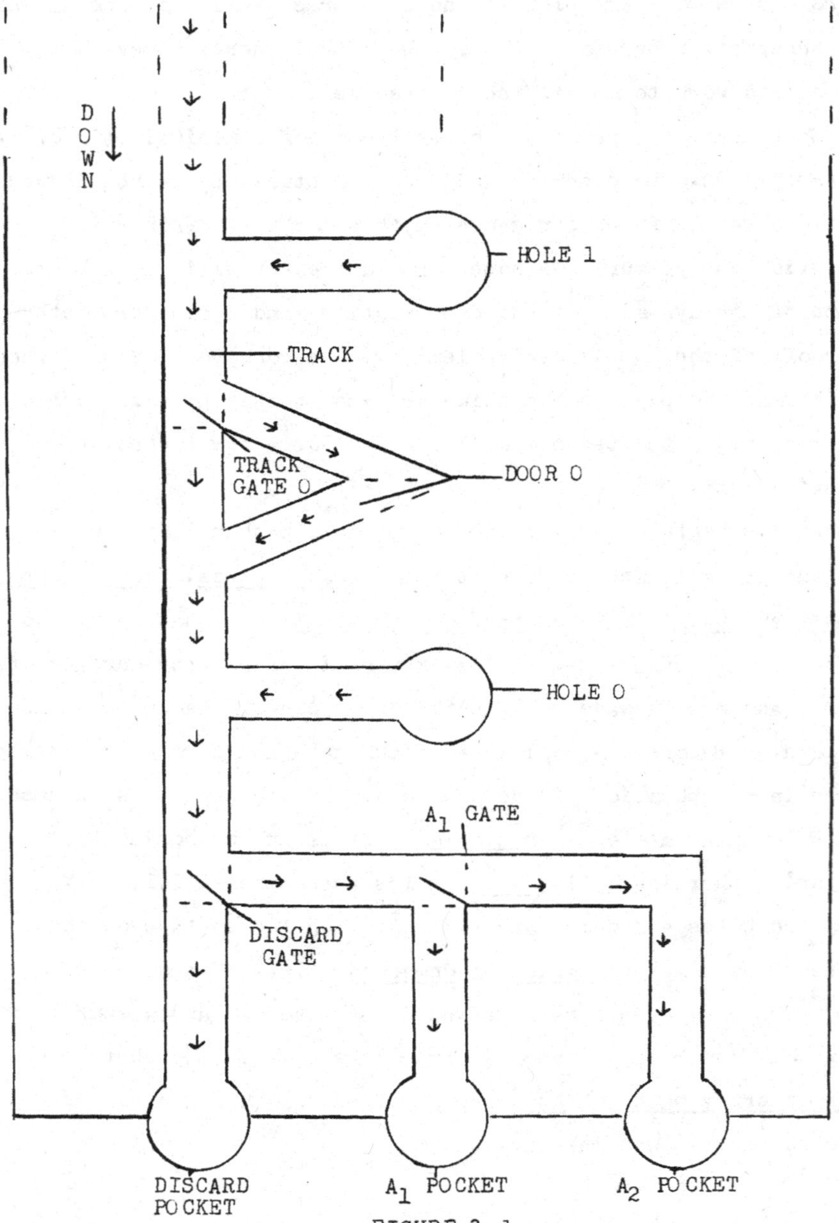
DOWN
HOLE 1
TRACK
TRACK GATE 0
DOOR 0
HOLE 0
A_1 GATE
DISCARD GATE
DISCARD POCKET
A_1 POCKET
A_2 POCKET

FIGURE 3.1

A game on this machine is played in a sequence of moves 0,1,2, ... of order-type ω.

Rules for a game:

Rule 3.1: Before the game begins, each hole has an infinite supply of unmarked balls of order-type ω and no marked balls, and there are no balls on the surface of the machine or in any pocket.

Rule 3.2: No unmarked ball ever gets onto the surface of the machine.

Rule 3.3: A move has only finitely many components. The following are the allowable components for any move n :

A. An unmarked ball sitting on top of a hole is marked.

B. A marked ball sitting on top of a hole is released onto the surface of the machine, and rolls until it ends up in front of a closed door or in a pocket.

C. A ball sitting in front of a closed door is released (by opening the door) and rolls until it ends up in front of another closed door or in a pocket.

Rule 3.4: Any move n must proceed via one of the sequences of components below.

A. An unmarked ball p is marked at move n via Rule 3.3.A. All marked balls of lower priority than p at move n and not already in a pocket are sent to the discard pocket via Rule 3.3.B and Rule 3.3.C. (For each such ball, gates and doors are set so that the ball will roll into the discard pocket.)

B. p is a marked ball sitting on top of hole i at the beginning of move n. All marked balls of lower priority than p at move n and not already in a pocket are sent to the discard pocket via Rules 3.3.B and 3.3.C. The discard gate is then closed as are all doors below p, all gates below p are set in some way, and p is set rolling via Rule 3.3.B.

C. p is a marked ball sitting in front of door i (which is closed) at the beginning of move n. All marked balls of lower priority than p at move n and not already in a pocket are sent to the discard pocket via Rules 3.3.B and 3.3.C. The discard gate is then closed as are all doors below p, all gates below p are set in some way, and p is set rolling via Rule 3.3.C.

In all three cases, we say ball p <u>generates</u> move n.

<u>Rule 3.5</u>: If a ball originating from hole i gets into pocket A_1 or pocket A_2 during move n, no ball in hole i which is unmarked at the end of move n can be marked at any move $m > n$.

<u>Rule 3.6</u>: Gates must be arranged in such a way so that at no time are two different balls prevented from rolling, by the same door.

We note that the rules of the game are consistent. The only possible conflict would be the requirement of Rule 3.3 that a move have only finitely many components and the requirement in Rule 3.4 that all balls satisfying a certain property be discarded. But all balls discarded in Rule 3.4 are marked, and by Rule 3.4 at most one ball can be marked during a given move. Hence only finitely many applications of Rules 3.3.B and 3.3.C are made during a given move.

We now prove some lemmas about any game played according to the above rules.

<u>Lemma 3.1</u>: Each ball generates only finitely many moves.

<u>Proof:</u> A ball p can generate moves only via Rule 3.4. If it does so via Rule 3.4.A, it was previously unmarked and now becomes marked, so this can happen only once. If it does so via Rule 3.4.B,

it is released from a hole and never again returns to any hole, so this can happen only once. And if it does so via Rule 3.4.C, it is resting before a door at the beginning of move n, and is further down the surface of the machine (either before another door or in a pocket) at the end of move n. Once a ball is in a pocket, it can no longer generate any moves. Since there are only finitely many doors below a given door, only finitely many such moves can be generated by ball p.

Lemma 3.2: Let i and n be given. Then at the end of move n, there are at most i marked balls on the surface of the machine which originated from hole i.

Proof: Each such ball must be in front of a door below hole i at the end of move n. There are only i such doors, and by Rule 3.6, at most one ball is in front of each door.

Lemma 3.3: Let i and n be given. Assume that a ball from hole i goes into pocket A_1 or pocket A_2 before move n. Then there are only finitely many moves $m \geq n$ such that a ball originating from hole i generates move m.

Proof: By Rule 3.5, no ball originating from hole i can be marked at any move $m \geq n$ if it is unmarked at the beginning of move n. Furthermore, if ball p generates move m, then ball p has been marked by the end of move m. Hence only finitely many balls originating from hole i can generate any move. The result now follows from Lemma 3.1.

Lemma 3.4: Let i be given. Then only finitely many moves can be generated by balls originating from hole i.

Proof: By induction on i. Assume that the lemma is true for all $j < i$. Then there is a move m_o such that no ball originating from any hole j, $j < i$, generates any move m, $m \geq m_o$. Hence by Rule 3.4, if ball p originates from hole i and is discarded at move m, $m \geq m_o$, move m must be generated by a ball originating from hole i which was marked before ball p was marked. We assume that no ball originating from hole i goes into pocket A_1 or pocket A_2, else the proof is immediate from Lemma 3.3.

Let p_1 be the marked ball of highest priority originating from hole i such that p_1 is on the surface of the machine or on top of hole i at some move m, $m \geq m_o$. If p_1 does not exist, then we are done since unmarked balls cannot generate moves. So assume that p_1 exists. Note that by choice of m_o and p_1, ball p_1 can never be discarded. By Lemma 3.1, ball p_1 generates only finitely many moves, and since it does not go into any pocket, it must end up either on top of hole i for all moves after some given move, or in front of some fixed door for all moves after some given move. Hence there is a move m_1 such that p_1 does not generate any move m, $m \geq m_1$.

We continue in this way defining p_j and m_j, $j = 2, \ldots, i+1$ by increasing the subscript j of m_j and p_j consecutively by 1 in the above paragraph. If some p_j is undefined along the way, then we are done. So assume p_{i+1} is defined. Then balls $p_1, \ldots, p_i$ are on the surface of the machine for all moves m, $m \geq m_i$. (Note that exactly one ball is on top of hole i at the end of any given move.) By Lemma 3.2, ball p_{i+1} can never end up on the surface of the machine, and we have assumed that it never goes into a pocket. Hence ball p_{i+1} can never be released from hole i, and so no balls originating from hole i can be marked after move m_{i+1}. The proof now follows from Lemma 3.1, since only marked balls can generate moves.

<u>Lemma 3.5</u>: Let $\mathcal{D}$ be a finite collection of doors, all of which open infinitely often. Then there are infinitely many moves m such that for all doors $d \in \mathcal{D}$, no ball is stopped in front of door d at the end of move m.

<u>Proof</u>: Let n be given. We show that there is a move m, $m \geq n$, such that for all $d \in \mathcal{D}$, no ball is stopped in front of door d at the end of move m. Let p be the ball of highest priority which

generates some move k, $k \geq n$. Note that by Lemma 3.4 and the definition of priorities, p must exist. (Move n is generated by a marked ball and there are only finitely many marked balls originating from the hole from which p originated or from holes lower down, by Lemma 3.4.) Let m be the last move $\geq n$ which p generates. We show that m has the desired property.

We first note that by choice of p , ball p can never be discarded. By choice of m, no ball on the surface of the machine at the end of move m can have lower priority than ball p . Assume that some ball q is stopped in front of door d, $d \in \mathcal{D}$, at the end of move m. Then q = p , or ball q has higher priority than ball p . Door d can open during move k , $k > m$, only to discard ball q, or if ball q generates move k . By choice of p and m, no move k, $k > m$ can be generated by q. But if q is discarded at stage $k > m$, then move k is generated by a ball of higher priority than ball q, hence of higher priority than ball p, contradicting the choice of p. Hence door d never opens after move m. But this contradicts the hypothesis that door d opens infinitely often. This completes the proof of the lemma.

We are now ready to present the recursion theory necessary to construct a minimal pair of r.e. sets. Following this, we describe a specific game on the pinball machine, devised to construct a minimal pair.

(3.7) Kleene Enumeration Theorem: There is a recursive enumeration of partial recursive functionals Ψ_0, Ψ_1, ..., and of finite functionals Ψ_0^s, Ψ_1^s, ... such that

(A) For all $i < \omega$ and $A \subseteq \omega$, if $\lambda x\, \Psi_i(A,x)$ is total, then it is recursive in A ;

(B) If f is recursive in A , then there is an $i < \omega$ such that for all x , $f(x) = \Psi_i(A;x)$;

(C) If A is a finite set, define $P(A, x, i, s) \Longleftrightarrow \Psi_i^s(A,x)$ is defined. Then $P(A, x, i, s)$ is recursive;

(D) Let $A \subseteq \omega$, and let $A_0, A_1, \ldots$ be an increasing sequence of finite sets such that $\bigcup_{i<\omega} A_i = A$. Assume that $\Psi_j(A;x)$ is defined. Then there is an s such that for all $t \geq s$, $\Psi_j(A;x) = \Psi_j^t(A_t;x)$;

(E) If $\Psi_e^s(A;x) = y$, then $\Psi_e^t(A;x) = y$ for all $t > s$; and

(F) If $\Psi_e^s(A;x) = y$ and $B(z) = A(z)$ for all $z < s$, then $\Psi_e^s(B;x) = y$.

(3.8) Pairing function: There is a recursive one-one correspondence between elements of ω and ordered pairs of elements of ω.

(3.9) From (3.7) and (3.8) we get an enumeration theorem for all pairs of partial recursive functionals. Let this enumeration be $\{<\Phi_i, \Theta_i> : i < \omega\}$, and let the corresponding enumeration of finite functionals be $\{<\Phi_i^t, \Theta_i^t> : i < \omega, t < \omega\}$.

We construct two r.e. sets, A_1 and A_2. For all s, let A_i^s be the set of x such that balls marked x, j, A_i for some j are placed in pocket A_i before move s, for $i = 1$ or 2. By Rule 3.4, for $i = 1$ or 2 and all s, A_i^s is finite.

For all $e < \omega$ and $s < \omega$, define $L(e,s)$ = the greatest $z \leq s$ such that for all $x < z$, $\Phi_e^s(A_1^s; x) = \Theta_e^s(A_2^s; x)$; and $M(e,s) = \max\{L(e,t) : t \leq s\}$.

In this way, we are assigning to door e the task of insuring that if $\Phi_e(A_1) = \Theta_e(A_2) = f$, then if f is total, f must be recursive.

To guarantee that neither A_1 nor A_2 is recursive, we have a recursive set of requirements $\{\Psi_e(\emptyset) \neq A_i : e < \omega, i = 1,2\}$, which we assign via a recursive one-one correspondence to holes.

We say that ball p <u>requires attention at move</u> n if one of the following holds:

(3.10) Ball p sits in front of a door, say door e, and $L(e,n) = M(e,n)$;

(3.11) Ball p is unmarked sitting on top of a hole, say hole i, and no ball originating from hole i has yet been placed in pocket A_1 or pocket A_2 ;

(3.12) Ball p is marked, sitting on top of a hole, say hole i, the marking is x, i , $\Psi_e(\emptyset) \neq A_k$ is the requirement corresponding to hole i, and $\Psi_e^n(\emptyset ; x)$ is defined.

Finally, we say that hole i has higher priority than hole j if $i < j$.

The construction:

Move n: Find the hole of highest priority such that a ball from that hole requires attention at move n. (Note that such a hole must exist since by Rule 3.4, only finitely many balls have been marked and finitely many balls have reached pocket A_1 or pocket A_2 before move n, hence some ball must require attention via (3.11).) Let this hole be hole i. Let ball p be the ball of highest priority originating from hole i which requires attention at move n, if such a ball exists. Otherwise, ball p will be the unmarked ball sitting on top of hole i. Adopt Case I, Case II, or Case III below, according as ball p requires attention via (3.10), (3.11), or (3.12).

Case I: We perform the move specified by Rule 3.4.C generated by ball p. To discard balls, all track gates are open, as is the discard gate, and all doors are closed before each component of the move. After balls are discarded, set the gates below the position of ball p as follows: Track gate j is closed if there is no ball now sitting in front of door j. (Thus ball p will be forced to stop in front of door j if it reaches track gate j.) Otherwise, track gate j is open. The discard gate is closed. The A_1 gate is open if ball p is marked r,s,A_1 for some r and s, and closed otherwise.

<u>Case II</u>: Follow the move specified by Rule 3.4.A, generated by p. Mark p with n,i where p is on top of hole i.

<u>Case III</u>: Follow the move specified by Rule 3.4.B, generated by p. After balls are discarded, mark ball p with A_1, where ball p is sitting on top of hole i, if either $\Psi_e^n(\emptyset,x) = 0$ where $\Psi_e(\emptyset) \neq A_1$ corresponds to hole i and p is marked with x,i, or if $\Psi_e^n(\emptyset ; x) \neq 0$ where $\Psi_e(\emptyset) \neq A_2$ corresponds to hole i and p is marked with x,i. Otherwise, mark ball p with A_2. Set all gates as in Case I.

From the enumeration theorem, it is not hard to see that all decisions are made effectively, and all searches are bounded, hence A_1 and A_2 are r.e. We note that this game obeys all the rules.

We now show that A_1 and A_2 are a minimal pair of r.e. sets.

<u>Lemma 3.6</u>: Neither A_1 nor A_2 is recursive.

<u>Proof</u>: Assume to the contrary that A_1 is recursive. (A similar proof holds for A_2.) Then by (3.7.B), there is an e such that for all x, $\Psi_e(\emptyset; x) = A_1(x)$. Let $\Psi_e(\emptyset) \neq A_1$ correspond to hole i. By Lemma 3.4, there is a move n such that for all $m \geq n$, no ball from hole i or a hole of higher priority generates move m. Hence there is a fixed ball p sitting on top of hole i at all moves $m \geq n$. If ball p were to require attention at move $m \geq n$, its failure to generate move m can only be caused by a ball originating from hole i or a hole of higher priority than hole i generating move m. Hence ball p never requires attention at any move $m \geq n$.

Assume that ball p is marked with x,i before move n. (If not, it can never be marked.) Then since ball p never requires attention, $\Psi_e^m(\emptyset,x)$ is undefined for all $m \geq n$. By (3.7.D),

$\Psi_e(\emptyset\ ;\ x)$ is undefined. This is impossible since $A_1(x)$ is defined and $\Psi_e(\emptyset\ ;\ x) = A_1(x)$. Hence ball p can never be marked. Ball p's failure to require attention can only be due to the fact that there is a ball q which originated from hole i and rolled into pocket A_1 or pocket A_2 before move n. Let q be marked by $y, i, A_k, k = 1$ or 2.

Assume $k = 1$. Then for some $r < n$, $\Psi_e^r(\emptyset\ ;\ y) = 0$. By (3.7.E), $\Psi_e^t(\emptyset\ ;\ y) = 0$ for all $t \geq r$, hence by (3.7.D), $\Psi_e(\emptyset\ ;\ y) = 0$. But $A_1(y) = 1$, hence $\Psi_e(\emptyset) \neq A_1$, contradicting our assumption. But if $k = 2$, then $\Psi_e^r(\emptyset\ ;\ y) \neq 0$ for some $r < n$, so by (3.7.E) and (3.7.D), $\Psi_e(\emptyset\ ;\ y) \neq 0$. But $A_1(y) = 0$ again yielding a contradiction. We must therefore conclude that A_1 is not recursive.

<u>Lemma 3.7:</u> If f is total and recursive in both A_1 and A_2, then f is recursive.

<u>Proof:</u> Assume that f is total and recursive in both A_1 and A_2. Then there is an e such that $f = \Phi_e(A_1) = \Theta_e(A_2)$. Fix such an e. Let n be the first move such that

(3.13) If m is a move and p a ball such that ball p generates move m and ball p originates from hole i for some $i \leq e$, then $m < n$;

(3.14) If m is a move and i a door such that $i \leq e$, door i opens only finitely often, and door i opens at move m, then $m < n$.

Note that n exists by Lemma 3.4.

To calculate $f(x)$, look for the first move $k > n$ such that

(3.15) $L(e,k) > x$; and

(3.16) For all $i \leq e$, if door i opens infinitely often, then there is no ball resting in front of door i at the end of move $k-1$.

By definition of $L(e,k)$ and (3.7.D), (3.15) holds for all sufficiently large k. By Lemma 3.5, (3.16) holds for infinitely many k. Hence such a $k > n$ must exist. Note that $\Phi_e^k(A_1^k; x) = \Theta_e^k(A_2^k ; x) = y$ for some y. We claim that $f(x) = y$. Once this claim is verified, we will have produced an effective way to calculate $f(x)$ for all x, so f will be recursive. By (3.7.D) it suffices to show that for all $s \geq k$,

(3.17) $\Phi_e^s (A_1^s ; x) = y$ or $\Theta_e^s (A_2^s ; x) = y$.

The proof is by induction on the set of moves $\geq k$. The assumption is true, by choice of k, for $s = k$. Assume that $j \geq k$ and we have shown (3.17) for $s = j$. We now prove (3.17) for $s = j + 1$. If no ball is placed in pocket A_1 or pocket A_2 during move j, then $A_1^{j+1} = A_1^j$ and $A_2^{j+1} = A_2^j$, so we are done by (3.7.E). Note that by Rule 3.4, at most one ball is placed in either of pocket A_1 and pocket A_2 during move j. By symmetry, we can assume that ball p is placed in pocket A_1 during move j. Since $j \geq k \geq n$, by (3.13), the hole from which p originated, say hole i, is such that $i > e$. Hence p must roll down the track to track gate e before going into A_1. So there is a move, say u, such that either p stops in front of door e during move u, or p rolls past track gate e during move u.

Assume first that p stops in front of door e during move u. Note that $u \geq k$, since if $u < k$, then at move k, ball p would be stopped in front of door r for some $r \leq e$. Door r cannot open finitely often by (3.14), and cannot open infinitely often by (3.16) yielding a contradiction. Hence there is a move v such that $k < u < v \leq j$ such that door e opens releasing p during move v. Thus by the construction, $\Phi_e^v (A_1^v ; x) = \Theta_e^v (A_2^v ; x)$ and both must equal y by the induction hypothesis. Since ball p is not discarded, no ball of higher priority than that of ball p can generate any move w where $v \leq w < j$, hence no such ball can be placed in pocket A_1 or pocket A_2 at any such move w. All balls of lower priority than that of ball p at move v and not already in a pocket, are discarded during move v. Hence if a ball goes into pocket A_1 or pocket A_2 at move w, $v \leq w < j$ this ball must be marked after move v, and hence is marked a,b with $a \geq v$. By (3.7.E) and (3.7.F), we must have $\Phi_e^j(A_1^j ; x) = Ⓗ_e^j(A_2^j ; x) = y$. But then at move j, p is placed in A_1 alone, so $Ⓗ_e^{j+1} (A_2^{j+1} ; x) = Ⓗ_e^j (A_2^j ; x) = y$ by (3.7.E), since $A_2^{j+1} = A_2^j$.

We must now consider the case where p bypasses door e during move u. Then by the construction, there is a ball q of higher priority than p at move u which is sitting in front of door e at the end of move $u-1$. We again note that $u \geq k$. Door e cannot open after move u and before $j + 1$, else q would generate this move, or a ball of higher priority than q would generate this move, and p would be discarded. Let ball q stop in front of door e during move v. Since q generates move v and ball p is marked and not in a pocket after move v, we must have $v < u$. $k \leq v$ by the same reasoning that showed that $k \leq u$. Note that (3.17) is satisfied for $s = v$ by induction. Since ball q is not discarded before move $j + 1$, no ball of higher priority than q can generate any move between v and $j + 1$, hence no such ball is placed in pocket A_1 or pocket A_2 at move w for any w such that $v \leq w < j + 1$. If a ball is placed in one of these pockets at such a move w, it must have been marked after move v, since at the end of move v, there are no marked balls not in a pocket and of lower priority than q. Hence if the ball is marked a,b, then $a \geq v$. From (3.7.F) and (3.7.E), we must conclude that (3.17) is satisfied for $s = j + 1$.

This concludes the proof of the lemma.

<u>Theorem 3.8</u>: There exists a minimal pair of r.e. sets.

Proof: We take A_1 and A_2 abpve. By Lemma 3.7, if $f \leq_T A_1$ and $f \leq_T A_2$, then f is recursive. Assume $A_1 \leq_T A_2$. Then since $A_1 \leq_T A_1$, by Lemma 3.7, A_1 is recursive, contradicting Lemma 3.6. So $A_1 \not\leq_T A_2$. A similar argument shows $A_2 \not\leq_T A_1$.

SECTION 4: GENERALIZING THE MINIMAL PAIR

Let α be an admissible ordinal, and let $A \subseteq \alpha$ and $B \subseteq \alpha$. We say that A and B are a minimal pair of α - r.e. sets if A and B are α - r.e., $A \not\leq_\alpha B$, $B \not\leq_\alpha A$, and for all $C \subseteq \alpha$, if $C \leq_\alpha A$ and $C \leq_\alpha B$, then C is α-recursive. Lerman and Sacks [8] have constructed minimal pairs for many admissible ordinals. There are still some admissible ordinals α, however, for which the existence of a minimal pair is still open.

We discuss only the case where $\alpha^* < \alpha$. (The case where $tp2(\alpha) < \alpha$ follows in a similar way with a slightly more complicated set of rules for a game.)

First, let us discuss the necessary modifications of the pinball machine. There will be more requirements, so we may need a longer machine. The length of the machine is important in proving the analogs of Lemma 3.1 through Lemma 3.4; and for $\alpha^* < \alpha$, we only know how to get these proofs if the machine has length α^*. (We remark at this point that if $\alpha^* < \alpha$, α^* is the greatest α-cardinal.) Fortunately, all requirements can be suitably indexed by an indexing of length α^*.

If $\alpha^* > \omega$, then we have another problem. For once a ball is released from hole ω, we would like it to stop in front of the first door along the way which has no ball in front of it. But the doors are not well-ordered in this direction, so this will not work. We must therefore let the ball roll all the way down and then start working its way up the surface of the machine. This, however, invalidates the proof of Lemma 3.7. For in that proof, we use the fact that a finite amount of information governs the action of all balls which are ever below a given door. By letting the balls roll up the surface of the machine, we now need an unbounded amount of information to tell us how balls act once they pass a given door.

We circumvent this problem by requiring that balls are subject to rolling up the surface of the machine and stopping in front of doors ω many times in succession, before they are allowed into the A_1 or A_2 pocket. In this way, each injury to the equality $\Phi_e^{\sigma}(A_1^{\sigma}; x) = \Theta_e^{\sigma}(A_2^{\sigma}; x)$ is followed by a repair of this injury. Furthermore, the number of alternations of the side of the equality which is injured, is finite. Thus we can prove the equivalent of Lemma 3.7.

Limit moves cause a problem because the position of a ball may not be well-defined at a limit move; it may roll during a set of moves cofinal with the limit ordinal we are considering. Hence at limit moves, we will look at all balls without a specified position at the beginning of the move, pick one, set a position for it based on the past history of its movement, and discard all other balls without fixed positions.

Finally, once a ball passes through a door, a decision must be made whether the ball is to continue up , or can now go down. For balls from hole i should only stop at doors below hole i. Hence after each door, we put a gate determining the direction the ball is to follow. We call such a gate a <u>return gate</u> and place one such gate below each hole except hole 0. A picture of the pinball machine used for the construction appears in Figure 4.1.

A game on the machine is played in a sequence of moves $\{i : i < \alpha\}$, where α is an admissible ordinal.

Rules for A Game:

Rule 4.1: Before the game begins, each hole has a supply of unmarked

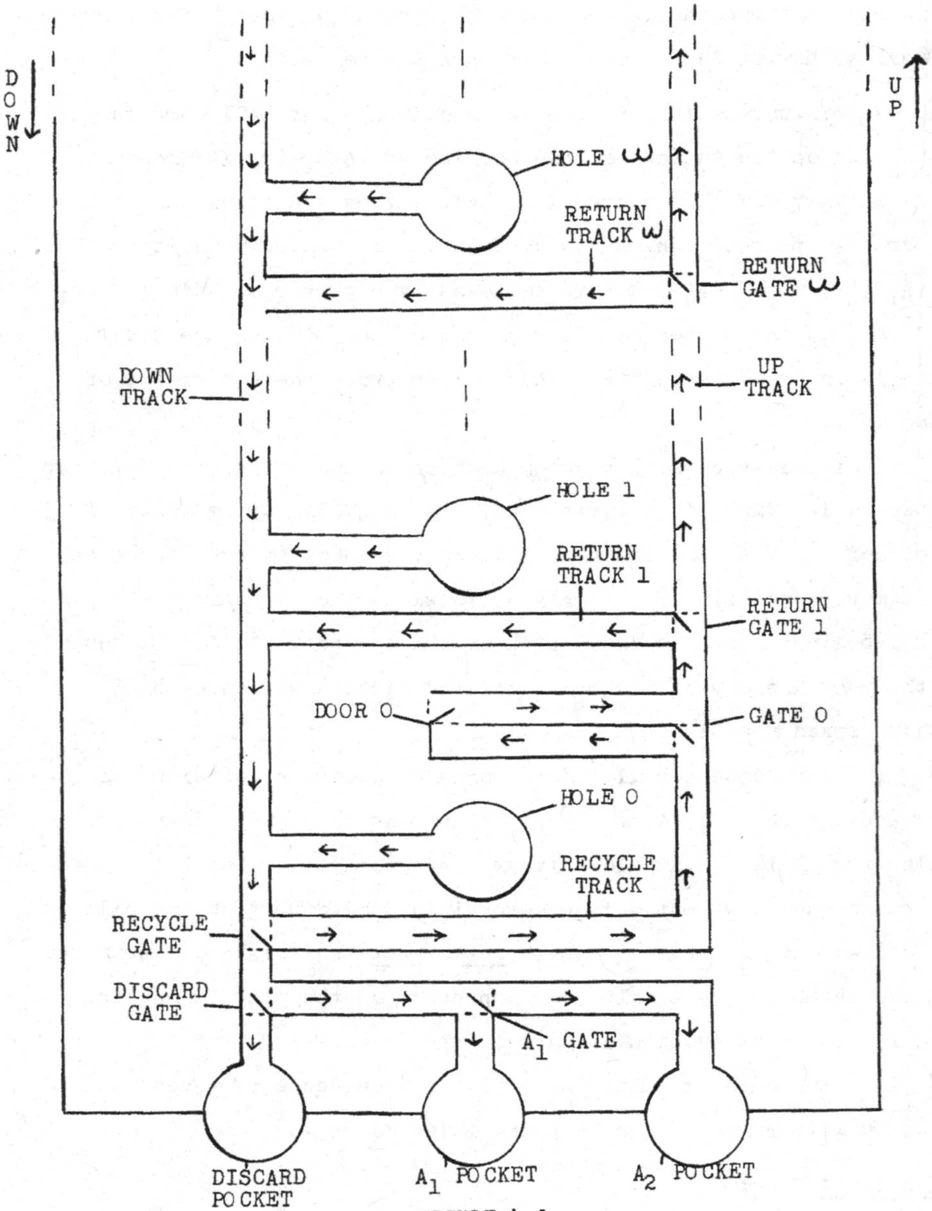
DOWN
UP
HOLE ω
RETURN TRACK ω
RETURN GATE ω
DOWN TRACK
UP TRACK
HOLE 1
RETURN TRACK 1
RETURN GATE 1
DOOR 0
GATE 0
HOLE 0
RECYCLE TRACK
RECYCLE GATE
DISCARD GATE
A1 GATE
DISCARD POCKET
A1 POCKET
A2 POCKET

FIGURE 4.1

balls of order-type α and no marked balls. There are no balls on the surface of the machine or in any pocket.

Rule 4.2: No unmarked ball ever rolls on the surface of the machine.

Rule 4.3: A move consists of an α-finite set of components. The following are the allowable components for moves:

A. An unmarked ball, sitting on top of a hole, is marked.

B. A marked ball, sitting on top of a hole, is released onto the surface of the machine, and rolls until it ends up in front of a closed door or in a pocket.

C. A ball sitting in front of a closed door is released (by opening the door), and rolls until it ends up in front of a closed door or in a pocket.

D. A ball which has no fixed position going into the move, is given a fixed position in front of a closed door or in a pocket.

Rule 4.4: Move n must proceed via one of the sequences described below:

A. An unmarked ball p is marked at move n. All balls now marked and of lower priority than p at this given point, and not already in a pocket, are discarded.

B. p is the marked ball sitting on top of hole i. All marked balls of lower priority than p at this given point, and not already in a pocket, are discarded. All doors are now closed, as is the discard gate. Gates are then set so that ball p will reach some final position, and not roll forever. Ball p is then released onto the surface of the machine and allowed to roll until it reaches a final position.

C. p is a marked ball sitting in front of closed door i. All marked balls of lower priority than p at this given point, and not already in a pocket, are discarded. All doors are then closed, as is the discard gate. Gates are then set so that ball p will reach some final position. Door i is then opened, and ball p is allowed to roll until it reaches a final position.

D. n is a limit ordinal. p is the marked ball of highest priority with no fixed position on the surface of the machine at the beginning of move n. All other marked balls with no fixed position at the beginning of move n are discarded, as are all marked balls of lower priority than p at move n, and not already in a pocket. All doors are then closed. If there is a place on the surface of the machine which is the limit of the positions occupied by p at previous moves in order of occupation, p is placed at this limit and allowed to roll after gates are set. Otherwise, p is placed in pocket A_1 or pocket A_2.

In each case, we say that ball p generates move n.

Rule 4.5: If a ball from hole i rolls into pocket A_1 or pocket A_2 at move m, then no ball from hole i becomes marked at any move after move m.

Rule 4.6: At the end of any given move, each door can impede the progress of at most one ball.

Rule 4.7: The game is uniquely determined from a finite set of instructions.

Rule 4.8: A ball originating from hole i can never be in any position above hole i.

Rule 4.9: The order-type of the set of moves during which a given ball p rolls through the recycle track is at most ω.

We leave it to the reader to verify that the rules are consistent.

We next prove some lemmas about a typical game. These proofs will be sketchy. We refer the reader to [8] for a rigorous proof of these lemmas.

Remark: The basic tool used in the proof of many of the lemmas below is the following: Let $R(i, x)$ be an α-recursive relation such that $R(i, x)$ holds if we find out during some move of the game that $R(i, x)$ is true. Let γ be a regular α-cardinal, and let I be an α-finite set of α-cardinality $< \gamma$. Assume that for all $i \in I$, the order-type of $\{x : R(i, x)\}$ is $< \gamma$. Then the order-type of $\mathcal{R} = \bigcup_{i \in I} \{x : R(i, x)\}$ is $< \gamma$. For were it $\geq \gamma$, then at some move of the game, we would have γ or more elements in $\mathcal{R}$ by that move, contradicting the assumption that γ is a regular α-cardinal.

Lemma 4.1: Let p be a marked ball, and let γ be a regular infinite α-cardinal. Then p generates only α-finitely many moves. Furthermore, if $i < \gamma$, $\gamma > \omega$, and p originates from hole i, then p generates $< \gamma$ many moves.

Proof: By Rule 4.8 and Rule 4.9, p can only cycle through the doors below hole i, ω many times. Thus as in the proof of Lemma 3.1, the order-type of the moves generated by p is at most $1 + i \cdot \omega + 1$ which is α-finite. If $\gamma > \omega$, then $1 + i \cdot \omega + 1 < \gamma$.

Let p be a marked ball originating from hole i. We say that p has order β at move n, if at move n the order-type of the balls originating from hole i of higher priority than p (under the priority ordering), and on the surface of the machine at the beginning of move n, is β. (Note that the order of p is always the same while p is on the surface of the machine.)

Lemma 4.2: Let i, n, and k be given such that no ball originating from below hole i generates any move m,

$n \leq m < k \leq \alpha$. Let γ be a regular infinite α-cardinal, and let $\beta < \gamma$. Then only α-finitely many moves m, $n \leq m < k$, can be generated by a ball from hole i of order β. Furthermore, if $\gamma > \omega$, then the order-type of such moves is $< \gamma$.

Proof: By induction on $\{\delta : \delta \leq \beta\}$. Assume that the lemma is true for all $\lambda < \delta$. If $\gamma = \omega$, then δ is finite, and the moves generated by balls of order $\lambda < \delta$ form an α-finite set (the finite union of α-finite sets). After all such moves, there can be only one ball of order δ until move k, (there is no way to discard this ball after it is marked). The result now follows from Lemma 4.1.

Now assume $\gamma > \omega$. Then by the Remark and by induction, the order-type of all moves m, $n \leq m < k$, generated by balls from hole i of order $\lambda < \delta$ is $< \gamma$. Between any two such moves, there is at most one ball of order δ from hole i. Hence there are $< \gamma$ many balls of order δ from hole i between move n and k. By Lemma 4.1, each such ball generates $< \gamma$ many moves. The Remark now yields the desired result.

Lemma 4.3: Let i and γ be given such that $i < \gamma$, and γ is an α-cardinal. Then at the end of each move, there are $< \gamma$ many balls on the surface of the machine originating from hole i.

Proof: Immediate from Rule 4.8, Rule 4.7, and the fact that γ is an α-cardinal.

Lemma 4.4: Let i, n, and k be given such that no ball originating from below hole i generates any move m, $n \leq m < k \leq \alpha$. Let γ be a regular infinite α-cardinal, and let $i < \gamma$. Then balls from hole i can generate only α-finitely many moves between n and k. If $\alpha > \omega$, then the order-type of the set of such moves is $< \gamma$.

Proof: If $\gamma = \omega$, then by Lemma 4.3, there are at most $i + 1$ orders of balls, and the result follows from Lemma 4.2.

Assume $\gamma > \omega$. Suppose that we have a move m, $n \leq m < k$, such that the order-type of the moves generated by balls from hole i and between n and m is $\geq \gamma$. Let $\beta \leq \gamma$ be the least order such that at least γ many such moves are generated by balls of order $< \beta$. Note that β cannot be $< \gamma$ by Lemma 4.2 and the Remark. But if $\beta = \gamma$, then at move m, there must be γ many balls from hole i on the surface of the machine (there must be a last ball of each order $< \gamma$, else $\beta < \gamma$). But this contradicts Lemma 4.3. Thus the lemma is proven.

Lemma 4.5: For all $i < \alpha^*$, there is a move n such that no ball from any hole $j \leq i$ generates any move $m \geq n$.

Proof: Fix i. Note that the result follows from Lemma 4.4 for $i < \omega$. Let $\gamma \leq \alpha^*$ be a regular infinite α-cardinal such that $i < \gamma$ and assume $\gamma > \omega$. (Note that α^* is an α-cardinal, so it is either regular, or the limit of regular α-cardinals.) By induction, we can show that for $j \leq i$, $< \gamma$ many moves are generated by balls originating from hole j. (The result will then follow from the Remark.) For by the Remark and by induction, we have for $j \leq i$, $< \gamma$ many moves are generated by balls from holes below hole j. By Lemma 4.4, between any two such consecutive moves, balls from hole j generate $< \gamma$ many moves. Hence by the Remark, the induction hypothesis is true for j.

We now present the recursion theory used to construct a minimal pair of α - r.e. sets. We then describe the game to construct a minimal pair.

(4.10) Enumeration Theorem: There is an α-recursive enumeration of partial α-recursive functionals $\{ \Psi_i : i < \alpha \}$ and α-finite functionals $\{ \Psi_i^s : i < \alpha, s < \alpha \}$ such that

A. For all $i < \alpha$ and α - r.e. $A \subseteq \alpha$, if $\lambda x \Psi_i(A ; x)$ is total, then it is weakly α-recursive in A;

B. If f is α-recursive in A, then there is an $i < \alpha$ such that for all x, $f(x) = \Psi_i(A ; x)$;

C. If A is α-finite, define $P(A,x,i,s) \Longleftrightarrow \Psi_i^s(A ; x)$ is defined. Then $P(A,x,i,s)$ is α-recursive;

D. Let A be α-r.e., and let $\{A_i : i < \alpha\}$ be an increasing α-recursive sequence of α-finite subsets of A such that $\bigcup_{i<\alpha} A_i = A$. Assume that $\Psi_j(A ; x)$ is defined. Then there is an s such that for all $t \geq s$, $\Psi_j^t (A_t ; x) = \Psi_j(A ; x)$;

E. If A is α-r.e. and $\Psi_e^s(A ; x) = y$, then $\Psi_e^t(A ; x) = y$ for all $t \geq s$;

F. If $\Psi_e^s(A ; x) = y$ and $B(z) = A(z)$ for all $z \leq s$, then $\Psi_e^s(B ; x) = y$.

(4.11) Pairing function: There is a one-one α-recursive correspondence between pairs of ordinals $< \alpha$, and ordinals $< \alpha$.

(4.12) Double Enumeration Theorem: Get $\{ < \Phi_i, \Theta_i > : i < \alpha\}$ and $\{ < \Phi_i^s, \Theta_i^s > : i < \alpha, s < \alpha\}$ as in Section 3.

We assume we are given a one-one α-recursive $f : \alpha \longrightarrow \alpha^*$. We construct two α-r.e. sets A_1 and A_2 and define $L(e,s)$ and $M(e,s)$ as in Section 3. We order all requirements $\{ \Psi_e(\emptyset) \neq A_i : e < \alpha, i=1,2\}$ into an α-recursive sequence of order-type α, and assign the βth such requirement to hole $f(\beta)$ at all moves $m \geq \beta$. Before move β, no requirement is assigned to hole $f(\beta)$. Note that if $\gamma < \alpha^*$, then there is a move n such that if a requirement is assigned to hole i, $i < \alpha$, it is assigned before move n. We associate requirement $L(e,y) = M(e,y)$ with door $f(e)$ at all moves m, $m \geq e$.

We say that ball p requires attention at move n if one of

the following holds:

(4.13) Ball p is marked, originating from hole i, n is a limit ordinal, and p has no fixed position at the beginning of move n.

(4.14) Ball p is marked sitting in front of closed door r at the beginning of move n, $L(e,n) = M(e,n)$, and $f(e) = r$, $e \leq n$.

(4.15) Ball p is unmarked sitting on top of hole i, no ball from hole i has yet been placed in pocket A_1 or pocket A_2, and a requirement has been associated with hole i by move n.

(4.16) Ball p is marked sitting on top of hole i, the marking is $x, i, \Psi_e(\emptyset) \neq A_k$, $k = 1$ or 2, has been associated with hole i by move n, and $\Psi_e^n(\emptyset\ ;\ x)$ is defined.

We say that hole i has higher priority than hole j if $i < j$.

The construction:

Move n: Find the hole of highest priority such that a ball from that hole requires attention at move n. (Note that the top ball in hole $f(n)$ requires attention via (4.15).) Let this hole be hole i. Let ball p be the ball of highest priority originating from hole i which requires attention at move n if such a ball exists. Otherwise, ball p will be the unmarked ball sitting on top of hole i. Adopt Case I, Case II, Case III, or Case IV below according as ball p requires attention through (4.13), (4.14), (4.15), or (4.16) above.

Case I: We perform a move given by Rule 4.4.D. Discard balls and close all doors. Let $S = \{i_j : j < k \leq n\}$ enumerate the doors that ball p has stopped in front of, in the order that ball p has stopped in front of them. If the sequence S has no limit, then ball p has been previously marked x, i, A_v for $v = 1$ or 2. Place p in pocket A_v and go to the next move.

Assume that the sequence S has a limit. Return gate i is closed, and all other return gates are open. Gate j is closed if there is no ball in front of door j and is open otherwise. Note that the limit of S is no higher than return gate i. If there is a $j < i$ such that gate j is closed, close the recycle, discard, and A_1 gates. If no such j exists, the discard gate is closed, the recycle gate is open, and the A_1 gate is set to let ball p roll into pocket A_1 if and only if ball p is marked x,i,A_1. Now place ball p at the limit of S, and allow it to roll until it stops.

Case II: We follow Rule 4.4.C.
Discard all balls which must be discarded, and close the discard gate and all doors. Gates are set as in Case I, where ball p originated from hole i. Door r is then opened, and ball p is allowed to roll until it stops.

Case III: Follow Rule 4.4.A. After balls are discarded, mark ball p with n,i.

Case IV: Follow Rule 4.4.B. Ball p is further marked with A_1 if $\Psi_e(\emptyset) \neq A_v$ is associated with hole i and ball p is already marked x,i, and $\Psi_e^n(\emptyset\ ;\ x) = 0$ and $v = 1$, or $\Psi_e^n(\emptyset;\ x) \neq 0$ and $v = 2$. Otherwise, ball p is further marked with A_2. Doors and gates are now set as in Case I, and ball p is released onto the surface of the machine.

If $A_i = \{x :$ ball p is in pocket A_i and is marked $x,j,A_i\}$, then A_i is α-r.e.

The following lemmas complete the proof.

Lemma 4.6: Neither A_1 nor A_2 is recursive.

Proof: As in the proof of Lemma 3.6, using Lemma 4.5 instead of Lemma 3.4.

Lemma 4.7: If f is α-recursive in both A_1 and A_2, then f is α-recursive.

Proof: The proof is similar to that of Lemma 3.7. We give a rough sketch of the differences here.

If $f \leq_\alpha A_1$ and $f \leq_\alpha A_2$, then there is an e such that $f = \Phi_e(A_1) = \Theta_e(A_2)$. Also, $L(e,y) = M(e,y)$ is assigned to door j for some j. Let n be a move such that no ball originating from a hole below door j, generates any move m, $m \geq n$. To calculate $f(x)$, look for the least $m \geq n$ such that $L(e,m) = M(e,m) > x$. Then $\Phi_e^m(A_1^m ; x) = \Theta_e^m(A_2^m ; x) = f(x)$. This is because at most one side of the computation of $\Phi_e^s (A_1^s ; x) = \Theta_e^s(A_2^s ; x)$ is injured at any given move. Every time that the injured side alternates, the previous injury has been repaired, and the ball causing the new injury has higher priority than all previous balls which injured such a computation after move m. Since "higher priority than" is a well-ordering, the number of alternations must be finite, else we would have an infinite decreasing sequence of ordinals. Since each injury is followed by a repair before the next injury is allowed, we must have $\Phi_e(A_1 ; x) = \Phi_e^s(A_1^s ; x)$ or $\Theta_e(A_2 ; x) = \Theta_e^s(A_2^s ; x)$ for all sufficiently large $s < \alpha$, concluding the proof.

Theorem 4.8: A_1 and A_2 are a minimal pair of α-r.e. sets.

Proof: As in the proof of Theorem 3.8.

SECTION 5: CONCLUSION

We conclude by describing the status of minimal pairs of α-r.e. sets. Such pairs have been constructed for all admissible ordinals α such that either $tp2(\alpha)$ is less than the greatest α-cardinal, or there is no Σ_2 projection from α into a smaller ordinal. If α is the limit of α-cardinals, such a pair has also been constructed.

The following case is open, and we conjecture that minimal pairs exist for such admissible α: There is a greatest α-cardinal, α is Σ_2 projectible into this greatest α-cardinal, but $tp2(\alpha)$ is bigger than this greatest α-cardinal.

REFERENCES

1. G. Driscoll, Contributions to metarecursion theory, Ph.D. Thesis, Cornell University, Ithaca, N.Y., 1965.

2. R. M. Friedberg, Three theorems on recursive enumeration, Journal of Symbolic Logic 23 (1958), 309-316.

3. __________, Two recursively enumerable sets of incomparable degrees of unsolvability, Proceedings of the National Academy of Science, U.S.A., 43(1957), 236-238.

4. C. Jockusch and R. I. Soare, A minimal pair of π_1^0 classes, Journal of Symbolic Logic 36(1971), 66-78 .

5. A. H. Lachlan, Lower bounds for pairs of recursively enumerable degrees, Proceedings of the London Mathematical Society (3) 16(1966), 537-569.

6. __________, The priority method for the construction of recursively enumerable sets, These Proceedings.

7. M. Lerman, On suborderings of the α-recursively enumerable α-degrees, to appear.

8. ________ and G. E. Sacks, On minimal pairs of α-recursively enumerable α-degrees, to appear.

9. H. Rogers, Jr., Theory of recursive functions and effective computability, McGraw-Hill, New York, 1967.

10. G. E. Sacks, Post's problem, admissible ordinals, and regularity, Transactions of the American Mathematical Society 124(1966), 1-23.

11. ________ and S. Simpson, The Friedberg Muchnik theorem in α-recursion theory, to appear.

12. C. E. M. Yates, A minimal pair of recursively enumerable degrees, Journal of Symbolis Logic 31(1966), 159-168.

XI. ABSTRACT COMPUTABILITY VERSUS ANALOG-GENERABILITY

(a survey)

Marian Boykan Pour-El[1]

The exact formulation of the concept of abstract computability is well-known to every mathematical logician. He is familiar with the pioneering work of Turing, Herbrand-Gödel, Church, Markov and others [16, 7, 1, 10]. He is aware that Turing's approach, via the Turing Machine, provides a theoretical basis for existing digital computers.

In this paper we investigate the relation between abstract computability and the general-purpose analog computer. An analog computer is a device in which the variables vary continuously. The paper is divided into three sections.

In the first section we present a definition of what it means for a function of a real variable to be generated by a general-purpose analog computer. The definition is formulated in terms of a simultaneous set of non-linear differential equations. It encompasses functions generable by existing general-purpose analog computers.

The second section is concerned with motivation. The motivation will be of three kinds. The first is conceptual. The reader will see how a study of analog-generable functions is related to traditional foundational problems - e.g. impredicative definitions. The second is technical. Since the basic definition is phrased in terms of a set of non-linear differential equations we would expect to use methods and techniques of classical analysis and differential equations -together

[1] Supported in part by N.S.F. GP 8829. The results were presented at a joint seminar on complex variables, differential equations and mathematical logic (Fall 1970) at the University of Minnesota. It is a pleasure to thank my colleagues, Professors J. Eagon, Y. Sibuya and J. Nitsche for valuable conversations. Thanks also to Professor J. Shepherdson for his hospitality at the University of Bristol at which some of this work was done. An abstract of the results appears in the Notices of the American Mathematical Society, February 1970.

with recursion theory. Indeed, this proves to be the case. The third - and most obvious - concerns itself with the relation between analog generability and computer science.

The last section of this paper is devoted to statements of results and brief remarks concerning proofs.

Throughout this work we assume that our functions have continuous derivatives - although the results hold under weaker assumptions. Indeed piecewise continuity of the derivative suffices for most of our applications.

§1 THE DEFINITION

In order to motivate the definition - and to show that it encompasses existing general-purpose analog computers - it is useful to consider the structure of present-day analog computers. We proceed as follows. First we discuss existing analog computers. This is followed by a discussion of a preliminary definition - which will be in terms of black boxes, feedback etc. (It should be clear from this account that present-day general-purpose analog computers are included under the preliminary definition.) We shall be led to a definition which is essentially graph-theoretic in character. Finally we relate our preliminary definition to the final definition which, as we have already remarked, involves non-linear differential equations.

Existing Analog Computers

It is often said that a digital computer is any device in which the variables vary discretely. Thus, on this definition, a drop of water issuing from a faucet at regular intervals may be conceived of as a digital computer. The literature on digital computers is more restrictive than that! Similarly, although an analog computer may be conceived of as any device in which the variables vary continuously, the literature on analog computers is more restrictive than that. What does the literature on analog computers conceive an analog computer to be? Clarence Johnson, in his well-known textbook on Electronic Analog Techniques divides the class of existing analog computers into two main categories. In the first category we have the so-called special-purpose devices. They include the planimeter, the wind tunnel and many other special-purpose machines. In this work, we will not be concerned with these very special-purpose devices. Rather our attention will be focused upon the general-purpose analog computer. So our question may be rephrased: What does the literature on analog computers conceive a general-purpose analog computer to be? Essentially, there are two main types. The first is the so-called electronic differential analyzer - often referred to as the electronic analog computer - and sometimes just referred to as the analog computer, as this is the main analog device. The second is the mechanical differential analyzer. Let us discuss briefly the construction of an electronic analog computer. Then we will make a few remarks relating this discussion to the mechanical differential analyzer. In this paper we assume that all components of our devices are perfect.

The functions generated by an electronic analog computer are functions of time: their values are usually measured in terms of voltage. Each electronic analog computer is composed of a finite number of black boxes or units. What are the units? There are essentially four kinds - integrator, adder, constant multiplier and variable multiplier. The integrator is a one-input, one-output device with a setting for initial conditions: if we feed in $u(t)$ as input and set the initial condition $e(a)$ at time $t = a$, our output will be $\int_a^t u(t)dt + e(a)$. The adder is a two-input, one-output device: if $u(t)$ and $v(t)$ are fed in as inputs then $u(t) + v(t)$ is the output. For each rational constant k, we have a constant multiplier$_k$. This is a one-input, one-output device: if we input $u(t)$ we obtain $ku(t)$ as output. Finally there is the variable multiplier, a two-input, one-output device: if we input $u(t)$ and $v(t)$ we obtain $u(t) \cdot v(t)$ as output.

The electronic analog computer is constructed from a finite number of these units. This is done by connecting the units with wires or plugs. Feedback is, of course, allowed. Interconnections and their relation to feedback as well as other matters will be considered in greater detail - in a more general context - when we discuss our preliminary definition of a general-purpose analog computer. The reader interested in the hardware required to realize an electronic analog computer is referred to the standard texts,[5], [6], [8], [14], [15].

The functions generated by a mechanical differential analyzer are also functions of time. Naturally they are not measured in volts. Nevertheless, the mechanical differential analyzer contains essentially the same units subject to the following modification. First, integration is slightly more general. Second, the hardware which realizes this analyzer is mechanical or electro-mechanical, rather than electronic. (Since this paper is not concerned with hardware this is not an essential difference for us.) A detailed account of the structural features of a mechanical differential analyzer may be found in [2].

Preliminary Definition

The preliminary definition is motivated by the description of the differential analyzers discussed above. Hence we will focus attention on the following units. (Note that our functions are not necessarily functions of time.) We will be concerned with functions of one variable only although our discussion is obviously applicable to functions of several variables.

1. Integrator: A two-input, one-output device with a setting for initial conditions. If $u(x)$ and $v(x)$ are inputs, we obtain as output $\int_{x_0}^{x} u(x)dv(x) + C$ - where C is a constant depending on the initial conditions and $\int_{x_0}^{x} u(x)dv(x)$ is the Riemann-Stieltjes integral.

2. Constant multiplier: For each real constant k, there is a device with one input and one output. If $u(x)$ is the input, then $ku(x)$ is the output.

3. Adder: A two-input,one-output device. If $u(x)$ and $v(x)$ are inputs, then $u(x) + v(x)$ is the output.

4. Variable multiplier: A two-input, one-output device. If $u(x)$ and $v(x)$ are inputs, then $u(x) \cdot v(x)$ is the output.

5. Constant function: A one-input, one-output device. If $u(x)$ is the input, then $C_1(x)$ - where $C_1(x) \equiv 1$ - is the output.

A general purpose analog computer is constructed by interconnecting a finite number of these units. We require - as in practice - that two inputs and two outputs can never be interconnected. We say that $\mathfrak{A}$ is a general-purpose analog computer if $\mathfrak{A}$ is a collection of $n-1$ units $U_2,\ldots,U_n$ which are interconnected so that each input is connected to at most one output.[2] Thus the reader might envision a portion of the computer as follows. (Assume the input terminals are on the left and the output terminals are on the right.)

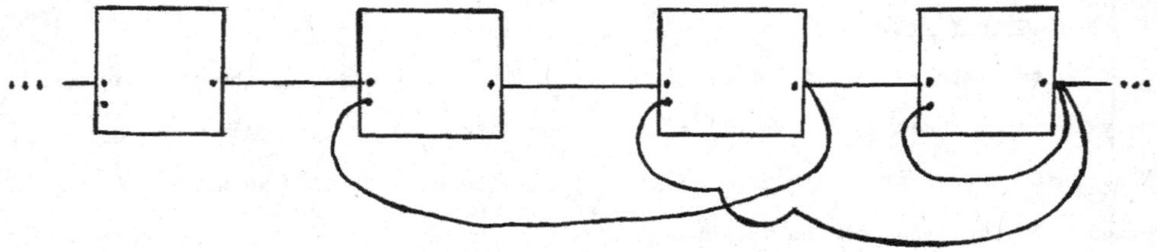

Note that feedback, which may be conceived of as a form of continuous recursion is permitted! Without feedback this study would degenerate into an uninteresting academic exercise. Incidentally it is easy to describe feedback in terms of the units of an analog computer. We say that a general-purpose analog computer $\mathfrak{U}$ - with $n-1$ units - has feedback if it is not possible to enumerate the units of $\mathfrak{U}$ - viz. $U_2,\dots,U_n$ - in such a way that if the output of U_i is connected to an input of U_j then $i<j$. Denote the output of U_i by y_i. We say that the function $u(x)$ is generated by $\mathfrak{U}$ on I if we can prescribe initial conditions to the integrators of $\mathfrak{U}$ at $x=a$ for some $a \in I$ so that if x is applied to every input not connected to an output, then for some i $(2 \le i \le n)$, we have $y_i(x) \equiv u(x)$ on I. We say that $u(x)$ is generable by an analog computable on I if there is some $\mathfrak{U}$ such that $u(x)$ is generated by $\mathfrak{U}$ on I.

This concludes our preliminary sketch of "general-purpose analog computer" and "function generated by a general-purpose analog computer." The reader will note

---

[2] The reason for this rather peculiar notation is the following. It is convenient to refer to the output of U_i as $y_i(x)$. As we will see in connection with our fundamental definition it is convenient to use the following notation: $y_o(x) \equiv 1$, $y_1(x) \equiv x$. Thus we index our units beginning with 2.

that the definitions given above are essentially graph-theoretic in nature. We invite him to translate these concepts into rigorous graph-theoretic definitions for himself.

Remark 1.

The variable multiplier can be eliminated. This is because

$$u(x)\cdot v(x) = \int_a^x u(x)dv(x) + \int_a^x v(x)du(x) + u(a)\cdot v(a)$$

The Fundamental Definition

We now give our fundamental definition and relate it to the preliminary account given above.

Notation: Define y_0 and y_1 by $y_0(x) \equiv 1$, $y_1(x) \equiv x$. Let I, J be closed bounded intervals with non-empty interiors. Let c_{ijk} be real numbers.

Definition.[3] The function $y(x)$ is generated by a general-purpose analog computer on I if there exists a set of functions $y_2(x),\ldots,y_n(x)$ and a set of initial conditions $y_i(a) = y_i^*$ - where $a \in I$ - such that

(1) $\{y_2,\ldots,y_n\}$ is the unique solution on I of a set of differential equations of the form

$$\frac{dy_2}{dx} = \sum_{i,j=0}^{n} c_{ij2} y_i \frac{dy_j}{dx}$$

$$\vdots$$

$$\frac{dy_n}{dx} = \sum_{i,j=0}^{n} c_{ijn} y_i \frac{dy_j}{dx}$$

[3] This definition - with its "domain of generation" - appears to differ considerably from the approach used in [13]. We have found this change in approach necessary because we believe that there is a serious gap in the short proof which appears on the top of p. 343 of [13]. For the sake of brevity the details concerning the gap - and the necessity for a better proof - will be omitted. They are discussed at length in the author's paper Abstract Computability and its Relation to the General-Purpose Analog Computer, which gives a complete account of the results surveyed here. We merely remark that our work is considerably more involved than that sketched in [13]. As indicated in section 3, it uses, among other things, the Cauchy-Peano existence theorem for differential equations and some facts concerning the degree of transcendence of transcendental field extensions.

<u>satisfying the initial conditions</u>

(2) For some i such that $2 \leq i \leq n$, $y(x) \equiv y_i(x)$ on I.

(3) $\langle a, y_2^*, \ldots, y_n^* \rangle$ possesses a "domain of generation" - to be explained next.

The use of computers in the real world suggests that each point $\langle x^{**}, u_2^{**}, \ldots, u_n^{**} \rangle$ sufficiently close to $\langle a, y_2^*, \ldots, y_n^* \rangle$ should provide initial conditions for a unique solution of the equations of (1) on some interval I^{**}. This is because the initial conditions represent initial settings on the integrators. (In practice the operator is allowed to vary the initial conditions on an analog computer slightly.) More precisely the following is required. There are closed intervals $J_1, \ldots, J_n$ - with non-empty interiors-such that

a. $\langle a, y_2^*, \ldots, y_n^* \rangle$ is an interior point of $J_1 \times J_2 \times \ldots \times J_n$

b. whenever $\langle x^{**}, u_2^{**}, \ldots, u_n^{**} \rangle \in J_1 \times J_2 \times \ldots \times J_n$ there exists a set of functions $\{u_2, \ldots, u_n\}$ such that

(i) $u_i(x^{**}) = u_i^{**}$ for $i = 2, \ldots, n$

(ii) $\{u_2, \ldots, u_n\}$ satisfies the equations of (1) on some I^{**} for which $x^{**} \in I^{**}$.

(iii) $\{u_2, \ldots, u_n\}$ is <u>locally unique</u> - i.e. unique on I^{**} and on every subinterval of I^{**} containing x^{**}.

<u>$J_1 \times J_2 \times \ldots \times J_n$ is called a domain of generation of $\langle a, y_2^*, \ldots, y_n^* \rangle$ with respect to the equations of (1) and is denoted by D_g</u>.

The definition given above is very closely related to the preliminary "graph theoretic" definition discussed earlier. In fact we state the following lemma [cf. 13].

<u>Lemma</u>. If a function $y(x)$ is generable on I in the "graph-theoretic" sense it is generable in the sense of the above definition.

The proof is very simple. Let us give a brief sketch. By the remark given above we can assume that the only units are the adder, the constant multiplier, the constant function $C_1(x)$ and the integrator. Let $U_2,\ldots,U_n$ be an enumeration of all integrators of the analog computer.[2] Let the output of the ith integrator be y_i. Recall that $y_0(x)=1$ and $y_1(x)=x$. Suppose $y(x)$ is the output of the kth integrator - i.e. $y=y_k$ where

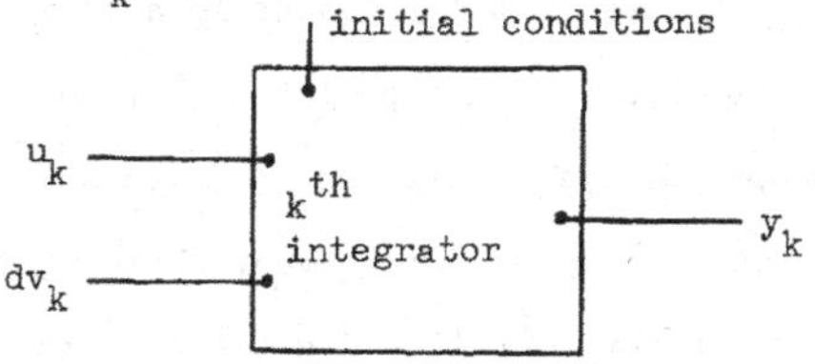

Then

$$y_k(x) = \int_a^x u_k dv_k + C_k$$

What is the most general form of u_k and v_k ? Since the only units in the analog computer are adders, constant multipliers, the constant function $C_1(x)$ and integrators, $u_k(x) = \sum_{i=0}^{n} a_{ik}y_i$ and $v_k(x) = \sum_{j=0}^{n} b_{jk}y_j$ for suitable reals a_{ik} and b_{jk}. Thus

$$y_k(x) = \int_a^x \left(\sum_{i=0}^{n} a_{ik}y_i\right) d\left(\sum_{j=0}^{n} b_{jk}y_j\right) + C_k$$

i.e.

$$\frac{dy_k}{dx} = \sum_{i,j=0}^{n} c_{ijk}y_i \frac{dy_j}{dx}$$

where $c_{ijk} = a_{ik}b_{jk}$. It is easy to extend this result to cover the output of any unit. For suppose y is the output of a unit of an analog computer $\mathfrak{A}$. Then since $y(x) = \int_a^x 1.dy + y(a)$, y can be considered as the output of an integrator in an analog computer $\mathfrak{A}'$ obtained by adding the integrator associated with $\int 1.dy$ and suitably altering the connections.

This concludes the first section. We have given our fundamental definition - which was our goal.

§2 MOTIVATION

There are three reasons why the study of functions generated by general-purpose analog computers appears to be of considerable interest. The first is conceptual. This research may be regarded as a first step in a study of an alternative approach to the concept of computability - an approach motivated by existing analog computers. Thus eventually we hope to do for existing analog computers what Turing did for digital computers. But much more is involved. Note that the distinguishing feature of an analog computer - as opposed to a digital computer - is that the variables change continuously. Thus on this approach we can expect computable function of a real variable to be the fundamental concept explicated. Computable functions from natural numbers to natural numbers will be a derived concept. This is, of course, a reversal of the procedure employed in connection with existing formulations of abstract computability. Existing formulations follow along the traditional foundational development of mathematics - from 0 and the successor through functions from N to N to definitions of real number and function of a real variable. They proceed by effectivizing the corresponding classical constructions. The consequences of the aforementioned reversal - for the foundations of mathematics and for research in higher order computability as studied by logicians - appear to be of considerable interest. We cite two examples. First, the role of impredicative definitions under this reversal. In this connection, the reader may wish to consider the least upper bound principle for real numbers. Second, the formulation of Church's thesis intrinsic for this context and its relation to research in higher order recursion theory. (Indeed, our work on analog generability may be conceived of as another approach to higher order computability - e.g. computability over the continuum.) We do not wish to convey the impression that we are against the traditional approach in any of its formulations. This approach is exceedingly fruitful. Furthermore, it satisfies man's desire for simplicity and order. Our point is merely this. It may be possible to conceive that nature generates a computable function of a real variable directly and not necessarily by approximation as in the traditional approach. We believe it is of some interest to investigate this phenomenon. Indeed, we may even go further and consider the following. Under certain circumstances the continuum provides an excellent foundational explanation, while under other circumstances discreteness is more useful. The idea of "discrete - continuum duality" is reminiscent of the concept of "wave - particle duality" prevalent in modern physical theory. Whether such a foundational approach will be fruitful for mathematical research remains to be seen. Perhaps future investigation will provide an answer. The author realizes that these considerations may be unpalatable to some readers. At all events the results of this paper are independent of any philosophical considerations - which we now bring to an abrupt halt.

Our second reason for studying analog-generable functions is technical: we believe the techniques employed here may be of independent interest. Recall that our fundamental definition is phrased in terms of a simultaneous set of non-linear differential equations. Hence, to obtain our results, we have found it necessary to combine the theorems and methods of proof of classical analysis and differential equations with those of recursion theory. Work by Pólya, Hölder, Cauchy-Peano, Weierstrass and others find application here. Some of the theorems and techniques employed are well-known; others are not. In some instances - e.g. in the use of Pólya's theorem and the Cauchy-Peano existence theorem for differential equations - we employ the original result. In other instances, an effective - i.e. recursion-theoretic - version of the result is used. Of course, some algebraic results are also evident. But this is hardly a novelty in logical investigations. Perhaps further research will broaden the interrelationships between recursion theory and mathematical logic on the one hand and classical analysis and differential equations on the other.

Third, and most obvious, is the following. We believe it is important for those who use analog computers to investigate - from the point of view of abstract computability - the class of functions their devices generate. Such a study will provide insight into the theoretical limitations of their machines. In addition such an investigation is of importance to those who use digital and hybrid machines. For, in recent years, experts in digital computers have been concerned with hybrids- i.e. computers combining the features of both analog and digital machines - in an attempt to maximize the advantages of both.

§3 RESULTS

The relation between analog generability and abstract computability is made clear by the following results.

Let us refer to a function of a real variable as entire if it is the restriction of an entire function to the real line. Let I and J be closed bounded intervals with non-empty interiors. (In fact we assume that all of our intervals have non-empty interiors.) Then we have

1. There exist entire functions f having the following properties

   a. f is computable - in the sense of recursive analysis [3]

   b. No restriction of f to a closed bounded interval of the real line can be generated by a general-purpose analog computer.

2. Let $f(x) = \Sigma a_i x^i$ be an entire function. Suppose for some closed bounded interval I, $f \upharpoonright I$ can be generated by a general-purpose analog computer. Then there exist a finite number of a_i such that, on every closed bounded interval, f is computable relative to these a_i.

Result 2 may be restated in the following form as it reflects a local rather than a global phenomenon. In 2', f need not be entire. As in 2, the a_i of 2' are real numbers. However, c may be chosen to be rational.

2'. Suppose for some I, $f \upharpoonright I$ can be generated by a general-purpose analog computer. Then there is a $J \subseteq I$ such that

   a. $f \upharpoonright J$ is expressible as $\Sigma a_i (x-c)^i$, the restriction of a function of a complex variable analytic in a region including J.

   b. $f \upharpoonright J$ is computable relative to a finite number of the a_i.

We cite the functions $\sum_{n=1}^{\infty} \left(\frac{x^n}{n^{(n^3)}} \right)$ and $\frac{1}{\Gamma(x)}$ as examples for 1. It is interesting to note that these functions cannot be generated by an analog computer even if we assume - as we do here - that the analog computer can generate every real number.

Actually result 1 is a consequence of a characterization of the analog generable functions in terms of solutions of algebraic differential polynomials. By an algebraic differential polynomial we mean an expression of the form

$$\sum_{i=0}^{p} a_i\, x^{m_{oi}}\, y^{n_{oi}} (y')^{n_{1i}} \cdots \left(y^{(k_i)}\right)^{n_{k_i i}} .$$

where a_i is a real number and m_{oi} , $n_{oi}, \ldots, n_{k_i i}$ are non-negative integers. To prove result 1 we first prove the following theorem: If $f \upharpoonright I$ can be generated by an analog computer there is a $J \subsetneq I$ such that on J, f satisfies an algebraic differential polynomial. The proof of this theorem uses among other results the Cauchy-Peano existence theorem together with some simple facts concerning the degree of transcendence of transcendental field extensions. To obtain result 1 we then apply a theorem of Pólya or of Hölder.

The characterization of analog-generable functions in terms of algebraic differential polynomials is carried out by proving a theorem which is essentially a converse to the theorem quoted in the previous paragraph: If f satisfies an algebraic differential polynomial on I , then for some $J \subsetneq I$, $f \upharpoonright J$ can be generated by a general-purpose analog computer. Our proof makes use of something stronger than the Cauchy-Peano existence theorem. In particular we require the Lipschitz condition in order to obtain unique solutions on a "domain of generation".

As a consequence of the above-mentioned characterization we see that the class of analog generable functions is exceedingly large. For example, many well-known special functions which arise as solutions of algebraic differential polynomials can be generated by an analog computer on every I for which they are defined - e.g. Bessel functions etc. This takes us far beyond the elementary functions - rational functions, trigonometric functions, e^x , log x - which are, of course, included. The reader is invited to consult the literature on special functions for additional examples.

We will not comment in detail upon the proofs of results 2 and 2' - except to state that they are involved and quite different from the proofs outlined above. They do, however, use the characterization of analog generable functions in terms of algebraic differential polynomials in one direction - viz. from analog generable functions to algebraic differential polynomials. One interesting sidelight of our work is that it provides criteria which allow us to conclude immediately that many well-known functions are computable in the sense of recursive analysis. For entire functions a criterion (which is actually proved in the proof of 2) is the following. Suppose $\sum a_i x^i$ satisfies an algebraic differential polynomial. Suppose further that each a_i is a computable real. Then $\sum a_i x^i$ is computable on every closed bounded interval I. Thus, for example, e^x, $\sin x$, $\cos x$, etc. are computable functions on every closed bounded interval. An analogous criterion arises from 2'. We leave the precise statement to the reader.

Thus we conclude our summary. As we have seen, the work is a curious mixture of classical analysis, differential equations, recursion theory, algebra and computer science. Numerous open problems can be formulated. Perhaps the reader may wish to try his hand at some of them.

Bibliography

1. Church, A. An unsolvable problem in elementary number theory. American Journal of Mathematics vol. 58 (1936) pp. 345-363.

2. Crank, J. The Differential Analyser. Longmans, Green & Co. 1947, viii + 137 pp.

3. Grzegorczyk, A. On the definitions of computable real continuous functions. Fundamenta Mathematicae vol. 44 (1957) pp. 61-71.

4. Hölder, O. Ueber die Eigenschaft der Gamma function keiner algebraischen Differentialgleichung zu genügen. Mathematische Annalen vol. 28 (1887) pp. 1-13.

5. Jackson, A. Analog Computation. McGraw Hill Book Co. Inc. (1960) xiv + 652 pp.

6. Johnson, C. Analog Computer Techniques. McGraw Hill Book Co., Inc. (1963) xii + 315 pp.

7. Kleene, S.C. Introduction to Metamathematics. D. van Nostrand Co., (1952) x + 550 pp.

8. Korn, G.A., and Korn, T.M. Electronic Analog and Hybrid Computers. McGraw-Hill Book Co., New York (1964) xxiii + 584 pp.

9. Lacombe, D. Extension de la notion de fonction récursive aux fonctions d'une ou plusieurs variables réelles. C.R. Acad. Sci. Paris, I t.240 (1955) pp. 2478-2480, II, t.241 (1955) pp. 13-14, III t.241 (1955) pp. 151-153.

10. Markov, A. The Theory of Algorithms (Russian) Trudy Mathematicheskogo Institutaimeni V.A. Steklova, vol. 38 pp. 176-189.

11. Pólya, G. Über das Anwachsen von ganzen Funktionen die einer Differentialgleichung genügen. Vierteljahrsschrift der Naturforsch. Gesellschaft in Zürich vol. 41 (1916) pp. 531-545.

12. Pólya, G. Zur Untersuchung der Grössenordnung ganzer Funktionen, die einer Differentialgleichung genügen. Acta Mathematica vol. 42 (1920) pp. 309-316.

13. Shannon, C. Mathematical theory of the differential analyzer. Journal of Mathematics and Physics vol. 20 (1941) pp. 337-354.

14. Soroka, W. Analog Methods in Computation and Simulation. McGraw-Hill Book Co. New York (1954) xii + 390 pp.

15. Tomovic, R. and Karplus, W.J. High Speed Analog Computers. Wiley and Sons, New York (1962) xi + 255 pp.

16. Turing, A. On computable numbers with an application to the Entscheidungsproblem. Proceedings of the London Mathematical Society ser. 2, vol. 42 pp. 230-265; vol. 43, pp. 544-546.

Department of Mathematics
University of Minnesota
MINNEAPOLIS, MN 55455/USA

XII. INFINITARY COMBINATORICS

E.M. Kleinberg

Massachusetts Institute of Technology

§1 We begin, appropriately, by dismissing the phrase "infinitary combinatorics". For, in mathematics, the word "combinatorial" traditionally applies to things which are anything but infinite. The fact remains, however, that it would somehow seem appropriate to at times describe the proof of a theorem about infinite sets as being "combinatorial". Most frequently, a proof's containing a "combinatorial argument" means that it involves some variation of the pigeon-hole principle, a notion equally at home in finite or infinite contexts. (The pigeon-hole principles notes that if you try to place too many items in too few places, then at least one of the places will contain many items. For example, if five thousand pigeons divided themselves among two pigeon-holes you could be certain that at least one of the holes contained more than one pigeon. An infinite example of the principle might be that any function from the integers into a finite set must be constant on an infinite set.)

There is, however, another sense of "combinatorial" natural to arbitrary contexts. This is the sense which might be meant when describing a proof of a theorem in set theory as being "combinatorial" rather than "model theoretic". One would interpret this as saying that the proof hinged upon some special facts about sets as opposed to some general technique involving logical formulas and models.

The importance of this sense of "combinatorial", that meaning "pure set theoretic", cannot be overstressed. For it is mainly the knowledge of specific properties of sets that usually serves to establish all but the most basic results of any given area.

In these lectures we examine certain set theoretic properties inspired by the pigeon-hole principle. We shall study their relationship to other well known set theoretic notions, as well as indicate how they may be used with striking success in proving various theorems concerning real numbers.

§2 To be perfectly accurate, we shall be examining a concept for extending the pigeon-hole principle originated by F.P. Ramsey in 1929. His discovery has turned out to have such a profound impact in set theory that it almost warrants being categorized as a principle itself.

Ramsey's idea goes as follows: in its simplest infinite form the pigeon hole principle yields that if an infinite set is partitioned into a finite number of pieces, then one of the pieces will be infinite. But what if we partitioned the collection of two-element subsets of an infinite set into a finite number of pieces? Surely one of the pieces will be infinite, but Ramsey noticed much more. He proved that there must exist some infinite subset of the original infinite set all of whose two-element subsets were contained in one of the pieces. This result is not so unnatural as it might, at first, appear. Let us restate it, this time using some notation: if n is any integer and x is any set, we denote by $[x]^n$ the collection of n-element subsets of x. Now the pigeon-hole principle yields that if x is any infinite set and F is any function from $[x]^1$ into a finite set, then there must be an infinite subset y of x such that F is constant on $[y]^1$. Ramsey's theorem just allows one to replace the 1 with a 2 ; if x is any infinite set and F is any function from $[x]^2$ into a finite set then there exists an infinite subset y of x such that F is constant on $[y]^2$.

A few minutes thought should tell one that Ramsey's generalization is by no means trivial to prove. As we hope to show, eventually, the idea it contains is both important and profound.

Let us begin by accurately stating and proving Ramsey's theorem.

2.1 Theorem (F.P. Ramsey 1929 [1]): Let x be an infinite set, let n and k be positive integers, and assume that $[x]^n$ is partitioned[(1)] into k disjoint pieces, $A_1, A_2, \ldots, A_k$. Then there is an infinite subset y of x such that for some i, $[y]^n \subseteq A_i$.

Remark: In Ramsey's original statement of his theorem he lists as an assumption, the axiom of choice. As it turns out, this assumption is only needed to show that x has a countable subset. We shall return to this point following the theorem's proof.

Proof: { Since the proof of this theorem for x the set of nonnegative integers and $n = k = 2$ conveys the essence of the full argument[(2)] we shall begin by sketching a proof of this case: let ω denote the set of nonnegative integers and suppose that $[\omega]^2 = A \cup B$ is a given partition. We build a binary branching tree with a nonnegative integer at each node as follows: we place 0 at the base of the tree. Let the left immediate successor of 0 be the least integer n such that $\{0,n\}$ is in A (if no such integer n exists 0 will have no immediate left successor) and let the right immediate successor of 0 be the least integer m such that $\{0,m\}$ is in B (same convention if none exists). Let n_1 and m_1 be the left and right immediate successors of 0 respectively. Now let the left (right) immediate successor of n_1 be the least integer n' out of those n such that $\{0,n\}$ is in A such that $\{n_1,n'\}$ is in A (B) and let the left (right) immediate successor of m_1 be the least integer m' out of those m such that $\{0,m\}$ is in B such that $\{m_1,m'\}$ is in A (B). Continuing in this way we build our desired tree.

(1) If Q is any set, a partition of Q is any collection of pairwise disjoint subsets of Q whose union is Q.

(2) It, in fact, contains the main idea behind the proof of every positive result concerning partition relations. We shall comment on this later.

Now since this tree is infinite and yet has only binary branching there must be an infinite branch σ through it. Furthermore, by our construction, if n is in the branch σ then either for every successor n' of n in σ $\{n,n'\}$ is in A or for every successor n' of n in σ $\{n,n'\}$ is in B. Call any such n an A point or a B point accordingly. We thus see that if n and n' are two A (B) points in σ, $\{n,n'\}$ is in A (B). Since σ is infinite it contains infinitely many A points or infinitely many B - in either case we are done.}

We now give a detailed and precise proof of Ramsey's theorem. We first assume $k = 2$ and proceed by induction on n. If $n = 1$ the theorem is clearly true and so suppose the theorem has been established for $n = m$ and that $[x]^{m+1} = A \cup B$ is a given partition. We define by induction on i members a_i of x and infinite subsets C_i of x: C_o is x itself and a_o is a member of C_o (pick one by AC[(1)]); at stage $i+1$, let C_{i+1} be an infinite subset z of $C_i - \{a_i\}$ such that $[z]^m \subseteq \{\{b_1,\ldots,b_m\} | \{a_i,b_1,\ldots,b_m\} \in A\}$ or $[z]^m \subseteq \{\{b_1,\ldots,b_m\} | \{a_i,b_1,\ldots,b_m\} \in B\}$ (we are simply using the inductive hypothesis on the partition $\{\{b_1,\ldots,b_m\} | \{a_i,b_1,\ldots,b_m\} \in A\} \cup \{\{b_1,\ldots,b_m\} | \{a_i,b_1,\ldots,b_m\} \in B\}$ of $[C_i - \{a_i\}]^m$), and let a_{i+1} be a member of C_{i+1} (again, pick one) Now the collection of a_i has just the property of our branch σ from the sketch. Namely, for any a_i, $\{\{a_i,a_{i_1},\ldots,a_{i_m}\} | i<i_1<\ldots<i_m\} \subseteq A$ or $\{\{a_i,a_{i_1},\ldots,a_{i_m}\} | i<i_1<\ldots<i_m\} \subseteq B$ - call a_i an A point or a B point accordingly. Thus if y_A is the collection of A points and y_B is the collection of B points, $[y_A]^{m+1} \subseteq A$ and $[y_B]^{m+1} \subseteq B$. Since either y_A or y_B is infinite we have established the theorem for $n = m+1$ and $k = 2$. By induction we have thus established the theorem for any n and $k = 2$. It is now a simple matter to finish the proof by induction on k. Namely suppose the result has been

(1) **"AC" is an abbreviation for "the axiom of choice".**

established for $k = \ell$, and suppose that $[x]^n = A_1 \cup \dots \cup A_{\ell+1}$ is a given partition. Consider the partition of $[x]^n$ into the two pieces A_1 and $A_2 \cup \dots \cup A_{\ell+1}$. By the case already established there is an infinite subset y of x such that either $[y]^n \subseteq A_1$ or $[y]^n \subseteq A_2 \cup \dots \cup A_{\ell+1}$. If $[y]^n \subseteq A_2 \cup \dots \cup A_{\ell+1}$ the inductive hypothesis yields an infinite subset z of y such that for some i, $2 \leqslant i \leqslant \ell+1$, $[z]^n \subseteq A_i$. In either case we have found an infinite subset of x all of whose n-element subsets lie in some A_i, $1 \leqslant i \leqslant \ell+1$. Ramsey's theorem is thus established. □

Remark: In the proof just given, we diverged from our sketch in that rather than construct a tree from which a branch is then chosen, we directly constructed the branch. Although it is somewhat more complicated, one can use the full tree approach in proving Ramsey's theorem for all n. When we examine generalizations of Ramsey's idea to uncountable cardinals however, the full tree proof is virtually necessary. To see just this sort of argument one might examine our proof of theorem 4.2.

Remark: Since Ramsey's original statement of his theorem noted his use of the axiom of choice, some words are in order concerning this point. Our proof seemed to use the axiom of dependent choice. But it is easy to see that countable choice suffices. Indeed, using countable choice one can show that any infinite set contains a countable subset and so one easily sees that countable choice reduces proving Ramsey's theorem to considering only the case $x = \omega$. Since ω is well-ordered our proof handles this case with no use of the axiom of choice whatever (the inductive hypothesis must be that the desired infinite subset of x can be found "canonically" - in this way the sequence of C_i's can be chosen without choice). A more elegant way to see that Ramsey's theorem for $x = \omega$ is provable without the axiom of choice is to simply observe that its statement is Π^1_2 in Kleene's hierarchy and so by Shoenfield's absoluteness theorem, it holds in L (the constructible universe of Gödel) iff it holds in fact. But we know it true in L as the axiom of choice is available there.

Now what about the possibility of proving the full Ramsey's theorem using no choice at all? It cannot be done. Indeed, let weak choice be the assertion that given any infinite collection of finite sets there is a choice function picking one element from each of intinitely many of them. Weak choice is independent of ZF[(1)] and yet we have

2.2 Theorem (see [11]): Ramsey's theorem implies weak choice.

Proof: {We first sketch the proof as follows: given a collection of finite sets it is easy to partition the two-element subsets of its union such that the set given us by the conclusion of Ramsey's theorem is either a subset of one of the original finite sets or it intersects any of them on at most one point. Since this set is infinite the second alternative happens and so we can now easily construct a choice function.} Let Q be a given infinite collection of nonempty finite sets. If infinitely many members of Q have a member in common we are easily done, and so let us assume not. Let Q* denote the union of the members of Q and let us partition $[Q^*]^2$ into the pieces A and B by having $\{p,q\}$ be a member of A iff for some set z in Q, p and q are both members of z. Now by Ramsey's theorem there is an infinite subset P of Q* such that either $[P]^2 \subseteq A$ or $[P]^2 \subseteq B$.

(1) This result is due to Cohen. See [3].

Claim: $[P]^2 \subseteq B$. (Proof: Suppose $[P]^2 \subseteq A$. Let $t \in P$. Then for any s in P, t and s must both lie in a common member of Q. Since there are infinitely many s in P this means that either t lies in infinitely many members of Q (which we are assuming it doesn't) or some member of Q is infinite (which it cannot be). This contradiction yields the claim. ⊠) Now by the claim, any two distinct members of P must lie in distinct members of Q and so the function which sends any member z of Q to $z \cap P$ is our desired choice function. □

As it turns out, weak choice is only one of the many choice principles implied by Ramsey's theorem.

Remark: Someone interested in recursion theory might ask the question as to just how "effective" Ramsey's theorem can be. Specifically, consider the case $n = k = 2$ and $x = \omega$. Then if we are given a recursive partition of $[x]^2$ into two pieces, what can we say about possible infinite sets y given us by Ramsey's theorem? We simply cite a result of Jockusch ([10]): given any recursive partition of $[x]^2$ into two pieces one can always find a Π^0_2 set y satisfying the conclusion of Ramsey's theorem. However, this is best possible, i.e., there exists a recursive partition of $[x]^2$ into two pieces for which no Σ^0_2 set satisfies the conclusion of Ramsey's theorem.

§ 3 Let us set the stage for our future work by discussing some preliminaries and notation. We shall work in Zermelo-Fraenkel set theory (ZF) which we take to not include the axiom of choice (AC). However, until we reach §§8b , AC will be freely used without explicit mention. We will assume the reader to be familiar with elementary set theory and in particular with the notion of ordinal.

Cardinals and cardinalities will always be considered in the sense of Von Neumann, that is, as ordinals onto which smaller ordinals cannot be mapped. Lower case Greek letters will always denote ordinals and $<$ will denote the usual well-ordering (the membership relation) of them. As usual, $\overline{A}$, $\overline{\overline{A}}$, $\cap A$, $\cup A$, 2^A , $f \upharpoonright A$, and $f''A$ (for f a function whose domain contains A as a subset) will denote the order-type of A(1), the cardinality of A(1), the intersection of A , the union of A , the power set of A , the restriction of f to A , and the range of f on A , respectively.

Recall that for α an ordinal, the <u>cofinality of α</u> , <u>$cf(\alpha)$</u> , is the least ordinal which can be mapped onto an unbounded subset of α . $\alpha (\geq \omega)$ is <u>regular</u> if $cf(\alpha) = \alpha$. A cardinal κ is said to be a <u>strong limit cardinal</u> (AC) if $\gamma < \kappa$ implies $\overline{\overline{2^\gamma}} < \kappa$.

§4 If x and y are sets, a partition of x into y pieces will be viewed as simply a function from x into y(2). If z is a set of ordinals and α is an ordinal $[z]^\alpha$ denotes the collection of α-sequences from z , that is, the collection of subsets of z of order-type α(3).

Of the many partition relations to be considered here we begin with the following: let λ denote a limit ordinal, α an arbitrary ordinal, and γ a cardinal. Then we denote by $\lambda \rightarrow (\lambda)^\alpha_\gamma$ (4) the

---

(1) This may not be defined if A fails to possess a well-ordering.

(2) If F maps x into y , F is identified with the partition $\{F^{-1}\{z\} \mid z \in y\}$ of x .

(3) We will often leave it to the context to say when one ordinal is assumed less than another, one set contained in another, and so forth. Here, for instance, it is assumed $\alpha \leq \overline{z}$.

(4) This notation is due to Erdös and Rado.

partition relation <u>for every partition of the α-sequences from λ into γ pieces there exists a subset C of λ of order-type λ such that all α-sequences from C lie in the same piece of the partition.</u> Using our earlier notation, $\lambda \to (\lambda)^{\alpha}_{\gamma}$ can be restated <u>for every function F from $[\lambda]^{\alpha}$ into γ there exists a subset C of λ such that $\overline{C} = \lambda$ and $\overline{\overline{F''[C]^{\alpha}}} = 1$</u> [(1)]. The set C mentioned in this definition is said to be <u>homogeneous for the partition F with respect to</u> $\lambda \to (\lambda)^{\alpha}_{\gamma}$. We will often simply refer to it as being "homogeneous". We will further adhere to tradition by deleting the subscript in this notation when it would be 2, that is, $\lambda \to (\lambda)^{\alpha}_{2}$ will be abbreviated $\lambda \to (\lambda)^{\alpha}$.

Let us now begin with a brief general discussion of this partition relation. For the case $\alpha = 1$, the relation conveys the traditional pigeon hole idea. Of course, if γ is too large the relation fails here. Clearly $\omega \to (\omega)^{1}_{\omega}$ is false. On the other hand, $\omega \to (\omega)^{1}_{15}$ is true. Indeed, it is easy to verify that if λ is an infinite cardinal, then the cofinality of λ exceeds γ, if and only if $\lambda \to (\lambda)^{1}_{\gamma}$.

Ramsey's idea comes into play when we consider the relation $\lambda \to (\lambda)^{\alpha}_{\gamma}$ when α is larger than 1. For example, $\omega \to (\omega)^{2}$ is true - it is the instance of Ramsey's theorem whose proof we sketched prior to presenting the full proof. More generally, $\omega \to (\omega)^{n}_{m}$ for <u>every</u> pair of positive integers n and m - this again is just an instance of Ramsey's theorem.

What about the relation $\lambda \to (\lambda)^{2}_{\gamma}$ for $\lambda > \omega$? Is it ever true? A moment's thought should tell one that Ramsey's theorem does not readily supply the answer, even for the simplest such case, $\lambda \to (\lambda)^{2}$.

---

(1) We shall be concerning ourselves primarily with only two types of partition relations - this one and another which will be discussed in §5- §7. There are, however, many other partition relations which have been studied and which are of interest. To learn about these one should refer to [5].

Indeed, as we shall soon see, it is impossible to prove within ZF[1] (or even within ZF + AC) that any ordinal $\lambda > \omega$ satisfies $\lambda \to (\lambda)^2$. Our approach in establishing this will be to show that any $\lambda > \omega$ satisfying $\lambda \to (\lambda)^2$ must be a "large cardinal" - more specifically, the existence of any $\lambda > \omega$ satisfying $\lambda \to (\lambda)^2$ would imply the consistency of ZF + AC and so by Godel's theorem, one could not in ZF + AC alone establish the existence of such a λ[2]. It is here, then, that we get our first glimpse of the fact that Ramsey's idea transcends the simple pigeon-hole idea - all infinite cardinals κ satisfy $\kappa \to (\kappa)^1$ yet only "large cardinals" κ can satisfy $\kappa \to (\kappa)^2$. Let us now make all of this more precise.

A cardinal κ is said to be <u>strongly inaccessible</u> if 1) $\kappa > \omega$, 2) κ is regular, and 3) κ is a strong limit. (Note: this definition requires AC). Strongly inaccessible cardinals have been studied in set theory for quite some time. It is well known that if κ is strongly inaccessible then $R(\kappa)$, the collection of sets of rank less than κ, forms a model for ZF, and that M_κ, the collection of sets constructible by order $<\kappa$, is a model for ZF + AC. Hence ZF + AC + "there exists a strongly inaccessible cardinal" $\vdash$ "ZF + AC is consistent". Thus by Gödel's second theorem, one cannot prove in ZF + AC that a strongly inaccessible cardinal exists. In ZF + AC + "$\exists\lambda>\omega(\lambda \to (\lambda)^2)$", however, one can:

4.1 <u>Theorem</u> (Erdos - see [7])[3]: Assume that λ is an ordinal

---

(1) We are implicitly assuming throughout that ZF (and hence, by a result of Gödel, ZF + AC) is consistent.

(2) Even though ZF + AC is not rich enough to prove the existence of a $\lambda > \omega$ satisfying $\lambda \to (\lambda)^2$, this does not imply that no such λ exists. Indeed, several well known set theoretic axioms do imply the existence of such λ. We shall examine this question in §5.

(3) Since these notes are not intended to be an historical account, I have made no special attempt to discover, for each of the results to

greater than ω such that $\lambda \to (\lambda)^2$. Then λ is a strongly inaccessible cardinal.

Proof: We first show that λ is a regular cardinal. To do this it will suffice to show that $cf(\lambda) = \lambda$, for it is clear that the cofinality of any ordinal is a cardinal. So suppose not, i.e., assume $\gamma = cf(\lambda) < \lambda$ - we will derive a contradiction. Let g map γ in an order-preserving manner onto an unbounded subset of λ, let[(1)] $F: [\lambda]^2 \to 2$ be given by

$$F(\{\alpha,\beta\}_<) = \begin{cases} 0 & \text{if for some } \eta < \gamma \\ & \alpha < g(\eta) < \beta \\ 1 & \text{otherwise} \end{cases},$$

and let C be homogeneous for F with respect to $\lambda \to (\lambda)^2$.
Claim: $F''[C]^2 = \{0\}$ (Proof: let α be the least member of C. As the range of g on γ is unbounded in λ let η be a member of γ such that $g(\eta) > \alpha$. Finally, as C has order-type λ, it must be unbounded in λ and so let β be a member of C larger than $g(\eta)$. Thus $\{\alpha,\beta\}_< \in [C]^2$ and $F(\{\alpha,\beta\}_<) = 0$. As C is homogeneous, we must have that $F''[C]^2 = \{0\}$. ⊠) Now let f map C into γ by setting, for any α in C, $f(\alpha)$ equal to the least η in γ such that $g(\eta) > \alpha$. By our claim it is easy to see that f is order-preserving and so, as $\bar{\bar{C}} = \lambda$, we contradict the fact that γ is less than λ.

---

be presented, to whom it is due. However, I will attribute theorems when I happen to know their source and so I apologize to those whose work is cited without reference.

(1) Let us indicate finite sets of ordinals as $\{\alpha_1,\alpha_2,\ldots,\alpha_n\}_<$ when we want it understood that $\alpha_1 < \alpha_2 < \ldots < \alpha_n$.

We now show that λ is a strong limit cardinal and thus complete our proof. Assume that λ is not a strong limit cardinal. Let γ, then, be less than λ, and such that $2^{\gamma} \geq \lambda$, and let us assume that γ is the least such ordinal. Let g map λ 1-1 into 2^{γ}, and let $F: [\lambda]^2 \to 2$ be given by

$$F(\{\alpha,\beta\}) = \begin{cases} 0 & \text{if, at the least place } g(\alpha) \text{ and } g(\beta) \text{ differ (when viewed as characteristic functions), } g(\alpha) \text{ is } 0 \text{ and } g(\beta) \text{ is } 1 \\ 1 & \text{otherwise} \end{cases}$$

Let C be homogeneous for F (with respect to $\lambda \to (\lambda)^2$). We now simply examine $\{g(\alpha) | \alpha \in C\}$. It is easy to see that for any $\xi < \gamma$ there exists an ordinal δ_ξ less than λ such that if α and β are any two members of C larger than δ_ξ, then $g(\alpha) \upharpoonright \xi = g(\beta) \upharpoonright \xi$. This is because 1) γ is the least ordinal below λ whose power set has cardinality at least λ, and 2) C is homogeneous for F. Since, now, γ is less than λ and λ is regular, the sup of all δ_ξ for $\xi < \gamma$, δ, is less than λ. But if α and β are members of C larger than δ, $g(\alpha)\upharpoonright\xi = g(\beta)\upharpoonright\xi$ for every $\xi < \gamma$, i.e., $g(\alpha) = g(\beta)$. This contradicts the fact that g was chosen to be 1-1. The proof of this theorem is now complete. □

In line with our earlier discussion we immediately have

<u>Corollary</u>: In ZF + AC one cannot prove the existence of an uncountable cardinal λ satisfying $\lambda \to (\lambda)^2$.

<u>Remark</u>: Since instances of $\lambda > \omega$ satisfying $\lambda \to (\lambda)^2$ cannot be found in ZF + AC alone, two questions are immediately suggested.

First of all, do any "reasonable" set theoretic axioms imply the existence of **uncountable** λ satisfying $\lambda \to (\lambda)^2$, and second of all, are the other relations of the form $\lambda \to (\lambda)^{\alpha}_{\gamma}$ any stronger than $\lambda \to (\lambda)^2$? We will comment on the first question in §5. The second we shall deal with here.

We first note that for $\gamma \geq \lambda$, any relation of the form $\lambda \to (\lambda)^{\alpha}_{\gamma}$ is trivially false (in ZF). For the partition which sends any α-sequence from λ to its least member can have no homogeneous set. Now when α is infinite, all relations of the form $\lambda \to (\lambda)^{\alpha}_{\gamma}$ contradict the axiom of choice. Such relations will be dealt with in section 8. We are thus left to consider relations of the form $\lambda \to (\lambda)^{n}_{\gamma}$ where $\lambda > \omega > n > 0$ and $\gamma < \lambda$. These turn out to all be equivalent:

4.2 Theorem: For any cardinal λ, $\lambda \to (\lambda)^2$ iff for every $n < \omega$ and $\gamma < \lambda$, $\lambda \to (\lambda)^{n}_{\gamma}$.

Proof: This proof closely parallels the original one of Ramsey. We are going to find our desired homogeneous sets as branches through the appropriate trees. Let us thus begin with some terminology: a (well-founded) tree is a partially ordered set such that the predecessors of any point are well-ordered by the ordering. The order of any point in a tree is the order-type of its set of predecessors and the αth level of a tree (for α an ordinal) is simply the set of members of the tree of order α. A branch of a tree is a linearly ordered subset. Finally, a cardinal κ is said to have the tree property if every tree of cardinality κ all of whose levels have cardinality less than κ has a branch of cardinality κ.

Now although it is clear that ω has the tree property (indeed this was a key fact we used in our proof of Ramsey's Theorem) it is by no means immediate that any uncountable cardinals have the tree property. In fact, our approach to proving this theorem will be to show that if $\lambda \to (\lambda)^2$ then λ has the tree property, and if λ has the tree property (and is strongly inaccessible) then $\lambda \to (\lambda)^n_\gamma$ for any $n < \omega$ and $\gamma < \lambda$. We would thus have $\lambda \to (\lambda)^2$ implying $\lambda \to (\lambda)^n_\gamma$ for any $n < \omega$ and $\gamma < \lambda$. The reverse direction is, of course, trivial.

Let us, then, proceed.

4.2.1 <u>Lemma</u>: If $\lambda \to (\lambda)^2$ then λ has the tree property.

<u>Proof of Lemma</u>: Let $\mathcal{T} = \langle T, <_T \rangle$ be a given tree of cardinality λ. By isomorphic translation we may clearly assume that T is λ itself. We extend $<_T$ to a total ordering $<^*$ of λ by $\alpha <^* \beta$ iff_{df} $\alpha <_T \beta$ or if $\beta \not<_T \alpha$ and, at the first level of $\mathcal{T}$ at which α and β have distinct predecessors α' and β', respectively, $\alpha' < \beta'$. We now use $\lambda \to (\lambda)^2$. Namely, let $F: [\lambda]^2 \to 2$ be given by $F(\{\alpha,\beta\}_<) = 0$ iff $\alpha <^* \beta$, and let C be homogeneous for F. Then using C we can define a subset B of λ by $B =_{df} \{\eta < \lambda \mid \eta$ has λ many $<_T$-successors in $C\}$. Claim: B is a size λ branch through $\mathcal{T}$ (Proof: To see that B has cardinality λ we need only show that each level of $\mathcal{T}$ has a point in B. Let us consider the δth level. Since C has cardinality λ and each level of $\mathcal{T}$ has cardinality less than λ, the regularity of λ (recall that $\lambda \to (\lambda)^2$) tells us that λ many points in λ have order at least δ and hence (again by regularity) that some point on the δth level has λ many $<_T$-successors in C. Thus some point on the δth level lies in B. Now why is B a branch? Well, if not, let α and β be two incomparable members of B, let δ be the least level at which α and β have distinct predecessors, and let $\eta_1 < \eta_2$ be predecessors of

α and β (not necessarily respectively) of order $\aleph$. Then since η_1 and η_2 both have λ many $<_T$-successors in C, we can find points $\xi < \tau < \sigma$ in C such that $\eta_1 <_T \xi$, $\eta_1 <_T \sigma$, and $\eta_2 <_T \tau$. But now $F(\{\xi,\tau\}_<) = 0 \neq 1 = F(\{\tau,\sigma\}_<)$, and so we contradict the homogeneity of C. Thus B must be a branch and so the claim follows. ⊠) This claim completes the proof of our lemma. ⊠

4.2.2 Lemma: If λ has the tree property and is strongly inaccessible then for any $n < \omega$ and $\gamma < \lambda$, $\lambda \to (\lambda)^n_\gamma$.

Proof of Lemma: **We prove our lemma by induction on** n: **if** $n = 1$, **then** $\lambda \to (\lambda)^n_\gamma$ **for any** $\gamma < \lambda$ **easily, for** λ **is** regular. Assume that $n = m+1$ and that we have already established $\lambda \to (\lambda)^m_\gamma$ for every $\gamma < \lambda$. Let $\gamma < \lambda$ and let $F: [\lambda]^{m+1} \to \gamma$ be a given partition. We then build (inductively by levels) a tree $\mathcal{T} = \langle \lambda, <_T \rangle$ as follows[1]: the first m levels of $\mathcal{T}$ are simply $0 <_T 1 <_T \ldots <_T m-1$. Starting with level $m+1$, now, our construction of $\mathcal{T}$ becomes much more complex. Let us sketch the idea: every time we now place an ordinal x at some position in $\mathcal{T}$ we will also define a set C_x of eventual $<_T$-successors of x. **The main idea, (*), behind** the sets C_x is that every member of $C_x \cup \{x\}$ behaves the same with respect to where F sends $m+1$-element subsets consisting of it added to m $<_T$-predecessors of x. Let us be specific: let $C_{m-1} = \lambda - m$. Suppose x has been placed in the tree, C_x the eventual $<_T$-successors of x have been defined, and we now wish to find the immediate $<_T$-successors of x. With an eye to making (*) correct we first break up C_x as follows: let x^* denote the set of $<_T$-predecessors of x, and for any u in $[x^*]^m$ and any $\eta < \gamma$, let $Q^u_\eta = \{\alpha \in C_x \mid F(u \cup \{\alpha\}) = \eta\}$. Then for any function f from

[1] Our tree $\mathcal{T}$ will be constructed such that given $\alpha_1 <_T \alpha_2 < \ldots <_T \alpha_m$, $F(\{\alpha_1, \alpha_2, \ldots, \alpha_m, \beta\})$ will be the same for all $<_T$ successors β of α_m.

$[x^*]^m$ into γ we define a set P_f by $P_f = \bigcap_{u\in[x^*]^m} Q^u_{f(u)}$.
If, now, y is the least member of P_f for some f, we let y be an immediate successor of x and define C_y to be $P_f - \{y\}$. (It is helpful to note that the P_f are pairwise disjoint subsets of C_x and that although various P_f may be empty, $\bigcup_f P_f = C_x$. It is also important to note that we have preserved (*). **Namely, for any f, if η and ξ are members of P_f and if $u\in[x^*]^m$, then $F(u\cup\{\eta\}) = F(u\cup\{\xi\}) = f(u)$). We have thus shown how to arrive at the $\alpha+1^{st}$ level of $\mathcal{T}$ given the** α^{th} level. Finally, suppose δ is a limit ordinal and that we have defined the αth level of $\mathcal{T}$ for each α less than δ. Then for each path p through $\mathcal{T}$ of length δ (i.e. for each linearly ordered subset of $\mathcal{T}$ containing one point from precisely each level less than δ) look at $\bigcap_{x\in p} C_x$. If it is empty, then p has no immediate successor on level δ. Otherwise let p have precisely one immediate successor on the δth level, namely the least member of $\bigcap_{x\in p} C_x$. Points arising in this way now form the δth level of .
Finally, if y is the successor of p on level δ, let $C_y =_{df} \bigcap_{x\in p} C_x - \{y\}$. We have thus inductively defined our tree $\mathcal{T} = \langle\lambda, <_T\rangle$.

Claim: $\mathcal{T}$ is a tree of cardinality λ all of whose levels have cardinality less than λ. (Proof: $\mathcal{T}$ clearly has cardinality λ since its universe is λ itself. It is easy to check by induction that each level of $\mathcal{T}$ has cardinality less than λ. Clearly the 0th level of $\mathcal{T}$ does; if the αth level does, the since each point x on the αth level gets at most as many immediate successors as there are functions from $[x^*]^m$ into γ, i.e., at most $2^{\overline{\overline{[x^*]^m}}\times\gamma}$ immediate successors, the fact that λ is strongly inaccessible immediately gives that there are fewer than λ points on the $\alpha+1^{st}$ level; finally, if δ is a limit ordinal and each level below the δth has cardinality less than λ, let ξ be the sup of these cardinalities.

$\xi < \lambda$ as λ is regular. Now there are no more points on the δth level than there are paths in $\mathcal{T}$ of length δ and there are at most ${}^{\delta}\xi$ (1) of these. But $\overline{\overline{{}^{\delta}\xi}} \leq 2^{\xi \times \delta} < \lambda$ as λ is strongly inaccessible, and so the δth level has cardinality less than λ. Thus the claim is proved. ⊠) Now since λ has the tree property, let $B \subseteq \lambda$ be a branch in $\mathcal{T}$ of cardinality λ. Then by our construction of $\mathcal{T}$, given any $\{\eta_1, \eta_2, \ldots, \eta_m\}_< \in [B]^m$, $F(\{\eta_1, \eta_2, \ldots, \eta_m, \beta\})$ is the same for every β in B larger than η_m. Let us, then, define a partition G of $[B]^m$ into γ pieces by having G send any $\{\eta_1, \ldots, \eta_m\}_<$ to $F(\{\eta_1, \ldots, \eta_m, \beta\})$, $\beta > \eta_m$, $\beta \in B$. Then our inductive hypothesis, $\lambda \to (\lambda)^m_\gamma$, presents us with a subset C of B of cardinality λ homogeneous for this partition G of $[B]^m$ (2). It is now immediate that C is homogeneous for our original partition F. For if $\{\alpha_1, \ldots, \alpha_{m+1}\}_<$ and $\{\beta_1, \ldots, \beta_{m+1}\}_<$ are two members of $[C]^{m+1}$, $F(\{\alpha_1, \ldots, \alpha_{m+1}\}_<) = G(\{\alpha_1, \ldots, \alpha_m\}) = G(\{\beta_1, \ldots, \beta_m\}) = F(\{\beta_1, \ldots, \beta_{m+1}\})$. Thus our lemma is proved. ⊠

Since we know that for $\lambda > \omega$ $\lambda \to (\lambda)^2$ implies that λ is strongly inaccessible, the two preceding lemmas give that for any cardinal λ, $\lambda \to (\lambda)^2$ implies that for every $n < \omega$ and $\gamma < \lambda$, $\lambda \to (\lambda)^n_\gamma$. We have thus proved our theorem. □

(1) Recall that if x and y are sets, ${}^{x}y$ denotes the collection of functions from x into y.

(2) Since our partition G is of $[B]^m$ rather than $[\lambda]^m$ the partition relation $\lambda \to (\lambda)^m_\gamma$ may not seem to apply. However, $\overline{\overline{B}} = \lambda$ and so if h is a bijection from λ onto B and $G' : [\lambda]^m \to \gamma$ is given by $G'(\{\alpha_1, \ldots, \alpha_n\}) = G(\{h(\alpha_1), \ldots, h(\alpha_n)\})$, $\lambda \to (\lambda)^m_\gamma$ applies to G' and yet any set homogeneous for G' is homogeneous for G. This general fact, that of being able to find appropriate homogeneous sets for partitions of $[x]^\alpha$ where the cardinality of x satisfies the appropriate partition relation, will be used by us repeatedly without explicit mention.

§ 5 We now take up the question of whether or not any "reasonable" set theoretic axioms imply the natural extension of Ramsey's Theorem just considered, namely that for some $\lambda > \omega$, $\lambda \to (\lambda)^2$. Since we know that any uncountable λ satisfying $\lambda \to (\lambda)^2$ is strongly inaccessible, we ought to look at so called "large cardinal" axioms to find such a λ. What we shall do is to look at one of the most well-known of these axioms[1].

An uncountable cardinal κ is said to be measurable if there exists a nontrivial two-valued κ-additive measure on κ, that is, if there exists a function $\mu : 2^\kappa \to \{0,1\}$ such that

(i) $\mu(\{\alpha\}) = 0$ for any $\alpha < \kappa$

(ii) $\mu(\kappa) = 1$

(iii) if A and B are disjoint subsets of κ then $\mu(A \cup B) = \mu(A) + \mu(B)$

(iv) if $\{A_\alpha | \alpha < \gamma < \kappa\}$ is a collection of subsets of κ such that $\mu(A_\alpha) = 0$ for all $\alpha < \gamma$, then $\mu(\bigcup_{\alpha<\gamma} A_\alpha) = 0$[2].

The axiom "there exists a measurable cardinal" has been extensively studied starting in the 1930's with its first proponent, S. Ulam. It is said to be a "large cardinal" axiom for as is well-known, measurable cardinals are strongly inaccessible. More recently, this axiom has been shown to imply generalizations of Ramsey's Theorem. In particular, it entails just the sort of result we want:

5.1 Theorem: If κ is a measurable cardinal then $\kappa \to (\kappa)^2$.

Proof: Using lemma 4.2.2, this theorem is extremely easy to prove. We need only show that any measurable cardinal is strongly inaccessible

---

(1) As it turns out, γ being strongly inaccessible does not imply that $\gamma \to (\gamma)^2$. We examine such negative results in §7 .

(2) (iii) and (iv) can easily be combined into one natural condition.

and has the tree property. So suppose κ is a measurable cardinal and let μ be a non-trivial two-valued κ-additive measure on κ.

Claim: κ is strongly inaccessible. (Proof: By definition, $\kappa > \omega$. If κ were not regular, we would have $\kappa = \bigcup_{\xi<\eta} \alpha_\xi$ where $\eta < \kappa$ and each $\alpha_\xi < \kappa$. But for any $\xi < \eta$, $\mu(\alpha_\xi) = \mu(\bigcup_{\beta<\alpha_\xi} \{\beta\}) = 0$ as μ is κ-additive and $\mu(\{\beta\}) = 0$ for each $\beta < \kappa$. Thus, again by the κ-additivity of μ, $\mu(\kappa) = \mu(\bigcup_{\xi<\eta} \alpha_\xi) = 0$, a contradiction. Thus κ is regular. Finally suppose that κ is not a strong limit cardinal and let $\gamma < \kappa$ be such that $2^\gamma \geqslant \kappa$. Now let $f: \kappa \xrightarrow[1-1]{} 2^\gamma$ and let us think of members of 2^γ as the associated characteristic functions, that is, as maps from 2^γ into $\{0,1\}$. Then clearly for every $\eta < \gamma$, $[f(\alpha)](\eta)$ is the same for almost every α, that is, for every $\eta < \gamma$, there is ϵ_η in $\{0,1\}$ such that $\mu(\{\alpha<\kappa | [f(\alpha)](\eta) = \epsilon_\eta\}) = 1$. Let, for each $\eta < \gamma$, $C_\eta = \{\alpha<\kappa | [f(\alpha)](\eta) = \epsilon_\eta\}$. Then since μ is κ additive and $\mu(C_\eta) = 1$ for each $\eta < \gamma$, $\mu(\bigcap_{\eta<\gamma} C_\eta) = 1$. Let α and β be distinct members of $\bigcap_{\eta<\gamma} C_\eta$. Then for every $\eta < \gamma$, $[f(\alpha)](\eta) = [f(\beta)](\eta)$ and so $f(\alpha) = f(\beta)$. This contradicts the fact that f is 1-1. We have thus established our claim. ☒)

Claim: κ has the tree property. (Proof: Suppose that $\mathcal{T} = \langle\kappa,<_T\rangle$ is a tree of cardinality κ all of whose levels have cardinality less than κ. (By isomorphic translation, we may assume the domain of $\mathcal{T}$ to be κ itself.) Then let B be the set of points in κ whose $<_T$-successors have μ-measure 1, that is,

$B =_{df} \{\alpha<\kappa | \mu(\{\eta<\kappa | \alpha<_T\eta\}) = 1\}$.

Claim: B is a branch in $\mathcal{T}$ of cardinality κ (Proof: To see that B has cardinality κ we need only show that every level of $\mathcal{T}$ has a point in B. Let us consider the δth level. Since each level of $\mathcal{T}$ has cardinality less than κ, the κ-additivity of μ tells us that almost every member of κ has order at least δ and hence (again by additivity) that some point on level δ has measure 1 $<_T$- successors. Now why is B linearly ordered under $<_T$? Well, if α and β are incomparable with respect to $<_T$, the sets of $<_T$-successors of α and of β are disjoint. Thus if B were not a branch, there would be two disjoint sets of μ-measure 1, an absurdity. Thus our claim is established.☒) Lemma 4.2.2 now gives our theorem. □

A very natural and important question comes to mind following this theorem. Namely, we know that if κ is measurable then any partition of $[\kappa]^2$ into two pieces has a homogemeous set of cardinality κ. Does it necessarily have a homogeneous set of measure 1 ? As it turns out, the answer to this question is, in general, no. However, for certain types of measures one can always find homogeneous sets of measure 1:

Definition: Let κ be a measurable cardinal and μ a nontrivial two-valued κ-additive measure on κ. Then μ is said to be normal if for any function $f: \kappa \to \kappa$, if $\mu(\{\alpha | f(\alpha) < \alpha\}) = 1$, then for some $\alpha_0 < \kappa$, $\mu(\{\alpha | f(\alpha) = \alpha_0\}) = 1$. Another way to say this is "μ is normal if any function from κ into κ which is less than the identity almost everywhere (with respect to μ) is constant almost everywhere".

As we shall see shortly, normal measures are extremely useful. Thus we are fortunate to have the following theorem:

Theorem: Let κ be any measurable cardinal. Then there exists a normal measure on κ.

Proof: Let ν be any nontrivial two-valued κ-additive measure on κ. We say that a function f from κ into κ is incompressible if f is not constant almost everywhere with respect to ν (abbrev: a.e.ν) and yet the only functions less than f a.e.ν are constant a.e.ν. It is routine to check (and we leave it to the reader) that if f is incompressible then the function $\mu: 2^{\kappa} \to \{0,1\}$ given by $\mu(x) = \nu(f^{-1}(x))$ is a nontrivial two-valued κ-additive normal measure on κ. Thus we need only show that an incompressible function exists. Now if the identity function is incompressible (i.e., if ν is normal) we are done. If it isn't, there is a function f_1 less than the identity a.e.ν which is not constant a.e.ν. If f_1 is incompressible we are done. Otherwise there is a function f_2 less than f_1 a.e.ν which is not constant a.e.ν. If f_2 is incompressible we are done. Otherwise we go on like this. After finitely many steps we must stop, i.e., some f_n must be incompressible. For otherwise, let $Q_n = \{\alpha | f_{n+1}(\alpha) < f_n(\alpha)\}$. Then as ν is κ additive, $\nu(\bigcap_{n<\omega} Q_n) = 1$, and so for any α in $\bigcap_{n<\omega} Q_n$, $f_1(\alpha), f_2(\alpha), f_3(\alpha), \ldots$ is an infinitely descending chain of ordinals, an absurdity. We have thus shown there must exist an incompressible function and so we are done. □

Now what good are normal measures? Well, for a start, they always give us homogeneous sets of measure 1:

5.2 Theorem (Rowbottom): Assume that κ is a measurable cardinal and that μ is a normal measure on κ. Then for any $n < \omega$ and $\gamma < \kappa$, any partition $F: [\kappa]^n \to \gamma$ has a homogeneous set of μ-measure 1.

Proof: If one cared to look at the proof of 4.2.2 he would easily see that if one could find a branch of μ-measure 1, he could find a homogeneous set of μ-measure 1 (assuming the inductive hypothesis modified to require μ-measure 1 homogeneous sets). But it is easy (though tricky) to see that the branch found in the proof of 5.1 has μ-measure 1. For suppose not. Then the function f from κ to κ given by $f(\alpha) =_{df}$ "the largest β such that $\beta \leqslant_T \alpha$ and $\beta \in B$" is less than the identity a.e.μ. Hence as μ is normal, there is some α_0 less than κ such that $\mu(\{\alpha | f(\alpha) = \alpha_0\}) = 1$. But if $\mu(\{\alpha | f(\alpha) = \alpha_0\}) = 1$, an immediate $<_T$-successor of α_0 in $\{\alpha | f(\alpha) = \alpha_0\}$, say β, must have μ-measure 1 $<_T$-successors in $\{\alpha | f(\alpha) = \alpha_0\}$. Having μ-measure 1 $<_T$-successors makes $\beta \in B$ yet as $f(\beta) = \alpha_0 < \beta$, $\beta \notin B$. We thus have a contradiction and so B must have μ-measure 1. Thus the theorem is proved. $\square$

Remark: We now give an even easier proof of this theorem (this time from scratch) which more closely parallels Ramsey's original argument. Namely, given a $\gamma < \kappa$ we proceed by induction on n, $n = 1$ being trivial (as μ is κ-additive). Suppose the theorem is true for $n = m$ and that $F: [\kappa]^{m+1} \to \gamma$ is a given partition. For each α, $\alpha < \kappa$, let C_α be of measure 1 and homogeneous for the partition of $[\kappa - (\alpha+1)]^m$ into γ pieces which sends any $\{\eta_1,\ldots,\eta_m\}$ to $F(\{\alpha,\eta_1,\ldots,\eta_m\})$. Now let C be the set of α less than κ such that $\alpha \in C_\beta$ for every $\beta < \alpha$. This set C is known as the diagonal intersection of the C_α's. It has many of the properties of the full intersection with the following highly desired additional property: Claim: $\mu(C) = 1$. (Proof: Since there are κ many C_α, the κ-additivity of μ says nothing about the measure of the intersection of the C_α even though each C_α has measure 1. That is why this claim is quite surprising. What we use to prove it is the normality of μ. In particular, let $f: \kappa \to \kappa$ be given by $f(\alpha) =_{df}$ "the least β such that $\alpha \notin C_\beta$." Then if $\mu(C) = 0$, $\mu(\{\alpha | f(\alpha) < \alpha\}) = 1$ and so for some $\alpha_0 < \kappa$, $\mu(\{\alpha | f(\alpha) = \alpha_0\}) = 1$. But $f(\alpha) = \alpha_0$ implies $\alpha \notin C_{\alpha_0}$ and so

$\{\alpha | f(\alpha) = \alpha_o\} \cap C_{\alpha_o} = \emptyset$. Since there cannot be two disjoint sets of μ-measure 1 , we have a contradiction. ⊠) Now if $\eta_1 < \eta_2 < \ldots < \eta_{m+1}$ are in C , then $\{\eta_2, \ldots, \eta_{m+1}\} \in C_{\eta_1}$, and so for any η in C , there is an ordinal $g(\eta) < \gamma$ such that $\{F(\{\eta\} \cup x) | x \in [C-(\eta+1)]^m\} = \{g(\eta)\}$. Since $\mu(C) = 1$, $\gamma < \kappa$, and μ is κ-additive, there exists a subset D of C of μ-measure 1 on which g is constant. This D is clearly homogeneous for F . □

Now theorem 5.2 tells us that the existence of a measurable cardinal entails an extension of Ramsey's theorem which is potentially even stronger than the existence of a $\lambda > \omega$ such that $\lambda \to (\lambda)^2$. Namely, if κ is measurable, then we know that given any partition $F: [\kappa]^n \to \gamma$ we can find a set of cardinality κ which is homogeneous for it. But what's more, given fewer than κ such partitions, we can find one set which is homogeneous, simultaneously, for all of them:

5.3 Theorem (Erdős-Hajnal): Assume that κ is a measurable cardinal, δ is a cardinal less than κ , and that for $\alpha < \delta$, n_α is a positive integer, γ_α is a cardinal less than κ and $F_\alpha: [\kappa]^{n_\alpha} \to \gamma_\alpha$ is a given partition. Then there exists a set C of cardinality κ such that for every $\alpha < \delta$, $\overline{\overline{F_\alpha''[C]^{n_\alpha}}} = 1$.

Proof (Rowbottom): Let μ be a normal measure on κ . By theorem 5.2 let, for each $\alpha < \delta$, C_α be a set of μ-measure 1 homogeneous for F_α. Then as μ is κ-additive, $\mu(\bigcap_{\alpha<\delta} C_\alpha) = 1$ and so $\overline{\overline{\bigcap_{\alpha<\delta} C_\alpha}} = \kappa$. Clearly $\bigcap_{\alpha<\delta} C_\alpha$ is homogeneous simultaneously for each F_α. □

The property of measurable cardinals given in 5.3 turns out to be of extreme use in set theory. We shall see an example of this in the next section.

§6 The main point of this section is to apply the generalization of Ramsey's Theorem indicated in 5.3 to proving a result about sets of real numbers. Specifically we shall prove D.A. Martin's result that if a measurable cardinal exists then every analytic set of reals is determinate.

Let us begin with a discussion of the notion of determinateness: we shall think of real numbers as functions from ω into ω, i.e., as members of ${}^{\omega}\omega$. Now given a set $A \subseteq {}^{\omega}\omega$, an associated game G_A is played as follows: two players, I and II, move alternately writing at each turn a nonnegative integer. In this way they build a real number, i.e., a sequence of nonnegative integers. Player I wins a given play of the game if the real number produced is a member of A - player II wins otherwise. A <u>winning strategy</u> for G_A is a function f from finite sequences from ω into ω such that a player making only moves as indicated by f (his move following any initial play $a_1, a_2, \ldots, a_n$ should be $f(\langle a_1, \ldots, a_n \rangle)$) wins any play of G_A. Clearly both players can never simultaneously possess a winning strategy for a given game. A set of reals A is said to be <u>determinate</u> if there is a winning strategy (for I or for II) for G_A.

The axiom of determinateness (AD) is the assertion that every set of reals is determinate. Using the axiom of choice it is fairly easy to put together a nondeterminate set simply by diagonalizing over all strategies. Thus the axioms of determinateness and choice contradict one another. This is unfortunate as the axiom of determinateness has many wonderful and important consequences[1]. However, if various specific sets turn out to be determinate, one can still establish interesting related results. It is toward showing specific sets determinate that we now concern ourselves.

We begin by citing the following well-known result of Gale and Stewart.

6.1 <u>Theorem:</u> Every **open** set of reals is determinate.

<u>Proof:</u> Let A be a given open set. To see that A is determinate we shall simply use the fact that any real which is in A is so on the basis of an initial segment, i.e., if f is in A then for some n in ω, any member of ${}^{\omega}\omega$ agreeing with f on integers $< n$ is also in A. Namely, suppose that player I does not have a winning strategy. Then we define a strategy for II as follows[2]: at any given move, II should write the least nonnegative integer so that after he does so, player I still has no winning strategy. It is easy to see inductively that II can always do this, for if at some move he can't, I already had a winning strategy which he used in making his previous move.

Claim: This is a winning strategy for player II. (Proof: Suppose II plays as dictated by this strategy and loses nonetheless. Let $r = \langle a_1, a_2, a_3, \ldots \rangle$ be the associated play. Then since $r \in A$, for some n, any real of the form $\langle a_1, \ldots, a_n, b_{n+1}, \ldots \rangle$ is a member of A. But this means that starting from his first move after the nth player I had a winning strategy (namely, "do anything") and this

---

(1) As an example, AD implies that every set of reals is Lebesgue measurable. Other applications of AD will be mentioned in §8.

(2) Rather than explicitly define a function for this strategy, we shall simply describe it informally. Our description, however, will clearly correspond to a function from finite sequences from ω into ω.

contradicts the fact that II played in such a way that at no initial point did player I have a winning strategy. ☒) This claim proves the theorem.[1] □

As it turns out, much better results are possible. Until recently, Morton Davis' theorem that every $\underset{=}{\Sigma}_3^0$ ($G_{\delta\sigma}$) set of reals is determinate was the best result known. However, Paris has recently shown that every $\underset{=}{\Sigma}_4^0$ set of reals is determinate.

This result of Paris provides one of the most compelling arguments for the study of large cardinals in infinitary combinatorics. We see this by tracing the history of Paris' result: Morton Davis found a method for pushing the Gale-Stewart result for open games up the hierarchy of sets and in so doing he established $\underset{=}{\Sigma}_3^0$ - determinateness. Then just recently Martin found a method for reducing the question of whether or not a particular set of reals is determinate to the question of whether or not some much simpler set is determinate. The key step in Martin's method involves the use of some sort of variation of Ramsey's theorem and the main applications Martin derived from his method used "large cardinal" generalizations of Ramsey's theorem such as those we considered in previous sections. As an example, under the assumption of the existence of a cardinal having the property 5.3 indicates measurable cardinals to have, Martin showed that arbitrary Borel sets (analytic sets, in fact) were as good as open sets relative to the question of determinateness. Unfortunately, this proof uses an assumption which cannot be proved in ZF and so people tried a finer approach, namely, to attempt a Martin-style analysis of sets somewhat less complex than arbitrary Borel sets (hoping to need only a provable (in ZF) partition property) and then use their new found simplicity in conjunction with Davis' method, to climb up

(1) Note that a simple modification of this proof shows that all closed sets are determinate.

the hierarchy. Martin used this approach to prove $\underset{=}{\Sigma}_4^0$ determinateness, but he still needed an independent partition property (though a weaker one than that of 5.3)[1] to succeed. Finally, Paris showed how to succeed here using a provable partition property.

The point is this: by considering general partition relations (in particular, ones not provable in ZF) Martin discovered a method for successfully dealing with questions of determinateness. The method, however, remains valid for use in ZF alone - for specific applications you need only find an appropriate provable partition relation to use. It is the simplicity and elegance that large cardinal axioms bring to set theory that makes the discovery of such general methods feasible.

Let us now examine Martin's method. Specifically, we shall consider his original theorem here:

6.2 Theorem (Martin [16]): Assume that there exists a measurable cardinal. Then every analytic set of reals is determinate.

Proof: Since we wish this proof to be understandable by those with no detailed knowledge of analytic sets, we shall leave for a remark a discussion of some facts about analytic sets we plan to use. In particular, we shall use the fact that there exists a countable linearly ordered set $(S,<)$ which we can list $s_1, s_2, s_3, s_4, s_5, \ldots$ (not in the ordering $<$) with the following property: let A be a given analytic set. Then there is a way to successively assign to each s_n a 0 or a 1 such that given two players engaged in a play of G_A, by the nth move of their play $s_1, s_2, \ldots, s_n$ have already received their 0 or 1, and if S^* is the set of those s_n which eventually receive a 1 during the play, I wins the play iff S^* is not

(1) What he needed was the existence of a $\lambda > \omega$ satisfying $\lambda \to (\lambda)^2$.

well-ordered under $<$. This is the only fact about analytic sets we need in order to proceed[1]. Let κ be a measurable cardinal. Now suppose A is a given analytic set. We wish to show that A is determinate. What we shall, in fact, do is to define an auxillary game G'_A which is open[2] and yet which is equivalent to G_A in the sense that if player I (II) has a winning strategy for G'_A , then player I (II) has a winning strategy for G_A. This is Martin's reduction of analytic games to open games which we mentioned earlier.

The game G'_A is played as follows (it is not the usual sort of game described earlier): Players I and II move alternately - at each of his moves, player I writes a nonnegative integer - at each of his moves, say the nth , player II writes an integer and also writes an ordinal α_n less than κ (κ is our measurable cardinal). We shall think of this α_n as being assigned to s_n - indeed, we define a play of G'_A to be a win for player II iff the members of S receiving 1 were assigned ordinals in an order-preserving way, i.e., iff for any two members s_n and s_m of S which received a 1 , $s_n < s_m$ iff $\alpha_n \epsilon\, \alpha_m$.

Now G'_A is clearly an open game, i.e., whenever player I wins, he has already won by some initial stage of play. Thus a trivial modification of our proof of Theorem 6.1 tells us that there is a winning strategy for G'_A (A strategy for G'_A for player I (II) here would be a function from pairs of finite sequences of nonnegative integers and of ordinals less than κ (from finite sequences of nonnegative integers) into nonnegative integers (into pairs of nonnegative integers and of ordinals less than κ)).

Claim: If player II has a winning strategy for G'_A then he has one

---

(1) For those who do not wish to wait for our later remark which elaborates on this fact, let us note here that (s,<) is the Kleene-Brouwer ordering on finite sequences of integers, and that we are simply using the Kleene-Brouwer analysis of Σ^1_1 sets.

(2) In a general setting, a game is "open" if it is such that I wins a play iff he has already won by some initial stage of the play.

for G_A. (Proof: Let g be a winning strategy for player II for $G_A^!$. Then the first coordinate of g, i.e., the function f such that for any x $f(x)$ is the first element of the pair $g(x)$, must be a winning strategy for II for G_A. For given any play of G_A with II using f, the set of those s_n which receive a 1, S^*, can be assigned ordinals in an order-preserving way (the ordinal moves of g for the play do this) and since the ordinals are well-ordered, S^* must also be well-ordered (under $<$). Thus by our fact about analytic sets, the sequence constructed is not in A and so II wins. ⊠)

Claim: If player I has a winning strategy for $G_A^!$ then he has one for G_A. (Proof: (This claim is much more difficult to prove than the preceeding one. Indeed, the previous claim should not have been surprising - when II wins $G_A^!$ not only is a sequence produced which fails to be in A - II well-orders the members of S receiving 1 as they appear. On the other hand, I winning $G_A^!$ only means that he messed up II's attempt to well-order the members of S a receiving 1 - it doesn't mean that they aren't well-ordered. This is where the notion of partition and homogeneous set come into play. To convert a strategy for I for G_A' to one for G_A, we must make guesses as to what II is writing for his ordinal moves - we shall guess that he writes ordinals from an appropriate homogeneous set. That way, even <u>after</u> a given play there will be no way to assign ordinals order-preservingly to those s_n's receiving a 1 for as the set is homogeneous the behavior of one sequence of guessed ordinals is just like that of any other.) Let us now make all of this specific: let f be a strategy for I for $G_A^!$. For each finite sequence x of nonnegative integers let $s_{i_1}, s_{i_2}, \ldots, s_{i_m}$ be those members of S among $s_1, s_2, \ldots, s_n$ (where n is the length of x) which receive a 1 during an initial play of G_A beginning with x, and let F_x be a partition of $[\kappa]^m$ into ω pieces given by

$F_x(\{\alpha_1,\ldots,\alpha_m\}) = f(x,\langle\alpha_{j_1},\ldots,\alpha_{j_m}\rangle)$, where $\langle\alpha_{j_1},\ldots,\alpha_{j_m}\rangle$ is an arrangement of $\{\alpha_1,\ldots,\alpha_m\}_<$ satisfying $\alpha_{j_p} \varepsilon \alpha_{j_q}$ iff $s_{i_p} < s_{i_q}$. Now by theorem 5.3 let C be a subset of κ of cardinality κ homogeneous for each partition F_x, i.e., C has cardinality κ and is such that for each finite sequence x (and associated m) $\overline{\overline{F_x''[C]^m}} = 1$. We are now in a position to define a strategy $\hat{f}$ for player I for G_A. $\hat{f}$ is simply f using guesses (for II's ordinal plays) from C, i.e., for any finite sequence x, let $\hat{f}(x)$ be the unique integer which is the range of F_x on $[C]^m$ (m as above). It is easy, now, to see that $\hat{f}$ is a winning strategy for player I for G_A. For suppose not, and let the sequence r be the result of a play of G_A in which I uses $\hat{f}$ and yet where $r \notin A$. Then S^*, the set of s_i's which receive a 1 during this play, is well-ordered under $<$, and hence one can associate in an order-preserving way an ordinal in C with each member of S^*. (Since S^* is countable its order-type is less than $\aleph_1$ and hence less than the order-type of C - that is why we can make this association using only ordinals in C). Let α_i be the ordinal associated in this way with s_i. Then if we consider a play of $G_A^!$ where I uses his strategy f and II writes integers as he did in forming r and writes α_i for any move where s_i receives a 1 (and writes the ordinal 0 otherwise), the resulting play of $G_A^!$ is a win for II since it produces r for the integer sequence and the order-preserving association $s_i \longleftrightarrow \alpha_i$ with the s_i's receiving a 1 for the ordinal sequence. But as f is a winning strategy for I for $G_A^!$, this play cannot be a win for II. This contradiction yields our claim. ☒) Martin's theorem is thus proved. □

<u>Remark:</u> For those interested in the details associated with the linearly ordered set $\acute{S}$ used in the above proof, we shall describe it now: let A be a given $\underset{=}{\Sigma_1^1}$ set. Then we know that there is a recursive predicate R and parameter u such that for any f,

$f \in A$ iff $\exists h \forall n R(\bar{h}(n), \bar{f}(n), \bar{u}(n))$ (f, h, and u denote functions from ω into ω and for any such function g, $\bar{g}(n)$ denotes (a code for) the finite sequence $\langle g(0), g(1), \ldots, g(n-1) \rangle$). Let s, r, and t be three finite sequences of integers of length n. Then s is said to be unsecured with respect to R, r, and t, if for any $m \leqslant n$ and initial segments s_m, r_m, and t_m of s, r, and t, respectively, of length m, $R(s_m, r_m, t_m)$. Now the set $(S, <)$ used in the proof of 6.2 is just the collection of finite sequences of nonnegative integers under the Kleene-Brouwer ordering: $s < t$ iff s is a proper extension of t or, at the first place s and t differ, t exceeds s. The members of S which receive a 1 (0) at a stage of play which has produced the sequence x of length n are just those which are unsecured (secured) with respect to R, x, and $\bar{u}(n)$. The Kleene-Brouwer theorem states that for any given $\underset{\sim}{\Sigma}^1_1$ set A, a real is in A iff the associated set of unsecured sequences is not well-ordered under the Kleene-Brouwer ordering (see [22]).

<u>Remark</u>: Since $\underset{\sim}{\Sigma}^1_1$-determinateness is false in L, the universe of constructible sets, one cannot hope to prove it in ZFC alone. However, the assertion that every Borel set of reals is determinate may be provable in ZFC. What is more, a result of Friedman tells one that any successful proof of Borel determinateness would, most likely, use some sort of property possessed by "large" sets (such as an appropriate partition property). For Friedman has constructed models for set theories allowing arbitarily large countable ranks and yet in which there are nondeterminate Borel sets.

<u>Remark</u>: In this section we have indicated applications of partition properties to questions of determinateness. There are, however, other notable applications to set theory as well as to model theory.

For example, Rowbottom [23] used partition properties to establish improved Lowenheim-Skolem theorems (so called strong two cardinal theorems) and then used these to infer that the constructible universe is small. For example, he showed that if one assumes the existence of a cardinal κ satisfying the conclusion of 5.3, then there are only countably many constructible real numbers. Further (and deeper) limitations of the constructible universe (under the assumption of the existence of such a κ) were found by Silver and Solovay - for these results the reader should refer to [24] and [27], respectively.

Now in model theory, partition properties, as mentioned just above, can be used to prove improved Lowenheim-Skolem theorems. But there are many other uses. For example one can use partition properties to prove the existence of end extensions of various models, and also to produce models with nontrivial automorphisms. For more information here the reader should refer to [2] and [18].

Of course, one should not forget Ramsey's original application of his theorem to decision theory [1].

§7 It seems appropriate to take a closer look at the special assumption used in proving 6.2. In particular, we did not really use the fact that there existed a measurable cardinal but only the fact that there existed a cardinal having the property 5.3 showed measurable cardinals to have. Let us call this partition property of cardinals $P(\kappa)$: $P(\kappa)$ iff given fewer than κ partitions of the form $F:[\kappa]^n \to \gamma$, $\gamma < \kappa$, there exists a set of size κ which is homogeneous for all of them. The problem we concern ourselves with here is the following: how does the assumption "there exists a $\kappa > \omega$ such that $P(\kappa)$" compare in strength to the assumptions "there exists a measurable cardinal" and "there exists a $\kappa > \omega$ such that

$\kappa \to (\kappa)^2$"? As it turns out, it is strictly between them. The two theorems we wish to establish here are:

7.1 <u>Theorem</u> (Rowbottom [23]): If κ is a measurable cardinal then there exists a cardinal γ, $\omega < \gamma < \kappa$, such that $P(\gamma)$.

7.2 <u>Theorem</u> (Reinhardt [21]): If $\kappa > \omega$ and $P(\kappa)$ then there exists a cardinal γ, $\omega < \gamma < \kappa$, such that $\gamma \to (\gamma)^2$.

The results we shall actually prove are much stronger than these. However, by using essentially the same argument with which the corollary to 4.1 was established (that which turns on Gödel's theorem that one cannot prove the consistency of any sufficiently rich system within itself) it is relatively routine to see that 7.1 and 7.2 give that the three assumptions we are considering are strictly nested by strength.

<u>Notation</u>: Let κ, λ, and γ be cardinals. Then we denote by $\kappa \to (\lambda)^{<\omega}_{\gamma}$ the assertion: <u>for every partition</u> $F: \bigcup_{n<\omega} [\kappa]^n \to \gamma$ <u>there exists a subset C of κ of cardinality λ such that for each $n < \omega$</u>, $F''[C]^n = 1$. As usual C is said to be homogeneous for F with respect to $\kappa \to (\lambda)^{<\omega}_{\gamma}$; also again, the subscript will be dropped when it would otherwise be 2.(1)

7.1.1. <u>Lemma</u>: If $\kappa \to (\kappa)^{<\omega}$ then $\kappa \to (\kappa)^{<\omega}_{\lambda}$ for every $\lambda < \kappa$.

<u>Proof of lemma</u>: Say $F: \bigcup_{n<\omega} [\kappa]^n \to \lambda$ is given. Define $G: \bigcup_{n<\omega} [\kappa]^{2n} \to 2$ by $G(\{\alpha_1, \dots, \alpha_{2n}\}_<) = 0$ iff $F(\{\alpha_1, \dots, \alpha_n\}_<) = F(\{\alpha_{n+1}, \dots, \alpha_{2n}\})$ and let C be homogeneous for G with respect to $\kappa \to (\kappa)^{<\omega}$. Then $G''[C]^{2n} = \{0\}$ for every n since, as $\kappa > \lambda$

(1) Note that $\omega \to (\omega)^{<\omega}$ is false. The map $F: \bigcup_{n<\omega} [\omega]^n \to 2$ given by $F(\{\alpha_1, \dots, \alpha_n\}_<) = 0$ iff $n \in \{\alpha_1, \dots, \alpha_n\}$ is easily seen to be a counterexample.

there are $\alpha_1 < \alpha_2 < \ldots < \alpha_{2n}$ in C such that $F(\{\alpha_1, \ldots, \alpha_n\}) = F(\{\alpha_{n+1}, \ldots, \alpha_{2n}\})$.

Claim: C is homogeneous for F with respect to $\kappa \to (\kappa)^{<\omega}_{\lambda}$. (Proof: Say $\{\alpha_1, \ldots, \alpha_n\}_<$ and $\{\beta_1, \ldots, \beta_n\}_<$ are members of $[C]^n$. Let $\{\delta_1, \ldots, \delta_n\}_<$ be a member of $[C]^n$ such that $\alpha_n < \delta_1$ and $\beta_n < \delta_1$. Then $G(\{\alpha_1, \ldots, \alpha_n, \delta_1, \ldots, \delta_n\}_<) = 0 = G(\{\beta_1, \ldots, \beta_n, \delta_1, \ldots, \delta_n\}_<)$ whence $F(\{\alpha_1, \ldots, \alpha_n\}) = F(\{\delta_1, \ldots, \delta_n\}) = F(\{\beta_1, \ldots, \beta_n\})$. ☒) This claim yields the lemma. ☒

7.1.2 Lemma: Assume $\kappa > \omega$. Then $\kappa \to (\kappa)^{<\omega}$ iff $P(\kappa)$.

Proof of Lemma: Clearly $P(\kappa)$ implies $\kappa \to (\kappa)^{<\omega}$. Conversely, suppose that $\delta < \kappa$ and that for $\alpha < \delta$, $F_\alpha: [\kappa]^{n_\alpha} \to \gamma_\alpha$ are given partitions. Let $\gamma = \bigcup_{\alpha<\delta} \gamma_\alpha$. Let $G: \bigcup_{n<\omega} [\kappa]^n \to {}^{\delta}\gamma$ by

$$[G(\{\alpha_1, \ldots, \alpha_n\}_<)](\alpha) = \begin{cases} F_\alpha(\{\alpha_1, \ldots, \alpha_n\}) & \text{if } n_\alpha = n \\ 0 \text{ otherwise} . \end{cases}$$

Since $\kappa \to (\kappa)^{<\omega}$, κ is strongly inaccessible and so $\overline{\overline{{}^{\delta}\gamma}} \leq 2^{\overline{\overline{\delta \times \gamma}}} < \kappa$. Thus ${}^{\delta}\gamma$ may be identified with $\overline{\overline{{}^{\delta}\gamma}} < \kappa$, and so let C be homogeneous for G with respect to $\kappa \to (\kappa)^{<\omega}_{\overline{\overline{{}^{\delta}\gamma}}}$.

Claim: C is homogeneous for each F_α (Proof: Say $\{\alpha_1, \ldots, \alpha_{n_\alpha}\}$ and $\{\beta_1, \ldots, \beta_{n_\alpha}\}$ are members of $[C]^{n_\alpha}$. Then $F_\alpha(\{\alpha_1, \ldots, \alpha_{n_\alpha}\}) = [G(\{\alpha_1, \ldots, \alpha_{n_\alpha}\})](\alpha) = [G(\{\beta_1, \ldots, \beta_n\})](\alpha) = F_\alpha(\{\beta_1, \ldots \beta_n\})$ ☒

This completes the proof of 7.1.2. ☒

Theorem 7.1 is clearly a consequence of the following much stronger fact:

7.3 <u>Theorem</u>: Assume that κ is a measurable cardinal and that μ is a normal measure on κ. Then

$$\mu(\{\lambda<\kappa \mid \lambda \to (\lambda)^{<\omega}\}) = 1 .$$

<u>Proof of 7.3</u>: Let $A = \{\lambda < \kappa \mid \lambda \to (\lambda)^{<\omega}\}$ and suppose that $\mu(A) = 0$. For each $\lambda \in \kappa - A$ let $F_\lambda : \bigcup_{n<\omega} [\lambda]^n \to 2$ be a partition having no set homogeneous with respect to $\lambda \to (\lambda)^{<\omega}$. Let $G: \bigcup_{n<\omega} [\kappa]^n \to 2$ by $G(\{\alpha_1,\ldots,\alpha_n\}_<) = F_{\alpha_n}(\{\alpha_1,\ldots,\alpha_{n-1}\}_<)^{(1)}$ and let C be homogeneous for G with respect to $\kappa \to (\kappa)^{<\omega}$ and such that $\mu(C) = 1$ (using 5.2) Claim: For some λ in $C\cap(\kappa-A)$, $\overline{\overline{C\cap\lambda}} = \lambda$. (Proof: Suppose not. Let $f: \kappa \to \kappa$ by $f(\alpha) =_{df} \overline{\overline{C\cap\alpha}}$. Then since $\mu(C) = 1$ and $\mu(\kappa-A) = 1$, $\mu(C\cap(\kappa-A)) = 1$ and so $f(\alpha) < \alpha$ for almost every α, i.e. $\mu(\{\alpha \mid f(\alpha) < \alpha\}) = 1$. By the normality of μ, then, there is some α_0 less than κ such that $\mu(\{\alpha \mid f(\alpha) = \alpha_0\})$. Thus $\overline{\overline{C\cap\alpha}} = \alpha_0$ for unboundedly many α in κ. This is absurd as $\overline{\overline{C}} = \kappa$ and κ is strongly inaccessible. ☒) So suppose $\lambda \in C\cap(\kappa-A)$ is such that $\overline{\overline{C\cap\lambda}} = \lambda$. Claim: $C\cap\lambda$ is homogeneous for F_λ with respect to $\lambda \to (\lambda)^{<\omega}$. (Proof: Say $\{\alpha_1,\ldots,\alpha_n\}_<$ and $\{\beta_1,\ldots,\beta_n\}_<$ are in $[C\cap\lambda]^n$. Then $F_\lambda(\{\alpha_1,\ldots,\alpha_n\}) = G(\{\alpha_1,\ldots,\alpha_n,\lambda\}) = G(\{\beta_1,\ldots,\beta_n,\lambda\}) = F(\{\beta_1,\ldots,\beta_n\})$. ☒) Since $C\cap\lambda$ is homogeneous for F_λ with respect to $\lambda \to (\lambda)^{<\omega}$, we contradict the definition of F_λ. This proves the theorem. □

We now consider theorem 7.2. It is easy to see that if $\kappa > \alpha > \beta > \omega$ and if $\kappa \to (\alpha)^{<\omega}$ then there exists a γ less than κ such that $\gamma \to (\beta)^{<\omega}$. One simply uses the main argument just given in proving 7.3. Thus what we shall prove is really stronger than 7.2:

(1) If F_{α_n} is not defined (i.e., if $\alpha_n \in A$) , we let $G(\{\alpha_1,\ldots,\alpha_n\})=0$.

7.4 Theorem[1]: Assume $\kappa \to (\omega)^{<\omega}$. Then there exists a cardinal γ, $\omega < \gamma < \kappa$, such that $\gamma \to (\gamma)^2$.

Proof of 7.4: Let κ be the least λ satisfying $\lambda \to (\omega)^{<\omega}$. Note that by the same argument used in establishing 7.1.1, we have that $\kappa \to (\omega)_3^{<\omega}$.

Claim: Given countably many partitions $F_i\colon \bigcup_{n<\omega} [\kappa]^n \to 3$ there is a set C homogeneous simultaneously for each with respect to $\kappa \to (\omega)_3^{<\omega}$. (Proof: Simply let $G\colon \bigcup_{n<\omega} [\kappa]^n \to 3$ be given by

$$G(\{\alpha_1,\dots,\alpha_n\}_<) = \begin{cases} F_i(\{\alpha_1,\dots,\alpha_j\}) & \text{if } n = 2^i 3^j \\ 0 & \text{otherwise}. \end{cases}$$

Then any set homogeneous for G with respect to $\kappa \to (\omega)_3^{<\omega}$ is clearly homogeneous for each F_i with respect to $\kappa \to (\omega)_3^{<\omega}$. ☒)

Now let $\mathcal{J}$ be any countable set of functions from $\bigcup_{n<\omega} [\kappa]^n$ into κ containing the coordinate functions $(F_i(\{\alpha_1,\dots,\alpha_n\}) = \alpha_i)$ and closed under all possible ways of composing such functions (eg: if $f(\{x,y\})$, $g(\{x,y,z\})$ and $h(\{x\})$ are in $\mathcal{J}$, then so is $k(\{x,y,z\}) =_{df} f(\{g(\{x,y,h(\{x\})\},z\}))$. Given any pair (f,g) of functions in $\mathcal{J}$ we associate a partition $F_{fg}\colon \bigcup_{n<\omega} [\kappa]^n \to 3$ defined by

$$F_{fg}(\{\alpha_1,\dots,\alpha_n\}_<) = \begin{cases} 0 & \text{if } f(\{\alpha_1,\dots,\alpha_n\}) < g(\{\alpha_1,\dots,\alpha_n\}) \\ 1 & \text{if } f(\{\alpha_1,\dots,\alpha_n\}) = g(\{\alpha_1,\dots,\alpha_n\}) \\ 2 & \text{if } f(\{\alpha_1,\dots,\alpha_n\}) > g(\{\alpha_1,\dots,\alpha_n\}) \end{cases}$$

Now let C be homogeneous, simultaneously for each F_{fg} with respect to $\kappa \to (\omega)_3^{<\omega}$, and let X denote the closure of C under $\mathcal{J}$, i.e., $X =_{dn} \bigcup_{f\in\mathcal{J}} f'' \bigcup_{n<\omega} [C]^n$[2]

[1] This theorem is due to Silver who first proved it using model theoretic methods (See [24]). The proof we give here is entirely combinatorial (see [9]).

[2] Note that as $\mathcal{J}$ itself is closed under function composition, X is closed under action by functions in $\mathcal{J}$

Let $\alpha_1, \alpha_2, \ldots$ be an enumeration of C in increasing order. Then we can define a map $j : X \to X$ by $j(f\{\alpha_{i_1}, \ldots, \alpha_{i_n}\}_<) =_{df} f(\{\alpha_{i_1+1}, \ldots, \alpha_{i_n+1}\})$. It is routine to see that j is well-defined as given - simply use the homogeneity of C. Since $j(\alpha_1) = \alpha_2$, j is not the identity function. Let γ be the least ordinal in X moved by j[(1)].

Our plan is to show that $\gamma > \omega$ and that $\gamma \to (\gamma)^2$. Notice that we have not put any special requirements as to what specific functions appear in $\mathcal{J}$. This we do as our proof proceeds. That is, by the end of our proof we will have described countably many functions such that given any $\mathcal{J}$ containing them, the associated γ as defined above will be larger than ω and satisfy $\gamma \to (\gamma)^2$. Hopefully this roundabout approach will make the proof clearer.

It is easy to see what $\mathcal{J}$ should contain to guarantee $\gamma > \omega$. Namely, for each $\alpha \leqslant \omega$, let $f_\alpha : \bigcup_{n<\omega} [\kappa]^n \to \kappa$ be given by "$f_\alpha(x) = \alpha$ for every x in $\bigcup_{n<\omega} [\kappa]^n$." Then if each f_α is in $\mathcal{J}$, we must have $\gamma > \omega$ for, for $\alpha < \omega$, $j(\alpha) = j(f_\alpha(\{\alpha_1\})) = f_\alpha(\{\alpha_2\}) = \alpha$.

Before proceeding let us note a simple fact which will be of great use to us: suppose that $\{\beta_1, \ldots, \beta_n\} \in [X]^n$, $f \in \mathcal{J}$, and $f(\{\beta_1, \ldots, \beta_n\}) < \gamma$. Then $f(\{\beta_1, \ldots, \beta_n\}) = j(f\{\beta_1, \ldots, \beta_n\}) = f(\{j(\beta_1), \ldots j(\beta_n)\})$. To see this simply unwind the definition of j and recall that j moves no ordinal less than γ.

---

(1) Observe that $\gamma < j(\gamma) < jj(\gamma) < \ldots$ For suppose $\gamma = f(\{\alpha_{i_1}, \ldots, \alpha_{i_n}\})$. Then if $f(\{\alpha_{i_1+1}, \ldots, \alpha_{i_n+1}\}) < f(\{\alpha_{i_1}, \ldots, \alpha_{i_n}\})$ the homogeneity of C would tell us that
$f(\{\alpha_{i_1}, \ldots, \alpha_{i_n}\}) > f(\{\alpha_{i_1+1}, \ldots, \alpha_{i_n+1}\}) > f(\{\alpha_{i_1+2} \ldots, \alpha_{i_n+2}\}) > \ldots$
and so we would have an infinite descending chain of ordinals, an impossibility. Thus $\gamma < j(\gamma)$. By the homogeneity of C, $\gamma < j(\gamma) < jj(\gamma) < \ldots$

Now what functions should we include in $\mathcal{J}$ to guarantee γ to be regular? Well, for each β in κ let $g_1 : [\kappa]^1 \to \kappa$ be given by $g_1(\{\beta\}) =_{df}$ "the cofinality of β" and for each λ less than κ let t_λ map $g_1(\{\lambda\})$ onto an unbounded subset of λ. We now define functions g_2 and g_3 from $[\kappa]^2$ into κ by $g_2(\{\beta_1,\beta_2\}_<)$ $=_{df}$ "the least β such that $t_{\beta_1}(\beta) \neq t_{\beta_2}(\beta)$" and $g_3(\{\beta_1,\beta_2\}_<) =_{df} t_{\beta_1}(g_2(\{\beta_1,\beta_2\}_<))$. Then if we now require g_1, g_2, and g_3 to be in $\mathcal{J}$, γ must be regular. We see this as follows: suppose not. Then $g_1(\{\gamma\}) < \gamma$ and so by our simple fact, $g_1(\{\gamma\}) = g_1(\{j(\gamma)\})$ $= g_1(\{jj(\gamma)\})$. Similarly, $g_2(\{\gamma, j(\gamma)\}_<) = g_2(\{j(\gamma), jj(\gamma)\}_<)$ and $g_3(\{\gamma, j(\gamma)\}_<)$ $= g_3(\{j(\gamma), jj(\gamma)\})$. But by definition, $g_3(\{\gamma, j(\gamma)\}) = t_\gamma(g_2(\{\gamma, j(\gamma)\}))$ $\neq t_{j(\gamma)}(g_2(\{\gamma, j(\gamma)\})) = t_{j(\gamma)}(g_2(\{j(\gamma), jj(\gamma)\}) = g_3(\{j(\gamma), jj(\gamma)\})$, a contradiction. Thus γ must be regular.

In a similar way we can add functions to $\mathcal{J}$ to make γ a strong limit cardinal: for each β in κ, let $h_1(\{\beta\})$ be the least α such that $\overline{\overline{2^\alpha}} \geq \beta$ if such an α exists, and 0 otherwise. Let, for $\beta < \gamma$, ℓ_β map $2^{h_1(\{\beta\})}$ into β, ℓ_β being onto if $\overline{\overline{2^{h_1(\{\beta\})}}} \geq \beta$, and define $h_2 : [\kappa]^2 \to \kappa$ by

$$h_2(\{\beta_1,\beta_2\}_<) =_{df} \begin{cases} \text{the least } \alpha < \beta_1 \text{ such that for some} \\ x \subseteq h_1(\{\beta_1\}),\ \ell_{\beta_1}(x) = \alpha \neq \ell_{\beta_2}(x) \\ \text{if } h_1(\{\beta_1\}) = h_1(\{\beta_2\}) > 0 \\ \\ 0 \text{ otherwise.} \end{cases}$$

Let us assume fixed a well-ordering of the bounded subsets of κ and, for $\beta_1 < \beta_2 < \kappa$ satisfying $h_1(\{\beta_1\}) = h_1(\{\beta_2\}) > 0$, let $s(\{\beta_1,\beta_2\})$ be the least $x \subseteq h_1(\{\beta_1\})$ such that $\ell_{\beta_1}(x) =$ $= h_2(\{\beta_1,\beta_2\})$ and $\ell_{\beta_2}(x) \neq h_2(\{\beta_1,\beta_2\})$. Now if for $\beta_1 < \beta_2 < \beta_3 < \kappa$ $h_1(\{\beta_1\}) = h_1(\{\beta_2\}) = h_1(\{\beta_3\})$ and if $h_2(\{\beta_1,\beta_2\}) = h_2(\{\beta_2,\beta_3\})$, then $s(\{\beta_1,\beta_2\}) \neq s(\{\beta_2,\beta_3\})$. So let h_3 and h_4 mapping $[\kappa]^3$ into κ be given by

$$h_3(\{\beta_1,\beta_2,\beta_3\}_<) =_{df} \begin{cases} \text{the least } \alpha \text{ in } s(\{\beta_1,\beta_2\}) - s(\{\beta_2,\beta_3\}) \text{ if} \\ s(\{\beta_1,\beta_2\}) - s(\{\beta_2,\beta_3\}) \neq \emptyset \\ \\ h_1(\{\beta_1\}) \text{ otherwise} \end{cases}$$

$$h_4(\{\beta_1,\beta_2,\beta_3\}_<) =_{df} \begin{cases} \text{the least } \alpha \text{ in } s(\{\beta_2,\beta_3\}) - s(\{\beta_1,\beta_2\}) \\ \text{if } s(\{\beta_2,\beta_3\}) - s(\{\beta_1,\beta_2\}) \neq \emptyset \\ \\ h_1(\{\beta_1\}) \text{ otherwise.} \end{cases}$$

Claim: If h_1,h_2,h_3, and h_4 are in $\mathcal{J}$, then γ is a strong limit cardinal. (Proof: Suppose not. Then $h_1(\{\gamma\}) < \gamma$ and so $h_1(\{\gamma\}) = h_1(\{j(\gamma)\}) = h_1(\{jj(\gamma)\}) = h_1(\{jjj(\gamma)\}) < \gamma$. Since $h_2(\{\gamma,j(\gamma)\})$, $h_3(\{\gamma,j(\gamma),jj(\gamma)\})$ and $h_4(\{\gamma,j(\gamma),jj(\gamma)\})$ are all less than γ we get $h_2(\{\gamma,j(\gamma)\}) = h_2(\{j(\gamma),jj(\gamma)\})$ $= h_2(\{jj(\gamma), jjj(\gamma)\})$, $h_3(\{\gamma,j(\gamma), jj(\gamma)\}) = h_3(\{j(\gamma),jj(\gamma),jjj(\gamma)\})$, and $h_4(\{\gamma,j(\gamma), jj(\gamma)\}) = h_4(\{j(\gamma),jj(\gamma),jjj(\gamma)\})$. But this is impossible since either h_3 or h_4, say h_3, satisfies "$h_3(\{\gamma,j(\gamma),jj(\gamma)\}) \notin s(\{j(\gamma),jj(\gamma)\})$ and $h_3(\{j(\gamma),jj(\gamma),jjj(\gamma)\}) \in s(\{j(\gamma),jj(\gamma)\})$. This contradiction yields the claim. ⊠)

Finally we wish to exhibit functions which, when members of $\mathcal{J}$, guarantee γ to satisfy $\gamma \to (\gamma)^2$. Actually we shall exhibit functions to guarantee γ having the tree property - by our previous work here and 4.2.2, this will suffice to establish our theorem. So for each α less than κ, let us choose a tree $T_\alpha = \langle \alpha, <_\alpha \rangle$ of cardinality $\overline{\overline{\alpha}}$ all of whose levels have cardinality less than $\overline{\overline{\alpha}}$ and yet which, if α fails to have the tree property, has no branch of cardinality $\overline{\overline{\alpha}}$ (For any such α, if $\beta < \alpha$ let $T_\alpha \upharpoonright \beta$ denote T_α restricted to points of order less than β. We may (and will) assume without loss of generality that for each α less than κ and each regular cardinal β less than α, the domain of $T_\alpha \upharpoonright \beta$ consists only of ordinals less than β. To see this simply note that in starting with an arbitrary T_α and replacing it with a desired tree conforming to this additional requirement, it may be necessary to carefully add new levels as well as to renumber nodes.) Now suppose $\beta_1 < \beta_2$ are regular cardinals less than κ such that β_1 fails to have the tree property. Then $T_{\beta_1} \neq T_{\beta_2} \upharpoonright \beta_1$ for since β_2 is regular, $T_{\beta_2} \upharpoonright \beta_1$ has a branch of cardinality β_1. With this in mind let us define functions k_1 and k_2 from $[\kappa]^2$ into κ by

$$k_1(\{\beta_1,\beta_2\}_<) =_{df} \begin{cases} \text{The least } \alpha < \beta_1 \text{ such that there is an } \eta < \beta_1 \text{ satisfying } \eta <_{\beta_1} \alpha \text{ and } \eta \not<_{\beta_2} \alpha \text{ if there is such an } \alpha \\ 0 \text{ otherwise} \end{cases}$$

$$k_2(\{\beta_1,\beta_2\}_<) =_{df} \begin{cases} \text{the least } \eta < \beta_1 \text{ such that } \eta <_{\beta_1} k_1(\{\beta_1,\beta_2\}) \text{ and } \eta \not<_{\beta_2} k_1(\{\beta_1,\beta_2\}) \text{ if there is such an } \eta \\ 0 \text{ otherwise.} \end{cases}$$

Claim: If $\mathcal{J}$ contains the functions $f_0, f_1, \ldots, f_n, \ldots f_\omega$, g_1, g_2, g_3, h_1, h_2, h_3, h_4, k_1 and k_2, then $\gamma > \omega$ and $\gamma \to (\gamma)^2$. (Proof: Considering our earlier work, we need only verify that the strongly inaccessible γ has the tree property. Now since γ is regular, $g_1(\{\gamma\}) = \gamma$ and so by the homogeneity of C it is easy to verify that $g_1(\{j(\gamma)\}) = j(\gamma)$. Thus $\gamma < j(\gamma)$ are both regular. Now since $k_1(\{\gamma, j(\gamma)\})$ and $k_2(\{\gamma, j(\gamma)\})$ are both less than γ, $k_1(\{\gamma, j(\gamma)\}) = k_1(\{j(\gamma), jj(\gamma)\})$ and $k_2(\{\gamma, j(\gamma)\}) = k_2(\{j(\gamma), jj(\gamma)\})$. But as γ fails to have the tree property,

$k_2(\{\gamma, j(\gamma)\}) \nless_{j(\gamma)} k_1(\{\gamma, j(\gamma)\})$ yet $k_2(\{j(\gamma), jj(\gamma)\}) <_{j(\gamma)} k_1(\{j(\gamma), jj(\gamma)\})$. This contradiction yields the claim. ☒)

This last claim completes the proof of the theorem. For let $\mathcal{J}$ be the closure under function composition of the set consisting of all coordinate functions from $\bigcup_{n<\omega} [\kappa]^n$ into κ in addition to the functions mentioned in this last claim. Then the associated $\gamma < \kappa$ satisfies $\gamma > \omega$ and $\gamma \to (\gamma)^2$. □

Remark: We have not considered the axiom "there exists a strongly inaccessible cardinal." As it turns out, this axiom is weaker than even "there exists a $\kappa > \omega$ such that $\kappa \to (\kappa)^2$." Although a proof of this is much easier than that given for 7.4, it is by no means trivial (see [18]).

Remark: There do exist "positive" results concerning the relative strengths of seemingly different large cardinal assumptions. See [29] for the equiconsistency of two "measure theoretic" large cardinal axioms, and see [13] for the equiconsistency of two "partition property" large cardinal axioms.

§ 8 We now consider infinite exponent partition relations, in particular, those of the form $\kappa \to (\kappa)^\alpha_\gamma$ where $\alpha \geqslant \omega$. As mentioned earlier, the existence of a cardinal κ satisfying any such relation contradicts the axiom of choice, and so it might be best to make several remarks here as to why one might be interested in studying them.

First of all, suppose that some interesting consequence, Φ, happens to follow if one assumes the existence of a cardinal satisfying some infinite exponent partition relation. Then with the intention of finding an outright proof of Φ, one can try two basic and potentially feasible approaches.

The first is to carefully analyze the use of the partition relation in the proof of Φ with the hope of finding a weakened (and provable) relation which also establishes Φ. (It often happens that some simple and elegant assumption suggests an approach to proving a theorem, and that at some later time, an appropriate modification is found whereby that basic approach still serves to establish the result but without the need for the special assumption.) As alluded to earlier, this general sort of thing happened when Paris eliminated, Martin's use of a large cardinal axiom in proving $\underset{\sim}{\Sigma}^0_4$ determinateness.

The second potential approach to finding an outright proof of Φ would be to try to find (in ZF+ AC , say) a model satisfying the partition relation (and hence Φ itself). If one could then show that the particular Φ at hand is "absolute" he would have established Φ as a theorem of ZF+AC. For example, if one were able to find in ZF+AC a model satisfying a partition relation which implies Borel determinateness, the above approach would yield an outright proof in ZF+AC of Borel determinateness.

It is in section 8a that we consider aspects of the feasibility of these two approaches. In particular, we deal there with the existence of provable (in ZF+AC) instances of weakened infinite exponent partition relations, and with the existence of models for ZF satisfying pure infinite exponent partition relations.

There exist other equally good reasons for pursuing a study of infinite exponent partition relations but these are less amenable to brief discussion. Hence, they will not be examined here.

However, as a final remark, let us mention that there exists g great beauty inherent in a study of such relations. This is especially true when dealing with these relations in their pure form as is done in section 8b .

To begin, let us recall what $\kappa \to (\kappa)^{\alpha}_{\gamma}$ means when $\alpha \geqslant \omega$. In particular, recall that under consideration are partitions of α-sequences of ordinals (as opposed to partitions of cardinality $\bar{\bar{\alpha}}$ sets of ordinals). Indeed, if α is an ordinal and x is a set of ordinals, $[x]^{\alpha}$ denotes the collection of α-sequences from x , that is, the collection of subsets of x of order-type α . Then for $\kappa > \gamma$ cardinals and $\alpha \leqslant \kappa$ an ordinal, $\kappa \to (\kappa)^{\alpha}_{\gamma}$ denotes <u>for every function $F:[\kappa]^{\alpha} \to \gamma$ there exists a set $C \subseteq \kappa$ such that $\bar{\bar{C}} = \kappa$ and $\overline{\overline{F''[C]^{\alpha}}} = 1$.</u>

Let us first examine some negative results:

8.1 <u>Theorem</u> (Rado- see [6]):Assume the axiom of choice. Then for any $\kappa \geqslant \omega$, $\kappa \not\to (\kappa)^{\omega}$.

<u>Proof</u>: Let $\kappa \geqslant \omega$ be given. We shall produce a partition $F:[\kappa]^{\omega} \to 2$ which does not even have an infinite homogeneous set.

Let us define the binary relation $\sim$ on $[\kappa]^{\omega}$ by $x \sim y$ iff_{df} x and y are identical modulo initial segments, i.e., iff x less some initial segment is identical with y less some initial segment. $\sim$ is easily seen to be an equivalence relation and so let us choose one representative from each of the resulting equivalence classes. We now define $F:[\kappa]^{\omega} \to 2$ as follows: given $x \in [\kappa]^{\omega}$, let $\hat{x}$ be the representative of the equivalence class in which x falls and let n be the length of the smallest initial segment s of $\hat{x}$ such that $\hat{x}-s$ is a final segment of x . We now let $F(x)$ equal 0 iff n is even. Now suppose that C is an infinite subset of κ . Let $x \in [C]^{\omega}$ and let x_1 be a final segment of x which agrees with some final segement of $\hat{x}$. Then x_1 and $x_1-\{\cap x_1\}$ are both members of $[C]^\omega$ and yet $F(x_1) \neq F(x_1-\{\cap x_1\})$. Thus C is not homogeneous for F . As C was arbitrary, F has no infinite homogeneous set. □

Remark: For one interested in various weak forms of the axiom of choice there is the question of just how little choice is needed to refute the existence of infinite exponent partition relations. The proof just given above shows that $\kappa \not\to (\kappa)^\omega$ follows from the existence of a well-ordering of $[\kappa]^\omega$. For $\kappa > \omega$ we can do much better: let AC_κ , well-ordered choice of length κ , be the assertion that for every collection of nonempty sets indexed by κ there exists a choice function. Then we have the following result:

8.2 Theorem ([12]): Let κ be an uncountable cardinal. Then AC_κ implies $\kappa \not\to (\kappa)^\omega$.

Proof ([14]): For each α less than κ of cofinality ω let x_α be an ω-sequence whose sup is α (1). Let $F:[\kappa]^\omega \to 2$ be defined as follows: given $x \in [\kappa]^\omega$ let $\alpha = \cup x$. Then we compare x and x_α . If for every n the nth member of x exceeds the nth member of x_α we let $F(x) = 0$. Otherwise $F(x) = 1$. Now suppose that C is homogeneous for F with respect to $\kappa \to (\kappa)^\omega$. (We will find a contradiction.) Then clearly $F''[C]^\omega = \{0\}$, for any ω-sequence from κ has a subsequence of type ω which F sends to 0 . Say x and y are members of $[C]^\omega$ such that $\cup x < \cap y$. For each $n < \omega$ let z_n be the ω-sequence from C consisting of the first n members of x adjoined to y . Then as we must have $F(z_n) = 0$

(1) This is our use of AC_κ .

for each n, it easily follows from the definition of F that $\cup x \geq \cup y$, a contradiction. □

The assumption that κ be uncountable in 8.2 is necessary, for as we shall remark later, AC_ω and $\omega \to (\omega)^\omega$ are consistent with one another.

The remainder of this section is organized into two main subsections. The first concerns itself with a weakened version of the infinite exponent partition relation $\omega \to (\omega)^\omega$ which can be proven outright in ZF and which has several useful applications. The second subsection deals exclusively with various consequences of the existence of full infinite exponent partition relations. In particular we shall show that uncountable cardinals satisfying infinite exponent partition relations must be measurable.

§§8a A weakened version of $\omega \to (\omega)^\omega$.

We weaken $\omega \to (\omega)^\omega$ through a topological approach. Specifically, let us consider the standard topology of $[\omega]^\omega$, namely the topology whose basic open sets are determined by finite sequences of integers: the basic open set determined by the finite sequence s consists of those members of $[\omega]^\omega$ whose initial segment is s. As usual, the Borel sets are those which are members of the smallest σ-field containing the open sets. A partition $F:[\omega]^\omega \to 2$ is said to be open if $F^{-1}\{0\}$ is open.

Now the main result we concern ourselves with here is that the assertion of $\omega \to (\omega)^\omega$ is true for open partitions, i.e.

8.3 Theorem (N. Williams): Assume $F:[\omega]^\omega \to 2$ is open. Then there exists an infinite subset C of ω such that $F''[C]^\omega = 1$.

Before giving a proof of this result let us compare it with Ramsey's Theorem. In particular, for $n < \omega$ let T_n be the topology on $[\omega]^\omega$ whose basic open sets are determined by finite sequences of nonnegative integers of length less than n. Let us say that the members of T_n are $(\text{open})_n$ sets. Then theorem 8.3 with "open" replaced by "$(\text{open})_n$" for any n is a simple consequence of Ramsey's theorem[(1)]. For if $F:[\omega]^\omega \to 2$ is $(\text{open})_n$, let $G: \bigcup_{i<n} [\omega]^i \to 2$ by $G(\{m_1,\ldots,m_i\}_<) = 0$ iff any member of $[\omega]^\omega$ with initial segment $\langle m_1,\ldots,m_i\rangle$ is sent by F to 0. Let C be homogeneous for G with respect to $\omega \to (\omega)^{<n}$. Then $\overline{\overline{F''[C]^\omega}} = 1$ for if x and y are in $[C]^\omega$ and $F(x) = 0$, some initial segment of x of length $i < n$, say x_i, is such that any member of $[\omega]^\omega$ with x_i as an initial segment is sent to 0 by F. Let y_i be the initial segment of length i of y. Then since $0 = G(x_i) = G(y_i)$, the definition of G tells us that $F(y) = 0$.

Now just as this discussion has shown theorem 8.3 for $(\text{open})_n$ partitions to correspond naturally to $\omega \to (\omega)^{<n}$, the full theorem would seem to correspond naturally to $\omega \to (\omega)^{<\omega}$. But as we saw on page 33, $\omega \to (\omega)^{<\omega}$ is false. This makes 8.3 all the more surprising.

Proof of Theorem 8.3: We must begin with some definitions: let $\mathcal{P}$ denote the set of all pairs $\langle s,S\rangle$ where $s(S)$ is a finite (infinite) subset of ω and $\bigcup s < \bigcap S$, and let us consider $\mathcal{P}$ to be partially ordered by $\leqslant$ given by "$\langle s,S\rangle \leqslant \langle t,T\rangle$ iff$_{df}$ $s \subseteq t$, $T \subseteq S$, and $t-s \subseteq S$." A subset x of ω is said to satisfy a member $\langle s,S\rangle$ of $\mathcal{P}$ iff$_{df}$ $s \subseteq x \subseteq s \cup S$. For $\langle s,S\rangle \in \mathcal{P}$, $2^{\langle s,S\rangle}$ denotes the collection of those subsets of ω satisfying $\langle s,S\rangle$. Now suppose $P \subseteq [\omega]^\omega$. Then $\langle s,S\rangle$ is homogeneous for P iff$_{df}$ $2^{\langle s,S\rangle} \subseteq P$ or $2^{\langle s,S\rangle} \subseteq P^C$

(1) $\omega \to (\omega)^{<n}$ for $n<\omega$ denotes "for every $F: \bigcup_{i<n} [\omega]^i \to 2$ there exists an infinite subset C of ω such that for every $i < n$, $\overline{\overline{F''[C]^i}} = 1$." Clearly Ramsey's theorem implies $\omega \to (\omega)^{<n}$ for any $n < \omega$. For given $F: \bigcup_{i<n} [\omega]^i \to 2$, let C_0 be ω and, inductively, let C_{i+1} be an infinite subset of C_i homogeneous for $F \restriction [C_i]^i$. Then C_n is clearly homogeneous for F.

(P^c denotes $[\omega]^\omega - P$, the complement of P) - $\langle s,S\rangle$ is <u>completely inhomogeneous</u> for P iff$_{df}$ for every $\langle t,T\rangle$ such that $\langle s,S\rangle \lesssim \langle t,T\rangle$, $\langle t,T\rangle$ is not homogeneous for P.

Now suppose that $F:[\omega]^\omega \to 2$ is open. Let $P = F^{-1}\{0\}$. Then clearly no $\langle s,S\rangle$ in $\mathcal{P}$ is completely inhomogeneous for P. For say $\langle s,S\rangle$ is completely inhomogeneous for P. Then as it is not homogeneous there exists an x in $2^{\langle s,S\rangle}$ such that $x \in P$. As P is open, x is in P on the basis of an initial segment of x, and so, for some initial segment $t \supseteq s$ of x, $2^{\langle t,\omega-t\rangle} \subseteq P$. Thus $\langle t,S-t\rangle \gtrsim \langle s,S\rangle$ and $2^{\langle t,S-t\rangle} \subseteq P$ contradicting the complete inhomogeneity of $\langle s,S\rangle$.

Thus our theorem follows from the following

8.3.1 <u>Lemma</u>: Suppose $P \subseteq [\omega]^\omega$ and $\langle s,S\rangle \in \mathcal{P}$. Then there exists an infinite subset S' of S such that $\langle s,S'\rangle$ is either homogeneous or completely inhomogeneous for P.

<u>Proof of Lemma</u>: Let $s_1, s_2, \ldots$ be an enumeration of all finite subsets of ω such that if $n < m$, $s_n \subseteq \cup s_m$. (The usual lexicographic ordering is such an enumeration). We inductively define for $n < \omega$ subsets S_n of ω as follows: $S_0 = S$; suppose S_n has been defined; if s_n is not a subset of S_n, let $S_{n+1} = S_n$. Suppose s_n is a subset of S_n. We consider three cases: (i) if for some $T \subseteq S_n - \cup s_n$ $\langle s \cup s_n, T\rangle$ is homogeneous for P, pick such a T and let $S_{n+1} = ((\cup s_n + 1) \cap S_n) \cup T$; (ii) if there is no T as in (i) but there is a $T \subseteq S_n - \cup s_n$ such that $\langle s \cup s_n, T\rangle$ is completely inhomogeneous for P, pick such a T and let $S_{n+1} = ((\cup s_n + 1) \cap S_n) \cup T$; (iii) if there is no T as in cases (i) or (ii), let $S_{n+1} = S_n$.

Now let $S' = \bigcap_{n<\omega} S_n$. Then it is easy to see that S' is infinite for whenever we thinned out S_n in going to S_{n+1}, s_n is guaranteed to be a subset of S'.

Claim: $<s,S'>$ is either homogeneous or completely inhomogeneous for P . (Proof: Let us first observe the following: suppose $s_m \subseteq S'$ is such that when s_m came up in our inductive construction of the S_n , case (iii) was encountered; then for some $i_o < \omega$, for any j in S' larger than i_o , $s_m \cup \{j\}$ encountered case (iii) when it came up for consideration. This is true since , for example, if case (ii) came up for $s_m \cup \{j\}$ for infinitely many j in S', let T be the set of such j which are larger than $\cup s_m$. Then $<s \cup s_m, T>$ is completely inhomogeneous for P and as $T \subseteq S_{m+1}$, we contradict our assumption that s_m encountered case (iii) . This proves the observation. Let us now consider what happens when $s_m = \emptyset$ comes up in the construction of the S_n . If cases (i) or (ii) are encountered then clearly $<s,S'>$ is either homogeneous or completely inhomogeneous (respectively) for P . Suppose case (iii) comes up. Then by our observation, case (iii) is encountered when $\{j\}$ is considered for all j in S' beyond some point. Again by our observation, for any such j , $\{j,k\}$ encounters case (iii) for every k in S' beyond some point. Continuing in this way we see that we can thin out S' to get an infinite subset S" of S' such that any finite subset of S" encountered case (iii) when it came up for consideration in the inductive construction of the S_n . Now since we assumed $\emptyset$ encountered case (iii) , $<s,S''>$ is not completely inhomogeneous for P and hence there exists $<t,T> \geq <s,S''>$ such that $<t,T>$ is homogeneous for P . Thus t-s encountered case (i) when it was considered. But as $t-s \subseteq S''$ we have a contradiction. Thus, the claim is proved. ☒) This completes the proof of the lemma. ☒

Thus the theorem is established. □

<u>Remark:</u> It is quite easy to extend this theorem to show that Borel partitions of $[\omega]^{\omega}$ have homogeneous sets. ($F: [\omega]^{\omega} \to 2$ is Borel

if $F^{-1}\{0\}$ is Borel). Let us do this now:

8.4 Theorem (Prikry-Galvin [8]). Assume $F: [\omega]^{\omega} \to 2$ is Borel. Then there exists an infinite subset C of ω such that $\overline{F''[C]^{\omega}} = 1$.

Proof: Let $P \subseteq [\omega]^{\omega}$. Then P is said to be completely Ramsey if for any $\langle s,S\rangle$ in $\mathcal{P}$ there exists an infinite subset S' of S such that $\langle s,S'\rangle$ is homogeneous for P. Theorem 8.3 (its proof, actually) showed that every open $P \subseteq [\omega]^{\omega}$ is completely Ramsey and clearly the collection of completely Ramsey subsets of $[\omega]^{\omega}$ is closed under complementation. Thus we will have established 8.4 if we can show that the collection of completely Ramsey sets is closed under countable intersection. So suppose $\{P_i\}_{i<\omega}$ is a collection of completely Ramsey sets, and suppose $\bigcap_{i<\omega} P_i$ is not completely Ramsey. We will find a contradiction: let $\langle s,S\rangle$ be such that for no infinite $S' \subseteq S$ is $\langle s,S'\rangle$ homogeneous for $\bigcap_{i<\omega} P_i$. By lemma 8.3.1, let S' be an infinite subset of S such that $\langle s,S'\rangle$ is completely inhomogeneous for $\bigcap_{i<\omega} P_i$. Thus it is immediate that if for some i $\langle t,T\rangle \geqslant \langle s,S'\rangle$ is homogeneous for P_i, $2^{\langle t,T\rangle} \subseteq P_i$ (for if we had $2^{\langle t,T\rangle} \subseteq P_i{}^{C} \subseteq (\bigcap_{i<\omega} P_i)^{C}$, $\langle t,T\rangle$ would be homogeneous for $\bigcap_{i<\omega} P_i$). We now use this fact iteratively as follows: let, as P_1 is completely Ramsey, $S_1 \subseteq S'$ be such that $2^{\langle s,S_1\rangle} \subseteq P_1$; let a_1 be the least member of S_1 and let $S_2 \subseteq S_1 - \{a_1\}$ be such that $2^{\langle t,S_2\rangle} \subseteq P_2$ for every t such that $s \subseteq t \subseteq s \cup \{a_1\}$ (we can find S_2 as follows: let $S_2^0 \subseteq S_1 - \{a_1\}$ be such that $2^{\langle s,S_2^0\rangle} \subseteq P_2$ and let $S_2^1 \subseteq S_2^0$ be such that $2^{\langle s\cup\{a_1\},S_2^1\rangle} \subseteq P_2$. We then let $S_2 = S_2^1$.) Let a_2 be the least member of S_2 and let $S_3 \subseteq S_2$ be such that $2^{\langle t,S_3\rangle} \subseteq P_3$ for every t such that $s \subseteq t \subseteq s \cup \{a_1,a_2\}$. Continuing in this way we generate an infinite set $T = \{a_1,a_2,a_3,a_4,a_5,\ldots\} \subseteq S'$. But it is easy to see that $2^{\langle s,T\rangle} \subseteq P_i$ for every i, i.e., that

$2^{<s,T>} \subseteq \bigcap_{i<\omega} P_i$, i.e., that $<s,T>$ is homogeneous for $\bigcap_{i<\omega} P_i$. But $<s,T> \geq <s, S'>$ and so we have contradicted the complete inhomogeneity of $<s,S'>$. □

Even 8.4 can be extended. As it turns out, analytic (Σ_1^1) partitions have homogeneous sets (in ZF+AC this is best possible) and, under the assumption of the existence of a measurable cardinal, Σ_2^1 partitions have homogeneous sets. These results are due to Silver ([25]).

Remark: There exist various applications of these weakened versions of $\omega \to (\omega)^\omega$. As an example, in [26] theorem 8.4 is used to construct a set of integers with no subset of strictly higher Turing degree. What is perhaps for us a more interesting example is Mathias' proof of the consistency of $\omega \to (\omega)^\omega$. In particular, let us consider forcing(1) (relative to a given standard model M of ZF+AC) with the partial ordering $\mathcal{P}$ used in the proof of 8.3 . A set x of nonnegative integers is said to be Mathias-generic over M if x is the unique set satisfying each condition in some M-generic filter on $\mathcal{P}$. Then using the idea in the proof of 8.4, it is fairly easy to show that every infinite subset of a Mathias generic set is Mathias generic. Using this fact it follows that $\omega \to (\omega)^\omega$ is true in Solovay's model of [28] in which every set of reals in Lebesgue measurable. (Note that AC_ω is also true in this model.)

Remark: It is possible to weaken $\kappa \to (\kappa)^\omega$ for any κ similarly to the way we weakened it for $\kappa = \omega$. In this way it is possible to exhibit, for $\kappa > \omega$, a naturally weakened version of $\kappa \to (\kappa)^\omega$ which is equivalent to $\kappa \to (\kappa)^{<\omega}$ (See [15]).

(1) For background on forcing one might refer to [28].

§§8b Partition Relations which yield Measurable Cardinals

In this subsection we show that uncountable cardinals satisfying various infinite exponent partition relations are measurable. What is initially of great interest here is due to the fact that virtually all of the well-known consequences of the existence of measurable cardinals have since been shown to follow from much weaker partition relations. What is more, the use of these weaker partition relations has produced even more refined consequences. Thus having a partition relation so strong as to imply the existence of measurable cardinals should be a very rich and elegant tool indeed. And besides the beautiful internal theory of such relations, they may be applied to more conventional mathematics in any of a number of ways as mentioned earlier.

Now keeping in mind 8.1 and 8.2 we shall work in ZF (without any form of the axiom of choice). Note that although the concept of "strongly inaccessible cardinal" does not really make sense in ZF without AC, all of the other set theoretic notions we have discussed so far do. However, various of the theorems we proved earlier needed AC. In particular, the theorem that every measurable cardinal has a normal measure does not necessarily remain valid in ZF alone. Thus the fact that we shall show that the cardinals in question, besides just being measurable, possess normal measures, deserves special mention. The result we are after is the following:

8.5 <u>Theorem</u> ([12]): Assume that κ is an uncountable cardinal and that $\kappa \rightarrow (\kappa)^{\omega}_{\gamma}$ for every $\gamma < \kappa$. Then κ is measurable and there exists a normal measure on κ.

So that we shall see most clearly the use of the partition relation in establishing this theorem, we shall present its proof as a series of lemmas.

Let us proceed: let κ be a cardinal. A filter on κ is collection $\mathcal{J}$ of subsets of κ satisfying

(i) $\emptyset \notin \mathcal{J}$

(ii) if $A \in \mathcal{J}$ and $A \subseteq B$ then $B \in \mathcal{J}$

(iii) if $A \in \mathcal{J}$ and $B \in \mathcal{J}$ then $A \cap B \in \mathcal{J}$.

Let $\mathcal{J}$ be a filter on κ. Then $\mathcal{J}$ is said to be nonprincipal if for no subset A of κ do we have $\mathcal{J} = \{B \subseteq \kappa | B \supseteq A\}$. $\mathcal{J}$ is said to be κ-complete if the intersection of any fewer than κ members of $\mathcal{J}$ is a member of $\mathcal{J}$. Finally, $\mathcal{J}$ is said to be an ultrafilter if for any $A \subseteq \kappa$, $A \in \mathcal{J}$ or $\kappa - A \in \mathcal{J}$.

Now it is well known (and trivial to verify) that for any cardinal κ there is a 1-1 correspondence between nontrivial two-valued κ-additive measures on κ and nonprinciple κ-complete ultrafilters on κ. Indeed, if μ is a nontrivial two-valued κ-additive measure on κ then the subsets of κ of μ-measure 1 form a nonprinciple κ-complete ultrafilter on κ and if $\mathcal{J}$ is a nonprinciple κ-complete ultrafilter on κ, then the map $\mu: 2^{\kappa} \to 2$ given by $\mu(A) = 1$ iff $A \in \mathcal{J}$ is a nontrivial two-valued κ additive measure on κ.

Now suppose that κ is any uncountable regular cardinal. Note that by 4.1, any uncountable cardinal κ satisfying $\kappa \to (\kappa)^{\alpha}$ where $\alpha \geq 2$ must be regular. (Our proof of this part of 4.1 did not use AC). A subset Q of κ is said to be ω-closed if the sup of any ω-sequence from Q is a member of Q, that is, if $x \in [Q]^{\omega}$ implies $\cup x \in Q$. As before, Q is unbounded (in κ) if the sup of Q is κ, i e., if $\cup Q = \kappa$. Now let E_{ω} be the collection of those subsets of κ which contain (as subset) an ω-closed unbounded subset of κ, i.e., let $E_{\omega} = \{A \subseteq \kappa |$ for some ω-closed unbounded subset Q of κ, $Q \subseteq A\}$. (If κ satisfies a sufficiently strong partition lation, E_{ω} will turn out to be an ultrafilter corresponding to a nontrivial two valued κ-additive normal measure on κ.) It is easy to verify using no assumption on κ (other than its regularity) that E_ω is a nonprinciple filter on κ : E_ω clearly satisfies clauses (i) and (ii) of the definition of filter and is clearly nonprinciple. To verify clause (iii), suppose that A_1 and A_2 are two members of E_ω and let Q_1 and Q_2 be ω-closed unbounded subsets of A_1 and A_2, respectively. We will be done if we could show that $Q_1 \cap Q_2$ were itself ω-closed and unbounded. Well, it is clearly ω-closed. To see that it is unbounded, suppose that $\nu < \kappa$. Let α_1 be the least member of Q_1 larger than ν[1], let α_2 be the least member of Q_2 larger than α_1, let α_3 be the least member of Q_1 larger than α_2, let α_4 be the least member of Q_2 larger than α_3, let α_5 be the least member of Q_1, larger than α_4, and so forth. Then $\{\alpha_1, \alpha_3, \alpha_5, \ldots\}$ and $\{\alpha_2, \alpha_4, \alpha_6 \ldots\}$ are ω-sequences from Q_1 and Q_2 respectively which have the same sup, say α. Since Q_1 and Q_2 are both ω-closed, $\alpha \in Q_1 \cap Q_2$. Since $\alpha > \nu$ and since ν was chosen arbitrarily, $Q_1 \cap Q_2$ must be unbounded.

We now investigate the additional special properties of E_ω which résult when we assume κ to satisfy various infinite exponent partition relations. A basic fact which we shall use repeatedly is the following: if x is any subset of κ, let $(x)_\omega$ denote the collection of sups of ω-sequences from x, i.e., $(x)_\omega =_{df} \{\cup p \mid p \in [x]^\omega\}$.

8.5.1 <u>Fact</u>: If x is any unbounded subset of κ then $(x)_\omega$ is an ω-closed unbounded subset of κ. (Proof of 8.5.1: Suppose $p \in [(x)_\omega]^\omega$. Let us view p as a function from ω into κ. For any n less than ω, $p(n) < p(n+1)$, and so, as $p(n+1)$ is the sup of an ω-sequence from x, there must be a β in x such that $p(n) < \beta < p(n+1)$. Let $q:\omega \to x$ be given by $q(i) =_{df}$ "the least

(1) Since Q_1 and Q_2 are sets of ordinals, they are well-ordered.

β in x such that $p(i) < \beta < p(i+1)$". (Notice that we did not use AC in defining q). Then $q \in [x]^\omega$ and since, for each n, $p(n) < q(n) < p(n+1)$, we must have $\cup p = \cup q$. As $q \in [x]^\omega$, $\cup p = \cup q \in (x)_\omega$ and so, as p was arbitrary, $(x)_\omega$ must be ω-closed. It is simple to see that $(x)_\omega$ is unbounded in κ. For if $\nu < \kappa$, $\overline{\overline{x-\nu}} = \kappa$ and so for any p in $[x-\nu]^\omega \subseteq [x]^\omega$, $\cup p > \nu$. ⊠)

8.5.2 <u>Lemma</u>: If $\kappa \to (\kappa)^\omega$ then E_ω is an ultrafilter.

<u>Proof of 8.5.2</u>: Suppose $A \subseteq \kappa$. We must show that either A or $\kappa - A$ contains an ω-closed unbounded subset of κ. Let $F:[\kappa]^\omega \to 2$ be given by "$F(p) = 0$ iff $\cup p \in A$", and let C be homogeneous for F with respect to $\kappa \to (\kappa)^\omega$. Now clearly if $F''[C]^\omega = \{0\}$, $(C)_\omega \subseteq A$, and if $F''[C]^\omega = \{1\}$, $(C)_\omega \subseteq \kappa - A$. Thus as $(C)_\omega$ is ω-closed and unbounded we are done. ⊠

8.5.3 <u>Lemma</u>: If $\kappa \to (\kappa)^\omega_\alpha$ then the intersection of any α members of E_ω is a member of E_ω.

<u>Proof of 8.5.3</u>: Suppose that sets A_β for $0 < \beta < \alpha$ are members of E_ω. We must show that $\bigcap_{0<\beta<\alpha} A_\beta$ contains an ω-closed unbounded set. Let $F: [\kappa]^\omega \to \alpha$ be given by

$$F(p) = \begin{cases} \text{the least } \beta \text{ such that } \cup p \notin A_\beta \text{ if} \\ \qquad\qquad \cup p \notin \bigcap_{0<\beta<\alpha} A_\beta \\ 0 \quad \text{otherwise}, \end{cases}$$

and let C be homogeneous for F with respect to $\kappa \to (\kappa)^\omega_\alpha$.
Claim: $F''[C]^\omega = \{0\}$. (Proof: Suppose not, i.e., suppose $0 < \beta < \alpha$ and $F''[C]^\omega = \{\beta\}$. Then $(C)_\omega \cap A_\beta = \emptyset$, and since $(C)_\omega$ is a member of E_ω, we contradict our earlier result that E_ω is a filter on κ. ⊠)

Now since $F''[C]^\omega = \{0\}$, $p \in [C]^\omega$ implies $\cup p \in \bigcap_{0<\beta<\alpha} A_\beta$, i.e., $(C)_\omega \subseteq \bigcap_{0<\beta<\alpha} A_\beta$. Since $(C)_\omega$ is ω closed and unbounded, we are done. ☒

8.5.4. Lemma: Assume that E_ω is a κ-complete ultrafilter on κ and and let μ_ω be the corresponding nontrivial two-valued κ-additive measure on κ . Then if $\kappa \to (\kappa)^\omega$, μ_ω is normal.

Proof of 8.5.4: Suppose $g:\kappa \to \kappa$ and $\mu_\omega(\{\alpha | g(\alpha)<\alpha\}) = 1$. Let A be an ω-closed unbounded subset of $\{\alpha | g(\alpha) < \alpha\}$, define $F:[A]^\omega \to 2$ by "$F(p) = 0$ iff $g(\cup p)$ is less than the least member of p" , and let C be homogeneous for F with respect to $\kappa \to (\kappa)^\omega$.
Claim: $F''[C]^\omega = \{0\}$ (Proof: Let $p \in [C]^\omega$. Since $C \subseteq A$ and since A is ω-closed , $g(\cup p) < \cup p$. Let q be the sequence consisting of those members of p larger than $g(\cup p)$. Then q is an ω-sequence and $g(\cup q) = g(\cup p)$. Thus $F(q) = 0$, and since $q \in [C]^\omega$, $F''[C]^\omega = \{0\}$. ☒) Now let α_0 be the least member of C . Then $g''(C)_\omega \subseteq \alpha_0$, for if $p \in [C]^\omega$, $\{\alpha_0\} \cup p \in [C]^\omega$, and since $F''[C]^\omega = \{0\}$, $F(\{\alpha_0\} \cup p) = 0$ and so $g(\cup p) < \alpha_0$. But $g''(C)_\omega \subseteq \alpha_0$ means that $(C)_\omega = \bigcup_{\beta<\alpha_0} g^{-1}\{\beta\}$ and since μ_ω is κ additive, $\alpha_0 < \kappa$, and $\mu_\omega((C)_\omega) = 1$, $\mu_\omega(g^{-1}\{\beta\}) = 1$ for some $\beta < \alpha_0$, i.e., for some $\beta < \alpha_0$, $\mu_\omega(\{\alpha | g(\alpha) = \beta\}) = 1$. Since g was arbitrary we have shown that μ_ω is normal. ☒

Proof of Theorem 8.5: Assume that $\kappa \to (\kappa)^\omega_\gamma$ for every $\gamma < \kappa$. Then by 4.1 , κ is regular. By lemmas 8.5.2, 8.5.3, and 8.5.4, E_ω defines a nontrivial, two-valued, κ-additive normal measure on κ . □

Remark: In 8.5 , the assumption "$\kappa \to (\kappa)^\omega_\gamma$ for every $\gamma < \kappa$" can be replaced with "$\kappa \to (\kappa)^{\omega+\omega}$" . For by essentially the same argument as that given in 7.1.1, $\kappa \to (\kappa)^{\omega+\omega}$ implies that $\kappa \to (\kappa)_\gamma^\omega$ for every $\gamma < \kappa$.

Remark: Suppose that λ is any limit ordinal less than κ and that $\kappa \to (\kappa)_\gamma^\lambda$ for every $\gamma < \kappa$. Then by considering the notion of "λ-closed unbounded set" in place of that of "ω-closed unbounded set" a development as given in proving 8.5 yields a nontrivial two-valued κ-additive normal measure μ_λ on κ : $\mu_\lambda(A) = 1$ iff A contains a λ-closed unbounded set. The general relation between the μ_λ is as follows:

Theorem ([12]): Suppose that λ and γ are limit ordinals less than κ . Then $\mu_\lambda = \mu_\gamma$ iff $cf(\lambda) = cf(\gamma)$.

This result is independent of any consideration of partition relations - it is simply a fact about the functions μ_λ and μ_γ . For a proof of this theorem see [12] .

Remark: The axiom of determinateness implies the existence of various infinite exponent partition relations. For example, Martin has shown that $\aleph_1 \to (\aleph_1)^{\aleph_1}$ is a consequence of AD . It is in fact the case that virtually every result of the form "AD implies that ...is a measurable cardinal" proceeds by first showing that AD implies ... to satisfy the appropriate infinite exponent partition relation. For more information on AD see Kunen's lecture in these proceedings and also [19].

I wish to thank Richard Shore, Fred Abramson, Barbara Jarvik, and Sy Friedman for their help in preparing the final draft of this manuscript.

REFERENCES

[1] F. P. Ramsey, On a problem of formal logic, Proceedings of the London Mathematical Society (2), Vol. 30 (1930), pp. 264-286.

[2] C. C. Chang and H. J. Keisler, Model Theory, to appear.

[3] P. J. Cohen, Set Theory and the Continuum Hypothesis, W. A. Benjamin, 1966.

[4] P. Erdős and A. Hajnal, On the structure of set mappings, Acta Mathematica Academiae Scientiarum Hungaricae, Vol. 9 (1958), pp. 111-131.

[5] P. Erdős, A. Hajnal, and R. Rado, Partition relations for cardinal numbers, Acta Mathematica Academiae Scientiarum Hungaricae, Vol. 16 (1965), pp. 93-196.

[6] P. Erdős and R. Rado, Combinatorial theorems on classification of subsets of a given set, Proceedings of the London Mathematical Society (3), Vol. 2 (1952), pp. 417-439.

[7] P. Erdős and A. Tarski, On families of mutually exclusive sets, Annals of Mathematics, Vol. 44 (1943), pp. 315-329.

[8] F. Galvin and K. Prikry, to appear.

[9] J. Henle and E. M. Kleinberg, A combinatorial proof of a combinatorial theorem, to appear.

[10] C. Jockusch, Ramsey's theorem and recursion theory, The Journal of Symbolic Logic, to appear.

[11] E. M. Kleinberg, The independence of Ramsey's theorem, The Journal of Symbolic Logic, Vol. 34 (1969), pp. 205-206.

[12] E. M. Kleinberg, Strong partition properties for infinite cardinals, The Journal of Symbolic Logic, Vol. 35 (1970), pp. 410-428.

[13] E. M. Kleinberg, The equiconsistency of two large cardinal axioms, to appear.

[14] E. M. Kleinberg and J. Seiferas, On infinite exponent partition relations and well-ordered choice, to appear.

[15] E. M. Kleinberg and R. A. Shore, On large cardinals and partition relations, The Journal of Symbolic Logic, Vol. 36 (1971), pp. 305-307.

[16] D. A. Martin, Measurable cardinals and analytic games, Fundamenta Mathematicae, Vol. 66 (1970), pp. 287-291.

[17] A. R. D. Mathias, Doctoral dissertation, Peterhouse, Cambridge, England.

[18] M. Morley, Partitions and models, Proc. Summer School in Logic, Leeds 1967, Springer, Berlin 1968, pp. 109-158.

[19] J. Mycielski, On the axiom of determinateness, Fund. Math., Vol. 53, (1963-64), pp.

[20] J. B. Paris, ZF $\vdash \Sigma_4^0$ determinateness, The Journal of Symbolic Logic, to appear.

[21] W. Reinhardt and J. Silver, On some problems of Erdös and Hajnal, Notices of the American Mathematical Society, Vol. 12 (1965), p. 723.

[22] H. Rogers, Jr., Theory of recursive functions and effective computability, McGraw Hill, 1967.

[23] F. Rowbottom, Some strong axioms of infinity incompatible with the axiom of constructibility, Annals of Mathematical Logic, Vol. 3 (1971), pp. 1-44.

[24] J. Silver, Some applications of model theory in set theory, Doctoral dissertation, University of California, Berkeley, California, 1966.

[25] J. Silver, Every analytic set is Ramsey, The Journal of Symbolic Logic, Vol. 35 (1970), pp. 60-64.

[26] R. Soare, Sets with no subset of higher degree, The Journal of Symbolic Logic, Vol. 34 (1969), pp. 53-56.

[27] R. M. Solovay, A nonconstructible Δ_3^1 set of integers, Transactions of the American Mathematical Society, Vol. 127 (1967), pp. 50-75.

[28] R. M. Solovay, A model for set theory in which every set of reals is Lebesgue measurable, Annals of Mathematics, Vol. 92 (1970), pp. 1-56.

[29] R. M. Solovay, Real-valued measurable cardinals, Proc. of Symposia in Pure Mathematics, Vol. XIII(1), American Mathematical Society 1971, pp. 397-428.

XIII. The Maximum Sum of a Family of Ordinals

John A. H. Anderson

§0 Preliminaries

General Notation. Small latin and greek letters denote ordinals unless otherwise stated. We assume ordinals have been defined so that each is the set of all smaller ordinals. The notation $f : \alpha \to \beta$ is thus appropriate when α and β are ordinals; we write f(i) or, more frequently, if to denote the image of i under f and $f[i]$ to denote the image set of the set i under f.

Cardinals and initial ordinals are identical, infinite cardinals being written in the form ω_p, but no cardinals operation are used. We denote the cardinal of the set A by $|A|$.

The order type of a set A of ordinals, ordered by inclusion, is denoted tp(A). The interval notation $[\alpha,\beta)$ is used for $\{i \mid \alpha \leq i < \beta\}$. We use $(x_i \mid i \in A)$ for an indexed family of ordinals or sets with index set A; $\{x_i\}$ $(i < \alpha)$ to denote a sequence of ordinals of type α; $\{x_i\}$ $(i \in A)$ to denote a sequence where A is a set of ordinals and the order of the sequence is that imposed by the natural order of A; the sum of the sequence is denoted $\sum(i < \alpha)x_i$ or $\sum(i \in A)x_i$.

We use k^+ to denote k + 1 and k^- to denote k - 1 if k has a predecessor and k if it has not; cf (α) is the smallest ordinal k for which there exists a strictly increasing function $f : k \to \alpha$ whose image set $f[k]$ is cofinal in α.

Special Notation. Given a sequence $\{x_i\}(i < \delta)$ and an ordinal β such that $|\beta| \leq |\delta|$ we use $S(\beta, \{x_i\}\ (i < \delta))$ to denote $\{\sum(i < \beta)x_{i\Gamma} \mid \Gamma : \beta \to \delta$ is an injection$\}$ and $s(\beta, \{x_i\}\ (i < \delta))$ to denote $\sup S(\beta, \{x_i\}(i < \delta))$. These symbols are abbreviated to $S(\beta)$ and $s(\beta)$ where the sequence referred to is apparent. We define $L(\{x_i\}(i < \delta))$ to be $\min\{\ell : s(\ell, \{x_i\}\ (i < \delta)) \in S(\ell, \{x_i\}\ (i < \delta))\}$. This may be abbreviated to L where the sequence is apparent. The Standard Form of an ordinal δ means the form $\sum(i < m)\omega_{ip}q_i + r$ where $m < \omega$, $0p > 1p > \ldots > m^{-}p$, $1 \leq |q_i| \leq \omega_{ip}$ for $i < m$ and $r < \omega$. This is rendered $\sum(i \leq m)\omega_{ip}q_i$ when δ is a limit number.

The general problem is to evaluate $s(\beta, \{x_i\}(i < \delta))$ in terms of finitely many parameters obtainable from β and $\{x_i\}(i < \delta)$ without considering different arrangements of the sequence, and find appropriate criteria equivalent to $s(\beta, \{x_i\}(i < \delta)) \in S(\beta, \{x_i\}(i < \delta))$. This is a generalisation of the problem posed by Dan Dushnik [1]. The problem is solved here when $\{x_i\}(i < \delta)$ can be arranged as a non-decreasing sequence whose type is a limit number.

§1 On the Smallest Ordinal β such that $s(\beta) \in S(\beta)$

Theorem 1. Let $\{x_i\}$ $(i < \delta)$ be a non-decreasing sequence of non-zero ordinals of type $\delta \geq 1$. Put $t = cf(\delta)$ and let $\Gamma : t \to \delta$ be an order preserving injection such that $\Gamma[t]$ is a cofinal subset of δ. Then $\sum(i < t)x_{i\Gamma} = s(t) \in S(t)$.

CASE 1. $t = 1$. Then $\delta^- < \delta$ and $\Gamma : 1 \to \delta$ must have $\Gamma(0) = \delta^-$ in order that $\Gamma[1]$ be cofinal in δ. The conclusion of the theorem follows.

CASE 2. $t \neq 1$. Then t is infinite. Let $\gamma : t \to \delta$ be an arbitrary injection. We define an order preserving injection $f : t \to t$ such that $i\gamma \leq if\Gamma$ for each $i < t$. Let us suppose that we have defined if for each $i < k$, where $k < t$, such that $\{if\}$ $(i < k)$ is strictly increasing and $i\gamma \leq if\Gamma$ for each $i < k$. Since $\Gamma[t]$ is cofinal in δ there exists $\xi < t$ such that $\Gamma(\xi) \geq \gamma(k)$. Now, $k < t$ and t is regular so that $\sup f[k] < t$. Put $\eta = \max\{\xi, \sup(f|k|) + 1\}$. Define kf to be η. Then $f : t \to t$ is defined by recursion to be order preserving and $i\gamma \leq if\Gamma$ for each $i < t$. Hence $\sum(i < t)x_{i\gamma} \leq \sum(i < t)x_{if\Gamma} \leq \sum(i < t)x_{i\Gamma}$, for $\{x_{if\Gamma}\}(i < t)$ is a subsequence of $\{x_{i\Gamma}\}(i < t)$. The conclusion of the theorem follows.

In the next theorem we evaluate $L(\{x_i\}(i < \delta))$ and $s(p)$ for $1 \leq p \leq L$ in terms of $\sup\{x_i \mid i < \delta\}$. We quote first a lemma from a previous paper [2].

Lemma 1. Let $\{x_i\}(i < t)$ be a non-decreasing sequence of non-zero ordinals of type a regular infinite ordinal t. Put $\lambda = \sup\{x_i \mid i < t\}$. Then $\sum(i < t)x_i$ is λ if $\lambda = t^\alpha$ for some $\alpha \geq 1$ and λ is not attained; otherwise it is λt.

Theorem 2. Let $\{x_i\}$ $(i < t)$ be a non-decreasing sequence of non-zero ordinals of type $\delta \geq 1$. Put $\lambda = \sup\{x_i \mid i < \delta\}$ and $t = cf(\delta)$.

CASE 1. λ is attained. Then $L = 1$ and $s(L) = \lambda$.

CASE 2. λ is not attained.

CASE 2A. $\lambda = t^\alpha$ for some $\alpha \geq 1$. Then $L = t$ and $s(p) = \lambda$ for $1 \leq p \leq L$.

CASE 2B. $\lambda \neq t^\alpha$ for any $\alpha \geq 1$. Then $L = \omega$, $s(L) = \lambda\omega$, and $s(p) = \sup\{x_i p \mid i < \delta\}$ for $1 \leq p < L$.

CASE 1. λ is attained. Then there exists $k < \delta$ such that $x_k = \lambda$ and we may define $\Gamma : 1 \to \delta$ by $\Gamma(0) = x_k$. Evidently $s(1) = \lambda$ $S(1)$.

CASE 2. $\lambda \notin \{x_i \mid i < \delta\}$. Then $t = cf(\lambda) \geq \omega$.

Hence $\lambda \geq t$ and there exists $\alpha \geq 1$ such that $t^\alpha \leq \lambda < t^{\alpha+1}$.

CASE 2A. $\lambda = t^\alpha$. Let $\Gamma : t \to \delta$ be an order preserving injection such that $\Gamma[t]$ is a cofinal subset of δ. By Theorem 1, $s(t) = \sum(i < t)x_{i\Gamma} \in S(t)$.

Since $\Gamma[t]$ is cofinal in δ, $\sup\{x_{i\Gamma} \mid i < t\} = \sup\{x_i \mid i < \delta\}$.

Hence Lemma 1 yields $s(t) = \lambda$.

Let p be such that $1 \leq p < t$. Let $\gamma : p \to \delta$ be an injection. Then $\sup \gamma[p] < \delta$ and since $\Gamma[t]$ is cofinal in δ there exists $\xi < t$ such that $\sup \gamma[p] < \xi\Gamma < \delta$. Then $\sum(i < p)x_{i\gamma} < \sum(\xi < i < t)x_{i\Gamma} = \lambda$. But $s(p) \geq s(1) = \lambda$. Hence $s(p) = \lambda \notin S(p)$ for $1 \leq p < t$. So $L = t$ and $s(L) = \lambda$.

CASE 2B. $t^{\alpha} < \lambda < t^{\alpha+1}$. Put $\lambda = t^{\alpha}b + c$ where $b < t$ and $c < t^{\alpha}$. We first prove that $s(\omega) = \lambda\omega \in S(\omega)$.

Subcase (i). $c \neq 0$. Then there exists $\xi < \delta$ such that $x_{\xi} \geq t^{\alpha}b$. Since $cf(\delta) \geq \omega$, $|[\xi,\delta)| \geq \omega$ and so there exists an injection $\Gamma : \omega \to [\xi,\delta)$.

Then $t^{\alpha}b\omega \leq \sum(i < \omega)x_{i\Gamma} \leq \lambda\omega = (t^{\alpha}b + c)\omega = t^{\alpha}b\omega$.

Thus $\sum(i < \omega)x_{i\Gamma} = \lambda\omega$. For any injection $\gamma : \omega \to \delta$ we have $\sum(i < \omega)x_{i\gamma} \leq \lambda\omega$ so $s(\omega) = \lambda\omega \in S(\omega)$.

Subcase (ii). $c = 0$. Since $t^{\alpha} < \lambda < t^{\alpha+1}$ we have $b \geq 2$. Were $b = b^{-}$ then would $cf(\lambda) = cf(t^{\alpha}b) = cf(b) \leq b < t$, contradiction. Hence $1 \leq b^{-} < b$. Thus there exists $\xi < \delta$ such that $x_{\xi} \geq t^{\alpha}b^{-}$. and an injection $\Gamma : \omega \to [\xi,\delta)$. Then $t^{\alpha}b^{-}\omega \leq \sum(i < \omega)x_{i\Gamma} \leq \lambda\omega = t^{\alpha}b\omega$. Now if $2 \leq b < \omega$ then $b^{-}\omega = \omega = b\omega$ while if $b \geq \omega$ then $b^{-} \geq \omega$ and $b\omega = (b^{-} + 1)\omega = b^{-}\omega$ by absorption. Thus $b^{-}\omega = b\omega$ whence $\sum(i < \omega)x_{i\Gamma} = \lambda\omega$ and so, as before, $s(\omega) = \lambda\omega \in S(\omega)$.

Next, let p be such that $1 \leq p < \omega$. We prove that $s(p) = \sup\{x_i p \mid i < \delta\} \notin S(p)$. Let $\gamma : p \to \delta$ be any injection. Since $p < cf(\delta)$ and $\lambda \notin \{x_i \mid i < \delta\}$, there exists ℓ with $\sup \gamma[p] < \ell < \delta$ and $x_{\ell} > \sup\{x_{i\gamma} \mid i < p\}$.

Let $\Gamma : p \to [\ell,\delta)$ be an injection; then $\sum(i < p)x_{i\gamma} < x_{\ell}p \leq \sum(i < p)x_{i\Gamma}$.

It follows that $s(p) \notin S(p)$. From the construction of ℓ given γ and Γ given ℓ it also follows that $S(p) = \sup\{x_i p \mid i < \delta\}$. Hence $L = \omega$ and the theorem is proved.

Note concerning $\sup\{x_i p \mid i < \delta\}$. It is not difficult to calculate from the form $\lambda = t^{\alpha}b + c$ of CASE 2B that $\sup\{x_i p \mid i < \delta\} = \lambda p$ if $c \neq 0$ or if $c = 0$ and b is infinite. Unfortunately, if $c = 0$ and b is finite then $\sup\{x_i p \mid i < \delta\} = t^{\alpha}(b^{-}p + 1)$ and there seems no obvious way of eliminating the representation of λ. We need not, however, mention the x_i.

§2. On $s(\beta)$ when β is a Multiple of L

Lemma 2. Let $\{x_i\}(i < \delta)$ be a non-decreasing sequence of non-zero ordinals of type $\delta \geq 1$. Let β be an ordinal such that $|\beta| \leq |\delta|$. Let $\beta = \tau\kappa$ for a pair of ordinals τ, κ. Then $s(\beta) \leq s(\tau)\kappa$.

Let $u \in S(\beta)$ and let $\Gamma: \beta \to \delta$ be an injection such that $u = \sum(i<\beta)x_{i\Gamma}$. Then $\sum(i < \beta)x_{i\Gamma} = \sum(j < \kappa)\sum(\tau j \leq i < \tau j^+)x_{i\Gamma}$. For each $j < \kappa$, $\sum(\tau j \leq i < \tau j^+)x_{i\Gamma} \in S(\tau)$ so $\sum(\tau j \leq i < \tau j^+)x_{i\Gamma} \leq s(\tau)$. The conclusion of the lemma follows.

Theorem 3. Let $\delta = \delta^- > 0$. Put δ in Standard Form; $\delta = \sum(i \leq m)\omega_{ip}q_i$. There exist ω_{mp} disjoint cofinal subsets of δ, each of type $cf(\delta)$.

Put $y = \sum(i < m)\omega_{ip}q_i$. Any cofinal subset of δ contains an element of $[y,\delta)$ and since $|[y,\delta)| = \omega_{mp}$ there are at most ω_{mp} such subsets.
CASE 1. $q_m < q^-_m$. Put $z = y + \omega_{mp}q^-_m$ so that $\delta = z + \omega_{mp}$. There exists ω_{mp} disjoint subsets of cardinality ω_{mp}, call this family $\{A_i \mid i \in \omega_{mp}\}$. The A_i must be of type ω_{mp} and cofinal in δ. Let $f_i : cf(\delta) \to A_i$ be an order preserving injection such that $f_i[cf(\delta)]$ is cofinal in A_i for each $i \in \omega_{mp}$; such exist since $cf(\delta) = cf(\omega_{mp})$. Then $\{f_i[cf(\delta)] \mid i < \omega_{mp}\}$ is the required family.
CASE 2. $q_m = q^-_m > 0$. For each $k < \omega_{mp}$ define B_k be $\{y + \omega_{mp}\, j + k \mid j < q_m\}$. The family $\{B_k : k < \omega_{mp}\}$ is disjoint and each element of it is cofinal in δ and of type q_m. Since $cf(\delta) = cf(q_m)$ we can define order preserving injections f_k, $f_k : cf(\delta) \to B_k$, such that $f_k[cf(\delta)]$ is cofinal in B_k for each $k < \omega_{mp}$. Then $\{f_k[cf(\delta)] \mid k < \omega_{mp}\}$ is the required family.

Theorem 4. Let $\{x_i\}(i < \delta)$ be a non-decreasing sequence of non-zero ordinals of type a limit ordinal δ. Put $t = cf(\delta)$ and put δ in Standard Form;

$$\delta = \sum(i \leq m)\omega_{ip}q_i .$$

Let A be any cofinal subset of δ of type t and let $\Gamma : t \to A$ be the order preserving bijection. Then $L(\{x_{i\Gamma}\}(i < t)) = L(\{x_i\}(i < \delta))$ and for each $r \leq L$, $s(r, \{x_i\}\ (i < t)) = s(r,\{x_i\}(i < \delta))$.

Now $\sup\{x_{i\Gamma} \mid i < t\} = \sup\{x_i \mid i < \delta\}$ and of course $cf(t) = t = cf(\delta)$. Hence by Theorem 2 it is sufficient to prove that $\sup\{x_{i\Gamma} \mid i < t\}$ is attained iff $\sup\{x_i \mid i < \delta\}$ is attained. But this later is evident.

Theorem 5. Let $\{x_i\}(i < \delta)$ be a non-decreasing sequence of non-zero ordinals of type δ, where the Standard Form of δ is $\omega_p q$, $1 \leq q \leq \omega_{p+1}$. Let β be such that $|\beta| \leq |\delta|$ and put $\beta = LQ + r$ where Q and r are ordinals and $r < L$. Then $s(\beta) = s(L)Q + s(r)$ and $s(\beta) \in S(\beta)$ iff $r = 0$.

There exists a family $\{A_i | i \leq Q\}$ of disjoint sets cofinal in δ, since $|Q| \leq \omega_p$, by Theorem 3, each of type $cf(\delta)$. By theorem 4, for each $j \leq Q$, $L(\{x_i\}(i \in A_j)) = L(\{x_i\}(i < \delta))$ and for each $r \leq L$, $s(r, \{x_i\}(i \in A_j)) = s(r, \{x_i\}(i < \delta))$. There exist $j\Gamma : L \to A_j$ such that $\sum(i < L)x_{i(j\Gamma)} = s(L)$ for each $j < Q$. Stringing these sequences together to form a sequence of type LQ, we can define $\Gamma : LQ \to \delta$ such that $\sum(i < LQ)x_{i\Gamma} = s(L)Q$. Using Lemma 2, $s(LQ, \{x_i\}(i < \delta)) = s(L)Q$ $\in S(LQ, \{x_i\}(i < \delta))$.

CASE 1. $r = 0$. Then $s(r) = 0$ and the conclusions of the theorem follow.
CASE 2. $r \neq 0$. Then $s(r) \notin S(r)$ since $r < L$. Let $\gamma : \beta \to \delta$ be an arbitrary injection. Then $\sum(i < \beta)x_{i\gamma} = \sum(i < LQ)x_{i\gamma} + \sum(LQ \leq i \leq \beta)x_{i\gamma}$. Firstly, $\sum(i < LQ)x_{i\gamma} \leq \sum(i < LQ)x_{i\Gamma}$. Secondly, $\sum(LQ \leq i < \beta)x_{i\gamma} \in S(r)$ and $s(r) \notin S(r)$ so there exists $\Gamma : [LQ,\beta) \to A_Q$ such that $\sum(LQ \leq i < \beta)x_{i\gamma} < \sum(LQ \leq i < \beta)x_{i\Gamma}$.
Then $\Gamma : \beta \to U(j \leq Q)A_j$ is an injection $\Gamma : \beta \to \delta$ and $\sum(i < \beta)x_{i\gamma} < \sum(i < \beta)x_{i\Gamma}$. Hence $s(\beta) \notin S(\beta)$.

It is evident that for any injection $\gamma : \beta \to \delta$
$\sum(i < \beta)x_{i\gamma} = \sum(j < Q) \sum(Lj \leq i < Lj^+)x_{i\gamma} + \sum(LQ \leq i < \beta) x_{i\gamma}$ $\leq s(L)Q + s(r)$. Also, given b such that $s(L)Q \leq b < s(L)Q + s(r)$ we can put $a = s(L)Q + \sigma$ where $\sigma < s(r)$; there must exist an injection $\Gamma : [LQ,\beta) \to A_Q$ such that $\sum(LQ \leq i < \beta)x_{i\Gamma} > \sigma$ since $s(r) = \sup S(r)$ so $\sum(i < \beta)x_{i\Gamma} \geq a$. This concludes the proof.

§3 The General Problem of $s(\beta, \{x_i\}\ (i < \delta)$ when δ is a Limit Number

Theorem 6. Let $1 \leq \sigma_o \leq \cdot \leq \sigma_n$ where $n < \omega$. Let $\delta = \omega_{op} + \ldots + \omega_{np}$ where $op > \ldots > np$. Put $I_k = [\sum(i < k)\omega_{ip}, \sum(i \leq k)\omega_{ip})$ for $k \leq n$. Define $\{y_i\}(i < \delta)$ by $y_i = \sigma_k$ if $i \in I_k$; we use the abbreviated forms of s and S for this sequence. Let ω^a be a prime component such that

$$|\delta| \geq |\omega^a| \geq |\delta \setminus \omega_{op}|.$$

Then (i) $s(\omega^a) = \max\ (\{\sigma_o\omega^a\} \cup \{\sigma_i\ \omega_{ip+1} \mid 1 \leq i \leq n\})$

and (ii) $s(\omega^a)\quad S(\omega^a)$ iff $s(\omega^a) = \sigma_o\omega^a$.

We observe that the theorem holds if $n = 0$ and proceed by finite induction on n.

Put $F = \{\sigma_o\omega^a\} \cup \{\sigma_i\ \omega_{ip+1} \mid 1 \leq i \leq n\}$.

Now; $\sigma_1\omega_{1p+1}, \ldots, \sigma_n\omega_{np+i}$ are distinct since $1p > \ldots > np$. Hence we can define Θ uniquely by $1 \leq \Theta \leq n$ and $\sigma_\Theta\ \omega_{\Theta p+1} = \max\{\sigma_i\omega_{ip+1} \mid 1 \leq i \leq n\}$.

CASE 1. $\sigma_o\ \omega^a \neq \max F$.

(1) We prove that $s(\omega^a) = \sigma_\Theta\ \omega_{\Theta p+1} \notin S(\omega^a)$. If $d < \sigma_\Theta\ \omega_{\Theta p+1}$ then there exists $e < \omega_{\Theta p+1}$ such that $d < \sigma_\Theta\ e$. Define injections $\Gamma : e \to I_\Theta$ and $\Gamma : [e, \omega^a) \to I_o$, then $d < \sum(i < \beta)x_{i\Gamma} \in S(\omega^a)$.

(2) Hence $s(\omega^a) \geq \sigma_\Theta\ \omega_{\Theta p+1}$.

Next, let $\gamma : \omega^a \to \delta$ be an arbitrary injection.

(3) We show $\sum(i < \omega^a)x_{i\gamma} < \sigma_\Theta\ \omega_{\Theta p+1}$.

Put $E = \{i \mid i < \omega^a\ \&\ i\Gamma \in I_\Theta\}$.

(4) Let $\phi : \eta \to E$ be the order preserving bijection; $\eta < \omega_{\Theta p+1}$.
By the induction hypothesio, $\sum(i \in \omega^a \setminus E)y_{i\gamma} \leq$
$s(\omega^a, \{y_i\}(i \in \delta \setminus I_\Theta)) = \max (F \setminus \{\sigma_\Theta \omega_{\Theta p+1}\}) < \sigma_\Theta\ \omega_{\Theta p+1}$.

(5) Put $\sum = \sum(i \in \omega^a \setminus E)y_{i\Gamma}$ and let $\xi < \omega_{\Theta p+1}$ be such that $\sum < \sigma_\Theta \xi$.
Then $\sum(i < \omega^a)y_{i\Gamma} \leq \sum(i < \eta)(\sum + y_{i\phi\Gamma}) + \sum \leq \sum(i < \eta)(\sigma_\Theta\xi + \sigma_\Theta) + \sigma_\Theta\xi <$
$\sigma_\Theta(\xi + 1)(\eta + 1) < \sigma_\Theta\ \omega_{\Theta p+1}$, see (4) and (5).
This proves (3). From (2) and (3) follows (1).

CASE 2. $\sigma_o\omega^a = \max F$. We prove that

(6) for any injection $\Gamma : \omega^a \to \delta$, $\sum(i < \omega^a)y_{i\Gamma} = \sigma_o\omega^a$.

(7) Evidently $\sum(i < \omega^a)y_{i\Gamma} \geq \sigma_o\omega^a$.

CASE 2A. $cf(\omega^a) \geq \omega_{\Theta p+1}$. Define E, η and ϕ as in CASE 1. Then $\eta < cf(\omega^a)$ and there exists $h < \omega^a$ such that $\{i\Gamma \mid h \leq i < \omega^a\} \subseteq \delta \setminus I_\Theta$.

(8) Evidently $\sum(h \leq i < \omega^a)y_{i\Gamma} = \sigma_o\omega^a$ and it will suffice to show that
$\sum(i < h)y_{i\Gamma} < \sigma_o\ \omega^a$. Since $\sigma_\Theta < \sigma_\Theta\ \omega_{\Theta p+1} \leq \omega^a$, there exists $\zeta < \omega^a$ with $\sigma_\Theta < \sigma_o\zeta$.

By the induction hypothesis $\sum(i < h\ \&\ i\Gamma \notin I_\Theta)y_{i\Gamma} <$
$\sum(i < \omega^a\ \&\ i\Gamma \notin I_\Theta)y_{i\Gamma} \leq s(\omega^a, \{y_i\}(i \in \delta \setminus I_\Theta)) = \sigma_o\omega^a$.
Putting $\sum = \sum(i < h\ \&\ i\Gamma \notin I_\Theta)y_{i\Gamma}$ we have $\sum < \sigma_o\xi$ for some $\xi < \omega^a$.
Now $\sum(i < h)y_{i\Gamma} \leq$
$\sum(i < \eta)(\sum + y_{i\phi\Gamma}) + \sum \leq \sigma_o \sum(i < \eta)(\xi + \zeta) + \sigma_o\ \xi <$
$\sigma_o \sum(i \leq \eta)(\xi + \zeta) = \sigma_o \sup \{(\xi + \zeta)i \mid i \leq \eta\} < \sigma_o\omega^a$
since $\xi + \zeta < \omega^a$ and $\eta < \omega_{\Theta p+1} \leq cf(\omega^a) \leq \omega^a$. With (8),
this concludes the proof of CASE 2A.

CASE 2B. $cf(\omega^a) < \omega_{\Theta p+1}$. Put $t = cf(\omega^a)$. Then there exists $b \geq 1$ and $c < \omega_{\Theta p+1}$ such that $\omega^a = \omega^b_{\Theta p+1}\ t^c$.

CASE 2B1. $c = 0$. Then $\omega^a = \omega^b_{\Theta p+1}$ and $b^- = b$ since $cf(\omega^a) = t < \omega_{\Theta p+1}$. Also $cf(b) = t$. By the hypothesis of CASE 2, $\sigma_\Theta\, \omega_{\Theta p+1} \leq \sigma_o\, \omega^b_{\Theta p+1}$. Since $b^- = b$ we must have $\sigma_\Theta\, \omega_{\Theta p+1} < \sigma_o \omega^b_{\Theta p+1}$ and so there exists $\xi < b$ such that $\sigma_\Theta\, \omega_{\Theta p+1} < \sigma_o\, \omega^\xi_{\Theta p+1}$.

Let $\{ib\}(i < t)$ be a strictly increasing sequence cofinal in b such that $\xi \leq ob$ and $(ib)^- < ib$ for each $i < t$. Then for each $i < t$, $\omega^{ib}_{\Theta p+1}$ is a prime component, $|\omega^{ib}_{\Theta p+1}| \geq \omega_{\Theta p+1}$,

$cf(\omega^{ib}_{\Theta p+1}) = \omega_{\Theta p+1}$ and $\sigma_o\ \omega^{ib}_{\Theta p+1} =$

$\max(\{\sigma_o \omega^{ib}_{\Theta p+1}\} \cup \{\sigma_j\ \omega_{jp+1} \mid 1 \leq j \leq n\})$.

By CASE 2A, $s(\omega^{ib}_{\Theta p+1}) = \sigma_o\ \omega^{ib}_{\Theta p+1} < \sigma_o \omega^a$.

Were $\sum(i < \omega^a) y_{i\Gamma} > \sigma_o \omega^a$ then there would exist $\mu < \omega^a$ such that $\sum(i < \mu) y_{i\Gamma} = \sigma_o \omega^a$, this because $\sigma_o \omega^a$ is a prime component. But there exists $j < t$ such that $\mu < \omega^{jb}_{\Theta p+1}$. Hence $\sum(i < \mu) y_{i\Gamma} \leq s(\omega^{jb}_{\Theta p+1}) < \sigma_o \omega^a$, contradiction.

CASE 2B2. $c \neq 0$. Now $\sigma_\Theta\ \omega_{\Theta p+1} \leq \sigma_o\ \omega^a = \sigma_o\ \omega^b_{\Theta p+1}\ t^c$. But $t^c < \omega_{\Theta p+1}$, whence $\sigma_\Theta\ \omega_{\Theta p+1} \leq \sigma_o\ \omega^b_{\Theta p+1}$. We apply CASE 2A if $cf(\omega^b_{\Theta p+1}) \geq \omega_{\Theta p+1}$ and CASE 2B1 if $cf(\omega^b_{\Theta p+1}) < \omega_{\Theta p+1}$ to obtain $s(\omega^b_{\Theta p+1}) = \sigma_o\ \omega^b_{\Theta p+1} \in S(\omega^b_{\Theta p+1})$, since

$\sigma_o \omega^b_{\Theta p+1} = \max(\{\sigma_o \omega^b_{\Theta p+1}\} \cup \{\sigma_i \omega_{ip+1} \mid 1 \leq i \leq n\})$.

Lemma 2 yeilds $s(\omega^a) \leq s(\omega^b_{\Theta p+1}) t^c = \sigma_o \omega^b_{\Theta p+1}\ t^c = \sigma_o \omega^a$. With (7) this completes the proof of the theorem.

Theorem 7. Let $\{x_i\}(i < \delta)$ be a non-decreasing sequence of non-zero ordinals of type a limit number δ. Put δ in Standard Form;

$$\delta = \sum(i \leq m)\omega_{ip}q_i .$$

Let $|\omega^a| > |\delta \setminus \omega_{op}q_o|$. We use the abbreviated forms $s(\omega^a)$ and $S(\omega^a)$ only when refering to $\{x_i\}(i < \delta)$. For each $k \leq m$, put $I_k = \left[\sum(i < k)\omega_{ip}q_i, \sum(i \leq k)\omega_{ip}q_i\right)$, put $\lambda_k = \sup\{x_i \mid i \in I_k\}$ and $\ell_k = L\{x_i\}(i \in I_k)$.

(1) CASE 1. $\ell_o > \omega^a$. Then $s(\omega^a) = \max(\{\lambda_o\} \cup \{\lambda_i\omega_{ip+1} \mid 1 \leq i \leq m\}) \notin S(\omega^a)$.

CASE 2. $\ell_o \leq \omega^a$. Put $\omega^a = \ell_o Q$. Put $\sigma_o = s(\ell_o, \{x_i\}(i < \omega_{op}q_o))$. (We can calculate σ_o from Theorem 5).

(2) Then $s(\omega^a) = \max(\{\sigma_o Q\} \cup \{\lambda_i\omega_{ip+1} \mid 1 \leq i < m\})$ and

$s(\omega^a)$ $S(\omega^a)$ iff $\sigma_o Q = \max(\{\sigma_o Q\}$ $\{\lambda_i \omega_{ip+1} \mid 1 \leq i \leq m\})$

If $m = 0$ then (1) and (2) follow from Theorem 2 and 4. We proceed by induction on m. Let $m \geq 1$. Since $op > \ldots > mp$ we can define Θ uniquely by $1 \leq \Theta \leq m$ and $\lambda_\Theta \omega_{\Theta p+1} = \max\{\lambda_i \omega_{ip+1} \mid 1 \leq i \leq m\}$. For any $\xi < \omega_{\Theta p+1}$, $s(\omega^a) \geq s(\ell_\Theta \xi, \{x_i\}(i \in I_\Theta)) = s(\ell_\Theta, \{x_i\}(i \in I_\Theta))\xi \geq \lambda_\Theta \xi$.

(3) Hence $s(\omega^a) \geq \lambda_\Theta \omega_{\Theta p+1}$.

Let $\gamma : \omega^a \to \delta$ be an arbitrary injection. Put $E = \{i \mid i\gamma \in I_\Theta\}$ and let $\phi : \eta \to E$ be the order preserving bijection. Put $\sum = \sum(i \in \omega^a \setminus E)x_{i\gamma}$.

CASE 1. $\ell_o > \omega^a$. To prove (1), it is sufficient, by (3), to show that $s(\omega^a) < \lambda_\Theta \omega_{\Theta p+1}$. By the induction hypothesis $\sum \leq \max(\{\lambda_o\} \cup \{\lambda_i \omega_{ip+1} \mid 1 \leq i \leq m$ and $i \neq \Theta\} < \lambda_\Theta \omega_{\Theta p+1}$. Hence $\sum < \lambda_\Theta \xi$ for some $\xi < \omega_{\Theta p+1}$ and

$\sum(i < \omega^a)x_{i\gamma} \leq \sum(i < \eta)(\sum + x_{i\phi\gamma}) + \sum$

$\leq \sum(i < \eta)(\lambda_\Theta \xi + \lambda_\Theta) = \lambda_\Theta(\xi + 1)(\eta + I) < \lambda_\Theta \omega_{\Theta p+1}$.

CASE 2. $\ell_o \leq \omega^a$.

CASE 2A. $\lambda_\theta \; \omega_{\theta p+1} > \sigma_o Q$. We prove that $s(\omega^a) = \lambda_\theta \omega_{\theta p+1} \notin S(\omega^a)$. By the induction hypothesis $\sum \leq s(\omega^a, \{x_i\}(i \in \delta \setminus I_\theta)) =$
$\max(\{\sigma_o Q\} \cup \{\lambda_i \omega_{ip+1} \mid 1 \leq i < m \;\; \& \; i \neq \theta\})$
$< \lambda_\theta \omega_{\theta p+1}$, so $\sum < \lambda_\theta \xi$ for some $\xi < \omega_{\theta p+1}$. Thus
$\sum(i < \omega^a) x_{i\gamma} \leq \sum(i < \eta)(\lambda_\theta \xi + \lambda_\theta) + \lambda_\theta < \lambda_\theta \omega_{\theta p+1}$.
With (3), we have proved (2) in this case.

CASE 2B. $\lambda_\theta \omega_{\theta p+1} \leq \sigma_o Q$. Then $\sigma_o Q =$
$\max(\{\sigma_o Q\} \cup \{\lambda_i \omega_{ip+1} \mid 1 \leq i \leq m\})$.
We prove that $s(\omega^a) = \sigma_o Q \in S(\omega^a)$. Were $Q < \omega_{1p+1}$ then would
$\lambda_\theta \omega_{\theta p+1} \geq \lambda_1 \omega_{1p+1} = (\lambda_1 \omega)\omega_{1p+1} \geq (\lambda_o \omega)\omega_{1p+1} \geq \sigma_o \omega_{1p+1}$ (Theorem 2)
$> \sigma_o Q$, contradiction.

(4) Hence $Q \geq \omega_{1p+1} > |\delta \setminus \omega_{op} q_o|$ and from the definition of Q, Q is a prime component. We define the sequence $\{y_i\}(i < \delta)$ by $y_i = \lambda_k$ if $i \in I_k$. Now $1 \leq \lambda_o \leq \ldots \leq \lambda_m$ and of course there is a bijection which will arrange $\{y_i\}(i < \delta)$ so as to satisfy the hypothesis of Theorem 6.

(5) Hence $s(Q, \{y_i\}(i < \delta)) = \max(\{\lambda_o Q\} \cup \{\lambda_i \omega_{ip+1} \mid 1 \leq i \leq m\})$
$= \lambda_o Q \in S(Q, \{y_i\}(i < \delta))$.

(6) We show there exists an injection $\Gamma : Q \to \delta$ such that $\sum(i < \omega^a) x_{i\gamma} \leq \sum(i < Q) y_{i\Gamma}$. If $\ell_o < \omega_{1p+1}$ then $\omega^a \geq \omega_{1p+1}$ yields $\ell_o \omega^a = \omega^a$ whence $Q = \omega^a$ and we simply put $\Gamma = \gamma$ since $x_i \leq y_i$ for each $i < \delta$. Let now $\ell_o \geq \omega_{1p+1}$. Then $\sum(i < \omega^a) x_{i\gamma} = \sum(j < Q) \sum(\ell_o j \leq i < \ell_o j^+) x_{i\gamma}$. Put $J_j = [\ell_o j, \ell_o j^+)$ for each $j < Q$. We divide these intervals into $m + 1$ different kinds. Put $P_o = \{j \mid j < Q \; \& \; \gamma[J_j] \subseteq I_o\}$. Then for each $j \in P_o$, $\sum(i \in J_j) x_{i\gamma} \leq s(\ell_o, \{x_i\}(i \in I_o)) = \lambda_o$ by Theorem 2, since $\ell_o \geq \omega_{1p+1} > \omega$.

(7) There exists a bijection $f: (m+1) \to (m+1)$ such that $of = o$ and $\lambda_{of} < \lambda_{1f}\omega_{1fp+1} < \ldots < \lambda_{mf}\,\omega_{mfp+1}$. Define P_k for k such that $1 \leq k \leq m$ by $P_k = \{j \mid j < Q$ and $\gamma[J_j] \cap I_{kf} \neq 0$ and $\gamma[J_j] \cap I_{if} = 0$ for $k \leq i \leq m\}$. Since the family $\{\gamma[J_j] \mid y < Q\}$ is disjoint, $|P_k| \leq \omega_{kfp} = |I_{kf}|$.

For each $j \in P_k$, $\gamma[J_j] \subseteq \bigcup(i \leq k)I_{if} = \sum(i \leq k)\omega_{ip}q_i$ so $\sum(\ell_o j \leq i < \ell_o j^+)x_{i\gamma} \leq s(\ell_o\{x_i\}(i < \sum(n \leq k)\omega_{np}q_n))$.

Now, the evaluation of $s(\ell_o,\{x_i\}(i < \sum(n \leq k)\omega_{np}q_n))$ falls under CASE 2A which yields $s(\ell_o,\{x_i\}(i < \sum(n \leq k)\omega_{np}q_n)) = \lambda_{kf}\,\omega_{kfp+1} \nleq S(\ell_o,\{x_i\}(i < \sum(n \leq k)\omega_{np}q_n))$, see (7).

(8) Hence there exists $\xi_j < \omega_{kfp+1}$ such that $\sum(\ell_o j \leq i < \ell_o j^+)x_{i\gamma} \leq \lambda_{kf}\xi_j$. The above applies to k with $1 \leq k \leq m$. If $k = 0$ then $j \in P_o$ yields $\sum(i \in J_j)x_{i\gamma} \leq \lambda_o = \lambda_{of}$ and we take $\xi_j = 1$. Then (8) applies for all $k \leq m$.

(9) We now define an injection $\Gamma : \sum(j < Q)\,\xi_j \to \delta$ such that $\sum(i < \omega^a)x_{i\gamma} \leq \sum(i < \sum(j < Q)\xi_j)\,y_{i\Gamma}$.

We can partition I_{kf} into ω_{kfp} disjoint sets each of cardinality ω_{kfp}. Now $|P_k| \leq \omega_{kfp}$ and $|\xi_j| \leq \omega_{kfp}$ for each $j \in P_k$. It follows that we can define injections $j\Gamma: \xi_j \to I_{kf}$ for each $j \in P_k$ such that $\{j\Gamma[\xi_j] \mid j \in P_k\}$ is a disjoint family of subsets of I_{kf}. For each $j < Q$, $j \in P_k$ for some unique $k \leq m$ and we thus have $j\Gamma$ defined for $j < Q$ such that the sequence indicated by the sum $\sum(j < Q)\sum(i < \xi_j)y_{i(j\Gamma)}$ yields the injection Γ required by (9). Observe only that $\sum(i \in J_j)x_{i\gamma} \leq \lambda_{kf}\xi_j = \sum(i < \xi_j)y_{i(j\Gamma)}$ where k is such that $j \in P_k$

The domain of Γ is $\sum(j < Q)\ \xi_j$.

Since $\xi_j \geq 1$ for each $j < Q$, $\sum(j < Q)\xi_j \geq Q$.

Define ξ to be sup $\{\xi_j \mid j < Q\}$.

Now, $\xi_j = 1$ if $j \in P_o$ and $\xi_j < \omega_{kfp+1} \leq \omega_{1p+1}$ for $j \in P_k$ where $1 \leq k \leq m$.

Also $|\{\xi_j \mid j < Q \text{ and } \xi_j > 1\}| < \omega_{1p+1}$.

Thence $\xi < \omega_{1p+1}$. Therefore $\sum(j < Q)\xi_j \leq \sum(j < Q)\xi =$ $\xi Q = Q$ by absorbtion, see (4).

Hence the domain of Γ is actually Q.

This proves (6). From (5) it now follows that

$\sum(i < \omega^a)x_{i\gamma} \leq \lambda_o Q \leq \sigma_o Q$.

By Theorem 5, $s(\omega^a) \geq s(\omega^a, \{x_i\}(i < \omega_{op}q_o)) =$

$s(\ell_o Q, \{x_i\}(i < \omega_{op}q_o)) = \sigma_o Q \in$

$S(\omega^a, \{x_i\}(i < \omega_{op}q_o)) \subseteq S(\omega^a)$.

Thence (2) follows.

This concludes the proof of the theorem.

Theorem 8. Let $\{x_i\}(i < \delta)$ be a non-decreasing sequence of non-zero ordinals of type a limit ordinal δ. Let β be such that $|\beta| \leq |\delta|$. Put δ into Standard Form,

$$\delta = \sum(i \leq m)\omega_{ip}q_i.$$

Let $m \geq 1$. Divide β by ω_{1p+1} to obtain $\beta = \omega_{1p+1}\alpha_o + \alpha_1$ where $\alpha_1 < \omega_{1p+1}$ and put $\alpha_o = \omega_{1p+1}\alpha_o'$. We use the abbreviated forms of s and S when referring to $\{x_i\}(i < \delta)$. Then $s(\beta) \in S(\beta)$ iff

(i) $s(\alpha_o) = s(\alpha_o, \{x_i\}\ (i < \omega_{op}q_o)) \in S(\alpha_o, \{x_i\}(i < \omega_{op}q_o)$

and (ii) $s(\alpha_1) = s(\alpha_1, \{x_i\}(\omega_{op}q_o \leq i \leq \delta)) \in$ $S(\alpha_1\{x_i\}(\omega_{op}q_o \leq i < \delta))$.

If (i) and (ii) hold, it easily follows that $s(\beta) \in S(\beta)$. Conversely, let $s(\beta) \in S(\beta)$.

(1) We first show that $s(\alpha_1) = s(\alpha_1, \{x_i\}(\omega_{op}q_o \leq i < \delta))$.

(2) Evidently $s(\alpha_1) \geq s(\alpha_1, \{x_i\}(\omega_{op}q_o \leq i < \delta))$ since
$S(\alpha_1) \supseteq S(\alpha_1, \{x_i\}(\omega_{op}q_o \leq i < \delta))$.
Let $\gamma : \alpha_1 \to \delta$ be an arbitrary injection.
Let $\{A_i \mid i < q_1\}$ be ω_{1p} disjoint cofinal subsets of
$[\omega_{op}q_o, \omega_{op}q_o + \omega_{1p}q_1)$.
Define $\Gamma : \alpha_1 \to \delta$ as follows.
If $i\gamma \in [\omega_{op}q_o + \omega_{1p}q_1, \delta)$ put $i\Gamma = i\gamma$;
otherwise there exists $ia \in A_i$ such that
$x_{i\gamma} \leq x_{ia}$ and we put $i\Gamma = ia$. Then Γ is an injection and
$\sum(i < \alpha_1)x_{i\gamma} \leq \sum(i < \alpha_1)x_{i\Gamma}$ whence with (2), (1) is proved.

(3) It also follows that $s(\alpha_1) \in S(\alpha_1)$ iff
$s(\alpha_1, \{x_i\}(\omega_{op}q_o \leq i < \delta)) \in S(\alpha_1, \{x_i\}(\omega_{op}q_o \leq i < \delta))$
We now consider several cases.
CASE 1. $\beta < \omega_{1p+1}$. Then $\alpha_o = 0$ and $\alpha_1 = \beta$. We observe that (i) holds trivially and that (ii) follows from (1) and (3).

CASE 2. $\beta \geq \omega_{1p+1}$. Put $\alpha_o = \omega^{oa} + \omega^{1a} + \ldots + \omega^{na}$ where $n < \omega$ and $oa \geq 1a \geq \ldots \geq na$.
By definition of $\alpha_o, \omega^{ia} \geq \omega_{1p+1}$ for $i \leq n$.

(4) Let $\mu \leq n$ be such that for each $i < \mu$ $s(\omega^{ia}) \in S(\omega^{ia})$. Put $z = \sum(i < \mu)\omega^{ia}$.

(5) We prove that $s(z) = s(z, \{x_i\}(i < \omega_{op}q_o)) \in S(z, \{x_i\}(i < \omega_{op}q_o))$.
Put $\ell_o = L(\{x_i\}(i < \omega_{op}q_o))$.

By Theorem 7, $\ell_o \leq \omega^{ia}$ for each $i < \mu$.

Put $\omega^{ia} = \ell_o Q_i$ and $\sigma_o = s(\ell_o, \{x_i\}(i < \omega_{op} q_o))$.

Then $s(\omega^{ia}) = \sigma_o Q_i$ for each $i < \mu$.

Thus $s(z) \leq \sum(i < \mu)\ s(\omega^{ia}) = \sigma_o \sum(i < \mu) Q_i$.

But Theorem 5 gives $s(z, \{x_i\}(i < \omega_{op} q_o)) =$

$s(\ell_o \{x_i\}(i < \omega_{op} q_o)) \sum(i < \mu) Q_i = \sigma_o \sum(i < \mu) Q_i$

$\in S(z, \{x_i\}(i < \omega_{op} q_o))$ whence (5).

(6) Next, suppose (4) holds but $s(\omega^{\mu a}) \notin S(\omega^{\mu a})$.

We shall deduce a contradiction.

Define I_j and λ_j for each $j \leq m$ by $I_j = [\sum(i < j)\omega_{ip} q_i, \sum(i \leq j)\omega_{ip} q_i)$ and

$\lambda_j = \sup\{x_i \mid i \in I_j\}$. Applying Theorem 7 in the light of (6),

$s(\omega^{\mu a}) = \max\{\lambda_i \omega_{ip+1} \mid 1 \leq i \leq m\}$.

(7) Define θ by $\lambda_\theta \omega_{\theta p+1} = \max\{\lambda_i \omega_{ip+1} \mid 1 \leq i \leq m\}$; this defines θ uniquely. Then $s(\omega^{\mu a}) = \lambda_\theta \omega_{\theta p+1}$.

(8) By hypothesis there exists an injection $\Gamma : \beta \to \delta$ such that

$s(\beta) = \sum(i < \beta) x_{i\Gamma}$.

By (5) there exists an injection $\gamma : z \to \omega_{op} q_o$ such that

$s(z) = \sum(i < z) x_{i\gamma}$.

By (6) and (7), $\sum(z \leq i < z + \omega^{\mu a}) x_{i\Gamma} < \lambda_\theta \omega_{\theta p+1}$.

For each j, $\mu \leq j \leq n$, $\omega^{ja} \leq \omega^{\mu a}$ so

$\sum(\sum(i < j)\omega^{ia} \leq k < \sum(i \leq j)\omega^{ia}) x_{i\Gamma} < \lambda_\theta \omega_{\theta p+1}$

Also $\alpha_1 < \omega_{\theta p+1} \leq \omega^{\mu a}$ so $\sum(\alpha_o \leq k < \beta)\ x_{i\Gamma} < \lambda_\theta \omega_{\theta p+1}$.

Hence by absorption $\sum(z \leq i < \beta) x_{i\Gamma} < \lambda_\theta\ \omega_{\theta p+1}$.

So $\sum(i < \beta) x_{i\Gamma} = \sum(i < z) x_{i\Gamma} + \sum(z \leq i < \beta) x_{i\Gamma}$

$< \sigma_o \sum(i < \mu) Q_i + \lambda_\theta \xi$ for some $\xi < \omega_{\theta p+1}$. Now $s(z) = s(z, \{x_i\}(i < \omega_{op} q_o))$

and $s(tp[z, \delta)) = \lambda_\theta \omega_{\theta p+1} \notin S(\omega^{\mu a})$.

Partition $\omega_{op}q_o$ into ω_{op} disjoint cofinal subsets. Put the first element of each subset into a set T. Put the resulting sets into a sequence $\{A_i\}(i < \beta)$. We use $\{A_i\}(i < z)$ to define $\gamma : z \to \delta$ such that $\sum(i < z)x_{i\gamma} = \sigma_o\sum(i < \mu)Q_i$, where $i\gamma \in A_i$ for each $i < z$. We know there exists an injection $g : [z,\beta) \to \delta$ such that $\sum(z \leq i < \beta)x_i g \geq \lambda_\theta \xi$. Define $\gamma : [z,\beta) \to \delta$ by $i\gamma = ig$ if $ig \geq \omega_{op}q_o$ and, if $ig < \omega_{op}q_o$, let $i\gamma$ be any element of A_i such that $x_{ig} \leq x_{i\gamma}$. Then $\gamma : \beta \to \delta$ is an injection and $\sum(i < \beta)x_{i\Gamma} < \sum(i < \beta)x_{i\gamma}$, contradicting (8). Recalling (4) and (6), this proves (i).

(9) Note. It follows from the above that, in the situation described in (4) and (6), $s(\beta) = \sum(i \leq \mu)\ s(\omega^{ia})$.

We now prove (ii).

(10) Suppose (ii) is false; in view of (1) we have

$s(\alpha_1, \{x_i\}(\omega_{op}q_o \leq i < \delta)) \notin S(\alpha_1, \{x_i\}\ (\omega_{op}q_o \leq i < \delta))$.

Let $s(\beta) = \sum(i < \beta)x_{i\Gamma} = \sum(i < \alpha_o)x_{i\Gamma} + \sum(\alpha_o \leq i < \beta)x_{i\Gamma}$ where Γ is an injection.

By (i) there exists an injection $\gamma : \alpha_o \to \omega_{op}q_o$ such that $\sum(i < \alpha_o)x_{i\Gamma} \leq \sum(i < \alpha_o)x_{i\gamma}$. By (1) there exists an injection $\gamma : [\alpha_o, \beta) \to [\omega_{op}q_o, \delta)$ such that $\sum(\alpha_o \leq i < \beta)x_{i\Gamma} < \sum(\alpha_o \leq i < \beta)x_{i\gamma}$. Hence $\gamma : \beta \to \delta$ is an injection and $\sum(i < \beta)x_{i\Gamma} < \sum(i < \beta)x_{i\gamma}$, contradiction. This concludes the proof of the theorem.

§4 Conclusions

We can now summarize the method of determining, for a given non-decreasing sequence of non-zero ordinals $\{x_i\}(i < \delta)$ of type a limit ordinal δ and given β with $|\beta| \leq |\delta|$, the value of $s(\beta)$ and also whether $s(\beta) \in S(\beta)$.

Put δ into Standard Form; $\delta = \sum(i \leq m)\omega_{ip}q_i$. Let $I_k = |\sum(i < k)\omega_{ip}q_i, \sum(i \leq k)\omega_{ip}q_i)$ for each $k \leq m$. Put β in the form $\beta = \sum(1 \leq i \leq m)\omega_{ip+1}\kappa_i + \kappa_{m+1}$ where $\kappa_i < \omega_{ip+1}$ for $1 \leq i \leq m$, which is obtained by repeated division. Put $\alpha_{i-1} = \omega_{ip+1}\kappa_i$ for $1 \leq i \leq m$ and $\alpha_m = \kappa_{m+1}$. Applying Theorem 8 a total of m times we obtain the following statesment; $s(\beta) \in S(\beta)$iff

† $s(\alpha_j,\{x_i\}(i \in I_j)) \in S(\alpha_j, \{x_i\}(i \in I_j))$ for each $j \leq m$. The validity of the m+1 statements † is determined by Theorem 5 using Theorem 2 to find the appropriate value of L.

In the event that † holds, it is evident that $s(\beta) = \sum(j \leq m) s(\alpha_j)$.

In the event that † does not hold, let ξ be $\min\{j \mid s(\alpha_j) \notin S(\alpha_j)\}$. Put $\alpha_\xi = \omega^{oa} + \omega^{1a} + \dots + \omega^{na}$ where $n < \omega$ and $oa \geq \dots \geq na$. Let μ be $\min\{j \mid j \leq n \ \& \ s(\omega^{ja}) \notin S(\omega^{ja})\}$.

CASE 1. $\omega^{\mu a} \neq 1$. Then

$s(\beta) = \sum(j < \xi)\ s(\alpha_j) + \sum(j \leq \mu)\ s(\omega^{ja})$.

CASE 2 . $\omega^{\mu a} = 1$. Then

$s(\beta) = \sum(j < m)\ s(\alpha_j) + \sum(j < \mu)\ s(\omega^{ja}) + s(\overline{n+1} - \mu)$.

The evaluation of $s(\omega^{ja})$ is an application of Theorem 7 to the sequence $\{x_i\}(i \in \bigcup(\xi \leq k \leq m)I_k)$ while the evaluation of $s(\overline{n+1} - \mu)$ is an application of Theorem 2 to $\{x_i\}(i \in I_m)$.

The calculation is readily explained.

CASE 1. $\omega^{\mu a} \neq 1$. If $\xi \neq m$ then $|\omega^{\mu a}| \geq \omega_{\xi^+ p+1}$ and by Theorem 7, $s(\omega^{\mu a})$

is a prime component. If $\xi = m$ then $s(\omega^{\mu a}) = s(\omega^{\mu a}, \{x_i\}\ (i \in I_m))$ and by Theorem 5 $L(\{x_i\}(i \in I_m)) > \omega^{\mu a} \geq \omega$, whence by Theorem 2, $s(\omega^{\mu a})$ is again a prime components. By (9) of Theorem 8 the result follows.

CASE 2. $\omega^{\mu a} = 1$. Then $\xi = m$ and $\omega^{ja} = 1$ for $\mu \leq j \leq n$ so
$\beta = \sum(i < m)\alpha_i + \sum(i < \mu)\omega^{ia} + (\overline{n+1} - \mu)$.

References

[1] Dan Dushnik, Maximal Sums of Ordinals, Proc. Amer. Math. Soc., 1947.

[2] John A.H. Anderson, The Minimum Sum of an Arbitrary Family of Ordinals, Jour. Lond. Math. Soc. to appear.

Department of Math.
Lancaster University
LANCASTER / ENGLAND

XIV. EFFECTIVE IMPLICATIONS BETWEEN THE "FINITE" CHOICE AXIOMS

J.H. Conway

If n is a positive integer, axiom [n] asserts that every collection of n-element sets has a choice function. We are concerned with the many surprising implications between these axioms for varying values of n. The subject has a long history, started by Tarski's observation that [2] and [4] are equivalent, and concluded in one sense by a recent paper of Gauntt, who gives an algorithm for settling any such question. Almost all of the theory prior to this was created by Mostowski.

Unfortunately, Gauntt's algorithm is complicated, and cannot easily be used to give us a general idea of which implications to expect. In this paper we show how the questions can be illuminated by some old and new results from finite group theory, and most strikingly, using Thompson's list of minimal simple groups (a result stronger than the celebrated Feit-Thompson theorem!). These theorems are used to prove a weakened form of a conjecture of Mostowski, and enable us to show that the original conjecture holds in many cases outside the range of previous techniques.

After stating and proving our main theorems, we discuss some of their applications, and in particular, enquire what can be said about the famous test case:

Do [3] & [5] & [13] imply [15]?

In particular, we show that there is no "effective" implication here, but that [3] & [5] & [13] do (effectively) imply a weakened form of [15] which in turn (ineffectively) implies that [15] holds for well-ordered collections of sets, and so of course for countable collections of sets. It should be noted that these remarks (and indeed the whole paper) were written in ignorance of Gauntt's techniques which have settled this test case. Nevertheless, the remarks on effectivity of the implications seem interesting for their own sake, and so have been retained.

It should perhaps also be noted that an earlier (unpublished) version of this paper was in error owing to the effects of a misreading of one of Mostowski's theorems. This error had (by accident) exactly the same effect as equating the two concepts of effective and ineffective implications. However, the error did not affect the sufficiency conditions which are the main improvements on published results. As a striking example, we use the methods to show that Mostowski's conditions suffice for implications towards [1050].

Notation. Let (a,b,c,...) be an ordered k-tuple of positive integers (in fact we shall usually tacitly suppose that these integers are all ≥ 2). Then $\langle a,b,c,\ldots\rangle$ will denote the additive semigroup of integers generated by a,b,c,... - that is to say, the closure of the set $\{a,b,c,\ldots\}$ under addition. The word *semigroup* will always mean this kind of semigroup. An ordered k-tuple (A,B,C,...) of disjoint sets A,B,C,..., of respective cardinals a,b,c,... will be called an *(a,b,c,...)-set.* Informally, we think of (A,B,C,...) as the set $A \cup B \cup C \cup \ldots$ given together with the partition into A,B,C,..., and we call the members of $A \cup B \cup C \cup \ldots$ the *entries* of (A,B,C,...). The axiom [a,b,c,...] asserts that any collection of (a,b,c,...)-sets has a *choice function* f - that is to say, a function whose typical value f(A,B,C,...) is an entry of (A,B,C,...). (This is slightly different from the usual definition of choice function, but obviously axiom [n] in the new sense is equivalent to [n] in the old sense.)

It turns out that if the semigroups $\langle a,b,c,\ldots\rangle$ and $\langle \alpha,\beta,\gamma,\ldots\rangle$ are equal, then the axioms [a,b,c,...] and $[\alpha,\beta,\gamma,\ldots]$ are equivalent. When we have proved this, we can use the notation [S], where S is an semigroup, to denote any one of the mutually equivalent axioms $[a,b,c,\ldots]$ for which $S = \langle a,b,c,\ldots \rangle$. Accordingly we can speak of semigroup axioms. We shall also speak rather loosely of S-sets, meaning $(a,b,c,\ldots)$-sets when $S = \langle a,b,c,\ldots \rangle$. A prime refinement of S (or of the corresponding axiom $[S]$) is any semigroup $T \supseteq S$ which is generated by prime numbers (or the corresponding axiom $[T]$). We similarly define prime power refinements.

If G is any finite group (all groups will be finite), we use $\langle G \rangle$ for the semigroup generated by the indices in G of all proper subgroups of G, and $[G]$ for the corresponding axiom. When G permutes a set X, we write x^{π} for the image of $x \in X$ under $\pi \in G$, and we extend the action of G in the obvious way by defining (for instance) $\{a,b,c,\ldots\}^{\pi} = \{a^{\pi},b^{\pi},c^{\pi},\ldots\}$. We shall say that an object U is invariant under G if $U^{\pi} = U$ for all $\pi \in G$.

Statements of theorems

We can now use this language to state our theorems, together with some results of Mostowski which we have added for comparison. (Actually, we have extended Mostowski's results to refer to semigroup axioms $[S]$ rather than just axioms $[n]$, but his proofs extend obviously.)

I. If $\langle \alpha,\beta,\gamma,\ldots \rangle \subseteq \langle a,b,c,\ldots \rangle$, then

$$[\alpha,\beta,\gamma,\ldots] \Rightarrow [a,b,c,\ldots] .$$

II. Let $S_1,\ldots,S_N,S$ be semigroups. Then for the implication

$$[S_1] \,\&\, [S_2] \ldots \&\, [S_N] \Rightarrow [S] \qquad (*)$$

to hold it suffices that every group G with $\langle G \rangle \supseteq S$ should have some subgroup H for which $\langle H \rangle \supseteq$ some S_i.

III. (Mostowski) : For (*) to hold it is necessary that every abelian group G with $\langle G \rangle \supseteq S$ should have some subgroup H with $< H > \supseteq$ some S_i (actually, we might take $H = G$).

IV. The condition of II is necessary and sufficient for effective implication (written $\Rightarrow_+$) - that is, for the existence of a formula

$$F(f_1,\ldots,f_N) = f$$

which defines a choice function f for the class of all S-sets whenever we substitute such choice functions for the classes of all S_1-sets, ... , S_N-sets for the letters $f_1,\ldots,f_N$.

V. If p,q,r,... are primes, then [S] is effectively implied by the conjunction of [p],[q],[r],... and all the refinements of [S] in which the generators are all prime to pqr ... .

VI. Axiom [S] is effectively implied by the conjunction of the axioms [2,5] and [3,4] and all the prime-power refinements [S*] of [S]. In almost all cases we can omit the conjuncts [2,5] and [3,4] and many of the [S*] - for instance we need only consider those whose least generator is prime.

VII. For the implication (*) to hold it suffices that each of the semigroups <2,5>, <3,4>, and the prime power refinements of S should contain some S_i. We can weaken these requirements exactly as in VI.

IIX. (Mostowski) : For (*) to hold it is necessary that each prime refinement of S should contain some S_i. Mostowski's conjecture is that this condition suffices (in the case [S] = [n]).

Remarks on the theorems

The result I is trivial, and justifies the use of the [S] notation. II is in principle due to Mostowski - given choice functions for the classes of all S_1-sets,...,S_N-sets we give an explicit rule which determines an entry of any given S-set. III is included for comparison with II. The necessity part of IV is included for its own interest only, since the main concern of the paper is with sufficiency conditions. We shall later discuss in a particular case the relations between the notions of implication and effective implication.

Parts V - IIX are efficient tools for deciding the status of any given putative implication. Tarski proves a theorem equivalent to V (in the case when [S] = [n]), but by a completely different method. It is in part VI that we use the Thompson classification of minimal simple groups. From this we deduce that any insoluble group G has a subgroup H with $\langle H\rangle \supseteq \langle 2,5\rangle$ or $\langle H\rangle \supseteq \langle 3,4\rangle$, so that with [2,5] and [3,4] as hypotheses we ensure that the group G in II is soluble, and can use the rich variety of theorems on subgroups of soluble groups. In particular cases we can usually locate all the 'essential' [S*] very easily. The cases [S] = [64] and [S] = [15] are discussed in detail near the end of the paper. Part VII is an immediate corollary of VI, in a form which enables us to decide the status of (*) very easily in almost all cases. In many cases the conditions produced by VII are also necessary for ineffective implication, using Mostowski's result IIX, which we have added for comparison with VII. Note that our result VII yields Mostowski's conjecture with the word "prime" replaced by "prime power".

After our proofs of these results we give a table of necessary and sufficient conditions for the effective implication

$$[S_1] \;\&\; [S_2] \;\&\ldots\&\; [S_N] \Rightarrow_+ [n]$$

for all $n \leq 55$ and for $n = 64$ (a particularly interesting case) as deduced from the condition VII. In all but 7 cases, the conditions obtained are Mostowski's necessary conditions for (ineffective) implication, so that Mostowski's conjecture is verified for most n in the range of the table. The same techniques can be used to verify the conjecture for many larger values of n - a fairly easy case is $n = 1050$.

<u>Proof of I.</u> To explain the method we first deduce [2] from [4]. We 'manufacture' a (4)-set

$$(\{(a,0),(a,1),(b,0),(b,1)\}) = (\{a,b\})^*$$

from any (2)-set $(\{a,b\})$. Then given any collection of (2)-sets $(\{a,b\})$ we apply axiom [4] to choose entries in each of the corresponding (4)-sets $(\{a,b\})^*$, and then ignore the 'tags' 0,1.

In general, if $X = (A,B,C,\ldots)$ is an $(a,b,c,\ldots)$-set, and $\alpha = aa' + bb' + cc' + \ldots \in \langle a,b,c,\ldots\rangle$, we can 'manufacture' an (α)-set

$$(X_\alpha) = (A \times [m,m+a') \cup B \times [n,n+b') \cup C \times [p,p+c') \cup \ldots),$$

where $[u,v)$ denotes the interval of integers $u \leq t < v$, and the numbers $m,n,p,\ldots$ increase sufficiently rapidly. If the numbers $\beta,\gamma,\ldots$ are also in $\langle a,b,c,\ldots\rangle$, we can similarly manufacture a (β)-set (X_β), an (γ)-set $(X_\gamma),\ldots,$ and so an $(\alpha,\beta,\gamma,\ldots)$-set

$$X^* = (X_\alpha, X_\beta, X_\gamma, \ldots) .$$

We now obtain a choice function g for any collection of $(a,b,c,\ldots)$-sets X from such a function f for the corresponding collection of sets X^* by defining $g(X) = x$ whenever $f(X^*) = (x,t)$, t an integer.

<u>Proof of II.</u> Let $S = \langle a,b,c,\ldots\rangle$, and let $X = (A,B,C,\ldots)$ be an S-set. A <u>group on X</u> means a group G of permutations π of the entries of X such that $A^\pi = A$, $B^\pi = B, \ldots$. Now given any group H on X, and a subgroup K of H of index α, we can construct some (α)-sets (X_α) invariant under H as follows.

Let $t = (x,y,z,\ldots)$ be an ordered tuple for which $x,y,z,\ldots$ are the entries of X in some order, without repetitions. (So t is invariant only under the trivial group on X.) Define the set

$t^K = \{ t^\kappa \mid \kappa \in K \}$. Then plainly t^K is invariant under precisely the subgroup K of H (ie., is invariant under K but not under any larger subgroup of H.) Now by applying the permutations θ of H to t^K we obtain only α distinct sets $t^{K\theta}$, for the set $t^{K\theta}$ depends only on the coset $K\theta$ of K in H to which θ belongs. Our (α)-set (X_α) is taken to be that one whose entries are the sets $t^{K\theta}$. Plainly (X_α) is invariant under H, while any entry of (X_α) is invariant precisely under some conjugate of K in H.

Now let $S_i = \langle\alpha,\beta,\gamma,\ldots\rangle \subseteq \langle H\rangle$. Then from sets $(X_\alpha),(X_\beta),\ldots$ constructed as above we can manufacture an S_i-set X^* with the property that X^* is invariant under H, while any entry of X^* is invariant precisely under some proper subgroup K of H (the subgroup depending on the entry).

Let us construct all such sets X^* corresponding to all possible t and all the semigroups $S_1,\ldots,S_N$ appearing in II, and select appropriate choice functions $f_1,\ldots,f_N$. Let G be the group of all permutations π on X for which

$$f_i(X^{*\pi}) = f_i(X^*)^\pi$$

for all appropriate X^* and f_i. <u>We assert that G has a fixed point among the entries of X.</u>

If not, let the orbits of G on A have sizes $a_1,a_2,\ldots,a_m$, those on B have sizes $b_1,b_2,\ldots,b_n$, etc. Then the numbers $a_i,b_j,\ldots$ are all greater than 1 and so indices of <u>proper</u> subgroups of G, so we conclude that $S = \langle a,b,c,\ldots\rangle \subseteq \langle G\rangle$. It follows from the condition of II that for some S_i and some subgroup H of G we have $S_i \subseteq \langle H\rangle$.

Let $f_i(X^*) = x^*$, say. Then $f_i(X^{*\pi}) = f_i(X^*)^\pi = x^{*\pi}$ for all $\pi \in G$, by the definition of G, while $X^{*\pi} = X^*$ for all $\pi \in H$. But taking $\pi \in H$, $\pi \notin K$, where K is precisely the (proper!) subgroup of H which leaves x^* invariant, we get

$$x^* = f(X^*) = f(X^{*\pi}) = x^{*\pi} \neq x^*,$$

a contradiction.

We now give a rule which specifies a particular fixed entry x of X, which will be the chosen entry $f(X)$.

We construct a standard $(a,b,c,\ldots)$-set

$$([m,m+a), [n,n+b), [p,p+c), \ldots) = X_o, \text{ say,}$$

where the integers $m,n,p,\ldots$ are $0,a,a+b,\ldots$. Lexicographically order all permutations π on X_o in the standard way, and then order all groups G_o on X_o by making G_o precede H_o if and only if the first permutation in their symmetric difference is in G_o. Now any 1-1 correspondence ϕ between the entries of X and those of X_o which takes $A \to [m,m+a)$, $B \to [n,n+b)$, $C \to [p,p+c)$,... will take our group G on X to a 'similar' group G_o on X_o . If two such maps ϕ,ψ take G to the same group G_o, then $\phi\psi^{-1} \varepsilon G$.

Let G_o be the lexicographically first group on X_o which is of the form G^ϕ, where G is the group on X which we have just proved to have a fixed point. Then G_o has a fixed point among the entries of X_o. Let x_o be the lexicographically first such fixed point. Then the equation

$$x_o = x^\phi$$

determines an entry x of X fixed under G, and x is independent of the choice of ϕ since if also $G^\psi = G_o$, then $\phi\psi^{-1} \varepsilon G$, so $x^{\phi\psi-1} = x$, so $x^\phi = x^\psi = x_o$.

So supposing the conditions of II and given choice functions f_i corresponding to the axioms $[S_i]$ of II, we have given a rule which specifies an entry x of an arbitrary S-set X, or in other words, produced a choice function f for the class of all S-sets. This shows that the condition of II suffices for the implication (*), and indeed that then this implication is effective.

It will be noted that we are using the 'class forms' of our semigroup axioms. However to show that the 'set forms' of the $[S_i]$ effectively imply the set form of the axiom [S] given condition II, it suffices to remark that at no time in the proof of [S] do we need to apply any $[S_i]$ to more than a set of S_i-sets.

Proof of IV. We have just established the sufficiency. For the necessity, we give less detail. Let X be an S-set, whose elements are to be thought of as individuals (Urelemente). Let V be the class which contains these entries, sets of them, sets of these, and so on, to every ordinal depth. Then any permutation on X extends naturally to a permutation on V. Let G be an abstract group such that $\langle G\rangle \supseteq S$ but no subgroup H of G has $\langle H\rangle \supseteq S_i$ for any $i = 1,2,\ldots,N$.

Then we can define an action of G on X which has no fixed entry, since each of the numbers a,b,c,... is a sum of indices of proper subgroups of G. From now on we identify G with the resulting permutation group.

Let $Y = (A',B',\ldots)$ be any S_i-set, and let H be the subgroup of G which leaves Y invariant. We claim that H has a fixed point among the entries of Y. If not, the typical entry y of Y is invariant under precisely a proper subgroup K of H, and has [H:K] images which are all in the same part A' or B' or ... of Y. But this implies that $\langle H\rangle \supseteq S_i$.

Now under G, Y has [G:H] images Y^π. We now use the axiom of choice (!) to prove the existence of a choice function f_i invariant under G. Let G act on the class of all S_i-sets Y, and select (using the axiom of choice) one representative Y from each G-orbit, and one of the entries y of Y which is fixed under the subgroup H of G that leaves Y invariant. Define $f_i(Y^\pi) = y^\pi$ for each $\pi \in G$. Since for each $\pi \in H$ we have $Y^\pi = Y, y^\pi = y$, this is consistent.

Supposing the condition of II fails, we have shown that there exist choice functions $f_1,\dots,f_N$ for the classes of all S_i-sets $(i = 1,2,\dots,N)$ which are invariant under a group G which acts on X with no fixed entry in X. If we had

$$[S_1] \& \dots \& [S_N] \Rightarrow_+ [S]$$

there would therefore be a choice function

$$f = F(f_1,\dots,f_N)$$

for the class of all S-sets. But then

$$f^\pi = F(f_1{}^\pi,\dots,f_N{}^\pi) = F(f_1,\dots,f_N) = f$$

for all $\pi \in G$, so that f would also be invariant under G, whence

$$f(X) = f(X^\pi) = f(X)^\pi$$

for all $\pi \in G$. This is a contradiction, since $f(X)$ is some entry of X, and no entry of X is fixed under G.

<u>Proof of V.</u> A group G whose order is divisible by p has a subgroup H of order p, so that $\langle H\rangle = \langle p\rangle$, and the condition of II is satisfied for all such G. But if a group G has order prime to pqr..., its orbits on the entries of any S-set X define a refinement of S whose generators are prime to pqr... .

<u>Proof of VI.</u> Every insoluble group G has subgroups M and N, with N normal in M, so that $M^* = M/N$ is one of Thompson's list of <u>minimal simple groups.</u> We can check from this list that each of the possible M^* contains a subgroup H^* of one of the three types :

i) dihedral of order 6

ii) dihedral of order 10

iii) alternating on 4 letters.

So H^* has subgroups K^* of indices 2 & 3, 2 & 5, 3 & 4, in the three cases. The inverse image H of H^* in the canonical projection $M \to M^*$ has therefore subgroups K (inverse images of the K^*) of the same indices, so that we have $\langle H\rangle \supseteq \langle 2,5\rangle$ <u>or</u> $\langle H\rangle \supseteq \langle 3,4\rangle$ (note that $\langle 2,3\rangle \supseteq \langle 2,5\rangle$.). It follows that if [2,5] and [3,4] are among the $[S_i]$ in II, then the corresponding group G is soluble, as of course are all its subgroups.

Now for a soluble group H the maximal subgroups have prime-power index, and in fact a theorem of Galois shows that any subgroup of maximal order has prime index. It follows that any semigroup $\langle H\rangle$ for a soluble group H is prime-power generated, the least generator being prime. (One also has, for instance, that any non-prime generator is congruent to 1 modulo some prime generator.) By replacing H by one of its Hall subgroups if necessary, we can suppose all the generators of H appear in some refinement of S.

Moreover, if no generator of S is 2,3 or 5 (or 1, when [S] is trivial), then each generator of S belongs to both $\langle 2,5\rangle$ and $\langle 3,4\rangle$, so that each of [2,5] and [3,4] is a prime-power refinement of [S], and can be omitted from the hypotheses of VI, while if [S] is [2], [3], or [5], then [S] is a prime refinement of [S] and VI is vacuous, even with [2,5] and [3,4] omitted.

These remarks establish all of VI, and VII is an easy corollary.

<u>Explanation and construction of the table</u>

The table shows for each $n \leq 55$, and for $n = 64$, a list of the axioms [S] which are <u>essential</u> for the effective implication to [n] in the sense that in order that

$$[S_1] \mathbin{\&} \ldots \mathbin{\&} [S_n] \Rightarrow_+ [n]$$

it is necessary and sufficient that each of the essential S should be included in at least one of the S_i. In the unstarred cases all the essential [S] are in fact prime generated, so that by Mostowski's

Table. Axioms [S] essential for the effective deduction of [n]

The meaning of the table is mostly explained in the text. (*) or (**) indicates cases where condition IIX is not sufficient for $\Rightarrow_+$. See the note on the case n = 69 for further explanation of the entry (**).

n	Essential [S] to $\Rightarrow_+$ [n]:
p	[p] (p prime)
4	[2]
6	[2],[3]
8	[2],[3,5]
9	[3],[2,7]
10	[2],[5],[3,7]
12	[2],[3],[5,7]
14	[2],[7],[3,11]
15(*)	[3],[5],[2,13],[2]or[7,8]
16	[2],[3,13],[5,11]
18	[2],[3],[5,13],[7,11]
20	[2],[5],[3,17],[7,13]
21(*)	[3],[7],[2,19],[5,11],[2]or[5,16]
22	[2],[11],[3,19],[5,17]
24	[2],[3],[5,19],[7,17],[11,13]
25	[5],[2,23],[3,19],[3,11],[7,11]
26	[2],[13],[3,23],[7,19],[5,11]
27	[3],[2,23],[5,11],[5,17],[7,13]
28(*)	[2],[7],[3,19],[5,23],[11,17],[5]or[3,25]
30	[2],[3],[5],[7,23],[11,19],[13,17]
32	[2],[3,29],[5,17],[13,19],[5,11],[7,11]
33	[3],[11],[2,31],[5,23],[7,13],[7,19]
34	[2],[17],[3,31],[5,29],[7,13],[11,23]
35	[5],[7],[2,31],[3,29],[11,13]
36	[2],[3],[5,31],[7,29],[13,23],[17,19]
38	[2],[19],[3,29],[5,23],[5,11],[7,31]
39(*)	[3],[13],[2,37],[5,29],[5,17],[5,11,23],[7,11],[11,17],[2]or[7,8]
40(*)	[2],[5],[3,37],[7,19],[7,13],[11,29],[17,23],[3]or[13,27]
42	[2],[3],[7],[5,37],[11,31],[13,29],[19,23]
44	[2],[11],[3,41],[5,29],[5,17],[13,31]
45	[3],[5],[2,43],[7,31],[7,19],[11,23],[11,17],[13,19]
46	[2],[23],[3,43],[5,41],[7,11],[7,13],[11,13],[17,29]
48	[2],[3],[5,43],[7,41],[11,37],[17,31],[19,29]
49	[7],[2,47],[3,43],[3,23],[5,29],[5,17],[5,11],[11,19], [13,23],[13,17,19]
50	[2],[5],[3,47],[7,43],[11,17],[13,37],[19,31]
51	[3],[17],[2,47],[5,41],[5,17,29],[5,23],[7,37],[7,13,31], [11,29],[11,17,23],[13,19]
52(*)	[2],[13],[7]or[3,49],[11,41],[5,47],[3,43],[3,23],[7,31], [7,19],[23,29]
54	[2],[3],[5,29],[7,47],[11,43],[13,41],[17,37],[23,31]
55(**)	[5],[11],[2,53],[7]or[3,49],[3,43],[3,23],[7,41], [7,19,29],[2]or[7,8],[13,29],[13,19,23],[17,19]
64	[2],[3,61],[5,59],[11,53],[17,47],[23,41],[7,43], [7,19],[7,13,37],[7,13,31],[13,19]

result IIX they are in fact essential for the <u>ineffective</u> deduction of [n]. (But in the starred cases we find alternatives for some of the [S] - this will be explained later by reference to a particular case.) Of course we do not regard [S] as essential if it is a consequence of some other (obviously <u>more</u> essential!) [S] in the list.

We can best further explain the construction and the meaning of this table by examining the hardest case n = 64. Since all the listed [S] in this case are prime generated, and since no one of the corresponding semigroups S includes another, Mostowski's condition IIX tells us that the conjunction of all the listed axioms is not equivalent to the conjunction of any proper subset of them - in other words, no axiom of the list may be omitted.

It is our aim to prove that any further prime-power generated refinement [S] of [64] is implied by some axiom of the list.

Since [2] is in the list, we can suppose that each generator q of S is odd. We can also suppose that each generator q of S is itself essential in the sense that 64 is not in the semigroup T obtained by omitting q from the generators of S, for otherwise $[T] \Rightarrow [S]$. We now consider the expression of 64 as a sum (with repetitions) of generators of S.

The expression of 64 as a sum of two primes are

$$3 + 61,\ 5 + 59,\ 11 + 53,\ 17 + 47,\ 23 + 41$$

which all correspond to axioms of the list. No one of these primes p can be a generator of any other S, for then $S \supseteq < p, 64-p >$, and so $[p,64-p] \Rightarrow [S]$. So the only permissible generators in a further S are the primes 7, 13, 19, 29, 31, 37, 43, and the prime powers 9,25, 49, 27.

For q = 9, 25, 49 the only primes p with $q \equiv 1$ (modulo p) are 2 and 3, which are not permitted, so these q cannot occur.

Similarly, if 27 is a generator, so is 13, and the corresponding partition of 64 refines 27 + 13 + 24. But there is no partition of 24 into permitted prime powers. We have now shown that all the required S are prime generated, so that Mostowski's condition holds for n = 64.

Any further axiom must correspond to a splitting of 64 into at least four permitted primes (allowing repetitions), the least of which is at most 64/4 and so either 7 or 13. If 7 occurs at least twice we have a refinement of 7+7+50. But the splittings of 50 into two permitted primes are 7+43, 13+37, which are coped with by the listed axioms [7,43] and [7,13,37]; while in a splitting into four or more primes the least is at most 50/4 and so must be 7, and [S] is again implied by [7,43].

If 7 and 13 both occur the partition is a refinement of

$$7 + 13 + 44$$

in which 44 can only be split into two permitted primes, and in fact the only possibility is 13 + 31, corresponding to the listed axiom [7,13,31].

If 7 occurs without 13 we have a refinement of 7 + 57 in which 57 is split into at least three primes all $\geq$ 19, namely the unique refinement 7 + 19 + 19 + 19, corresponding to [7,19]. Finally, if 7 does not occur, we find ourselves successively refining

13 + 51 through 13 + 13 + 38 to 13 + 13 + 19 + 19,

corresponding to the last listed axiom [13,19].

We have therefore shown that

$$\text{"}[S_1] \,\&\ldots\&\, [S_N] \Rightarrow_{+} [64]$$

holds if and only if for each of the listed axioms [S] the semigroup

S includes some one of $S_1,\ldots,S_N$". This is the meaning of the last line in our table.

As an example where Mostowski's condition is insufficient for the effective implication, we now take $n = 15$. Here the refinements [3],[5] and [2,13] are plainly essential, so we can suppose that no one of 3^a, 5^b, 2, 13 is a generator of any further S, leaving only the possibilities 4,7,8,11. Since the least generator can be supposed prime, 4 is disallowed, and plainly the only permissible refinement is [7,8].

So we deduce from VI (or VII) that

$$[3] \& [5] \& [2,13] \& [7,8] \Rightarrow_+ [15],$$

the first three being essential by Mostowski's criterion. Now the only group of order $2^a 7^b$ on 15 letters with $\langle G\rangle = \langle 7,8\rangle$ is a group of order 56 consisting of an elementary abelian group of order 8 extended by an automorphism of order 7. Each subgroup H of G has $\langle H\rangle \subseteq \langle 2\rangle$ or $\langle H\rangle \subseteq \langle 7,8\rangle$, and both $\langle 2\rangle$ and $\langle 7,8\rangle$ occur, so that by IV we find :

"For $[S_1] \& \ldots \& [S_N] \Rightarrow_+ [15]$ it is necessary and sufficient that there exist subscripts h,i,j,k such that

$$S_h \subseteq \langle 3\rangle,\ S_i \subseteq \langle 5\rangle,\ S_j \subseteq \langle 2,13\rangle, \text{ and } (S_k \subseteq \langle 2\rangle \ \underline{\text{or}}\ \langle 7,8\rangle)."$$

The other starred entries in the table are to be interpreted similarly.

<u>How to use the table</u>

For what n does $[n] \Rightarrow [8]$?

For this it is necessary and sufficient that each of the semigroups $\langle 2\rangle$ and $\langle 3,5\rangle$ include $\langle n\rangle$ - that is, that n be expressible in both forms $2a$ and $3b + 5c$ $(a,b,c \geq 0)$. The n of form $3b + 5c$ are 3,5,6,8,9,10 and all larger numbers (continue by adding 3s to 8,9,10), and so we conclude that $[n] \Rightarrow [8]$ only if [n] is an even one of these, i.e., if and only if n is an even number other than 2 or 4.

Similarly [n] $\Rightarrow$ [6] only if n is a multiple of 6, while [n] $\Rightarrow$ [9] if and only if n = 6 or any larger multiple of 3. Again, [3] & [7] $\Rightarrow$ [9], since the semigroup <3> contains 3, and <2,7> contains 7, but [3] & [5] $\not\Rightarrow$ [9] since neither 3 nor 5 is a member of <2,7>. The smallest n $\neq$ 16 for which [n] $\Rightarrow$ [16] is n

Note that in each case so far implications have been effective whenever they were valid.

Is it true that [3] & [5] & [13] $\Rightarrow$ [15]?

This is a famous question, the first test case of Mostowski's conjecture. Here we have elements in each of <3>, <5>, <2,13>, but no element in either of <2> or <7,8>, so certainly there is no <u>effective</u> implication. We shall discuss the possibilities for ineffective implications in a moment.

We should remark that Mostowski himself verified his conjecture for all n > 20 other than 15, so that the particular cases discussed above are none of them new. But it would be difficult to verify the conjecture for many n > 20 using only previously published ideas. It is a remarkable fact that there are so few doubtful cases with n $\leq$ 55 (only 15,21,28,39,40,52,55) and that the remaining problem is similar in each of these cases.

<u>The case n = 1050</u>

Mostowski's condition can be shown to suffice in the case n = 1050 by the following argument. Since $1050 = 2.3.5^2.7 < 11^3 = 1331$ we need only consider as non-prime generators the squares of the primes from 11 to 31. But for these primes p, all the prime divisors of p^2-1 are among 2,3,5,7,11, which we need not allow, since 1050 is contained in each of the prime-generated semigroups <2>, <3>, <5> <7>, <11,1039>.

<u>The case $n = 69$</u>

Mostowski's conjecture in its original form holds for $n = 69$, although our generalisation IIX does not. This is because of a curious number-theoretical accident. In fact, if we are concerned only with implications of the form

$$[m_1] \& \ldots \& [m_N] \Rightarrow [n],$$

as distinct from the more general implications

$$[S_1] \& \ldots \& [S_N] \Rightarrow [n],$$

then in the case $n = 69$ we can omit the terms $[2]$ <u>or</u> $[7,8]$, $[2]$ <u>or</u> $[5,16]$ from the extension of our table since

$m \in \langle 2,67\rangle \Rightarrow m \in \langle 2\rangle$ or $(m \in \langle 7,8\rangle \;\&\; m \in \langle 5,16\rangle)$.

A similar phenomenon happens when $n = 55$, the term $[2]$ <u>or</u> $[7,8]$ being then 'dominated' by $[2,53]$, but here we cannot deduce the Mostowski conjecture, since we still have the bad term $[7]$ <u>or</u> $[3,49]$. (It was a protracted consideration of such phenomena - first noted at $n = 55$ - that arrested further growth of the table!)

<u>Ineffective implications in the residual cases?</u>

We take

$$[3] \& [5] \& [13] \Rightarrow [15] \qquad (?)$$

as our example, since all the other cases in the range of our table are entirely similar. Select choice functions f_3, f_5, f_{13} corresponding to the three hypotheses.

The only case when we fail to define $f(X)$ (X being some 15-element set) is when the group of permutations leaving f_3, f_5, f_{13} invariant is isomorphic to the abstract group G_o of order 56 mentioned earlier. (Mostowski, of course, was aware of the relevance of this group.) So when we cannot define $f(X)$ we can define an action of G_o on X, so that G_o will have orbits of sizes 7 and 8 on X.

Now because G_o has a <u>unique</u> subgroup of order 8, it can transitively permute a 7-element set in essentially just one way. So if we are given fixed-point-free actions of G_o on two 15-element sets X and Y, we can determine a 1-1 correspondence between the two corresponding 7-element orbits $\overline{X}$ and $\overline{Y}$ of G_o in X and Y.

We can express the resulting axiomatic implication as

$$[3] \;\&\; [5] \;\&\; [13] \Rightarrow_{+} [15]_{(1 \;\underline{\text{or}}\; 7=7)} \qquad (**).$$

The new axiom $[15]_{(1 \;\underline{\text{or}}\; 7=7)}$ asserts that:

there is a function $f(X,Y)$ defined for all pairs X,Y of 15-element sets whose value is <u>either</u> an element of $X \cup Y$ <u>or</u> a 1-1 correspondenc between a 7-element subset of X and one of Y.

This implication (**) seems to be the strongest reasonable (?) effective approximation to the desired implication

$$[3] \;\&\; [5] \;\&\; [13] \Rightarrow [15].$$

We show for instance that $[15]_{(1 \;\underline{\text{or}}\; 7=7)} \Rightarrow [15]_W$, where $[15]_W$ asserts the existence of choice functions for <u>well-ordered</u> collections of 15-element sets. (This will be our only example of an <u>ineffective</u> implication.)

Let $\sum = \{ X_\beta \mid \beta \;\; \alpha \quad On \}$ be some well-ordered sequence of 15-element sets. Apply $[15]_{(1 \;\underline{\text{or}}\; 7=7)}$ to $\sum$, and let β denote the typical ordinal less than α.

If $f(X_\beta, X_\gamma) \in X_\beta$ for some ordinal $\gamma < \alpha$, define $g(X_\beta) = f(X_\beta, X_\gamma)$ for the least such γ. If not, but $f(X_\gamma, X_\beta) \in X_\beta$ for some $\gamma < \alpha$, define $g(X_\beta) = f(X_\gamma, X_\beta)$ for the least such γ. Ther for all β, γ for which $g(X_\beta)$ and $g(X_\gamma)$ are both undefined, $f(X_\beta, X_\gamma)$ is a 1-1 correspondence $f_{\beta\gamma}$ between 7-element subsets $\overline{X_\beta}$ and $\overline{X_\gamma}$ of X_β and X_γ. (These subsets might depend on both paramete β, γ but this is immaterial.)

If there is any β for which $g(X_\beta)$ has not yet been defined, <u>choose at random an ordering of</u> X_β. Then for any further X_γ with $g(X_\gamma)$ undefined, 'transfer' this ordering to $\overline{X_\gamma}$ using the 1-1 correspondence $f_{\beta\gamma}$, and define $g(X_\gamma)$ to be the first element of $\overline{X_\gamma}$ in this ordering, while $g(X_\beta)$ is to be the first element in the chosen ordering of X_β. Since this completes the definition of a choice function for $\sum$, we have verified $[15]_W$, but our deduction has been ineffective, since we have specified 15! choice functions rather than any particular one.

<u>Other problems and axioms</u>

It seems likely that 2 and 4 are the only distinct numbers for which the corresponding axioms are equivalent, but I do not think that anyone has proved this. It also seems likely that the only cases of $[m] \Rightarrow [n]$ with $m < n$ are $[2] \Rightarrow [4]$, $[6] \Rightarrow [8]$, and $[6] \Rightarrow [9]$. (It is easy to see that n cannot be square free.) Proofs of these conjectures could possibly be carried through using known additive properties of primes, although it might be necessary to assume the Goldbach conjecture.

It would be interesting to know whether or not there is a largest number n for which Mostowski's condition suffices for effective implications. We have already discussed the case $n = 1050$, and it seems very likely that with a table of primes and a few hours of hard work one could produce a much larger number. (Occasionally, one might need to make use of some slight knowledge of properties of soluble groups.) Of course, highly composite numbers are the most hopeful candidates. It is not at all clear whether or not there are infinitely many cases of this kind.

Finally, we note that the same methods are applicable to many generalisations of the basic axioms. An example which is not hard to prove, but is nevertheless quite surprising, is the equivalence

$$[2] \mathbin{\&} [3] \mathbin{\&} [7] \Leftrightarrow [10]_5 ,$$

where $[10]_5$ asserts the existence of a function f defined on the class of all 10-element sets X, f(X) being always a 5-element subset of X. The reader might like to verify this equivalence ab initio - the details are curious, but not difficult. I do not know whether, for instance, there is any larger n for which $[2n]_n$ is equivalent to a conjunction [a] & [b] &... of ordinary choice axioms, but there are certainly many strange implications to be found among these axioms $[m]_n$.

Dept. of PMMS
CAMBRIDGE / England

XV. ON DESCENDINGLY COMPLETE ULTRAFILTERS

KAREL PRIKRY

University of California, Los Angeles

Dedicated to Professor A. Mostowski

The notion of an α-descendingly complete (abbrev. α-d.c.) ultrafilter is due to Keisler. Descendingly complete ultrafilters were studied by Chang [1], G. V. and D. V. Chudnovsky [3], Kunen and Prikry [10], Jorgensen [5] and [6], and Silver.

In the present paper we present a theory of ultraproducts with respect to d.c. ultrafilters which is very similar to the theory of ultraproducts with respect to countably complete ultrafilters as developed in Keisler, Tarski [8]. Some of the results of the present paper were also obtained by the above mentioned authors. G. V. and D. V. Chudnovsky proved Theorems 1 and 3 by making use of Ulam and Hajnal matrices constructed in [15] and [4]. Compare also Kunen, Prikry [10]. Recently, Jorgensen [5] announced Proposition 4a. We proved this in September 1970 and most of the results of the present paper were obtained by February 1971.

In April 1971, Silver, using as a starting point the theory presented here, derived the existence of $0^{\#}$, thus advancing further the analogy between descendingly complete and countably complete ultrafilters. See Theorem 15 for the precise statement of Silver's theorem. The definition and properties of $0^{\#}$ can be found in Solovay [14]. Compare also Silver [13]. Silver has also obtained interesting results of a combinatorial nature about d.c. ultrafilters.

Further results about d.c. ultrafilters can be found in [12].

In the first part of the present paper we present results concerning cardinalities of ultraproducts and the length of decreasing sequences. We use these results to investigate further the structure of ultraproducts and the size of

The preparation of this paper was supported in part by NSF Grants #144-A966 and GP-27964.

cardinals carrying ultrafilters of various degrees of completeness. Finally, at the end of the paper, we combine some results of R. Jensen, J. Silver and the author to answer a question of Chang [1] under the hypothesis of $V = L$. We show, among other things, that, if $V = L$, then every uniform ultrafilter over $\omega_{\omega+1}$ is ω_n-descendingly incomplete for all $n \in \omega$.

I would like to thank C. C. Chang, whose suggestions led to many improvements and C. Richards, S. Shelah and J. Silver for their interest in this work.

Until further notice, let $\mathcal{U}$ be an ultrafilter over an arbitrary infinite set I.

The following definition is due to Keisler.

DEFINITION 1. (Chang [1]). $\mathcal{U}$ is α-descendingly incomplete (abbrev. α-d.i.) if there are sets $X_\xi \in \mathcal{U}$ $(\xi < \alpha)$ such that

$$(\forall \xi,\eta)(\xi < \eta < \alpha \rightarrow X_\xi \supseteq X_\eta) \text{ and } \bigcap \{X_\xi : \xi < \alpha\} = 0 .$$

$\mathcal{U}$ is α-descendingly complete (abbrev. α-d.c.) if $\mathcal{U}$ is not α-d.i.

It is easy to see that $\mathcal{U}$ is α-d.i. iff $\mathcal{U}$ is $cf(\alpha)$-d.i.

The following theorem, proved first by Chang [1] under the assumption $2^\lambda = \lambda^+$, was proved in [3] and [10] without any such hypotheses.

THEOREM 1. Let λ be a cardinal.

(a) If λ is regular and $\mathcal{U}$ is λ^+-d.i., then $\mathcal{U}$ is λ-d.i.

(b) If $\nu = cf(\lambda) < \lambda$ and $\mathcal{U}$ is λ^+-d.i., then either

(i) $\mathcal{U}$ is ν-d.i.,

or

(ii) There is some $\alpha < \lambda$ such that $\mathcal{U}$ is β-d.i. for all regular β such that $\alpha < \beta < \lambda$.

Also the following notion, closely related to "descendingly complete," is due to Keisler.

DEFINITION 2. Let α be a cardinal. $\mathcal{U}$ is α-decomposable (abbrev. α-dec.) if there are sets Y_ξ $(\xi < \alpha)$ such that

(i) $(\forall\xi,\eta)(\xi,\eta < \alpha \;\&\; \xi \neq \eta \rightarrow Y_\xi \cap Y_\eta = 0)$;

(ii) $\bigcup\{Y_\xi : \xi < \alpha\} = I$;

(iii) for all $S \subseteq \alpha$, if $|S| < \alpha$, then

$$\bigcup\{Y_\xi : \xi \in S\} \notin \mathfrak{U} .$$

If $\mathfrak{Y} = \{Y_\xi : \xi < \alpha\}$ satisfies (i), (ii), (iii), then $\mathfrak{Y}$ is called an α-decomposition for $\mathfrak{U}$.

$\mathfrak{U}$ is α-indecomposable (abbrev. α-indec.) if $\mathfrak{U}$ is not α-dec.

REMARK. If $\mathfrak{U}$ is α-dec., then $\mathfrak{U}$ generates a uniform ultrafilter $\mathfrak{V}$ on α in the following manner: For every $S \subseteq \alpha$, $S \in \mathfrak{V}$ iff $\bigcup\{Y_\xi : \xi \in S\} \in \mathfrak{U}$. If $\mathfrak{V}$ and $\mathfrak{U}$ are related in the above manner, then $\mathfrak{V}$ is said to be $\leq \mathfrak{U}$ in Rudin-Keisler ordering.

The following lemma is easy.

LEMMA 1. _For regular_ α, $\mathfrak{U}$ _is_ α-_dec. iff_ $\mathfrak{U}$ _is_ α-_d.i._ _For singular_ α, _if_ $\mathfrak{U}$ _is_ α-_dec., then_ $\mathfrak{U}$ _is_ α-_d.i._

We sketch a proof of the following proposition.

PROPOSITION 1. _Suppose that_ λ _is a singular strong_[†] _limit cardinal and_ $\nu = cf(\lambda)$. _Let_ $\mathfrak{U}$ _be_ ν-_dec. and_ μ-_dec. for arbitrarily large_ $\mu < \lambda$. _Then_ $\mathfrak{U}$ _is_ λ-_dec._

First we show

LEMMA 2. _Let_ $\mathfrak{U}$ _be as in Proposition 1. Then there are_

$$\mathfrak{A}_\xi = \{A^\xi_\rho : \rho \in \lambda_\xi\}$$

and λ_ξ $(\xi \in \nu)$ _such that_

[†] If $cf(\lambda) = \omega$, "strong" can be omitted. This becomes evident by examining the proof.

(i) $\mathfrak{A}_\xi$ <u>is a</u> λ_ξ<u>-decomposition for</u> $\mathfrak{U}$ $(\xi \in \nu)$.

(ii) λ_ξ $(\xi \in \nu)$ <u>are cardinals</u>; $\lambda_0 = \nu$; <u>if</u> $\varkappa < \nu$, <u>then</u> $\sup\{\lambda_\xi : \xi \in \varkappa\} < \lambda_\varkappa$; $\lim\{\lambda_\xi : \xi \in \nu\} = \lambda$.

(iii) <u>If</u> $\xi < \eta < \nu$, <u>then</u> $\mathfrak{A}_\eta$ <u>is a refinement of</u> $\mathfrak{A}_\xi$, <u>i.e. for every</u> $\rho \in \lambda_\eta$ <u>there is a</u> $\sigma \in \lambda_\xi$ <u>such that</u> $A^\eta_\rho \subseteq A^\xi_\sigma$.

<u>Proof</u>. We define $\mathfrak{A}_\xi$, λ_ξ by induction. Let $\lambda_0 = \nu$ and $\mathfrak{A}_0$ be any λ_0-decomposition for $\mathfrak{U}$. If $\mathfrak{A}_\xi$, λ_ξ are defined for $\xi < \varkappa < \nu$ as required, let $\sim$ be the canonical refinement of the equivalence relations given by $\mathfrak{A}_\xi$, $\xi < \varkappa$. Since λ is strong limit, $\sim$ has fewer than λ equivalence classes. Let $\lambda_\varkappa$ be chosen so that $\mathfrak{U}$ is $\lambda_\varkappa$-dec., $\lambda_\varkappa < \lambda$ and $\sim$ has fewer than $\lambda_\varkappa$-equiv. classes. Let $\mathfrak{B}$ be a $\lambda_\varkappa$-decomposition for $\mathfrak{U}$ and $\mathfrak{A}_\varkappa$ be the canonical refinement of $\sim$ and $\mathfrak{B}$.

<u>Proof of Proposition 1</u>. Let μ_ξ $(\xi \in \nu)$ be chosen as follows: $\mu_0 = \lambda_0 = \nu$. For all $\xi < \nu$, μ_ξ is regular and $\lim\{\lambda_\eta : \eta < \xi\} < \mu_\xi \leq \lambda_\xi$. We further define a sequence ν_ξ, $1 \leq \xi < \nu$ of ordinals $< \nu$. ν_ξ is the least ordinal such that $\sup\{\nu_\eta : \eta < \xi\} < \nu_\xi < \nu$ and $\mathfrak{A}_\xi$ splits

$$\bigcup \{A^0_\rho : \sup\{\nu_\eta : \eta < \xi\} < \rho < \nu_\xi\}$$

into at least μ_ξ equivalence classes. Such a ν_ξ exists because $\mathfrak{A}_\xi$ splits

$$\bigcup \{A^0_\rho : \sup\{\nu_\eta : \eta < \xi\} < \rho < \nu\}$$

(which is a set in $\mathfrak{U}$) into λ_ξ pieces, $\lambda_\xi \geq \mu_\xi > \nu$ and μ_ξ is regular. We now define a λ-decomposition $\mathfrak{A}$ for $\mathfrak{U}$ as follows: $\mathfrak{A}$ coincides with $\mathfrak{A}_\xi$, $1 \leq \xi < \nu$, on

$$\bigcup \{A^0_\rho : \sup\{\nu_\eta : \eta < \xi\} < \rho < \nu_\xi\} .$$

We claim that $\mathfrak{A}$ is a λ-decomposition for $\mathfrak{U}$. It is enough to show that if $S \subseteq \mathfrak{A}$ and $|S| = \mu < \lambda$, then

$$B = \bigcup \{A : A \in S\} \notin \mathfrak{U} .$$

Assume the contrary. Let ξ be such that $\mu_\xi > \mu$. Let

$$S' = \{A \in S : (\exists \rho)(\rho > \sup\{\nu_\eta : \eta < \xi\}, A \subseteq A^0_\rho)\} ,$$

$$R = \{B : B \in \mathfrak{A}_\xi \;\&\; (\exists A \in S')(A \subseteq B)\} .$$

It is easy to see that $\bigcup S' \in \mathfrak{U}$ and hence $\bigcup R \in \mathfrak{U}$. But $|R| \leq \mu < \mu_\xi \leq \lambda_\xi$ and we have a contradiction with $\mathfrak{A}_\xi$ being a λ_ξ-decomposition for $\mathfrak{U}$. Hence Proposition 1 is proved.

The following proposition is due to Carl Richards and the author.

PROPOSITION 2. <u>Let</u> λ <u>be an inaccessible cardinal,</u> $2^\lambda = \lambda^+$ <u>and</u> $\mathfrak{U}$ <u>be</u> μ-<u>dec. for arbitrarily large</u> $\mu < \lambda$. <u>Then</u> $\mathfrak{U}$ <u>is</u> λ-<u>dec</u>.

<u>Proof</u>. This is similar to Proposition 1 but somewhat easier. We consider μ-decompositions for arbitrarily large $\mu < \lambda$. If their canonical refinement has a subfamily of cardinality λ with union in $\mathfrak{U}$, we are done. Otherwise it is a λ^+-decomposition and we use Theorem 1.

PROPOSITION 3. (GCH). <u>If</u> λ <u>is the least cardinal such that</u> $\mathfrak{U}$ <u>is</u> λ-<u>indec</u>., <u>then</u> $\lambda = \omega$ <u>or</u> λ <u>is a successor cardinal</u>.

<u>Proof</u>. This follows easily from Propositions 1,2.

If $\mathfrak{U}$ is an ultrafilter over I, we let $\prod A_i/\mathfrak{U}$ denote the ultraproduct of $\{A_i : i \in I\}$ with respect to $\mathfrak{U}$. If $f \in \prod A_i$, then $[f]$ will denote the corresponding element of $\prod A_i/\mathfrak{U}$.

DEFINITION 3. For every cardinal μ, define by induction: $\mu^{(0)} = \mu$, $\mu^{(n+1)} = (\mu^{(n)})^+$.

LEMMA 3. <u>If</u> $\mathfrak{U}$ <u>is</u> μ^+-<u>indec. or</u> μ <u>is regular and</u> $\mathfrak{U}$ <u>is</u> μ-<u>indec., then</u> $\mathfrak{U}$ <u>is</u> $\mu^{(n)}$-<u>indec. for all</u> $n \in \omega$, $n \neq 0$.

<u>Proof</u>. By Theorem 1(a) and Lemma 1.

PROPOSITION 4. (GCH). *Let* μ *be a cardinal and* $\mathfrak{u}$ *be* μ^+*-indec. Then*

[1](a) $|\prod \mu/_{\mathfrak{u}}| \leq \mu^+$

[1](b) *For all* $n \geq 1$, $|\prod \mu^{(n)}/_{\mathfrak{u}}| \leq \mu^{(n)}$

(c) *For every ordinal* ν, $\prod \nu/_{\mathfrak{u}}$ *has no decreasing sequence of type* μ^{++}.

Proof. (a) Assume that $|\prod \mu/_{\mathfrak{u}_I}| > \mu^+$. Hence there are functions $f_\alpha \in \mu^I$, $\alpha < \mu^{++}$, such that for $\alpha,\beta \in \mu^{++}$, if $\alpha \neq \beta$, then $f_\alpha(i) \neq f_\beta(i)$ a.e., i.e. $\{i : f_\alpha(i) \neq f_\beta(i)\} \in \mathfrak{u}$. For every $i \in I$, let

$$A_i = \{\{\alpha,\beta\} : f_\alpha(i) \neq f_\beta(i)\} .$$

For every $A \subseteq [\mu^{++}]^2$, let

$$X_A = \{i : A_i = A\} .$$

This gives a partition of I into at most μ^{+++} equivalence classes. By Lemma 3, $\mathfrak{u}$ is μ^{++}-indec. and μ^{+++}-indec. Hence one can find sets A^ξ $(\xi \in \mu)$ such that

$$X = \bigcup \{X_{A^\xi} : \xi \in \mu\} \in \mathfrak{u} .$$

We claim that

$$(*) \qquad \bigcup \{A^\xi : \xi \in \mu\} = [\mu^{++}]^2 .$$

Indeed, for any $\{\alpha,\beta\} \in [\mu^{++}]^2$ there is some $i \in X$ such that $f_\alpha(i) \neq f_\beta(i)$. Hence $\{\alpha,\beta\} \in A_i$. But $i \in X_{A^\xi}$ for some $\xi \in \mu$. Hence $A^\xi = A_i$, i.e. $\{\alpha,\beta\} \in A^\xi$.

By applying the Erdös-Rado theorem to (*) we can find $Z \subseteq \mu^{++}$ such that $|Z| = \mu^+$ and $[Z]^2 \subseteq A^{\xi_0}$ for some $\xi_0 \in \mu$. Since $A^{\xi_0} = A_{i_0}$ for some $i_0 \in I$, we get $f_\alpha(i_0) \neq f_\beta(i_0)$ for all $\{\alpha,\beta\} \in [Z]^2$. This yields μ^+ distinct elements of μ - a contradiction.

(b) By our hypothesis and Lemma 3, $\mathfrak{u}$ is $\mu^{(n)}$-indec. for all $n \geq 1$. Hence it can be easily seen that for all $n \geq 1$,

[1] This was also proved by M. Jorgensen [5].

$$\prod \mu^{(n)}/_{u} = \bigcup_{\nu\in\mu^{(n)}} \prod \nu/_{u} .$$

The result now follows from (a).

(c) The proof is similar to (a). This time we choose f_α, $\alpha \in \mu^{++}$, so that $f_\alpha(i) < f_\beta(i)$ a.e. whenever $\beta < \alpha < \mu^{++}$. A_i, X_A, A^ξ, Z, ξ_0, i_0 are then defined accordingly. We now have $|Z| = \mu^+$ and for any $\alpha,\beta \in Z$, if $\beta < \alpha$, then $f_\alpha(i_0) < f_\beta(i_0)$. This gives an infinite decreasing sequence of ordinals - a contradiction.

Let us also state a modification of Proposition 4 which does not use GCH.

PROPOSITION 4'. Let μ be a cardinal and let u be λ-indec. for all λ such that $\mu < \lambda \leq 2^{(2^\mu)^+}$. Then

(a) $|\prod \mu/_u| \leq 2^\mu$.

(b) For all $n \geq 1$, $|\prod \mu^{(n)}/_u| \leq \mu^{(n)} \cdot 2^\mu$.

(c) For every ordinal ν, $\prod \nu/_u$ has no decreasing sequence of type $(2^\mu)^+$.

(d) If $\lambda \leq 2^{(2^\mu)^+}$, then $|\prod \lambda/_u| \leq \lambda^\mu$. Hence in particular $|\prod 2^\mu/_u| = 2^\mu$.

Proof. We shall comment only on (d). It follows from the indecomposability hypothesis that if

$$\lambda \leq 2^{(2^\mu)^+} ,$$

then

$$\prod \lambda/_u \subseteq \bigcup \left\{ \prod X/_u : X \in [\lambda]^{\leq\mu} \right\} .$$

Hence

$$|\prod \lambda/_u| \leq 2^\mu \cdot \lambda^\mu = \lambda^\mu ,$$

$$|\prod 2^\mu/_u| \leq (2^\mu)^\mu = 2^\mu$$

and (d) follows.

The following proof of Proposition 4(a) was found by S. Shelah.

Suppose that $|\prod \mu/_{u}| > \mu^+$ and let f_α, $\alpha \in \mu^{++}$, be as in the previous proof of Proposition 4(a). Let $\sim_\alpha$ be the equivalence relation on I given by

$$i \sim_\alpha j \text{ iff } f_\alpha(i) = f_\alpha(j) .$$

Let $\sim$ be the canonical refinement of $\sim_\alpha$, $\alpha \in \mu^{++}$. $\sim$ gives a partition of I into at most μ^{+++} pairwise disjoint sets. Hence $\leq \mu$ of them have union in u. Let A_ξ, $\xi \in \mu$, be such a family of equivalence classes of $\sim$. Clearly, for every $\xi \in \mu$ and $\alpha \in \mu^{++}$, f_α is constant on A_ξ. Hence we can define $g_\alpha \in \mu^\mu$ by $g_\alpha(\xi) = \rho$ iff f_α takes value ρ on A_ξ, $g_\alpha(\xi) = 0$ if $A_\xi = 0$. Since $\bigcup \{A_\xi : \xi \in \mu\} \in u$ and $f_\alpha \neq f_\beta$ a.e. whenever $\alpha \neq \beta$, $\alpha,\beta \in \mu^{++}$, it follows that $g_\alpha \neq g_\beta$ for $\alpha \neq \beta$, $\alpha,\beta \in \mu^{++}$. Hence $|\mu^\mu| \geq \mu^{++}$, contradicting GCH.

DEFINITION 3. An ordered set K has cofinality $\geq \lambda$, abbreviated $cf(K) \geq \lambda$, if for every $H \subseteq K$ such that $|H| < \lambda$, there is some $k \in K$ such that $k > h$ for all $h \in H$. $cf(K) = \lambda$ if $cf(K) \geq \lambda$ and not $cf(K) \geq \lambda^+$.

The following result plays a crucial role in the present paper.

PROPOSITION 5. (GCH). _Let_ μ _be a cardinal and_ u _be_ μ^+_-indec. Let_ ν _be any ordinal and_ $K \subseteq \prod \nu/_u$, $cf(K) \geq \mu^{++}$, _and_ K _be bounded from above. Then there is a least_ $h \in \prod \nu/_u$ _such that_ h _bounds_ K.

Proof. Suppose there is no such h. Then by Proposition 4(c) we can find a cardinal $\lambda \leq \mu^+$ and functions f_α, $\alpha \in \lambda$, such that

(i) $\alpha > \beta \rightarrow f_\alpha(i) < f_\beta(i)$ a.e.;

(ii) $[f_\alpha]$ bounds K for each $\alpha \in \lambda$;

and

(iii) for each f, if $[f]$ bounds K, then $[f_\alpha] < [f]$ for some $\alpha \in \lambda$.

Let $A_i = \{f_\alpha(i) : \alpha \in \lambda\}$. Then $|A_i| \leq \lambda \leq \mu^+$. Hence by Proposition 4(b), $|\prod A_i/_u| \leq \mu^+$. Let

$$F = \left\{[f] : [f] \in \prod A_i/_{\mathcal{U}} \ \& \ (\exists [k] \in K)([k] > [f])\right\} ,$$

i.e. F is the set of K-bounded elements of $\prod A_i/_{\mathcal{U}}$. Since $|F| \leq \mu^+$ and $cf(K) \geq \mu^{++}$, we can find $[k_0] \in K$ such that $[k_0] > [f]$ for all $[f] \in F$. For $\alpha \in \lambda$, let

$$I_\alpha = \{i : f_\alpha(i) > k_0(i)\} .$$

Clearly $I_\alpha \in \mathcal{U}$, for each $\alpha \in \lambda$. We define

$$h(i) = \min\{f_\alpha(i) : i \in I_\alpha\} .$$

$h(i)$ is defined a.e., since h is defined for all $i \in I_0$. Further $h(i) \leq f_{\alpha+1}(i)$ for all $i \in I_{\alpha+1}$. Hence $h(i) < f_\alpha(i)$ a.e., for all $\alpha \in \lambda$. Using (iii) we can therefore find $[k] \in K$ such that $[h] < [k]$. Note, however, that $[h] \in \prod A_i/_{\mathcal{U}}$, and thus $[h] \in F$. Hence $[h] < [k_0]$. On the other hand, it follows immediately from the definition of h that $h(i) > k_0(i)$ a.e., a contradiction.

The following proposition gives some further information of the kind stated in Proposition 4.

PROPOSITION 6. Let μ be a strongly inaccessible cardinal, $2^\mu = \mu^+$ and $\mathcal{U}$ be μ-indec. Then $|\prod \mu/_{\mathcal{U}}| = \mu$.

Proof. By Proposition 2, there is some $\delta < \mu$ such that $\mathcal{U}$ is λ-indec. for all λ such that $\delta < \lambda < \mu$. By μ-indec.,

$$\prod \mu/_{\mathcal{U}} = \bigcup_{\nu \in \mu} \prod \nu/_{\mathcal{U}} .$$

But by Proposition 4' and the above stated consequence of Proposition 2, we certainly have

$$|\prod \nu/_{\mathcal{U}}| < \mu$$

for all $\nu < \mu$ and the result follows.

The following lemma is proved similarly to Theorem 1.8, p. 241, [8]. We omit the proof.

LEMMA 4. Let $\mathcal{U}$ be an ultrafilter over I and A be an ordered set. Suppose that for some $[f] \in \prod A/_{\mathcal{U}}$,

(i) $[f] > [c_a]$ for each $a \in A$

(ii) for all $[g] \in \prod A/_{\mathcal{U}}$, if $[g] < [f]$, then $[g] < [c_a]$ for some $a \in A$.

Define $\mathcal{V} \subseteq \mathcal{P}(A)$ by

$$X \in \mathcal{V} \longleftrightarrow f^{-1}(X) \in \mathcal{U} .$$

Then

(a) $\mathcal{V}$ is an ultrafilter on A.

(b) In $\prod A/_{\mathcal{V}}$, $[id]$ is the least element $> [c_a]$ for each $a \in A$.

(c) For each cardinal λ, if $\mathcal{U}$ is λ-indec., so is $\mathcal{V}$.

REMARK. We have used $[c_a]$ is two different ways. Firstly, it was the equivalence class of the constant function a with respect to $\mathcal{U}$, secondly with respect to $\mathcal{V}$.

DEFINITION 4. If A is an ordered set and $\mathcal{V}$ is an ultrafilter over A such that the conditions (a) and (b) of Lemma 4 are satisfied, then $\mathcal{V}$ is said to be normal.

THEOREM 2. Let κ be a cardinal and $\mathcal{U}$ a uniform ultrafilter over κ such that one of the following holds:

(a) There is a cardinal μ such that $\mathcal{U}$ is μ^+-d.c.

$$2^{\mu} = \mu^+, \quad 2^{\mu^+} = \mu^{++}, \quad 2^{\mu^{++}} = \mu^{+++}, \quad cf(\kappa) > \mu^+ .$$

(b) There is a cardinal μ such that $\mathcal{U}$ is μ^+-d.c., $2^{\mu^{(n)}} < \lim\{\mu^{(n)} : n \in \omega\}$ for all $n \in \omega$, and $cf(\kappa) > cf(2^{\mu})$.

(c) There is a cardinal μ such that $\mathcal{U}$ is ν-indec. for all ν,

$\mu < \nu < 2^{(2^\mu)^+}$, and $cf(\kappa) > cf(2^\mu)$.

In each of the cases (a), (b), (c), there is a normal ultrafilter $\mathcal{V}$ over κ satisfying (a), (b), (c), respectively. Moreover, for all λ, if $\mathcal{U}$ is λ-indec., so is $\mathcal{V}$.

REMARK. In (a), the assumption about cardinals will hold if GCH holds and κ is regular and $> \mu^+$.

Proof of Theorem 2. In each case we take

$$K = \{[c_\gamma] : \gamma \in \kappa\}$$

and apply Proposition 5. Although we stated Proposition 5 under the assumption of GCH only, we can see that it will still hold in each of the cases (a), (b), (c). Then we use Lemma 4.

LEMMA 5. Let $\mathcal{U}$ be normal over κ and κ regular. Then every closed cofinal subset $C \subseteq \kappa$ belongs to $\mathcal{U}$. Hence the set of limit ordinals $< \kappa$ belongs to $\mathcal{U}$. And if κ is a limit cardinal, the set of limit cardinals $< \kappa$ belongs to $\mathcal{U}$.

This is well known.

PROPOSITION 7. Let $\mathcal{U}$ be normal over κ and κ regular. For every limit ordinal $\alpha < \kappa$, let S_α be a cofinal subset of α of type $cf(\alpha)$, $S_\alpha = \{0\}$ otherwise. Then

$$|\prod S_\alpha/\mathcal{U}| \geq \kappa .$$

Proof. If $[f] \in \prod S_\alpha/\mathcal{U}$, then $f(\alpha) < \alpha$ a.e. Hence by normality, $[f] \leq [c_\beta]$ for some $\beta \in \kappa$. If

$$|\prod S_\alpha/\mathcal{U}| < \kappa ,$$

then by regularity, there will be a $\gamma \in \kappa$ such that $[f] \leq [c_\gamma]$ for all $[f] \in \prod S_\alpha/\mathcal{U}$. On the other hand,

$$A = \{\alpha : \alpha \text{ limit } \& \, \alpha > \gamma\} \in \mathcal{U} .$$

For each such α we can find $\rho_\alpha \in S_\alpha$, $\rho_\alpha > \gamma$. Define $f(\alpha) = \rho_\alpha$ for $\alpha \in A$ and $f(\alpha) = 0$ otherwise. Then

$$[f] \in \prod S_\alpha/\mathcal{U}$$

and $[f] > [c_\gamma]$. This is a contradiction.

PROPOSITION 8. Let $\mathcal{U}$ be normal over κ and κ regular.

(a) Suppose $\kappa = \lambda^+$, where λ is singular, and

$$|\prod \nu/\mathcal{U}| < \kappa$$

for each $\nu < \lambda$. Then

$$\{\alpha : \alpha \in \kappa \, \& \, cf(\alpha) > \nu\} \in \mathcal{U}$$

for each $\nu < \lambda$.

(b) Suppose κ is a strongly inaccessible cardinal and

$$|\prod \nu/\mathcal{U}| < \kappa$$

for each $\nu < \kappa$. Then

$$\{\alpha : \alpha < \kappa \, \& \, \alpha \text{ strongly inacc.}\} \in \mathcal{U} .$$

Also, every closed subset of κ contains a strongly inaccessible cardinal. Hence κ is strongly Mahlo.

Proof. Assume the contrary. Note that in the case (b),

$$A = \{\alpha : \alpha \text{ is a strong limit cardinal}\} \in \mathfrak{U} ,$$

this set being a closed cofinal subset of κ. Form sets S_α as in Proposition 7. In the case (a) there is some $\nu < \lambda$ such that $|S_\alpha| \leq \nu$ a.e. In the case (b) there is a $\nu < \kappa$ such that $|S_\alpha| \leq \nu$ a.e. Otherwise $cf(\alpha)$ would be unbounded on every set of α's in $\mathfrak{U}$, and $cf(\alpha) < \alpha$. This contradicts normality. Hence in either case,

$$|\prod S_\alpha/_\mathfrak{U}| < \kappa .$$

This contradicts Proposition 7.

If $C \subseteq \kappa$ is closed and unbounded, then $C \in \mathfrak{U}$ by Lemma 5. Hence $C \cap A \in \mathfrak{U}$, giving the last part of (b).

THEOREM 3. (a) (GCH). _Let_ κ _be an inaccessible cardinal which is not Mahlo and_ $\mathfrak{U}$ _a uniform ultrafilter over_ κ. _Then there is some_ $\alpha < \kappa$ _such that_ $\mathfrak{U}$ _is_ λ-_dec. for all_ λ $(cf(\lambda) > \alpha, \lambda < \kappa)$.

(b) _Let_ κ _be a strongly inaccessible cardinal which is not Mahlo. Then for every_ $\alpha < \kappa$ _there is some_ λ _such that_ $\alpha < \lambda < \kappa$ _and_ $\mathfrak{U}$ _is_ λ-_dec_.

G. V. and D. V. Chudnovsky [3] showed the following theorem.

THEOREM 3.1. _Let_ κ _be a strongly or weakly inaccessible cardinal which is not in the_ ω-_th Mahlo class of strongly or weakly inaccessible cardinals respectively. Let_ $\mathfrak{U}$ _be a uniform ultrafilter over_ κ. _Then there is some_ $\alpha < \kappa$ _such that_ $\mathfrak{U}$ _is_ β-_d.i. for all regular_ $\beta > \alpha, \beta < \kappa$. _Even stronger, either_ $\mathfrak{U}$ _is_ α-_d.i. for all regular_ $\alpha < \kappa$, _or_ $\mathfrak{U}$ _is_ (α,κ)-_regular for some_ $\alpha < \kappa$.

Compare also our Theorems 6 and 7. Assuming GCH our method gives the same result as the simpler version of Chudnovsky's theorem. Chudnovsky used a result of Hajnal [4].

Proof of Theorem 3. (a) We assume the contrary. Let $\mathcal{U}$ be a uniform ultrafilter over κ such that $\mathcal{U}$ is λ-indec. for arbitrarily large regular $\lambda < \kappa$. By Theorem 2(a) and Lemma 4(c), there is a $\mathcal{V}$ over κ which is normal and λ-indec. for the same λ as $\mathcal{U}$. It is now easily seen, e.g. from Propositions 4,6, that

$$|\prod \mu/_{\mathcal{U}}| < \kappa$$

for all $\mu < \kappa$. Hence κ is Mahlo by Proposition 8 - a contradiction.

(b) is similar.

DEFINITION 5. $S \subseteq \kappa$ is said to be stationary if for every closed unbounded $C \subseteq \kappa$, $C \cap S \neq 0$.

DEFINITION 6. κ is 0-Mahlo if κ is strongly inaccessible. κ is $(\alpha+1)$-Mahlo if it is α-Mahlo and

$$\{\lambda : \lambda < \kappa \text{ and } \lambda \text{ is } \alpha\text{-Mahlo}\}$$

is stationary. If α is limit, then κ is α-Mahlo if κ is β-Mahlo for all $\beta < \alpha$.

Some of the following results are due to J. Silver and are included with his permission.

LEMMA 6. (J. Silver). Let $\mathcal{U}$ be a ν-indec. ultrafilter over κ, ν regular and $< \kappa$. Let $\delta \in \kappa$ and $cf(\delta) = \nu$. Further let δ_ξ, $\xi \in \nu$, be a strictly increasing sequence converging to δ. Then $[c_{\delta_\xi}]$ converges to $[c_\delta]$ in $\prod \kappa/_{\mathcal{U}}$.

Proof. Suppose not. Then there is some f such that

$$[c_{\delta_\xi}] < [f] < [c_\delta], \quad \xi \in \nu .$$

Define

$$A_\xi = \{\rho : \delta_\xi \leq f(\rho) < \delta_{\xi+1}\}, \quad \xi \in \nu .$$

Then A_ξ's are pairwise disjoint, and

$$\bigcup \{A_\xi : \xi \in \nu\} \in \mathcal{U} ,$$

since $f(\rho) < \delta$ a.e. Further if $S \subseteq \nu$ and $|S| < \nu$, then

$$A = \bigcup \{A_\xi : \xi \in S\} \notin \mathcal{U} .$$

(Suppose $A \in \mathcal{U}$. By regularity of ν, there is some $\eta \in \nu$ such that $\eta > \sigma$ for all $\sigma \in S$. Clearly, if $\rho \in A$, then

$$f(\rho) \leq \delta_\eta < \delta_{\eta+1} .$$

Hence $f(\rho) < \delta_{\eta+1}$ a.e., i.e.

$$[f] < [c_{\delta_{\eta+1}}] .$$

This contradicts the choice of f.) This shows that $\{A_\xi : \xi \in \nu\}$ is a ν-decomposition for $\mathcal{U}$. This contradicts our hypothesis and the lemma is proved.

LEMMA 7. (J. Silver). Let κ be regular and $\mathcal{U}$ normal over κ. Further suppose that for some regular $\nu < \kappa$, $\mathcal{U}$ is ν-indec. Let $C \subseteq \prod \kappa/_{\mathcal{U}}$ be closed and unbounded below $[id]$. Let

$$D = \{\delta : [c_\delta] \in C\} .$$

Then

(a) $|D| = \kappa$,

(b) D is closed under limits of sequences of type ν.

Proof. (b) follows immediately from Lemma 6. To prove (a), it is sufficient to show that for every $\alpha \in \kappa$, there is a $\delta \in D$, $\delta > \alpha$. It is easy to get sequences $[f_\xi]$, δ_ξ, $\xi \in \nu$ with the following properties:

$$[c_\alpha] < [f_\xi] \in C, \quad \xi \in \nu ,$$

$$[f_\xi] \leq [c_{\delta_\xi}], \quad \xi \in \nu ,$$

$$[f_{\xi+1}] > [c_{\delta_\xi}], \quad \xi \in \nu .$$

Let $\delta = \lim\{\delta_\xi : \xi \in \nu\}$. By Lemma 6, $[c_{\delta_\xi}]$, $\xi \in \nu$, converge to $[c_\delta]$. Hence $[f_\xi]$, $\xi \in \nu$, also converge to $[c_\delta]$. Hence $[c_\delta] \in C$, i.e. $\delta \in D$.

THEOREM 4. (J. Silver and the author). _Let_ $\mathfrak{U}$ _be normal over_ κ, κ _regular and_ $S \subseteq \kappa$ _be stationary in_ κ. _Further suppose that for all_ $\alpha \in S$, $\mathfrak{U}$ _is_ $cf(\alpha)$-_indec. Then_

$$\{\beta : S \cap \beta \text{ is stationary in } \beta\} \in \mathfrak{U} .$$

Proof. Suppose not. Hence

$$B = \{\beta : S \cap \beta \text{ is not stationary in } \beta\} \in \mathfrak{U} .$$

For each $\beta \in B$, let $C_\beta \subseteq \beta$, C_β be closed and unbounded in β, of type $cf(\beta)$ and $C_\beta \cap S = 0$. Let $C = \prod C_\beta/_{\mathfrak{U}}$. S being stationary, it is non-empty and therefore for some reg. $\nu < \kappa$, $\mathfrak{U}$ is ν-indec. It follows from Lemma 7, that

$$D = \{\delta : [c_\delta] \in C\} ,$$

(i) has cardinality κ, and

(ii) is closed under limits of sequences of type $cf(\alpha)$ for all $\alpha \in S$.

Let $\overline{D}$ be the closure of D. Since S is stationary, there is some $\alpha \in S \cap \overline{D}$. We shall show that $\alpha \in D$. Suppose not. Hence there is an increasing sequence $\delta_\xi \in D$, $\xi \in cf(\alpha)$, converging to α. By (ii), it follows that $\alpha \in D$, since $\alpha \in S$. This contradiction shows that $\alpha \in D$. Therefore $[c_\alpha] \in C$, which gives $\alpha \in C_\beta$ for almost all β. Hence for some $\beta \in B, \alpha \in C_\beta$. But $\alpha \in C_\beta \cap S$, $\beta \in B$ are contradictory. This completes the proof of the theorem.

THEOREM 5. _Let_ κ _be a regular cardinal, and_ $\mathfrak{U}$ _a normal ultrafilter over_ κ. _Let_ $C \subseteq \prod \kappa/_{\mathfrak{U}}$ _be closed and unbounded below_ $[id]$. _Then one of the following holds_:

(i) $\{\delta : [c_\delta] \in C\} \in \mathfrak{U}$

(ii) $\{\delta : \mathfrak{U}$ _is_ $cf(\delta)$-_dec._$\} \in \mathfrak{U}$.

Proof. Assume $\neg$(i). Hence

$$E = \{\delta : [c_\delta] \notin C\} \in \mathfrak{U} .$$

For every $\delta \in E$, let $h(\delta)$ = some $[f]$, where $[f] < [c_\delta]$, $[f] \geq [g]$ for all $[g] \in C$. Let

$$F = \{\delta : \delta \in E \ \& \ (\exists \eta)(h(\delta) \leq [c_\eta] < [c_\delta])\} .$$

We claim that $F \notin \mathfrak{U}$. Assume $F \in \mathfrak{U}$. Define $H(\delta)$ = some η such that

$$h(\delta) \leq [c_\eta] < [c_\delta] ,$$

for all $\delta \in F$. By normality, H is bounded below the identity on some $G \in \mathfrak{U}$. This clearly means that C is bounded below the identity, a contradiction. Hence $F \notin \mathfrak{U}$. Hence for almost all δ, there is no γ such that

$$h(\delta) < [c_\gamma] < [c_\delta] .$$

This implies that for almost all δ, $\mathfrak{U}$ is cf(δ)-decomposable. This gives (ii). For the proof of this last step see Lemma 6.

THEOREM 6. *Assume* GCH. *Let* κ *be strongly inaccessible and* $\mathfrak{U}$ *over* κ *be* λ*-indec. for arbitrarily large reg.* $\lambda < \kappa$. *Then* κ *is in the* ω*-th Mahlo class*.

Proof. We need to show that for all $n \in \omega$, κ is in the n-th Mahlo class. We show this for each $n \in \omega$, by induction on κ. For $n = 1$, the result follows from Theorem 3a. Assume the result true for n and all κ. We can assume that $\mathfrak{U}$ is normal, and $\{\delta : \delta \text{ is } (n-1) \text{ at Mahlo}\} \in \mathfrak{U}$.

Case I.

$$D = \{\delta : \delta \text{ is } (n-1)\text{st Mahlo } \& \ \mathfrak{U} \text{ is } \delta\text{-dec.}\} \in \mathfrak{U} .$$

Let $A = \{\nu : \mathfrak{U} \text{ is } \nu\text{-indec.}, \nu \text{ reg.}\}$. Then $|A| = \kappa$. Let

$$B = \{\beta : |\beta \cap A| = \beta\} .$$

B is closed and unbounded in κ. Hence $D \cap B \in \mathcal{U}$. If $\delta \in D \cap B$, then δ carries an ultrafilter which is ν-indec. for arbitrarily large $\nu < \delta$. Since $\delta < \kappa$, δ is n-th Mahlo by the induction hypothesis. Thus we have shown that

$$\{\delta : \delta < \kappa \ \& \ \delta \text{ is n-th Mahlo}\} \in \mathcal{U} .$$

<u>Case II</u>. Suppose Case I fails. By Theorem 5, if $C \subseteq \prod \kappa/_{\mathcal{U}}$ is closed and unbounded below $[id]$, then

$$\{\delta : [c_\delta] \in C\} \in \mathcal{U} .$$

We proceed by contradiction and assume

$$\{\delta : \delta \text{ is n-th Mahlo}\} \notin \mathcal{U} .$$

Hence we can associate to almost every $\delta \in \kappa$ a subset $C_\delta \subseteq \delta$ such that C_δ is closed and unbounded in δ and contains no cardinals of the $(n-1)$st Mahlo class. Now

$$C = \prod C_\delta/_{\mathcal{U}}$$

is closed and unbounded below $[id]$. Hence $\{\delta : [c_\delta] \in C\} \in \mathcal{U}$, i.e.

$$\{\delta : \delta \text{ is not } (n-1)\text{st Mahlo}\} \in \mathcal{U} ,$$

a contradiction.

This concludes the proof of Theorem 6.

The next theorem is due to G. V. and D. V. Chudnovsky [3] and in the present framework to J. Silver.

THEOREM 7. (J. Silver). <u>Let</u> κ <u>be a strongly inaccessible cardinal and</u> $\mathcal{U}$ <u>over</u> κ <u>be</u> λ-<u>indec. for all sufficiently large</u> $\lambda < \kappa$. <u>Then</u> κ <u>is in the</u> ω-<u>th Mahlo class</u>.

This can be proved in the same way as Theorem 6, except that we use the versions of theorems and lemmas that did not depend on GCH.

However we can also give a faster proof as follows. By Theorem 3(b), κ is in the 1-st Mahlo class. Hence $\{\delta : \delta < \kappa$ and δ is strongly inaccessible$\}$ is stationary. It is easy to see that we can apply Theorem 4 and get $\{\delta : \delta$ is in 1-st Mahlo class$\} \in \mathfrak{u}$. Hence this set is stationary and the rest is clear.

The following lemma is due to J. Ketonen.

LEMMA 8. (Ketonen [9]). _Let_ $\mathfrak{u}$ _be an ultrafilter over_ κ, $A_\rho \subseteq \kappa$ _for_ $\rho \in \kappa$, $|A_\rho| < \lambda < \kappa$. _Suppose that there are_ $[f_\xi] \in \prod A_\rho/_\mathfrak{u}$ _and a strictly increasing sequence_ $\gamma_\xi \in \kappa$ $(\xi \in \nu)$, _with the following properties_:

$$[c_{\gamma_\xi}] \leq [f_\xi] < [c_{\gamma_{\xi+1}}], \quad \xi \in \nu .$$

Then there are sets $E_\xi \in \mathfrak{u}$ $(\xi \in \nu)$, _such that any_ λ _of them have empty intersection_. ($\mathfrak{u}$ _satisfying this last condition is said to be_ (λ,ν)_-regular_. Cf. Keisler [7].)

Proof. Let $E_\xi = \{\rho : \gamma_\xi \leq f_\xi(\rho) < \gamma_{\xi+1}\}$ $(\xi \in \nu)$. If $\rho \in E_\xi \cap E_\eta$ and $\xi \neq \eta$, then clearly $f_\xi(\rho) \neq f_\eta(\rho)$. If $\rho \in E_\xi$, $\xi \in S$, $|S| = \lambda$, then

$$R = \{f_\xi(\rho) : \xi \in S\} \subseteq A_\rho$$

(we can assume without loss of generality that for all ξ, ρ, $f_\xi(\rho) \in A_\rho$), and $|R| = \lambda$. Hence $|A_\rho| \geq \lambda$, a contradiction.

LEMMA 9. _Let_ κ _be a regular cardinal_, $\lambda < \kappa$, $\lambda \leq \nu \leq \kappa$, $\mathfrak{u}$ _not_ (λ,ν)-_regular_, $|A_\rho| < \lambda$ _for each_ $\rho \in \kappa$, $A_\rho \subseteq \kappa$. _Let_

$$\mathfrak{A} = \left\{[f] : [f] \in \prod A_\rho/_\mathfrak{u} \;\&\; (\exists \delta < \kappa)([f] < [c_\delta])\right\} .$$

Then there is some $\gamma < \kappa$ _such that_ $[f] < [c_\gamma]$ _for all_ $[f] \in \mathfrak{A}$.

Proof. Assume the contrary. It is easy to construct sequences $[f_\xi]$, γ_ξ $(\xi \in \nu)$ satisfying the conditions of Lemma 8. Hence $\mathfrak{u}$ is (λ,ν)-regular, a contradiction.

LEMMA 10. Let $\varkappa$ be regular, $\mathfrak{U}$ be normal over $\varkappa$, $\lambda < \varkappa$, $A_\rho \subseteq \rho$ for each $\rho < \varkappa$, $|A_\rho| < \lambda$, and $\prod A_\rho/\mathfrak{U}$ be unbounded below [id]. Then $\mathfrak{U}$ is $(\lambda,\varkappa)$-regular.

Proof. Let $\mathfrak{A}$ be defined as in Lemma 9. By $A_\rho \subseteq \rho$ and the normality of $\mathfrak{U}$, we get

$$\mathfrak{A} = \prod A_\rho/\mathfrak{U} .$$

Now apply Lemma 9 with $\nu = \varkappa$.

THEOREM 8. Let $\varkappa$ be a regular limit cardinal, $\mathfrak{U}$ be normal over $\varkappa$ and $(\lambda,\varkappa)$-nonregular for all $\lambda < \varkappa$. Then every closed unbounded subset of $\varkappa$ contains a regular limit cardinal. If $\varkappa$ is strongly inaccessible, then every closed unbounded subset of $\varkappa$ contains a strongly inaccessible cardinal.

Proof. Since $\mathfrak{U}$ is normal, every closed unbounded subset of $\varkappa$ belongs to $\mathfrak{U}$. Hence the set of limit cardinals belongs to $\mathfrak{U}$, and if $\varkappa$ is strongly inaccessible, the set of strong limit cardinals belongs to $\mathfrak{U}$.

We need to show that the set of regular cardinals belongs to $\mathfrak{U}$. Assume the contrary. Hence the set of singular cardinals is in $\mathfrak{U}$. For every singular $\alpha < \varkappa$, let $A_\alpha \subseteq \alpha$, A_α be closed and unbounded in α, of type $\mathrm{cf}(\alpha)$. Now the function $f(\alpha) = \mathrm{cf}(\alpha)$ is pressing down a.e. Hence there is a set $X \in \mathfrak{U}$ such that f is bounded below $\varkappa$ on X. Hence there is some cardinal $\lambda < \varkappa$ such that for all $\alpha \in X$, $\mathrm{cf}(\alpha) < \lambda$. Hence $|A_\alpha| < \lambda$ a.e. By Lemma 10 and $(\lambda,\varkappa)$-nonregularity, $\prod A_\alpha/\mathfrak{U}$ is bounded below [id]. Say $[c_\beta]$ bounds $\prod A_\alpha/\mathfrak{U}$. On the other hand $\alpha > \beta$, for almost all α. For each such α we can find $\rho_\alpha \in A_\alpha$, $\rho_\alpha > \beta$. Then defining

$$g(\alpha) = \rho_\alpha ,$$

$$[g] \in \prod A_\alpha/\mathfrak{U}, \quad [g] > [c_\beta] ,$$

a contradiction. This proves the theorem.

COROLLARY. <u>Let $\varkappa$ be a regular limit cardinal, $\mathfrak{U}$ be normal over $\varkappa$ and λ-d.c. for arbitrarily large regular $\lambda < \varkappa$. Then the conclusion of Theorem 8 holds.</u>

THEOREM 9. <u>Let $\varkappa = \nu^+$, $\lambda \leq \nu$ and $\mathfrak{U}$ be normal over $\varkappa$. If $\{\alpha : cf(\alpha) < \lambda\} \in \mathfrak{U}$, then $\mathfrak{U}$ is $(\lambda,\varkappa)$-regular, and hence $(\mu,\varkappa)$-regular for all $\mu \geq \lambda$, $\mu < \varkappa$.</u>

<u>Proof</u>. For every α such that $cf(\alpha) < \lambda$, let $A_\alpha \subseteq \alpha$, be closed unbounded in α and of type $cf(\alpha)$. It is easy to see that the hypothesis of Lemma 10 is satisfied. Hence $\mathfrak{U}$ is $(\lambda,\varkappa)$-regular.

COROLLARY. <u>Under the hypotheses of Theorem 9, $\mathfrak{U}$ is μ-d.i. for all regular $\mu \geq \lambda$, $\mu < \varkappa$.</u>

The following theorem is a special case of a result due to Chang and Keisler. We include a proof for the convenience of the reader. Compare e.g. Chang [2], p. 97.

THEOREM 10. (Chang-Keisler). <u>Let $\mathfrak{U}$ over $\varkappa$ be $(\lambda,\varkappa)$-regular, where λ is a strong limit cardinal. Then $|\prod \lambda/_{\mathfrak{U}}| \geq 2^\varkappa$.</u>

<u>Proof</u>. Let $S_\alpha \in \mathfrak{U}$ $(\alpha \in \varkappa)$, and any λ of S_α's have empty intersection. This is by $(\lambda,\varkappa)$-regularity. For $x \in \varkappa$, let

$$R_x = \{\alpha : x \in S_\alpha\} .$$

For $A \subseteq \varkappa$, define

$$f_A(x) = A \cap R_x .$$

Let $A \neq B$. Hence for reasons of symmetry we can assume that for some α, $\alpha \in A$, $\alpha \notin B$. Now if $x \in S_\alpha$, then $\alpha \in R_x$ and therefore

$$\alpha \in f_A(x) = A \cap R_x$$

$$\alpha \notin f_B(x) = B \cap R_x .$$

Hence for all $x \in S_\alpha$,

$$f_A(x) \neq f_B(x) .$$

Thus $[f_A] \neq [f_B]$. This shows that

$$(*) \qquad |\prod \mathcal{P}(R_x)/_{\mathfrak{U}}| \geq 2^\kappa .$$

Since λ is strong limit and $|R_x| < \lambda$ for all x, it follows that $|\mathcal{P}(R_x)| < \lambda$ for all x. Hence (*) implies

$$|\prod \lambda/_{\mathfrak{U}}| \geq 2^\kappa .$$

The following theorems give some further information of the kind given in Theorem 1(b).

THEOREM 11. (GCH). Let $\kappa = \lambda^+$ where λ is singular, $cf(\lambda) = \nu < \lambda$. Let $\mathfrak{U}$ be uniform over κ. Then either

(i) $\mathfrak{U}$ is λ-dec., or

(ii) There is some $\alpha < \lambda$ such that $\mathfrak{U}$ is β-d.i. for all regular β such that $\alpha < \beta < \lambda$.

Proof. Assume that (ii) fails. Hence there are arbitrarily large regular $\beta < \lambda$ such that $\mathfrak{U}$ is β-d.c. By Theorem 2 and Lemma 4(c) we can assume that $\mathfrak{U}$ is normal. Since λ is singular and $\kappa = \lambda^+$, we trivially have

$$\{\alpha : cf(\alpha) < \lambda\} \in \mathfrak{U} .$$

Hence by Theorem 9, $\mathfrak{U}$ is (λ, κ)-regular. Hence by Theorem 10,

$$(*) \qquad |\prod \lambda/_{\mathfrak{U}}| \geq 2^\kappa = \kappa^+ .$$

We claim that there is an $f \in \lambda^\kappa$ such that f takes exactly λ distinct values on each set from $\mathfrak{U}$. Assume the contrary. Hence each $f \in \lambda^\kappa$ is equiv. mod $\mathfrak{U}$ to a g such that g takes fewer than λ values on a set from $\mathfrak{U}$. Hence

$$\prod \lambda/_{\mathcal{U}} \subseteq \bigcup_{X \in [\lambda]^{<\lambda}} \prod X/_{\mathcal{U}} .$$

It follows, from Propositions 4, 6, that

$$|\prod \mu/_{\mathcal{U}}| < \kappa$$

for all $\mu < \lambda$. Further, by GCH,

$$|[\lambda]^{<\lambda}| = \lambda^+ = \kappa .$$

Hence

$$|\prod \lambda/_{\mathcal{U}}| \leq \kappa \cdot \kappa = \kappa .$$

This contradicts (*). Hence our claim is established. But the existence of an f as in the claim is equivalent to the λ-decomposability of $\mathcal{U}$. Hence (i) follows and the theorem is proved.

Without assuming GCH, we can prove the following by a similar argument.

THEOREM 12. Let $\kappa = \lambda^+$ where λ is singular strong limit cardinal, $cf(\lambda) = \nu < \lambda$. Suppose that $\lambda^{\lambda} < 2^{\kappa}$. Let $\mathcal{U}$ be uniform over κ. Then either

(i) $\mathcal{U}$ is λ-dec., or

(ii) $\mathcal{U}$ is β-dec. for arbitrarily large $\beta < \lambda$.

In the next theorem we consider further the situation described in Theorem 11 in the special case $cf(\lambda) = \omega$.

THEOREM 13. (GCH). Let $\kappa = \lambda^+$, $cf(\lambda) = \omega < \lambda$ and κ be less than the first measurable cardinal. Let $\mathcal{U}$ be a uniform ultrafilter over κ. Then $\mathcal{U}$ is λ-decomposable.

Proof. Since κ is less than the first measurable cardinal, $\mathcal{U}$ is clearly ω-decomposable. Thus if $\mathcal{U}$ is μ-dec. for arbitrarily large $\mu < \lambda$, then $\mathcal{U}$ is λ-dec. by Proposition 1. On the other hand, if there is some $\beta < \lambda$ such that $\mathcal{U}$ is α-indec. for all α, $\beta < \alpha < \lambda$, then $\mathcal{U}$ is λ-dec. by Theorem 11.

DEFINITION 7. An ultrafilter $\mathcal{U}$ over κ is said to be indecomposable if it satisfies the following condition: for all $\lambda < \kappa$, if $\lambda \neq \omega$, then $\mathcal{U}$ is λ-indec.

THEOREM 14. Assume GCH. Let κ be a cardinal such that some $\mathcal{U}$ over κ is indecomposable. Then either κ is a strongly inaccessible cardinal of at least the ω-th Mahlo class, or $\mathrm{cf}(\kappa) = \omega$.

Proof. If κ is strongly inaccessible, then by Theorem 6, κ is in the ω-th Mahlo class. If κ is singular and $\mathcal{U}$ is uniform over κ, then trivially, $\mathcal{U}$ is $\mathrm{cf}(\kappa)$-dec. Hence if $\mathrm{cf}(\kappa) \neq \omega$, we have the desired decomposition. If $\kappa = \lambda^+$, where λ is regular and $\mathcal{U}$ is uniform over κ, the existence of the desired decomposition follows from Theorem 1. Otherwise we use Theorem 11.

The following result is due to J. Silver and the proof will appear elsewhere.

THEOREM 15. Suppose that for some strongly inaccessible cardinal κ there is an indecomposable ultrafilter $\mathcal{U}$ over κ. Then $0^{\#}$ exists.

It is a well known theorem of D. Scott that if κ is a measurable cardinal and GCH holds below κ, then $2^{\kappa} = \kappa^+$. We propose to show

THEOREM 16. Let κ be a strongly inaccessible cardinal such that some $\mathcal{U}$ over κ is indecomposable. If GCH holds below κ, then $2^{\kappa} = \kappa^+$.

Proof. We can assume that $\mathcal{U}$ is normal. By normality,

$$\prod_{\alpha\in\kappa} \alpha/_{\mathcal{U}} = \bigcup_{\alpha\in\kappa} \prod_{i\in\kappa} \alpha/_{\mathcal{U}} .$$

Hence by the previous results,

$$\left|\prod_{\alpha\in\kappa} \alpha/_{\mathcal{U}}\right| = \kappa .$$

It follows that if for each $\alpha \in \kappa$, $|\rho_\alpha| = |\alpha|$, then

$$\left|\prod_{\alpha\in\kappa} \rho_\alpha/\mathfrak{U}\right| = \kappa ,$$

and therefore

$$\left|\prod_{\alpha\in\kappa} |\alpha|^+/\mathfrak{U}\right| \leq \kappa^+ .$$

We shall now establish a 1-1 map F of $P(\kappa)$ into

$$\prod_{\alpha\in\kappa} |\alpha|^+/\mathfrak{U} .$$

This will show that $|P(\kappa)| \leq \kappa^+$ and therefore $2^\kappa = \kappa^+$.

We construct an F as above by the same method that is used to prove Scott's theorem. For every cardinal $\alpha < \kappa$, let

$$\{a_\xi^\alpha : \xi < \alpha^+\}$$

be a 1-1 enumeration of $P(\alpha)$. If $a \in P(\alpha)$, then $\mathrm{ind}_\alpha(a)$ is by definition the unique ξ such that $a = a_\xi^\alpha$.

If $X \subseteq \kappa$, we define

$$f_X(\alpha) = \mathrm{ind}_\alpha(X \cap \alpha) .$$

Since $\{\alpha : \alpha \text{ is a cardinal}\} \in \mathfrak{U}$, f_X is defined almost everywhere on κ. Hence we can set

$$F(X) = [f_X] .$$

It is clear that if $X \neq Y$, then f_X, f_Y can agree only on a set of power $< \kappa$. Hence F is 1-1 and the theorem is proved.

THEOREM 17. Let κ be strongly inaccessible and some uniform $\mathfrak{U}$ over κ

be indecomposable. Then Kurepa's hypothesis fails for $\varkappa$.

REMARK. Kurepa's hypothesis for $\varkappa$ is the following statement: There is a family $\mathfrak{F} \subseteq P(\varkappa)$ such that $|\mathfrak{F}| = \varkappa^+$ and for all $\alpha < \varkappa$, $|\mathfrak{F} \upharpoonright \alpha| \leq \alpha$, where

$$\mathfrak{F} \upharpoonright \alpha = \{X \cap \alpha : X \in \mathfrak{F}\} .$$

Proof of Theorem 17. Let $\mathfrak{F} \subseteq P(\varkappa)$ and $|\mathfrak{F} \upharpoonright \alpha| \leq |\alpha|$ for all $\alpha < \varkappa$. For each $\alpha \in \varkappa$, let $\{a_\xi^\alpha : \xi \in \alpha\}$ be an enumeration of $\mathfrak{F} \upharpoonright \alpha$. We can assume that $\mathfrak{U}$ is normal. If $X \in \mathfrak{F}$, we define

$$f_X(\alpha) = \mathrm{ind}_\alpha(X \cap \alpha)$$

where $\mathrm{ind}_\alpha(a)$ is defined as in the proof of Theorem 16. Since for every $X \in \mathfrak{F}$, $f_X(\alpha) < \alpha$, we have

$$[f_X] \in \prod_{\alpha \in \varkappa} \alpha/_\mathfrak{U}$$

and if $X \neq Y$, then $f_X \neq f_Y$ a.e. Hence

$$\left|\prod_{\alpha \in \varkappa} \alpha/_\mathfrak{U}\right| \leq \varkappa$$

implies $|\mathfrak{F}| \leq \varkappa$. Thus we have shown: If $\mathfrak{F} \subseteq P(\varkappa)$ and for all $\alpha \in \varkappa$, $|\mathfrak{F} \upharpoonright \alpha| \leq |\alpha|$, then $|\mathfrak{F}| \leq \varkappa$. This means that Kurepa's hypothesis fails for $\varkappa$.

REMARK. The proof actually shows that if there is a normal $\mathfrak{U}$ over $\varkappa$ such that

$$\left|\prod_{\alpha \in \varkappa} \alpha/_\mathfrak{U}\right| \leq \varkappa ,$$

then Kurepa's hypothesis fails for $\varkappa$.

Finally, we shall discuss the problem of the existence of descendingly complete ultrafilters in the light of the axiom of constructibility.

Since the existence of $0^{\#}$ implies $V \neq L$, it follows by Silver's Theorem 15, that if $V = L$ and κ is inaccessible, then there is no indec. $\mathcal{U}$ over κ.

Next, Jensen [16], §5, 6, proved the following theorem:

THEOREM 18. Assume $V = L$. Let κ be regular and not weakly compact. Then there is an $S \subseteq \kappa$ such that

(i) for every $\alpha \in S$, $cf(\alpha) = \omega$,

(ii) S is stationary in κ,

(iii) $S \cap \alpha$ is not stationary in α for all $\alpha < \kappa$.

In a discussion, he asserted that his proof can be modified to give

THEOREM 19. (R. Jensen). Assume $V = L$. Let κ, λ be regular, $\omega \leq \lambda < \kappa$, and κ not weakly compact. Then there is an $S \subseteq \kappa$ such that

(i) for every $\alpha \in S$, $cf(\alpha) = \lambda$,

(ii) S is stationary in κ,

(iii) $S \cap \alpha$ is not stationary in α for all $\alpha < \kappa$.

We can now prove the following theorem:

THEOREM 20. (R. Jensen, J. Silver and the author). Assume $V = L$. Let $\kappa > \omega$ be regular and not weakly compact. Let $\mathcal{U}$ be uniform over κ. Then $\mathcal{U}$ is λ-d.i. for all regular $\lambda \leq \kappa$.

Proof. ω-incompleteness is a well known theorem of D. Scott. The case $\lambda = \kappa$ is trivial.

Let $\omega < \lambda < \kappa$, λ regular, and suppose that $\mathcal{U}$ is λ-d.c. By Theorem 1(a), $\mathcal{U}$ is $\lambda^{+}, \lambda^{++}, \ldots$-d.c. Hence clearly $\lambda^{+}, \lambda^{++}, \ldots$ are all $< \kappa$. By Theorem 2(a), there is a $\mathcal{V}$ over κ which is normal and λ-d.c. Now let $S \subseteq \kappa$ be as in Theorem 19. By Theorem 4,

$$\{\beta : S \cap \beta \text{ is stationary in } \beta\} \in \mathcal{V} .$$

This clearly contradicts (iii). Hence Theorem 20 is proved.

Chang [1] raised the following problems:

I. If $\mathcal{U}$ is $\omega_{\omega+1}$-d.i., is $\mathcal{U}$

(a) ω_1-d.i.?

(b) ω_n-d.i. for all n?

II. If $\mathcal{U}$ is ω_{ω_1+1}-d.i., is $\mathcal{U}$

(a) ω_1-d.i.?

(b) ω_n-d.i. for all n?

A positive answer to both I, II, follows from Theorem 20, at least under the hypothesis that $V = L$. On the other hand, Chang's questions I', II' seem to be still open, even under the hypothesis that $V = L$. E.g., question I' is:

(I') If $\mathcal{U}$ is uniform over ω_ω, is $\mathcal{U}$

(a) ω_1-d.i.?

(b) ω_n-d.i. for all n?

Let us state without proof a result about indecomposable ultrafilters over singular κ.

THEOREM 21. (GCH). <u>Let</u> κ <u>be singular</u>, $cf(\kappa) = \omega$ <u>and</u> $\mathcal{U}$ <u>be a uniform indecomposable ultrafilter over</u> κ. <u>Then there is a uniform indecomposable</u> $\mathcal{V}$ <u>over</u> κ <u>such that for all</u> $g \in \kappa^\kappa$, <u>if</u> $g(\alpha) < \alpha$ <u>a.e., then</u> g <u>takes at most countably many values on a set in</u> $\mathcal{V}$.

Theorem 21 asserts that a uniform indecomposable ultrafilter over singular κ cnn be "normalized." A more general result concerning singular cardinals κ of $cf(\kappa) > \omega$ also holds.

Results concerning the consistency of the existence of indecomposable ultrafilters over singular κ can be found in [11].

REFERENCES

[1] Chang, C. C., Descendingly incomplete ultrafilters, Trans. Amer. Math. Soc. 126, 108-118 (1967).

[2] __________, Methods of constructing models, Sets, models and recursion theory, Edit. J. N. Crossley, Amsterdam 85-121 (1967).

[3] Chudnovsky, G. V. and Chudnovsky, D. V., Regularnye i ubyvajusche nepolnye ultrafiltry, Dokl. Akad. Nauk SSSR 198, 779-782 (1971).

[4] Hajnal, A., Ulam-matrices for inaccessible cardinals, Bull. Acad. Polon. Sci. XVII, 683-688 (1969).

[5] Jorgensen, M. A., Images of ultrafilters and cardinality of ultrapowers, Amer. Math. Soc. Notices 18, 826 (1971).

[6] ______________, Regular ultrafilters and long ultrapowers, Ibid., 928.

[7] Keisler, H. J., On cardinalities of ultraproducts, Bull. Amer. Math. Soc. 70, 644-647 (1964).

[8] Keisler, H. J. and Tarski, A., From accessible to inaccessible cardinals, Fund. Math. 53, 225-308 (1964).

[9] Ketonen, J., Everything you wanted to know about ultrafilters, Doctoral Dissertation, Univ. of Wisconsin, (1971).

[10] Kunen, K. and Prikry, K., On descendingly incomplete ultrafilters, to appear in J. Symb. Logic.

[11] Prikry, K., Changing measurable into accessible cardinals, Dissertationes Mathematicae (Rozprawy Matematyczne) LXVIII, 5-52 (1970).

[12] Prikry, K. and Shelah, S., On decomposability of families of sets, to appear.

[13] Silver, J. H., Some applications of model theory in set theory, Ann. Math. Logic (1) 3, 45-110 (1971).

[14] Solovay, R., A Δ_3^1 non-constructible set of integers, Trans. Amer. Math. Soc. 127, 50-75 (1967).

[15] Ulam, S., Zur Masstheorie in der allgemeinen Mengelehre, Fund. Math. 16, 140-150 (1930).

[16] Jensen, R. B., The fine structure of the constructible hierarchy, to appear in Annals of Math. Logic.

XVI. A MODEL FOR THE NEGATION OF THE AXIOM OF CHOICE[1]

Kenneth Kunen[2]
University of Wisconsin and
University of California

In analogy with Gödel's L, we consider here the model M (introduced in Chang [1] and called there C^{ω_1}), constructed by using the infinitary language $\mathcal{L}_{\omega_1\omega_1}$. Thus, $M_0 = 0$, $M_\gamma = \cup\{M_\alpha : \alpha < \gamma\}$ for γ a limit ordinal, and $M_{\alpha+1}$ is the set of all subsets of M_α definable over M_α from a countable sequence of elements of M_α by a formula in $\mathcal{L}_{\omega_1\omega_1}$. $M = \cup\{M_\alpha : \alpha \in ON\}$, where ON is the class of ordinals.

M is the least transitive model for ZF such that $ON \subseteq M$ and $M^\omega \subseteq M$. Furthermore (assuming that the axiom of choice, AC, holds in the universe, V), one can, following the procedure for L, represent elements of M by countable abstraction terms involving countably many ordinals. Hence, we see that M is also the least transitive model for ZF such that $ON^\omega \subseteq M$.

The method of Cohen implies immediately that one cannot prove in ZFC that AC holds in M; however, $V = L$ implies that $M = L \models AC$. Hence, the statement that $M \models AC$ is independent of ZFC. The purpose of this note is to prove that AC is false in M. Our proof will be formalizable in the system ZFC[3] + "there are at least ω_1 measurable cardinals".

For brevity, we write $u(\kappa, \mathcal{U})$ to mean that $\mathcal{U}$ is a κ-complete non-principal ultrafilter on the measurable cardinal κ. Scott [5] introduced the idea of using $\mathcal{U}$ to take an ultrapower of V. Follow-

---

1 This was not the subject of the author's talk at Cambridge.
2 Research partially supported by the A.P. Sloan Foundation and NSF Grant GP-23114.
3 The use of AC may in fact be eliminated by a more complicated argument.

ing the notation in [4], $Ult(V,\mathcal{U})$ is the transitive class isomorphic to $V^{\kappa}/\mathcal{U}$ and $i^{\mathcal{U}}: V \to Ult(V,\mathcal{U})$ is the canonical elementary embedding.

Since $Ult(V,\mathcal{U})$ is closed under countable sequences, $i^{\mathcal{U}}\restriction M$ is an elementary embedding from M into itself. Also, using the representation of elements of M by abstraction terms we see that

Lemma 1 (ZFC). For each $x \in M$, there is a countable set of ordinals, s, such that whenever $u(\kappa,\mathcal{U})$ and $\forall\alpha \in s[i^{\mathcal{U}}(\alpha) = \alpha]$, then $i^{\mathcal{U}}(x) = x$.

Lemma 1 may also be proved directly by transfinite induction on the construction of x, without using abstraction terms.

Lemma 2 (ZFC). For each ordinal α, there are at most finitely many measurable cardinals κ such that

$$\exists\mathcal{U}[u(\kappa,\mathcal{U}) \wedge i^{\mathcal{U}}(\alpha) \neq \alpha].$$

We defer the proof of Lemma 2, which uses iterated ultrapowers, until after we prove the main theorem. Lemmas 1 plus 2 give immediately

Lemma 3 (ZFC). For each $x \in M$, there are at most countably many measurable cardinals κ such that

$$\exists\mathcal{U}[u(\kappa,\mathcal{U}) \wedge i^{\mathcal{U}}(x) \neq x].$$

For any ordinal θ, let $[\theta]^{\omega}$ be the set of subsets of θ of order type ω. Note that $[\theta]^{\omega} \in M$.

Theorem (ZFC). If there are ω_1 measurable cardinals less than θ, then M contains no well-ordering of $[\theta]^{\omega}$.

Proof. If there were such a well-ordering, then there would be an ordinal ρ and a 1-1 function F in M from ρ onto $[\theta]^{\omega}$.

By Lemma 3, let $\kappa_0 < \kappa_1 < \dots < \kappa_n < \dots < \theta$ $(n < \omega)$, with $u(\kappa_n, \mathcal{U}_n)$ and $i^{\mathcal{U}_n}(F) = F$ for each n. Let $t = \{\kappa_n : n < \omega\}$. Say $t = F(\alpha)$. By Lemma 2, $i^{\mathcal{U}_n}(\alpha) = \alpha$ for some n, so $i^{\mathcal{U}_n}(t) = t$. Hence, $i^{\mathcal{U}_n}(\kappa_n) = \kappa_n$, which is impossible.

We return now to the proof of Lemma 2. We begin with some remarks on the products of ultrafilters. If $\mathcal{U}$ is an ultrafilter on I and $\mathcal{W}$ an ultrafilter on J, then $\mathcal{U} \overset{\rightarrow}{\times} \mathcal{W}$ and $\mathcal{U} \overset{\leftarrow}{\times} \mathcal{W}$ are the ultrafilters on $I \times J$ defined by

$$X \in \mathcal{U} \overset{\rightarrow}{\times} \mathcal{W} \leftrightarrow \{i \in I: \ \{j \in J: \ \langle i,j \rangle \in X\} \in \mathcal{W}\} \ \in \mathcal{U}$$

$$X \in \mathcal{U} \overset{\leftarrow}{\times} \mathcal{W} \leftrightarrow \{j \in J: \{i \in I: \langle i,j \rangle \in X\} \in \mathcal{U}\} \in \mathcal{W} \ .$$

There is in general no relation between $\mathcal{U} \overset{\rightarrow}{\times} \mathcal{W}$ and $\mathcal{U} \overset{\leftarrow}{\times} \mathcal{W}$. However, if $\mathcal{W}$ is $\overline{\overline{I}}^{+}$-complete, then it is easy to see that

$$\mathcal{U} \overset{\rightarrow}{\times} \mathcal{W} = \mathcal{U} \overset{\leftarrow}{\times} \mathcal{W} = \{X \subseteq I \times J: \exists U \in \mathcal{U} \ \exists \ W \in \mathcal{W} (U \times W \subseteq X)\} \ .$$

The notion $\overset{\rightarrow}{\times}$ is used in [4] to give a presentation of Gaifman's theory of iterated ultrapowers. Thus, if $\mathcal{U}$ and $\mathcal{W}$ are countably complete, then $\mathrm{Ult}(V, \mathcal{U} \overset{\rightarrow}{\times} \mathcal{W})$ can be identified with the model obtained by working <u>within</u> $\mathrm{Ult}(V, \mathcal{U})$ and taking the ultrapower of the universe (i.e. $\mathrm{Ult}(V, \mathcal{U})$) by $i^{\mathcal{U}}(\mathcal{W})$ (which, <u>in</u> $\mathrm{Ult}(V, \mathcal{U})$, is an ultrafilter on $i^{\mathcal{U}}(J)$). Hence, if $\mathcal{U}_n$ are countably complete ultrafilters on I_n $(n < \omega)$, then $\mathrm{Ult}(V, \mathcal{U}_0 \overset{\rightarrow}{\times} \dots \overset{\rightarrow}{\times} \mathcal{U}_n)$ can be identified with the ultrapower of $\mathrm{Ult}(V, \mathcal{U}_0 \overset{\rightarrow}{\times} \dots \overset{\rightarrow}{\times} \mathcal{U}_{n-1})$ by $i^{\mathcal{U}_0 \overset{\rightarrow}{\times} \dots \overset{\rightarrow}{\times} \mathcal{U}_{n-1}}(\mathcal{U}_n)$, taken within $\mathrm{Ult}(V, \mathcal{U}_0 \overset{\rightarrow}{\times} \dots \overset{\rightarrow}{\times} \mathcal{U}_{n-1})$. This gives a canonical elementary embedding, $\vec{j}_n$ from $\mathrm{Ult}(V, \mathcal{U}_0 \overset{\rightarrow}{\times} \dots \overset{\rightarrow}{\times} \mathcal{U}_{n-1})$ into $\mathrm{Ult}(V, \mathcal{U}_0 \overset{\rightarrow}{\times} \dots \overset{\rightarrow}{\times} \mathcal{U}_n)$. Furthermore, the direct limit of the $\mathrm{Ult}(V, \mathcal{U}_0 \overset{\rightarrow}{\times} \dots \overset{\rightarrow}{\times} \mathcal{U}_n)$ under the embeddings $\vec{j}_n$ is well-founded, as one can see by an obvious generalization of Theorem 3.6 of [4] (which discusses the case where all the $\mathcal{U}_n$ are the same; for a discussion of iterated ultrapowers by different ultrafilters, see [3]).

The notion $\overset{\leftarrow}{\times}$ was discussed by Keisler to give a treatment of iterated ultrapowers in model theory (see, e.g., [2]). $\mathrm{Ult}(V,\mathcal{U}\overset{\leftarrow}{\times}\mathcal{W})$ corresponds to taking $(\mathrm{Ult}(V,\mathcal{U}))^{J}/\mathcal{W}$, where one uses here the usual model-theoretic definition of ultrapower. We can form similarly $\mathrm{Ult}(V,\mathcal{U}_0\overset{\leftarrow}{\times}\ldots\overset{\leftarrow}{\times}\mathcal{U}_n)$, which will be well-founded if the $\mathcal{U}_i$ are all countably complete, and elementary embeddings $\overset{\leftarrow}{j}_n$ from $\mathrm{Ult}(V,\mathcal{U}_0\overset{\leftarrow}{\times}\ldots\overset{\leftarrow}{\times}\mathcal{U}_{n-1})$ into $\mathrm{Ult}(V,\mathcal{U}_0\overset{\leftarrow}{\times}\ldots\overset{\leftarrow}{\times}\mathcal{U}_n)$. Note that $\overset{\leftarrow}{j}_n$ is just the restriction of $i^{\mathcal{U}_n}$ to $\mathrm{Ult}(V,\mathcal{U}_0\overset{\leftarrow}{\times}\ldots\overset{\leftarrow}{\times}\mathcal{U}_{n-1})$. We may again take the direct limit of the $\mathrm{Ult}(V,\mathcal{U}_0\overset{\leftarrow}{\times}\ldots\overset{\leftarrow}{\times}\mathcal{U}_n)$ under the $\overset{\leftarrow}{j}_n$, but it need not be well-founded; in fact, it is not well-founded if there is an α such that $i^{\mathcal{U}_n}(\alpha) > \alpha$ for infinitely many n. (Thus, e.g., if all the $\mathcal{U}_n$ are the same and non-principal, this direct limit is never well-founded).

Now, if each $\mathcal{U}_n$ is $(\mathrm{Card}\,(I_0\times\ldots\times I_{n-1}))^+$-complete, then the products $\overset{\rightarrow}{\times}$, $\overset{\leftarrow}{\times}$ and the embeddings $\vec{j}_n$, $\overset{\leftarrow}{j}_n$ are the same, so the two direct limits are the same, and hence both well-founded. So, in this case, we have that for each α, $i^{\mathcal{U}_n}(\alpha) = \alpha$ for all but finitely many n. In particular, we cannot have an increasing sequence of measurable cardinals, $\langle\kappa_n : n < \omega\rangle$, with $u(\kappa_n,\mathcal{U}_n)$ and $i^{\mathcal{U}_n}(\alpha) > \alpha$ for each n. This proves Lemma 2.

A simpler proof of Lemma 2, which avoids the use of iterated ultrapowers, has been found by William Fleissner. Both our proof and his are indirect, in that they derive a contradiction from the negation of Lemma 2. We know of no way of explicitly defining from α (say, in terms of cardinal arithmetic) a finite set, s_α, such that

$$\forall\kappa\notin s_\alpha\,\forall\mathcal{U}[u(\kappa,\mathcal{U})\rightarrow i^{\mathcal{U}}(\alpha)=\alpha].$$

In conclusion, we remark that one cannot prove that AC fails in M using just one measurable cardinal. For, following the notation of [4], assume $V = L[\mathcal{U}]$, where $\mathcal{U}$ is a normal ultrafilter on the measurable cardinal κ, and let $a = \{i^{\mathcal{U}}_{0n}(\kappa): n < \omega\}$. Obviously, $L[a] \subseteq M$. Now, by the construction of the iterated ultrapower $\mathrm{Ult}_\omega(V,\mathcal{U})$ in [4], any ordinal α is of the form $(i^{\mathcal{U}}_{0\omega}(f))(a)$, where $f: [\kappa]^\omega \to ON$. Hence, any ω-sequence of ordinals, $\langle\alpha_m : m < \omega\rangle$ is of the form $\langle (i^{\mathcal{U}}_{0\omega}(G))(m,a): m < \omega\rangle$, where $G:\omega \times [\kappa]^\omega \to ON$. Since $i^{\mathcal{U}}_{0\omega}(G) \in \mathrm{Ult}_\omega(V,\mathcal{U}) \subseteq L[a]$, $ON^\omega \subseteq L[a]$, so $M = L[a] \models AC$.

The above paragraph easily generalizes to show that a countable sequence of measurable cardinals is also insufficient to conclude that $M \models \neg AC$.

Finally, the methods of this paper easily generalize to show that AC fails in Chang's C^{λ^+} for all regular $\lambda > \omega$. The proof will, of course, use the existence of λ^+ measurable cardinals.

REFERENCES

[1] C. C. Chang, Sets constructible using $L_{\kappa\kappa}$, A.M.S. Proceedings of Symposia in Pure Mathematics, Vol. 13 (1967 UCLA Summer Institute), Part I (1971) 1-8.

[2] C. C. Chang and H. J. Keisler, Model Theory, to appear.

[3] K. Kunen, Inaccessibility properties of cardinals, Doctoral dissertation, Stanford University, 1968.

[4] K. Kunen, Some applications of iterated ultrapowers in set theory, Annals Math. Logic 1 (1970) 179-227.

[5] D. Scott, Measurable cardinals and constructible sets, Bull. Acad. Polon. Sci. Sér. Sci. Math. Astronom. Phys. 9 (1961) 521-524.

XVII. FILTERS CLOSED UNDER MAHLO'S AND GAIFMAN'S OPERATION

K. GLOEDE

As pointed out by JENSEN [1967], the classical method for obtaining and strengthening the axioms of ZERMELO-FRAENKEL set theory can be described schematically as follows:

(I) Certain combinatorial properties true for the finite sets are extended to hold for the class of all sets. (In this way one can obtain the usual axioms of ZERMELO-FRAENKEL set theory with the exception of the axiom of infinity.)

(II) Given a suitable description Φ of V, the class of all sets, one assumes - by way of relativization - that there is a set a satisfying Φ. Or, alternatively and in combination with (I), a suitable property of ω is extended to hold for some uncountable cardinal. (Thus one obtains the axiom of infinity, axioms for the existence of a measurable cardinal, etc.)

(III) The methods described above are iterated transfinitely.

We do not intend to make steps (I) - (III) more explicit but shall restrict ourselves to MAHLO's method which is an illustrative example of this procedure (cf. LEVY [1960] and §2 of this paper). Moreover, though it seems to be impossible to give any precise meaning to the notion of a "large" cardinal, MAHLO's method has turned out to provide a suitable interpretation of the concept of a "large class" of ordinals. Starting with such a class A, e.g. the class of all ordinals, and applying MAHLO's operation to A, we obtain a subclass B of A consisting of ordinals which are large in the sense that they cannot be obtained by the iterated operation of taking fixed points of sequences of ordinals in A (for a precise statement cf. MAHLO's and GAIFMAN's papers). Iterating MAHLO's operation and regarding those ordinals which are not removed from A by these processes we get an approach to certain "large" cardinals in A which may be even further continued by introducing stronger principles. In order to make this approach work, we have (i) to give a precise meaning to the "large" classes with which we start and (ii) ensure that the application of MAHLO's operation applied to these classes does not break down, i.e. yields a non-empty, preferably again a "large" class. We have chosen to interprete "large" class by "element of a proper, non-trivial filter F", and our aim is to study such filters which are closed under MAHLO's operation (or even GAIFMAN's operation). In addition, F is required to be complete and normal in order to ensure that F is closed even under iterations of MAHLO's operation and diagonalization of these iterations. We also try to find various conditions which imply the existence of such filters. Our results are similar to and extend those of KEISLER-TARSKI [1964], LEVY [1967], GAIFMAN [1967] and SOLOVAY [1967a].

Although the author did not succeed to solve some problems which naturally arose out of the approach outlined above, and fully resumes the responsibility for these defects, he would like to express his deepest gratitude to Prof. G. H. Müller for his help and encouragement during and beyond the preparation of this paper.

§ 1. Preliminaries

We are working in von-NEUMANN-BERNAYS-GÖDEL set theory (NBG) with lower case set variables and upper case class variables. Besides the usual axioms of NBG (group of axioms A - D of GÖDEL [1940]) we assume the set-form of the axiom of choice, AC; $\vdash$ refers to this theory. Recall that NBG (with AC included) is a conservative extension of ZERMELO-FRAENKEL set theory with the axiom of choice (ZFC), hence any result proved for ZF-formulae can be obtained in ZFC. Φ, Ψ, ..., φ, ψ, ... denote formulae of NBG; the latter are used for formulae containing possibly free but no bound class variables (i.e. Π^1_0-formulae). We use the standard set-theoretical notations, in particular, 0 denotes the empty set, $\cup$ (union), $\cap$ (intersection), $\mathfrak{P}$ (power set), Rng (range), Dom (domain). Ft(F) means "F is a function", $F''A := \{y \mid \exists x \varepsilon A \ \langle x,y \rangle \varepsilon F\}$ $F \restriction A := F \cap (A \times V)$ and $R|A := R \cap (A \times A)$ denote restriction to A (as domain and field resp.). The unique value of F at a is denoted by F(a) if it exists, otherwise F(a) is the empty set.

Ordinals are denoted by greek letters $\alpha, \beta, \gamma, \ldots, \xi, \eta, \zeta, \ldots$ If F is a function with range included in On, the class of all ordinals, then $\sup_{\xi<\alpha} F(\xi) := \bigcup\{F(\xi) \mid \xi<\alpha\}$ is the least ordinal γ such that $\forall\xi<\alpha\ F(\xi) \leq \gamma$. Lim($\alpha$) means that α is a limit number, the smallest limit number being denoted by ω. Cardinals are identified with initial ordinals, the cardinality of a set a is denoted by $\bar{\bar{a}}$. α^+ is the least cardinal greater than α. The cumulative von-NEUMANN hierarchy is defined by $V_\alpha = \bigcup\{\mathcal{P}(V_\xi) \mid \xi<\alpha\}$.

We assume that a suitable <u>Gödelization</u> $\ulcorner\Phi\urcorner$ of the set of formulae Φ of NBG has been defined (in some cases formulae are identified with their resp. Gödel numbers) and also a suitable formalization of the notion of <u>satisfaction</u> $\langle a,\varepsilon\rangle \models \ulcorner\varphi\urcorner[a_0,\ldots,a_n]$ (where $a_0,\ldots,a_n \varepsilon a$) for first-order formulae φ (meaning that φ is a formula of ZF containing at most the free variables $v_0,\ldots,v_n$ and that φ is true in $\langle a,\varepsilon\rangle$ if $a_0,\ldots,a_n$ are assigned to $v_0,\ldots,v_n$ resp.). Similarly,
$\langle a,\varepsilon\rangle \models^2 \ulcorner\Phi\urcorner[A_0,\ldots,A_m,a_0,\ldots,a_n]$
is a suitable formalization of the statement " Φ is a formula of NBG with free class variables among $X_0,\ldots,X_m$, free set variables among $v_0,\ldots,v_n$, $A_0,\ldots,A_m \subseteq a$, $a_0,\ldots,a_n \varepsilon a$ and Φ is true in $\langle a,\varepsilon\rangle$ under the given assignment", where $\forall X$ is interpreted in $\langle a,\varepsilon\rangle$ by $\forall x \subseteq a$. (Replacing the free class variables $X_0,\ldots,X_m$ by relation constants $\dot{A}_0,\ldots,\dot{A}_m$, resp., we also write in this case $\langle a,\varepsilon,A_0,\ldots,A_m\rangle \models^2 \ulcorner\Phi\urcorner[a_0,\ldots,a_n]$.) $\langle a,\varepsilon\rangle \models^2 \ulcorner\Phi\urcorner$ for a formula Φ means $\langle a,\varepsilon\rangle \models^2 \ulcorner\Psi\urcorner$, where Ψ is the universal closure of Φ, similarly $\langle a,\varepsilon\rangle \models^2 \Gamma$ for a schema (set of formulae) Γ.

Φ^A denotes relativization of Φ to A, i.e. quantifiers $\forall X$, $\exists X$, $\forall x$, $\exists x$ occurring in Φ are replaced by $\forall x \subseteq A$, $\exists x \subseteq A$, $\forall x \varepsilon A$, $\exists x \varepsilon A$, resp. (renaming bound variables if necessary). Note that $\langle a,\varepsilon\rangle \models^2 \ulcorner\Phi\urcorner \leftrightarrow \Phi^a$ for a sentence Φ is provable in ZF, and similar for formulae containing free variables.

The end of a proof is denoted by $\square$.

§2 MAHLO's Principle

In his papers from 1911 - 1913, MAHLO describes two principles for obtaining large cardinal numbers. The first method starts with the sequence of regular cardinals, called π_0-numbers,[1] and proceeds by taking fixed points of this sequence thus arriving at the weakly inaccessible cardinals (π_1-numbers). In general, the $\pi_{\gamma+1}$-numbers are the fixed points of the sequence of π_γ -numbers, and for limit numbers λ , the π_λ -numbers are the ordinals which are π_α -numbers for each $\alpha<\lambda$. Here we shall be mostly concerned with the second principle which is known to be "stronger" than the iterated fixed point operation; the relationship between both these methods has been studied in detail by GAIFMAN [1967].

In describing MAHLO's (second) Principle it seems to be advantageous to use LEVY's approach via the concept of normal functions:

DEFINITION

$$Nft(F) :\leftrightarrow Ft(F) \wedge Dom\ F = On \wedge Rng\ F \subseteq On \wedge \forall \xi, \eta\ (\ \xi < \eta \rightarrow F(\xi) < F(\eta)) \wedge \forall \lambda\ (\ Lim(\lambda) \rightarrow F(\lambda) = \sup_{\xi<\lambda} F(\xi)\ )$$

"F is a normal function (defined for all ordinals)"

We assume familiarity with the main properties of normal functions. In particular, for each normal function F there is a normal function $\mathfrak{D}$ F enumerating the critical points of F (i.e. the ordinals α such that $\alpha = F(\alpha)$) in increasing order.

LEMMA 2.1

(i) $Nft(F) \rightarrow Nft(\mathfrak{D} F) \wedge \forall \xi \in Rng(\mathfrak{D} F)\ \ F(\xi) = \xi$

(ii) $Nft(F) \wedge Nft(G) \rightarrow Nft(F \circ G) \wedge \forall \xi\ (\ (F \circ G)(\xi) = \xi \rightarrow F(\xi) = \xi \wedge G(\xi) = \xi)$

DEFINITION

A is closed, unbounded $:\leftrightarrow\ \ A \subseteq On \wedge \bigcup A = On \wedge \forall x \subseteq A(\ x \neq 0 \rightarrow \bigcup x \in A)$

The closed, unbounded classes are just the ranges of normal functions:

---

1) We shall disregard the fact that MAHLO considers 0 as the least regular cardinal and start with ω .

LEMMA 2.2

A is closed, unbounded $\leftrightarrow$ $\exists F\ (\ Nft(F) \wedge A = Rng\ F\ )$

The concept of a closed, unbounded class of ordinals is for certain applications not a suitable interpretation of a "large" class of ordinals, since the limit of a sequence of regular numbers need not be regular (and hence the class of MAHLO numbers of any type is not closed). Therefore we consider the wider notion of a <u>dense</u> class:

DEFINITION

A is dense $:\leftrightarrow$ $\forall F\ (\ Nft(F) \rightarrow Rng(F) \cap A \neq 0\ )$

The corresponding relativized definitions are the following:

a is closed, unbounded in α $:\leftrightarrow$ $a \subseteq \alpha \wedge \bigcup a = \alpha \wedge \forall x \subseteq a(x \neq 0 \rightarrow \bigcup x \in a \cup \{\alpha\})$

a is dense in α $:\leftrightarrow$ $\forall x \subseteq \alpha$ (x is closed, unbounded in α $\rightarrow x \cap a \neq 0$)

A set a which is dense in α is also called <u>stationary</u> (in α), or, equivalently, α is called <u>a-MAHLO</u> (FODOR[1956,1966], MATHIAS [1968]).

From Lemma 2.1 and 2.2 we immediately have:

LEMMA 2.3

A is dense $\leftrightarrow$ $\forall F\ (\ Nft(F) \rightarrow \exists \xi \in A \quad F(\xi) = \xi\ )$

$\leftrightarrow$ $\forall X\ (\ X$ is closed, unbounded $\rightarrow X \cap A \neq 0)$

Dense classes are "large" classes of ordinals in the following sense:

LEMMA 2.4

(i) On is dense.

(ii) A is dense $\wedge$ $A \subseteq B \subseteq On$ $\rightarrow$ B is dense.

(iii) A is dense $\rightarrow$ $A - \alpha$ is dense.

(iv) A is not dense $\rightarrow On - A$ is dense.

(v) A is dense $\wedge$ $Nft(F) \rightarrow F''A$ is dense $\wedge$ $A \cap Rng(F)$ is dense,
in particular, each closed, unbounded class is dense.

(vi) $\bigcup_{\xi<\alpha} A_\xi$ is dense $\leftrightarrow$ $\exists \xi<\alpha\ A_\xi$ is dense .

The proof of (i) - (v) is straightforward using the previous lemmas. The proof of (vi) requires the following strong axiom of choice for classes:

Π^1_0-AC: $\forall x\, \exists Y\ \Phi(x,Y) \rightarrow \exists Y\, \forall x\ \Phi(x,Y_x)$ (Φ a Π^1_0-formula),

where sequences of classes are coded in the following way:

DEFINITION 2.5

$$A_a := \{x \mid \langle a,x\rangle \varepsilon A\} = A''\{a\}.$$

Proof of (vi): Assume that $\forall \xi<\alpha$ A_ξ is not dense. By the above axiom of choice (in fact, a somewhat weaker axiom suffices) there is a class B such that

$$\forall \xi<\alpha (\ B_\xi \text{ is closed, unbounded} \wedge B_\xi \cap A_\xi = 0\ ).$$

Let $B' := \bigcap_{\xi<\alpha} B_\xi$. This class is closed, unbounded and $B' \cap \bigcup_{\xi<\alpha} A = 0$, hence $\bigcup_{\xi<\alpha} A_\xi$ is not dense. The converse is obvious from (ii). □

REMARKS

1) The results of Lemma 2.4 relativize to κ provided κ is regular and uncountable; in this case the set form of the axiom of choice is sufficient to prove (vi). For more details consult the papers by FODOR and HAJNAL (see the bibliography).

2) The intersection of two dense classes need not be dense, whereas the sets which are closed, unbounded in κ generate a κ-complete filter on κ (cf. §5) (assuming again κ to be regular and uncountable).

We now turn to the definition of MAHLO's operation.

DEFINITION 2.6

$$\text{reg}(\alpha) :\leftrightarrow \text{Lim}(\alpha) \wedge \forall f\, \forall \xi<\alpha (\ \text{Ft}(f) \wedge \text{Dom } f = \xi \wedge \text{Rng } f \subseteq \alpha \rightarrow$$
$$\rightarrow \sup_{\eta<\xi} f(\eta)<\alpha\ ) \qquad \text{" } \alpha \text{ is \underline{regular}"}$$

$$\text{in}(\alpha) :\leftrightarrow \alpha>\omega \wedge \forall f\ (\ \text{Ft}(f) \wedge \text{Dom } f\, \varepsilon V_\alpha \wedge \text{Rng } f \subseteq V_\alpha \rightarrow \text{Rng } f\ \varepsilon V_\alpha\ )$$
$$\text{" } \alpha \text{ is (strongly) \underline{inaccessible}"}$$

$\text{Reg} := \{\xi \mid \text{reg}(\xi)\}$, $\text{In} := \{\xi \mid \text{in}(\xi)\}$, <u>Card</u> is the class of cardinals.

$$H_0(A) := \{\xi \mid \text{reg}(\xi) \wedge A \cap \xi \text{ is dense in } \xi\}$$

$$H(A) := \{\xi \mid \text{in}(\xi) \wedge A \cap \xi \text{ is dense in } \xi\}$$

H_0 and H are (variants of) MAHLO's operation and roughly the duals of the operation M of KEISLER-TARSKI [1964].

Applying H to the class of inaccessible cardinals and iterating this process one obtains the hierarchy of MAHLO numbers in the sense of BERNAYS [1962] and LEVY [1960]:

$$\mathrm{Ma}(\alpha,0) \leftrightarrow \mathrm{in}(\alpha) \leftrightarrow \alpha \in \mathrm{In} = H^{0}(\mathrm{In})$$

$$\mathrm{Ma}(\alpha,\gamma+1) \leftrightarrow \alpha \in H^{\gamma+1}(\mathrm{In}) = H(H^{\gamma}(\mathrm{In}))$$

$$\mathrm{Ma}(\alpha,\lambda) \leftrightarrow \alpha \in H^{\lambda}(\mathrm{In}) = \bigcap_{\xi<\lambda} H^{\xi}(\mathrm{In}) \quad \text{for } \lambda \text{ a limit number.}$$

$\mathrm{Ma}(\alpha,\gamma)$ is read as " α is a MAHLO number of degree γ " (BERNAYS) or " α is hyperinaccessible of type γ " (LEVY). The MAHLO numbers of degree γ in their original sense (i.e. MAHLO's ρ_{γ} -numbers) are the elements of $H_{0}^{1+\gamma}(\mathrm{Reg})$. In particular, an ordinal α in $H_0(\mathrm{Reg})$ is a π_0-number and is already weakly inaccessible, i.e. a π_1-number. But even more, α is a π_{γ} -number for each $\gamma<\alpha$, hence it is a π_{α} -number, but not the least such number. The exact process leading from the π_0-numbers to the ρ_0-numbers can be described in terms of GAIFMAN's operation (cf. §6).

It is well-known (cp. SHEPHERDSON [1951-1953]) that the existence of weakly inaccessible uncountable cardinals and hence the existence of ρ_0-numbers cannot be proved in ZFC. In this section we shall propose several principles which imply the existence of MAHLO numbers of high degree.

A simple axiom of this type is the following

<u>MAHLO's Principle</u> <u>M</u> : A is dense → H(A) is dense .

THEOREM 2.7

$$M \vdash \quad \forall n<\omega \ \{\xi \mid \mathrm{Ma}(\xi,n)\} \text{ is dense} .$$

The principle M allows us to deal with the successor stages in the iteration process generating the MAHLO numbers. Unfortunately, the intersection of dense classes need not be dense and may even be empty. So in order to iterate MAHLO's Principle at limit stages we introduce the following strengthening of M:

<u>M^+</u> $\quad \forall\xi<\alpha\ (A_{\xi} \text{ is dense}) \rightarrow \{\eta \mid \mathrm{in}(\eta) \wedge \alpha<\eta \wedge \forall\xi<\alpha\ A_{\xi}\cap\eta \text{ is dense in } \eta\}$ is dense .

THEOREM 2.8

$$M^+ \vdash \forall\gamma\,(\{\xi \mid Ma(\xi,\gamma)\} \text{ is dense}).$$

Proof by induction on γ : Since M^+ implies M, we need only consider the case of a limit number γ . By induction hypothesis,

$$A_\alpha := \{\xi \mid Ma(\xi,\alpha)\} \text{ is dense} \quad \text{for each } \alpha<\gamma .$$

Hence, by M^+,

$$\{\eta \mid in(\eta) \wedge \forall\xi<\gamma\; A_\xi \cap \eta \text{ is dense in } \eta\} \quad \text{is dense .}$$

However, this class is equal to

$$\{\eta \mid in(\eta) \wedge \forall\xi<\gamma\; \eta \in H(A_\xi)\} = \{\eta \mid in(\eta) \wedge \forall\xi<\gamma\; Ma(\eta,\xi+1)\}$$
$$= \{\eta \mid Ma(\eta,\gamma)\} \quad . \;\square$$

Note that $Ma(\alpha,\beta) \wedge \gamma<\beta \rightarrow Ma(\alpha,\gamma)$

and $Ma(\alpha,\beta) \rightarrow \beta\leq\alpha$.

This suggests to apply again MAHLO's first principle (the fixed point process) and introduce MAHLO numbers α which are MAHLO numbers of degree α , and iterate MAHLO's operation on these classes:

DEFINITION (<u>hyper-MAHLO numbers</u>)

$$HMa(\alpha,0) \leftrightarrow Ma(\alpha,\alpha)$$

$$HMa(\alpha,\gamma+1) \leftrightarrow \alpha \in H^{\gamma+1}(\{\xi \mid Ma(\xi,\xi)\})$$

$$HMa(\alpha,\lambda) \leftrightarrow \forall\xi<\lambda\; HMa(\alpha,\xi) \quad \text{for } \lambda \text{ a limit number.}$$

There is a further strengthening of M^+ which allows us to take "diagonal" intersections and hence obtain the existence of hyper-MAHLO numbers:

<u>M^*</u> : $\forall\eta\; A_\eta$ is dense $\rightarrow \{\xi \mid in(\xi) \wedge \forall\eta<\xi\; A_\eta\cap\xi \text{ is dense in } \xi\}$ is dense.

Obviously, M^* implies M^+ and hence $\{\xi \mid Ma(\xi,\gamma)\}$ is dense for each γ . Applying M^* to these classes we obtain:

THEOREM 2.9 Assuming M^*,

$\{\xi | Ma(\xi,\xi)\}$ is dense, $\{\xi | HMa(\xi,\gamma)\}$ is dense,

$\{\xi | HMa(\xi,\xi)\}$ is dense.

One may continue with this process, iterate the "hyper", take diagonal intersections, etc. A bound on these iterations is given by GAIFMAN's operation (see § 6).

REMARK 2.1o

1) Note that the predicates $Ma(\alpha,\gamma)$, $HMa(\alpha,\gamma)$, etc. are definable in ZF, hence we could derive the results of Theorem 2.7 - 2.9 from the respective schemata $\hat{M}$, $\hat{M}^+$ and $\hat{M}^*$, resp., which result from the corresponding axioms by replacing the free class variable A by a syntactical variable $\hat{A}$ ranging over class terms $\{x | \varphi(x,a,..)\}$ (for φ a formula of ZF set theory).

2) The principles M, M^+ and M (and the corresponding schemata) can be regarded as particular cases of a (second order) reflection principle; this will be shown in the next section. In this respect it is interesting to give a characterization of the hierarchy of MAHLO numbers in model-theoretical terms (cf. LEVY [196o]). First let us state the following absoluteness property of the predicate $Ma(\alpha,\gamma)$:

LEMMA 2.11 (BERNAYS [1962])

If α is (strongly) inaccessible, $\beta,\gamma<\alpha$, then:

$Ma(\beta,\gamma) \leftrightarrow (Ma(\beta,\gamma))^{V_\alpha}$.

The reason why LEVY and BERNAYS start from the inaccessibles (rather than from the regular or weakly inaccessible cardinals as in MAHLO's original paper) is the fact that the stronger concept of MAHLO number allows to apply the previous lemma to obtain:

THEOREM 2.12

$Ma(\alpha,0) \leftrightarrow \langle V_\alpha,\varepsilon\rangle \overset{2}{\models} NBG$

$Ma(\alpha,\gamma+1) \leftrightarrow \langle V_\alpha,\varepsilon,\gamma\rangle \overset{2}{\models} (NBG + \{\xi | Ma(\xi,\dot{\gamma})\}$ is dense$)$.

This is the interpretation of MAHLO's method which we had in mind in our introduction.

Let M_0 be defined as MAHLO's Principle replacing H by H_0:

M_0: A is dense $\rightarrow$ $H_0(A)$ is dense .

HAJNAL [1969] has defined the following property P :

DEFINITION

$$P(\alpha) \;:\leftrightarrow\; \exists x,y\,(\; x \subseteq y \wedge y \subseteq \alpha \cap \mathrm{Card} \wedge x \text{ is dense in } \alpha \;\wedge$$
$$\wedge\; y \text{ is closed, unbounded in } \alpha \wedge H_0(x) \cap \alpha \subseteq \alpha - y\;) .$$

LEMMA 2.13 If κ is (weakly) inaccessible:

$$P(\kappa) \leftrightarrow \exists x \subseteq \kappa\,(\; x \text{ is dense in } \kappa \;\wedge\; H_0(x) \cap \kappa \text{ is not dense in } \kappa\;) .$$

Proof: The implication from left to right is obvious. To prove the converse, assume that there is some $a \subseteq \kappa$ such that a is dense in κ, but $H_0(a) \cap \kappa$ is not dense in κ. Then there is some $b \subseteq \kappa$ such that b is closed, unbounded in κ and $H_0(a) \cap b = 0$. If κ is weakly inaccessible, $b_0 = b \cap \mathrm{Card}$ is closed, unbounded in κ, too. Let $a_0 = a \cap b$. Then a_0 is dense in κ by (v) of Lemma 2.4 (relativized to κ). Moreover, $H_0(a_0) \cap \kappa \subseteq H_0(a) \cap \kappa \subseteq \kappa - b \subseteq \kappa - b_0$. This proves $P(\kappa)$. □

COROLLARY 2.14

If κ is inaccessible:

$$\langle V_\kappa , \varepsilon \rangle \models^{2} M_0 \;\leftrightarrow\; \neg\, P(\kappa) \;.$$

This result schould be compared with Theorem 2 of HAJNAL [1969]. (Note that HAJNAL uses the operation M' which is the dual of the operation H_0.)

§ 3 Indescribability

The formulae of NBG are classified according to their prefix normal form as Π^1_n-, Σ^1_n- and Δ^1_n-formulae (identifying formulae which are equivalent in the theory NBG; for the corresponding classification of ZF-formulae cf. LEVY [1965]), i.e. $\Pi^1_o = \Sigma^1_o$ is the set of formulae of NBG containing no bound class variables, Π^1_{n+1} and Σ^1_{n+1} are the sets of formulae which are equivalent in NBG to formulae of the form $\forall X\, \Psi$ and $\exists X\, \Psi$, resp., where Ψ is in Σ^1_n and Π^1_n, resp. Δ^1_n is $\Pi^1_n \wedge \Sigma^1_n$.

The partial Π^1_n-reflection principle is the following schema:

Π^1_n-PR: $\quad \Phi(A_1,\ldots,A_n) \rightarrow \exists \xi \quad \Phi^{V_\xi}(A_1 \cap V_\xi, \ldots, A_n \cap V_\xi)$ for Φ any Π^1_n-formula containing at most the free class variables $A_1,\ldots,A_n$.

(Σ^1_n-PR is defined analogously.) The weak partial Π^1_n-reflection principle is the same schema with V_ξ replaced by ξ (weak Π^1_n-PR).

DEFINITION (Π^1_n-indescribable cardinals, HANF-SCOTT [1961], LEVY [1967])

$$\Pi^1_n\text{-ind}(\alpha) : \leftrightarrow (\langle V_\alpha, \varepsilon\rangle \models^2 \Pi^1_n\text{-PR}) \wedge \alpha > 0 \quad ,$$

$$\text{weakly } \Pi^1_n\text{-ind}(\alpha) : \leftrightarrow (\langle \alpha, \varepsilon\rangle \models^2 \text{weak } \Pi^1_n\text{-PR}) \wedge \alpha > 0 \ .$$

THEOREM 3.1

$$\Pi^1_o\text{-ind}(\alpha) \leftrightarrow \text{in}(\alpha) \ .$$

The proof from left to right is obvious from the definition of inaccessibility given in § 2, the converse is due to MONTAGUE-VAUGHT [1959]. □

On the other hand, weakly Π^1_o-ind$(\alpha) \leftrightarrow$ reg$(\alpha) \wedge \alpha > 0$ by LEVY [1967], and a weakly Π^1_1-indescribable cardinal can only be shown to be weakly inaccessible, but need not be strongly inaccessible.

For the following we need some strengthening of the above reflection principles.

THEOREM 3.2 $(n \geq 1)$

The schema Π^1_n-PR implies the following:

$$\Phi(A_1,..,A_n) \rightarrow \exists\xi\ (in(\xi) \wedge \Phi^{V_\xi}(A_1 \cap V_\xi,...,A_n \cap V_\xi))$$

for $\Phi \in \Pi^1_n$.

Proof: Let Ψ be the Π^1_1-sentence $\forall F(\ Ft(F) \wedge \exists x(x=Dom\ F) \rightarrow \exists x(x=Rng\ F))$ (i.e. the axiom of replacement). Ψ holds in NBG, and for limit numbers α :

$\langle V_\alpha, \varepsilon \rangle \models^2 \ulcorner\Psi\urcorner \leftrightarrow \Psi^{V_\alpha} \leftrightarrow in(\alpha)$. Hence we obtain the above schema by applying Π^1_n-reflection to the formula

$$\Phi(A_1,..,A_n) \wedge \Psi \wedge \forall x \exists y\ x\varepsilon y \wedge \exists\xi\ Lim(\xi)\ . \ \square$$

THEOREM 3.3 $(n \geq 1)$

Π^1_n -PR implies the following schema (for $\Phi \in \Pi^1_n$):

$$\Phi(A_1,..,A_n) \rightarrow \{\xi \mid in(\xi) \wedge \Phi^{V_\xi}(A_1 \cap V_\xi,...,A_n \cap V_\xi)\} \text{ is dense .}$$

Proof: Use the preceding theorem and the fact that

$$Nft(F) \wedge Lim(\alpha) \rightarrow [\ (Nft(F \cap V_\alpha))^{V_\alpha} \leftrightarrow \alpha = F(\alpha)\ ]\ . \ \square$$

It is known that a schema of complete reflection (for ZF-formulae) can be proved in ZF. A similar proof shows:

THEOREM 3.4

For any Π^1_0-formula $\varphi(A_1,..,A_n,a_1,..,a_m)$ with free variables as indicated there is a normal function F such that for any critical point α of F (i.e. $\alpha = F(\alpha)$):

$$\forall x_1...x_m \varepsilon V_\alpha\ (\ \varphi(A_1,..,A_n,x_1,...,x_m) \leftrightarrow \varphi^{V_\alpha}(A_1 \cap V_\alpha,...,A_n \cap V_\alpha,x_1,...,x_m)).$$

Proof:(By induction on the number of logical symbols occurring in φ :)

If φ does not contain any quantifier, F can be taken to be the identity. Now suppose φ is of the form $\exists x\ \psi(A_1,..,A_n,x,x_1,..,x_m)$. By ind. hyp. there is a normal function F such that for $\alpha = F(\alpha)$:

$$\forall x,x_1,..,x_m \varepsilon V_\alpha\ (\ \psi(A_1,..,A_n,x,x_1,..,x_m) \leftrightarrow \psi^{V_\alpha}(A_1 \cap V_\alpha,..,A_n \cap V_\alpha,x,x_1,..,x_m)).$$

Define functions G,H_1 by

$$G(a_1,\dots,a_m,\alpha) = \begin{cases} \mu\xi[\xi \geq \alpha \wedge \exists x \varepsilon V_\xi \ \psi(A_1,\dots,A_n,x,a_1,\dots,a_m)] & \text{if there exists such a } \xi, \\ \alpha & \text{otherwise.} \end{cases}$$

$H_1(\alpha) = \sup\{G(x_1,\dots,x_m,\alpha) \mid x_1,\dots,x_m \varepsilon V_\alpha\}$. Then for $\beta = G_1(\alpha)$:

$\forall x_1 \dots x_m \varepsilon V_\alpha \ (\exists x \ \psi(A_1,\dots,A_n,x,x_1,\dots,x_m) \leftrightarrow \exists x \varepsilon V_\beta \ \psi(A_1,\dots,A_n,x,x_1,\dots,x_m))$.

Let H_2 be the normal function enumerating the range of H_1 (which is closed, unbounded), and let $F_1 := F \circ H_2$. It is easy to see that F_1 is a normal function as required for φ . □

THEOREM 3.5

$$\Pi^1_1\text{-PR} \vdash M^* .$$

Proof: Let $\Phi(A,\eta)$ be the Π^1_1-formula

$\{x \mid \langle\eta, x\rangle \varepsilon A\}$ is dense .

Applying Π^1_1-PR to the Π^1_1-formula $\forall\eta \ \Phi(A,\eta)$ (via Theorem 3.3), we obtain:

$B := \{\xi \mid \text{in}(\xi) \wedge \forall\eta < \xi \ \Phi^{V_\xi}(A \cap V_\xi, \eta)\}$ is dense.

Since $B \subseteq \{\xi \mid \text{in}(\xi) \wedge \forall\eta < \xi \ (A_\eta \cap \xi$ is dense in $\xi)\}$,

the class on the r.h.s. is dense, too. □

COROLLARY 3.6

$$\Pi^1_1\text{-ind}(\alpha) \rightarrow \langle V_\alpha, \varepsilon\rangle \models^2 M^* .$$

For later applications, let M^*_o be the principle M^* with "inaccessible" being replaced by "regular", i.e.

$\underline{M^*_o}$: $\forall\eta \ A_\eta$ is dense $\rightarrow \{\xi \mid \text{reg}(\xi) \wedge \forall\eta < \xi \ A_\eta \cap \xi$ is dense in $\xi\}$ is dense.

Using the following coding of sequences of classes of ordinals (rather than Def. 2.5):

$$A_\alpha := \{\xi \mid P(\xi, \alpha) \varepsilon A\} ,$$

where P is GÖDEL's pairing function $P: On \times On \xrightarrow[\text{onto}]{1:1} On$,

M^*_o is a statement on ordinals and classes of ordinals only. Similar to

Theorem 3.5 and Cor. 3.6 one obtains:

THEOREM 3.5.a

$$\text{weak-}\Pi^1_1\text{-PR} \quad \vdash \quad M^*_o \ .$$

COROLLARY 3.6.a

$$\text{weakly } \Pi^1_1\text{-ind}(\alpha) \quad \rightarrow \quad \langle \alpha, \varepsilon \rangle \models^2 M^*_o \ .$$

REMARK 3.7

1. JENSEN [1969] has recently shown that for regular cardinals α :

$$\Pi^1_1\text{-ind}(\alpha) \leftrightarrow \forall x \subseteq \alpha \ (\ x \text{ is dense in } \alpha \rightarrow H_o(x) \neq 0) \ ,$$

assuming $V = L$, hence under this assumption, the converse of Cor. 3.6 is true.(Cf. however the remark following Theorem 5.9.)

2. Let $\hat{M}^*$ be the schema obtained from M^* by restricting the free class variable A to range only over class terms $\{x| \ \varphi(x,a,..)\}$ for φ a formula of ZF, and suppose there is a Π^1_1-indescribable cardinal, α_o **denoting the** least such cardinal. Note that $\hat{M}^*$ is a Σ^1_2-schema (i.e. each instance of $\hat{M}^*$ is a Σ^1_2-formula) which is true in $\langle V_{\alpha_o}, \varepsilon \rangle$. Since any Π^1_1-indescribable cardinal is Σ^1_2-indescribable, an argument due to LEVY [1967] (for any φ $\{\xi<\alpha_o | \langle V_\xi, \varepsilon \rangle \models^2 M^*_\varphi\}$ (M^*_φ is $\hat{M}^*$ corr. to φ) is in LEVY's filter $F_L(\Pi^1_1)$,cf.5.5.1) shows that there is a $\beta < \alpha_o$ such that $\langle V_\beta, \varepsilon \rangle \models^2 \hat{M}^*$, hence the schema $\hat{M}^*$ is properly weaker than Π^1_1-reflection (assuming consistency), in contradistinction to JENSEN's result mentioned above.

Finally a remark on strengthening the axiom of choice using second-order reflection principles. BERNAYS [1962] has shown that Π^1_1-reflection implies GÖDEL's strong axiom of choice E. Similarly, consider the following strong axiom of choice:

$$\Pi^1_1\text{-AC:} \quad \forall x \exists X \ \Phi(X,x) \ \rightarrow \exists Y \forall x \ \Phi(Y_x,x) \qquad (\ \Phi \text{ a } \Pi^1_1\text{-formula}) \ .$$

THEOREM 3.8

$$\Pi^1_3\text{-PR} \ \vdash \ \Pi^1_1\text{-AC} \ .$$

Proof: Assume that for some Π^1_1-formula Φ we have

$$\forall x \exists X \ \Phi(X,x) \wedge \forall Y \exists x \neg \ \Phi(Y_x,x) \ .$$

Since this statement is a Π^1_3-formula, reflection gives the existence of some limit number α such that

(i) $\forall x \varepsilon V_\alpha \ \exists X \subseteq V_\alpha \ \Phi^{V_\alpha}(X,x) \ \wedge \forall Y \subseteq V_\alpha \ \exists x \varepsilon V_\alpha \ \ \Phi^{V_\alpha}(Y_x,x)$.

However, the usual axiom of choice for sets, AC, gives the existence of a function f such that $\mathrm{Dom}\ f = V_\alpha \wedge \mathrm{Rng}\ f \subseteq \mathcal{P}(V_\alpha) \wedge \forall x \varepsilon V_\alpha \ \Phi^{V_\alpha}(f(x),x)$.

This contradicts the second part of (i). $\square$

Similarly, Π^1_2-reflection implies the axiom of choice (Π^1_0-AC) used in the proof of Lemma 2.4.(vi).

§ 4 TAKEUTI's generalization of MAHLO's Principle

Throughout this section we assume axiom M^* of §2 and the Π^1_1-AC of §3. We shall give an interpretation of the concept of a "<u>nodical class</u>" (cf. TAKEUTI [1969]):

DEFINITION

$$\mathfrak{N}(A) :\leftrightarrow \exists X \ (X \text{ is dense} \wedge \{\xi \mid in(\xi) \wedge X \cap \xi \text{ is dense in } \xi\} \subseteq A) ,$$

$$\text{i.e. } \exists X \ (X \text{ is dense } \wedge \ H(X) \subseteq A) .$$

THEOREM 4.1

1. $\neg \mathfrak{N}(0)$
2. $\mathfrak{N}(On)$
3. $\mathfrak{N}(A) \wedge A \subseteq B \ \rightarrow \ \mathfrak{N}(B)$
4. $In = \{\xi \mid in(\xi)\} \subseteq A \ \rightarrow \ \mathfrak{N}(A)$
5. $\mathfrak{N}(A) \ \rightarrow \ \mathfrak{N}(A-\alpha)$
6. $\mathfrak{N}(A) \ \rightarrow$ A is dense
7. $\forall \xi \ \mathfrak{N}(A_\xi) \ \rightarrow \ \mathfrak{N}(\{\xi \mid \forall \eta < \xi \quad \xi \ \varepsilon A_\eta\})$
8. $\forall \xi < \alpha \ \mathfrak{N}(A_\xi) \ \rightarrow \ \mathfrak{N}(\bigcap_{\xi<\alpha} A_\xi)$
9. $Nft(F) \ \rightarrow \ \mathfrak{N}(\mathrm{Rng}\ F)$, i.e. A is closed, unbounded $\rightarrow \mathfrak{N}(A)$.
10. $\mathfrak{N}(\{\xi \mid in(\xi) \wedge \alpha < \xi \wedge \forall x \varepsilon V_\xi \quad (\varphi(x,A) \leftrightarrow \varphi^{V_\xi}(x, A \cap V_\xi))\})$

for any Π^1_0-formula φ with free variables as indicated.

Proof: 2) - 5) are obvious. 1) and 6) use M^* :If X is dense, so is H(X).

(7) is proved as follows: By assumption and Π^1_1-AC there is a class X such that for all α : X_α is dense $\wedge$ $H(X_\alpha) \subseteq A_\alpha$. By M^*,

B:= $\{\xi \mid in(\xi) \wedge \forall \eta < \xi \quad X_\eta \cap \xi$ is dense in $\xi\}$ is dense. It remains to show: $H(B) \subseteq \{\xi \mid \forall \eta < \xi \quad \xi \in A_\eta\}$: Let $\gamma \in H(B)$, i.e. $in(\gamma) \wedge B \cap \gamma$ is dense in γ. Let $\beta < \gamma$ and a be closed, unbounded in γ . a', the set of limit points of a (cf. the proof of 5.1o), is closed, unbounded in γ . Since $B \cap \gamma$ is dense in γ:

$\exists \xi (\beta < \xi < \gamma \wedge \xi \in a' \cap B)$, i.e.

$\exists \xi (\beta < \xi < \gamma \wedge a \cap \xi$ is closed, unb. in $\xi \wedge \forall \eta < \xi \; X_\eta \cap \xi$ is dense in ξ).

Hence $a \cap X_\beta \neq 0$. This proves: $\gamma \in H(X_\beta) \subseteq A_\beta$ for all $\beta < \gamma$. □

8) follows from 7). If A is closed, unbounded, then A is dense by 2.4.(v), and $H(A) \subseteq A$, since A is closed. This proves 9). Finally, 1o) follows from 9) and Theorem 3.4. □

Theorem 4.1 shows that - assuming M^* and Π^1_1-AC - one can define a property $\mathcal{M}$ applying to classes such that $\mathcal{M}$ has properties similar to $\mathcal{G}^o$ of TAKEUTI [1961] and satisfies the axioms of NTT except that 1o) of the preceding theorem holds for Π^1_o-formulae $\varphi(A,a)$ only.

§ 5 Filters closed under MAHLO's operation

The following results could be formalized in a theory which extends NBG and allows the use of such concepts as classes of classes; however, in order to keep technical problems to a minimum, we use relativization to some ordinal κ . Therefore we need only assume the axioms of ZFC; κ is always supposed to be a regular and uncountable cardinal.

DEFINITION (Variants of MAHLO's operation restricted to subsets of κ)

For $a \subseteq \kappa$:

$h_o(a) = \{\xi < \kappa \mid reg(\xi) \wedge a \cap \xi$ is dense in $\xi\}$

$h(a) = \{\xi < \kappa \mid in(\xi) \wedge a \cap \xi$ is dense in $\xi\}$

$h_s(a) = \{\xi < \kappa \mid (\xi = 0 \vee Lim(\xi)) \wedge a \cap \xi$ is dense in $\xi\}$ (SCOTT)

$h_G(a) = \{\xi \mid \xi \in a \wedge a \cap \xi$ is dense in $\xi\}$ (GAIFMAN)

REMARK: For $a \subseteq \kappa$: $h(a) = h_o(a) \cap In = h_s(a) \cap In \wedge h_G(a) = h_s(a) \cap a$.

DEFINITION: For functions f,g:

$$f \leq g :\leftrightarrow \text{Dom } f = \text{Dom } g \wedge \forall x \varepsilon \text{Dom } f \ (f(x) \subseteq g(x)) \ .$$

REMARK: $h \leq h_o \leq h_s \wedge h_G \leq h_s$

DEFINITION

$$G := \{x \subseteq \kappa \mid \exists y \subseteq \kappa \ (y \text{ is dense in } \kappa \wedge h(y) \subseteq x)\} ,$$

$$G_o := \{ x \subseteq \kappa \mid \exists y \subseteq \kappa \ (y \text{ is dense in } \kappa \wedge h_o(y) \subseteq x)\} .$$

Note that $G \subseteq G_o$. We want to show that these sets are examples of filters closed under MAHLO's operation, provided κ is "large".

A filter on κ is a non-empty set $F \subseteq \mathcal{P}(\kappa)$ satisfying $\forall x \varepsilon F \ \forall y \ (x \subseteq y \subseteq \kappa \rightarrow y \varepsilon F) \wedge \forall x,y \varepsilon F \ \ x \cap y \ \varepsilon \ F$. F is proper iff $0 \notin F$, non-trivial iff $\forall \xi < \kappa \ \ \kappa - \{\xi\} \ \varepsilon F$, α-complete iff for every $\gamma < \alpha$ and every sequence $\langle a_\xi \mid \xi < \gamma \rangle$: $\forall \xi < \gamma \ \ a_\xi \ \varepsilon \ F \rightarrow \bigcap_{\xi < \alpha} a_\xi \ \varepsilon \ F$. F is normal iff it is closed under diagonal intersection, i.e. for every sequence $\langle a_\xi \mid \xi < \kappa \rangle$:

$$\forall \xi < \kappa \quad a_\xi \ \varepsilon \ F \rightarrow \{\xi < \kappa \mid \forall \eta < \xi \ \ \xi \ \varepsilon a_\eta \} \ \varepsilon \ F \ .$$

Note that a normal filter on κ is κ-complete, therefore a normal filter is sometimes called strongly complete. Usually this condition is given in terms of incompressible functions, in fact one can prove the following

LEMMA

Let F be a proper, non-trivial and κ-complete filter on κ . Then F is normal iff F satisfies the following condition:

If g is a function $\kappa \rightarrow \kappa$ which is nowhere constant (w.r.t. F) , i.e. $\forall \xi < \kappa \ \ \{\eta < \kappa \mid g(\eta) \neq \xi \} \ \varepsilon \ F$, then $\{\xi < \kappa \mid \ \xi \leq g(\xi)\} \ \varepsilon \ F$.

(I.e. $I \restriction \kappa$ is incompressible on κ (w.r.t.F).)

We shall not use this condition and hence omit the proof of this lemma.

THEOREM 5.1

Let κ be an inaccessible cardinal such that $\langle V_\kappa , \varepsilon \rangle \models^2 M^*$. Then:

(1) G is a proper, non-trivial and κ-complete filter on κ .

(2) $\forall \xi < \kappa \quad a_\xi \ \varepsilon \ G \rightarrow \{\xi < \kappa \mid \forall \eta < \xi \ \ \xi \ \varepsilon a_\eta \} \varepsilon \ G$, i.e. G is normal.

(3) $\{\xi < \kappa \mid \text{Lim}(\xi)\} \ \varepsilon \ G$.

(4) $\{\xi<\kappa \mid in(\xi) \wedge \forall x \varepsilon V_\xi \ (\varphi^{V_\kappa}(x,a) \leftrightarrow \varphi^{V_\xi}(x,a \cap V_\xi))\} \ \varepsilon\ G$

for any $a \subseteq \kappa$, φ a ZF-formula.

The same holds for G_o.

Proof: This is just the relativized version of Theorem 4.1.◻ In fact, in order to obtain (1) - (3), κ need only be regular, and by the same arguments as before:

COROLLARY 5.2

If κ is a regular cardinal satisfying $\langle \kappa, < \rangle \models^2 M_o^*$, then G_o is a filter on κ satisfying (1) - (3) of 5.1.

Note that the assumptions of Theorem 5.1 cannot be weakened in general, since JENSEN's result mentioned in Remark 3.7 shows that for κ regular: G is proper iff κ is Π_1^1-indescribable, and both conditions are equivalent to the assumption that $\langle V_\kappa, \varepsilon \rangle \models^2 M^*$, assuming $V = L$.

THEOREM 5.3

Let F be a filter on κ satisfying (1) - (3) of 5.1. Then F satisfies:

(5) $a \ \varepsilon\ F \rightarrow$ a is dense in κ,

(6) $b \subseteq \kappa \wedge$ b is closed, unbounded in $\kappa \rightarrow b \ \varepsilon\ F$.

Proof of (6): Let b be closed, unbounded in κ, and let $f: \kappa \rightarrow b$ be the function enumerating b in increasing order (note that we always assume that κ is regular). For $\gamma<\kappa$ define

$a_\gamma = \{\xi<\kappa \mid f(\gamma)<\xi\}$. Since F is non-trivial and κ-complete, $\forall \xi<\kappa\ a_\xi \ \varepsilon\ F$. Hence, by the normality of F, $a = \{\xi<\kappa \mid \forall \eta<\xi \ \ \xi \ \varepsilon\ a_\eta\} \ \varepsilon\ F$.

By (3), $c = a \cap \{\xi<\kappa \mid Lim(\xi)\} \ \varepsilon\ F$. Let $\gamma \ \varepsilon\ c$. Then $Lim(\gamma) \wedge \forall \xi<\gamma\ f(\xi)<\gamma$, therefore $f(\gamma) = \gamma$, since b is closed. In particular, $\gamma \ \varepsilon\ b$. This proves $c \subseteq b$, and hence $b \ \varepsilon\ F$.

Proof of (5): Let $a \ \varepsilon\ F$, b be closed, unbounded in κ. By (6), $b \in F$ whence $a \cap b \in F$ and $a \cap b \neq 0$ since F is proper. This proves that a is dense in κ. ◻

COROLLARY 5.4

If κ is inaccessible and $\langle V_\kappa, \varepsilon\rangle \models^2 M^*$, then G is closed under MAHLO's operation: $a \in G \rightarrow h(a), h_G(a) \in G$.

Similarly, if κ is regular and $\langle\kappa,<\rangle \models^2 M_o^*$, then G_o is closed under h_o and h_G.

5.5. <u>APPLICATIONS</u>

1. LEVY [1967] has defined a filter

$F = F(\Pi^1_1) = \{x \subseteq \kappa \mid x \text{ is } \Pi^1_1\text{-enforceable at } \kappa\}$.

If κ is Π^1_1-indescribable, F satisfies (1) - (3) of 5.1, and $\{\xi<\kappa \mid in(\xi)\} \varepsilon F$.

Moreover, a is dense in κ $\rightarrow h(a) \varepsilon F$:

Let a be dense in κ. Then $\langle\kappa, \varepsilon, a\rangle \models^2 \Phi$ for some Π^1_1 - sentence Φ which says that a is dense. Therefore $b = \{\xi<\kappa \mid \langle\xi, a\cap\xi\rangle \models^2 \Phi\}$ is Π^1_1-enforceable at κ, hence $h(a) = b \cap \{\xi<\kappa \mid in(\xi)\} \varepsilon F$. □

As a consequence, $G \subseteq F$, and F is closed under MAHLO's operation h (and h_G) by (5) of Theorem 5.3.

2. Let F_o be the filter generated by the closed, unbounded subsets of κ, $F_o := \{x \subseteq \kappa \mid \exists y \subseteq \kappa\ (y \text{ is closed, unbounded in } \kappa \wedge x \subseteq y\}$. Since κ is regular and uncountable, F_o is a proper, non-trivial and normal filter on κ, $F_o \subseteq G$ by (6) of Theorem 5.3 (if the assumptions of this theorem are satisfied). However, $\{\xi<\kappa \mid reg(\xi)\}$ is not closed and not in F_o, in particular, F_o is <u>not</u> closed under MAHLO's operation h_o. From Cor. 5.7 below it follows that F_o is not κ-saturated, a result due to SOLOVAY [1967 b].

For the following, a filter F on κ is called <u>α-saturated</u> (for α a cardinal) iff for every family $E \subseteq \mathfrak{P}(\kappa)$ which satisfies

(i) $\forall x,y \varepsilon E\ (x \neq y \rightarrow x \cap y = 0)$ (i.e. the elements of E are pairwise disjoint)

and (ii) $\forall x \varepsilon E\ \ \kappa - x \notin F$ (i.e. $E \subseteq \mathfrak{P}(\kappa) - I$, where I is the dual ideal), E has cardinality $< \alpha$.-From SOLOVAY [1967a] we cite the following

THEOREM 5.6 (SOLOVAY)

Let $\kappa > \omega$ be regular and $F \subseteq \mathfrak{P}(\kappa)$ be a proper, non-trivial, normal and κ-saturated filter on κ. Then F satisfies

(i) $\{\xi<\kappa \mid reg(\xi)\} \in F$ and

(ii) $\kappa - a \notin F \wedge a \subseteq \kappa \rightarrow h_o(a) \in F$.

COROLLARY 5.7

Assuming the hypothesis of Theorem 5.6, F satisfies

(7) a is dense in $\kappa \rightarrow h_o(a) \in F$ and

(8) $a \in F \rightarrow h_o(a) \in F$,

i.e. $G_o \subseteq F$ and F is closed under MAHLO's operation h_o.

The proof of (7) is a slight modification of SOLOVAY's proof of (ii): The set F used in SOLOVAY's proof is closed, unbounded in κ ,hence $a \cap F \neq 0$ if a is dense in κ . Now (8) follows from (5) of 5.3 and (7). □

REMARK 5.8

1. The assumptions of Theorem 5.6 hold if κ is real-valued measurable, in fact F can be taken to be $\aleph_1$-saturated in this case (cf. SOLOVAY [1967a]). Note that (7) implies $G_o \subsetneq F$, in particular G_o is proper; however, $V \neq L$ in this case (see below).

2. SOLOVAY and KUNEN (cf. KUNEN [1970]) have proved the following result: If κ carries a proper, non-trivial, κ-complete and κ-saturated (or only κ^+-saturated) filter, then κ carries a normal filter having the same properties, and κ is measurable in some subuniverse of V (more exactly, in L[u,z] for two sets u,z), in particular, $V \neq L$. On the other hand, the existence of a Π^1_1-indescribable cardinal is compatible with V = L . Hence, assuming the existence of such a cardinal and V = L, if α_o is the least Π^1_1-indescribable cardinal, then by 5.4 α_o carries a proper, non-trivial and normal filter closed under h_o, but not an α_o^+-saturated filter having these properties.-On the other hand, for κ-saturated filters we have the following converse of Cor. 5.4:

THEOREM 5.9

Let F be a proper, non-trivial, normal and κ-saturated filter on κ . Then $\langle\kappa, <\rangle \models^2 M_o^*$.

Proof: Suppose $\forall\xi<\kappa$ a_ξ is dense in κ . Then by (7) of 5.7:

$\forall\xi<\kappa \quad h_o(a_\xi) \varepsilon F$. Hence

$b = \{\zeta<\kappa | \forall\eta<\xi \quad \xi \varepsilon h_o(a_\eta)\} \quad \varepsilon F$ by normality. Moreover, by the definition of h_o: $b \subseteq \{\xi<\kappa | \operatorname{reg}(\xi) \wedge \forall\eta<\xi \ a_\eta \cap \xi$ is dense in $\xi\} \quad \varepsilon F$, in particular, by (5) of 5.3, the set on the r.h.s. is dense in κ . □

As a consequence, under the assumptions of Theorem 5.9 (e.g. if κ is real-valued measurable) $\langle\kappa,<\rangle$ satisfies M_o^* but need not be Π_1^1-indescribable (SOLOVAY [1967c] proves that ZFC +" $2^{\aleph_0}$ is real-valued measurable" is consistent if ZFC is consistent with the existence of a (2-valued) measurable cardinal), so JENSEN's result (mentioned in Remark 3.7) need not hold without any additional assumption, in fact, CH is badly violated if $2^{\aleph_0}$ is real-valued measurable.

Remark 5.8.1 and 5.5.1 suggest that G is minimal among several proper, non-trivial and normal filters closed under MAHLO's operation. We shall now give a characterization of G which shows that G is most naturally related to MAHLO's operation (we do not know whether G is really minimal among all those filters).

LEMMA 5.1o

(i) $\operatorname{reg}(\alpha) \wedge h(a)\cap\alpha$ is dense in $\alpha \rightarrow a\cap\alpha$ is dense in α ,

(ii) $h(h(a)) \subseteq h(a)$ for $a \subseteq \kappa$.

Proof: Let $c \subseteq \alpha$ be closed, unbounded in α and let c' be the set of limit points of c, i.e. $c' = \{\xi<\alpha | \xi = \bigcup(c\cap\xi)\}$.

If α is regular, c' is again closed, unbounded in α . Therefore, if $h(a)\cap\alpha$ is dense in α , then $\gamma \varepsilon c' \cap h(a)$ for some $\gamma<\alpha$. Since $\gamma \varepsilon c'$, $c\cap\gamma$ is closed, unbounded in γ, and $a\cap\gamma$ is dense in γ , since $\gamma \varepsilon h(a)$. Therefore $c\cap a \neq 0$. This proves (i). (ii) is a simple consequence of (i). □

LEMMA 5.11

If F is a normal filter on κ , then so is $\{x \subseteq \kappa | \exists y\varepsilon F \quad h(y) \subseteq x\}$.

The proof is straightforward.- As an application to G we have:

THEOREM 5.12

Let κ be inaccessible and assume $\langle V_\kappa, \varepsilon\rangle \models^2 M^*$.

Then $G = \{x \subseteq \kappa \mid \exists y \varepsilon G \ h(y) \subseteq x\}$.

Proof: Suppose, $a \subseteq \kappa \wedge \exists y \varepsilon G \ h(y) \subseteq a$. Since the elements of G are dense in κ, $a \varepsilon G$. To prove the converse, let $a \varepsilon G$. Then $h(b) \subseteq a$ for some b which is dense in κ. By (ii) of 5.1o, $h(h(b)) \subseteq h(b)$. Since $h(b) \varepsilon G$, $\exists y \varepsilon G \ h(y) \subseteq h(b)$ whence $\exists y \varepsilon G \ h(y) \subseteq a$. □

During the Cambridge Summer School in Logic, A. MÁTÉ informed us that he had obtained similar results though by different methods which resulted from a generalization of the notion of incompactness. For an (uncountable) cardinal α, let $L_{\alpha\alpha}$ be the (infinitary) language of KARP [1964]. Roughly speaking, $L_{\alpha\alpha}$ allows to take conjunctions and disjunctions over less than α formulae and strings of quantifiers of length less than α, atomic formulae are assumed to be of the form $u=v$ and $u \varepsilon v$ only. If α and γ are cardinals, α is called <u>(α, γ)-compact</u> iff every set S of sentences of the language $L_{\alpha\alpha}$ such that $\overline{\overline{S}} \leq \alpha$ and every subset $E \subseteq S$, $\overline{\overline{E}} < \alpha$, has a (two-valued) model S has a Boolean-valued model (A,B) where A is the set of individuals and B is the complete Boolean algebra of truth values such that B satisfies the γ-chain condition and the pair (A,B) satisfies the Maximum Principle of SCOTT-SOLOVAY [1967]. In particular, α is $(\alpha,2)$-compact iff α is weakly compact, and if α is inaccessible, α is weakly compact iff α is Π^1_1-indescribable (a proof is given in SILVER [1966]).

Using our notation, MÁTÉ's results can be stated as follows (α is assumed to be inaccessible):

(1) If α is (α, α)-compact, then $\langle V_\alpha, \varepsilon\rangle \models^2 M^*$

(MÁTÉ [1971], Theorem 7.9)

(2) If α carries a proper, non-trivial, normal and α-saturated filter, then α is (α, α)-compact.

As a consequence,

(3) If α carries a proper, non-trivial, normal and α-saturated filter, then $\langle V_\alpha, \varepsilon\rangle \overset{2}{\models} M^*$.

This result corresponds to Theorem 5.9 (which was proved without using Boolean-valued models) whereas Theorem 5.1 proves the converse except that the required filter need not be α-saturated (in fact, as pointed out above, this additional property cannot be proved in general).

§ 6 GAIFMAN's Operation

A strong generalization of MAHLO's operation has been given by GAIFMAN [1967], in fact, the operation h^∇ is stronger than operations obtained from taking iterated applications and diagonal intersections of these, starting with h. - For the following, let κ again be regular and uncountable.

DEFINITION

$Ft^2(f) :\leftrightarrow \; Ft(f) \wedge \text{Dom } f \subseteq \mathcal{P}(\kappa) \wedge \text{Rng } f \subseteq \mathcal{P}(\kappa)$.

LTF_o is a class of functions containing the "local thinning functions" of GAIFMAN; we define it to be the set of f satisfying

$Ft^2(f) \wedge \forall x,y\,(\, y \subseteq x \,\varepsilon\, \text{Dom } f \rightarrow y \,\varepsilon\, \text{Dom } f\,) \wedge$

$\wedge\, \forall \xi < \kappa \;\; \forall x \varepsilon \text{Dom } f \;(\; f(x \cap (\xi+1)) = f(x) \cap (\xi+1))$.

The reason we choose this wider concept was to include h and h_o among these functions, in fact:

REMARK 6.1

The functions h, h_o, h_s and h_G (defined in §5) are in LTF_o.

DEFINITION

f is <u>monotone</u> $:\leftrightarrow Ft(f) \wedge \forall x,y \varepsilon \text{Dom } f \;(\; x \subseteq y \rightarrow f(x) \subseteq f(y))$.

REMARK 6.2

The functions h, h_o, h_s and h_G are monotone.

We now turn to operations defined on functions and sequences of functions:

DEFINITION 6.3

Let f be a function, $\langle f_\xi \mid \xi < \alpha\rangle$ be a sequence of functions.

1. $\bigcap\langle f_\xi \; \xi<\alpha\rangle = \bigcap_{\xi<\alpha} f_\xi$ is defined to be the function g such that

$\mathrm{Dom}\, g = \bigcap_{\xi<\alpha} \mathrm{Dom}\, f_\xi$, and for $a\varepsilon \mathrm{Dom}\, g$:

$g(a) = \bigcap_{\xi<\alpha} f_\xi(a)$.

(Though the "intersection" of functions defined above is not to be confused with the set-theoretical intersection (as in the last two applications), we hope that no confusion will arise.)

2. $\langle f_\xi \mid \xi<\alpha\rangle^D$ is the following diagonal function g:

$\mathrm{Dom}\, g = \bigcap_{\xi<\alpha} \mathrm{Dom}\, f_\xi$, and for $a \varepsilon \mathrm{Dom}\, g$:

$$g(a) = \begin{cases} \{\xi<\kappa \mid \xi \,\varepsilon f_\xi(a)\} & \text{if } \alpha \geq \kappa, \\ \{\xi<\alpha \mid \xi \,\varepsilon f_\xi(a)\} \cup \bigcap_{\xi<\alpha} f_\xi(a) & \text{if } \alpha<\kappa. \end{cases}$$

3. The iterated functions f^α are defined by recursion on α :

$f^0 = I\restriction \mathrm{Dom}\, f$ (I is the identity function)

$f^{\gamma+1} = f\circ f^\gamma$, $f^\lambda = \bigcap_{\xi<\lambda} f^\xi$ for λ a limit number.

$f^\Delta := \langle f^\xi \mid \xi<\kappa\rangle^D$

4. $\langle f_\xi \mid \xi<\alpha\rangle$ is decreasing $:\leftrightarrow \forall \xi,\eta \; (\xi<\eta<\alpha \rightarrow f_\eta \leq f_\xi)$,

$\langle f_\xi \mid \xi<\alpha\rangle$ is continously decreasing $:\leftrightarrow \langle f_\xi \mid \xi<\alpha\rangle$ is decreasing $\wedge$

$\wedge \; \forall\lambda \; (\mathrm{Lim}(\lambda) \wedge \lambda<\alpha \rightarrow f_\lambda = \bigcap_{\xi<\lambda} f_\xi)$.

5. $A^a := \{f \mid \mathrm{Ft}(f) \wedge \mathrm{Dom}\, f = a \wedge \mathrm{Rng}\, f \subseteq A\}$.

LEMMA 6.4

Suppose $\alpha \leq \kappa \wedge \forall\xi<\alpha \; f_\xi \,\varepsilon\, \mathrm{LTF}_0$. Then $\bigcap_{\xi<\alpha} f_\xi \,\varepsilon \mathrm{LTF}_0$ and $\langle f_\xi \mid \xi<\alpha\rangle^D \;\varepsilon\; \mathrm{LTF}_0$.

Moreover, the operations defined in 1.-3. above preserve the property of being monotone.

EXAMPLE:

$h^\gamma(\kappa\cap \mathrm{In}) = h_G(\kappa\cap \mathrm{In}) = \{\xi<\kappa \mid \mathrm{Ma}(\xi,\gamma)\}$,

$h^\Delta(\kappa\cap \mathrm{In}) = h_G(\kappa\cap \mathrm{In}) = \{\xi<\kappa \mid \mathrm{Ma}(\xi,\xi)\}$.

GAIFMAN's operation is defined as follows:

DEFINITION 6.5 (GAIFMAN [1967])

1) If $\alpha \leq \kappa \wedge Ft^2(f)$, then <u>$Q(\alpha, f)$</u> is the least set $E \subseteq \{x \mid Ft^2(f)\}$ satisfying the following conditions:

   (i) $I \upharpoonright \text{Dom } f \; \varepsilon \; E \quad \wedge \quad f \; \varepsilon \; E$,

   (ii) $g \; \varepsilon \; E \rightarrow f \circ g \; \varepsilon \; E$,

   (iii) $\forall \xi < \alpha \quad \forall y \; \varepsilon \; E^{\xi} \quad \bigcap y \; \varepsilon \; E$ ($\bigcap$ as defined in 6.4.1),

   (iv) $g \; \varepsilon \; E^{\beta} \wedge g$ is continuously decreasing $\rightarrow g^{D} \; \varepsilon \; E$,

   where $\beta = \alpha$, if $\text{Lim}(\alpha)$, $\beta = \alpha + 1$ otherwise.

2) If $\alpha < \kappa \wedge Ft^2(f)$, $J_{\alpha f}$, $J'_{\alpha f}$ and f^{∇} are defined by

$$J_{\alpha f}(a) = \bigcap \{g(a) \mid g \; \varepsilon \; Q(\alpha, f)\} \quad ,$$

$$J'_{\alpha f}(a) = \bigcap \{g(a \cap (\alpha + 1)) \mid g \; \varepsilon \; Q(\alpha, f)\} \quad \text{and}$$

$$f^{\nabla}(a) = \langle J_{\xi f} \mid \xi < \alpha \rangle^{D}(a) = \{\xi < \kappa \mid \xi \; \varepsilon \; J_{\xi f}(a)\}$$

   for $a \; \varepsilon \; \text{Dom } f$.

For later applications we need the following lemmata the proofs of which are straightforward in most cases or can be found in GAIFMAN [1967]:

LEMMA 6.6

1) If $\alpha \leq \kappa \wedge f \; \varepsilon \; LTF_o$, then $Q(\alpha, f) \subseteq LTF_o$ and all the members of $Q(\alpha, f)$ have the same domain (viz. Dom f). Moreover, $Q(\alpha, f)$ is closed under composition of functions.

2) If $\alpha < \kappa \wedge f \; \varepsilon \; LTF_o \wedge a \; \varepsilon$ Dom f , then $\alpha \; \varepsilon \; J_{\alpha, f}(a) \leftrightarrow \alpha \; \varepsilon \; J'_{\alpha f}(a)$.

3) $f \; \varepsilon \; LTF_o \rightarrow f^{\nabla} \; \varepsilon \; LTF_o$,

   $f \; \varepsilon LTF_o \wedge \alpha \leq \kappa \wedge g \; \varepsilon \; Q(\alpha, f) \rightarrow f^{\nabla} \leq g$.

In particular, $h^{\nabla} \leq h^{\Delta}$, $h^{\nabla} \leq (h^{\Delta})^{\Delta}$, etc. Similarly, if f^{β} is obtained from h by applying the operation $^{\Delta}$ β-times (taking intersections at limit stages in order to obtain a continously decreasing sequence), then $h^{\nabla} \leq f^{\Delta}$, etc. Hence one can continue taking iterations, diagonal intersections, etc. As long as the sequences thus obtained are continously decreasing, the resulting function g is still $\geq h^{\nabla}$ (so h^{∇} is "stronger" than g). It should be noted, however, that $\langle J_{\xi f} \mid \xi < \kappa \rangle$ need not be continously decreasing, and this fact indicates that the operation h^{∇} amounts to a real "jump" beyond the operations considered above (consult GAIFMAN's paper for more details; we shall return to this question at the end of this paper).

LEMMA 6.7

1) $a \subseteq \kappa \wedge \alpha \leq \kappa \rightarrow h_G(a \cap \alpha) = h_G(a) \cap \alpha \wedge$

$$\wedge \forall g \varepsilon Q(\alpha, h_G)\ (\ g(a \cap \alpha) = g(a) \cap \alpha\ ),$$

2) $\forall x \subseteq \kappa\ \ \forall g \varepsilon Q(\kappa, h_G)\ \ \ g(x) \subseteq x$.

Next we consider the problem of the existence of filters closed under GAIFMAN's operation h^{∇} or h_G^{∇}. It is obvious that such filters should be κ-complete and normal (in view of (iii) and (iv) in Def. 6.5.1). For the remaining part of this section, let F be a normal filter on κ such that $\{\xi < \kappa \mid Lim(\xi)\} \varepsilon F$.

PROPOSITION 6.8

If F is closed under $f \varepsilon LTF_o$, then F is closed under each $g \varepsilon Q(\alpha, f)$ for each $\alpha \leq \kappa$.

The proof is by induction on $g \varepsilon Q(\alpha, f)$ according to the cases (i) - (iv) in the definition of $Q(\alpha, f)$. (Henceforth we shall simply say "by induction on $g \varepsilon Q(\alpha, f)$".) □

PROPOSITION 6.9

If κ is inaccessible and F is closed under $f \varepsilon LTF_o$, then F is closed under $J'_{\alpha f}$ for each $\alpha < \kappa$.

Proof: If κ is inaccessible and $\alpha < \kappa$, then

$\{ g \restriction \mathfrak{P}(\alpha + 1) \mid g \in Q(\alpha, f) \}$ has power $< \kappa$, so the result follows from Lemma 6.6.1 and the κ-completeness of F. □

Note that by 2) of Lemma 6.6 we could conclude that F is closed under f^{∇} if it is closed under diagonal intersection of the $J'_{\alpha f}$'s , and this is true if this sequence is continously decreasing (which need not be true in general as mentioned above).-By Cor. 5.4:

COROLLARY 6.10

If κ is inaccessible and $\langle V_\kappa, \varepsilon\rangle \models^2 M^*$, then G is closed under $J'_{\alpha h}$ and $J'_{\alpha h_G}$ for each $\alpha<\kappa$. Similar for M_o^*, G_o and h_o.

LEMMA 6.11

Suppose α is a limit number, $\alpha<\kappa$, $f \in LTF_o$ and $a \in \mathrm{Dom}\, f$. Then

$$\alpha \in J_{\alpha f}(a) \leftrightarrow \forall g \in Q(\alpha,f) \quad \alpha \in g(a)$$

$$\leftrightarrow \alpha \in a \wedge \forall g \in Q(\alpha,f)(\alpha \in g(a) \rightarrow \alpha \in f(g(a)))\,.$$

Proof: The first part is obvious from the definitions, similarly the second "$\rightarrow$". To prove the converse, show by induction on $g \in Q(\alpha,f)$ that

$$g \in Q(\alpha,f) \rightarrow \alpha \in g(a)\,. \quad \square$$

COROLLARY 6.12

Suppose α is a limit number, $\alpha<\kappa$ and $a \subseteq \kappa$. Then: $\alpha \in J_{\alpha,h_G} \leftrightarrow$

$$\leftrightarrow \alpha \in a \wedge \forall g \in Q(\alpha,h_G)[\alpha \in g(a) \rightarrow g(a)\cap\alpha \text{ is dense in } \alpha\,].$$

The following Theorem shows that under additional assumptions G can be shown to be closed under h_G^∇:

THEOREM 6.13

Assume $\langle\kappa,<\rangle \models^2 M_o^*$, $G_o \subseteq F$ and

$$\{\xi<\kappa \mid reg(\xi) \wedge \langle\xi,<\rangle \models^2 M_o^*\} \in F\,. \text{ Then:}$$

$$\forall x \in G_o \quad h_G^\nabla(x) \in F\,.$$

Proof: Let $a \in G_o$. Then there is a set $a' \subseteq \kappa$, a' is dense in κ and $h_o(a') \subseteq a$. First we prove:

(i) $b = \{\xi<\kappa \mid a'\cap\xi$ is dense in $\xi \wedge h_o(a'\cap\xi)\cap\xi \subseteq a\cap\xi\} \in G_o$:

By Cor. 5.4: $h_o(a'), h_o(h_o(a')) \in G_o$. This implies:

$$\{\xi<\kappa \mid a'\cap\xi \text{ is dense in } \xi \wedge h_o(a')\cap\xi \text{ is dense in } \xi\} \in G_o\,.$$

Since $h_o(a') \subseteq a$, we have $h_o(a')\cap\xi \subseteq a\cap\xi$, and this proves (i).

Using the additional assumption of our theorem, we have:

$$c = a\cap\{\xi<\kappa \mid reg(\xi) \wedge \langle\xi,<\rangle \models^2 M_o^*\} \in F\,.$$

(ii) $\alpha \in b \cap c \wedge g \in Q(\alpha, h_G) \wedge \alpha \in g(a) \rightarrow \alpha \in h_G(g(a))$.

Outline of proof: Assuming the hypothesis of (ii), $a' \cap \alpha$ is dense in α . Let Q_α be $Q(\alpha, h_G \upharpoonright \mathfrak{P}(\alpha))$ in the sense of V_α (i.e. in the definition of $Q(\kappa, h_G)$ given in Def. 6.5, κ is replaced by α and h_G by $h_G \upharpoonright \mathfrak{P}(\alpha)$). By induction on the cases (i) - (iv) in the def. of Q_α and using Lemma 6.7 one shows that $g \upharpoonright \mathfrak{P}(\alpha) \in Q_\alpha$. Since $a' \cap \alpha$ is dense in α and $h_o(a' \cap \alpha) \cap \alpha \subseteq a \cap \alpha$, we have $a \cap \alpha \in G_\alpha$ where again G_α is defined as G_o with κ being replaced by α, h_o by h_o relativized to α . Now α is regular and $\langle \alpha, < \rangle \overset{2}{\models} M_o^*$, so by Prop. 6.8 (in the sense of V_α), we have $g(a \cap \alpha) \in G_\alpha$, in particular, $g(a \cap \alpha) = g(a) \cap \alpha$ is dense in α , and this proves (ii).

Finally, from (ii) and Cor. 6.12 we obtain: $\alpha \in b \cap c \rightarrow \alpha \in J_{\alpha, h_G}(a)$.

Since $b \cap c \in F$, this completes the proof of our theorem. □

COROLLARY 6.14

Assume $\langle V_\kappa, \varepsilon \rangle \overset{2}{\models} M^*$ and $G \subseteq F$ and F satisfies

$\{\xi < \kappa \mid in(\xi) \wedge \langle V_\xi, \varepsilon \rangle \overset{2}{\models} M^*\} \in F$. Then:

$\forall x \in G \ \ \overset{\nabla}{h}_G(x) \in F$.

For our applications, we need some further lemmas:

LEMMA 6.15

$a \subseteq \kappa \rightarrow h_G(h(a)) \cap a \subseteq h_G(a)$.

Proof: Let $\alpha \in a \cap h_G(h(a))$. Then

$\alpha \in h(a) \wedge h(a) \cap \alpha$ is dense in α , hence

$a \cap \alpha$ is dense in α by 2) of Lemma 5.1o, therefore $\alpha \in h_G(a)$. □

COROLLARY 6.16

$a \subseteq \kappa \wedge \alpha \leq \kappa \rightarrow \forall g \in Q(\alpha, h_G) \ \ g(h(a)) \cap a \subseteq g(a)$.

The proof is again by induction on $g \in Q(\alpha, h_G)$ using Lemma 6.15. □

As a consequence,

PROPOSITION 6.17

$$a \subseteq \kappa \;\rightarrow\; \overset{\nabla}{h}_G(h(a)) \cap a \subseteq \overset{\nabla}{h}_G(a) \;\wedge\; \overset{\nabla}{h}_G(h_o(a)) \cap a \subseteq \overset{\nabla}{h}_G(a) .$$

LEMMA 6.18

1) If $f \varepsilon LTF_o$ is monotone, so is $\overset{\nabla}{f}$.

2) If $f, g \varepsilon LTF_o$ and f or g is monotone, then $f \leq g \rightarrow \overset{\nabla}{f} \leq \overset{\nabla}{g}$.

The proof can be found in GAIFMAN [1967].

COROLLARY 6.19

1) $\overset{\nabla}{h}_G \leq \overset{\nabla}{h}_s$

2) $\forall x \subseteq \kappa \;\; \overset{\nabla}{h}_G(x) \subseteq \overset{\nabla}{h}_s(x) \cap x$

3) $\forall x \subseteq \kappa \cap Reg \;\; \overset{\nabla}{h}_G(x) \subseteq \overset{\nabla}{h}_o(x)$

4) $\forall x \subseteq \kappa \cap In \;\; \overset{\nabla}{h}_G(x) \subseteq \overset{\nabla}{h}(x)$

As a further consequence:

COROLLARY 6.2o

If F is closed under $\overset{\nabla}{h}_G$, then:

1) F is closed under $\overset{\nabla}{h}_o$, if $\kappa \cap Reg \;\; \varepsilon \;\; F$,

2) F is closed under $\overset{\nabla}{h}$, if $\kappa \cap In \;\; \varepsilon \;F$.

6.21 APPLICATIONS

1. Suppose $\langle \kappa, < \rangle \models^2 M_o^*$. If $\{\xi < \kappa \mid reg(\xi) \wedge \langle \xi, < \rangle \models^2 M_o^*\} \;\varepsilon\; G_o$,

then G_o is closed under $\overset{\nabla}{h}_G$ and $\overset{\nabla}{h}_o$, similarly for M^*, G and h.
In particular, if $\{\xi < \kappa \mid \Pi_1^1\text{-ind}(\xi)\} \;\varepsilon\; G$ (e.g. if $\langle V_\kappa, \varepsilon, G\rangle$ is a model of TAKEUTI's system NTT with (1o) of Theorem 4.1 holding at least for a suitable Π_2^1-formula Φ such that

$$in(\alpha) \rightarrow (\;\Pi_1^1\text{-ind}(\alpha) \leftrightarrow \langle V_\alpha, \varepsilon\rangle \models^2 \Phi\;)$$) , then G is closed under $\overset{\nabla}{h}_G$ and $\overset{\nabla}{h}$ (assuming that κ is Π_1^1-indescribable).

2. If κ is weakly Π_2^1-indescribable, LEVY's filter

$$F = F(\Pi_2^1) = \{x \subseteq \kappa \mid x \text{ is } \Pi_2^1\text{-enforceable at } \kappa\}$$ (cf. 5.5.1)

satisfies: $G_o \subseteq F(\Pi_1^1) \subseteq F(\Pi_2^1) = F$ and

$$\{\xi < \kappa \mid \text{weakly} - \Pi_1^1\text{-ind}(\xi)\} \;\varepsilon\; F .$$

Hence by (6) of 5.3:

$$\begin{aligned} a \in F \quad &\rightarrow \quad a \text{ is dense in } \kappa \\ &\rightarrow \quad h_o(a) \in G_o \\ &\rightarrow \quad h_G^{\nabla}(h_o(a)) \in F \quad \text{by 3.6.a and 6.13} \\ &\rightarrow \quad h_G^{\nabla}(a) \in F \quad \text{by 6.17.} \end{aligned}$$

Therefore F is a proper, non-trivial and normal filter on κ closed under h_G^{∇} and h_o^{∇}. Similarly, if we drop the restriction "weakly" and replace G_o by G, h_o by h.

3. Let κ be a cardinal carrying a proper, non-trivial, normal and $\aleph_1$-saturated filter F. We have proved in ZFC that for any such cardinal κ :

$\langle \kappa, < \rangle \models^2 M_o^*$ (Theorem 5.9). Since M_o^* is a Π_2^1-sentence (over the ordinals), using a result of KUNEN one can prove in ZFC:

$$\{\xi < \kappa \mid \langle \xi, < \rangle \models^2 M_o^*\} \in F \quad \text{for any such } \kappa, F.$$

Moreover, $G_o \subseteq F$ by (7) of Cor. 5.7. Using Theorem 6.13 and the same arguments as in the case of LEVY's filter, F is closed under h_G^{∇} and h_o^{∇}. Note that this result applies in particular to a real-valued (or 2-valued) measurable cardinal and its normal measure. For the case of 2-valued measurable cardinals cf. KEISLER [1962].

We conclude with some remarks bearing on the question whether the operation h_G^{∇} can be regarded as the supremum of all the iterated diagonal operations starting with h_G. For this purpose we propose the following

DEFINITION

α is an M-limit $:\leftrightarrow \operatorname{reg}(\alpha) \wedge \exists F \subseteq \mathcal{P}(\alpha)$ (F is a proper, non-trivial and normal filter on α $\wedge\ \forall x \varepsilon F\ \ h_G(x) \in F \wedge \{\xi < \kappa \mid \operatorname{Lim}(\xi)\} \in F$).

If α is an M-limit, α can be regarded as an ordinal such that $\mathcal{P}(\alpha)$ contains a subset of sets "large" in α and closed under MAHLO's operation h_G and the operations obtained from h_G by taking iterations and diagonal intersections. Note that by our results, a (weakly) Π_1^1-indescribable cardinal is an M-limit. We do not know whether it is consistent to assume the converse or whether there is an M-limit strictly less than the first Π_1^1-indescribable cardinal. On the other hand, using Remark 3.7.2, if α is Π^1_1-indescribable, then there is a $\beta < \alpha$ satisfying $\hat{M}^*$ and hence there is a proper, non-trivial filter on β which is β-complete and normal with respect to sequences first-order definable in V_β and closed under MAHLO's operation restricted to subsets of β which are first-order definable in V_β .

DEFINITION

For $a \subseteq \kappa$, $\alpha \leq \kappa$:

$$F(a, \alpha) := \{x \leq \kappa \mid \exists g \varepsilon Q(\alpha, h_G) \quad g(a) \cap \alpha \subseteq x\} .$$

THEOREM 6.22

For α a limit number, $\alpha < \kappa$, $0 \notin a \subseteq \kappa$: $F(a, \alpha)$ is a non-trivial, α-complete and normal filter on α which is closed under h_G.

Proof: Let F be $F(a, \alpha)$. Obviously, $\alpha \varepsilon F$, $a \cap \alpha \varepsilon F$, and

$$\forall x \varepsilon F \ \forall y \leq \alpha \ (x \subseteq y \rightarrow y \varepsilon F) .$$

F is normal: Suppose $\forall \xi < \alpha \quad a_\xi \varepsilon F$. Then there is a sequence $\langle g_\xi \mid \xi < \alpha \rangle$ such that $\forall \xi < \alpha \ g_\xi(a) \cap \alpha \subseteq a_\xi \quad \wedge \quad \forall \xi < \alpha \ g_\xi \varepsilon Q(\alpha, h_G)$.

Define a sequence $\langle f_\xi \mid \xi < \alpha \rangle$ by recursion as follows:

$$f_0 = g_0 \quad , \quad f_{\gamma+1} = g_\gamma \circ f_\gamma \qquad \text{for} \quad \gamma < \alpha \quad ,$$

$$f_\lambda = \bigcap_{\xi < \lambda} f_\xi \qquad \text{for } \lambda \text{ a limit number } < \alpha .$$

Then $\forall x \subseteq \kappa \quad f_{\gamma+1}(x) = g_\gamma(f_\gamma(x)) \subseteq f_\gamma(x)$ by 6.7. Therefore the sequence $\langle f_\xi \mid \xi < \alpha \rangle$ is continously decreasing and $\forall \xi < \alpha \quad f_\xi \varepsilon Q(\alpha, h_G)$.

It remains to show: (i) $f(a \cap \alpha) \subseteq \{\xi < \alpha \mid \forall \eta < \xi \quad \xi \varepsilon a_\eta\}$ where $f := \langle f_\xi \mid \xi < \alpha \rangle^D \varepsilon Q(\alpha, h_G)$. In fact, if $\beta \varepsilon f(a \cap \alpha)$ then

$$\beta \varepsilon f_\beta(a \cap \alpha) \qquad (\text{if } \alpha = \kappa \quad \text{or} \quad \beta < \alpha \wedge \alpha < \kappa \ ) \text{ or}$$

$$\beta \varepsilon \bigcap_{\xi < \alpha} f_\xi \ (a \cap \alpha)$$ (if $\alpha < \kappa$). In the first case, for all $\gamma < \beta$:

$$\beta \varepsilon f_{\gamma+1}(a) \cap \alpha = g_\gamma(f_\gamma(a)) \cap \alpha \subseteq g_\gamma(a) \cap \alpha \subseteq a_\gamma \quad , \text{ hence } \beta \varepsilon a_\gamma ,$$

in the second case: $\forall \xi < \alpha \ \beta \varepsilon f_\xi(a) \cap \alpha$, hence $\beta \varepsilon \alpha \wedge \beta \varepsilon f_\beta(a \cap \alpha)$, and therefore $\forall \xi < \beta \quad \beta \varepsilon a_\xi$ as in the first case. This proves (i).

F is non-trivial: Let $b \in F \wedge \beta < \alpha$. Then $g(a \cap \alpha) \subseteq b$ for some $g \varepsilon Q(\alpha, h_G)$.

Since $0 \notin a$ (note that 0 is never removed from a by operations $g \varepsilon Q(\alpha, h_G)$)

$h_G^{\beta+1}(g(a)) \subseteq g(a) - \beta$, and hence $b - \beta \in F$.

Finally, F is closed under h_G : Let $b \in F$, i.e. $g(a) \cap \alpha \subseteq b$ for some $g \in Q(\alpha, h_G)$. Then $g' = h_G \circ g \in Q(\alpha, h_G) \wedge g'(a) \cap \alpha = h_G(g(a) \cap \alpha) \subseteq$
$\subseteq h_G(b)$, therefore $h_G(b) \in F$. □

THEOREM 6.23

Suppose $\alpha < \kappa$, α a limit number, and $0 \notin a \subseteq \kappa$.

If $\alpha \in J_{\alpha\, h_G}(a)$, then $F(a, \alpha)$ is a proper, non-trivial and normal filter on α closed under h_G.

Proof: Let $b \in F = F(a, \alpha)$. Then $g(a \cap \alpha) \subseteq b$ for some $g \in Q(\alpha, h_G)$. Since $\alpha \in J_{\alpha\, h_G}(a) \wedge h_G \circ g \in Q(\alpha, h_G)$, wehave $\alpha \in h_G \circ g\,(a)$, i.e.
$g(a \cap \alpha) = g(a) \cap \alpha$ is dense in α , and so is b, in particular $b \neq 0$. □

COROLLARY 6.24

If $\alpha < \kappa \wedge \alpha \in J_{\alpha\, h_G}(a)$ for some $a \subseteq \kappa \cap \mathrm{Reg}$, then α is an M-limit.

We interprete this result as follows: Let κ be a regular cardinal such that for some $a \subseteq \kappa \cap \mathrm{Reg}$, $h_G(a)$ is non-empty. Then there is an M-limit below κ , in fact, each $\alpha \in h_G(a)$ is an M-limit. If we regard any M-limit α as a bound for the operations obtained from MAHLO's operation h_G by repeatedly taking iterations and diagonal intersections, the Cor. 6.24 defines sucha limit in terms of a larger cardinal. Moreover, these results indicate that the operation h_G^{∇} involves a real "jump" beyond these limits which is obviously due to the fact that the sequence of $J_{\alpha\, h_G}$'s is not continously decreasing.

BIBLIOGRAPHY

P. Bernays

[1962] Zur Frage der Unendlichkeitsschemata in der axiomatischen Mengenlehre, Essays ded. to A.A. Fraenkel, pp.3-49, North-Holland Publ. Co. 1962

G. Fodor

[1956] Eine Bemerkung zur Theorie der regressiven Funktionen, Acta Sci. Math. (Szeged) 17 (1956), pp. 139-142

[1966] On stationary sets and regressive functions, ibid. 27 (1966), pp. 1o5-11o

[1966] On a process concerning inaccessible cardinals I, ibid. 27 (1966), pp. 111-124

G. Fodor and A. Hajnal

[1967] On regressive functions and α -complete ideals, Bull. Acad. Pol. Sci. 15 (1967), pp.427-432

G. Fodor and A. Máté

[1969] Cardinals inaccessible with respect to a function defined on pairs of cardinals, Acta Sci. Math. (Szeged) 3o (1969), pp.1o7-111

H. Gaifman

[1967] A generalization of Mahlo's method for obtaining large cardinal numbers, Israel J. Math. 5 (1967), pp.188-2oo

K. Gödel

[194o] The consistency of the axiom of choice and of the generalized continuum hypothesis with the axioms of set theory (Annals of Math. Studies No. 3), Princeton, New Jersey, 194o

A. Hajnal

[1969] Ulam matrices for inaccessible cardinals, Bull. Acad. Pol. Sci. 17 (1969), pp. 683-688

W. Hanf

[1964] Incompactness in laguages with infinitely long expressions, Fund. Math. 53 (1964), pp.3o9-324

W. Hanf and D. Scott

[1961] Classifying inaccessible cardinals (abstract), Notices AMS 8 (1961), p.445

R.B. Jensen

[1967] Große Kardinalzahlen, Notes of lectures given at the Mathematisches Forschungsinstitut Oberwolfach (Germany), 1967

[1969] Souslin's hypothesis = weak compactness in L (abstract), Notices AMS 16 (1969), p.842

C. Karp

[1964] Languages with expressions of infinite length, North- Holland Publ. Co. 1964, XIX+183 pp.

H.J. Keisler

[1962] Mahlo's operation and the existence of α -complete prime ideals (abstract), Notices AMS 9 (1962), p.339

H.J. Keisler and A. Tarski

[1964] From accessible to inaccessible cardinals, Fund. Math. 53 (1964), pp. 225-308, Corrections ibid. 57 (1965), p.119

H.J. Keisler and F. Rowbottom

[1965] Constructible sets and weakly compact cardinals (abstract), Notices AMS 12 (1965), pp. 373-374

K. Kunen

[1970] Some applications of iterated ultrapowers in set theory, Annals of Math. Logic 1 (1970), pp. 179-227

A. Levy

[1960] On the principles of reflection in axiomatic set theory, Pacific J. Math. 10 (1960), pp.223-238

[1965] A hierarchy of formulas in set theory, Memoirs of the AMS No. 57, Providence, Rhode Island, 1965

[1967] The sizes of the indescribable cardinals, Proc. U.C.L.A. Summer Institute on Set Theory 1967, to appear

P. Mahlo

[1911-1913] Über lineare transfinite Mengen; Zur Theorie und Anwendung der ρ_0-Zahlen, Berichte über die Verhandlungen der Königl. Sächs. Gesellschaft der Wissenschaften zu Leipzig, Math.-Phys. Klasse 63 (1911), pp. 187-225; ibid. 64 (1912), pp. 108-112; ibid. 65 (1913), pp. 268-282

A. Máté

[1971] Incompactness in infinitary languages with respect to Boolean-valued interpretations, University of Szeged (1971), 55 pp. (mimeographed)

A.R.D. Mathias

[1968] A survey of recent results in set theory, Proc. U.C.L.A. Summer Institute on Set Theory 1967, to appear

R. Montague and R. Vaught

[1959] Natural models of set theory, Fund. Math. 47 (1959), pp. 219-242

D. Scott and R.M. Solovay

[1967] Boolean-valued models for set theory, Proc. U.C.L.A. Summer Institute on Set Theory 1967, to appear

J.C. Shepherdson

[1951-1953] Inner models for set theory, Journ. Symb. Logic 16 (1951), pp. 161-190; ibid. 17 (1952), pp. 225-237; ibid. 18 (1953), pp. 145-167

J. Silver

[1966] Some applications of model theory in set theory, Thesis, Univ. of California, Berkeley, 1966 , 110 pp.

R.M. Solovay

[1967a] Real-valued measurable cardinals, Proc. U.C.L.A. Summer Institute on Set Theory 1967, to appear

[1967b] Solution of a problem of Fodor and Hajnal, ibid.

[1967c] If ZF+MC+AC is consistent, then ZF+AC+ $2^{\aleph_0}$ is r.v.m. is consistent, ibid.

G. Takeuti

[1961] Axioms of infinity of set theory, Journ. Math. Soc. Japan 13 (1961), pp. 220-233

G. Takeuti

[1969] The universe of set theory, in: Foundations of Mathematics, Symposium papers comm. the 60th birthday of K. Gödel, pp. 74-128, Springer-Verlag, 1969

R. Vaught

[1963] Indescribable cardinals (abstract), Notices AMS 10 (1963), p. 126

University of Heidelberg
69 Heidelberg/Germany

XVIII. ON CHROMATIC NUMBER OF GRAPHS AND SET SYSTEMS

P. Erdős and A. Hajnal

Mathematical Institute, Hungarian Academy of Sciences

B. Rothchild

University of California, Los Angeles

§ 1. Introduction

The aim of this note is to correct a mistake the first two authors made in [1]. Set theoretical notation will be standard otherwise we will use the graph theoretical notation described in [1].

One of the main results of [1] Corollary 5.6 states that every graph of chromatic number $> \omega$ contains complete bipartite graphs $[i, \omega_1]$ for $i < \omega$.

At the end of our paper in 12.1 we claimed the following generalization of this result for uniform set systems.

If $\mathcal{H} = \langle h, H \rangle$ is a uniform set system with $\chi(\mathcal{H}) = k$, $2 \leq k < \omega$, $\beta \geq \omega$ then either the colouring number of $\mathcal{H}$ is at most β or there is $H' \subset H$, $|H'| = \beta^+$ such that $|\bigcap H'| \geq k-1$. For $k = 2$ this is a trivial consequence of e.g. [1] 5.6. In [1] we omitted the "proof". We verified the result for $\alpha(\mathcal{H}) = \alpha = \beta^+$ and thought that the induction method described in [1] § 4 yields the general result. However, this is not true and we are going to state a correct version of the theorem.

To have a brief notation we introduce a relation $R(\alpha, \beta, \gamma, k, i)$.

Definition. Let α, β, γ be cardinals, $2 \leq k < \omega$, $1 \leq i < k$. $R(\alpha, \beta, \gamma, k, i)$ is said to hold if for every uniform set system $\mathcal{H} = \langle h, H \rangle$, with $\alpha(\mathcal{H}) = \alpha$,

$\kappa(\mathcal{H}) = k$ either there is $H' \subset H$, $|H'| = \gamma$ with $|\bigcap H'| \geq i$ or $Chr(\mathcal{H}) \leq \beta$.

The false theorem claimed that $R(\alpha, \beta, \beta^+, k, k-1)$ holds for $\beta \geq \omega$, $2 \leq k < \omega$, and for every α

Theorem 1. Let $\beta = \omega_\xi$, $3 \leq k < \omega$. Then

$R(\alpha, \beta, \beta^+, k, k-1)$ holds for $\alpha = \beta^+$

$R(\alpha, \beta, \beta^+, k, 2)$ holds for $\alpha \leq \omega_{\xi+k-2}$

Put $exp_0(\beta) = \beta$, $exp_{k+1}(\beta) = 2^{exp_k(\beta)}$ for $k < \omega$

On the other hand we have

Theorem 2. Let $\beta = \omega_\xi$, $3 \leq k < \omega$, $2 \leq i \leq k-1$ and put $\alpha = (exp_{k-i}(\beta))^+$

Then $R(\alpha, \beta, 2, k, i)$ is false.

If we now denote by $f(\beta, k, i)$ the minimal α, for which $R(\alpha, \beta, 2, k, i)$ is false for $\beta \geq \omega$, $3 \leq k < \omega$ $2 \leq i \leq k-1$ then assuming G.C.H. we obtain the following

Corollary:

$$f(\beta, k, 2) = \omega_{\xi+k-1}$$
$$f(\beta, k, k-1) = \omega_{\xi+2}$$

$$f(\beta, k, i) \leq \omega_{\xi+k-i+1} \quad \text{for} \quad 2 \leq i \leq k-1$$

We have an example to show that the upper estimate is not best possible.

Theorem 3. Assume $k = \binom{\ell}{t}$, $\alpha = (exp_{t-1}(\beta))^+$, $k > 1$ $i = \binom{\ell-1}{t} + 1$. Then $R(\alpha, \beta, 2, k, i)$ is false.

As a corollary if G.C.H. is assumed and $\beta = \omega_\xi$ we get

$$f\left(\beta, k, \binom{\ell-1}{t}+1\right) \le \omega_{\xi+t}$$

E.g. in case $k=6$, $\ell=4$ we get

$$f(\beta, 6, 4) \le \omega_{\xi+2}$$ while $k-i+1=3$

Note that assuming G.C.H. the simplest unsolved problem is the following:

Does $R(\omega_2, \omega, 2, 5, 3)$ hold?

We can not determine $f(\beta, k, i)$ for other values of , however with Galvin the first two authors have a number of similar but more complicated results giving sharp upper estimations for the chromatic number of uniform set systems not containing certain types of finite subsystems. These will appear in a forthcoming triple paper of Erdős, Hajnal and Galvin.

Before we turn to the proofs we mention a few other problems which led us to discover our mistake.

Let β be a cardinal. Let $H_0(\beta)$ be the minimal γ, for which there is a partition I of length γ of $\mathcal{P}(\beta)$, i.e. $\mathcal{P}(\beta) = \bigcup_{\nu<\gamma} I_\nu$ such that no I_ν contains three different sets, A, B, C with $A \cup B = C$.

The function H_0 was introduced by Hanson for finite β. Hanson proved

(1) $c\sqrt{\beta} < H_0(\beta) \le \frac{\beta}{2}+2$ for $\beta < \omega$

A theorem of Erdős and Komlós implies that in (1) we have $\frac{\beta}{4} < H_0(\beta)$ as well.

For $\beta \ge \omega$, G. Elekes proved recently that $H_0(\omega) > \omega$ See [2]. For more problems arising here see [3].

Later Erdős considered the following similar problem.

Let $H_1(\beta)$ be the smallest γ for which there is $\mathcal{P}(\beta) = \bigcup_{\nu<\gamma} I_\nu$ such that there are no distinct $A, B, C, D \in \mathcal{P}(\beta)$ in the same I_ν satisfying

$$(2) \qquad A \cup B = C \qquad , \; A \cap B = D$$

For finite β a theorem of Erdős and Kleitman gives

$$c_1 \beta^{1/4} < H_1(\beta) < c_2 \beta^{1/2}$$

Meditation shows that $R(2^\beta, \gamma, 2, 4, 3)$ implies $H_1(\beta) \le \gamma$ hence, by the false "theorem", $H_1(\beta) \le \omega$ for every β. Investigation of $H_1(\beta)$ led us finally to the simple proof of Theorem 2.

It is worth to remark that by the above consideration and by Theorem 1 we have

$$(3) \qquad 2^\beta = \beta^+ \quad \text{implies} \quad H_1(\beta) = \beta \quad \text{for} \quad \beta \ge \omega$$

$H_1(\beta)$ will be studied in the forthcoming Erdős-Galvin-Hajnal paper as well.

§ 2. Proofs

Proof of Theorem 1. Let $\mathcal{H} = \langle h, H \rangle$ be a uniform set system with $\alpha(\mathcal{H}) = \alpha$, $\kappa(\mathcal{H}) = k$ and assume $2 \le i \le k-1$, $\beta < \alpha$ and

$$(4) \qquad |\bigcap H'| < i \qquad \text{for} \quad H' \subset H, \; |H'| = \beta^+$$

It follows easily from the Lemmas stated in [1] § 4 that then there is a sequence B_ξ, $\xi < \alpha$ of disjoint subsets of α, satisfying the following conditions

(5) $|B_\xi| < \alpha$ for $\xi < \alpha$; $\bigcup_{\xi<\alpha} B_\xi = h$

(6) If $C_\xi = \bigcup_{\eta<\xi} B_\eta$, $X \in H$, $|X \cap C_\xi| \geq i$

then $X \subset C_\xi$ for $\xi < \alpha$.

To prove statement (i) of Theorem 1. let $\alpha = \beta^+$, $i = k-1$ Then, by (5), $|B_\xi| \leq \beta$ for $\xi < \alpha$. Then there are sets D_ν, $\nu < \beta$ such that $h = \bigcup_{\nu<\beta} D_\nu$ and $|D_\nu \cap B_\xi| \leq 1$ for $\nu < \beta$, $\xi < \alpha$. Then, by (6), $X \not\subset D_\nu$ for $X \in H$ otherwise there is a maximal ξ with $X \cap B_\xi = \{u\}$ for some u, and $X - \{u\} \subset C_\xi$, $|X - \{u\}| = k-1$ implies $u \in C_\xi$ a contradiction. Hence $Chr(\mathcal{H}) \leq \beta$

To prove part (ii) we apply induction on α. We assume that (ii) is true for every $\mathcal{H}'$ with $\alpha(\mathcal{H}') < \alpha$, $k(\mathcal{H}') = \ell$, $3 \leq \ell < \omega$ Since $R(\beta, \beta, \dots)$ is true we may assume $\beta < \alpha \leq \omega_{\xi+k-2}$ and that (4), (5) and (6) hold with $i = 2$ and we have to prove $Chr(\mathcal{H}) \leq \beta$.

By part (i) we may assume $k > 3$.

For $X \in H$ let $\xi(X) = \max\{\xi < \alpha : B_\xi \cap H \neq \emptyset\}$

$$H_\xi = \{X \in H : \xi(X) = \xi\}; \quad H = \bigcup_{\xi<\alpha} H_\xi$$

Then, by (6), $|X \cap B_\xi| \geq k-1 \geq 3$ for $X \in H_\xi$. Hence there is a uniform set system $\mathcal{H}_\xi = \langle B_\xi, \hat{H}_\xi \rangle$ such that

(7) $\alpha(\mathcal{H}_\xi) = |B_\xi| \le \omega_{\xi+k-3}$ for each $\xi < \alpha$, $\varkappa(\mathcal{H}_\xi) = k-1$

and there is $Y \in \hat{H}_\xi$, $Y \subset X$ for $X \in H_\xi$

Then obviously $|\bigcap H'| < 2$ for $H' \subset \hat{H}_\xi$, $|H'| = \beta^+$

Applying the induction hypothesis for the set systems $\mathcal{H}_\xi$ we get that there are sets

$$D_{\xi,\nu} \subset B_\xi \quad ; \quad \bigcup_{\nu<\beta} D_{\xi,\nu} = B_\xi$$

such that $Y \not\subset D_{\xi,\nu}$ for $Y \in \hat{H}_\xi$, $\xi < \alpha$, $\nu < \beta$.

Put $D_\nu = \bigcup_{\xi<\alpha} D_{\xi,\nu}$. Then $\bigcup_{\nu<\beta} D_\nu = H$

Let $\nu < \beta$, $X \in H$. Then $X \in H_\xi$ for some $\xi < \alpha$

Then by (7) there is $Y \in \hat{H}_\xi$, $Y \subset X$

Since $Y \not\subset D_{\xi,\nu}$, $X \not\subset D_\nu$. Hence

$$\mathrm{Chr}(\mathcal{H}) \le \beta .$$

<u>Proof of Theorem 2</u>. Put

$$h = [\alpha]^{k-i+1} \quad , \quad 2 \le k-i+1 < k$$

Let $X \in [\alpha]^k$, $X = \{x_0, \ldots, x_{k-1}\}$, $x_0 < \ldots < x_{k-1}\}$

We define $Z(X) \in [h]^k$ by

$$Z(X) = \Big\{ \{u_0^j, \ldots, u_{k-i}^j\} \in h : j < k \text{ and } u_\nu^j = x_{j+\nu \bmod k} \text{ for } \nu \le k-i \Big\}$$

i.e. $Z(X)$ consists of the k intervals of length $k-i+1$ of X considered in the cyclical order $x_0 \prec \cdots \prec x_{k-1} \prec x_0$.
By $k-i+1<k$, $|Z(X)|=k$.

Put $H = \{Z(X) : X \in [\alpha]^k\}$

$\mathcal{H} = \langle h, H\rangle$. Then $\alpha(\mathcal{H}) = \alpha = (\exp_{k-i}(\beta))^+$

Let now $X \neq Y \in [\alpha]^k$, then there are $x \in X-Y$, $y \in Y-X$. Hence

$$|Z(X)-Z(Y)| \geq k-i+1, \quad |Z(Y)-Z(X)| \geq k-i+1$$

and thus

$$|Z(X)\cap Z(Y)| \leq \frac{2k-2(k-i+1)}{2} = i-1$$

Thus $H' \subset H$, $|H'|=2$ implies $|\bigcap H'| < i$

We prove $Chr(\mathcal{H}) > \beta$.

Let $h = \bigcup_{\nu<\beta} D_\nu$ be a partition of h. Then by $h = [\alpha]^{k-i+1}$ and as a corollary of the Erdős-Rado theorem $(\exp_{k-i}(\beta))^+ \to (\beta^+)^{k-i+1}_\beta$ there is $X \subset \alpha$, $|X|=k$ which is homogeneous, i.e. there is $\nu<\beta$ such that $[X]^{k-i+1} \subset D_\nu$. But then $Z(X) \subset D_\nu$, $Z(X) \in H$ for this X. Hence $\mathcal{H}$ has the properties to show that $R(\alpha,\beta,2,k,i)$ is false.

As to the proof of Theorem 3, take

$$h = [\alpha]^t, \quad H = \{[X]^t : X \in [\alpha]^\ell\}, \quad \mathcal{H} = \langle h, H\rangle$$

It follows quite similarly as in the proof of Theorem 2 that $\mathcal{H}$ disproves $R(\alpha,\beta,2,k,i)$

References

[1] P. Erdős and A. Hajnal, On chromatic number of graphs and set systems, Acta Math. Acad. Sci. Hung. 17(1966)61-99.

[2] G. Elekes, On a partition property of infinite subsets of a set. Periodica Math. Hung. to appear.

[3] P. Erdős and A. Hajnal, Unsolved and solved problems in set theory, Proceedings of the Berkeley Symposium 1971, to appear.

XIX. COUNTABLE MODELS OF SET THEORIES

Harvey Friedman
Department of Philosophy, Stanford University[1]

This paper is a contribution to the study of the structure of countable models of set (or class) theories. The emphasis here is on those questions which, in our opinion, are the most mathematically natural, and have the most attractive solutions. The structure of uncountable models of set theories has not been amenable, as of now, to the kind of systematic treatment given here; we mention several open questions about them in this Introduction. (One sort of question can be continually asked in any context in which good results are obtained for countable structures. Namely, can we replace isomorphism by $\mathcal{L}_{\infty,\omega}$-equivalence?

All of the terminology used in this Introduction to describe the results presented in the paper are fully explained in Section 1. We describe some of our results below, with special attention paid to the consequences they have for the three familiar theories first-order arithmetic (Z_1), 2nd-order arithmetic (Z_2), and Zermelo-Fraenkel set theory without the axiom of choice (ZF).

The principal classes of well-founded models considered in the paper are admissible and power admissible sets and systems, and power reflective systems (all under the ϵ-relation). It is trivial that the first coordinates of the admissible and power admissible systems are just the admissible and power admissible sets respectively. It is shown in Section 2 that the first coordinates of countable power reflective systems are exactly the countable power admissible sets. It is also shown in Section 2 that every countable admissible set is contained in a power admissible set with the same ordinals. Do these latter two results have an analogue for the uncountable case?

In Section 2, it is shown that the standard set of every nonstandard admissible set (hence of every nonstandard model of ZF) is an admissible set. (This result was known by several investigators.) However, the converse is false: although the hereditarily hyperarithmetic sets form an admissible set, they do not form the

[1] Research partially supported by NSF 29254.

standard set of any nonstandard admissible set. We do, however, prove a weak converse. Let K be the class of standard sets of nonstandard admissible sets. We prove in Section 2 that every countable admissible set is contained in a countable element of K with the same ordinals. Let K(ZF) be the class of standard sets of nonstandard models of ZF. We prove, also in Section 2, that every countable admissible set is contained in a countable element of K(ZF) with the same ordinals. Do these theormes have an analogue in the uncountable case? Can we characterize, at least, the countable elements of K or K(ZF)? One precise question: are the countable elements of K, or K(ZF) the class of countable well-founded models of some set theory?

The situation is neater for power admissibility. We show in Section 2 that the standard set of any nonstandard power admissible set (hence of every nonstandard model of ZF) is a power admissible set. (J. Barwise privately informed the authors that the standard set of any model of ZF is a power admissible set.) <u>And</u> every countable power admissible set is the standard set of some nonstandard power admissible set. We also prove in Section 2 that every countable power admissible set which is an initial segment of some model of ZF is the standard set of some nonstandard model of ZF. Do these theorems have analogues in the uncountable case?

Turning to standard systems, in Section 2 we show that the standard system of any nonstandard power admissible set (hence any nonstandard model of ZF) is a power reflective system. Conversely, every countable power reflective system is the standard system of some nonstandard power admissible set. In addition, every countable power reflective system, the first coordinate of which is an initial segment of some model of ZF, is the standard system of some nonstandard model of ZF. Do these theorems have analogues in the uncountable case?

Section 3 is devoted to a study of the order type of the ordinals in countable nonstandard admissible sets. The study naturally breaks into three cases. The first case is when the standard ordinal of the countable nonstandard admissible set is greater than ω. In this case, the order type is completely determined by the standard ordinal, and is simply $\alpha \times (1 + Q)$, where Q denotes the order type of the rationals. An obvious consequence of results of Section 2 is that the countable admissible ordinals are exactly the standard ordinals of countable nonstandard admissible sets, and also exactly the standard ordinals of countable nonstandard models of ZF. So the order types of the ordinals in countable ω-models of ZF are exactly the same as the order types of countable nonstandard admissible sets, and are given by $\alpha(1 + Q)$, for countable admissible ordinals $\alpha > \omega$. Does this have an analogue in the uncountable case? For instance, do models of ZF and ZF + $\exists$ inaccessible cardinal have the same order types of ordinals? (For countable models, they do.)

The second case is when the standard ordinal of the countable nonstandard admissible set is ω, and the axion of infinity fails. Such nonstandard admissible sets correspond exactly to nonstandard models of Z_1. The order type is well known to be $\omega + ((-\omega + \omega) \times Q)$, where $-\omega$ is the negative integers under <. Does this have an analogue in the uncountable case? For instance, are the order types of nonstandard models of Z_1 the same as the order types of nonstandard models of the true sentences of arithmetic?

The third and most delicate case is when the countable nonstandard admissible set has standard ordinal ω and the axiom of infinity holds. In this case, the order type of the ordinals is explicitly calculated in terms of the standard system. Now since the hereditarily finite sets always form an initial segment of models of any of the theories considered, we have, by the results of Section 2, a characterization of the standard systems of nonstandard admissible sets with standard ordinal ω, as well as of non ω-models of ZF, as those countable power reflective (HF,C). Hence, once again, the order types of the ordinals in countable nonstandard admissible sets with standard ordinal ω are the same as the order types of the ordinals in countable non ω-models of ZF. Do these results have analogues in the uncountable case?

We mention a problem about models of Z_1. We can define the standard system of a model M of Z_1 as (ω,C), where C is the set of all $\{n \in \omega$: M $\models$ "nth prime divides a"$\}$, for objects <u>a</u> in M. Interpreting the results on standard systems in terms of recursion theory yields: (ω,C) is the standard system of some countable nonstandard model of Z_1 if and only if i) C is countable and nonempty

ii) if $y_1, \ldots, y_n \in C$ and x is recursive in $(y_1, \ldots, y_n)$, then $x \in C$
iii) any infinite tree of 0's and 1's recursive in some element of C has an infinite path recursive in some element of C. (For a closely related problem with the same answer, and which inspired the work on standard systems here, see Scott [3].) But what are the C such that (ω, C) is the standard system of some non-standard model of Z_1? Are the standard systems of nonstandard models of Z_1 the same as the standard systems of nonstandard models of the true sentences of arithmetic?

In Section 4 we give a necessary and sufficient condition for one countable nonstandard power admissible set satisfying $\Sigma^S(P)$-Sep to be isomorphic to an initial segment of a second. The condition is that i) they have the same standard set ii) any $\Sigma^S(P)$ sentence true about an element of the standard set in the first model is also true about that element in the second model. Since every countable nonstandard model of ZF is a countable nonstandard power admissible set satisfying $\Sigma^S(P)$-Sep, these same two conditions are necessary and sufficient for countable nonstandard models of ZF.

In the case of models with standard ordinal ω, an examination of the proof yields an even more elegant theorem: we can drop $\Sigma^S(P)$-Sep. We have that one nonstandard power admissible set with standard ordinal ω is isomorphic to an initial segment of a second if and only if every $\Sigma^S(P)$-sentence true in the first is true in the second. This has consequences both for ZF and Z_1. One countable non ω-model of ZF is isomorphic to an initial segment of a second if and only if every $\Sigma^S(P)$-sentence true in the first is true in the second. Since models of Z_1 correspond to models of set theory in which the axiom of infinity fails, clearly $\Sigma^S(P)$ and Σ are essentially the same. So we obtain: one countable nonstandard model of Z_1 is isomorphic to an initial segment of a second if and only if every Σ_1^0 sentence true in the first is true in the second.

These theorems above, of course, have no content if the two models are isomorphic, for then trivially one is isomorphic to an initial segment of the other. So we prove that i) every countable nonstandard model of ZF is isomprhic to a proper initial segment of itself ii) every countable nonstandard model of Z_1 is isomorphic to a proper initial segment of itself. We do not have analogues to any of the theorems of Section 4 for the uncountable case.

In Section 5 we consider models (HF, ϵ, C) of Σ^1_∞ - CA. Thus each such (HF, ϵ, C) is a model of Morse-Kelley without infinity. We prove that to every such (HF, ϵ, C) there is a $D \subset C$, $D \neq C$ with (HF, ϵ, D) satisfying Σ^1_∞ - CA. Stated in the language of analysis, we prove that every ω-model of Z_2 has a proper ω-submodel of Z_2. What about uncountable ω-models of Z_2? For instance, does every uncountable ω-model of Z_2 have an uncountable proper ω-submodel of Z_2? Does every countable well-founded model of Morse-Kelley have a proper submodel with the same sets?

Section 6 deals with well-founded models only. It is well known that L is a transitive class containing all ordinals, satisfying ZF, and has a first-order axiomatization that is categorical relative to ordinals. In other words, there is a sentence φ of set theory such that φ holds in L and any two transitive sets with the same ordinals satisfying φ are equal. This is one of the most useful properties of L. Is there any other class sharing the above properties? In Section 6 we answer this in the negative. In doing so, we need to prove the following for "many" countable ordinals α: if T is a recursive set theory such that $ZF + T + V \neq L$ has a transitive model with ordinal α, then $ZF + T + V \neq L$ has more than one transitive model with ordinal α. Is this statement true for all countable ordinals α, and in particular for the ordinal of the minimum transitive model of ZF?[2]

Section 6 ends with a rather remarkable proof that any countable transitive set satisfying $ZF + V \neq HOD$ is elementary equivalent to some other countable transitive set with the same ordinals. Remarkable in that the method one would expect to need - forcing - seems completely powerless here.

---

[2] We know the statement is true for all countable ordinals and all recursive T which prove the existence of $0^{\#}$; for then the α have $V(\alpha) \cap L(n(\alpha)) = L(\alpha)$.

SECTION 1. PRELIMINARIES

In this section we present the basic axiom systems, and state without proof those facts used about them throughout the paper.

We first introduce the language in which we will write axiomatic set theories, which we call LST.

DEFINITION 1.1. The parameters of LST are written a_n, $0 \leq n$. The variables of LST are written v_n, $0 \leq n$. The atomic formulae of LST are written $a = b$, $a \varepsilon b$, where a, b are parameters of LST.

DEFINITION 1.2. If λ is a string of symbols, $\underline{a}$ a parameter, v a variable, then λ_v^a is the string of symbols obtained by replacing each occurrence of a by v in λ.

DEFINITION 1.3. The formulae of LST are given by i) all atomic formulae are formulae ii) $(\sim\varphi)$ is a formula if φ is iii) $(\varphi \& \psi)$ is a formula if φ, ψ are iv) $(\forall v)(\varphi_v^a)$ is a formula if φ is, and $\underline{a}$ is a parameter occurring in φ, v a variable not occurring in φ. The sentences of LST are the formulae of LST with no parameters. A set theory is a set for formulae of LST.

Thus $(\forall v)(\varphi_v^a) = (\forall v)(\psi_v^b)$ iff $a = b$ and $\varphi = \psi$.

DEFINITION 1.4. The s-structures are (B,E), where B is a nonempty set and E is a binary relation on B. E serves as the interpretation of ε.

DEFINITION 1.5. Let $(\varphi \rightarrow \psi)$ be $(\sim(\varphi \& (\sim\psi)))$, $(\varphi \vee \psi)$ be $((\sim\varphi) \rightarrow \psi)$, $(\varphi \longleftrightarrow \psi)$ be $((\varphi \rightarrow \psi) \& (\psi \rightarrow \varphi))$, $(\exists v)(\varphi_v^a)$ be $(\sim(\forall v)((\sim\varphi_v^a)))$. Let $(\exists! v)(\varphi_v^a)$ be $(\exists v)(\varphi_v^a \& (\forall w)(\varphi_w^a \rightarrow w = v))$.

We next introduce the language in which we will write axiomatic class theories.

DEFINITION 1.6. The parameters of LCT are written a_n, and a_n^1, $0 \leq n$. The variables of LCT are written v_n, and v_n^1, $0 \leq n$. The a_n, v_n are said to have sort $\underline{0}$; a_n^c, v_n^c have sort $\underline{1}$. The atomic formulae of LCT are written $a = b$, $c \varepsilon d$, where a, b are parameters of the same sort, c is a parameter of sort 0, d a parameter. Unless specified otherwise, if no superscript appears, then the object is presumed to be of sort 0. The <u>formulae</u> of LCT are then defined as in Definition 1.3.

DEFINITION 1.7. The <u>c-structures</u> are given by (B,E,C), where (B,E) is an s-structure and C is a nonempty set of subsets of B. C serves as the domain of individuals of sort 1 and ϵ serves as ε between B and C. We take this definition of c-structure since we will be dealing only with theories containing extensionality.

DEFINITION 1.8. We carry over Definition 1.5 to LCT.

We now define four important classes of formulae.

DEFINITION 1.9. The <u>Δ_0^s-formulae</u> are the formulae of LST given by i) each atomic formula of LST is Δ_0^s ii) $(\sim\varphi)$ is Δ_0^s if φ is Δ_0^s iii) $(\varphi \& \psi)$ is Δ_0^s if φ, ψ are Δ_0^s iv) $(\forall v)(v \varepsilon a \rightarrow \varphi_v^b)$ is Δ_0^s if φ is Δ_0^s, written $(\forall v \varepsilon a)(\varphi_v^b)$.

DEFINITION 1.10. The <u>Σ^s-formulae</u> are the formulae of LST given by i) each Δ_0^s-formula is Σ^s ii) $(\varphi \& \psi)$, $(\varphi \vee \psi)$ are Σ^s if φ, ψ are Σ^s iii) $(\forall v \varepsilon a)(\varphi_v^b)$ is Σ^s if φ is Σ^s iv) $(\exists v)(\varphi_v^a)$ is Σ^s if φ is Σ^s.

DEFINITION 1.11. The <u>Δ_0^c-formulae</u> are the formulae of LCT given by i) $a = b$, $c \varepsilon d$ are Δ_0^c, where d may have sort 1 ii) $(\sim\varphi)$ is Δ_0^c if φ is Δ_0^c iii) $(\varphi \& \psi)$ is Δ_0^c if φ, ψ are Δ_0^c iv) $(\forall v)(v \varepsilon a \rightarrow \varphi_v^b)$ is Δ_0^c if φ is Δ_0^c.

DEFINITION 1.12. The <u>Σ^c-formulae</u> are the formulae of LCT given by i) each Δ_0^c-formula is Σ^c ii) $(\varphi \& \psi)$, $(\varphi \vee \psi)$ are Σ^c if φ, ψ are iii) $(\forall v \varepsilon a)(\varphi_v^b)$ is Σ^c if φ is Σ^c iv) $(\exists v)(\varphi_v^a)$ is Σ^c if φ is Σ^c.

DEFINITION 1.13. <u>Adm^s</u> is the set theory given by 1. <u>Ext^s</u>. $(\forall v)(v \varepsilon a \longleftrightarrow v \varepsilon b) \rightarrow a = b$. 2. <u>Found</u>. $(\exists v)(\varphi_v^a) \rightarrow (\exists v)(\varphi_v^a \& (\forall w \varepsilon v)((\sim\varphi_w^a)))$, for formulae φ of LST. 3. <u>Pair</u>. $(\exists v)(\forall w)(w \varepsilon v \longleftrightarrow (w = a \vee w = b))$. 4. <u>Union</u>. $(\exists v)(\forall w)(w \varepsilon v \longleftrightarrow (\exists z \varepsilon a)(w \varepsilon z))$. 5. <u>$\Delta_0^s$-Sep</u>. $(\exists v)(\forall w)(w \varepsilon v \longleftrightarrow (w \varepsilon a \& \varphi_w^b))$, for Δ_0^s-formulae φ.
6. <u>Δ_0^s-Coll</u>. $(\forall v \varepsilon a)(\exists w)(\varphi_{vw}^{bc}) \rightarrow (\exists z)(\forall v \varepsilon a)(\exists w \varepsilon z)(\varphi_{vw}^{bc})$.

We now introduce the class theory Adm^c.

DEFINITION 1.14. <u>Adm^c</u> is the class theory given by 1. Ext^s. 2. <u>Ext^c</u>. $(\forall v)(v \varepsilon a^1 \longleftrightarrow v \varepsilon b^1) \rightarrow a^1 = b^1$. 2. Found. 3. Pair. 4. Union. 5. <u>Δ_o^c-Sep</u>. $(\exists v)(\forall w)(w \varepsilon v \longleftrightarrow (w \varepsilon a \& \varphi_w^b))$, for Δ_o^c-formulae φ. 6. <u>Δ_o^c-Coll</u>. $(\forall v \varepsilon a)(\exists w)(\varphi_{vw}^{bc}) \rightarrow (\exists z)(\forall v \varepsilon a)(\exists w \varepsilon z)(\varphi_{vw}^{bc})$, for Δ_o^c-formulae φ. 7. <u>Δ^c-CA</u>. $(\forall w)(\varphi_w^b \longleftrightarrow (\sim\psi_w^b)) \rightarrow (\exists v^1)(\forall w)(w \varepsilon v^1 \longleftrightarrow \varphi_w^b)$, where φ, ψ are Σ^c-formulae.[3]

DEFINITION 1.15. Take <u>Δ^s-Sep</u> to be $(\forall v)(\varphi_v^b \longleftrightarrow (\sim\psi_v^b)) \rightarrow (\exists v)(\forall w)(w \varepsilon v \longleftrightarrow w \varepsilon a \& \varphi_w^b))$, for Σ^s-formulae φ, ψ. Take <u>Σ^s-Coll</u> to be $(\forall v \varepsilon a)(\exists w)(\varphi_{vw}^{bc}) \rightarrow (\exists z)(\forall v \varepsilon a)(\exists w \varepsilon z)(\varphi_{vw}^{bc})$, for Σ^s-formulae φ.

THEOREM 1.1. $Adm^s \vdash \Sigma^s$-Coll and $Adm^s \vdash \Delta^s$-Sep.

DEFINITION 1.16. Take <u>Δ^c-Sep</u> to be $(\forall v)(\varphi_v^b \longleftrightarrow (\sim\psi_v^b)) \rightarrow (\exists v)(\forall w)(w \varepsilon v \longleftrightarrow (w \varepsilon a \& \varphi_w^b))$, for Σ^c-formulae φ, ψ. Take <u>Σ^c-Coll</u> to be $(\forall v \varepsilon a)(\exists w)(\varphi_{vw}^{bc}) \rightarrow (\exists z)(\forall v \varepsilon a)(\exists w \varepsilon z)(\varphi_{vw}^{bc})$, for Σ^c-formulae φ.

THEOREM 1.2. $Adm^c \vdash \Sigma^c$-Coll and $Adm^c \vdash \Delta^c$-Sep.

THEOREM 1.3. To every $(B,E) \models Adm^s$ there is a C with $(B,E,C) \models Adm^c$. Consequently, Adm^c is a conservative extension of Adm^s; i.e., $Adm^s \vdash \varphi$ iff $Adm^c \vdash \varphi$, for $\varphi \in$ LST.

DEFINITION 1.17. Let <u>Ord(a)</u> be the formula $(\forall x \varepsilon a)(\forall y \varepsilon x)(y \varepsilon a) \& (\forall x \varepsilon a)(\forall y \varepsilon a)(x \varepsilon y \vee y \varepsilon x \vee x = y)$. Introduce 0 as a defined constant of Adm^s given by $(\forall x)(\sim x \varepsilon 0)$. Introduce the defined operators to Adm^s: $\{a,b\}$, $\langle a,b\rangle$, $a \cup b$ given by $(\forall x)(x \varepsilon \{a,b\} \longleftrightarrow (x = a \vee x = b))$, $\langle a,b\rangle = \{a,\{a,b\}\}$, $(\forall x)(x \varepsilon a \cup b \longleftrightarrow (x \varepsilon a \vee x \varepsilon b))$. Introduce the defined constants to Adm^s: i) $\overline{0} = 0$ ii) $\overline{n+1} = n \cup \{n\}$.

THEOREM 1.4. $Adm^s \vdash (Ord(a) \& Ord(b)) \rightarrow (a \varepsilon b \vee b \varepsilon a \vee a = b)$.

[3] From some points of view it may be better to have foundation for all formulae in LCT. Here it does not matter since we will only consider well-founded models of Adm^c.

THEOREM 1.5. There are Σ^s-formulae φ, ψ with exactly two parameters a, b such that the following are provable in Adm^s: i) $\varphi \longleftrightarrow (\sim\psi)$ ii) $(\exists! x)(\varphi^b_x)$ iii) $\varphi \rightarrow \mathrm{Ord}(b)$ iv) $\varphi \rightarrow (\forall x)(x \in b \longleftrightarrow (\exists y \in a)(\exists z)(\varphi^{ab}_{yz}\ \&\ (x \in z \vee x = z)))$.

DEFINITION 1.18. In view of Theorem 1.4, we can introduce the operation <u>Rk</u>, for rank, to Adm^s, by $\mathrm{Rk}(a) = b \longleftrightarrow \varphi$.

We now introduce the class theory PAdm^c, for "power admissible."

DEFINITION 1.19. <u>PAdm^c</u> consists of i) Adm^c. ii) <u>P^s</u>. $(\exists v)(\forall w) w \in v \longleftrightarrow (\forall z)(z \in w \rightarrow z \in a)$. iii) <u>$P^c$</u>. $(\exists v^1)(\forall w)(w \in v^1 \longleftrightarrow (\exists x)(\exists y)(w = \langle x,y\rangle\ \&\ (\forall z)(z \in y \longleftrightarrow (\forall v)(v \in z \rightarrow v \in x))))$.

Finding a set theory which corresponds to PAdm^c as Adm^s corresponds to PAdm^s leads to the definition of $\Delta^s_o(P)$-formulae and $\Sigma^s(P)$-formulae.

DEFINITION 1.20. The <u>P-terms</u> are given by i) each parameter of LST is a P-term ii) P(t) is a P-term if t is. The <u>atomic P-formulae</u> are given by $s = t$, or $s \in t$ for P-terms s, t.

DEFINITION 1.21. The <u>pseudo $\Delta^s_o(P)$-formulae</u> and the <u>pseudo $\Sigma^s(P)$</u>-formulae are defined exactly as the Δ^s_o-formulae and Σ-formulae in Definitions 1.9 and 1.10, except that atomic P-formulae are used instead of atomic formulae.

DEFINITION 1.22. The <u>$\Delta^s_o(P)$-formulae</u> and the <u>$\Sigma^s(P)$-formulae</u> are the formulae of LST (without P), that are the obvious translations into the usual language of set theory generated by the defining axiom $a \in P(b) \longleftrightarrow (\forall v)(v \in a \rightarrow v \in b)$.

DEFINITION 1.23. Take PAdm^s to be the set theory consisting of i) Adm^s ii) P^s iii) <u>$\Delta^s_o(P)$-Coll.</u> $(\forall v \in a)(\exists w)(\varphi^{bc}_{vw}) \rightarrow (\exists z)(\forall v \in a)(\exists w \in z)(\varphi^{bc}_{vw})$, for $\Delta^s_o(P)$-formulae φ.

THEOREM 1.6. To every $(B,E) \models \mathrm{PAdm}^s$ there is a C with $(B,E,C) \models \mathrm{PAdm}^c$. Consequently, PAdm^c is a conservative extension of PAdm^s; i.e., $\mathrm{PAdm}^s \vdash \varphi$ iff $\mathrm{PAdm}^c \vdash \varphi$, for $\varphi \in$ LST.

DEFINITION 1.24. Take $\underline{\Delta^s(P)\text{-Sep}}$ to be $(\forall v)(\varphi^b_v \longleftrightarrow (\sim\psi^b_v)) \rightarrow (\exists v)(\forall w)(w \in v \longleftrightarrow (w \in a \;\&\; \varphi^b_w))$, for $\Sigma^s(P)$-formulae φ, ψ. Take $\underline{\Sigma^s(P)\text{-Coll}}$ to be $(\forall v \in a)(\exists w)(\varphi^{bc}_{vw}) \rightarrow (\exists z)(\forall v \in a)(\exists w \in z)(\varphi^{bc}_{vw})$, for $\Sigma^s(P)$-formulae φ.

THEOREM 1.7. $PAdm^s \vdash \Sigma^s(P)$-Coll and $PAdm^s \vdash \Delta^s(P)$-Sep.

DEFINITION 1.25. We add the defined operator $\underline{P^s}$ to $PAdm^s$, by $P^s(a) = b \longleftrightarrow (\forall v)(v \in b \longleftrightarrow (\forall x \in v)(x \in a))$, and the defined constant $\underline{P^c}$ to $PAdm^c$, by $P^c = a^1 \longleftrightarrow (\forall w)(w \in a^1 \longleftrightarrow (\exists x)(w = \langle x, P^s(x)\rangle))$.

THEOREM 1.8. There are Σ^c-formulae φ, ψ with exactly three parameters a, b, c^1 such that the following are provable in $PAdm^c$ i) $\varphi^{c^1}_{P^c} \longleftrightarrow (\sim\psi^{c^1}_{P^c})$ ii) $\varphi^{c^1}_{P^c} \longleftrightarrow (Ord(a) \;\&\; Rk(b) \in a)$.

DEFINITION 1.26. In view of Theorem 1.6, we introduce the defined operator V to $PAdm^c$, and by Definition 1.20, also to $PAdm^s$, given by $(Ord(a) \rightarrow (\forall x)(x \in V(a) \longleftrightarrow Rk(x) \in a)) \;\&\; (\sim Ord(a) \rightarrow V(a) = \emptyset)$.

DEFINITION 1.27. We now describe a form of reflection, $\underline{\Pi^1_1 \text{ - Rfn}}$: $(\forall x^1)(\exists y)(\varphi^{a^1b}_{x^1y}) \rightarrow (\exists z)(Ord(z) \;\&\; (\forall x^1)(\exists y \in V(z))(\varphi^{a^1b}_{x^1y}))$, where φ is a Δ^c_0-formula. We only use Π^1_1-Rfn in conjunction with $PAdm^c$.

The relation between $PAdm^c$ and $PAdm^c + \Pi^1_1$-Rfn will be dealt with in the paper.

THEOREM 1.9. (Normal Form.) Let $0 \leq n$. There is a Δ^s_0-formula φ with exactly the parameters $a_0, \ldots, a_n, a_{n+1}$ such that for all Σ^s_0-formulae ψ with exactly the parameters $a_1, \ldots, a_n$ there is a $k \in \omega$ such that $Adm^s \vdash (\exists x)(\varphi^{a_0 a_{n+1}}_{\bar{k}\, x}) \longleftrightarrow \psi$.

We have also the analogue for $PAdm^s$.

THEOREM 1.10. Let $0 \leq n, m$. There is a $\Delta^s_0(P)$-formula φ with exactly the parameters $a_0, \ldots, a_n, a_{n+1}, a^1_1, \ldots, a^1_m$ such that for all $\Sigma^s(P)$-formulae ψ with exactly the parameters $a_1, \ldots, a_n$, there is a $k \in \omega$ such that $PAdm^s \vdash (\exists x)(\varphi^{a_0 a_{n+1}}_{\bar{k}\, x}) \longleftrightarrow \psi$.

DEFINITION 1.28. HF is the set of all hereditarily finite sets. HC is the set of all hereditarily countable sets. Inf is the axiom of infinity, which reads $(\exists v)(\mathrm{Ord}(v) \;\&\; (\forall a \in v)(\exists b \in v)(a \in b))$.

DEFINITION 1.29. Let $(B,E) \models \mathrm{Adm}^s$. Then the standard part of (B,E) is $\{b \in B:$ no $f: \omega \to B$ has $(\forall n)(E(f(n+1),f(n))) \;\&\; f(0) = b\}$. If Y is the standard part of (B,E), then clearly (Y,E) is an extensional well-founded structure, and hence there is a unique transitive set X, called the standard set of (B,E), and a unique isomorphism j from (Y,E) onto $(X,\in)$. We will always assume for $(B,E) \models \mathrm{Adm}^s$, that $X = Y$ and j is the identity. Let $(B,E) \models \mathrm{PAdm}^s$. Then the standard system of (B,E) is (A,C), where A is the standard set of (B,E) and $C = \{x \cap A: (\exists b \in B)(x = \{a: E(a,b)\})\}$. Whenever $y = A \cap \{a: E(a,b)\}$, we say that b represents y in (B,E), or b is a representative of y in (B,E).

DEFINITION 1.30. An admissible set is a transitive set A such that $(A,\in) \models \mathrm{Adm}^s$. An admissible system is a pair (A,C), where A is a transitive set and the c-structure $(A,\in,C) \models \mathrm{Adm}^c$. An admissible ordinal is an ordinal of the form $A \cap \mathrm{On}$, for some admissible set A. A power admissible set is a transitive set A such that $(A,\in) \models \mathrm{PAdm}^s$. A power admissible system is a pair (A,C), where A is a transitive set such that $(A,\in,C) \models \mathrm{PAdm}^c$. A power reflective system is a power admissible system (A,C) such that $(A,\in,C) \models \Pi^1_1\text{-Rfn}$.

DEFINITION 1.31. A nonstandard admissible set is a $(B,E) \models \mathrm{Adm}^s$ such that (B,E) is not well-founded. A nonstandard power admissible set is a $(B,E) \models \mathrm{PAdm}^s$ such that (B,E) is not well-founded.

DEFINITION 1.32. Let $(B_1,E_1) \models \mathrm{PAdm}^s$, $(B_2,E_2) \models \mathrm{PAdm}^s$. Then (B_1,E_1) is an initial segment of (B_2,E_2) if and only if i) $B_1 \subset B_2$ ii) $(\forall x,y \in B_1)(E_1(x,y) \longleftrightarrow E_2(x,y))$ iii) $(\forall x \in B_1)(\forall y)([(B_2,E_2) \models \mathrm{Rk}(y) \in \mathrm{Rk}(x)] \to y \in B_1)$.

THEOREM 1.11. Suppose (B_1,E_1) is an initial segment of (B_2,E_2). Then every $\Sigma^s(P)$-formula true of some elements of B_1 in (B_1,E_1) is also true about those same elements in (B_2,E_2).

DEFINITION 1.33. Let $(B,E) \models \mathrm{Adm}^s$. Then x is called standard in (B,E) iff x is in the standard set of (B,E); x is called a standard ordinal in (B,E) iff x is in the standard set of (B,E) and $(B,E) \models \mathrm{Ord}(x)$. The standard ordinal of (B,E) is the supremum of the standard ordinals in (B,E).

For each admissible set A Barwise [1] introduces the infinitary language $\mathcal{L}_A$, done in a natural way so that each formula in an element of A, and for all admissible systems (A,C), the set of formulae of $\mathcal{L}_A$ is in C. In addition, if $A \subset B$, B an admissible set, then $\mathcal{L}_A = \mathcal{L}_B \cap A$. We do not need the whole of $\mathcal{L}_A$: just two fragments tailor made for our purposes. We will use & instead of $\wedge$, and v for ~ & ~.

DEFINITION 1.34. Let A be an admissible set. Then LST_A is the $\mathcal{L}_A$ of Barwise [1] with only ε, $=$ and no constants. LST_A^+ is the ${}_A$ of Barwise [1] with only ε, $=$, and the constants c_n, $n < \omega$. A theory in LST_A is a set of sentences of LST_A. A theory in LST_A^+ is a set of sentences of LST_A^+. A theory (sentence) is consistent iff it has a model. Write $T \vdash \varphi$ iff every model of T is a model of φ. When dealing with theories in LST_A we use models (B,E); with theories in LST_A^+ we use models $(B,E,d_n)_{n<\omega}$.

We now give a form of the compactness Theorem 2.13 of Barwise [1] tailor made for our purposes.

THEOREM 1.12. (Compactness.) Let (A,C) be a countable admissible system, and let $T \in C$ be a theory in LST_A^+. Suppose that every $T_0 \subset T$ with $T_0 \in B$ is consistent. Then T is consistent. Similarly for LST_A.

DEFINITION 1.35. Let A be an admissible set. An A-supercomplete theory T is a theory in LST_A^+ satisfying 1) every $\varphi \in T$ is consistent 2) for all sentences φ in LST_A^+, $\varphi \in T$ or $\sim\varphi \in T$ 3) for all sentences $\&(\Gamma)$ in LST_A^+, if $\Gamma \subset T$, then $\&(\Gamma) \in T$ 4) for all sentences $\forall x\varphi(x)$ in LST_A^+, if $\varphi(c_n) \in T$ for all $n < \omega$, then $\forall x\varphi(x) \in T$.

THEOREM 1.13. If T is A-supercomplete, then T has a model in which every element of the domain is the interpretation of some constant.

Note that LST is a sublanguage of LST_A for all admissible sets A.

THEOREM 1.14. Let A be an admissible set. Then there is a function f such that

1) for all sets a, f(a) is a formula φ_a of some LST_A with only one parameter
2) for all a, Ext $\vdash (\exists ! x)(\varphi_a(x))$ & $(\forall x)(\varphi_a(x) \rightarrow (\forall y)(y \varepsilon x \longleftrightarrow v(\{\varphi_b(y): b \in a\})))$
3) for all admissible systems (A,C) we have $f \restriction A \in C$.

DEFINITION 1.36. In view of Theorem 1.11, introduce defined constants c_a, for every set a, to the theory {Ext}, with the defining axioms $\varphi_a(c_a)$, and such that, for all admissible systems (A,C), we have $\{\langle a,c_a\rangle: a \in A\} \in C$.

DEFINITION 1.37. In Section 4 we will have use for the theory $\Sigma^s(P)$-Sep, given by $(\exists v)(\forall w)(w \varepsilon v \longleftrightarrow (w \varepsilon a \ \& \ \varphi^b_w))$, for $\Sigma^s(P)$-formulae φ. In Section 5 we will have use for the theory Σ^1_∞ - CA, given by $Ext^1 + (\exists v^1)(\forall w)(w \varepsilon v^1 \longleftrightarrow \varphi^b_w)$, for formulae φ of LCT. We will also use the theory Σ^1_∞ - AC, given by $Ext^1 + (\forall w)(\exists v^1)(\varphi^{ab}_{wv^1}) \rightarrow (\exists x^1)(\forall w)(\exists v^1)(\varphi^{ab}_{wv^1} \ \& \ (\forall y)(\langle w,y\rangle \varepsilon x^1 \longleftrightarrow y \varepsilon v^1))$, for formulae φ of LCT.

SECTION 2. STANDARD SYSTEMS OF NONSTANDARD ADMISSIBLE SETS

We first establish a basic fact about standard sets.

THEOREM 2.1. The standard set of a nonstandard admissible set is an admissible set.

Proof: Let (B,E) be a nonstandard admissible set, and let A be the standard set of (B,E). Then Ext^s, Found, Pair, Union, and Δ_0^s-Sep hold in (A,ϵ) since they hold in (B,E), A is transitive, and $(A,\epsilon) = (A,E)$. To verify Δ_0^s-Coll in (A,ϵ), let $\varphi(v,w)$ be Δ_0^s with all parameters replaced by elements of A, $x \in A$, and $(A,\epsilon) \models (\forall v \in x)(\exists w)(\varphi(v,w))$. Then $(B,E) \models (\forall v \in x)(\exists w)(\varphi(v,w))$, and let $b \in B$ have $(B,E) \models (\forall v \in x)(\exists w \in b)(\varphi(v,w))$, by Δ_0^s-Coll in (B,E). Then for some ordinal c in (B,E) we have $(B,E) \models (\forall v \in x)(\exists w)(\mathrm{Rk}(w) \in c \;\&\; \varphi(v,w))$. Choose c to be the E-least ordinal in B such that the above holds. Now clearly the above holds for all nonstandard ordinals c of (B,E), and hence c must be standard; i.e., $c \in A$. Now by Δ^s-Coll in (B,E), choose $d \in B$ with $(B,E) \models (\forall v \in x)(\exists w)(\mathrm{Rk}(w) \in c \;\&\; w \in d \;\&\; \varphi(v,w))$. By Δ^s-Sep in (B,E), choose e with $(B,E) \models (\forall v \in x)(\exists w)(\mathrm{Rk}(w) \in c \;\&\; w \in e \;\&\; \varphi(v,w)) \;\&\; (\forall v \in e)(\mathrm{Rk}(v) \in c)$. Then $e \in A$ and we are done.

Not every (countable) admissible set is the standard set of some nonstandard admissible set; e.g., the least admissible set with an infinite element.

However, we can prove the following weak converse:

THEOREM 2.2. To every countable admissible set A there is a nonstandard admissible set whose standard set contains A and whose standard ordinal is $A \cap \mathrm{On}$. More generally, let (A,C) be an admissible system, $T \in C$, T a theory in LST_A. Suppose that some $(B,E) \models T + \mathrm{Adm}^s$ has standard set containing A. Then some nonstandard admissible set satisfying T has standard set containing A and standard ordinal $A \cap \mathrm{On}$.

Proof: Let T_1 be $T \vdash \mathrm{Adm}^s + \{(\exists v)(v = \bar{x}):\ x \in A\}$. Then $T_1 \in C$, T_1 is a theory in LST_A, and T_1 is consistent. We must prove that T_1 has a nonstandard model whose standard ordinal is $A \cap \mathrm{On}$. It is easily seen that it suffices to construct an A-supercomplete theory T_2 such that 1) $T_1 \subset T_2$ 2) $(\bar{\alpha}\ \varepsilon\ c_o) \in T_2$ for all $\alpha \in A$ 3) for each c_n such that $(\bar{\alpha}\ \varepsilon\ c_n) \in T_2$ for all $\alpha \in A$, there is a c_m such that $(\bar{\alpha}\ \varepsilon\ c_m) \in T_2$, for all $\alpha \in A$, and $(c_m\ \varepsilon\ c_n) \in T_2$.

Enumerate the sentences of LST_A^+ by $\varphi_o, \varphi_1, \varphi_2, \ldots$. We define a sequence of consistent theories T_2^n, $0 \le n$. Take $T_2^o = T_2 + \{\bar{\alpha}\ \varepsilon\ c_o\colon\ \alpha \in A\}$. This is consistent by the compactness theorem. Suppose T_2^{3n} has been defined. Take $T_2^{3n+1} = T_2^{3n} + \varphi_n$ if consistent; $T_2^{3n} + \sim\varphi_n$ otherwise. Take T_2^{3n+2} as follows: if $\sim\varphi_n \in T_2^{3n+1}$ and $\varphi_n = \&(\Gamma)$ then $T_2^{3n+2} = T_2^{3n+1} + \sim\psi$ for some $\psi \in \Gamma$ with $T_2^{3n+1} + \sim\psi$ consistent; if $\sim\varphi_n \in T_2^{3n+1}$ and $\varphi_n = (\mathrm{V}v)(\psi(v))$ then $T_2^{3n+2} = T_2^{3n+1} + \sim\psi(c_m)$ for some c_m not appearing in T_2^{3n+1}; $T_2^{3n+2} = T_2^{3n+1}$ otherwise. Take T_2^{3n+3} as follows: if for some $\alpha \in B$, $T_2^{3n+2} + \sim(\bar{\alpha}\ \varepsilon\ c_n)$ is consistent, take $T_2^{3n+3} = T_2^{3n+2} + \sim(\bar{\alpha}\ \varepsilon\ c_n)$. If not, then by the compactness theorem, $T_2^{3n+2} + c_m\ \varepsilon\ c_n + \{\bar{\alpha}\ \varepsilon\ c_m\colon\ \alpha \in B\}$ is consistent, where c_m does not appear in T_2^{3n+2}, $n \neq m$, and take T_2^{3n+3} to be that. Finally, take $T_2 = \bigcup_n T_2^n$.

If we want to strengthen Theorem 2.2 to give a sufficient condition for being a standard set, we must consider power admissible sets.

THEOREM 2.3. The standard set of a power admissible set is a power admissible set. Every countable power admissible set is the standard set of some nonstandard power admissible set. More generally, let (A,C) be a power admissible system, $T \in C$, T a theory in LST_A. Suppose that $(A,\in)$ is an initial segment of some (B,E) satisfying $PAdm^S + T$. Then some nonstandard power admissible set satisfying T has standard set A.

<u>Proof</u>: The first statement is a special case of what we prove later about standard systems. We will obtain the third statement from Theorem 2.2. Consider the auxiliary theory $T' = T + PAdm^S + \{(\forall y)(Rk(y)\ \varepsilon\ \bar{\alpha} \longleftrightarrow v(\{y = \bar{x}\colon\ x \in A \cap V(\alpha)\}))\colon\ \alpha \in A\}$. Then $T' \in C$ and T' is a theory in LST_A, and so apply Theorem 2.2.

We now consider standard systems.

THEOREM 2.4. The standard system of every nonstandard power admissible set is a power reflective system.

Proof: Let (B,E) be a nonstandard power admissible set, (A,C) the standard system of (B,E). We first verify that $(A,\epsilon,C) \models \mathrm{Adm}^c$. Note that Ext^c holds since $C \subset \mathbb{P}(A)$. Ext^s, Found, Pair, Union hold since A is a transitive set, $(A,\epsilon,C) = (A,E,C)$, and they hold in (B,E). We have $(A,\epsilon,C) \models \Delta_0^c$-Sep by replacing parameters from C by representatives in (B,E). Also $(A,\epsilon,C) \models \Delta_0^c$-Coll by replacing parameters from C by representatives from (B,E) and employing the same argument as that for Theorem 2.1.

Now for Δ^c - CA, let $(A,\epsilon,C) \models (\forall w)((\exists v)(\varphi(v,w) \longleftrightarrow \sim(\exists v)(\psi(v,w)))$, where φ, ψ contain parameters from A, C and φ, ψ are Δ_0^c. Let φ^*, ψ^* be the result of replacing the parameters from C by representatives from (B,E). Note that for all ordinals $\alpha \in A$, $(B,E) \models (\forall w \in V(\alpha))(\sim((\exists v)(\varphi^*(v,w) \;\&\; v \in V(\alpha)) \;\&\; (\exists v)(\psi^*(v,w) \;\&\; v \in V(\alpha))))$. Then for some nonstandard ordinal β in (B,E) we have $(B,E) \models (\forall w \in V(\beta))(\sim((\exists v)(\varphi^*(v,w) \;\&\; v \in V(\beta)) \;\&\; (\exists v)(\psi^*(v,w) \;\&\; v \in V(\beta))))$. Fix $x \in B$ such that $(B,E) \models (\forall w)(w \in x \longleftrightarrow (w \in V(\beta) \;\&\; (\exists v \in V(\beta))(\varphi^*(v,w))))$, by Δ_0^s-Sep in (B,E). Then finally take $v = \{a \in A\colon\ E(a,x)\}$.

We leave it to the reader to verify P^s and P^c in (A,ϵ,C).

To check $(A,\epsilon,C) \models \Pi_1^1$ - Rfn, assume $(A,\epsilon,C) \models (\forall z)(\mathrm{Ord}(z) \rightarrow (\exists x^1)(\forall y \in V(z))(\varphi(x^1,y))$, where φ contains parameters from A, C and φ is Δ_0^c. Let φ^* be the result of replacing the parameters from C by representatives from (B,E). Then for all ordinals $\alpha \in A$ we have $(B,E) \models (\exists x)(\forall y \in V(\alpha))(\varphi^*(x,y))$. Hence for some nonstandard ordinal β in (A,E) we have $(B,E) \models (\exists x)(\forall y \in V(\beta))(\varphi^*(x,y))$. Finally choose $x^1 = \{a \in A\colon\ E(a,x)\}$.

THEOREM 2.5. Every countable power reflective system is the standard system of some nonstandard power admissible set. More generally, let (A,C) be a countable power reflective system, $T \in C$, T a theory in LST_A. Suppose that A is an initial segment of some (B,E) satisfying $\mathrm{PAdm}^s + T$. Then there is a nonstandard power admissible set satisfying T whose standard system is (A,C).

Proof: Let (A,C) be a countable power reflective system, $T \in C$ satisfying hypotheses. Let $T_1 = T + \mathrm{PAdm}^s + \{(\exists v)(v = \overline{x})\colon\ x \in A\} + \{(\forall v)(\mathrm{Rk}(v) \in \overline{\alpha} \longleftrightarrow v \in \overline{A \cap V(\alpha)})\colon\ \alpha \in A\}$. Then T is consistent. It suffices to find an A-supercomplete theory T_2 satisfying 1) $T_1 \subset T_2$ 2) for each $Y \in C$ there is a c_k such that for each $\alpha \in A$ we have $(\forall x)((x \varepsilon c_k \;\&\; Rk(x) \varepsilon \bar{\alpha}) \longleftrightarrow x \varepsilon \overline{Y \cap V(\alpha)}) \in T_2$ 3) for each c_k there is a $Y \in C$ such that for each $\alpha \in A$ we have $(\forall x)((x \varepsilon c_k \;\&\; Rk(x) \varepsilon \bar{\alpha}) \longleftrightarrow x \varepsilon \overline{Y \cap V(\alpha)}) \in T_2$ 4) for each c_k such that for each $x \in A$, $(\bar{x} \varepsilon c_k) \in T_2$, there is a c_m such that for each $x \in A$, $(\bar{x} \varepsilon c_m)$ & $(c_m \varepsilon c_k) \in T_2$.

Enumerate the sentences of LST_A^+ by $\varphi_o, \varphi_1, \varphi_2, \ldots,$ and enumerate the elements of C by $Y_o, Y_1, Y_2, \ldots$. We define a sequence of consistent theories T_2^n, $0 \leq n$ such that each $T_2^n \in C$, so that the compactness theorem can be applied to each T_2^n. Define $T_2^o = T_1$. Suppose that $T_2^{5n} \in C$ has been defined and is consistent. Take $T_2^{5n+1} = T_2^{5n} + \varphi_n$ if consistent; $T_2^{5n} + \sim\varphi_n$ otherwise. Take T_2^{5n+2} as follows: if $\sim\varphi_n \in T_2^{5n+1}$ and $\varphi_n = \&(\Gamma)$ then $T_2^{5n+2} = T_2^{5n+1} + \sim\psi$ for some $\psi \in \Gamma$ with $T_2^{5n+1} + \sim\psi$ consistent; if $\sim\varphi_n \in T_2^{5n+1}$ and $\varphi_n = (\forall x)(\psi(x))$ then $T_2^{5n+2} = T_2^{5n+1} + \sim\psi(c_m)$ for some c_m not appearing in T_2^{5n+1}; $T_2^{5n+2} = T_2^{5n+1}$ otherwise. Take T_2^{5n+3} as follows: if for some $x \in A$, $T_2^{5n+2} + \sim(\bar{x} \varepsilon c_n)$ is consistent, take $T_2^{5n+3} = T_2^{5n+2} + \sim(\bar{x} \varepsilon c_n)$. If not, then by compactness, $T_2^{5n+2} + c_m \varepsilon c_n + \{\bar{x} \varepsilon c_m\colon x \in A\}$ is consistent, where c_m is the first constant not appearing in T_2^{5n+2}, $n \neq m$, and take T_2^{5n+3} to be that. Take $T_2^{5n+4} = T_2^{5n+3} \cup \{(\forall x)((x \varepsilon c_m \;\&\; Rk(x) \varepsilon \bar{\alpha}) \longleftrightarrow x \varepsilon \overline{Y \cap V(\alpha)})\colon \alpha \in A\}$, where c_m is the first constant not appearing in T_2^{5n+3}. By compactness, T_2^{5n+4} is consistent. Take T_2^{5n+5} as follows. Note that for each $\alpha \in A$ there is a $y \in A \cap V(\alpha + 1)$, and hence a $y \in C$ such that $T_2^{5n+4} + (\forall x)((x \varepsilon c_n \;\&\; Rk(x) \varepsilon \bar{\alpha}) \longleftrightarrow x \varepsilon \bar{y})$ is consistent. We will be done if we can find $y \in C$ such that $T_2^{5n+4} + \{(\forall x)((x \varepsilon c_n \;\&\; Rk(x) \varepsilon \bar{\alpha}) \longleftrightarrow x \varepsilon \overline{y \cap V(\alpha)})\colon \alpha \in A\}$ is consistent, setting T_2^{5n+5} to be that. Suppose not. Then by Π_1^1 - Rfn in $(A, \in, C)$, there must be an $\alpha \in A$ such that for all $y \in C$, there is a β and an inconsistency proof, both in $A \cap V(\alpha)$, of $T_2^{5n+4} + (\forall x)((x \varepsilon c_n \;\&\; Rk(x) \varepsilon \bar{\beta}) \longleftrightarrow x \varepsilon \overline{y \cap V(\beta)})$. But this is a contradiction.

We now present a theorem about sets and systems only.

THEOREM 2.6. Every countable admissible set is contained in a power admissible set with the same ordinals. To every countable power admissible set A there is a C

such that (A,C) is a power reflective system.

Proof: By Theorem 2.2, every countable admissible set is contained in the standard set of some nonstandard power admissible set. By Theorem 2.4, this standard set is power admissible. For the second part, by Theorem 2.4, it suffices to prove that every countable power admissible set is the standard set of some nonstandard power admissible set. But this is clear from Theorem 2.2, using the appropriate T.

SECTION 3. THE ORDINALS IN NONSTANDARD ADMISSIBLE SETS

Let (B,E) be a nonstandard admissible set. The theory of the ordinals in (B,E) naturally breaks into three cases. The first case we consider is if the standard ordinal of (B,E) is $> \omega$. In this case we will calculate the order type (of the ordinals) of (B,E) solely in terms of the standard ordinal of (B,E). The second case is if the standard set of (B,E) is HF, and the axiom of infinity fails in (B,E). This determines the order type of (B,E), and is essentially due to Henkin. The third case is if the standard set of (B,E) is HF and the axiom of infinity holds in (B,E). Here we calculate the order type of (B,E) in terms of the standard system (HF,Y) of (B,E).

THEOREM 3.1. Let T be a recursive theory, $Adm^s \subset T$, T has a model with uncountably many standard ordinals. Then the ordinals of countable nonstandard models of T are exactly the countable admissible ordinals.

Proof: Using Theorem 2.2, it suffices to note that for each countable admissible ordinal α, there is a nonstandard admissible set satisfying T whose standard set contains an admissible set with ordinal α. This is clear, using the fact that α is an admissible ordinal iff $L(\alpha) \models Adm^s$.

Until further notice, we fix (B,E) to be a nonstandard admissible set whose standard ordinal α is $> \omega$. We let $\sim$ be the following equivalence relation on the ordinals of (B,E): $a \sim b$ iff either the interval (a,b) or (b,a) is well-ordered by E.

Recall the Cantor normal form theorem which asserts that every ordinal $\alpha > 0$ is uniquely of the form $(\omega^{\alpha_0} \times n_0) + \cdots + (\omega^{\alpha_k} \times n_k)$, where $\alpha_0 > \alpha_1 > \cdots > \alpha_k$, and each $n_j \in \omega - \{0\}$. The Cantor normal form thoerem can be stated and proved within Adm^s + Inf. Since the integers are standard in (B,E), the normal form of an ordinal of (B,E) must have finite length.

LEMMA 3.2.1. Every equivalence class of $\sim$ under E has a least element.

Proof: Take α to be an ordinal of (B,E). We must prove that there is a least β such that (β,α) is well-ordered by E. Within (B,E), write $\alpha = (\omega^{\alpha_o} \times n_o) + \cdots + (\omega^{\alpha_k} \times n_k)$, $\alpha_o > \alpha_1 > \cdots > \alpha_k$, each $n_j \in \omega - \{0\}$. Take β to be the result of deleting all terms in the above normal form in which α_j is < standard ordinal of (B,E). Then clearly $\beta \sim \alpha$. Now since it is a theorem of Adm^s that any tail of an ordinal ω^γ is of the same length as ω^γ, we clearly have that β is the least element of the equivalence class of α.

LEMMA 3.2.2. The equivalence classes of $\sim$ under E are densely ordered by E.

Proof: Straightforward, and left to the reader; use Lemma 3.2.1.

LEMMA 3.2.3. Every equivalence class is of length the standard ordinal of (B,E).

Proof: Let α be the standard ordinal of (B,E). Assume β is the least element of the equivalence class $[\beta]$. Consider the function $f(\gamma) = \beta + \gamma$ in (B,E). Note that $f \restriction \beta$ is one-one, order preserving, and has range an interval, with left endpoint β, in (B,E). This range is included in $[\beta]$, and has no sup in (B,E), since the standard ordinals have no sup in (B,E). Hence this range must be exactly $[\beta]$.

LEMMA 3.2.4. There is a least equivalence class but no greatest equivalence class.

Proof: The standard ordinals constitute the least equivalence class. Note that if α is nonstandard, then the interval $(\alpha, \alpha + \alpha)$ is not well-ordered by E. So, since (B,E) is a nonstandard admissible set, there is no greatest equivalence class.

THEOREM 3.2. Let (B,E) be a countable nonstandard admissible set with standard ordinal $\alpha > \omega$. Then the order type of the ordinals of (B,E) is $\alpha \times (1 + Q)$, where Q is the ordering of the rationals.

Proof: Immediate from Lemmas 3.2.1-3.2.4.

We now come to the case of a nonstandard admissible (B,E), with $(B,E) \models \sim$ Infinity, and (consequently) the standard ordinal of (B,E) is ω. Here we let $\sim$ be the equivalence relation on the ordinals of (B,E) given by $a \sim b$ iff a and b are separated by only finitely many points in (B,E).

LEMMA 3.3.1. The equivalence classes are densely ordered by E.

Proof: We can do arithmetic within (B,E), and since there are no limit ordinals in (B,E), the ordinals act like the natural numbers. Thus if α, β are not equivalent, then the equivalence class of $(\alpha + \beta)/2$ lies strictly in between those of α, β.

LEMMA 3.3.2. The equivalence classes of nonstandard ordinals of (B,E) are each isomorphic to $(-\omega + \omega)$ under E.

Proof: This is a consequence of the fact that there are no limit ordinals in (B,E), which means that every nonstandard ordinal in (B,E) has an immediate predecessor.

LEMMA 3.3.3. There is a least equivalence class but no greatest.

Proof: Similar to Lemma 3.2.4.

THEOREM 3.3. Any countable nonstandard admissible set not satisfying Infinity has order type $\omega + ((-\omega + \omega) \times Q)$.

We now come to the most delicate case: (B,E) has standard ordinal ω and satisfies Infinity. We again employ the Cantor normal form theorem in (B,E). We have that each nonzero ordinal α of (B,E) is of the form $(\omega^{\alpha_0} \times n_0) + \cdots + (\omega^{\alpha_k} \times n_k)$ in (B,E), where k is a possibly nonstandard integer, $\alpha_0 > \cdots > \alpha_k$, each n_j is a nonzero integer of (B,E). Here the all important equivalence relation $\sim$ on ordinals is defined by $\alpha \sim \beta$ iff, aside from terms of the form $(\omega^{\alpha_j} \times n_j)$ for α_j standard, α and β have identical Cantor normal forms in (B,E). Note that the equivalence classes form intervals in the ordinals of (B,E) under E.

We now wish to calculate the order type of each equivalence class $[\alpha]$ under $\sim$ when ordered by E.

LEMMA 3.4.1. If $\alpha \sim \beta$, then the Cantor normal form of α and β differ at only finitely many places.

<u>Proof</u>: Use Foundation in (B,E).

LEMMA 3.4.2. The integers of (B,E) have order type $\omega + ((-\omega + \omega) \times Q)$.

<u>Proof</u>: See Theorem 3.3.

Now given $[\alpha]$ ordered by E, we can describe the situation explicitly by first considering the function $f^{\alpha}: \omega \to \omega^*$, where ω^* is the set of all integers of (B,E), by $f^{\alpha}(n) = k$ if $(\omega^n \times k)$ occurs in the Cantor normal form of α in (B,E); 0 if no $(\omega^n \times k)$ occurs. We then order the set $f^{\alpha +}$ of all functions $g: \omega \to \omega^*$ which differ from f^{α} at only finitely many places, by $g < h$ iff $E(g(n),h(n))$, where n is the greatest argument at which g, h differ.

LEMMA 3.4.3. $([\alpha],E)$ is isomorphic to $(f^{\alpha +},<)$.

<u>Proof</u>: Obvious from the way Cantor normal forms are compared for size.

DEFINITION 3.1. Let $f: \omega \to \omega \cup \{\infty\}$. We say that f is <u>in the standard system of (B,E)</u> if and only if f is in, in the usual sense, if the ∞'s are replaced by some specific element of HF - ω; say, $\{\{\emptyset\}\}$. We associate, to each $f: \omega \to \omega \cup \{\infty\}$, a linear ordering L(f) given by $\sum_{-\infty}^{0} ((\omega + ((-\omega + \omega) \times Q))^n \times f(n)$ if $f(n) \neq \infty$; $(\omega + ((-\omega + \omega) \times Q))^{n+1}$ if $f(n) = \infty) + \sum_{1}^{\infty} (\omega + ((-\omega + \omega) \times Q))^n$.

DEFINITION 3.2. To each ordinal α of (B,E) we associate the function $f_{\alpha}: \omega \to \omega \cup \{\infty\}$ given by $f_{\alpha}(n) = k$ if $(\omega^n \times k)$ occurs in the Cantor normal form of α in (B,E) and k is standard; ∞ if k is nonstandard; 0 if no such term occurs.

LEMMA 3.4.4. $([\alpha],E) \approx L(f_{\alpha})$.

<u>Proof</u>: $(f^{\alpha +},<) \approx L(f_{\alpha})$ can be seen by inspection, using Lemma 3.4.2.

We now define an equivalence relation on equivalence classes, given by $[\alpha] \sim [\beta]$ iff f^{α} and f^{β} differ at only finitely many places. We let $[[\alpha]]$ be the equivalence class of $[\alpha]$.

LEMMA 3.4.5. Each $[[\alpha]]$ is dense among the $[\beta]$. I.e., if $[\beta] < [\gamma]$, then $[\beta] < [\delta] < [\gamma]$ for some $[\delta] \in [[\alpha]]$.

Proof: Take the largest exponent on which the Cantor normal forms of β and γ differ, say $(\omega^\rho \times n)$ for β, $(\omega^\rho \times m)$ for γ where n may of course be 0. It is clear that $n < m$ and ρ is nonstandard. Let $\rho' < \rho$ be nonstandard, and set β^* to be the result of deleting all terms in the Cantor normal form of β with exponents $< \rho'$. Set α^* be the result of deleting all terms in the Cantor normal form of α with exponents $\geq \rho'$. Then set $\delta = \beta^* + \omega^{\rho'} + \alpha^*$.

DEFINITION 3.3. Let Y be a countable collection of countable linear orderings. We define a new linear ordering, $\text{mix}(Y)$. Let $f\colon Q \to Y$ be any function from the rationals onto Y such that $f^{-1}(y)$ is dense in Q for all $y \in Y$. Take the domain of $\text{mix}(Y)$ to be $\{(q,x)\colon q$ is rational and x is in the domain of $f(q)\}$. Take the $<$ of $\text{mix}(Y)$ to be $(q,x) < (r,y)$ if $q < r$ or ($q = r$ and $x < y$ in $f(q)$). It can be seen that $\text{mix}(Y)$ is independent of the choice of f.

LEMMA 3.4.6. The order thpe of the nonstandard ordinals of (B,E) is the mix of the order types of the equivalence classes $[\alpha]$ under E, for nonstandard α.

Proof: Clear by Lemma 3.4.5.

THEOREM 3.4. Let (B,E) be a countable nonstandard admissible set with standard ordinal ω. Then the order type of the ordinals of (B,E) under E is given by $(\sum_1^\infty (\omega + ((-\omega + \omega) \times Q))^n) + \text{mix}\{L(f) \mid f\colon \omega \to \omega \cup \{\infty\}$ and f is in the standard system of $(B,E)\}$.

Proof: In view of Lemmas 3.4.4, 3.4.5, 3.4.6, we only need to remark that there is a least $[\alpha]$, and that is $[0]$, and its order type under E is $\sum_1^\infty (\omega + ((-\omega + \omega) \times Q))^n$.

COROLLARY 3.4.1. Let T_1, T_2 be recursive theories in LST, $\text{Adm}^s + \text{INF} \subset T_1$, $\text{Adm}^s + \text{Inf} \subset T_2$. Then the order types of the ordinals of countable models of T_1

with standard ordinal ω are the same as the order types of the ordinals of countable models of T_2 with standard ordinal ω.

<u>Proof</u>: Immediate from Theorems 2.4, 2.5, and 3.4.

SECTION 4. INITIAL SEGMENTS OF NONSTANDARD POWER ADMISSIBLE SETS

We give a necessary and sufficient condition for one countable nonstandard power admissible set satisfying $\Sigma^s(P)$-Sep to be isomorphic to an initial segment of a second.

DEFINITION 4.1. Let (B_1,E_1), (B_2,E_2) be nonstandard power admissible sets with standard sets A. Let $j\colon B_1 \to B_2$ have finite domain $\{x_1, \ldots, x_n\}$. Then j is $\Sigma^s(P)$-preserving iff for every $\Sigma^s(P)$-formula $\varphi(a_o, \ldots, a_n)$, $x \in A$, with $(B_1,E_1) \models \varphi(x, x_1, \ldots, x_n)$, we have $(B_2,E_2) \models \varphi(x, j(x_1), \ldots, j(x_n))$.

THEOREM 4.1. Let (B_1,E_1), (B_2,E_2) be two countable nonstandard power admissible sets such that both satisfy $\Sigma^s(P)$-Sep. Then (B_1,E_1) is isomorphic to an initial segment of (B_2,E_2) if and only if they have the same standard system and the empty function is $\Sigma^s(P)$-preserving from (B_1,E_1) into (B_2,E_2).

Proof: Let (B_1,E_1), (B_2,E_2) be as in hypotheses. Necessity. Let (A_1,C_1), (A_2,C_2) respectively be the standard systems of (B_1,E_1) and (B_2,E_2). Assume that (B_1,E_1) is an initial segment of (B_2,E_2). Clearly $A_1 = A_2$, $C_1 \subset C_2$. Let $x \in C_2$, and let x^* represent x in (B_2,E_2). Let α be a nonstandard ordinal of (B_1,E_1). Choose $y \in B_2$ with $(B_2,E_2) \models y = x^* \cap V(\alpha)$. Then $y \in B_1$ and y represents x in (B_1,E_1); i.e., $y \in C_1$. So $C_2 \subset C_1$, and hence $C_1 = C_2$. It is clear that the empty function is $\Sigma^s(P)$-preserving.

Sufficiency. Let (A,C) be the standard system of (B_1,E_1), (B_2,E_2). We construct an isomorphism ρ from (B_1,E_1) into (B_2,E_2) such that the range of ρ is an initial segment of (B_2,E_2). We construct ρ in stages ρ_n, where each ρ_n is $\Sigma^s(P)$-preserving. We can obviously set ρ_o = empty function. Let $\{x_i\}_{i<\omega}$ be an enumeration of B_1, and $\{y_i\}_{i<\omega}$ be an enumeration of B_2. Suppose ρ_{2n} has been defined. Let $\mathrm{Dom}(\rho_{2n}) = \{z_1, \ldots, z_{k-1}\}$, and $\rho_{2n}(z_i) = w_i$. We wish to find w_k such that $\rho_{2n} \cup \{\langle x_n, w_k\rangle\}$ is $\Sigma^s(P)$-preserving. Let $\varphi(a, a_o, \ldots, a_k)$ be a $\Sigma^s(P)$-formula such that for each $\Sigma^s(P)$-formula $\psi(a_o, \ldots, a_k)$ there is an m with $\mathrm{PAdm}^s \vdash \varphi^a_{\underline{m}} \longleftrightarrow \psi$. Now for each $\alpha \in A$ let $Y_\alpha = \{\langle m,x\rangle\colon\ x \in A \cap V(\alpha)$ & $(B_1,E_1) \models \varphi(m,x,z_1, \ldots, z_{k-1},x_n)\}$. Let $Y = \bigcup_\alpha Y_\alpha$. It is clear that $Y \in C$. Let

y^* represent Y in (B_2,E_2). Note that for each $\alpha \in A$ we have $(B_1,E_1) \vdash (\exists v)(\forall\langle m,x\rangle \in Y_\alpha(\varphi(m,x,z_1, \ldots, z_{k-1},v))$. Hence for each $\alpha \in A$ we have $(B_2,E_2) \models (\exists v)(\forall\langle m,x\rangle \in y^* \cap V(\alpha))(\varphi(m,x,w_1, \ldots, w_{k-1},v))$. Hence for some w_k we have $(B_2,E_2) \models (\forall\langle m,x\rangle \in Y_\alpha)(\varphi(m,x,w_1, \ldots, w_k))$ for all $\alpha \in A$, and set $\rho_{2n+1} = \rho_{2n} \cup \{\langle x_n,w_k\rangle\}$.

Now suppose ρ_{2n+1} has been defined. Let $\mathrm{Dom}(\rho_{2n+1}) = \{z_1, \ldots, z_k\}$, $\rho_{2n+1}(z_i) = w_i$. If for no w_i does $(B_2,E_2) \vdash \mathrm{rk}(y_n) \leq \mathrm{rk}(w_i)$, then set $\rho_{2n+1} = \rho_{2n}$. If there is such a w_i, then we assume for convenience that $(B_2,E_2) \models \mathrm{rk}(y_n) \leq \mathrm{rk}(w_k)$, and we will find z_{k+1} such that $\rho_{2n+1} \cup \{\langle z_{k+1},y_n\rangle\}$ is $\Sigma^s(P)$-preserving. For each $\alpha \in A$ let $Z_\alpha = \{\langle m,x\rangle: x \in A \cap V(\alpha)\ \&\ (B_2,E_2) \models \sim\varphi(m,x,w_1, \ldots, w_k,y_n)\}$. Let $Z = \bigcup_\alpha Z_\alpha$. It is clear that $Z \in C$. Let z^* represent Z in (B_1,E_1). Note that for each $\alpha \in A$ we have $(B_2,E_2) \models (\exists v)(\mathrm{Rk}(v) \leq \mathrm{Rk}(w_k)\ \&\ (\forall\langle m,x\rangle \in Z_\alpha)(\sim\varphi(m,x,w_1, \ldots, w_k,v)))$. Hence for each $\alpha \in A$ we have $(B_1,E_1) \models (\exists v)(\mathrm{Rk}(v) \leq \mathrm{Rk}(w_k)\ \&\ (\forall\langle m,x\rangle \in z^* \cap V(\alpha))(\sim\varphi(m,x,z_1, \ldots, z_k,v)))$. Hence for some z_{k+1} we have $(B_1,E_1) \models (\forall\langle m,x\rangle \in Z_\alpha)(\sim\varphi(m,x,z_1, \ldots, z_{k+1}))$, for all $\alpha \in A$, and set $\rho_{2n+2} = \rho_{2n+1} \cup \{\langle z_{k+1},y_n\rangle\}$.

In case the standard ordinals of (B_1,E_1), (B_2,E_2) are ω, we have the following sharper version.

THEOREM 4.2. Let (B_1,E_1), (B_2,E_2) be two countable nonstandard power admissible sets with standard ordinal ω. Then (B_1,E_1) is isomorphic to an initial segment of (B_2,E_2) iff every $\Sigma^s(P)$-sentence true in (B_1,E_1) is also true in (B_2,E_2).

<u>Proof</u>: A routine examination of the proof of Theorem 4.1 reveals that the full $\Sigma^s(P)$-Sep is not used; only $\Sigma^s(P)$-Sep applies to separating <u>standard</u> sets. In this case, this comes for free, since the standard sets are exactly the hereditarily finite sets.

Note that Theorems 4.1 and 4.2 have no content if (B_1,E_1) and (B_2,E_2) are isomorphic. So we prove two theorems to cover this interesting case.

THEOREM 4.3. Let (B,E) be a countable nonstandard power admissible set such that for some nonstandard ordinal α, $(B,E) \models V(\alpha)$ is an elementary submodel with respect to $\Sigma^s(P)$-formulae of V. Then (B,E) is isomorphic to a proper initial segment of itself.

Proof: Clearly by hypotheses we have $(B,E) \vdash \Sigma^s(P)$-Sep. Let $B_2 = \{b: (B,E) \models b \in V(\alpha)\}$. Then (B_2,E) is a countable nonstandard power admissible set such that the empty function is $\Sigma^s(P)$-preserving from (B,E) into (B_2,E). Hence applying Theorem 4.1, we are done.

Clearly, if $(B,E) \models \sim$Inf, then Theorem 4.3 has no content.

LEMMA 4.4.1. Every countable nonstandard power admissible set satisfying ~Inf is a proper initial segment, as well as an elementary submodel, of some countable nonstandard admissible set.

Proof: The nonstandard power admissible sets satisfying ~Inf are the same as the nonstandard admissible sets satisfying ~Inf as well as the nonstandard models of ZFC + ~Inf, and they correspond in the obvious way to nonstandard models of first-order arithmetic. So we simply quote the result of MacDowell and Specker [2] to the effect that every model of first-order arithmetic has a proper elementary end extension.

THEOREM 4.4. Every countable nonstandard power admissible set satisfying ~Inf is isomorphic to a proper initial segment of itself.

Proof: Let (B_2,E_2) be the countable nonstandard admissible set satisfying ~Inf. By Lemma 4.4.1, let (B_2,E_2) be a proper initial segment of (B_1,E_1), where (B_1,E_1) is a countable nonstandard power admissible set satisfying the same $\Sigma^s(P)$-sentences as (B_2,E_2). Since both (B_1,E_1), (B_2,E_2) have standard ordinal ω, apply Theorem 4.2 to obtain an initial segment (B_3,E_3) of (B_2,E_2) isomorphic to (B_1,E_1). Hence (B_3,E_3) is a proper initial segment of (B_1,E_1), and we are done.

SECTION 5. SUBMODELS OF Σ^1_∞ - CA

In this section we consider models (HF,ϵ,C) of Σ^1_∞ - CA.

DEFINITION 5.1. Let $(HF,\epsilon,C) \models \Sigma^1_\infty$ - CA. Then $\alpha \in C$ is a definable element of (HF,ϵ,C) if and only if there is a formula $\varphi(a)$ of LCT such that $(HF,\epsilon,C) \models (\forall v)(v \in \alpha \longleftrightarrow \varphi(v))$.

LEMMA 5.1.1. There is a Σ^1_2-formula $\varphi(a^1,b^1)$ such that for any $(HF,\epsilon,C) \models \Sigma^1_\infty$ - CA there is an $(HF,\epsilon,D) \models (\Sigma^1_\infty$ - AC + φ defines a well-ordering on D), with $D \subset C$.

Proof: φ is the well known Σ^1_2 formulation of (a^1 is constructible and b^1 is constructible and a^1 occurs before b^1 in the constructible hierarchy).

LEMMA 5.1.2. For any $(HF,\epsilon,C) \models \Sigma^1_\infty$ - CA there is an $E \subset C$ such that $(HF,\epsilon,E) \models \Sigma^1_\infty$ - AC and every element of E is definable in (HF,ϵ,E).

Proof: First pass to (HF,ϵ,D) as in Lemma 5.1.1. Next let E = set of all definable elements of (HF,ϵ,D). Then (HF,ϵ,E) is an elementary submodel of (HF,ϵ,D), and we are done.

DEFINITION 5.2. Let LCT^+ be LCT augmented with constants c_x, $x \in HF$ and c^1_n, $n \in \omega$. Let f be a one-one recursive function from ω onto the sentences of LCT^+, and g be a one-one recursive function from ω onto the formulae $\varphi(a)$ of LCT. A special theory T is a set of sentences of LCT^+ such that i) the universal closure of every axiom of Σ^1_∞ - CA is in T ii) $(\forall v)(v \in c^1_n \longleftrightarrow g(n)(v)) \in T$, for each $n \in \omega$ iii) $\varphi \in T$ iff $\sim\varphi \notin T$ iv) $(\varphi \,\&\, \psi) \in T$ iff $\varphi \in T$ and $\psi \in T$ v) $\forall v\, \varphi(v) \in T$ iff $\varphi(c_x) \in T$ for all $x \in HF$ vi) $\forall v^1 \varphi(v^1) \in T$ iff $\varphi(c^1_n) \in T$ for all $n \in \omega$ vii) $(c_x \in c_y) \in T$ iff $x \in y$. For special theories T let $|T|$ be (HF,ϵ,C), where $C = \{\{x \in HF\colon (c_x \in c^1_n) \in T\}\colon n \in \omega\}$.

LEMMA 5.1.3. $|\;|$ is one-one and onto the $(HF,\epsilon,C) \models \Sigma^1_\infty$ - CA with all elements definable.

Proof: Left to the reader.

Until Theorem 5.1, we fix (HF,ϵ,C) to be a model of Σ^1_∞ - AC. We now define the all important tree of finite sequences of elements of C.

DEFINITION 5.3. Let $\underline{Y}$ be the set of all finite sequences of theories in LCT^+, $\langle T_o, \ldots, T_{n-1}\rangle$, satisfying 1) T_o = the set of universal closures of axioms of Σ^1_∞ - CA + $\{(\forall v)(v \in c^1_k \longleftrightarrow g(k)(v)\colon\ k \in \omega\}$. 2) T_{n-1} does not contain a sentence and its negation. 3) for each $2k + 1 < n$, one of the following holds: i) $f(k) = \sim\varphi$ and $(T_{2k+1} = T_{2k} + \varphi$, or $T_{2k+1} = T_{2k} + \sim\varphi)$ ii) $f(k) = c_x \in c_y$ and $T_{2k+1} = T_{2k} + c_x \in c_y$, if $x \in y$; $T_{2k+1} = T_{2k} + \sim c_x \in c_y$, if $x \notin y$ iii) $f(k) = c_x \in c^1_n$ or $c^1_m = c^1_r$, and $(T_{2k+1} = T_{2k} + \varphi$, or $T_{2k+1} = T_{2k} + \sim\varphi)$ iv) $f(k) = (\varphi \ \&\ \psi)$ and $(T_{2k+1} = T_{2k} + (\varphi \ \&\ \psi) + \varphi + \psi$, or $T_{2k+1} = T_{2k} + \sim(\varphi \ \&\ \psi) + \sim\varphi$, or $T_{2k+1} = T_{2k} + \sim(\varphi \ \&\ \psi) + \sim\psi)$ v) $f(k) = \forall v\varphi(v)$ and $(T_{2k+1} = T_{2k} + \forall v\varphi(v) + \{\varphi(c_x)\colon x \in HF\}$, or $T_{2k+1} = T_{2k} + \sim\forall v\varphi(v) + \sim\varphi(c_x)$ for some $x \in HF)$ vi) $f(k) = \forall v^1\varphi(v^1)$ and $(T_{2k+1} = T_{2k} + \forall v^1\varphi(v^1) + \{\varphi(c^1_n)\colon\ n \in \omega\}$, or $T_{2k+1} = T_{2k} + \sim\forall v^1\varphi(v^1) + \sim\varphi(c^1_n)$ for some $n \in \omega)$. 4) for each $2k + 2 < n$, $T_{2k+2} = T_{2k+1} + \{c_x \in c^1_k\colon\ x \in \alpha\} + \{\sim c_x \in c^1_k\colon\ x \in \alpha\}$, for some $\alpha \in C$. For each $\langle T_o, \ldots, T_{n-1}\rangle \in Y$, let $Y_{\langle T_o,\ldots,T_{n-1}\rangle} = \{\langle T_n, \ldots, T_{n+r-1}\rangle\colon\ \langle T_o, \ldots, T_{n+r-1}\rangle \in Y\}$. An <u>infinite path through</u> $\underline{Y}$ is a function $h\colon\ \omega \to C$ with $(\forall n)(\langle h(0), \ldots, h(n-1)\rangle \in Y)$.

LEMMA 5.1.4. If $\langle T_o, \ldots, T_{n-1}\rangle \in Y$, then each $T_i \in C$. If h is an infinite path through Y, then $\bigcup Rug(h) = T$ is a special theory and $|Rng(h)|$ is a submodel of (HF,ϵ,C). If h_1, h_2 are distinct infinite paths through Y, then $\bigcup Rug(h_1) \neq \bigcup Rug(h_2)$.

<u>Proof</u>: Left to the reader.

LEMMA 5.1.5. The tree Y can be described with class quantifiers within (HF,ϵ,C); however, it is too "big" to be described as an object in (HF,ϵ,C). Y has an infinite path (perhaps not an object in (HF,ϵ,C)).

We assume the reader is familiar with the formalization within (HF,ϵ,C) of the concept of a well-founded tree; if a tree is well-founded within (HF,ϵ,C), then it may not really be well-founded. When considering trees in (HF,ϵ,C), usually the nodes are elements of HF. However, here we will consider trees whose nodes are elements of C, and which are sufficiently "small" (unlike Y) to be considered as objects in (HF, ϵ, C), using the standard coding devices. We can compare the lengths of well-founded trees coded in (HF, ϵ, C) by comparison maps coded within (HF, ϵ, C).

We wish to prove the existence of $D \subset C$, $D \neq C$, with (HF, ϵ, D) $\models \Sigma^1_\infty$ - CA. By the lemmas, it is enough to prove that Y has at least two infinite paths.

The key idea is to look at the lengths in (HF, ϵ, C) of subtrees of Y that are "small enough" to be coded as objects in (HF, ϵ, C). (A subtree of Y is a nonempty subset of Y closed under subsequences.) We call these subtrees "small subtrees." If $y \in Y$, we let Y_y be the subtree (not necessarily small) of extensions of y in Y.

LEMMA 5.1.6. Suppose that every ordering on HF coded in C satisfied to be a well-ordering in (HF, ϵ, C) is truly a well-ordering. Then there is a $D \subset C$, $D \neq C$, with (HF, ϵ, D) $\models \Sigma^1_\infty$ - CA.

Proof: The hypothesis of absoluteness of well-foundedness guarantees absoluteness of Σ^1_1 sentences. Now the existence of a D with (HF, ϵ, D) $\models \Sigma^1_\infty$ - CA is a true Σ^1_1 sentence, and so is true in (HF, ϵ, C). So there must be a D, coded as an object in (HF, ϵ, C) with (HF, ϵ, D) $\models \Sigma^1_\infty$ - CA. Now by a simple diagonalization, $D \neq C$.

LEMMA 5.1.7. The following holds in (HF, ϵ, C): "if every small subtree of Y is well-founded, and there is a bound on the lengths of the small subtrees of Y, then to each $y \in Y$ there is a small subtree of Y_y with maximum length."

Proof: Argue within (HF, ϵ, C): under the hypotheses there clearly is a sup of the lengths of the small subtrees of Y_y. To show that this sup is realized as a length of a small subtree of Y_y, choose, by Σ^1_∞ - AC, a sequence of small subtrees of Y_y whose lengths are cofinal with the sup. Since the union of small subtrees is again a small subtree, we are done.

LEMMA 5.1.8. Some small subtree of Y is not well-founded (though perhaps well-founded in (HF, ϵ, C)).

<u>Proof</u>: Suppose not. <u>Case 1</u>. The hypothesis of Lemma 5.1.7 fails. Then the only possibility is that, within (HF, ϵ, C), there are arbitrarily long small subtrees of Y. But then the hypothesis of Lemma 5.1.6 holds. Hence Y has an infinite path which is an object in (HF, ϵ, C), from the proof of Lemma 5.1.6. So this infinite path will constitute a small subtree of Y which is not well-founded. <u>Case 2</u>. The hypothesis of Lemma 5.1.7 holds. Then define the function F from Y into the (genuine) ordinals given by F(y) is the ordinal of the small subtree of Y_y of maximum length. A moments reflection reveals that this is an order-preserving map, and so Y is a well-founded tree.

So both cases lead to a contradiction.

LEMMA 5.1.9. Any non well-founded tree which is an object in (HF, ϵ, C) and satisfied to be well-founded in (HF, ϵ, C) has more than one infinite path.

<u>Proof</u>: Assume the negation and define, within (HF, ϵ, C), $f(0) = \langle\ \rangle$, $f(n+1) =$ node extending $f(n)$ on the next level, with highest ordinal in the tree. It is easily seen that f is defined on all of ω (and is the unique infinite path through the tree), which is a contradiction.

LEMMA 5.1.10. There is a $D \subset C$, $D \neq C$, with $(HF, \epsilon, D) \models \Sigma^1_\infty$ - CA.

<u>Proof</u>: By Lemma 5.1.8, there is a non well-founded small subtree of Y. If this subtree is not satisfied to be well-founded in (HF, ϵ, C), then we obtain D just as in Lemma 5.1.6. If this subtree is satisfied to be well-founded in (HF, ϵ, C), then by Lemma 5.1.9, Y will have at least two infinite paths, which is enough by Lemmas 5.1.3 and 5.1.4.

THEOREM 5.1. If $(HF, \epsilon, C) \models \Sigma^1_\infty$ - CA, then there is a $D \subset C$, $D \neq C$, with $(HF, \epsilon, D) \models \Sigma^1_\infty$ - CA.

<u>Proof</u>: Lemma 5.1.10 established this under the assumption that $(HF, \epsilon, C) \models \Sigma^1_\infty$ - AC. But this is good enough by Lemma 5.1.1.

SECTION 6. CATEGORICITY RELATIVE TO ORDINALS

In this section we consider well-founded models only. Unless stated otherwise, if we write (N,ϵ) we assume that N is a transitive class (not necessarily a set).

DEFINITION 6.1. Let T be a theory in LST, α an ordinal. Then T is <u>α-categorical</u> if and only if $(\forall M,N)(((M,\epsilon) \models T \ \& \ (N,\epsilon) \models T \ \& \ M \cap On = N \cap On = \alpha) \rightarrow M = N)$. T is <u>On-categorical</u> if and only if $(\forall M,N)(((M,\epsilon) \models T \ \& \ (N,\epsilon) \models T \ \& \ On \subset M \ \& \ On \subset N) \rightarrow M = N)$.

DEFINITION 6.2. Let T be a theory in LST, α an ordinal, φ a sentence of LST. Then $T \vdash_\alpha \varphi$ iff for all M with $(M,\epsilon) \models T$, $M \cap On = \alpha$, we have $(M,\epsilon) \models \varphi$. Also $T \vdash_{On} \varphi$ iff for all M with $On \subset M$, $(M,\epsilon) \models T$, we have $(M,\epsilon) \models \varphi$.

THEOREM 6.1. $Adm^s + V = L$ is On-categorical, and α-categorical for all α.

Proof: Classical.

We will need the following lemmas for the results of this section.

LEMMA 6.2.1. $Adm^s \vdash (\forall x)(x \subset On \rightarrow x \in L) \rightarrow V = L$. $ZF \vdash (\forall x)(x \subset On \rightarrow x \in HOD) \rightarrow V = HOD$.

Proof: Assume hypothesis. Then prove $(\forall x)(x \in L)$ by induction on the rank of x. Similarly for HOD.

LEMMA 6.2.2. Let (A,C) be a countable admissible system, and let $T \in C$ be a theory in LST_A, $\beta \in A$, $x \subset \beta$, $x \notin A$. If $Adm^s + T$ is consistent, then there is a model of $Adm^s + T$ whose standard set does not contain the element x.

Proof: Straightforward use of supercomplete theories together with the fact that validity of sentences of LST_A is Σ on (A,ϵ).

DEFINITION 6.3. Let <u>$n(\alpha)$</u> be the least ordinal greater than α which is admissible.

THEOREM 6.2. Let T be a recursive theory in LST, α a countable admissible ordinal such that $V(\alpha) \cap L(n(\alpha)) = L(\alpha)$. Then $Adm^s + T$ is α-categorical if and only if $Adm^s + T \vdash_\alpha V = L$.

Proof: If $Adm^s + T \vdash_\alpha V = L$, then clearly $Adm^s + T$ is α-categorical.

For the converse, we employ the lemmas. Let $(M,\epsilon) \models Adm^s + T + V \neq L$, $M \cap On = \alpha$. By Lemma 6.2.1, let $\beta < \alpha$, $x \subset \beta$ with $(M,\epsilon) \models \sim x \in L$. Then $x \notin L(\alpha)$. Hence $x \notin L(n(\alpha))$. Let T^* be the theory $Adm^s + T + V \neq L + (\forall x)(Ord(x) \longleftrightarrow v(\{x = \bar{\gamma}: \gamma < \alpha\}))$. Let $(L(n(\alpha)),C)$ be any admissible system. Then $T^* \in C$, and so by Lemma 6.2.2, there is a model of T^* whose standard set does not contain the element x. Hence this model of T^* must be some $(N,\epsilon) \models T$, $N \cap On = \alpha$, and $M \neq N$.

LEMMA 6.3.1. Let Y be a transitive class containing all ordinals. Then there is a countable ordinal α with $V(\alpha) \cap L(n(\alpha)) = L(\alpha)$, and an (M,ϵ) elementary equivalent to (Y,ϵ) with $M \cap On = \alpha$.

Proof: First choose a cardinal κ so that $(Y \cap V(\kappa),\epsilon)$ is an elementary submodel of (Y,ϵ). By Godel's work on GCH, $V(\kappa) \cap L(n(\kappa)) = L(\kappa)$. Consider the structure $(V(n(\kappa)), Y \cap V(\kappa),\epsilon)$, where $Y \cap V(\kappa)$ is taken as a constant. Choose a countable elementary substructure of $(V(n(\kappa)), Y \cap V(\kappa),\epsilon)$, and since this is well-founded, it is isomorphic to some (Z,c,ϵ), where Z is transitive. Then (c,ϵ) will be the desired (M,ϵ).

THEOREM 6.3. Suppose T is a recursive theory in LST such that $Adm^s + T$ is α-categorical for all countable α. Then $Adm^s + T \vdash_{On} V = L$.

Proof: Assume not $Adm^s + T \vdash_{On} V = L$. By Theorem 6.2 it is enough to prove that there is some countable ordinal α with $V(\alpha) \cap L(n(\alpha)) = L(\alpha)$, and a model $(M,\epsilon) \models Adm^s + T + V \neq L$ with $M \cap On = \alpha$. But this comes immediately from Lemma 6.3.1.

DEFINITION 6.4. Let $(M,\epsilon) \models ZF$, $M \cap On = \alpha$. Take $\leq_n$ to signify "elementary substructure with respect to Σ_n formulae." Take T(M) to be the following theory in LST_M: $ZF + \{(\exists x)(x = \bar{\beta}): \beta < \alpha\} + \{V(\bar{\beta}) \leq_n V(\bar{\gamma}): \beta, \gamma < \alpha$ and $(M \cap V(\beta),\epsilon) \leq_n (M \cap V(\gamma),\epsilon)\} + \{V(\bar{\delta}) \models \psi: \delta < \alpha$ and $(M \cap V(\delta),\epsilon) \models \psi\}$. Let Df(M) be those subsets of M that are definable in (M,ϵ). Take HOD(M) to be the elements of

M that are satisfied to be hereditarily ordinal definable in (M,ϵ).

LEMMA 6.4.1. Let $(M,\epsilon) \models ZF$. Then $(HOD(M), Df(HOD(M)))$ is an admissible system, and $T(M) \in Df(HOD(M))$.

Proof: Left to the reader.

LEMMA 6.4.2. Let M be countable with $(M,\epsilon) \models ZF$, $M \cap On = \alpha$. Let $(B,E) \models T(M)$, with standard set A. Then $(A \cap V(\alpha),\epsilon)$ is elementary equivalent to (M,ϵ).

Proof: Let M, α, B, E, A be given. Note that since $(M,\epsilon) \models ZF$, for each n there is a strictly increasing sequence of ordinals $B_{n,m}$, $m < \omega$, with supremum α, such that $(M \cap V(\beta_{n,m}),\epsilon) <_n (M \cap V(\beta_{n,m+1}),\epsilon) <_n (M,\epsilon)$. Hence $(A \cap V(\beta_{n,m}),\epsilon) <_n (A \cap V(\beta_{n,m+1}),\epsilon)$. Now let $(M,\epsilon) \models \psi$. We want to prove $(A \cap V(\alpha),\epsilon) \models \psi$. Assume ψ is Σ_n. Then each $(M \cap V(\beta_{n,m}),\epsilon) \models \psi$, each $m < \omega$. Hence each $(A \cap V(\beta_{n,m}),\epsilon) \models \psi$. So $(A \cap V(\alpha),\epsilon) \models \psi$.

THEOREM 6.4. Let α be a countable ordinal, T an arbitrary theory in LST. Suppose ZF + T is α-categorical. Then $ZF + T \vdash_\alpha V = HOD$.

Proof: Suppose not $ZF + T \vdash_\alpha V = HOD$. Let $(M,\epsilon) \models ZF + T + \sim V = HOD$, $M \cap On = \alpha$. It is enough to produce an (N,ϵ), $N \cap On = \alpha$, $N \neq M$, elementary equivalent to (M,ϵ). By Lemma 6.4.1, we can apply Lemma 6.2.2 to the theory $T(M)$, which has the model M. By Lemma 6.2.1, let $\beta < \alpha$ such that some $x \subset \beta$ has $x \notin M - HOD(M)$. By Lemma 6.2.2, let $(B,E) \models T(M)$, with standard set A, and $x \notin A$. The by Lemma 6.4.2, $(A \cap V(\alpha),\epsilon)$ is elementary equivalent to (M,ϵ). It is clear that $A \cap V(\alpha) \neq M$, and $A \cap V(\alpha) \cap On = \alpha$.

REFERENCES

[1] J. Barwise, Infinitary Logic and Admissible Sets, JSL, 34 (1969), 226-252.

[2] R. MacDowell and E. Specker, Modelle der Arithmetik, Infinistic Methods, Proc. Sympos. on the Foundations of Math. at Warsaw, 1959, Oxford, 1961, p. 257-263.

[3] D. Scott, Algebras of Sets Binumerable in Complete Extensions of Arithmetic, Proc. Sympos. Pure Math., Vol. 5, pp. 117-121, AMS, 1962.

Errata

Introduction, p. 4, ℓ. 12: "set ii)" should read "system ii)."

Introduction, p. 4, ℓ. 21, 23, 27: "second." should read "second, and they have the same standard system."

Section 1, p. 1, ℓ. 16: Delete.

Section 4, p. 2, ℓ. 21: "in (B_2, E_2)." should read "in (B_2, E_2), and they have the same standard system."

Section 4, p.1, ℓ. 10: "empty function" should read "identity function on the common standard part."

XX. DESCRIPTIVE SET THEORY IN $L_{\omega_1\omega}$

Robert Vaught

University of California
Berkeley, California

Descriptive set theory deals with Borel and projective sets in metrizable spaces and especially in the spaces ω^ω and 2^ω. This 'classical' work goes back to Borel, Lebesgue, Lusin, Souslin, Sierpinski, and others. The canonical reference is Kuratowski [11]. About ten years ago, Scott [26], Lopez-Escobar [5], and others showed that some of the same results, notably the first separation theorem, can be obtained in a stronger, more general form as part of the model theory of the language $L_{\omega_1\omega}$. (This language is like the ordinary first order predicate logic except that countably long disjunctions are allowed.) However, there was a kind of gap, since a group of classical results closely related to the first separation theorem, and even the best-known proof of the latter, involving the ordinals less than ω_1, were not extended to $L_{\omega_1\omega}$.

Recently, Moschovakis [20-23] has discovered that some of the best-known classical arguments can be generalized if one replaces the operation (A) (cf. [11]) by a certain infinite game (see §3 below), which had been studied earlier by Svenonius [28]. However, Moschovakis did not work in $L_{\omega_1\omega}$ (except briefly in a joint paper [3] with Barwise and Gandy). We will show here that when Moschovakis' methods are applied to $L_{\omega_1\omega}$, the gap described above is filled.

For an $L_{\omega_1\omega}$-sentence σ, or for a second-order $L_{\omega_1\omega}$-sentence of the form $\exists \underline{S}\sigma(\underline{R},\underline{S})$, $\forall \underline{S}\sigma(\underline{R},\underline{S})$, $\exists \underline{T}\forall \underline{S}\sigma(\underline{R},\underline{S},\underline{T})$, etc., the class of all models of the sentence is called elementary, $\underset{\sim}{\exists}_1$, $\underset{\sim}{\forall}_1$, $\underset{\sim}{\exists}_2$, etc., respectively, (imitating the classical Borel, $\underset{\sim}{\Sigma}^1_1$, $\underset{\sim}{\Pi}^1_1$, $\underset{\sim}{\Sigma}^1_2$). However, we are concerned only with the induced classes of countable models. We show that over countable models, in full analogy with the classical results ([11]):

The author is grateful to Professor John Addison for encouragement at a time when no results had yet been obtained.

The author was partially supported by National Science Foundation grant GP-24352.

(i) *The reduction principle holds for* $\underset{\sim}{\forall}_1$ *and for* $\underset{\sim}{\exists}_2$.

(ii) *Every* $\underset{\sim}{\exists}_2$ *class is the union of* $\aleph_1$ *elementary classes*.

(iii) *A 'covering theorem' holds, yielding a new proof of the Lopez-Escobar interpolation theorem*.

(Some of these results are nearer to what is in the work of Moschovakis [20-23] than others. In particular, this applies to the reduction principle for $\underset{\sim}{\forall}_1$.)

Scott [25, 26] proved that the isomorphism type of any countable structure is *elementary*, over countable models. (Throughout, all structures are assumed to have countably many relations or operations.) His result did not appear to be in analogy with any classical result. However, the type of $\mathfrak{A}$ is clearly $\underset{\sim}{\exists}_1$. Hence Scott's theorem is an immediate consequence of (ii) (since an isomorphism type is minimal). Indeed, Scott's theorem is clearly equivalent to (ii) for $\underset{\sim}{\exists}_1$, with $\aleph_1$ omitted.

As Ryll-Nardzewski remarked (see Scott [26]), results about *elementary*, $\underset{\sim}{\exists}_1$, etc., *over countable models*, can be rephrased so as to lie back in descriptive set theory. An *invariant* subset X, say of $2^{\omega\times\omega}$ is one such that $R \in X$ and $(\omega,R) \cong (\omega,R')$ implies $R' \in X$. Then, roughly speaking, $\underset{\sim}{\exists}_1$ = invariant $\underset{\sim}{\Sigma}^1_1$ (etc. for $\underset{\sim}{\forall}_1, \underset{\sim}{\exists}_2, \ldots$); moreover, by the interpolation theorem, *elementary* = invariant Borel. In this style, (ii), for example, says that an invariant $\underset{\sim}{\Sigma}^1_2$ set is the union of $\aleph_1$ *invariant* Borel sets.

Descriptive set theory was improved in another direction, starting over twenty years ago, by Kleene, Mostowski, and others, in what is sometimes called effective descriptive set theory. A few years ago, Barwise [2], using the Kripke-Platek 'admissible' sets, showed that the effective version could be carried over to $L_{\omega_1\omega}$. Even more was accomplished. Firstly, the admissible sets refined the effective theory even without passing to $L_{\omega_1\omega}$. Secondly, the passage to $L_{\omega_1\omega}$ had the benefit (as Barwise showed) of including the ordinary logic as just one part of the whole picture. Addison (see e.g. [1]) had earlier stressed that all these results should be unified. Now Barwise could give a single proof of at least four known first separation (or interpolation) theorems (Lusin, Kleene, Craig, Lopez-Escobar), and of a new one over *any* admissible set. Moreover, Kreisel [10] and

Barwise [2] developed a new compactness theorem, extending the one for ordinary logic.

Moschovakis' work is also in the effective vein. It is presented in a new framework differing from that of Kripke-Platek-Barwise, though some interconnections appear in Barwise-Gandy-Moschovakis [3].

Most of our results will be established in effective versions. In particular, our new proof of interpolation will give 'Barwise interpolation'.

We also obtain a result in the model theory of ordinary (finite) logic. (Svenonius had introduced the 'game' in finite logic in [28].) In 1962, Addison [1] raised the question whether the reduction theorem holds for cPC (complements of PC classes, in Tarski's sense). Silver [27] showed the answer is negative over uncountable models. Addison [1] had pointed out that over finite models the question is related to the difficult Scholz-Asser spectrum problems. We show below that

(iv) <u>the reduction principle holds for</u> cPC <u>over exactly denumerable models, and for</u> cPC_{δ} <u>over countable models</u>.

The main results are in §2-4, especially in §4. In §5 we show that the methods can also be used to give proofs of the Barwise compactness and completeness (i.e. 5.2) theorems -- avoiding the proof theory initiated by Karp [6].

§1. <u>Preliminaries</u>. The language $L_{\omega_1\omega}$ has variables $v_0, v_1, \ldots$ (and occasionally individual constants), n-ary relation symbols for $n \in \omega$ (denoted by 'R','S', etc.), and the symbols $\approx$, $\sim$, $\forall$, and $\bigwedge$ (conjunction). (First order) formulas of $L_{\omega_1\omega}$ are formed as in ordinary logic with the added proviso that, if X is a countable set of formulas, then $\bigwedge_{\varphi \in X} \varphi$ is also a formula. 'θ', 'φ', 'ψ', 'δ' denote formulas of $L_{\omega_1\omega}$.

Countable sets of relation symbols will be denoted by '$\underline{R}$', '$\underline{S}$', etc. An $\underline{R}$-structure is a pair $\mathfrak{A} = (A, R^{\mathfrak{A}})_{R \in \underline{R}}$ where $R^{\mathfrak{A}} \subseteq A^n$ if R is n-ary.

We also consider (prenex) second-order sentences over $L_{\omega_1\omega}$ (denoted by capital letters 'Φ', 'Ψ', etc.) such as $\exists \underline{S}\sigma$, $\forall \underline{S}\sigma$, $\exists \underline{T}\forall \underline{S}\sigma$, etc. The usual first or second order formulas of ordinary, finitary logic are just those in which no infinite conjunction (or quantification) occurs; they are called finite formulas.

Let $H_\omega(H_{\omega_1})$ be the set of all hereditarily finite (countable) sets. It is sometimes desirable to determine $L_{\omega_1\omega}$ more completely. In Barwise [2] it is exactly specified, as a certain subset of H_{ω_1}. Barwise' definition can obviously be extended, by putting $\forall \underline{S}\Phi = (9,\underline{S},\Phi)$, to include our second order formulas above.

Mod Φ is the class of all models of Φ and Mod' Φ the class of all countable models of Φ. Let $\mathcal{A}$ be any set or class. A class $\mathcal{K}$ of $\underline{R}$-structures is called $\mathcal{A}$-elementary, $\exists_1(\mathcal{A})$, $\forall_1(\mathcal{A})$, $\exists_2(\mathcal{A})$, etc., if $\mathcal{K}$ is of the form Mod Φ where $\Phi \in \mathcal{A}$ and Φ is of the form σ, $\exists \underline{S}\sigma$, $\forall \underline{S}\sigma$, $\exists \underline{T}\forall \underline{S}\sigma$, etc., respectively. If $\mathcal{A}$ is the universe, we drop it but write ('boldface') elementary, $\underset{\sim}{\Sigma}_1$, etc. If $\mathcal{A}$ is the class of all finite sentences, we drop it altogether. Thus elementary = EC and $\exists_1$ = PC (in Tarski's sense). As is usual, to say that a formula Φ is, e.g., $\underset{\sim}{\exists}_1$ can mean either that Mod Φ is $\underset{\sim}{\exists}_1$ or that Φ is literally of the form $\exists \underline{S}\sigma$, depending on the context. Incidentally it is easy to see that <u>over infinite models</u> the notions $\exists_1(\mathcal{A})$, etc. (for $\mathcal{A}$, say, admissible) are unchanged if only quantifications $\exists \underline{R}$, $\forall \underline{R}$ for <u>finite</u> $\underline{R}$ are permitted.

Recall that K is EC_δ if it is a countable intersection of EC's. A sentence $\varphi(\underline{R})$ is called EC_d if it is of the form $\bigwedge_n \sigma_n$ where the function $\lambda n\sigma_n$ is recursive. To justify the term 'recursive' we assume whenever 'EC_d' is used that $\underline{R} \subseteq H_\omega$. As usual, Φ is PC_δ if it is of the form $\exists \underline{S}\Psi$ where Ψ is EC_δ. If in addition Ψ is EC_d then Φ is called PC_d or $\exists_1^d$. The negation of Φ is cPC_d or $\forall_1^d$, etc.

There is a variant of PC in which "additional universe" is allowed as well as additional relations. $\mathcal{K}$ is $PC'(PC'_\delta, PC'_d)$ if it is the set of all $\mathfrak{A} = (A, R^{\mathfrak{A}})_{R \in \underline{R}}$ such that for some $A' \supseteq A$, some $\mathcal{S}$,

$$(A', A, \underline{R}^{\mathfrak{A}}, \mathcal{S}) \in \text{Mod } \Psi, \text{ where } \Psi \text{ is } EC(EC_\delta, EC_d).$$

Our principal interest is in classes Mod' Φ rather than Mod Φ, but a separate hierarchy notation seems unnecessary. We simply say, e.g., "over countable models, Ψ is $\underset{\sim}{\exists}_1$" to mean that Mod' Ψ = Mod' Φ, for some $\Phi \in \underset{\sim}{\exists}_1$. In a similar way, we could say "over all models of the form $(\omega, 0, 1, \ldots, \underline{R})$, Ψ is elementary, $\exists_1$, etc." and capture exactly the meaning of (classical) Borel, Σ_1^1, etc.

$\vdash \Phi$ (or $\vdash' \Phi$) means Φ holds in all (or all countable) models. $\Phi \equiv \Psi$ ($\Phi \equiv' \Psi$) means Φ and Ψ are equivalent in all (countable) models. (Note that, by the Löwenheim-Skolem theorem for $L_{\omega_1 \omega}$, $\vdash \Phi$ and $\vdash' \Phi$ coincide if Φ is $\forall_1$.)

Recall the result (coming from Makkai [14], Craig-Vaught [4], Kleene [9]):

THEOREM 1.1. Suppose R is finite. Then $PC_d' = PC'$, and, over infinite models, $PC_d = PC$.

An important theorem of Makkai [7] is:

THEOREM 1.2. (Makkai) $PC_\delta' = PC_\delta$ and $PC_d' = PC_d$.

In effective matters, an understanding of the notions 'admissible', '$\mathfrak{a}$-r.e.', etc., as in [2], will be assumed. The notion 'primitive recursive (p.r.) set function or relation' of [5] will also be used. These notions appear throughout, but §2-4 can be read for their non-effective content by just ignoring all parts concerning effective notions. For us, all admissible sets $\mathfrak{a}$ may as well be assumed subsets of H_{ω_1}, and a p.r. set function F can be considered on (and hence to) H_{ω_1}.

The simplest admissible sets are H_ω, H_{ω_1}, and $L(\omega_1^c)$. (ω_1^c is the first non-recursive ordinal; $L(\alpha)$ is the family of sets Gödel-constructible before level α.) Certain simple functions, like $F(x) = \{x\}$, and certain simple generating methods yields the p.r. set functions. Any admissible set in p.r. closed (i.e., closed under each p.r. set function). The restriction to an admissible $\mathfrak{a}$ of a p.r. relation is $\mathfrak{a}$-recursive.

§2. Normal form and $\mathcal{L}$-logic

This section contains some simple lemmas whose purpose is to reduce certain questions concerning $L_{\omega_1 \omega}$ to questions concerning finitary logic; it could be avoided at the cost of having to give various arguments twice.

Normal forms for $\underset{\sim}{\exists}_1$-sentences are well-known. For example, it is well-known that $\underset{\sim}{\exists}_1$ is the same as 'PC$_\delta$ in ω-logic', a fact which in essence goes back to Lopez-Escobar [13]. These normal forms would suffice for the 'non-effective' results in §4, but not for the 'effective'. A slightly different normal form is needed having the property that the normal form of Φ belongs to an admissible set $\mathcal{a}$ if Φ does.

In ω-logic one considers that each structure $(A,\mathcal{R})$ has tacked onto it an ω-part, giving $(A \cup \omega, A, \omega, +, \cdot, \mathcal{R})$, say. (In weak second order logic one also tacks on another part $A^{\underset{\sim}{\omega}}$ and appropriate relations.)

Suppose an arbitrary but fixed, structure $\mathcal{L} = (B, W^{\mathcal{L}})_{W \in \underline{W}}$ is given. (For us, below, $\underline{W}$ will be finite and B countable.) Consider any structure $\mathfrak{A}$, of a (different) type $\underline{R}$ such that the set $\underline{R}$ is a subset of B. Put $\mathfrak{A} + \mathcal{L} = (A \cup B, A, B, W^{\mathcal{L}}, R\ell_n)$ $(W \in \underline{W}, n < \zeta)$ -- A and B being considered disjoint. This is the 'cardinal sum', except that the relations $R^{\mathfrak{A}}$ have been replaced by relations $R\ell_n$ defined by:

$$R\ell_n(R, a_0, \ldots, a_{n-1}) \text{ iff } R \in \underline{R},\ R \text{ is n-ary, and } R^{\mathfrak{A}}(a_0, \ldots, a_{n-1}).$$

(In the sense of an admissible set $\mathcal{a}$, the set $\underline{R}$ might appear uncountable; while the type of $\mathfrak{A} + \mathcal{L}$ is clearly 'recursive', if $\underline{W}$ is.) $\zeta (\leq \omega)$ is taken as the smallest ordinal such that $\underline{R}$ contains n-ary relations only for $n < \zeta$. This ensures that the type of $\mathfrak{A} + \mathcal{L}$ is finite if $\underline{R}$ and $\underline{W}$ are finite (in which case one could also just use $\underline{R}$ instead of the $R\ell_n$).

A sentence Φ concerning structures of the form $\mathfrak{A} + \mathcal{L}$ ($\mathcal{L}$ fixed) can be regarded as in the '$\mathcal{L}$-logic' of structures $\mathfrak{A}$. That is, one is interested not in Mod Φ but in $\text{Mod}^{\mathcal{L}} \Phi = \{\mathfrak{A} : \mathfrak{A} + \mathcal{L} \in \text{Mod } \Phi\}$. (Closely related, though different notions, over arbitrary $\mathcal{L}$, are involved in Morley [19] ('$\mathcal{L}$-model') and in Moschovakis [20-23].

We will see that (2.1) an $\underset{\sim}{\exists}$-sentence Φ about the $\mathfrak{A}$'s can be put in a finite or PC$_d$ normal form Φ^* in $\mathcal{L}$-logic, for a suitable $\mathcal{L}$, 'near' to Φ; and (2.2) sentences of $\mathcal{L}$-logic can always be translated back into sentences concerning only the $\mathfrak{A}$'s.

Since A and B are always considered disjoint, the ordinary logic of structures $\mathfrak{A}+\mathfrak{B}$ can alternatively and conveniently be written as a two-sorted logic -- with variables (denoted by 'x','y') ranging (in $\mathfrak{A}+\mathfrak{B}$) over A and variables ('s','t') ranging over B (and $x \approx s$ not allowed). Even in considering $\underset{\sim}{\exists}_1$, etc., in $\mathfrak{B}$-logic it is clearly equivalent to consider only sorted (new) relations $S(a^{(m)};b^{(n)})(a_i \in A, b_i \in B)$. ($x^{(n)}$ always denotes an n-termed sequence $(x_0,\ldots,x_{n-1})$.)

THEOREM 2.1. <u>With each</u> Φ <u>can be correlated a structure</u> $\mathfrak{B}_\Phi$ <u>of finite type and a sentence</u> Φ^* <u>so that, for any</u> $\mathfrak{A}$,

$$\mathfrak{A} \models \Phi \quad \underline{\text{iff}} \quad \mathfrak{A}+\mathfrak{B}_\Phi \models \Phi^*.$$

<u>Moreover</u>, (a) <u>If</u> Φ <u>is</u> $\underset{\sim}{\exists}_n$ $(\underset{\sim}{\forall}_n)$, <u>then</u> Φ^* <u>is</u> $\exists_n^d$ $(\forall_n^d)$.

(b) <u>The function</u> $F(\Phi) = (\mathfrak{B}_\Phi, \Phi^*)$ <u>is a p.r. set function</u>.

By passing to the weak second order logic of $\mathfrak{A}+\mathfrak{B}_\Phi$ one could (preserve 2.1 and 2.2 and) make Φ^* finite. (This would also bring out more clearly the relationship with Moschovakis' work.) But the notation is a little more complicated, and for our purposes there seems to be no gain.

<u>Proof</u>. First consider an $L_{\omega_1\omega}$-sentence $\varphi(\underline{R})$. We take $B = \omega \cup \text{Sub}$ (the set of all subformulas of φ). After forming a suitable structure $\mathfrak{B}$ on B, we take for φ^* the sentence

$$(\exists K,d)(\text{Sat}(K,d) \wedge K_0(\bar{\varphi}))$$

where (d is only technical and) Sat says that 'K is the satisfaction relation for subformulas of φ'.

The (familiar) details are as follows: For convenience we can clearly assume that any atomic formula, not involving $\approx$, which occurs in φ is of the form $Pv_0\ldots v_{k-1}$ or, simply, P. Over B define the following relations: $\text{Neg}(\theta,\psi)$ iff ψ is $\sim\theta$. $\text{Con}(\theta,\psi)$ iff θ is a conjunct of ψ. $\text{Uq}(k,\theta,\psi)$ iff ψ is $\forall v_k\theta$. $\text{Eq}(j,k,x)$ iff ψ is $v_j \approx v_k$. $\text{Pred}(m,\psi)$ iff ψ is m-ary and $\psi \in \underline{R}$. Now put $\mathfrak{B}_\varphi = (B,\omega,0,\text{Sc},\text{Sub},\text{Neg},\text{Con},\text{Uq},\text{Eq},\text{Pred},\varphi)$. Write $\bar{k}$ for $\text{Sc}\ldots\text{Sc}\,\bar{0}$ (k times).

Let $\zeta'(\leq \omega)$ be the sup of all $i + 2$ such that v_i or some i-ary R occurs in φ. Thinking now of structures $\mathfrak{A} + \mathcal{L}_\varphi$, let d be a new individual constant and K_n be new n+1-ary relation symbols $(n < \zeta')$ (sorted in a way that will be obvious). Put $K = \{K_n : n < \zeta'\}$. (In the formula Sat, each sequence $(a_0,\ldots,a_{k-1})$ will be 'considered' as an infinite sequence $(a_0,\ldots,a_{k-1},d_0,d_0,d_0,\ldots)$.) The formula Sat(K,d) is as follows:

$$\bigwedge_{j,k,n<\zeta'} (\forall x^{(n)})(\forall s,t)\{[\mathrm{Neg}(s,t) \to (K_n(t,x^{(n)}) \leftrightarrow \sim K_n(s,x^{(n)}))]$$

$$\wedge\ [\mathrm{Con}(s,t) \to (K_n(t,x^{(n)}) \leftrightarrow \forall t'(\mathrm{Con}(t',t) \to K_n(t',x^{(n)})))]$$

$$\wedge\ [\mathrm{Uq}(\bar{k},s,t) \to (K_n(t,x^{(n)}) \leftrightarrow \forall y K_p(s,x_0,\ldots,x_{k-1},y,x_{k+1},\ldots))]$$

$$\wedge\ [\mathrm{Eq}(\bar{j},\bar{k},t) \to (K_n(t,x^{(n)}) \leftrightarrow x_j \approx x_k)]$$

$$\wedge\ [\mathrm{Pred}(\bar{j},t) \to (K_n(t,x^{(n)}) \leftrightarrow R\ell_j(t,x_0,\ldots,x_{j-1}))] \ .$$

(Here x_m means d if $m \geq n$; 'p' should be $\max(n,k+1)$.)

It is clear that φ^* and $\mathcal{L}_\varphi$ meet all the demands of the theorem.

Now suppose Φ is, say, $\exists \underline{S}\Psi(\underline{R},\underline{S})$, and for Ψ we already have $\mathcal{L}$ and Ψ^*. Then, trivially, $\mathfrak{A} \in \Phi$ if and only if $\mathfrak{A} + \mathcal{L} \models \exists(R\ell_n^{\underline{S}} : n \in \omega)\Psi^*(R\ell_n^{\underline{R}} \cup R\ell_n^{\underline{S}} : n \in \omega)$. So clearly 2.1 holds in general.

Theorem 2.2, below, is simply an extension of the well-known fact that ω-logic (or weak second-order logic) is 'included' in $L_{\omega_1\omega}$. Consider sorted structures of the form $\mathfrak{C} = (\mathfrak{A} + \mathcal{L}, S^{\mathfrak{C}})_{S \in \underline{S}}$, where each 'additional relation' $S^{\mathfrak{C}}$ is of sorted type $(j(S),k(S))$, i.e. of the form $S(a^{(j)};b^{(k)})$. Introduce new relation symbols S_b (for $S \in \underline{S}$, $b \in B^{\ell(S)}$) with k(S) places. Pass to structures $\mathfrak{C}^0$ having a constant $\bar{b}$ naming b, for each $b \in B$.

With each (two-sorted) formula φ (of $L_{\omega_1\omega}$ over structures like $\mathfrak{C}^0$), having no free B-variables, we correlate a formula $\varphi^{\mathcal{L}}$ (of type $\underline{R}$) as follows:

(a) If φ is $S(x_0,\ldots;\bar{b}_0,\ldots)$, $\varphi^{\mathcal{L}}$ is $S_b(x_0,\ldots)$. $x \approx y$ is unchanged, and any $W(\bar{b}_0,\ldots)$ is replaced by its truth value. $R\ell_n(\bar{b};x_0,\ldots)$ becomes $b(x_0,\ldots)$.

(b) The operator $\mathcal{L}$ passes through $\sim, \wedge$, and $(\forall x)$. (c) $(\forall s\theta)^{\mathcal{L}}$ is $\bigwedge_{b \in B} (\theta(\bar{b}))^{\mathcal{L}}$. The correlation is extended to $\underset{\sim}{\exists}_n$ (and similarly $\underset{\sim}{\forall}_n$) by stipulating that $(\exists \underline{T}\Phi)^{\mathcal{L}}$ is $\exists(T_b : T \in \underline{T}$ and $b \in B^{\ell(T)})\Phi^{\mathcal{L}}$. Obviously $\Phi^{\mathcal{L}}$ may not be finite when Φ is. However, it is clear that:

THEOREM 2.2 (a) In $\mathcal{L}$-logic, $\Phi \equiv \Phi^{\mathcal{L}}$

(b) If Φ is elementary, $\exists_n$, $\forall_n$, respectively, so is $\Phi^{\mathcal{L}}$

(c) The function $F(\mathcal{L},\Phi) = \Phi^{\mathcal{L}}$ is a p.r. set function.

§3. The Game. Infinite games have been applied in various ways in model theory, especially by Keisler (see, e.g., [7], [8]). Their special application to countable models was initiated, very early, by Svenovius [28]. Let M_n be a (2n)-ary relation symbol $(n \in \omega)$. Let Γ be the sentence

$$\forall x_0 \exists y_0 \forall x_1 \exists y_1 \ldots \forall x_n \exists y_n \ldots \bigwedge_n M_n(x_0,y_0,\ldots,x_{n-1},y_{n-1}).$$

Γ is not in any language above, but the meaning of $\mathfrak{A} \in \text{Mod}\,\Gamma$ is (clearly) to be that the 'y'-player has a winning strategy in the infinite game suggested by Γ. Perhaps an even simpler equivalent of Γ in the (slightly different) condition:

(1) $\exists(T_n : n \in \omega)\{T_0 \wedge \bigwedge_n(\forall x^{(n)},y^{(n)})[T_n(x_0,y_0,\ldots,x_{n-1},y_{n-1}) \rightarrow \forall x_n \exists y_n T_{n+1}(x_0,\ldots,y_n) \wedge M_n(x_0,\ldots,y_{n-1})]\}$

From (1) follows at once:

THEOREM 3.1 (Svenonius [28]). Γ is PC_d.

As Moschovakis (e.g., [23]) has pointed out, Γ can and should be regarded as a close relation of the operation (A), which is expressed by the formula:

$$\exists k_0 \exists k_1 \ldots \exists k_n \ldots \bigwedge_n M_n(k_0,\ldots,k_{n-1})$$

For example, 3.1 should be strengthened by observing that, in (1), each M_n occurs only positively. It then follows at once that, for any sequence Θ of formulas $\Theta_n(x_0,y_0,\ldots,x_{n-1},y_{n-1})$ $(n \in \omega)$, if each Θ_n is $\exists_1$, so is $\Gamma(\Theta)$ (obtained by ordinary substitution of Θ_n for M_n in $\Gamma(n \in \omega)$). This generalizes the classical result that applying the operation (A) to analytic sets always yields an analytic set.

Conversely, every analytic set can be obtained by applying the operation (A) to suitable closed sets. There is a corresponding result concerning $\underset{\sim}{\exists}_1$ classes, over countable models, and a game operation. It will be inferred from a special case, namely, the following basic theorem of Svenonius [28]:

THEOREM 3.2(a) (Svenonius). With each PC_δ-formula Φ can be correlated a sequence N^Φ of finite open formulas $N_n^\Phi(x_0,y_0,\ldots,x_{n-1},y_{n-1})$ $(n \in \omega)$ in such a way that $\vdash \Phi \rightarrow \Gamma(N^\Phi)$ and $\vdash' \Gamma(N^\Phi) \rightarrow \Phi$.

Moreover, (b) There is a recursive functional H such that $H(\lambda_k\sigma_k,\underline{S},n) = N_n^\Phi$, whenever $\Phi = \exists\underline{S}\bigwedge_n \sigma_n \in PC_\delta$.

Proof. A proof is in Svenonius [28] and another (at least for a weak second-order situation) in Moschovakis [22]; hence we shall only give an outline. Let Φ be $\exists\underline{S}\bigwedge_n \sigma_n(\underline{R},\underline{S})$ (each σ_n elementary). By Skolem normal form we can clearly assume each σ_n is of the form $(\forall u_0,\ldots,u_{k(n)-1})(\exists v)Q_n$, Q_n open. Make a list of all $(p;j_0,\ldots,j_{k(p)-1})$ in such a way that in the m-th entry each $j_i \leq m$. Let N_m' be $Q_p(x_{j_0},\ldots,x_{j_{k(p)-1}},y_m)$ where $(p,j_0,\ldots)$ is the m-th entry. The N_m' still contain S's, but for N_n^Φ we take a (finite open) $\underline{R}$-formula which is equivalent to: "there exist S's such that $\bigwedge_{m<n} N_m'(x_0,\ldots,x_m,y_m)$".

It is easy to prove that $\vdash \Phi \rightarrow \Gamma(N^\Phi)$. If $\mathfrak{A}$ is countable and $\mathfrak{A} \models \Gamma(N^\Phi)$, let the x-player exhaust the model and apply Konig's infinity lemma to obtain $\mathfrak{A} \models \Phi$.

The steps in the argument above are easily made effective, yielding (b).

Theorem 3.2, together with 2.1 and 2.2, yields a similar result for an arbitrary $\exists_1$-sentence Φ. In fact, we pass (by 2.1) to Φ^* (a PC_d-sentence); then (by 3.2) to $\Gamma(N^{\Phi^*})$, which is interpreted as in (1); and, finally, by 2.1, to: $(\Gamma(N^{\Phi^*}))^{\mathcal{L}_\Phi}$, which is denoted by Γ^Φ. - Clearly:

COROLLARY 3.3. If Φ is $\underset{\sim}{\exists}_1$, then Γ^Φ is $\exists_1$ and is equivalent to Φ in countable models.

In §4, using 2.1 and 2.2, we shall infer results about Γ^{Φ} directly from results about Γ. However, there would also be advantages in working directly with Γ^{Φ} or rather (2) below -- thus omitting or absorbing 2.1 and 2.2. It is easy to verify that Γ^{Φ} is equivalent to a ('game-theoretic') assertion of the form:

$$(2) \qquad \forall x_0 \bigwedge_{b_0 \in B} \exists y_0 \bigvee_{c_0 \in B} \forall x_1 \bigwedge_{b_1 \in B} \cdots \bigwedge_n P_n^{b_0 \ldots c_{n-1}}(x_0,\ldots,y_{n-1})$$

where each $P_n^{b_0\ldots}$ is finite open, and the whole function P depends primitive recursively on Φ and ω. Thus 3.3 asserts that every $\underset{\sim}{\exists}_1$ class over countable models can be obtained by applying (2) to suitable finite open formulas.

§4. Theory of Γ and Applications. The whole §4 will be in close analogy to the classical theory [11]. To begin with, the basic theory of the operation (A) (cf. [11]) can be imitated almost exactly for Γ:

DEFINITION 4.1 (a). The formulas $\delta^n_\alpha(x_0,y_0,\ldots,x_{n-1},y_{n-1})$ are defined recursively:

$$\sigma^n_0 \text{ is } \bigwedge_{i \leq n} M_i(x_0,y_0,\ldots,x_{i-1},y_{i-1})$$

$$\delta^n_{\alpha+1} \text{ is } \forall x_n \exists y_n\, \delta^{n+1}_\alpha$$

$$\delta^n_\alpha \text{ is } \bigwedge_{\beta<\alpha} \delta^n_\beta \text{ if } \alpha \text{ is a limit ordinal.}$$

We also write δ_α for δ^0_α.

(b) ρ_α is the sentence $\bigwedge_n (\forall x_0,y_0,\ldots,x_{n-1},y_{n-1})(\delta^n_\alpha \rightarrow \delta^n_{\alpha+1})$.

Obviously, if $\alpha < \omega_1$, then δ^n_α and ρ_α are formulas of $L_{\omega_1\omega}$ (and in any case of $L_{\infty\omega}$).

THEOREM 4.2. (a) $\vdash \delta^n_\alpha \rightarrow \delta^n_\beta$ if $\beta < \alpha$.

(b) $\vdash \Gamma \rightarrow \delta_\alpha$.

(c) $\vdash \rho_\alpha \wedge \delta_\alpha \rightarrow \Gamma$.

(d) $\vdash' \bigvee_{\alpha<\omega_1} \rho_\alpha$. (If A is finite, then $\mathfrak{A} \models \rho_\omega$.)

(e) $\Gamma \equiv' \bigwedge_{\alpha < \omega_1} \delta_\alpha$. (For finite A, $\mathfrak{A} \models \Gamma \leftrightarrow \bigwedge_k \delta_k$.)

(f) $\Gamma \equiv' \bigvee_{\alpha < \omega_1} (\rho_\alpha \wedge \delta_\alpha)$

(g) (d), (e), and (f) <u>hold in all models of power</u> $< \kappa$ <u>if</u> ω_1 <u>is replaced by the cardinal</u> $\kappa > \omega$.

<u>Proof</u> (just as for the operation (A)). (a) and (b) (in a version about δ_α^n) are easily proved by induction on α. In (c), assume $\mathfrak{A} \models \rho_\alpha \wedge \delta_\alpha$. Looking at form (1) of Γ, take $T_n = \delta_\alpha^n$. Since $\mathfrak{A} \models \delta_\alpha^n \rightarrow \alpha_{\alpha+1}^n$, it follows (by 4.1) that (1) holds for $\mathfrak{A}$, i.e., $\mathfrak{A} \models \Gamma$. (d) is clear as the sets δ_α^n in a model $\mathfrak{A}$ are decreasing and A is countable (resp., finite). (e) and (f) follow from (b), (d), and (c). The proof of (g) is entirely analogous. (d) and (e) will be improved in effective versions in 5.1.

Given any $\underset{\sim}{\exists}$-sentence Φ, write $\delta_\alpha^{n^\Phi}$ (or ρ_α^Φ) for $\delta_\alpha^n(N^{\Phi*})^{\mathcal{L}\Phi}$ (or $\rho_\alpha(N^{\Phi*})^{\mathcal{L}\Phi}$). Then, by §2 and §3, 4.2 <u>holds with</u> Γ, δ, ρ <u>replaced by</u> $\Gamma^\Phi, \delta^\Phi, \rho^\Phi$. (It seemed worthwhile, however, to emphasize that everything is contained in the pure theory of Γ.) We remark that

(3) <u>The function</u> $F(n,\Phi,\alpha) = \delta_\alpha^{n^\Phi}$ <u>is a p.r. set function.</u>

(3) is obvious except when Φ and α are finite. Then it depends on the observations that δ_k^n only depends on δ_j^m for $j + m = k + n$, and that $\delta_k(N^\Phi) = \delta_k(N_0^\Phi,\ldots,N_k^\Phi)$. Incidentally, if Φ is PC, passing to $\mathcal{L}$-logic is a waste of time, and $\delta_\alpha^\Phi \equiv \delta_\alpha(N^\Phi)$. For $\Phi \in PC_d$ we <u>define</u> $\delta_\alpha^\Phi = \delta_\alpha(N^\Phi)$, and similarly for Γ^Φ, ρ_α^Φ (see 4.5), etc.

The basic mathematical content of 4.2 is due to Moschovakis [20-23]. However, it is expressed there without using $L_{\omega_1\omega}$. Throughout, our references to Moschovakis' work (or [3]) will necessarily be somewhat vague, leaving a detailed comparison to the reader; this is because his work involves various notions not appearing here. (Incidentally, the reader should realize that Moschovakis studies several topics -- for example, non-denumerable models--which are not considered here at all, beyond 4.2(g) and 4.7.)

THEOREM 4.3. Over countable models, every $\exists_2$-class is the union of $\aleph_1$ elementary classes.

Proof. First let Φ be $\exists_1$. Then, by 4.2 (f) (and §2, §3), $\Phi \equiv' \bigvee_{\alpha<\omega_1} (\rho_\alpha^\Phi \wedge \delta_\alpha^\Phi)$, which establishes 4.3 for such Φ. Now $\forall_1$-sentences can be dealt with using 4.2 (d), and $\exists_2$ follow by arguing just as in Kuratowski [11]. Alternatively, once we have the case of $\exists_1$ we can infer the rest of 4.3 from the classical 4.3, since an invariant union of $\aleph_1$ non-invariant Borel sets can trivially be expressed as a union of $\aleph_1$ invariant Σ_1^1 sets.

As already mentioned in the introduction, 4.3 yields at once:

COROLLARY 4.4. (Scott [25]). The isomorphism type of a countable structure is elementary, over countable models.

Morley [18] proved that any $\exists_1$-class contains up to isomorphism countably many, $\aleph_1$, or $2^{\aleph_0}$ countable models. It may be remarked that his result extends at once to $\exists_2$-classes by an (unnecessary) use of 4.3, or by arguing from the classical 4.3 as above.

A very simple 'formalization' of δ can be made as follows. Let U, $<$, D^n $(n \in \omega)$ be new relation symbols.

DEFINITION 4.5. (a) Let $\zeta(U,<,c,D,M)$ be the conjunction of (the universal closures of) the following elementary formulas (where 'u', 'v' range over U; 'x', 'y' over A):

(i) '$(U,<)$ is an ordering with last element c'

(ii) $D^n(u;x_0,y_0,\ldots,x_{n-1},y_{n-1}) \leftrightarrow (\forall v<u)\forall x_n \exists y_n D^{n+1}(v;x_0,\ldots,y_n) \wedge \bigwedge_{i\le n} M_i(x_0,\ldots,y_{i-1})$

(b) Let ζ^+ be $\zeta \wedge D^0(c)$.

LEMMA 4.6. (a) Suppose $(U^{\mathfrak{A}},<^{\mathfrak{A}})$ is w.o. (well-ordered) with last element $c^{\mathfrak{A}}$. Then $\mathfrak{A} \models \zeta$ iff $D^{\mathfrak{A}}$ 'is' $\delta^{\mathfrak{A}}$ (i.e., in $\mathfrak{A}$, $D^n[w;a_0,\ldots]$ iff $\delta_\alpha^n[a_0,\ldots]$, where w is the α-th element of $(U,<)$).

(b) $\vdash \zeta^+ \wedge \sim \Gamma \rightarrow$ '(U,<) is w.o.'

Proof. (a) is trivial. For (b), if (U,<) is not w.o. (in a given model $\mathfrak{A}$ of ζ^+), then clearly, by 4.5 (ii), the y-player can win the game Γ (using a descending sequence $c > u_0 > u_1 > \ldots$).

As before, if Φ is Π_1, 4.5 and 4.6 extend themselves directly (via §2,3) to notions and results concerning Γ^Φ. In fact, put $\zeta^\Phi = \zeta(N^{\Phi *})^{\mathcal{L}\Phi}$ and similarly for $\zeta^{+\Phi}$. Then clearly:

4.6' 4.6(a), (b) hold with ζ, ζ^+, δ, Γ replaced by ζ^Φ, $\zeta^{+\Phi}$, δ^Φ, Γ^Φ, respectively.

A family $\mathcal{F}$ satisfies the reduction principle if for any K, L $\in \mathcal{F}$ there exist disjoint K' $\subseteq$ K and L' $\subseteq$ L such that K' $\cup$ L' = K $\cup$ L. (Also, K', L' 'reduce' K and L). Let M_n' (D_n') be new 2n-ary (2n+1-ary) relation symbols.

THEOREM 4.7 (a) $\sim \Gamma(M)$, $\sim \Gamma(M')$ are reduced, over infinite models, by the cPC_d-sentences $\Psi(M,M')$ and $\Psi'(M,M')$:

$$(\Psi) \sim \Gamma(M) \wedge (\forall U,<,c,D,D')[\zeta^+(U,<,c,D,M) \wedge \zeta(U,<,c,D',M') \rightarrow D'^0(c)]$$

$$(\Psi') \sim \Gamma(M') \wedge (\forall U,<,c,D,D')[\zeta(U,<,c,D,M) \wedge \zeta(U,<,c,D',M') \wedge (\forall u < c)D'^0(u) \rightarrow D^0(c)]$$

(b) Now read Ψ, Ψ' in the style of PC_d' as 'for any U, any <, D over A $\cup$ U, if $\zeta^+ \ldots$ '. Then Ψ, Ψ' are cPC_d' and reduce $\sim \Gamma(M)$, $\sim \Gamma(M')$ over all models.

Proof. By 4.2 (g), (e), in any model $\mathfrak{A}$ of Ψ of infinite power κ, $\sim \delta_\alpha$ holds for some $\alpha < \kappa^+$. By 4.6 (b), any ordering satisfying in $\mathfrak{A}$ the hypothesis ζ^+ is a well-ordering. Hence, by 4.6 (a), Ψ 'says' that $\sim \Gamma$ holds and $\mu\alpha \sim \delta_\alpha(M) \leq \mu\alpha \sim \delta_\alpha(M')$, if the latter exists. Ψ says the reverse, with strict <. Thus the reduction takes place, in the usual way. Since 4.6 clearly holds in a PC'-type interpretation, the same argument establishes (b).

THEOREM 4.8. Over countable models, the reduction principle holds:

(a) for the family $\underset{\sim}{\forall}_1$;

(b) for the family $\forall_1(\mathfrak{a})$, if $\mathfrak{a}$ is p.r. closed and $\omega \in \mathfrak{a}$;

(c) for the families cPC_δ and cPC_d and, over denumerably infinite models, for the family cPC.

Of course, 4.8 means that these families are formed for a given similarity type $\underline{R}$.

Proof. We first show (c). By 3.2, if $\Phi(\underline{R})$ is cPC_δ (resp., cPC_d) then $\Phi \equiv' \sim \Gamma(N^\Phi)$ where N^Φ is a (recursive) sequence of finite open $\underline{R}$-formulas. Similarly, if Φ' is given, $\Phi' \equiv' \sim \Gamma(N'^\Phi)$. Applying 4.7 (b), Φ, Φ' are reduced by $\Psi(N,N')$, $\Psi'(N,N')$ which are cPC'_δ (or cPC'_d). But, by Makkai's theorem 1.2, $cPC'_\delta = cPC_\delta$ and $cPC'_d = cPC_d$. For the family cPC it is clear that $\underline{R}$ may as well be assumed finite. Hence, by 1.1, over denumerably infinite models $cPC = cPC_d$, so (c) is established.

Now assume Φ, Φ' are $\underset{\sim}{\forall}_1$. By 2.1 we get $\Phi \equiv \Phi^*$ and $\Phi' \equiv \Phi'^*$ in $\mathcal{L}$-logic, where Φ^*, Φ'^* are cPC_d. (2.1 gives different $\mathcal{L}^\Phi$ and $\mathcal{L}^{\Phi'}$ but, for example, the two can simply be put together by taking for $\mathcal{L}$ the cardinal sum of $\mathcal{L}^\Phi$ and $\mathcal{L}^{\Phi'}$.) By (c), Φ^*, Φ'^* can be reduced by cPC_d-sentences Ψ, Ψ' (over countable models). Then, by 2.2, $\Psi^{\mathcal{L}}$ and $\Psi'^{\mathcal{L}}$ clearly reduce Φ, Φ' and are $\underset{\sim}{\forall}_1$. This proves (a), but in fact (by 2.1, 2.2, and 4.7) $\Psi^{\mathcal{L}}$ and $\Psi'^{\mathcal{L}}$ were obtained primitive recursively from (Φ,Φ',ω), and hence belong to $\mathfrak{a}$ (in (b)) if $\Phi, \Phi' \in \mathfrak{a}$.

(Incidentally, for cPC and in (a), (b), Makkai's theorem was clearly not needed.)

The use of Γ to obtain reduction principles is due to Moschovakis [20-23]. The reduction principle for $\underset{\sim}{\forall}_1$ and for $\forall_1(\mathfrak{a})$ for certain $\mathfrak{a}$ can be obtained from a reduction principle of his by a suitable translation. To deal with arbitrary $\mathfrak{a}$ the less familiar translation method in §2 was developed. The author did not realize that (c) holds until some months after the talk in Cambridge. As stated in the introduction, (c) for PC partly solves affirmatively a problem of Addison [1]. It is an open problem whether the reduction principle holds for cPC over all countable, or over all finite, models. As Addison [1] pointed out, the (general) Asser conjecture "PC = cPC over finite models" would trivially imply the reduction principle for cPC over finite models. (Asser considered the case $\underline{R} = 0$, but we are considering arbitrary $\underline{R}$.)

Moschovakis established the reduction principle in a stronger, 'coming up from below' way which is important for his work with 'inductively defined' classes over non-denumerable models. Still a different proof of a (closely related) reduction principle is given in [3] (after Theorem 3.1).

The reduction principle for $\mathcal{F}$ always implies the second separation principle [11] for the class $c\mathcal{F}$. As yet, the author knows no interesting application in model theory for the second separation principle.

The 'generalized reduction principle' ([11]) 'reduces' a list $\Phi_0,\ldots,\Phi_n,\ldots$ of $\forall_1$-sentences, instead of two. It is easy to modify the argument above to show that <u>the generalized reduction principle holds for</u> $\forall_1$ <u>(over countable models)</u>, improving 4.8 (a). The same method appears to fail for 4.8 (c). (In (b) there may be a question of formulation.)

THEOREM 4.9. <u>Over countable models the reduction principle holds for</u> $\exists_2$ <u>and for</u> $\exists_2(\mathfrak{a})$, <u>if</u> $\mathfrak{a}$ <u>is primitive recursively closed and</u> $\omega \in \mathfrak{a}$.

<u>Proof</u>. Modulo what we have done so far, the proof is just like the classical proof (cf. [11]). By §2, it will clearly be enough to reduce $\exists_2^d$-sentences Φ and Φ' over exactly denumerable models (since $\mathcal{L}$ has been taken infinite). Let Φ and Φ' be $\exists\underline{T}\forall\underline{S}\sigma(\underline{R},\underline{S},T)$ and $\exists\underline{T}'\forall\underline{S}'\sigma'(\underline{R},\underline{S}',\underline{T}')$. Then $\Phi \equiv' \exists\underline{T} \sim \Gamma(N(\underline{R},\underline{T}))$ and $\Phi' \equiv' \exists\underline{T}' \sim \Gamma(N'(\underline{R},\underline{T}'))$, for suitable $\underline{N}$, $\underline{N}'$. The reducing sentences are Ψ, Ψ', where Ψ, say, is

$$\text{'}\mu\alpha\exists\underline{T} \sim \delta_\alpha(N) \text{ exists, and } \leq \mu\alpha\exists\underline{T}' \sim \delta_\alpha(N') \text{ if the latter exists'}.$$

Precisely, Ψ is

$$(\exists U,<,c,\underline{T},\underline{D})\{'(U,<) \text{ is w.o.}' \wedge \zeta(U,<,c,\underline{D},N) \wedge \sim D^0(c)$$

$$\wedge (\forall \underline{S}',\underline{D}',u)[\zeta(U,<,c,\underline{D}',N') \wedge u < c \rightarrow D'^0(u)]\}.$$

The details of Ψ' and the rest of the argument imitate (in analogy) the proof of 4.7 and 4.8.

Over finite models, we do not see how to get in 4.9 an analogue of 4.8 (c), i.e., reduction for $\exists_2^d$. (Note that Ψ, above, is not $\exists_2^d$ as it stands.) On the other hand, the part of 4.8 (c) about denumerably infinite models has here a true by vacuous analogue. Indeed, because, roughly speaking, $(\omega,<)$ can be characterized, one sees trivially that, over infinite models and finite $\underline{R}$, $\exists_2 = \exists_2(L(\omega_1^c))$. (Compare [1], where the same argument is used, though erroneously.)

Denote the order type of $(U,<)$ by $\overline{(U,<)}$. If $\mathfrak{a}$ is admissible, Ord $\mathfrak{a}$ is the least ordinal $\notin \mathfrak{a}$.

THEOREM 4.10. _Suppose_ $\mathfrak{a}$ _is admissible and_ Φ _is in_ $\exists_1(\mathfrak{a})$. _If_ $\vdash \Phi \rightarrow {}'(U,<)$ _is w.o._', _then, for some_ $\beta <$ Ord $\mathfrak{a}$, $\vdash \Phi \rightarrow {}'\overline{(U,<)} < \beta'$.

4.10 in full generality is due to Barwise [2]. Various cases of it were proved by Morley [16, 17], Lopez-Escobar [13], Spector, and Lusin-Sierpinski! Note that the case $\mathfrak{a} = H_{\omega_1}$ follows directly from the Lusin-Sierpinski result about Σ_1^1, since an invariant Σ_1^1-class is certainly Σ_1^1. Several proofs of 4.10 are known, so for now we simply assume 4.10. Two proofs making use of Γ are indicated in §5.

THEOREM 4.11 ('_Covering_', _cf._ [11]). _Suppose_ Φ _and_ Ψ _are_ $\exists_1$ _and_ $\vdash \Phi \rightarrow \sim \Psi$. _Then, for some_ $\beta < \omega_1$, $\vdash \Phi \rightarrow \sim \delta_\beta^\Psi$. _Moreover, if_ $\mathfrak{a}$ _is admissible and_ $\Phi, \Psi \in \mathfrak{a}$, _then_ β _can be taken in_ $\mathfrak{a}$ _and hence_ $\delta_\beta^\Psi \in \mathfrak{a}$.

Proof. By 3.3 and our hypothesis, $\vdash \Phi \rightarrow \sim \Gamma^\Psi$. Hence Lemma 4.6'(b) implies that

$$(*) \qquad \vdash \Phi \wedge \zeta^{+\Psi} \rightarrow {}'(U,<) \text{ is w.o.}'.$$

Applying 4.10 gives:

$$\vdash \Phi \wedge \zeta^{+\Psi} \rightarrow \text{'}\overline{(U,<)} < \beta\text{'} \quad \text{for some} \quad \beta < \omega_1.$$

Hence, by 4.6'(a), $\vdash \Phi \rightarrow \sim \delta^{\Psi}_{\beta}$. To cover the case of finite models, the proof just given should be read (here and below) so as to allow U to be additional universe.

Now suppose $\Phi, \Psi \in \mathcal{A}$. If $\omega \in \mathcal{A}$, then (by §2, §3, 4.5) $\zeta^{+\Psi} \in \mathcal{A}$. Hence, by 4.10, we can take $\beta \in \mathcal{A}$. Consequently, $\delta^{\Psi}_{\beta} \in \mathcal{A}$ (since $\lambda\alpha\delta_{\alpha}$ is p.r.).

Only the case $\mathcal{A} = H_{\omega}$ remains. Now we ignore the passage to $\mathcal{L}_{\Psi}$ and Ψ^* and take $\zeta^{+\Psi}$ to be $\zeta^{+}(N^{\Psi})$. The left side of (*) can be written $(\exists \underline{R}, D, c)(\Phi \wedge \zeta^{+}(N^{\Psi}))$, which is PC_d and hence (by 1.1) PC'. Hence 4.10 can be applied, giving $\beta < \omega$. $\delta_{\beta}(N^{\Psi})$ is obviously finite.

As in the classical case, the covering theorem implies the strong first separation (interpolation) theorem, of Barwise, Lopez-Escobar, et al. Let $\mathcal{A}$ be admissible.

COROLLARY 4.12. If $\Phi(\underline{R}), \Psi(\underline{R})$ are $\underset{\sim}{\exists}_1$ (or $\exists_1(\mathcal{A})$) and $\vdash \Phi \rightarrow \sim \Psi$ then, for some $\sigma(\underline{R})$ (some $\sigma \in \mathcal{A}$), $\vdash \Phi \rightarrow \sigma$ and $\vdash \sigma \rightarrow \sim \Psi$.

For the finite case ($\mathcal{A} = H_{\omega}$), almost exactly 4.11 and its use to prove 4.12 are in Svenonius [28]. Moschovakis [22] discussed interpolation (or Souslin-Kleene) type theorems by means of Γ in a setting close to the general case, but only for classes of <u>elements</u> (rather than relations (=, here, structures)). For these his methods are somewhat different, and yield other results of interest.

Makkai [15] established that Novikov's generalized first separation principle holds for $\underset{\sim}{\exists}_1$. This result follows (as in [11]) from 4.12 and the <u>generalized</u> reduction principle for $\underset{\sim}{\forall}$, mentioned above (after 4.8).

It is possible to improve 4.11 in the following way. Suppose $\Psi(\underline{R})$ is $\underset{\sim}{\exists}_1$. By taking a little more care, we could have insured that <u>each</u> R (<u>in</u> $\underline{R}$) <u>occurs positively</u> (<u>negatively</u>) <u>in</u> δ^{Ψ}_{α} <u>only if it does so in</u> Ψ. To do this we first improve 2.1 in a similar way, by redefining Sat in a well-known, 'more positive' way. Next a similar improvement of 3.2 is obtained, by simply observing that Lyndon's interpolation theorem for propositional calculus answers that

N_n $(= \exists S \bigwedge_{m<n} N'_m)$ can be written in the desired way. Looking at the definition of δ_α (4.1) completes the proof. 4.11 now insures that in 4.12, σ has a positive (negative) occurrence of R only if $\sim\Psi$ does. The Lyndon-type version of 4.12 says also the same thing for Φ, but that result can in fact easily be derived from the one-sided result just mentioned.

Unfortunately, the above remarks are much less helpful than I had thought (in a earlier manuscript version of this paper). The reason is that many applications of the Lyndon-type interpolation theorem it is essential that the theorem be for logic <u>without equality</u>. In any case, the whole question of using Γ in connection with (various) preservation theorems has now been beautifully dealt with by Makkai (in this volume).

Just as in the classical case [11], the covering theorem was stated for a specific 'decomposition' of $\sim\Psi$, into $\bigvee_{\alpha<\omega_1} \sim\delta^\Psi_\alpha$. Various weaker statements can be made, such as: For any Ψ there exist θ_α $(\alpha < \omega_1)$ such that $\vdash \theta_\alpha \rightarrow \sim\Psi$ $(\alpha < \omega_1)$ and whenever $\vdash \Phi \rightarrow \sim\Psi$ then for some $\alpha < \omega_1$, $\vdash \Phi \rightarrow \theta_\alpha$. Moschovakis pointed out to the author that this statement can be rather easily be derived directly from the classical covering theorem together with the Lopez-Escobar interpolation theorem. To what extend the same may apply to various stronger weakenings of 4.11 is not clear.

The reduction principle for Π^1_1 is a consequence of a very strong result, the uniformization principle of Novikov-Kondo-Addison. The latter says, roughly, that if K(R,S) is Π^1_1 then there is a Π^1_1 L(R,S) such that $\exists SK \equiv \exists SL$ and $\exists SL \Rightarrow \exists! SL$. It is natural to ask whether the uniformization principle can also be extended to $L_{\omega_1\omega}$. Now, the literal generalization clearly fails, since S can never be determined in (A,R) beyond an (A,R) automorphism. What seems the most plausible generalization is a principle (about $\forall_1$) in which '$\exists!$' is just replaced by 'exactly one up to R-automorphism'. (Since (ω, Sc) has only one automorphism, this would include the old principle.) However, Dale Myers has recently shown that this proposed 'uniformization principle' cannot be established in ZF. For this and other related results, see [24].

§5. Γ and Admissible Sets. We will show that Γ can be used to give proofs of other results of Barwise concerning $L_{\omega_1\omega}$ and admissible sets. Since the results are not new, proofs will be give rather sketchily.

Henceforth $\mathfrak{A}$ is an admissible set $\neq H_\omega$ and $\alpha = \mathrm{Ord}\ \mathfrak{A}$ (The results also hold for H_ω but to adapt some of the proofs to H_ω we would need Makkai's 1.2; however, the (proof of the) latter already implies Godel's completeness theorem!)

THEOREM 5.1 (Improvement of 4.2(d),(e),(f)). If $\mathfrak{O}(\underline{M}) \in \mathfrak{A}$ then $\mathfrak{O} \models \rho_\alpha$. Hence, if $\mathcal{L}(\underline{R}) \in \mathfrak{A}$ and $\Phi(\underline{R})$ is $\exists_1(\mathfrak{A})$, then $\mathcal{L} \models \rho_\alpha^\Phi$.

Proof. It will suffice to show $\mathfrak{O} \models \delta_\alpha \to \delta_{\alpha+1}$ (as the same will clearly then hold for δ_α^n). Suppose $\mathfrak{O} \vdash \delta_\alpha$, i.e., $\mathfrak{O} \vdash \bigwedge_{\beta<\alpha} \delta_{\beta+1}$, or in other words:

$$(4) \qquad (\forall x_0 \in A) \bigwedge_{\beta<\alpha} (\exists y_0 \in A) \mathfrak{O} \models \delta_\beta^1[x,y].$$

The formula, $\mathfrak{O} \models \delta_{\alpha+1}$, which we want, can be written

$$(5) \qquad (\forall x_0 \in A)(\exists y_0 \in A) \bigwedge_{\beta<\alpha} \mathfrak{O} \models \delta_\beta^1[x,y].$$

The needed permutation of $\bigwedge$ and $\exists$ holds in all admissible sets (almost by definition), provided the complement of the matrix "$\mathfrak{O} \models \delta_\beta^1[x,y]$" is $\mathfrak{A}$-r.e. and, as here, $\vdash \delta_\beta^1 \to \delta_{\beta'}^1$, $(\beta > \beta')$). But satisfaction and $\lambda\beta\delta_\beta^1$ and hence our matrix are primitive recursive. The second statement follows from the first (by §2, §3).

Theorem 5.1 essentially follows from Theorem 1 of [3] and is related to earlier work of Moschovakis [20,21].

From 5.1 follows an improvement in our discussion of Scott's theorem (see 4.3, 4.4): Suppose $\mathfrak{O} \in \mathfrak{A}$ and Φ is the natural $\exists_1$ 'describing sentence' for $\mathfrak{O}$. Then $\Phi \equiv' \rho_\alpha^\Phi \wedge \delta_\alpha^\Phi$ (by 5.1 and 4.2). In 4.4 we had only 'for some α'.

Another consequence of 5.1 is:

COROLLARY 5.2. (Barwise [2]). The set of logically valid sentences belonging to $\mathfrak{A}$ is $\mathfrak{A}$-r.e.

<u>Proof</u>. By 5.1,

(6) $$\{(\mathfrak{A},\Psi):\Psi \text{ is } \forall_1,\ \Psi,\mathfrak{A} \in \mathcal{A},\ \text{and } \mathfrak{A} \models \Psi\} \text{ is}_J\ \mathcal{A}\text{-r.e.}$$

((6) holds because $\mathfrak{A} \models \Psi$ iff $(\exists\beta < \alpha)\mathfrak{A} \models \sim \tilde{\delta}_\beta^{\Psi}$, which is clearly $\mathcal{A}$-r.e.) Now for sentences $\sigma(\underline{S}) \in \mathcal{A}$, $\vdash \sigma$ iff $(\forall\gamma \leq \omega)\gamma \models \forall\underline{S}\sigma$, which (by (6)) is obviously $\mathcal{A}$-r.e.

THEOREM 5.3. (Compactness) (Barwise [2]). <u>If</u> $X \subseteq \mathcal{A}$ <u>is</u> $\mathcal{A}$-<u>r.e.</u> <u>and every</u> $\mathcal{A}$-<u>finite subset of</u> X <u>has a model, so has</u> X.

<u>Proof</u>. By introducing a unary predicate we can assume that every $\mathcal{A}$-finite subset of X has a model with universe ω. Let $\mathcal{A} \models Xa \leftrightarrow \exists sQ(a,s)$, where Q has only (ε-) bounded quantifiers, and a parameter a_0. Henceforth let 'w' range over transitive sets containing a_0 and ω. Now refer to the proof of 2.1. Given w, take $\mathfrak{A} = \omega$, $B = w$, form $\mathcal{L}$ as there and put $\bar{w} = (\mathcal{L},\varepsilon,\omega,a_0)$ (which is also $\mathfrak{A} + \mathcal{L}$). By our hypothesis,

$$(\forall w \in \mathcal{A})\bar{w} \models \exists K\psi;\ \text{when } \psi \text{ is } Sat(K,\bar{0}) \wedge (\forall x,y)(Q(x,y) \rightarrow K_0(x)).$$

Our desired conclusion is $\mathcal{A} \vdash \exists K\psi$. By inspection of Sat(§2) we see that ψ is a conjunction of sentences $\forall u^{(n)}P$ where P has only bounded quantifiers. Repeating the proof of 3.2 exactly, we see that consequently $\exists K\psi \equiv' \Gamma(N)$, where

(7) $$\vdash N_{n+1}(x_0,\ldots,y_n) \rightarrow y_n \in a_0 \cup \omega \cup x_0 \cup \ldots \cup x_n.$$

Thus the game is 'bounded' for the y-player.

Now let $\Gamma^n(x_0,\ldots,y_{n-1})$ be $\forall x_n \exists y_n \forall x_{n+1} \ldots \bigwedge_k N_k(x_0,\ldots,y_{k-1})$. We will show that $\bar{\mathcal{A}} \vdash \Gamma(N)$, as desired, by showing that (1) (preceding 3.1) holds for the following T_n $(n \in \omega)$: $T_n(x_0,\ldots,y_{n-1})$ iff $(\forall w \in \mathcal{A})(x_0,\ldots,y_{n-1} \in w \rightarrow \bar{w} \vdash \Gamma^n(x_0,\ldots,y_{n-1}))$. By hypothesis, T_0 holds and clearly $T_n \Rightarrow N_n$. It only remains to show that $T_n \Rightarrow \forall x_n \exists y_n T_{n+1}$. Now, by looking at the 'bounded' game Γ^n, one sees immediately that

(8) $$x_0,\ldots,y_{n-1} \in w \in w' \wedge w' \models \Gamma^n \Rightarrow w \models \Gamma^n.$$

By (6),

(9) $$"\bar{w} \not\models \Gamma^n(x_0,\ldots,y_{n-1})" \text{ is } \mathcal{A}\text{-r.e.}$$

Assume T_n, which can be written:

(10) $$\forall w(x_0,\ldots,y_{n-1} \in w \Rightarrow \bar{w} \models \forall x_n \exists y_n \Gamma^{n+1}).$$

In view of (7), (8), and (9) the quantifiers $\forall w$ and $\exists y_n$ in (10) can be permuted as in the proof of 5.1; and this yields $\forall x_n \exists y_n T_{n+1}$, as desired.

Now we return to the basic theorem, 4.10, on boundedness of Ξ-classes of well-orderings, and mention two 'pure Γ' proofs. The first consists simply in recalling that Barwise, in [2], has proved 4.10 using only Compactness (5.3 above). Alternatively, a proof of 4.10 can be given, using 5.1, which is analogous to the classical proof (cf. [11]) and Spector's proof, involving a 'universal' class, etc. We leave the details to the reader. (Some related arguments are in Moschovakis [20,23].)

In [2] there is a generalization of 5.3 involving the notions S-admissible and S-$\mathcal{A}$-r.e., where S is a relation over $\mathcal{A}$. The proof above works verbatim if "$\mathcal{A}$-r.e." in (9) is read "S-$\mathcal{A}$-r.e.".

A different generalization can be given for various effective results, including several in this paper. The possibility of such a generalization is 'known', but no details seem to be in print. For motivation, suppose that a countable structure $\mathcal{L} = (B,S)$ (S binary, say) is just given to us, and we want to consider $\mathcal{L}$-logic, or Morley's $\mathcal{L}$-models [19], or Moschovakis' theory [20-22] over $\mathcal{L}$. In all of these, the elements of $\mathcal{L}$ are given essentially as urelementen. Thus, in considering the $L_{\omega_1\omega}$ of $\mathcal{L}$-logic and 'admissible fragments' of it, we may lose something by insisting that $\mathcal{L} \in H_{\omega_1}$. For this reason, we start over, working in a set theory with many urelementen. The notions 'transitive', 'admissible', 'p.r. set function', etc. are modified, mostly in the most obvious way. For the purposes above, we could require that the set of urelementen in an 'admissible' $\mathcal{A}$ belongs to $\mathcal{A}$. One should also require that if $\mathcal{A}$ contains an infinite set, then $\omega \in \mathfrak{a}$. It seems very likely that under such a wider interpretation all the theorems in this paper (concerning admissible sets, etc.) remain valid. However the author has not done the detailed work, and there may be non-trivial questions of formulation to be settled.

Such a generalization is needed if one tries to compare Moschovakis' framework and results with those here. Had it been adopted in [3], the results there would also stand in a clearer relation to Moschovakis' work. Of course, it is very natural from an algebraic point of view. Hopefully such a study will soon be made. While writing I have heard that Barwise plans a paper on 'urelementen'.

BIBLIOGRAPHY

1. Addison, J., The theory of hierarchies, Logic, methodology, and philosophy of science, Proc. 1960 International Congress, Stanford 1962, pp. 26-37.

2. Barwise, J., Infinitary logic and admissible sets, J. Symbolic Logic vol. 34 (1969), pp. 226-252.

3. Barwise, J., Gandy, R., and Moschovakis, Y., The next admissible set, J. Symbolic Logic vol. 36 (1971), pp. 108-120.

4. Craig, W., and Vaught, R., Finite axiomatizability using additional predicater, J. Symbolic Logic vol. 23 (1958), pp. 289-308.

5. Jensen, R., and Karp, C., Primitive recursive set functions. Axiomatic set theory, Amer. Math. Soc. Proc. Symp. Pure Math., Providence, 1971, pp. 143-176.

6. Karp, C., Languages with Expressions of Infinite Length, Amsterdam, 1964.

7. Keisler, H. J., Finite approximations of infinitely long formulas, Proc. Berkeley Symposium on the Theory of Models, Amsterdam, 1965, pp. 158-169.

8. Keisler, H. J., Infinite quantifier and continuous games. Applications of Model Theory to Algebra, Analysis, and Probability, New York, 1969, pp.228-264.

9. Kleene, S., Finite axiomatizability of theories in the predicate calculus using additional predicate symbols. Memoirs Amer. Math. Soc. no. 10 (1952), pp. 27-68.

10. Kreisel, G., Model-theoretic invariants: Applications to recursive and hyper-arithmetic operations. Proc. Berkeley Symposium on the Theory of Models, Amsterdam, 1965, pp. 190-205.

11. Kuratowski, K., Topology I, New York-Warsaw, 1966.

12. Lopez-Escobar, E. G. K., An interpolation theorem for denumerably long formulas, Fund. Math. vol. 57 (1965), pp. 253-272.

13. Lopez-Escobar, E. G. K., On defining well-orderings, Fund. Math. vol. 59 (1966), pp. 13-21.

14. Makkai, M., On PC_Δ-classes in the theory of models, Publ. Math. Inst. Hungarian Acad. Sci. vol. 9 (1964), pp. 159-194.

15. Makkai, M., An application of a method of Smullyan to logics on admissible sets, Bull. Acad. Polon. des Sciences (Serie des sciences math.) vol. 17 (1969), pp. 341-346.

16. Morley, M., Omitting classes of elements. Proc. Berkeley Symposium on the Theory of Models, Amsterdam, 1965, pp. 265-273.

17. Morley, M., The Hanf number for ω-logic. J. Symbolic Logic vol. 32 (1967), p. 437.

18. Morley, M., The number of countable models. J. Symbolic Logic vol. 35 (1970), pp. 14-18.

19. Morley, M., The Lowenheim-Skolem theorem for models with standard part. Symposia Mathematica vol. v, Academic Press, London-New York, 1971, pp. 43-52.

20. Moschovakis, Y., Abstract first order computability I, II, Trans. Amer. Math. Soc. vol. 138 (1969), pp. 427-464, pp. 465-504.

21. Moschovakis, Y., Abstract computability and invariant definability, J. Symbolic Logic vol. 34 (1969), pp. 605-633.

22. Moschovakis, Y., The Suslin-Kleene Theorem for countable structures, Duke Math. J. vol. 37 (1970), pp. 341-352.

23. Moschovakis, Y., The game quantifier. Proc. Amer. Math. Soc. vol. 31 (1972), pp. 245-250.

24. Myers, D., The failure of the $L_{\omega_1\omega}$ analogue of Π^1_1 uniformization, Notices Amer. Math. Soc. vol. 19 (1972), p. A-330.

XXI. MODAL MODEL THEORY

C. C. Chang*

To the memory of Richard Montague

1. INTRODUCTION

We are all familiar with the usual modal operators $\Diamond$, for possibility, and $\Box$, for necessity. In any given model $\mathfrak{A}$ for modal logic, the meanings assigned to $\Diamond\varphi$ and $\Box\varphi$ are the following:

(1) $\mathfrak{A} \models \Diamond\varphi$ means: in $\mathfrak{A}$ it is possible that φ;

$\mathfrak{A} \models \Box\varphi$ means: in $\mathfrak{A}$ it is necessary that φ.

We propose to introduce a system of logic with modal operators like, for example, $\Diamond x$ and $\Box x$, where the meanings assigned to the formulas $\Diamond x\varphi(xyz\ldots)$ and $\Box x\varphi(xyz\ldots)$ are the following: Let $a,b,c,\ldots$ be elements of $\mathfrak{A}$. Then

(2) $\mathfrak{A} \models \Diamond a\varphi[abc\ldots]$ means: in $\mathfrak{A}$ the individual a finds it possible that $\varphi[abc\ldots]$;

$\mathfrak{A} \models \Box a\varphi[abc\ldots]$ means: in $\mathfrak{A}$ the individual a finds it necessary that $\varphi[abc\ldots]$.

We see that the impersonal "it" in (1) is replaced by the personal "a" in (2). As far as we know this seems to be a new approach in modal logic, and we shall develop in this paper the modest beginnings of a model theory for such systems. In the process, we shall arrive at some interesting definitions of models and satisfaction. There is nothing technically difficult in this paper. We have, however, raised some interesting problems, and the subject may be of interest to more philosophically inclined logicians.

The ideas in this paper owe much to a paper by Richard Montague [1] and to a Carnap Prize essay by Donald Victery [2]. I wished that I had made these investigations before Montague's death so that I could have a chance to discuss these matters with him.

---

*The research and writing of this paper was partially supported by NSF Grant GP-27964.

2. THE FORMAL SYSTEM

Let L be a set of symbols of the following sorts:

finitary predicate symbols, P, Q, R,...,

finitary function symbols, F, G, H,...,

individual constant symbols, $c_1, c_2, \ldots$,

the identity symbol =,

individual variable symbols, $v_0, v_1, v_2, \ldots$,

and in addition,

finitary modal operator symbols, N, M,... .

We shall now build up a formal system over L using the usual connectives $\neg$, $\wedge$, $\vee$, $\rightarrow$, $\leftrightarrow$, and the two quantifiers $\forall x$, $\exists x$. The formation rules are just a slight extension of the formation rules for first order logic. Terms of L are built up in the usual way from constants, variables, and function symbols. Atomic formulas are of the forms $t_1 = t_2$ and $P(s_1 s_2 \ldots s_n)$ where $t_1, t_2, s_1, s_2, \ldots, s_n$ are terms and P is an n-place predicate symbol. The set of all formulas of L is obtained from the set of atomic formulas by the following closure rules (only rule (iv) is new):

(i) every atomic formula is a formula;

(ii) if φ, ψ are formulas, then $\neg\varphi$, $\varphi \wedge \psi$, $\varphi \vee \psi$, $\varphi \rightarrow \psi$, $\varphi \leftrightarrow \psi$ are formulas;

(iii) if φ is a formula and x is a variable, then $(\forall x)\varphi$ and $(\exists x)\varphi$ are formulas;

(iv) if N is an n-place modal operator symbol, x is a variable, and $\varphi_1, \ldots, \varphi_n$ are formulas, then $Nx(\varphi_1 \ldots \varphi_n)$ is a formula;

(v) φ is a formula if and only if φ is generated from the rules (i)-(iv).

Every variable occurring in an atomic formula occurs free in that formula. If a variable occurs free in φ, then it occurs free in $\neg\varphi$; if it occurs free in φ or ψ, then it occurs free in $\varphi \wedge \psi$, $\varphi \vee \psi$, $\varphi \rightarrow \psi$, $\varphi \leftrightarrow \psi$. If a variable y occurs free in φ and $y \neq x$, then it occurs free in $\forall x\, \varphi$ and $\exists x\, \varphi$. The variable x always occurs free in the formula $Nx\ \varphi_1 \ldots \varphi_n$. A formula is a sentence if it has no free occurrences of variables within it. Note that even if $\varphi_1, \ldots, \varphi_n$ are sentences, the formula $Nx\ \varphi_1 \ldots \varphi_n$ is not a sentence. Given a formula φ of L, we follow the usual convention when we write $\varphi(xyz\ldots)$ to indicate that all of the free variables of φ are among those displayed within the parentheses. If we disregard the modal operators and the rule (iv), then of course every first order formula or sentence of L is a formula or sentence of L in the new sense.

As we are not interested in proving completeness theorems, this is about as much of the syntax of L as we would want to discuss. There remains the interesting questions of what gives rise to a model $\mathfrak{A}$ for L and how is satisfaction defined.

3. INTUITIVE MOTIVATIONS

The intended meaning of the formula $Nx\ \varphi_1 \ldots \varphi_n$ is that: "x finds it is N that $\varphi_1 \ldots \varphi_n$". Here N can be any adjective which describes some abstract quality. We give a list of examples: useful, helpful, possible, believable, likely, necessary, good, bad, consistent, beautiful, horrible, incredible, strange, etc. Rather than using the phrase "x finds it is", which I find quite suitable, the reader for his own convenience could substitute the phrases "x considers it as", "x thinks it is", or "x considers it to be", if it suits him better. At any rate, we hope that the idea is clear. As for the meaning of "$\varphi_1 \ldots \varphi_n$", we can for the moment assume that it stands for the conjunction of the statements φ_1 through φ_n. Generally, there will be a difference between $Nx\ \varphi_1 \ldots \varphi_n$ and $Mx\ (\varphi_1 \wedge \ldots \wedge \varphi_n)$, where M is a 1-place modal operator.

A possible model $\mathfrak{A}$ for L should consist of a non-empty set A (the universe of $\mathfrak{A}$) together with the intended interpretations of all the symbols in L. For the usual predicate, function, or constant symbols in L, we take their interpretations in $\mathfrak{A}$ to be the usual relations $P \subset A \times \ldots \times A$, functions $F: A^n \to A$, and distinguished elements $d_1, d_2, \ldots$ of A. For simplicity, let us assume N is a 1-place modal operator. The interpretation for N in $\mathfrak{A}$ offers many interesting possibilities. We should expect that the interpretation of N will carry with it the idea that individuals in A are making some sort of judgment in their world, namely $\mathfrak{A}$, based upon information supplied by the interpretation of N in $\mathfrak{A}$.

As an example, let's say that the set A is a roomful of people, a,b,c,... . Let P be a 2-place predicate so that its interpretation in $\mathfrak{A}$ is a subset $P \subset A \times A$. (Here we make no distinction between P and its interpretation in $\mathfrak{A}$.) Let us further assume that the intended meaning of P is that

P(ab) if and only if the individual a speaks to the individual b.

Thus, to find out all about P, we merely stand discreetly outside the room and observe (presumably over a period of time) who is or is not speaking to whom. Suppose now we give Nx the intended meaning:

"x finds it enjoyable".

Then in the model $\mathfrak{A}$ (the roomful of people A, the binary relation P, and some interpretation of N):

NaP(ab) holds if and only if a finds it enjoyable to speak to b.

Let us make the first observation that whether a finds it enjoyable to speak to b or not, it may happen to have no connection with whether indeed a speaks to b, that is whether

P(ab) holds in $\mathfrak{A}$.

Secondly, how should the individual a find out whether or not it is enjoyable for him to speak to b? Since this is a (somewhat tacky) social situation, we must give it a corresponding social analysis.

It is natural that the enjoyment of a speaking to b (from a's point of view) depends on both a and b. So, a's decision on whether or not he would enjoy speaking to b will then depend on who the person b is. One reasonable way for a to decide this question would be to observe who else in the room is speaking to b. Finally, of course, a's decision may also depend on the predicate P, so that his reaction in other situations may vary from situation to situation.

So, in an ideal world, we see that the question of

$$NaP(ab) \text{ holding in } \mathfrak{A}$$

could depend on at least four factors, namely,

$$a, b, P(xy), \text{ and } \{c \in A : P(cb)\}.$$

In general, then, given an arbitrary formula $\varphi(xyz\ldots)$ of L and individuals $a,b,c,\ldots$ in A, the question of whether or not

(1) $\mathfrak{A} \models Na\varphi[abc\ldots]$

depends on at least the following factors,

(2) $a,b,c,\ldots,\varphi(xyz\ldots)$, and $\{a' \in A : \mathfrak{A} \models \varphi[a'bc\ldots]\}$.

Observe that the set

$$C = \{a' \in A : \mathfrak{A} \models \varphi[a'bc\ldots]\}$$

of course depends both on $b,c,\ldots,$ and on φ.

We now make a great simplication. We shall henceforth assume that:

(*) The validity of (1) shall depend only on a and on the set C.

Under this assumption (*), the natural interpretation of N in the model $\mathfrak{A}$ is that of a function mapping A into sets of subsets of A. So,

$$N : A \to S(S(A))$$

and for each $a \in A$, $N(a) \subset S(A)$, and each set in $N(a)$ is called an acceptable set to a. We then say that

$$\mathfrak{A} \models Na\varphi[abc\ldots] \text{ iff } C \in N(a).$$

In particular,

$\mathfrak{A} \models NaP(ab)$ (a finds it enjoyable to speak to b) if and only if

$$\{c \in A : P(cb)\} \in N(a).$$

Note that our assumption (*) may lead one to suspect that generally speaking the people in room A are somewhat simple minded. For example, an individual $a \in A$ will decide the two different questions of

$$\mathfrak{A} \models NaP(ab) \quad \text{(a finds it enjoyable to speak to b)}$$

and

$$\mathfrak{A} \models NaP(ba) \quad \text{(a finds it enjoyable to be spoken to by b)}$$

by using the same set of acceptable sets $N(a)$. There are a number of convincing arguments why the validity of (1) should depend in some way on all of the factors in (2). (There are undoubtedly more factors than in (2) in real life.) For the purpose of this paper, we shall stick to the assumption (*), leaving open and untouched the possibility of more complicated interpretations of N. We might point out that one consequence of (*) is that the following sentence

$$\forall x(\varphi(x) \leftrightarrow \psi(x)) \rightarrow \forall x\,(Nx\varphi(x) \leftrightarrow Nx\psi(x))$$

is valid in every model for L. Our assumption (*) does not imply that the following is a valid sentence:

$$\forall x(\varphi(x) \rightarrow \psi(x)) \rightarrow \forall x\,(Nx\varphi(x) \rightarrow Nx\psi(x)).$$

Before we leave this section, let's give two more examples. The ternary predicate $Q(abc)$ can stand for:

a helps b to beat up c.

The 1-place modal operator N could range from "necessary" to "helpful" to "stupid". The quatenary predicate $R(abcd)$ can stand for:

a asks b to referee c's paper generalizing a result of d.

M then could be interpreted as: "natural", "tiresome", "outrageous", "amusing". One can have a lot of fun making up and reading various formulas made up of Q and N or R and M.

4. MODELS, SATISFACTION, AND TRUTH

A model $\mathfrak{A} = \langle A, \ldots \rangle$ for L consists of a non-empty set A together with an interpretation of each symbol in L. For a predicate, function, or constant symbol in L, we take their usual interpretations. Let N be an n-place modal operator symbol in L. Then the interpretation of N in $\mathfrak{A}$ is a function mapping A into $S(S(A)^n)$. So, for each $a \in A$, $N(a) \subset S(A) \times \ldots \times S(A)$ (n-times).

Let $\varphi(x_1 \ldots x_n)$ be a formula of L and let $a_1, \ldots, a_n \in A$. We define the satisfaction predicate

$$\mathfrak{A} \models \varphi[a_1 \ldots a_n]$$

in the usual way for formation rules (i) - (iii). Let N be an n-place modal operator and let $\varphi_1(x_1 \ldots x_n), \ldots, \varphi_m(x_1 \ldots x_n)$ and $a_1, \ldots, a_n \in A$ be given. We say that

$$\mathfrak{A} \models Nx_1(\varphi_1 \ldots \varphi_m)[a_1 a_2 \ldots a_n]$$

if and only if

$$\langle \{c_1 \in A : \mathfrak{A} \models \varphi_1[c_1 a_2 \ldots a_n]\}, \ldots, \{c_m \in A : \mathfrak{A} \models \varphi_m[c_m a_2 \ldots a_n]\} \rangle \in N(a_1).$$

It is built into the definition that if $a_1, a_2, \ldots, a_n$ and $b_1, b_2, \ldots, b_n$ are two sequences of elements of A which coincide on the free variables in $\varphi(x_1 \ldots x_n)$, then

$$\mathfrak{A} \models \varphi[a_1 \ldots a_n] \text{ iff } \mathfrak{A} \models \varphi[b_1 \ldots b_n].$$

A sentence φ of L is <u>true in</u> $\mathfrak{A}$, or <u>holds in</u> $\mathfrak{A}$, if $\mathfrak{A} \models \varphi$. A sentence φ is <u>valid</u> if $\mathfrak{A} \models \varphi$ for every model $\mathfrak{A}$ for L. Two models $\mathfrak{A}$ and $\mathfrak{B}$ are <u>elementarily equivalent</u>, in symbols $\mathfrak{A} \equiv \mathfrak{B}$, iff $\mathfrak{A} \models \varphi$ iff $\mathfrak{B} \models \varphi$ for every sentence φ of L. This completes the short list of definitions in this section.

Before going on to the next section, this is an appropriate place to point out briefly the differences in the paper [1] of Montague and this paper. Montague starts out with the same set L of predicate, function, and operator symbols. His formation rules for formulas of L is very much like ours except that in (the crucial) rule (iv) he only allows the formation of

(iv′) $N(\varphi_1 \ldots \varphi_n)$ if $\varphi_1, \ldots, \varphi_n$ are formulas.

A model $\mathfrak{A}$ for pragmatics (this is what he called the language L) has a universe $I \cup A$ composed of two parts I and A. An n-place predicate P is interpreted as an (n+1)-ary relation

$$P \subset I \times A^n.$$

An n-place function F is interpreted as an (n+1)-place function

$$F: I \times A^n \to A.$$

Finally, an n-place operator N is interpreted as a mapping

$$N: I \to S(S(I)^n).$$

He then defines for each formula $\varphi(x_1 \ldots x_n)$ of L and elements $i \in I$ and $a_1, \ldots, a_n \in A$ the satisfaction predicate

$$\mathfrak{A} \models \varphi_i[a_1 \ldots a_n]$$

by doing more or less the natural thing. That is,

$$\mathfrak{A} \models P_i(a_1 \ldots a_n) \text{ means } P(ia_1 \ldots a_n);$$
$$\mathfrak{A} \models F_i(a_1 \ldots a_n) = b \text{ means } F(ia_1 \ldots a_n) = b;$$
$$\mathfrak{A} \models (\varphi \wedge \psi)_i[a_1 \ldots a_n] \text{ iff } \mathfrak{A} \models \varphi_i[a_1 \ldots a_n] \text{ and } \mathfrak{A} \models \psi_i[a_1 \ldots a_n];$$

etc., through all the connectives; and

$$\mathfrak{A} \models (\forall x \varphi)_i\,[xa_2 \ldots a_n] \text{ iff for all } a \in A,\ \mathfrak{A} \models \varphi_i[aa_2 \ldots a_n];$$

similarly for $\exists x$; and, finally,

$$\mathfrak{A} \models (N\varphi_1 \ldots \varphi_m)_i[a_1 \ldots a_n] \text{ iff}$$

$$\langle\{j \in I : \mathfrak{A} \models \varphi_{1j}[a_1 \ldots a_n]\}, \ldots, \{j \in I : \mathfrak{A} \models \varphi_{mj}[a_1 \ldots a_n]\}\rangle \in N(i).$$

Montague showed in his paper that his models and his notion of satisfaction, when suitably restricted, will yield practically all known examples of modal and tense logics. (For instance, ordinary and generalized tense logics, standard and generalized modal logics, general and special deontic logics, Kripke's semantics for nonnormal modal logics, inductive logic, etc.) It seemed to me that, from the model-theoretic point of view, his separation of the universe into I and A and the consequent different roles that $\forall x$ and $N(\varphi_1 \ldots \varphi_n)$ take was unnatural. It is much better to treat all elements of $I \cup A$ on an equal footing. The small step I took resulted in the notions of model and satisfaction given above. It is clear that we can suitably restrict our models to give his models as special cases. Furthermore, by introducing two new predicates to stand for the two different parts of his universe, we can recapture his notion of satisfaction as well. Thus, it would appear that practically all known examples of modal and tense logics are special cases of the notions given here. At the moment it is not clear why this is supposed to be an advantage. However, it does make the model theory of the whole theory more interesting and natural appearing.

5. BEGINNING MODEL THEORY

Two models $\mathfrak{A} = \langle A, \ldots\rangle$ and $\mathfrak{B} = \langle B, \ldots\rangle$ for L are _isomorphic_ iff there is a one-to-one mapping f of A onto B which preserves all relations, functions, and constants, and in addition if $N^{\mathfrak{A}}$ and $N^{\mathfrak{B}}$ are the interpretations of an n-place operator N in $\mathfrak{A}$ and $\mathfrak{B}$, respectively, then for each $a \in A$

$$N^{\mathfrak{B}}(f(a)) = \{\langle f^{*}c_1, \ldots, f^{*}c_n\rangle : \langle c_1, \ldots, c_n\rangle \in N^{\mathfrak{A}}(a)\}.$$

The function f is an _isomorphism_ of $\mathfrak{A}$ onto $\mathfrak{B}$, and we write $\mathfrak{A} \cong \mathfrak{B}$ if there is such an isomorphism. Clearly, the relation $\cong$ is an equivalence. We also have very easily:

5.1 If $\mathfrak{A} \cong \mathfrak{B}$, then $\mathfrak{A} \equiv \mathfrak{B}$.

The converse of 5.1 is of course false. But we have that:

5.2 If $\mathfrak{A} \equiv \mathfrak{B}$ and $\mathfrak{A}$ is finite, then $\mathfrak{A} \cong \mathfrak{B}$.

Proof. Let $A = \{a_1,\ldots,a_n\}$. Since the usual identity sentences are in L, it follows that $B = \{b_1,\ldots,b_n\}$. There are altogether n! one-to-one mappings f of A onto B. If each such function f fails to be an isomorphism, it is because it fails to preserve some interpretation of some particular symbol in L. Whence there are a collection L′ of at most n! symbols in L such that no function f is an isomorphism with respect to L′. We show that this leads to a contradiction. To simplify the argument, let's suppose that L′ contains only a 2-place predicate P and a 2-place operator N. Let φ be a sentence of L′ which expresses the following facts about the model $\mathfrak{A}$:

There are n elements $a_1,\ldots,a_n$ such that:

diagram of $\{a_1,\ldots,a_n\}$ using identity and P;

for each element $a_i \in A$ and each pair $\langle C_1,C_2\rangle \in N(a_i)$, we have $Na_i(\bigvee_{x\in C_1} x = a_i, \bigvee_{x\in C_2} x = a_i)$; for each $a_i \in A$ and each pair $\langle C_1,C_2\rangle \notin N(a_i)$, we have $\neg Na_i (\bigvee_{x\in C_1} x = a_i, \bigvee_{x\in C_2} x = a_i)$.

Clearly φ holds in $\mathfrak{A}$. So it holds in $\mathfrak{B}$. Then it is easy to construct an isomorphism of $\mathfrak{A}$ onto $\mathfrak{B}$ which preserves P and N. ⊣

Let $\mathfrak{A}$ be a model for L and let N be an n-place operator symbol. Thus, for each $a \in A$,

$$N(a) \subset S(A) \times \ldots \times S(A) \qquad \text{n-times.}$$

Let $\varphi_i(xyz\ldots)$, $1 \leq i \leq n$, be any formulas of L and let $a,b,c,\ldots$ be arbitrary elements of A. We observe that the truth of

(1) $$\mathfrak{A} \models Nx\varphi_1 \ldots \varphi_n[abc\ldots]$$

depends on whether or not the n-tuple

(2) $$\langle C_1,\ldots,C_n\rangle \in N(a),$$

where

(3) $$C_i = \{a' \in A : \mathfrak{A} \models \varphi_i[a'bc\ldots]\}, \qquad 1 \leq i \leq n.$$

Exactly, how many times are we called upon to decide the truth of (1)? Precisely, $|L| + |A| + \omega$ number of times. From this we see that we only need the information supplied by at most $|L| + |A| + \omega$ n-tuples as in (2). This leads to the following definition. For each $a \in A$, we define the essential part $N^e(a)$ of $N(a)$ to be the subset of $N(a)$ consisting exactly of those n-tuples $\langle C_1,\ldots,C_n\rangle$ such that each C_i satisfies (3) for some formula $Nx\varphi_1 \ldots \varphi_n$ and some elements $b,c,\ldots$ in A. The essential part of N is the function N^e. The essential part of the model $\mathfrak{A}$ is the model $\mathfrak{A}^e$ which is exactly like $\mathfrak{A}$ except that each modal operator N is replaced by its essential part N^e. Note that $\mathfrak{A}^e$ is still a model for L. The following is obvious.

5.3 $\mathfrak{A} \equiv \mathfrak{A}^e$ and for each $a \in A$ and each N, $|N^e(a)| \leq |L| + |A| + \omega$.

A model $\mathfrak{A}$ is reduced iff $\mathfrak{A} = \mathfrak{A}^e$.

$\mathfrak{A}$ is said to be a submodel of $\mathfrak{B}$ (and $\mathfrak{B}$ an extension of $\mathfrak{A}$) if $\mathfrak{A}$ is a submodel of $\mathfrak{B}$ in all the usual respects and for each n-place operator N and each $a \in A$, each n-tuple in $N^{\mathfrak{A}}(a)$ is majorized by an n-tuple in $N^{\mathfrak{B}}(a)$, i.e., whenever $C_i \subset A$, $1 \leq i \leq n$, and $\langle C_1,\ldots,C_n\rangle \in N^{\mathfrak{A}}(a)$, then there are $D_i \subset B$, $1 \leq i \leq n$, such that $C_i = A \cap D_i$, and $\langle D_1,\ldots,D_n\rangle \in N^{\mathfrak{B}}(a)$. We write $\mathfrak{A} \subset \mathfrak{B}$ to mean that $\mathfrak{A}$ is a submodel of $\mathfrak{B}$. Note that $\subset$ is a reflexive, transitive, and antisymmetric relation. It follows that if $\mathfrak{A}$ is a submodel of $\mathfrak{B}$ in all the usual respects and for each $a \in A$ and N, we have $N^{\mathfrak{A}}(a) \subset N^{\mathfrak{B}}(a)$, then $\mathfrak{A} \subset \mathfrak{B}$. So, in particular, $\mathfrak{A}^e \subset \mathfrak{A}$. $\mathfrak{A}$ is said to be a <u>conservative submodel of</u> $\mathfrak{B}$ (and $\mathfrak{B}$ a <u>conservative extension of</u> $\mathfrak{A}$) iff $\mathfrak{A}$ is a submodel of $\mathfrak{B}$ and for each $a \in A$ and N,

$$\text{if } \langle D_1,\ldots,D_n\rangle \in N^{\mathfrak{B}}(a), \quad \text{then } \langle D_1 \cap A,\ldots,D_n \cap A\rangle \in N^{\mathfrak{A}}(a).$$

We use the symbols $\subset\subset$ for conservative extensions. The relation $\subset\subset$ is also reflexive, transitive, and anti-symmetric. If $\mathfrak{A} \subset\subset \mathfrak{B}$ and $A = B$, then $\mathfrak{A} = \mathfrak{B}$. Let

$$\mathfrak{A}_0 \subset \mathfrak{A}_1 \subset \ldots \subset \mathfrak{A}_n \subset \ldots, \qquad n \in \omega,$$

be a (simple infinite) chain of models for L. The union of the chain $\mathfrak{A} = \bigcup_n \mathfrak{A}_n$, is the model defined as follows: Again we skip over the usual parts of $\mathfrak{A}$. Let $a \in \bigcup_n A_n$ and for simplicity let N be a 1-place operator. Let m be any number such that $a \in A_m$. Consider all sequences

$$D_m \subset D_{m+1} \subset \ldots \subset D_n \subset \ldots$$

such that $D_n \in N^{\mathfrak{A}_n}(a)$, $n \geq m$, and such that $D_{n+1} \cap A_n = D_n$. Define $N^{\mathfrak{A}}(a)$ to be the set of all unions $\bigcup_n D_n$ of such sequences D_n, $n \geq m$, and for all possible m. It is clear that each member of the chain is a submodel of the union. Furthermore, if the $\mathfrak{A}_n$'s form a conservative chain, then the union is a conservative extension of each $\mathfrak{A}_n$. We record these facts as:

5.4 The union of a (conservative) chain of models $\mathfrak{A}_n$, $n \in \omega$, is a (conservative) extension of each $\mathfrak{A}_n$.

<u>Proof</u>. Straightforward.

It is easy to generalize the definition of $\bigcup_n \mathfrak{A}_n$ to unions of arbitrary chains $\mathfrak{A}_s$, $s \in S$, of models for L so that the result is an extension of each $\mathfrak{A}_s$. In general, it is impossible for an ω_1-union of a conservative chain to be a conservative extension of each member of the chain. To see this, Let $L = \{N\}$

where N is a 1-place modal operator. Let T be a Suslin tree. For each $\xi < \omega_1$, let A_ξ = the set of all $t \in T$ at level at most ξ. Then

$$A_0 \subset A_1 \subset \ldots \subset A_\xi \subset \ldots, \xi < \omega_1.$$

If $t \in A_\xi$, define $N^\xi(t) \subset S(A_\xi)$ by

> $N^\xi(t)$ = the set of all branches of A_ξ which contains t and contains an element at level ξ.

Let $\mathfrak{A}_\xi = \langle A_\xi, N^\xi \rangle$. Then it is clear that if $\xi \leq \eta$ then $\mathfrak{A}_\xi \subset\subset \mathfrak{A}_\eta$. Now $T = \cup_{\xi<\omega_1} A_\xi$ and there is no N^{ω_1} on T so that $\langle T, N^{\omega_1} \rangle$ is a conservative extension of each $\mathfrak{A}_\xi$. To see this, let $t \in A_0$. Since $N^0(t) \neq 0$, $N^{\omega_1}(t) \neq 0$. So let $B \in N^{\omega_1}(t)$. Then, if each $\mathfrak{A}_\xi \subset\subset \langle T, N^{\omega_1} \rangle$, we would have

$$B \cap A_\xi \in N^\xi(t) \quad \text{for all } \xi < \omega_1.$$

This implies that B is a branch of T of length ω_1.

We say that $\mathfrak{A}$ is an <u>elementary submodel</u> of $\mathfrak{B}$ and $\mathfrak{B}$ is an <u>elementary extension</u> of $\mathfrak{A}$ iff $\mathfrak{A} \subset \mathfrak{B}$ and for all formulas $\varphi(xyz\ldots)$ of L and all elements a,b,c,... in A,

$$\mathfrak{A} \models \varphi[abc\ldots] \text{ iff } \mathfrak{B} \models \varphi[abc\ldots].$$

We use the usual symbol $\mathfrak{A} \prec \mathfrak{B}$ to denote that $\mathfrak{A}$ is an elementary submodel of $\mathfrak{B}$. Clearly $\mathfrak{A}^e \prec \mathfrak{A}$. The relation $\prec$ is reflexive and transitive, and $\mathfrak{A} \prec \mathfrak{B}$ implies $\mathfrak{A} \equiv \mathfrak{B}$. There are, as far as we know, no simple natural extensions of the Tarski characterization for elementary submodels. However, the Tarski-Vaught theorem on elementary chains has a generalization in the following form.

5.5 Let $\mathfrak{A}_n$, $n < \omega$, be a conservative elementary chain of models, i.e., if $n \leq m$, then $\mathfrak{A}_n \prec \mathfrak{A}_{n+1}$ and $\mathfrak{A}_n \subset\subset \mathfrak{A}_{n+1}$. Then the union $\mathfrak{A} = \cup_n \mathfrak{A}_n$ is a conservative elementary extension of each $\mathfrak{A}_n$.

Proof. The proof proceeds just as in first order logic by induction on the formulas of L. The only possible difficulty will occur at formation rule (iv). Again, for simplicity, let N be a 1-place modal operator and let N^n be its interpretation in $\mathfrak{A}_n$ and N^ω be its interpretation in $\mathfrak{A}$. Let a,b,c,... be in A_n, and let $\varphi(xyz\ldots)$ be a formula of L. We wish to show that

$$(1) \qquad \mathfrak{A}_n \models Nx\varphi[abc\ldots] \text{ iff } \mathfrak{A} \models Nx\varphi[abc\ldots].$$

If $\mathfrak{A}_n \models Nx\varphi[abc\ldots]$, then the set

$$D_n = \{a' \in A_n : \mathfrak{A}_n \models \varphi[a'bc\ldots]\} \in N^n(a).$$

For each $m \geq n$, since $\mathfrak{A}_n < \mathfrak{A}_m$, the set

$$D_m = \{a' \in A_m : \mathfrak{A}_m \models \varphi[a'bc\ldots]\} \in N^m(a),$$

and furthermore, $D_{m+1} \cap A_m = D_m$. Whence, the set $D = \bigcup_{m \geq n} D_m \in N^\omega(a)$. By our inductive hypothesis on the formula φ,

$$(2) \qquad D = \{a' \in A : \mathfrak{A} \models \varphi[a'bc\ldots]\}.$$

Whence, $\mathfrak{A} \models Nx\varphi[abc\ldots]$. On the other hand, if $\mathfrak{A} \models Nx\varphi[abc\ldots]$, then let D be as in (2) and we have $D \in N^\omega(a)$. Since $\mathfrak{A}$ is a conservative extension of $\mathfrak{A}_n$, the set $D \cap A_n \in N^n(a)$. But, by induction,

$$D \cap A_n = \{a' \in A_n : \mathfrak{A}_n \models \varphi[a'bc\ldots]\}.$$

So $\mathfrak{A}_n \models Nx\varphi[abc\ldots]$. (1) is proved and the induction is complete. ⊣

The proof of 5.5 obviously generalizes to conservative elementary chains of arbitrary lengths provided that the union is a conservative extension of each member of the chain. This, however, as we have seen, is not an easy requirement to fulfill.

The following downward Löwenheim-Skolem theorem can either be proved directly by a version of the Tarski characterization of elementary submodels, or it can be proved by indirect means.

5.6 Let $\mathfrak{A}$ be a model for L. Let $\alpha = |A|$, let $|L| + \omega \leq \beta \leq \alpha$, and let X be a subset of A of power β. Then there is an elementary submodel $\mathfrak{B}$ of $\mathfrak{A}$ such that $X \subset B$ and $|B| = \beta$.

We shall now discuss the indirect proof of 5.6 using expansions and translations. Let $\mathfrak{A} = \mathfrak{A}^e$, and assume that $|L| + \omega \leq \alpha = |A|$. It follows that $|L| + \omega + |A| \leq |A|$. Since $\mathfrak{A} = \mathfrak{A}^e$, for each N and each $a \in A$, we have $|N(a)| \leq |A|$. Whence, we can attempt to replace the n-place operator N by one 1-place relation $\widetilde{N}$ and n 3-place relations $N_1^*, \ldots, N_n^*$, on A as follows: For each $a \in A$, $\widetilde{N}(a)$ iff $N(a) \neq 0$. To each sequence $\langle C_1, \ldots, C_n \rangle \in N(a)$, we assign an element (not necessarily unique) $b \in A$, with different sequences being assigned different b's. If $N(a) \neq 0$ every $b \in A$ is assigned. We define the 3-place relations N_i^* as follows: for all $c \in A$,

$$N_i^*(abc) \text{ iff } \langle C_1, \ldots, C_n \rangle \text{ is assigned to } b \text{ and } c \in C_i.$$

Let L^* be a first order language containing all the predicate, function, constant symbols of L and, in addition, the symbols $\widetilde{N}$, N_i^* for each $N \in L$. We say that L^* is the <u>expansion</u> of L. Given the model $\mathfrak{A}$ as above, let $\mathfrak{A}^*$ be the model obtained from $\mathfrak{A}$ by replacing all interpretations for modal operator symbols N by the relations $\widetilde{N}$, N_i^*, $1 \leq i \leq n$. We say that $\mathfrak{A}^*$ is the <u>expansion</u> of $\mathfrak{A}$. Obviously, $\mathfrak{A}^*$ is a model for L^*. The relationship between $\mathfrak{A}$ and $\mathfrak{A}^*$ will be made clear by the following translation.

For each formula φ of L, we define its <u>translation</u> φ^* a formula of L^* by induction on φ:

$$\varphi^* = \varphi \text{ if } \varphi \text{ is atomic;}$$

$$(\neg\varphi)^* = \neg\varphi^*, \ (\varphi \wedge \psi)^* = \varphi^* \wedge \psi^*, \text{ etc;}$$

$$((\forall x)\varphi)^* = (\forall x)\varphi^*, ((\exists x)\varphi)^* = (\exists x)\varphi^*;$$

$$(Nx\varphi_1 \ldots \varphi_n)^* = [(\exists y)\{\bigwedge_{1\leq i\leq n} (\forall z)(N_i^*(xyz) \leftrightarrow \varphi_i^*(\tfrac{x}{z}))\} \wedge \widetilde{N}(x)] \vee [x \neq x \wedge \neg\widetilde{N}(x)].$$

(In the last formula, y and z are new variables and $\varphi_i^*(\frac{x}{z})$ is obtained from φ_i^* by replacing all free occurrences of x by z.) A simple induction will establish the following:

5.7 Let $\varphi(xyz\ldots)$ be a formula of L and let $a,b,c,\ldots$ be in A. Then

$$\mathfrak{A} \models \varphi[abc\ldots] \text{ iff } \mathfrak{A}^* \models \varphi^*[abc\ldots].$$

Note that the second $\models$ sign is the classical satisfaction relation for models of first order logic. Note also that $\mathfrak{A} \equiv \mathfrak{B}$ does not necessarily imply that $\mathfrak{A}^* \equiv \mathfrak{B}^*$. Let $\mathfrak{B}^*$ be an elementary submodel of $\mathfrak{A}^*$ in the sense of L^*. We now reconstitute the interpretation of N on $\mathfrak{B}^*$ by defining: For $C_1,\ldots,C_n \subset B$, $a \in B$,

$$\langle C_1,\ldots,C_n\rangle \in N(a) \text{ iff for some } b \in B,\ C_i = \{c \in B : \mathfrak{B}^* \models N_i^*[abc]\},\ 1 \leq i \leq n.$$

Let $\mathfrak{B}$ be the model for L with the universe B and carrying with it the interpretations of all the symbols in L. Clearly, $\mathfrak{B}^*$ is the expansion of $\mathfrak{B}$. We now claim,

5.8 $\mathfrak{B}$ is an elementary submodel of $\mathfrak{A}$ in the sense of L.

<u>Proof</u>. It is easy to check that $\mathfrak{B} \subset \mathfrak{A}$. Let $\varphi(xyz\ldots)$ be a formula of L and let $a,b,c,\ldots$ be in B. Then

$$\mathfrak{B} \models \varphi[abc\ldots] \text{ iff } \mathfrak{B}^* \models \varphi^*[abc\ldots] \text{ iff}$$
$$\mathfrak{A}^* \models \varphi^*[abc\ldots] \text{ iff } \mathfrak{A} \models \varphi[abc\ldots].$$

$\dashv$

Since $|L| + \omega = |L^*| + \omega$, 5.6 follows from 5.8 by applying the classical downward Löwenheim-Skolem Theorem to the model $(\mathfrak{A}^e)^*$, and then use transitivity of $\prec$.

We next prove the compactness theorem and the upward Löwenheim-Skolem Theorem by using the same tricks of expansion and translation. We first define ultraproducts for models for L. Let $\mathfrak{A}_i$, $i \in I$, be models for L and let D be an ultrafilter over I. Let $f \in \Pi_{i\in I} A_i$ and we define in the usual way the equivalence class f_D of f modulo D. Let

$$\Pi_D A_i = \{f_D : f \in \Pi_{i\in I} A_i\}.$$

We shall now define the ultraproduct $\mathfrak{A} = \Pi_D \mathfrak{A}_i$ of the models $\mathfrak{A}_i$, $i \in I$, modulo the ultrafilter D. We take the usual interpretations of the predicate, function, and constant symbols in L. Let N be an n-place modal operator, and let N^i be its interpretation in the model $\mathfrak{A}_i$. The interpretation of N in the ultraproduct $\mathfrak{A}$ is defined as follows: Let $f \in \Pi_{i\in I} A_i$. For each $i \in I$, let $F_i \in N^i(f(i))$. We may write each F_i as

$$F_i = \langle C_1^i,\ldots,C_n^i\rangle.$$

Define, for $1 \leq j \leq n$,

$$C_{jD} = \{g \in \Pi_{i\in I} A_i : \{i \in I : g(i) \in C_j^i\} \in D\},$$

and

$$F_D = \langle C_{1D},\ldots,C_{nD}\rangle.$$

Finally, let

$$N^{\mathfrak{A}}(f_D) = \{F_D : F \in \Pi_{i\in I} N^i(f(i))\}.$$

5.9. Let $\mathfrak{A}_i$, $i \in I$, be models for L, D be an ultrafilter over I, $\varphi(xyz\ldots)$ be a formula of L, and $a,b,c,\ldots$ be in $\Pi_{i\in I} A_i$. Then

$$\Pi_D \mathfrak{A}_i \models \varphi[a_D b_D c_D \ldots] \text{ iff } \{i \in I : \mathfrak{A}_i \models \varphi[a(i)b(i)c(i)\ldots]\} \in D.$$

Proof. 5.9 can be proved by induction on φ and by using a direct computation based on the definition of ultraproducts. We shall give only an idea why the theorem is true (at least if each $|A_i| \geq |L| + \omega$) by using expansions and translations. First, let $\mathfrak{A}_i^*$, $i \in I$, be the expansions of $\mathfrak{A}_i$. Form the ordinary ultraproduct $\Pi_D \mathfrak{A}_i^*$. It is easily verified that

$$\Pi_D \mathfrak{A}_i^* \text{ is an expansion of } \Pi_D \mathfrak{A}_i.$$

Now, let $\varphi^*(xyz\ldots)$ be the translation of φ. Then

$$\Pi_D \mathfrak{A}_i \models \varphi[a_D b_D c_D \ldots] \text{ iff}$$

$$\Pi_D \mathfrak{A}_i^* \models \varphi^*[a_D b_D c_D \ldots] \text{ iff}$$

$$\{i \in I : \mathfrak{A}_i^* \models \varphi^*[a(i)b(i)c(i)\ldots]\} \in D \text{ iff}$$

$$\{i \in I : \mathfrak{A}_i \models \varphi[a(i)b(i)c(i)\ldots]\} \in D.$$

⊣

5.10 Let Σ be a set of sentences of L. Σ has a model if and only if every finite subset of Σ has a model.

<u>Proof</u>. Use 5.9 and the usual argument.

If all $\mathfrak{A}_i = \mathfrak{A}$, then the ultraproduct $\Pi_D \mathfrak{A}_i$ becomes the <u>ultrapower</u> $\Pi_D \mathfrak{A}$. The natural embedding $d : \mathfrak{A} \to \Pi_D \mathfrak{A}$ defined by

$$d(a) = (\text{the constant function } a)_D$$

is an elementary embedding of $\mathfrak{A}$ into $\Pi_D \mathfrak{A}$.

5.11 Let $\mathfrak{A}$ be an infinite model for L. Then $\mathfrak{A}$ has arbitrarily large elementary extensions.

<u>Proof</u>. Take index set I and ultrafilter D so that $|\Pi_D A| = |A|^{|I|}$.

⊣

We should mention that versions of 5.6, 5.10, and 5.11 were obtained for models of pragmatics by Victery in [2]. He, however, did not realize that they could be proved by considering $\mathfrak{A}^*$ and φ^*.

This beginning list of results in modal model theory probably gives the misleading impression that one can now keep on churning out new model-theoretic results one after the other, and, at worst, one can always fall back on $\mathfrak{A}^*$ and φ^* and reduce things to first order model theory. This is not so. The sample questions below are not answered. All of them have answers in first order logic, or if the interpretations of the model operators N are trivial or very simple. The point is to find the largest classes of models for L which are "natural" and in which some answers may be provided.

1. Is there a normal form theorem for the language L?

2. Are there preservation theorems for

   submodels,

   extensions,

   unions of chains,

   homomorphic images,

   direct products,

   reduced products, etc.?

3. Are there interpolation, consistency, and definability theorems for L?

4. How can one define saturated and special models? And are there any applications?

5. Is it true that two models for L are elementarily equivalent if and only if they have isomorphic ultrapowers? (We may have to restrict this question to models $\mathfrak{A} = \mathfrak{A}^e$.)

6. Finally, does any of this model theory contribute to a better understanding of modal logic?

REFERENCES

[1] Montague, Richard. Pragmatics, in Contemporary Philosophy, a survey, ed., Raymond Klibansky, Florence, 1968.

[2] Victery, Donald. Carnap Prize Essay, Department of Philosophy, UCLA, 1971.

University of California, Los Angeles

XXII. A PRESERVATION THEOREM FOR INTERPRETATIONS

K. Jon Barwise
University of Wisconsin, Madison

A question arose in connection with some work on the foundations of abstract logic[1] which seems to have been overlooked even for the usual first order predicate logic. It is best understood by means of some specific examples.

Let π be an interpretation[2] of number theory in ZF set theory, number theory being formulated in terms of 0, +, x; set theory in terms of $\in$. If a sentence φ in the language of set theory has the property that for all models $\mathfrak{A}$ of ZF, the truth of $\mathfrak{A} \models \varphi$ depends only on the integers of $\mathfrak{A}$, is there a sentence ψ of number theory such that $ZF \vdash (\varphi \leftrightarrow \psi^{\pi})$? Or suppose that the truth of $\mathfrak{A} \models \varphi$ depends only on $L^{\mathfrak{A}}$, the constructible sets of $\mathfrak{A}$, for each $\mathfrak{A} \models ZF$. Is there a θ in the language of set theory such that $ZF \vdash (\varphi \leftrightarrow \theta^{(L)})$?

THEOREM. Let π be an interpretation of a language L_0 in a theory T, T being formulated in a language L. For any sentence φ of L the following are equivalent:

(1) For some sentence ψ of L_0

$$T \vdash (\varphi \leftrightarrow \psi^{\pi}) ,$$

(2) For all models $\mathfrak{A}, \mathfrak{B}$ of T with ${}^{\pi}\mathfrak{A} = {}^{\pi}\mathfrak{B}$,

$$\mathfrak{A} \models \varphi \text{ iff } \mathfrak{B} \models \varphi .$$

---

[1]The results in abstract logic presented in Cambridge will appear in Abstract First Order Logic, a paper to appear in the Annals of Mathematical Logic. Our work on the foundations of abstract logic is in too tentative a state for publication in this volume.

[2]The reader is free to use his favorite definition of "interpretation". If he does not have a favorite we suggest the one given in Section 2. 7 of Enderton [3].

Research partially supported by grant NSF GP-27633.

Here ψ^{π} is the interpretation of ψ in L, ${}^{\pi}\mathfrak{A}$ is the natural structure for the language L_0 induced by π and $\mathfrak{A}$. This theorem answers the above questions, and a number of related ones, affirmatively.

COROLLARY. Let π be an interpretation of a complete theory T_0 of L_0 in an incomplete theory T of L. There are models $\mathfrak{A}$ and $\mathfrak{B}$ of T with ${}^{\pi}\mathfrak{A} = {}^{\pi}\mathfrak{B}$ but $\mathfrak{A} \not\equiv \mathfrak{B}$. In fact, for any φ independent of T there are models $\mathfrak{A}$ and $\mathfrak{B}$ of T with

$$\mathfrak{A} \models \varphi$$
$$\mathfrak{B} \models \neg\varphi$$
$${}^{\pi}\mathfrak{A} = {}^{\pi}\mathfrak{B}$$

For an application of this corollary let $T_0 = \mathrm{Th}(\langle \text{Ordinals}, < \rangle)$, T = ZFL, $L_0 = \{<\}$, $L = \{\in\}$, π the usual interpretation of T_0 in T. Thus there are models of ZFL with the same ordinals which are not elementary equivalent. (This is due to Rosenthal [5], his proof used the strong undecidability of ZF.)

The theorem follows from the following interpolation lemma, the proof of which is an application of Feferman's many-sorted interpolation theorem.

INTERPOLATION LEMMA. Let π_1, π_2 be interpretations of L_0 in theories T_1, T_2 in languages L_1, L_2 and suppose $\mathcal{K}_1 \cap \mathcal{K}_2 = \emptyset$, where

$$\mathcal{K}_i = \{ {}^{\pi_i}\mathfrak{A} : \mathfrak{A} \text{ is an } L_i\text{-structure which is a model of } T_i \} .$$

There is a sentence φ of L_0 such that $\mathcal{K}_1 \subseteq \mathcal{K}$, $\mathcal{K}_2 \cap \mathcal{K} = \emptyset$, where

$$\mathcal{K} = \{ \mathfrak{A} : \mathfrak{A} \text{ an } L_0\text{-structure, } \mathfrak{A} \models \varphi \} .$$

The usual interpolation theorem is a special case of the above where the interpretations are the identity maps of L_0 into L_1 and L_2.

To prove the theorem from the lemma let $T_1 = T \cup \{\varphi\}$, $T_2 = T \cup \{\neg\varphi\}$, $\pi_1 = \pi_2 = \pi$, $L_1 = L_2 = L$.

To prove the lemma we may assume that the nonlogical symbols of L_0, L_1, L_2 are disjoint. We introduce two new sorts of variables: a, b, c, ... (sort 1) and u, v, w, ... (sort 2), the old variables x, y, z, ... being sort 0. Since $\mathcal{K}_1 \cap \mathcal{K}_2 = \emptyset$ the following many sorted theory $S_{01} \cup S_{02}$ is inconsistent. S_{01} is the theory which consists of all $\varphi^{(1)}$ for $\varphi \in T_1$ and axioms which insure that the sort 0-structure $\mathfrak{A}$ of a model $(\mathfrak{A}, \mathfrak{B})$ of S_{01} is ${}^{\pi_1}\mathfrak{B}$ for the sort 1 structure $\mathfrak{B}$. (Here $\varphi^{(1)}$ is φ just rewritten using sort 1 variables.) Examples of the second kind of sentence in S_{01} are:

$$\forall x\, \exists a(x = a \wedge \sigma^{(1)}(a))$$

$$\forall a\, (\sigma^1(a) \rightarrow \exists x\, (x = a))$$

$$\forall x, y\, [R(x, y) \longleftrightarrow \exists a\, \exists b\, (\, x = a \wedge y = b \wedge \theta^{(1)}(a, b))]$$

where $\sigma(x)$ is the π_1-interpretation of $\forall$, R is a typical binary relation symbol of L_0 and $\theta(x,y)$ its π_1-interpretation. S_{02} is the theory which consists of all $\varphi^{(2)}$ for $\varphi \in T_2$ and axioms which insure that the sort 0-structure $\mathfrak{A}$ of a model $(\mathfrak{A}, \mathcal{L})$ of S_{02} is ${}^{\pi_2}\mathcal{L}$ for the sort 2 structure $\mathcal{L}$. By compactness there are sentences σ_{01}, a conjunction of members of S_{01}, and σ_{02}, a conjunction of members of S_{02}, such that

$$\sigma_{01} \rightarrow \neg\sigma_{02}$$

is valid. The Feferman interpolation theorem tells us that there is a sentence φ involving only symbols and sorts common to σ_{01} and σ_{02} such that $\sigma_{01} \rightarrow \varphi$ and $\varphi \rightarrow \neg\varphi$ are valid. The sentence φ has only sort 0 variables and symbols from L_0 so it is the desired sentence since $S_{01} \vdash \varphi$, $S_{02} \vdash \neg\varphi$ just means $\mathcal{K}_1 \subseteq$ Models of φ, $\mathcal{K}_2 \cap$ Models of $\varphi = \phi$. $\dashv$

If we go back to the example we gave above we see that we can do better using the interpolation lemma. If T_1 and T_2 are any consistent extension of ZFL then we can get models $\mathfrak{A}_1 \models T_1$, $\mathfrak{A}_2 \models T_2$ where $\mathfrak{A}_1$ and $\mathfrak{A}_2$ have the same ordinals. We can do better still if we turn to infinitary extension of the above results.

The theorem, corollary and lemma stated above carry over from first order predicate logic to any logic where one has the many-sorted interpolation theorem of Feferman and enough compactness to handle the theories involved; in particular, for any admissible fragment L_A of $L_{\omega_1\omega}$, A countable, with theories which are Σ_1 in X for some $X \subseteq A$ with $\langle A, \in, X\rangle$ admissible. Using this, and techniques from Barwise [1], one can show that given any countable model $\mathfrak{A}$ of ZF there are $2^{\aleph_0}$ countable end extensions $\mathcal{L}$ of $\mathfrak{A}$, $\mathcal{L} \models$ ZFL, such that no two are elementary equivalent but all of them have the same $\langle$ Ordinals, $<, +, \times\rangle$. (This last makes use of some results from section 4 of Chang [2].) This considerably improves a result in Rosenthal [5] to the effect that if ZF has an ω-model then ZFL has nonisomorphic models $\mathfrak{A}, \mathcal{L}$ with the same $\langle$ Ordinals, $<, +, \times\rangle$.

We have probably overemphasized set theory in this note. After all, interpretations are an old tool in logic, both in undecidability and relative consistency results. It would be nice to see applications of the above results to other theories.

REFERENCES

1. Barwise, K. J., Infinitary methods in the model theory of set theory, Logic Colloquium '69, Amsterdam 1971.

2. Chang, C. C., Infinitary properties of models generated from indiscernibles, Proceedings of 1967 Int'l Cong. Logic Math. and Phil. Sci., North Holland, 1968, pp. 9-21.

3. Enderton, H. B., A mathematical introduction to logic, New York and London, 1972.

4. Feferman, S., Lectures on proof theory, Proceedings of the summer school in logic, Leeds, 1967, Springer lecture notes in mathematics, Berlin, 1968.

5. Rosenthal, J., Relations not determining the structure of L, Pacific J. Math, 37 (1971), pp. 497-514.

XXIII. Vaught sentences and Lindström's regular relations

M. Makkai.*

Introduction. In this paper we generalize the results of our [14] to the language $L_{\omega_1, \omega}$. As [14] was based primarily on methods of Svenonius [20], the present paper is based on recent work of Robert Vaught [21] generalizing both Svenonius [20] and classical descriptive set-theoretic methods and results to the model theory of $L_{\omega_1, \omega}$.

Vaught considers infinitary sentences of the form:

$$\forall x_0 \bigwedge i_0 \in I \ \exists y_0 \bigvee j_0 \in I \ \cdots \ \forall x_n \bigwedge i_n \in I \ \exists y_n \bigvee j_n \in I \ \cdots$$
$$\bigwedge n < \omega \ N^{i_0 j_0 \cdots i_n j_n} (x_0, y_0, \ldots, x_n, y_n)$$

where I is a countable set and $N^{i_0 - j_n}$ is a formula of $L_{\omega_1, \omega}$ with at most $x_0, y_0, \ldots, x_n, y_n$ free, for any $\langle i_0, \ldots, j_n \rangle \in I^{\omega}$.(1) Such a sentence is called here a Vaught sentence. We note that we will slightly change the appearance of the ω-type prefix of Vaught sentences. It is easy to see that the class of models of a Vaught sentence is a Σ_1^1-class i.e. one defined by a Σ_1^1 formula of the form $\exists S_1 \cdots S_n \cdots \varphi$ where $S_1, \ldots, S_n, \ldots$ are second order variables and $\varphi \in L_{\omega_1, \omega}$.(2) Vaught's normal form theorem (see Corollary 3.3 [21]) is a partial converse of this; it says that any Σ_1^1 class has the same countable (including finite) members as the class of models of some Vaught sentence. A related result (though in a more special context not mentioning $L_{\omega_1, \omega}$) was obtained by Moschovakis [16].

We generalize Vaught's theorem to a normal form theorem for Σ_1^1 classes closed under a particular but arbitrary regular relation R.(3) As a special case, we obtain that, if we disregard uncountable structures, a

---

* The author's research was supported by the National Research Council of Canada.

$\underset{\sim}{\Sigma}_1^1$ class is closed under homomorphisms iff it is definable by a positive Vaught sentence, i.e. where the $N^{\underline{k}}$ are positive (actually, the $N^{\underline{k}}$ can be chosen to be finitary and quantifierfree in addition). We note, however, that this special case and others can be proved directly (though in essentials very similarly to our proof of the general theorem) without use of Lindström games (for the latter, see below). In the case of homomorphic images, the direct proof will be much simpler than the proof of the general theorem.

From the normal form theorem we derive, using results of Vaught, a general interpolation theorem for regular relations which in turn implies a general preservation theorem. (Corollaries 2.10, 2.11). We also obtain an apparently new result for homomorphisms and positive sentences of $L_{\omega_1,\omega}$ as a special case (Corollary 2.12).

As we will point out, our corollaries can be proved by other methods without use of the main theorem. The proofs via the main theorem are, however, still of interest as being "purely model-theoretic" contrasted to the other known proofs all having various degrees of a proof-theoretic character.

Vaught sentences are generalizations of Svenonius sentences. These are of the form

$$Q_0 v_0 Q_1 v_1 \cdots Q_n v_n \cdots \bigwedge_n N^n(v_0, \cdots, v_n) \tag{2}$$

where the N^n are finitary formulas. Svenonius' theorem [20] is the corresponding normalform theorem for finitary logic and it is equivalent to saying that if the $\underset{\sim}{\Sigma}_1^1$ class K is defined by $\exists \vec{S} \varphi$ where φ is a countable conjunction of finitary sentences (let such a K be called <u>finitary</u> $\underset{\sim}{\Sigma}_1^1$), then there is a Svenonius sentence whose countable models are exactly the countable elements of K. In [14] we gave a related normalform theorem for finitary $\underset{\sim}{\Sigma}_1^1$ classes closed under a given regular relation. Though strictly speaking the results of our earlier [14] are not consequences of the present ones, the proofs in [14] can be obtained as simplifications of the present ones. In some details, the present treatment improves [14], see e.g. the proof of 2.3 and 2.5 via 2.1.

In the formulation of some of our statements, we will include more refined versions involving admissible fragments of $L_{\omega_1, \omega}$ (see [1] or [9].

The reader not interested in admissible fragments can simply ignore all reference to them without impairing the arguments.

Finally, as a word of warning, we have to note that in many cases the general preservation theorem we obtain as Corollary 2.11 does not yield the familiar, natural syntactic forms of sentences we obtain by other methods, though in the cases considered by the author (e.g., $\mathfrak{A} R \mathfrak{B} \Leftrightarrow \mathfrak{B}$ is an endextension of $\mathfrak{A}$) the equivalence of the two syntactic classes can be shown in an elementary way.

We would like to express our thanks to Professor Robert Vaught for his valuable comments on a previous version of this paper, especially for pointing out that his improved version of his treatment, "working for arbitrary admissible fragments", can be extended to the situation in this paper.

§1. Preliminaries

Our terminology follows standard usage. A "formula" or "sentence" is meant to be one of $L_{\omega_1, \omega}$ if not said explicitly otherwise. A formula is assumed to have finitely many free variables. All formulas are formed using only v_n for $n < \omega$ as free or bound variables.

Infinitary sentences of the following form

$$\bigwedge_{i_0 \in I} \bigvee_{j_0 \in I} \bigwedge_{k_0 \in I} Q_0 v_0 \cdots \bigwedge_{i_n \in I} \bigvee_{j_n \in I} \bigwedge_{k_n \in I} Q_n v_n \cdots \bigwedge_{n < \omega} N^{i_0 j_0 k_0 \cdots i_n j_n k_n}(v_0, \ldots, v_n) \quad (1)$$

are called <u>Vaught sentences</u>; here I is a fixed countable set (see also footnote (1)), each Q_n is either $\forall$ or $\exists$, and for any $i_0, \dots, k_n \in I$, $N^{i_0 \dots k_n}$ is a formula (of $L_{\omega_1, \omega}$) with the free variables $v_0, \dots, v_n$ at most.

Let Φ be the Vaught sentence displayed in (1). The definition of when the structure $\mathfrak{A}$ satisfies Φ is suggested by the notation of Φ. In fact, this definition is obtained from that of the notion of satisfaction for sentences with quantifier prefixes of length ω by interpreting $\bigwedge_{k \in I}$ ($\bigvee_{k \in I}$) as a universal (existential) quantifier on the "variable" k ranging over I. For sentences with ω-type, or more general, prefixes see e.g. [8] or [14]. In particular, we will assume that it is understood what is meant by a <u>winning strategy for</u> "$\mathfrak{A} \models \Phi$", c.f. also [14].

Following Vaught [21], we define, for given Φ as in (1) the formulas

$$\varphi_\upsilon^{i_0 j_0 k_0 \dots i_{n-1} j_{n-1} k_{n-1}} (v_0, \dots, v_{n-1})$$

for any $n < \omega$, i_ℓ, j_ℓ, k_ℓ in I $(\ell < n)$ and for any ordinal $\upsilon < \omega_1$ by induction on υ as follows:

$$\varphi_0^{\emptyset} =_{df} \bigwedge \emptyset \quad \text{(identically true)}$$

($\emptyset$ is the empty sequence and the empty set of formulas)

$$\varphi_0^{i_0 \dots k_n} =_{df} N^{i_0 \dots k_n};$$

for $\underline{k} = i_0 j_0 k_0 \dots i_{n-1} j_{n-1} k_{n-1}$ with all the members of $\underline{k}$ in I ($\underline{k}$ may be the empty sequence), and for any $\upsilon < \omega_1$ and any limit ordinal

$\lambda < \omega_1$, we define

$$\varphi^k_{\upsilon+1} =_{df} \bigwedge_{i_n \in I} \bigvee_{j_n \in I} \bigwedge_{k_n \in I} Q_n v_n \varphi^{k\,i_n j_n k_n}_{\upsilon}$$

(here we used the abbreviation $\bigwedge_{k \in I} \psi^k$ for $\bigwedge\{\psi^k : k \in I\}$ and the other one with $\bigwedge$ replaced by $\bigvee$),

$$\varphi^k_\lambda =_{df} \bigwedge\{\varphi^k_\upsilon : \upsilon < \lambda\} .$$

Put $\varphi_\upsilon = \varphi^{\emptyset}_\upsilon$ $(\upsilon < \omega_1)$. We will refer to φ_υ as the υth ω_1, ω-approximation of Φ

Let us say that Φ is in the admissible set A if $\underline{N} = \langle N^{i_0 \cdots k_n} : i_0, \cdots, k_n \in I\rangle$ amd $\langle Q_n : n < \omega\rangle$ are in A . It is clear that in this case $\varphi_\upsilon \in L_A$ for every $\upsilon <$ Ord (A).

In the next lemma we state three facts concerning ω_1, ω-approximations; they were proved by Vaught along with other important related results (see 4.2(b), (e) and 4.10 in [21]). In 1.1, φ_υ is the υth ω_1, ω-approximation of the Vaught sentence Φ .

Lemma 1.1. (i) $\Phi \models \varphi_\upsilon$ for every $\upsilon < \omega_1$.

(ii) If ψ is a sentence in $L_{\omega_1, \omega}$ and $\Phi \models \psi$ (in fact, if every countable model of Φ is a model of ψ), then $\varphi_\lambda \models \psi$ for some $\lambda < \omega_1$. If, in addition, $\mathfrak{G}$ is an admissible subset of HC, $\Phi \in \mathfrak{G}$ and $\psi \in L_{\mathfrak{G}}$ then λ can be chosen $<$ Ord $(\mathfrak{G})$.

(iii) For any countable $\mathfrak{A}$, $\mathfrak{A} \models \Phi$ iff $\mathfrak{A} \models \bigwedge\{\varphi_\upsilon : \upsilon < \omega_1\}$.

Now we reproduce the notion of a regular relation between structures, introduced by Lindström [10]. Let $\langle P_n : n<\omega\rangle = p$ be a sequence of quantifier symbols $P_n = \forall$ or $P_n = \exists$ and let Γ be a countable set of formulas. Though in the original definition by Lindström the elements of Γ are finitary, we do not need this restriction here. We refer to $G = (p, \Gamma)$ as a Lindström game. We note that the original notion of a Lindström game is slightly more general (see also [14] where p as given here is called a simple prefix in contrast to more complex objects used in the general notion). However, the extension of our definitions and results to the general case presents only notational difficulties.

Let $\mathfrak{A}$ and $\mathfrak{B}$ similar structures appropriate for Γ. Let $I = \langle I_n : 1 \le n < \omega \rangle$ be a sequence of relations $I_n \subset A^n \times B^n$. We write $a_0, \cdots, a_{n-1} \, I_n \, b_0, \cdots, b_{n-1}$ for $\langle \langle a_0, \cdots, a_{n-1} \rangle, \langle b_0, \cdots, b_{n-1} \rangle \rangle \in I_n$. For the sake of conciseness in 1.2, let us define $\emptyset I_0 \emptyset$ to hold for the empty sequence $\emptyset$.

Definition 1.2. I is a winning strategy for "$\mathfrak{A}R(G)\mathfrak{B}$" if for every $n < \omega$ the following (i), (ii) and (iii) are true:

(i) If $P_n = \exists$ and $a_0, \cdots, a_{n-1} \, I_n \, b_0, \cdots, b_{n-1}$, then $\forall a \in A \, \exists b \in B \; a_0, \cdots, a_{n-1}, a \; I_{n+1} \; b_0, \cdots, b_{n-1}, b$,

(ii) If $P_n = \forall$ and $a_0, \cdots, a_{n-1} \, I_n \, b_0, \cdots, b_{n-1}$, then $\forall b \in B \, \exists a \in A \; a_0, \cdots, a_{n-1}, a \; I_{n+1} \; b_0, \cdots, b_{n-1}, b$,

(iii) If $\gamma(v_0, \cdots, v_n) \in \Gamma$ with the free variables among $v_0, \cdots, v_n$, then $a_0, \cdots, a_n \, I_{n+1} \, b_0, \cdots, b_n$ and $\mathfrak{A} \models \gamma[a_0, \cdots, a_n]$ imply $\mathfrak{B} \models \gamma[b_0, \cdots, b_n]$.

Given a similarity type τ containing all non-logical symbols in Γ, define a binary relation $R(G)$ (or more particularly, $R_\tau(G)$) between structures of type τ as follows:

$\mathfrak{A}R(G)\mathfrak{B} \Leftrightarrow_{df}$ there is a winning strategy for "$\mathfrak{A}R(G)\mathfrak{B}$".

We say that G is associated with R, a binary relation between structures of type τ, if $R \subseteq R(G)$ and for countable $\mathfrak{A}$ and $\mathfrak{B}$, $\mathfrak{A}R(G)\mathfrak{B}$ iff $\mathfrak{A}R\mathfrak{B}$. Finally, we call R regular if there is some Lindström game associated with R. R. Lindström [10] has shown that many relations for which preservation theorems have been proved are regular (e.g., all relations treated in [13] are regular). In [14], we pointed out that the relation of $\mathfrak{B}$ being an endextension of $\mathfrak{A}$ (c.f. Feferman-Kreisel [3]) is regular too. On the other hand, at least directly the results below do not give the natural preservation theorem in this case (c.f. Feferman [7], also [13]).

Digression.

In the rest of this § we would like to point out some facts in connection with Vaught sentences and regular relations. We will not use them in §2 , except for (1.7 and) 1.8, but they should indicate the importance of the notions concerned. On the other hand, these facts must be known to many people.

The first remark is due in a special case to S. Shelah [19] (see Lemma 5 in [19]) and it says briefly that the two person game associated with a Vaught sentence is determinate.[4] In other words, the negation of the Vaught sentence in (1) is logically equivalent to

$$\bigvee_{i_0} \bigwedge_{j_0} \bigvee_{k_0} \bar{Q}_0 v_0 \cdots \bigvee_{n < \omega} \neg N^{i_0 \cdots k_n} \tag{2}$$

where $\bar{Q}$ is $\forall$ or $\exists$ according to Q is $\exists$ or $\forall$. The proof of this is the same as that of Lemma 5 in [19]. This leads to the following:

Proposition 1.3. (i) Let Φ be a Vaught sentence and K the class of all models of Φ . Let K' be any $\sum_{\sim 1}^1$ class such that every countable structure in K' is in K . Then the whole of K' is contained in K .

(ii) Let the class of countable members of K be called the countable part of K . Let K_0 be the countable part of some $\sum_{\sim 1}^1$ class. Then there is a largest $\sum_{\sim 1}^1$-class K whose countable part is K_0 . In fact, K is axiomatized by some, and any, Vaught sentence the class of countable models of which is K_0 .

(iii) The classes defined by two Vaught sentences coincide if their countable members are the same.

Proof of (i). Assume that $\mathfrak{A} \in K' - K$. Then $\mathfrak{A}$ is the model of (2) and a Σ_1^1-formula σ defining K. The form of (2) allows us to use a downward Löwenhein-Skolem argument and to show that there is a countable model of (2) and σ, contradicting the hypothesis, q.e.d.

(ii) and (iii) follow from (i) and Vaught's theorem mentioned in the Introduction. The maximality property (ii) and the uniqueness property (iii) give "invariant" significance to Σ_1^1 classes defined by Vaught-sentences.

Next we turn to relations between structures. We consider a many-sorted version $L^{(m)}$ of $L_{\omega_1, \omega}$ with denumerable many different sorts of variables. Among these there are two distinguished sorts the variables of which are denoted by $a, a_1, \cdots$ and $b, b_1, \cdots$, respectively. For simplicity, there are no operation symbols in $L^{(m)}$. Each predicate symbol has a definite sort assigned to each of its places and there are enough predicate symbols of all possible finitary kinds. When forming formulas of $L^{(m)}$, only the right sort of variables may occupy any given place of a predicate symbol. It should be obvious how we define the syntax of $L^{(m)}$ precisely. As for the semantics, we only note that in every possible interpretation of a part of the sorts and predicates of $L^{(m)}$, i.e., in every many sorted structure, the range of every sort of variable is supposed to be nonempty, but otherwise it may be a completely arbitrary set.

We are going to consider many-sorted structures of the form $(\mathfrak{A}, \mathfrak{B}, F)$ where $\mathfrak{A}, \mathfrak{B}$ are structures of the ordinary similarity type τ. In this, the range of the variables $a, a_1, \cdots$ is $A = |\mathfrak{A}|$, that of

$b, b_1, \cdots$ is $B = |\mathfrak{B}|$. To every predicate P in $\tau^{(5)}$, there correspond two predicates in $(\mathfrak{A}, \mathfrak{B}, F)$, one with all places of sort "a" , the other of "b" ; the interpretation of the first one is $P^{\mathfrak{A}}$, of the second one $P^{\mathfrak{B}}$. F represents the interpretations of some additional sorts and predicates.

Definition 1.4. A relation R between structures of type τ is called $\underset{\sim}{\Sigma}_1^1$, if there is a many-sorted ω_1, ω-sentence Φ of $L^{(m)}$ such that $\mathfrak{A}R\mathfrak{B}$ iff there are appropriate F such that $(\mathfrak{A}, \mathfrak{B}, F)$ is a model of Φ . R is finitary $\underset{\sim}{\Sigma}_1^1$ ($f-\underset{\sim}{\Sigma}_1^1$) if in addition Φ is a conjunction of finitary sentences, and strictly finitary $\underset{\sim}{\Sigma}_1^1$ ($sf-\underset{\sim}{\Sigma}_1^1$) if Φ is a single finitary sentence.

There are a number of relations which are "naturally" $sf-\underset{\sim}{\Sigma}_1^1$, e.g. "$\mathfrak{B}$ is a homomorphic image of $\mathfrak{A}$", "$\mathfrak{B}$ is a direct factor of $\mathfrak{A}$", "$\mathfrak{B}$ is an endextension of $\mathfrak{A}$" . There is a slightly more general possible definition of $\underset{\sim}{\Sigma}_1^1$ ($f-\underset{\sim}{\Sigma}_1^1$, $sf-\underset{\sim}{\Sigma}_1^1$) relations in terms of relativized reducts and avoiding manysorted languages. But we think that the present definition is more natural.

We want to point out a connection between Svenonius sentences and the defining scheme of a regular relation. Let us call Ψ a two-sorted Svenonius sentence if it is of the form (2) in the Introduction but every v_n is either of sort "a" or of sort "b" , moreover every N^n is a finitary two-sorted formula appropriate for two-sorted structures of the form $(\mathfrak{A}, \mathfrak{B})$ where $\mathfrak{A}$ and $\mathfrak{B}$ are ordinary structures of type τ . We can define two-sorted Vaught sentences similarly. Now, the relation $R(G)$, for any Lindström game $G = (p, \Gamma)$ with all elements of Γ being finitary is defined by a two-sorted Svenonius sentence. E.g. if $p = \langle \exists\forall\exists\forall\cdots\rangle$,

this sentence is

$$\forall a_0 \exists b_0 \forall b_1 \exists a_1 \forall a_2 \exists b_2 \cdots \bigwedge_{\gamma \in \Gamma} [\gamma^a[a_0, \ldots, a_n] \to \gamma^b[b_0, \ldots, b_n]]$$

where $\gamma^a(\gamma^b)$ denotes the formula obtained from γ by changing every bound variable to sort "a" (sort "b"). Even if the elements of Γ are not finitary, R(G) is defined by a generalized two-sorted Svenonius sentence with the conjuncts in $L_{\omega_1, \omega}$, so a fortiori R(G) is defined by a two-sorted Vaught sentence.

Though we do not know how to derive the next proposition from the original forms of the Svenonius and Vaught normal-form theorems directly, by similar proofs we have

<u>Proposition 1.5.</u> Any $\underset{\sim}{\Sigma}^1_1$ (f-$\underset{\sim}{\Sigma}^1_1$) relation coincides with one defined by a two-sorted Vaught sentence (two-sorted Svenonius sentence) when both are restricted to countable structures.

The many-sorted analog of 1.3 can be proved similarly to 1.3.; we state only the consequence for regular relations.

<u>Proposition 1.6.</u> Suppose that the relation R is regular and the Lindström-game G is associated with R . Then $\bar{R}(G)$ is the largest $\underset{\sim}{\Sigma}^1_1$-relation which coincides with R for countable structures. Thus R(G) (though not G) is determined uniquely by R.

Note that the relations R(G) are reflexive and transitive. We do not know if every reflexive and transitive relation defined by a two-sorted Svenonius sentence is of the form R(G) for some Lindström-game G . It is possible that the construction of the relations R(G) should be generalized to have such a result. Using an argument as in the proof of 1.3(i), it is easy to see that the "Vaught extension" of a $\underset{\sim}{\Sigma}^1_1$ relation R, asserted to exist in 1.5., remains reflexive and transitive if R was so.

We wish to point out another way of looking at regular relations.

Definition 1.7. Given $G = (p, \Gamma)$, the class $F_{\infty,\omega}(G)$ of formulas of $L_{\infty,\omega}$ is defined as the least class F such that F contains Γ, F is closed under arbitrary conjunctions and disjunctions (provided the results have only finitely many free variables) and if $\varphi(v_0, \cdots, v_n) \in F$ with at most the free variables indicated then $P_n v_n \varphi \in F$.

Proposition 1.8. $\mathfrak{A} R(G) \mathfrak{B}$ iff for every sentence φ in $F_{\infty,\omega}(G)$, $\mathfrak{A} \models \varphi$ implies $\mathfrak{B} \models \varphi$.

The proof of 1.8 is completely analogous to the proof of Karp's Theorem [8] (see also [2]) (which is in fact a special case of 1.8, see Example 1 below).

1.8 suggests that perhaps a preservation theorem for $R(G)$ in $L_{\omega_1,\omega}$ can be proved with the $L_{\omega_1,\omega}$-sentences in $F_{\infty,\omega}(G)$. We have not been able to prove such a result. Instead, we introduce an auxiliary Lindström game $G' = (q, \Gamma')$ such that $R(G') = R(G)$ and we consider the $L_{\omega_1,\omega}$-sentences in $F_{\infty,\omega}(G')$; c.f. 2.2.

Digressing even farther from our main goal, we wish to point out an absoluteness property of Vaught sentences. We were informed of this by the talk by J. Barwise at the Summer School in Logic, Cambridge, 1971.

Proposition 1.9 (Barwise). Let M and N be models of ZF, N an endextension of M. Let Φ be a Vaught sentence in M, $\mathfrak{A}$ a structure in M. Then $M \models$ "$\mathfrak{A} \models \Phi$" iff $N \models$ "$\mathfrak{A} \models \Phi$".

The proof may be given by observing that the negation of "$\mathfrak{A} \models \Phi$" is equivalent to the well-foundedness of a certain relation defined in a simple way. Another method is to first show that $\mathfrak{A} \models \Phi$ is equivalent to saying that $\mathfrak{A}$ is a model of a certain simply definable proper class of sentences of $L_{\infty,\omega}$; this will imply the "if" part.

Now it is straightforward to imitate the statement and proof of Theorem 12 of Barwise [2] to get a result saying roughly that the Vaught normal form for any $\underset{\sim}{\Sigma}_1^1$ condition is the strongest possible absolute weakening of the given $\underset{\sim}{\Sigma}_1^1$ condition.

§2. The normal form theorem

Unless otherwise stated, all formulas and structures are of the fixed countable similarity type τ.

Perhaps the main point of the present paper is that there is a natural, simple syntactic category of Vaught sentences which are "obviously" preserved under a given relation R(G), very much like, say, positive sentences are preserved for homomorphic images.

Definition 2.1. The set V(G) of Vaught sentences associated with the Lindström game $G = (p, \Gamma)$, $p = \langle P_n : n < \omega\rangle$, consists of all Φ of the form

$$\bigwedge_{i_0 \in \omega} \bigvee_{j_0 \in \omega} \bigwedge_{k_0 \in \omega} P_0 v_0 \bigwedge_{i_1 \in \omega} \bigvee_{j_1 \in \omega} \bigwedge_{k_1 \in \omega} P_1 v_1 \cdots \bigwedge_n N^{i_0 j_0 k_0 \cdots i_n j_n k_n}(v_0,\cdots,v_n)$$

where the quantifier symbols P_n come from the p component of G and each $N^{i_0 j_0 k_0 \cdots i_n j_n k_n}(v_0,\cdots,v_n)$ is a finite disjunction of elements of Γ containing at most $v_0,\cdots,v_n$ free.

Proposition 2.2. Any Vaught sentence Φ in $V(G)$ is preserved under $R(G)$, i.e.

$$\mathfrak{A} \models \Phi \text{ and } \mathfrak{A}R(G)\mathfrak{B} \Rightarrow \mathfrak{B} \models \Phi .$$

Proof (in sketch). Assuming the left-hand side of the implication, we have two winning strategies S_1 and S_2 for "$\mathfrak{A} \models \Phi$" and "$\mathfrak{A}R(G)\mathfrak{B}$", respectively. We describe informally a winning strategy S_3 for "$\mathfrak{B} \models \Phi$". For the sake of illustration, suppose $P_0 = \forall$ and $P_1 = \exists$. As the first step in the specification of S_3, we let it associate with a given i_0 the j_0 which is associated with i_0 by S_1. Secondly, let $i_0 \in \omega$ (and thus $j_0 \in \omega$), $k_0 \in \omega$, $b_0 \in B$ (as v_0) and $i_1 \in \omega$ be given. By the strategy $S_2 = \langle I_0, \ldots \rangle$ for "$\mathfrak{A}R(G)\mathfrak{B}$", there is a_0 such that $a_0 I_0 b_0$ (see 1.2.(ii), note that $P_0 = \forall$). Let S_3 associate with $\langle i_0, k_0, b_0, i_1 \rangle$ the j_1 that is associated with $\langle i_0, k_0, a_0, i_1 \rangle$ by S_1. Thirdly, let the opponent play $k_1 \in \omega$ next. Let us look at the a_1 which is played by S_1 for $\langle i_0, k_0, a_0, i_1, k_1 \rangle$. By $S_2 = \langle I_0, I_1 \ldots \rangle$ there is $b_1 \in B$ such that $a_0, a_1 I_1 b_0, b_1$ (see 1.2.(i), note that $P_1 = \exists$). Let S_3 play b_1 as v_1. "E.t.c.". From the fact that S_3 thus described is indeed winning for "$\mathfrak{B} \models \Phi$", we show the part that we have $\mathfrak{B} \models N^{\underline{k}}[b_0, b_1]$ where $\underline{k} = i_0 j_0 k_0 i_1 j_1 k_1$. Since S_1 is winning, $\mathfrak{A} \models N^{\underline{k}}[a_0, a_1]$. Since S_2 is winning, and $N^{\underline{k}}$ has a special form related to Γ, and since $a_0, a_1 I\, b_0, b_1$ holds, $\mathfrak{A} \models N^{\underline{k}}[a_0, a_1] \Rightarrow \mathfrak{B} \models N^{\underline{k}}[b_0, b_1]$. Hence $\mathfrak{B} \models N^{\underline{k}}[b_0, b_1]$ as required.

Besides 2.2., elements of $V(G)$ have all their ω_1, ω-approximations preserved by $R(G)$ as is seen by 1.8 and

Proposition 2.3. Every ω_1, ω-approximation of an element of $V(G)$ belongs to $\Delta(G) =_{df} F_{\infty, \omega}(G) \cap L_{\omega_1, \omega}$.

Proof: by inspection.

Unfortunately, we could not prove a "preservation theorem" (see 2.11.) for $R(G)$ with only $V(G)$. Instead, we have to modify G to get some G' such that in particular $R(G) = R(G')$. Some preparations first.

Let $\langle s_n : n<\omega\rangle$ be a recursive enumeration of all nonempty finite sequences of natural numbers such that $s_n \subsetneq s_m$ implies $n<m$. We write $n \rho m$ if s_n is a proper initial segment of s_m, $n \bar{\rho} m$ if $n \rho m$ and there is no k such that $n \rho k$, $k \rho m$. The enumeration $\langle s_n : n<\omega\rangle$ and ρ are fixed throughout this § .

Let $\ell(n)$ be the (length of s_n) $- 1$. A sequence $i_0, \cdots, i_m$ is called a <u>path</u> if $\ell(i_0) = 0$ and $i_0 \bar{\rho} i_1 \bar{\rho} i_2 \cdots \bar{\rho} i_m$.

The relation ρ is an (irreflexive) partial tree-ordering the natural numbers, since $\underline{n} =_{df} \{m : m \rho n$ or $m = n\}$ is totally ordered by ρ . Note that a path $i_0, \cdots, i_m$ coincides with $\underline{i_m}$ indexed according to the ordering ρ . Note also that there are infinitely many natural numbers n on level 0 (i.e., $\ell(n) = 0$) and for each n , there are infinitely many ρ-successors of n (which are one level higher than n). In what follows, we will refer to the order ρ rather than $\langle s_n : n<\omega\rangle$.

Let p be a prefix $\langle P_n v_n : n<\omega\rangle$, Γ a countable set of formulas, G the Lindström game (p, Γ). G is fixed throughout the rest of the paper. Using ρ, we associate with G another Lindström game $G' = (q, \Gamma)$ such that $R(G) = R(G')$ as follows. Define $q = \langle Q_n : n<\omega\rangle$ by $Q_n = P_{\ell(n)}$. Define Γ' to be set of all formulas $\gamma(v_{i_0}, v_{i_1}, \cdots, v_{i_m})$ such that $\gamma(v_0, \cdots, v_m) \in \Gamma$, γ has only the free variables indicated and $i_0, \cdots, i_m$ is a path.

<u>Lemma 2.4.</u> $R(G) = R(G')$.

First suppose that $I = \langle I_n : 1 \leq n<\omega\rangle$ is a winning strategy for "$\mathfrak{A}R(G)\mathfrak{B}$". Define the winning strategy $I' = \langle I'_n : 1 \leq n<\omega\rangle$ for "$\mathfrak{A}R(G')\mathfrak{B}$" as follows. Let $a_0, \cdots, a_n I'_{n+1} b_0, \cdots, b_n$ hold if and only if for every path $i_0, \cdots, i_m$ with $i_m \leq n$, we have $a_{i_0}, \cdots, a_{i_m} I_{m+1} b_{i_0}, \cdots, b_{i_m}$. It is easy to check that I' is indeed winning for "$\mathfrak{A}R(G')\mathfrak{B}$". This proves that $R(G) \subseteq R(G')$.

Now let I' be a winning strategy for "$\mathfrak{A}R(G')\mathfrak{B}$". Let $\langle i_j : j<\omega\rangle$ be an infinite path, i.e., $i_0, \cdots, i_n$ a path for every n. Define $a_0, \cdots, a_m I_{m+1} b_0, \cdots, b_m$ to be true if and only if there are $a'_0, \cdots, a'_{i_m}, b'_0, \cdots, b'_{i_m}$ such that $a_j = a'_{i_j}$ and $b_j = b'_{i_j}$ for $j \leq m$ and $a'_0, \cdots, a'_{i_m} I'_{i_{m+1}} b'_0, \cdots, b'_{i_m}$. It is easy to check that $I = \langle I_m : 1 \leq m<\omega\rangle$ is winning for "$\mathfrak{A}R(G)\mathfrak{B}$", which proves $R(G') \subseteq R(G)$.

We define the set $\Delta'(G)$ of formulas of $L_{\omega_1, \omega}$ as $\Delta(G')$ (see 2.3). The definition can be given directly in terms of G as follows.

<u>Definition 2.5.</u> $\Delta(G)$ is the smallest set Δ such that

(i) $\Gamma' \subset \Delta$,

(ii) Δ is closed under finite or denumerably infinite conjunction and disjunction (but the conjunctions and disjunctions are to contain only finitely many free variables),

(iii) if $\varphi \in \Delta$, and φ does not contain v_m free for any m such that $m > n$, then $P_{\ell(n)} v_n \varphi \in \Delta$.

<u>Proposition 2.6.</u> Any sentence θ in $\Delta'(G)$ is preserved under $R(G)$.

2.6 follows from 2.4 and 1.8.

<u>Definition 2.7.</u> $V'(G) =_{df} V(G')$.

Observe that the prefix of an element in $V'(G)$ is fixed once $p = \langle P_k : k < \omega \rangle$ is fixed, hence the Vaught sentence Φ in Definition 2.1 applied to G' depends only on $\underline{N} = \langle N^{\underline{k}} \rangle_{\underline{k} \in \bar{I}}$ (where $\bar{I} = \bigcup \{ I^{3n+3} : n < \omega \}$) . We will denote Φ by $V(\underline{N})$.

Proposition 2.8. Any element Φ of $V'(G)$ is preserved under $R(G)$, i.e., $\mathfrak{A}R(G)\mathfrak{B}$ and $\mathfrak{A} \models \Phi$ imply $\mathfrak{B} \models \Phi$.

Proof: by 2.2. and 2.4.

For a class K of structures of type τ, and a relation R between structures of type τ, we write $C_R(K)$ to denote $\{\mathfrak{B} :$ for some $\mathfrak{A} \in K$, $\mathfrak{A}R\mathfrak{B}\}$. $K^{(\leq \aleph_0)}$ denotes the class of countable (understood always as including finite) structures in K; $C_R^{(\leq \aleph_0)}(K) =_{df} (C_R(K))^{(\leq \aleph_0)}$, $\mathrm{Mod}^{(\leq \aleph_0)}(\sigma) =_{df} (\mathrm{Mod}\ (\sigma))^{(\leq \aleph_0)}$.

Theorem 2.9. For any sentence $\varphi \in L_{\omega_1, \omega}$ there is a Vaught sentence in $V'(G)$ such that for $K = \mathrm{Mod}\ (\varphi)$,

$$C_{R(G)}(K) \subseteq \mathrm{Mod}\ \Phi$$

and $C_{R(G)}^{(\leq \aleph_0)}(K) = C_{R(G)}^{(\leq \aleph_0)}(K^{(\leq \aleph_0)}) = \mathrm{Mod}^{(\leq \aleph_0)}(\Phi)$.

If in addition $\mathfrak{a} \subset HC$, $\mathfrak{a}$ is admissible, $\varphi \in L_{\mathfrak{a}}$ and $G \in \mathfrak{a}$, then $\Phi = V(\underline{N})$ can be found in $\mathfrak{a}$.

Proof. In what follows, free variables will mostly be treated as individual constants. Therefore, if α is an assignment of elements of

$A = |\mathfrak{A}|$ to variables, we will say that $(\mathfrak{A}, \alpha)$ is a model of a set Σ of formulas if $\mathfrak{A} \models \varphi[\alpha]$ for every $\varphi \in \Sigma$. Also, for a set Σ of formulas or a single formula Σ and a formula φ, we mean by $\Sigma \models \varphi$ that every model $(\mathfrak{A}, \alpha)$ of Σ is a model of φ, i.e. $\Sigma \models \varphi$ is understood with "the free variables held constant".

In the following it is convenient to consider only the existential quantifier $\exists$ and the infinitary disjunction $\bigvee$ acting on countable (perhaps finite) sets of formulas, and the finitary connectives $\neg, \wedge, \vee, \rightarrow$ primitive.

Let φ be any sentence in $L_{\omega_1, \omega}$. We will describe how to obtain Φ of the theorem.

Let Θ be the smallest fragment containing φ and Γ, i.e. the least set Θ of formulas such that $\varphi \in \Theta$, $\Gamma \subset \Theta$, every atomic formula of τ is in Θ, Θ is closed under finitary logical operations $(\neg, \wedge, \vee, \rightarrow, \exists v_n)$ Θ is closed under (proper) substitution of variables for free variables, and, finally, every subformula of a formula in Θ is again in Θ. Clearly, Θ is countable.

We put $I = \not{\emptyset}^{\Theta}$. For any $\underline{k} = i_0 j_0 k_0 \cdots i_n j_n k_n$, sequence of elements of I, we first define $M^{\underline{k}}$ as follows.

Case 1. For every $m = 0, \cdots n$, i_m and k_m are formulas with at most $v_0, \cdots, v_{m-1}$ free, and in particular

$$i_m \text{ is of the form } \bigvee \Psi_m \text{ with } \Psi_m \neq \emptyset$$

and

$$k_m \text{ is of the form } \exists x_m \varphi_m .$$

Subcase 1 of Case 1. For every $m = 0, \cdots n$, j_m is a formula ψ_m which is an element of Ψ_m .

Under these conditions, define $M^{\underline{k}}$ to be the conjunction of all formulas

$$\bigvee\Psi_m \to \psi_m \quad \text{for } m \leq n$$

and

$$\exists x_m \varphi_m \to \varphi_m(v_m/x_m) \quad \text{for } m \leq n \text{ such that } Q_m = P_{\ell(m)} = \exists .$$

(v_m is substituted for the free occurrences of x_m).

Subcase 2 of Case 1. Subcase 1 does not hold. In this case, let

$$M^{\underline{k}} = \bigvee\emptyset \quad \text{(identically false)} .$$

Case 2. Case 1 does not hold. In this case, let

$$M^{\underline{k}} = \bigwedge\emptyset \quad \text{(identically true)} .$$

The second type of formulas included in $M^{\underline{k}}$ in Subcase 1 of Case 1 must be familiar to the reader as the "Henkin formulas" from Henkin's classical proof of the completeness theorem. The first type is the straightforward analog of Henkin formulas when we pass to infinitary disjunction from existential quantifier. We claim that $V(\underline{M})$, i.e.

$$\bigwedge i_0 \in I \bigvee j_0 \in I \bigwedge k_0 \in I \; Q_0 v_0 \cdots \bigwedge n < \omega \, M^{i_0 \cdots k_n}$$

is valid , i.e. true in every structure. The proof of this is straightforward and essentially "the same" as the proof that addition of Henkin sentences to consistent theories does not destroy consistency. We note that in this proof it is essential that all v_m introduced with the second kind of Henkin formulas are existentially quantified in the prefix of $V(\underline{M})$.

Now we return to the given sentence φ and we define

$$N^{\underline{k}}(v_0, \ldots, v_n) \qquad (\underline{k} = i_0, \ldots, k_n)$$

as the conjunction of all finite disjunctions (including the empty one) δ of elements of Γ' such that δ contains $v_0, \ldots, v_n$ free at most and

$$\varphi \wedge M^{\underline{k}}(v_0, \ldots, v_n) \models \delta(v_0, \ldots, v_n)$$

with the variables held constant. Put

$$\Phi = V(\underline{N}) .$$

We have

$$\varphi \wedge M^{\underline{k}}(v_0, \ldots, v_n) \models N^{\underline{k}}(v_0, \ldots, v_n) ,$$

which has the consequence that for any model $\mathfrak{A}$ of φ, any winning strategy for "$\mathfrak{A} \models V(\underline{M})$" is a winning strategy for "$\mathfrak{A} \models V(\underline{N})$". It follows by the validity of $V(\underline{M})$ that

$$\mathrm{Mod}\ (\varphi) \subseteq \mathrm{Mod}\ (\Phi) .$$

Clearly, $\Phi \in V'(G)$. By 2.8,

$$C_{R(G)}\ \mathrm{Mod}\ (\varphi) \subseteq C_{R(G)}\ \mathrm{Mod}\ (\Phi) \subseteq \mathrm{Mod}\ \Phi$$

and we have proved the first assertion of the theorem.

Having the last inclusion, we see that for showing the equalities of the theorem it is sufficient to prove

$$\mathrm{Mod}^{(\leq \aleph_0)}(\Phi) \subseteq C_{R(G)}\ \mathrm{Mod}^{(\leq \aleph_0)}(\varphi) .$$

Let $\mathfrak{B}$ be a countable model of Φ. Our goal is to construct a countable model $\mathfrak{A}$ or φ such that $\mathfrak{A}R(G)\mathfrak{B}$. We will use the hypothesis $\mathfrak{B} \models \Phi$ thus: we choose particular values for the "universally quantified variables"

i_n, k_n and x_n with $Q_n = \forall$, and we obtain certain values for the "existentially quantified variables" j_n and x_n with $Q_n = \exists$ to work with.(6)

For every $n < \omega$, let $i_n = \bigvee \Psi_n$ be such that $\{i_n : n < \omega\}$ is the set of all non-empty infinitary disjunctions in $\Theta = I$; we can choose the enumeration $\langle i_n : n < \omega \rangle$ so that i_n contains at most $v_0, \ldots, v_{n-1}$ free.(7)

Next we turn to fixing values for k_n, $n < \omega$.

Assume from now on that at least one of the quantifier symbols P_ℓ (i.e., at least one Q_n) is existential. In the contrary case, the proof is even simpler than below, in particular, the items $\exists x_\ell \psi_\ell$ and y_ℓ introduced below are not needed. Also, for the sake of uniformity, extend the ordering ρ onto $\omega \cup \{-1\}$ by defining $(-1)\rho n$ to be true for every $n \in \omega$. For $m \in \omega \cup \{-1\}$, let S_m be the set of ρ-successors of m, i.e. $S_m = \{m' : m\rho m'$ and there is no m'' with $m\rho m''$, $m''\rho m'\}$. Notice that each S_m is infinite. Put $\ell(-1) = -1$.

Let $\exists x_\ell \psi_\ell \in \Theta$ and the variable $y_\ell = v_{m_\ell}$ be defined for $\ell < \omega$ such that

(i) if v_m occurs in $\exists x_\ell \psi_\ell$, then $m < m_\ell$,

(ii) $m_i < m_\ell$ for $i < \ell$,

(iii) for every $\ell < \omega$, $P_{\ell(m_\ell)} = Q_{m_\ell} = \exists$,

(iv) for any $m \in \omega \cup \{-1\}$ such that $P_{\ell(m)+1} = \exists$,

and for any formula in Θ of the form $\exists x \psi$, there is ℓ such that m_ℓ is a ρ-successor of m (i.e. $m_\ell \in S_m$; hence also $\ell(m_\ell) = \ell(m)+1$) and $\exists x_\ell \psi_\ell = \exists x \psi$.

Notice that we need our assumption that there is m with $P_m = \exists$ to make (iii) hold. The construction is straightforward and it uses that each S_m is infinite.[8]

Now, we put

$$k_{m_\ell} = \exists x_\ell \psi_\ell$$

(notice that by (ii) $m_i \neq m_\ell$ for $i \neq \ell$) and put k_n for $n \notin \{m_\ell : \ell < \omega\}$ to be an arbitrary sentence $\exists x \psi$.

Finally, we assign a value $b_n \in B$ to every variable v_n with $Q_n = \forall$ such that we have

$$\{b_n : n \in S_m\} = B = |\mathfrak{B}| \tag{1}$$

for every $m \in \omega \cup \{-1\}$ with $P_{\ell(m)+1} = \forall$. Since for such an m , we have $Q_n = P_{\ell(m)+1} = \forall$ whenever $n \in S_m$ and S_m is infinite, moreover, since B is countable, such an assignment $v_n \mapsto b_n$ $(Q_n = \forall)$ is possible.

As mentioned above, the winning strategy for "$\mathfrak{B} \models \Phi$" provides elements $j_n \in \Theta = I$ for every $n < \omega$ and b_n for n such that $Q_n = \exists$, b_n assigned to v_n , such that we have

$$\mathfrak{B} \models N^{i_0 j_0 k_0 \cdots i_n j_n k_n}[b_0, \ldots, b_n] \tag{2}$$

for every $n < \omega$.

Looking at the definition of the formulas $M^{\underline{k}}$, we see that by (i), (ii) and (iii) now Case 1 holds for our particular $\underline{k} = i_0 j_0 k_0 \cdots i_n j_n k_n$, for any $n < \omega$. Because of (2), Subcase 2 of Case 1 cannot hold (since then $N^{\underline{k}}$ should contain as a conjunct the identically false empty disjunction). Hence Subcase 1 holds and

$$j_n = \psi_n \text{ is an element of } \Psi_n, \text{ for } n < \omega,$$

and $M^{\underline{k}}$ is defined as described in Subcase 1.

Let H_1 be the set of formulas

$$\bigvee \Psi_m \rightarrow \psi_m \quad \text{for } m < \omega$$

and

$$\exists x_\ell \psi_\ell \rightarrow \psi_\ell (v_{m_\ell} / x_\ell) \quad \text{for } \ell < \omega .$$

Notice that now $M^{i_0 \cdots k_n}$ is the conjunction of these latter formulas for $m \leq n$ and $m_\ell \leq n$.

Let H_2 be the set of all valid formulas in Θ. Let H_3 be the set of all formulas $\neg\gamma'(v_0, \cdots, v_n)$ with arbitrary $n < \omega$ such that $\gamma' \in \Gamma'$ and $\mathfrak{A} \models \neg\gamma'[b_0, \cdots, b_n]$. We claim that

$$T = \{\varphi\} \cup H_1 \cup H_2 \cup H_3$$

is <u>finite-propositionally consistent in</u> Θ in the sense that there is an assignment ξ assigning every element of Θ one of the truth values "true" and "false" such that ξ commutes with finitary Boolean operations and ξ makes every element of T true. In view of the compactness of finitary propositional logic, it suffices to show that every finite subset of T is finite-propositionally consistent in Θ . Note that every structure $(\mathfrak{A}, \alpha)$ is automatically a model of H_2 , and also, the formulas $M^{\underline{k}}$ are essentially the initial segments of H_1 . Hence it suffices to show that

$$M^{i_0 \cdots k_n}(v_0, \cdots, v_n) \cup \Gamma''$$

has a model $(\mathfrak{A}, \alpha)$ for every $n < \omega$ and every finite set Γ'' of formulas $\neg\gamma'(v_0, \cdots, v_n)$ with $\gamma' \in \Gamma'_n \cap H_3$. But if this failed then $\delta = \bigvee\{\gamma' : \neg\gamma' \in \Gamma''\}$ would be a conjunct of $N^{i_0 \cdots k_n}$, by the definition of $N^{\underline{k}}$, hence by (2), $\mathfrak{B} \models \delta[b_0, \cdots, b_n]$ which is obviously false.

Having the truth-assignment ξ on Θ , we can easily show that ξ in fact defines a model $(\mathfrak{A}, \alpha)$ of T . The model is defined to have the domain consisting of the terms of τ modulo equality in T and the basic operations and relation on $\mathfrak{A}$ are defined such that

$$\mathfrak{A} \models \theta[\alpha] \Leftrightarrow \xi(\theta) \text{ is "true"}$$

for every atomic formula θ . Then by induction on the formula $\theta \in \Theta$ we show that

$$\mathfrak{A} \models \theta[\alpha] \Leftrightarrow \xi(\theta) \text{ is "true"}.$$

This straightforward induction uses that elements of H_1 and certain validities in H_2 receive the value "true" by ξ . Note that $\mathfrak{A} \models \varphi$.

Let $a_n = \alpha(v_n)$. We claim that for every $m \in \omega \cup \{-1\}$ such that $P_{\ell(m)+1} = \exists$ we have

$$\{a_n : n \in S_m\} = A = |\mathfrak{A}| \quad . \tag{3}$$

Let a be an arbitrary element of A . Then $a = t^{\mathfrak{A}}[\alpha]$ for some term t . By (iv) above applied for $\exists x[t \approx x]$ (where x is a variable not occurring in t), there is k such that $\exists x_k \psi_k = \exists x[t \approx x]$ and $m_k \in S_m$. We have

$\theta = \exists x_k \psi_k \rightarrow \psi_k(y_k/x_k) \in H \subset T$, hence $\mathfrak{A} \models \theta[\alpha]$. It follows that $\mathfrak{A} \models (t \approx v_{m_k})[\alpha]$ (since $y_k = v_{m_k}$) , i.e. $a = a_{m_k}$, proving that $a \in \{a_n : n \in S_m\}$.

Next define $I_n \subset A^n \times B^n$ for $n \geq 1$ as follows. For $a'_0, \cdots, a'_{n-1} \in A$ and $b'_0, \cdots, b'_{n-1} \in B$, let $a'_0, \cdots, a'_{n-1} I_n b'_0, \cdots, b'_{n-1}$ hold iff there is a <u>path</u> $i_0, \cdots, i_{n-1}$ such that $a'_j = a_{i_j}$ and $b'_j = b_{i_j}$ for $j < n$. Put $I = \langle I_n : 1 \leq n < \omega \rangle$. We are going to verify 1.2 (i), (ii) and (iii).

Suppose $P_n = \exists$ and $a'_0, \cdots, a'_{n-1} I_n b'_0, \cdots, b'_{n-1}$, i.e. $a'_j = a_{i_j}$ and $b'_j = b_{i_j}$ for $j < n$ and for some path $i_0, \cdots, i_{n-1}$. Let $a \in A$. Since $\ell(i_{n-1}) + 1 = n$ and $P_n = \exists$, by (3) there is $i_n \in S_{i_{n-1}}$ such that $a = a_{i_n}$. Put $b = b_{i_n}$. Clearly, $i_0, \cdots, i_n$ is a path, hence by the definition of I_{n+1} , we have $a'_0, \cdots, a'_{n-1}, a I_{n+1} b'_0, \cdots, b'_{n-1}, b$, proving 1.2(i).

The verification of 1.2(ii) is similar; it uses (1).

Finally, let $\gamma(v_0, \cdots, v_{n-1}) \in \Gamma$, $\langle i_0, \cdots, i_{n-1} \rangle$ a path, $a'_j = a_{i_j}$, $b'_j = b_{i_j}$ for $j < n$. The formula $\gamma' = \gamma(v_{i_0}, \cdots, v_{i_{n-1}})$ belongs to Γ'_m for $m = i_{n-1} + 1$. To see 1.2(iii), we have to show that $\mathfrak{A} \models \gamma[a'_0, \cdots, a'_{n-1}]$ implies $\mathfrak{B} \models \gamma[b'_0, \cdots, b'_{n-1}]$. But this is equivalent to saying that $\mathfrak{B} \models \neg\gamma'[b_0, \cdots, b_m]$ implies $\mathfrak{A} \models \neg\gamma'[a_0, \cdots, a_m]$ which is true since $(\mathfrak{A}, \alpha)$ satisfies every element $\neg\gamma'$ of $H_3 \subset T$.

To sum up, we have shown that there is a winning strategy I for "$\mathfrak{A}R(G)\mathfrak{B}$", hence $\mathfrak{A}R(G)\mathfrak{B}$ holds. Since $\mathfrak{A}$ is a countable model of φ, by remarks made above this completes the proof of the theorem.

To have the Vaught sentence Φ belong to $\mathcal{C}$, a given admissible set containing φ and G, we have to be slightly more careful with the definition of $N^{\underline{k}}$. Since we want $\langle N^{\underline{k}}\rangle_{\underline{k}}$, a fortiori each $N^{\underline{k}}$, to belong to $\mathcal{C}$, the definition given in the above proof does not quite work, as "$\models$" is only $\mathcal{C}$-r.e. and not $\mathcal{C}$-rec.

First, notice that Θ in the proof is on element of $\mathcal{C}$. Secondly, let H_2 be the set of those validities in Θ which are <u>actually used</u> in the proof, i.e. those of the form $\psi \to \bigvee\Psi\ (\psi \in \Psi), \neg\bigvee 0$, $\varphi \to \exists x\varphi$, substitution instances of finitary Boolean tautologies and equality axioms. Notice that $H_2 \in \mathcal{C}$. Returning to the definition of $N^{\underline{k}}$ for fixed $\underline{k}$, let the modified $N^{\underline{k}}$ be the conjunction of those finite disjunctions $\delta(v_0, \cdots, v_n)$ of elements of Γ' such that for some finite subset H_2' of H_2, δ is a <u>finite-propositional consequence</u> in the obvious sense of $\{\varphi, M^{\underline{k}}\} \cup H_2'$. Observe that (i) $N^{\underline{k}} \in \mathcal{C}$ by using Δ_1^1-separation for $\mathcal{C}$ and the "recursiveness" of finitary propositional logic, (ii) in fact, $N^{\underline{k}}$ is an $\mathcal{C}$-recursive function of $\underline{k} \in \Theta^{\underline{\omega}}$, hence $\langle N^{\underline{k}}\rangle_{\underline{k}} \in \mathcal{C}$, (iii) the modified $N^{\underline{k}}$ contains as conjuncts at most those δ which enter in the original one and thus Mod $(\varphi) \subseteq$ Mod (Φ) remains true, and finally, (iv) the proof of the existence of the model $(\mathfrak{A}, \alpha)$ of T remains valid since in that proof we actually use only the presence of those δ which enter in the modified $N^{\underline{k}}$. This completes the proof of the supplement to the theorem.

Corollary 2.10. For any $\underset{\sim}{\Sigma}_1^1$ class K there is $\Phi \in V'(G)$ such that the relations of 2.9 hold.

Proof. Let K be the class of reducts to τ of all structures in $K' = Mod_{\tau'}(\varphi)$, $K = K' \mid \tau$, where $\tau' \supset \tau$, τ' is countable, and φ is an $L_{\omega_1, \omega}$ sentence appropriate for τ'. Let $R'(G) = R_{\tau'}(G)$ be the relation between structures of type τ' defined by G. Observe that $\mathfrak{A}'R'(G)\mathfrak{B}'$ iff $\mathfrak{A}' \mid \tau \; R(G)\mathfrak{B}' \mid \tau$. Hence by applying 2.9 for φ, τ', $R'(G)$ in place of φ, τ, $R(G)$, we obtain the desired conclusion.

A $\underset{\sim}{\Sigma}_1^1$-class of countable structures is a class which consists of the countable (including finite) members of a $\underset{\sim}{\Sigma}_1^1$-class. We say that K is closed under R among countable structures if for any countable $\mathfrak{A}$ and $\mathfrak{B}$, $\mathfrak{A} \in K$ and $\mathfrak{A}R\mathfrak{B}$ implies $\mathfrak{B} \in K$.

Corollary 2.11. Let R be a relation which coincides with R(G) for countable structures. Then for any K, the following are equivalent:

(i) K is a $\underset{\sim}{\Sigma}_1^1$-class of countable structures closed under R among countable structures,

(ii) K is the class of countable models of some Vaught sentence in $V'(G)$.

The fact that (ii) implies (i) follows from 2.8 and the fact that Vaught sentences define $\underset{\sim}{\Sigma}_1^1$-classes. Conversely, if (i) holds, then by the reflexivity of $R(G)$ we have $K = C_R^{(\leq \aleph_0)}(K)$. By 2.10, we conclude (ii).

Let us call K a $\underset{\sim}{\Delta}_0^1$-class (of countable structures) if K is the class of (countable) models of some sentence.

Corollary 2.12. Let R be a regular relation (in fact, let G be associated with R). For every $\underset{\sim}{\Sigma}_1^1$-class K there are $\aleph_1$ $\underset{\sim}{\Delta}_0^1$-classes K_υ $(\upsilon < \omega_1)$ each closed under R (in fact, each K_υ is defined by an element of $\Delta'(G)$) such that

(i) $C_R(K) \subseteq \bigcap_{\upsilon < \omega_1} K_\upsilon$,

(ii) $C_R^{(\leq \aleph_0)}(K) = C_R^{(\leq \aleph_0)}(K^{(\leq \aleph_0)}) = \bigcap_{\upsilon < \omega_1} K_\upsilon^{(\leq \aleph_0)}$,

and (iii) for any $\underset{\sim}{\Sigma}_1^1$ class K′ disjoint from $C_R(K)$ there is some $\upsilon < \omega_1$ such that $K' \cap K_\upsilon = 0$.

Proof. Given the $\underset{\sim}{\Sigma}_1^1$-class K, let $\Phi \in V'(G)$ be chosen according to 2.10. Let θ_υ be the υth ω_1, ω-approximation of Φ, and $K_\upsilon = \mathrm{Mod}(\theta_\upsilon)$. Then $\theta_\upsilon \in \Delta'(G)$ by 2.3, and K_υ is closed under R by 2.6. (i) follows from 2.10 and 1.1(i), (ii) from 2.10 and 1.1(iii). To see (iii), let K′ be the class of models of the $\underset{\sim}{\Sigma}_1^1$-sentence $\exists \bar{S}\psi$, ψ being a sentence appropriate for τ plus the S and let K′ be disjoint from $C_R(K)$. Since

$C_R^{(\leq \aleph_0)}(K) = \mathrm{Mod}^{(\leq \aleph_0)}(\Phi)$, the parenthetical hypothésis of 1.1(ii) holds for the present Φ and ψ . Thus (iii) follows from 1.1(ii).

Notice that 2.12 without the parenthetical phrases is a purely modeltheoretic result for regular relations, not involving any particular syntactic notion.

Let φ and ψ be sentences, possibly mentioning some non-logical constants beyond τ . We write $\varphi \models^R \psi$ if for any $\mathfrak{A}$ and $\mathfrak{B}$ appropriate for φ and ψ , resp., and both having types including τ , $\mathfrak{A} \models \varphi$ and $\mathfrak{A} \mid \tau R\, \mathfrak{B} \mid \tau$ imply $\mathfrak{B} \models \psi$.

The next corollary is just a reformulation of a part of 2.12.

<u>Corollary 2.13.</u> Let R be regular (and G associated with R). Then $\varphi \models^R \psi$ if and only if there is a sentence θ <u>of</u> τ such that $\varphi \models \theta \models^R \theta \models \psi$ (or, equivalently, there is $\theta \in \Delta'(G)$ such that $\varphi \models \theta \models \psi$) .(9) If, in addition, $\mathfrak{a}$ is admissible, $G \in \mathfrak{a}$ and $\varphi, \psi \in L_{\mathfrak{a}}$, then an appropriate θ can be found in $L_{\mathfrak{a}}$.

The addition concerning admissible sets follows from the corresponding additions to 2.9 and 1.1(ii).

<u>Example 1.</u> Let p be the alternating prefix $\langle \exists\forall\exists \cdots \rangle$, Γ the set of all atomic and negated atomic formulas of τ . Then $G = (p, \Gamma)$ is associated with the relation of isomorphism (in fact, by 1.8 R(G) is the relation of ∞, ω-equivalence; see [6], [7], [10], [2]) . Now 2.11 becomes the Craig-Lopez-Escobar interpolation theorem [11] and it's present proof is closely related to Vaught's proof [21]. Of course, if we are interested only in this case, or even if we want only 2.14 below, the present proof is unnecessarily complicated and we can avoid Lindström games altogether.

Example 2. Let p be as before and Γ the set of all atomic formulas of τ. Let $\mathfrak{A}R\mathfrak{B}$ hold iff $\mathfrak{B}$ is a homomorphic image of $\mathfrak{A}$. $G = (p, \Gamma)$ is associated with R(c.f. [10]). Clearly, every element of $\Delta(G)$ is positive (we consider the empty disjunction positive). Hence 2.13 reduces to the Lyndon-Lopez-Escobar theorem on homomorphisms [11] and 2.12 to

Corollary 2.14. For every $\underset{\sim}{\Sigma}_1^1$ class K of countable structures, the class Hom (K) of all homomorphic images of structures in K is axiomatised, among countable structures, by $\aleph_1$ positive $L_{\omega_1, \omega}$-sentences θ_υ, $\upsilon < \omega_1$, such that every $\underset{\sim}{\Sigma}_1^1$ class disjoint from Hom (K) is disjoint from Mod (θ_υ) for some $\upsilon < \omega_1$.

Example 3. Let R be defined by $\mathfrak{A} R \mathfrak{B} \Leftrightarrow \exists \mathfrak{C}\ \mathfrak{A} \simeq \mathfrak{B} \times \mathfrak{C}$. We designate as basic formulas all formulas $\alpha(x_0, \cdots, x_{n-1})$ and $\alpha(x_0, \cdots, x_{n-1}) \wedge \neg\alpha(y_0, \cdots, y_{n-1})$ if $\alpha(v_0, \cdots, v_{n-1})$ is atomic of the fixed similarity type, it has exactly the free variables shown, and $x_0, \cdots, x_{n-1}, y_0, \cdots, y_{n-1}$ are arbitrary variables substituted for $v_0, \cdots, v_{n-1}, v_0, \cdots, v_{n-1}$ respectively. We say that x and y have complementary occurrences in a basic formula of the second kind if $x = x_i$ and $y = y_i$ for some $i = 0, \cdots, n-1$. We also say that x and y have complementary occurrences in an arbitrary formula φ if φ has a subformula which is a basic formula of the second kind and in which x and y have complementary occurrences. Now, we define Δ to be the smallest set X of formulas such that X contains the basic formulas, is closed under $\forall$, $\wedge$ and $\vee$, and, finally, $\varphi \in X$ implies $\exists x \varphi \in X$ whenever there is no free y in $\exists x \varphi$ such that x and y have complementary occurrences in φ . What was shown in [13] regarding direct factors is equivalent to saying that a sentence is preserved under R iff it is logically equivalent to a sentence in Δ . We want to point out how this result can be derived from 2.13.

Let us first define a Lindström game G associated with R . Let $\langle\langle i_k, j_k\rangle : k \in \omega\rangle$ be a repetition-free indexing of $\omega \times \omega$. Let $p = \langle P_k : k \in \omega\rangle$ be any sequence of $\forall$'s and $\exists$'s such that (i) for every j, there are infinitely many k such that $j_k = j$ and $P_k = \forall$, (ii) there are infinitely many k such that $P_k = \exists$, and (iii) for any k such that $P_k = \exists$ we have $j_k \neq j_\ell$ for all $\ell < k$. Let Γ be the set of all formulas $\alpha(v_{k_0}, \cdots, v_{k_n})$, $\alpha(v_{k_0}, \cdots, v_{k_n}) \wedge \neg\alpha(v_{\ell_0}, \cdots, v_{\ell_n})$ where α is atomic and $j_{k_m} = j_{\ell_m}$ for $m = 0, \cdots, n$. It can be verified that $\Gamma = (p, G)$ is associated with R indeed.

It is easy to see that $\Delta'(G)$ is a subset of Δ as defined above. Hence, (the non-trivial part of) our assertion follows from 2.13.

We wish to make some more remarks. The first one is that the assertion of 2.12, and hence of 2.13, without the parenthetical phrase are true for a simply defined class of relations including the regular ones.

<u>Proposition 2.15.</u> Suppose that the relation R is $\underset{\sim}{\Sigma}^1_1$ (see Def. 1.4), it is reflexive and transitive. Then the assertions of 2.12 and 2.13 are true for R (without the parenthetical phrases).

<u>Proof.</u> First we show that 2.13 holds for R . Suppose $\varphi \models^R \psi$ and assume for simplicity that φ, ψ do not contain symbols beyond τ . By assumption, there is a many sorted sentence Φ of $L_{\omega_1, \omega}$ such that $\mathfrak{A} R \mathfrak{B}$ iff for some F, $(\mathfrak{A}, \mathfrak{B}, F) \models \Phi$. We claim that for any $\mathfrak{A}, \mathfrak{B}, \mathfrak{C}$, F, G we have

$$(\mathfrak{A}, \mathfrak{B}, F) \models \Phi \text{ and } \mathfrak{A} \models \varphi \Rightarrow [(\mathfrak{B}, \mathfrak{C}, G) \models \Phi \Rightarrow \mathfrak{C} \models \psi] .$$

Indeed, $(\mathfrak{A}, \mathfrak{B}, F) \models \Phi$ and $(\mathfrak{B}, \mathfrak{C}, G) \models \Phi$ imply $(\mathfrak{A}, \mathfrak{C}, H) \models \Phi$ for some H by the transitivity of R. Thus the claim follows from

$$(\mathfrak{A}, \mathfrak{C}, H) \models \Phi \text{ and } \mathfrak{A} \models \varphi \Rightarrow \mathfrak{C} \models \psi .$$

We now apply Feferman's interpolation theorem for many-sorted logic [4], or alternatively, the Craig-Lopez-Escobar interpolation theorem together with the remark in Footnote 2. It follows that there is an ordinary sentence θ in $L_{\omega_1, \omega}(\tau)$ such that

$$(\mathfrak{A}, \mathfrak{B}, F) \models \Phi \text{ and } \mathfrak{A} \models \varphi \Rightarrow \mathfrak{B} \models \theta_0$$

and
$$\mathfrak{B} \models \theta_0 \Rightarrow [(\mathfrak{B}, \mathfrak{C}, G) \models \Phi \Rightarrow \mathfrak{C} \models \psi]$$

or in other words,

$$\varphi \models^R \theta_0 \models^R \psi .$$

Repeating the argument, we obtain sentences $\theta_1, \theta_2, \cdots, \theta_n, \cdots$ of $L_{\omega_1, \omega}(\tau)$ such that

$$\varphi \models^R \theta_{n+1} \models^R \theta_n$$

for $n < \omega$. Now put $\theta = \bigwedge_{n<\omega} \theta_n$. We have $\theta \models^R \theta$, $\varphi \models^R \theta$ and $\theta \models^R \psi$. By the reflexivity of R, $\varphi_1 \models^R \varphi_2$ implies $\varphi_1 \models \varphi_2$. Hence we have $\varphi \models \theta \models^R \theta \models \psi$ as desired.

To obtain 2.12, apply Vaught's normal-form theorem mentioned in the Introduction to obtain a Vaught sentence Φ such that for any countable $\mathfrak{B}$,

$$\mathfrak{B} \models \Phi \Leftrightarrow \exists \mathfrak{A} \exists F [(\mathfrak{A}, \mathfrak{B}, F) \models \Phi \;\&\; \mathfrak{A} \in K] .$$

Let θ_0^υ be the ω_1, ω-approximations of $\Phi(\upsilon < \omega_1)$. In particular, we will have $\varphi \models^R \theta_0^\upsilon$ by 1.1(i). Now, construct, for any given $\upsilon < \omega_1$, θ^υ from θ_0^υ as θ is obtained from θ_0 in the above proof. It is then easy to see that $K_\upsilon = \mathrm{Mod}^{(\leq \aleph_0)}(\theta^\upsilon)$ will satisfy 2.12.

This proof gives an infinitary interpolant θ even if the formulas φ and ψ we start with are finitary. Thus we have the problem if the finitary analog of 2.15 is true for reflexive and transitive f-Σ_1^1 relations. In [10], the finitary analog of 2.13 (with a different $\Delta'(G)$) was shown to hold for relations $R(G)$ with Γ containing only finitary formulas; see also [14].

Finally, we mention that for relations $R = R(G)$, the full 2.13 with the syntactic specification $\theta \in \Delta(G)$ can also be proved by the methods of [13].

Actually, in this way we obtain 2.1 in a slightly stronger form, namely, instead of $(p, \Gamma) \in \underset{\sim}{A}$ we need only that p is an $\underset{\sim}{A}$-recursive function and Γ is an $\underset{\sim}{A}$-r.e. subset of $\underset{\sim}{A}$. Combining now the full 2.13 with 2.15, we obtain the full 2.12 at once. This completes showing how to circumvent the Theorem, if we are only interested in the corollaries.

The University of Manitoba, Winnipeg

and

The Mathematical Institute of the
Hungarian Academy of Sciences, Budapest.

Footnotes.

(1) I could always be taken to be ω unless we want applications to arbitrary admissible fragments (see [2], [9]). Then it turns out that the conjunctions $i_n \in I$ could be omitted too, and still the same results will hold.

(2) An equivalent and in most respects more natural definition of $\underset{\sim}{\Sigma}_1^1$ is that K is $\underset{\sim}{\Sigma}_1^1$ if it consists of the relativized reducts of the models of a sentence in $L_{\omega_1, \omega}$. More precisely, a class K of structures of the countable type τ is called $\underset{\sim}{\Sigma}_1^1$ if there are: a countable similarity type $\tau' \supseteq \tau$, a formula $\theta(x)$ of $L_{\omega_1, \omega}(\tau')$ with one free variable and a sentence φ of $L_{\omega_1, \omega}(\tau')$ such that $\mathfrak{A} \in K$ iff for some model $\mathfrak{A}'$ of φ of type τ' we have $\mathfrak{A} = (\mathfrak{A}' \mid \theta) \mid \tau$ where $\mathfrak{A}' \mid \theta$ is the substructure of $\mathfrak{A}'$ with domain $\{a \in A' : \mathfrak{A}' \models \theta[a]\}$ and $\mid \tau$ denotes taking reduct to the type τ. The equivalence of the two definitions is easily proved by the method of [12], proof of Theorem 2. Vaught's notation [21] for $\underset{\sim}{\Sigma}_1^1$ is "$\exists_1$".

(3) Regular relations were introduced by Lindström [10] to unify the treatment of many special preservation theorems. See also [14] and §1 below.

(4) This fact also was mentioned by Robert Vaught and Jon Barwise in their talks at the Summer School in Logic, Cambridge, 1971.

(5) For simplicity, we assume here that there are no operation symbols or individual constants in τ.

(6) Notice that here we use, instead of $\mathfrak{B} \models \Phi$, only the weaker statement obtained from Φ by rearranging the prefix of Φ so that every "universal quantifier" precedes every "existential one" (except that $\bigwedge n < \omega$ keeps its place).

(7) If Θ does not contain any formula of the form $\bigvee\Psi$ then, it turns out, no infinite disjunctions $\bigvee k_n$ are needed in the prefix of the Vaught sentence Φ to be constructed and if the elements of Γ are finitary, we get a Svenonius sentence considered in [12]. On the other hand, if there is no formula $\bigvee\Psi$ with $\Psi \neq 0$ in Θ, we can add $\bigvee\{v_0 \approx v_0\}$ to Θ so that all $\bigvee\Psi_n$ can be defined to be $\bigvee\{v_0 \approx v_0\}$.

(8) We use this opportunity to point out an oversight in [14]. In the analogous situation in the proof of 1.2 in [14], we constructed the enumeration $\langle s_n : n < \omega\rangle$ giving rise to ρ and the $\exists x_k \psi_k$ and y_k simultaneously in an unnecessarily complicated, though correct, way.

(9) Conversely, taking $\{K_\eta : \eta < 2^{\aleph_0}\}$ to be the collection of all classes $K' = \mathrm{Mod}\ (\theta)$ such that $\theta \in \Delta(G)$ and $K \subseteq K'$, the weakened version of the statement of 2.12 with $\aleph_1$ replaced by $2^{\aleph_0}$ follows from 2.13. To show (ii), one uses Scott's isomorphism theorem. This is to correct an oversight in [13] where we proved results of the sort of 2.10(ii) without the cardinality bound $\aleph_1$ but did not notice that they followed from our results of the sort of 2.13. We have seen the derivation of the weakened 2.12(ii) from 2.13 in [17] and [18].

[1] J. Barwise, Infinitary logic and admissible sets, J. Symbolic Logic 34(1969), 226-252.

[2] J. Barwise, Back and forth thru infinitary logic, to appear in a volume on model theory, edited by M. Morley.

[3] S. Feferman, G. Kreisel, Persistent and invariant formulas relative to theories of higher order, Bull. Amer. Math. Soc. 72(1966), 480-485.

[4] S. Feferman, Lectures on proof theory, in: Proceedings of the Summer School in Logic, Leeds, 1967, Lecture Notes in Math., No. 70, Springer Verlag, 1968, 1-107.

[5] S. Feferman, Persistent and invariant formulas for outer extensions, Compos. Math. 20(1968), 29-52.

[6] R. Fraissé, Sur l'extension aux relations de quelques propriétés des ordres, Ann. Sci. École Norm. Sup., 3iéme série, 71(1954), 363-388.

[7] C. Karp, Finite quantifier equivalence, in: The Theory of Models, North Holland, Amsterdam, 1965, pp. 407-412.

[8] H.J. Keisler, Infinite quantifiers and continuous games, in: Applications of Model Theory to Algebra, Analysis and Probability, ed. W.A.I. Luxemburg, New York, 1969, pp. 228-264.

[9] H.J. Keisler, The model theory of infinitary languages, North Holland, Amsterdam, 1971.

[10] P. Lindström, On relations between structures, Theoria (Lund), 1966, 172-185.

[11] E.G.K. Lopez-Escobar, An interpolation theorem for denumerably long formulas, Fund. Math. 57(1965), 253-272.

[12] M. Makkai, On PC_{Δ}-classes in the theory of models, Publ. Math. Inst. Hung. Acad. Sci. 9 Ser. A(1964), 159-194.

[13] M. Makkai, On the model theory of denumerably long formulas, J. Symbolic Logic 34(1969), 437-459.

[14] M. Makkai, Svenonius sentences and Lindström's theory on preservation theorems, to appear in Fund. Math.

[15] M. Makkai, Vaught sentences and regular relations, abstract to appear in Notices Amer. Math. Soc., February 1972.

[16] Y. Moschovakis, The Suslin-Kleene theorem for countable structures. Duke Math. J. 37(1970), 341-352.

[17] N. Motohashi, Model theory in a positive second order logic with countable conjunctions and disjunctions, preprint.

[18] B.F. Nebres, Doctoral Dissertation, Stanford University, Stanford, Calif.

[19] S. Shelah, On the number of non-almost isomorphic models of T in a power, Pacific J. Math. 36(1971), 811-818.

[20] L. Svenonius, On the denumerable models of theories with extra predicates, in: The Theory of Models, North Holland, Amsterdam, 1965, pp. 376-389.

[21] R. Vaught, Descriptive set theory in $L_{\omega_1, \omega}$, this volume.